AF679831

Food and Agriculture in Papua New Guinea

edited by R. Michael Bourke and Tracy Harwood

THE AUSTRALIAN NATIONAL UNIVERSITY

E PRESS

ANU
E PRESS

Published by ANU E Press
The Australian National University
Canberra ACT 0200, Australia
Email: anuepress@anu.edu.au
This title is also available online at: http://epress.anu.edu.au/food_agriculture_citation.html

National Library of Australia Cataloguing-in-Publication entry:

Food and agriculture in Papua New Guinea /
editors, R. Michael Bourke, Tracy Harwood.

Bibliography.
Includes index.
ISBN 9781921536601 (pbk)
ISBN 9781921536618 (pdf)
Agriculture--Papua New Guinea
Agriculture--Economic aspects--Papua New Guinea.
Agricultural productivity--Papua New Guinea.
630.9953

Suggested citation:
Bourke, R.M. and Harwood, T. (eds) (2009). Food and Agriculture in Papua New Guinea. ANU E Press, The Australian National University, Canberra.

http://rspas.anu.edu.au/lmg/

Design: Sam Highley, Clarus Design
Typeface: Minion Pro
Index: Tracy Harwood

Cover: Yasisime from Asirangka village in the Aiyura basin, Eastern Highlands Province, weeds a peanut plot. Peanut is grown for both home consumption and for sale locally. It is grown in rotation with sweet potato to maintain soil fertility. The plot on the left contains sweet potato, the staple food in the highlands, and other food crops. Casuarina trees in the background shade Arabica coffee, the most important cash crop in the highlands. Photo by Mike Bourke.

Contents

Foreword

It gives me great pleasure to contribute the Foreword to this impressive book.

I have long held the view that subsistence agriculture underwrites the PNG cash economy. The cash sector should supplement and complement but not replace the subsistence sector. Agriculture in PNG provides direct benefits to over 80% of our population. A strong subsistence sector and the *wantok* system provide surrogate social welfare support for many people. The resilience of the rural majority was seen recently when steep increases in food prices caused considerable distress in many parts of the world. However, most rural people in PNG were spared the worst impact of these rises because of their strong subsistence base and cash income from agricultural export crops.

Agriculture is also important because it is a central part of our traditional cultures and maintains our lifestyles. It also reduces poverty among rural people and lessens migration to towns. While the mining and petroleum sectors are presently important to the economy, agriculture uses renewable resources that will continue to be critical to the PNG economy long into the future.

I am attracted to many aspects of this important book. It contains information about the significance of subsistence food production, the role of export cash crops, and details on forestry and fisheries. I was particularly pleased to see the wide variety of subjects covered including, for example, population, land use, global climate change, animal husbandry, gender issues and access to services. I was also pleased to note that the authors have included information on addressing current constraints to the expansion of agriculture.

The book is written in language that is readily accessible to non-specialists. It will be a valuable teaching resource for senior high school and university students. I commonly hear advisers new to PNG saying that there is no information available on a particular topic, or that something that has been tried before should be tried again. This book makes that 'missing' information available in one place and describes the history of previous attempts to, for example, introduce new crops or bring about rural development.

As a former director of the PNG National Planning Office, it is clear to me that much of the information in this book will and should form the basis for planning further agricultural development and indeed form the basis for defining development policies for PNG. The authors have not been afraid to put forward their expert opinions on possible future trends. They have also constructively addressed some contentious issues in the interests of agricultural development in PNG.

I congratulate the authors, editors and The Australian National University for producing this book and warmly commend it.

Charles W. Lepani
Papua New Guinea High Commissioner
to Australia, Canberra
October 2008

Contributors

Most of this book was written by R. Michael Bourke and Bryant Allen. Some sections were contracted to specialists. Matthew Allen wrote 11 sections of Part 5.

Bryant Allen

Bryant Allen is a Senior Fellow with the Land Management Group, Research School of Pacific and Asian Studies (RSPAS), The Australian National University. He has carried out research in the Cook Islands in the 1960s and in rural Papua New Guinea (PNG) from the early 1970s to the present. He has worked at the University of PNG, the PNG Department of Environment and Conservation and the PNG National Research Institute. He has researched in many parts of rural PNG, in particular in the Dreikikir area of East Sepik and the Tari area of Southern Highlands. His main interests are in the sustainability of agricultural systems, in particular shifting cultivation systems; rural development; rural poverty; and food security. **Authored:** 1.1–1.7, 1.9–1.12, 1.14–1.15, 2.5, 3.6–3.12, 6.1–6.10.

Matthew Allen

Matthew Allen is a Postdoctoral Fellow with the Resource Management in Asia-Pacific Program, RSPAS, The Australian National University. Matthew has conducted research on various aspects of the societies and environments of Melanesia since 1997. His PhD research examined the relationships between the state, identity, resources and conflict in Solomon Islands. His research interests are resource conflict and identity politics in Melanesia in general and Solomon Islands in particular, with further interests in smallholder agriculture, food security and rural development. **Authored:** 5.2, 5.4–5.13.

Kate Barclay

Kate Barclay is a Senior Lecturer with the International Studies program at the University of Technology, Sydney. Kate researches the social and political aspects of fisheries in the Asia–Pacific region. She is particularly interested in the connections between economic development and national identity, and globalisation, using fisheries enterprises as case studies. **Authored:** 2.10.

R. Michael Bourke

Mike Bourke is an independent consultant and an Adjunct Senior Fellow with the Land Management Group, RSPAS, The Australian National University. He has formal training as an agricultural scientist and human geographer and has been continuously involved in research and development work in village agriculture in PNG since 1970. He has worked in most parts of rural PNG. His research interests include relationships between people, crops and the environment; intensification of land use; climate change; sustainability of agriculture; food security; rural poverty; rural development; and root crop agronomy. **Authored:** History of agriculture, 1.8, 1.13, 2.1, 2.2, 2.4, 3.1–3.5, 4.2–4.4, 5.1, 5.3, 5.15–5.20; and significant contribution to all sections in Part 5.

John Gibson

John Gibson is a Professor of Economics at the Waikato Management School, University of Waikato, Hamilton, New Zealand, and a Visiting Fellow at the Center for Global Development, Washington, DC. He has previously worked as an adviser with the PNG Department of Agriculture and Livestock and as a researcher with the PNG Institute of National Affairs. His research interests include poverty measurement; the design and analysis of household survey data; and economic development, especially in China and other Asian and Pacific economies. **Authored:** 2.7–2.9.

Tracy Harwood

Tracy Harwood is an editor with the Land Management Group, RSPAS, The Australian National University. She has a Bachelor of Applied Science in Ecology and Natural Resources from the University of Canberra and is interested in environmental sciences. She has been editing since 2002 and is a member of the Canberra Society of Editors and the Australian and New Zealand Society of Indexers.

Jean Kennedy

Jean Kennedy is an archaeologist and Visiting Fellow in the Department of Archaeology and Natural History, RSPAS, The Australian National University. She is interested in the prehistory and ethnobotany of perennial tropical crops of the south-west Pacific region. **Authored:** 2.3.

Andrew McGregor

Andrew McGregor is an agricultural economist specialising in the Pacific islands. He currently manages the Trade and Development Office, Fiji. Over the past 30 years Andrew has worked in most Pacific island countries, and also in the eastern Caribbean and Hawaii. **Authored:** 4.1, 5.14, 5.21; contributed to future prospects for export crops in Part 5.

Alan Quartermain

Alan Quartermain is Dean of the School of Natural Resources, University of Vudal, Rabaul, PNG. He trained as an agricultural scientist and animal geneticist. He has taught at the University of PNG and University of Vudal, was Chief Scientist at the PNG National Agricultural Research Institute, and works as a consultant for the Food and Agriculture Organization of the United Nations. His research interests include livestock production, and agricultural research and development issues in developing countries. **Authored:** 2.6.

Acknowledgements

The concept for this book was developed by Mike Bourke, Bryant Allen and John Gibson. The basic idea was presented to the Papua New Guinea National Agricultural Research Institute; Department of Agriculture and Livestock; and Department of National Planning and Monitoring. Staff at these institutions, including Naihuwo Ahai, Raghunath Ghodake, Valentine Kambori, Matthew Kanua, Roy Masamdu and Geoff Wiles, supported the idea, commented on the proposal and suggested additional material. Participants at a seminar at The Australian National University suggested inclusion of some other sections, as did Geoff Fox of AusAID.

Many people from the commercial sector, industry bodies and government departments in PNG provided data, as did some based in Australia and elsewhere. They include: Ganei Agodop, Joseph Aka, Will Akus, Stephen Althaus, Moses Alu, Ram Anatharaman, Theresa Arek, David Armour, David Bain, Felix Bakani, Valonna Baker, Sergei Bang, Paul Barker, Chew Kui Boo, Greg Bosa, Maurice Brownjohn, Guy Cameron, Paul Chatterton, John Cook, George Curry, Charles Dambui, Edwin Dawana, Robert DeBrouwere, Bob Densley, Frederic Dumortier, Warren Dutton, Geoff Fahey, Mathias Faure, Philip Franklin, Brad Furness, Chris Gard, Botu Gaupu, Anton Goonetilleke, John Goumei, Jim Greg, Marianne Grosclaude, Brian Gunn, Martin Gunther, Gaona Gwaibo, Peter Heywood, Neville Howcroft, Ross Humphries, Kasis Inape, Sharryl Ivahupa, Mike Jackson, Ron Jenkins, Mark Johnston, Vele Kagena, Patricia Kaiola, John Kaius, Aquina Kango, Pauka Karafa, Ram Karan, Henao Kari, Prescilla Kauni, John Kendiga, Jeff Kinch, Michael Knight, Gina Koczberski, Marcel Kuiper, Tom Kukhang, Lastus Kuniata, Hugh Laird, Leanne Lawrance, Phil Leahy, Malcolm Levett, Frank Lewis, Ishmel Libitino, David Loh, Tella Loie, Mark Lowe, Michael Lowe, Sue Lowe, Victor Maggio, John Mandich, Mike Manning, Anna Martin, David Mather, Leigh Matters, Wakasa Mecksaene, Rahman Mizanur, Jane Mogina, John Moxon, Jennifer Moyes, Kulala Mulung, John Nightingale, Norah Omot, Theresa Orim, Diane Ormsby, Ian Orrell, Dale Palmer, Nigel Parker, Sudari Pawiro, Andrew Pawley, John Petersen, Philip Pondikou, Kanawi Pouru, Gary Preston, Alan Quartermain, Sisiwaka Roah, Graeme Ross, Alan Rumsey, David Sali, Joselito Santiago, Wila Saweri, Brett Schofield, Reuben Sengere, Ian Sexton, Tim Sharp, Robert Shelton, Rob Small, John Sowei, Graeme Stowell, Willie Susuke, Ben Tara, Lauatu Tautea, Bruce Telfer, Stephen Tovita, Katie Tripp, Jack Tulvue, Ruth Turia, Roger Vallance, Michael Varapik, Phillip Vovola, Michael Waisime, Peter Walton, Tamzin Wardley, Nathan Wartovo, Jimmy Weiner, David Wilkin, Steve Woodhouse, Charles Yala, Andrew Yamanea and Sylvanius Yavisa.

Some sections were sent to specialists for comments. Those provided by Jack Golson on the history of agriculture were particularly comprehensive. Other reviewers were Chris Ballard, Kate Barclay, Paul Barker, Chew Kui Boo, John Burton, Brian Chung, George Curry, Peter Cusack, Warren Dutton, Colin Filer, Jim Fingleton, Phil Franklin, John Gibson, Anton Goonetilleke, Brian Gunn, Simon Haberle, Alfred Hartemink, Robin Hide, Braden Jenkin, Margaret Jolly, Gael Keig, Jeff Kinch, Lastus Kuniata, Hugh Laird, Joseph Lelang, Anthony Lewis, John McAlpine, Andrew McGregor, Mark Mosko, Ian Orrell, Alan Quartermain, Ian Sexton, Tim Sharp, Rob Small, John Spriggs, Bill Standish and Phillip Vovola.

Most of the information on village-sector agriculture was collected from hundreds of people from every district in PNG who willingly gave their time and immense knowledge about their food production and cash crop systems. One of the aims of this book is to explain the critical importance to PNG's national economy of village agriculture through the production of food, shelter and cash crops.

The book was produced by members of the Land Management Group, Research School of Pacific and Asian Studies, The Australian National University. As well as the editors, Laura Vallee (data organisation) and Bryant Allen (editing) made a significant input. Others who contributed included Matthew Allen, Karen Fisher, Quintin Gravatt, Amber Pares, Natalie Stuckings and Veerle Vlassak. Graphs and pie charts were generated in house. Maps were drawn by Kay Dancey, Cartography Unit, ANU. Additional figures were provided by Catherine Eadie, Philippe Rekacewicz and Robert Rohde. Design was by Sam Highley. Several members of the Canberra Society of Editors advised informally on some editing conventions. Funding was provided by the Research School of Pacific and Asian Studies at ANU, the Australian Government through AusAID, and by members of the Land Management Group from consulting income.

The assistance of all those named above is acknowledged with thanks. Responsibility for any errors or omissions rests with the authors of individual sections.

Acronyms and initialisms

ACIAR	Australian Centre for International Agricultural Research
ADB	Asian Development Bank
AGSYS	agricultural system
ANU	The Australian National University
AusAID	Australian Agency for International Development
AVHRR	Advanced Very High Resolution Radiometer
CBO	community-based organisation
CCI	Cocoa Coconut Institute of PNG
CIC	Coffee Industry Corporation Ltd
CIMC	Consultative Implementation and Monitoring Council
CLTC	Christian Leaders Training College
CMB	Copra Marketing Board
COPD	Community Oil Palm Development
CPI	consumer price index
CRES	Centre for Resource and Environmental Science (ANU)
CSIRO	Commonwealth Scientific and Industrial Research Organisation (Australia)
CU	census unit
DAL	Department of Agriculture and Livestock
DASF	Department of Agriculture, Stock and Fisheries
DEC	Department of Environment and Conservation
DHS	Demographic and Health Survey
DIS	delivered-in-store
DNPM	Department of National Planning and Monitoring
DPI	Department of Primary Industry
DSG	District Support Grant
DSIP	District Services Improvement Program
EBC	Evangelical Brotherhood Church Inc.
ENSO	El Niño Southern Oscillation
EU	European Union
FAO	Food and Agriculture Organization of the United Nations
FFB	fresh fruit bunch (oil palm)
FIMS	Forest Resource Information System
FPDA	Fresh Produce Development Agency Ltd
GIS	geographical information system
HIV/AIDS	human immunodeficiency virus/ acquired immune deficiency syndrome
IASER	Institute of Applied Social and Economic Research
IFTA	Insect Farming and Trading Agency
INA	Institute of National Affairs
IPCC	Intergovernmental Panel on Climate Change

ITTO	International Tropical Timber Organization
KIK	Kokonas Indastri Koporesen
LAES	Lowlands Agricultural Experiment Station, Keravat
LDC	Livestock Development Corporation Ltd
LMS	London Missionary Society
LSS	land settlement scheme
MASP	Mapping Agricultural Systems of Papua New Guinea Project
MP	member of parliament
MTDS	Medium Term Development Strategy
NAC	National Agricultural Council
NACS	National AIDS Council Secretariat
NADP	National Agriculture Development Plan
NAIS	National Agricultural Information System
NAQIA	National Agriculture Quarantine and Inspection Authority
NARI	National Agricultural Research Institute
NBPOL	New Britain Palm Oil Limited
NES	nucleus estate and smallholder (model)
NFA	National Fisheries Authority
NFB	National Forestry Board
NFS	National Forestry Service
NGO	non-government organisation
NIS	nut-in-shell
NNS	National Nutrition Survey
NPK	nitrogen–phosphorus–potassium fertiliser
NPO	National Planning Office
NSO	National Statistical Office of PNG
OPGA	Oil Palm Growers' Association
OPIC	Oil Palm Industry Corporation
OPRA	Oil Palm Research Association
ORD	Office of Rural Development
PDS	Provincial Data System
PNG	Papua New Guinea
PNGCC	Papua New Guinea Coconut Commodities
PNGFA	Papua New Guinea Forest Authority
PNGRIS	Papua New Guinea Resource Information System
POPA	Palm Oil Producers' Association
PRA	participatory rural appraisal
RAMS	Road Asset Management System
RHS	rural household survey
RIC	Rural Industries Council
RMU	resource mapping unit
RRA	rapid rural appraisal
SGS	Société Générale de Surveillance
SIB	Spice Industry Board
SOI	Southern Oscillation Index
SSTC	South Seas Tuna Corporation
TPC	Tactical Pilotage Chart
UN	United Nations
UNICEF	United Nations Children's Fund
US	United States of America
USDA	United States Department of Agriculture
VOP	village oil palm
WWF	(formerly) World Wide Fund for Nature

Botanical names

Staple foods

Common name	Scientific name
Banana	*Musa* cvs
Cassava	*Manihot esculenta*
Coconut	*Cocos nucifera*
Cordyline	*Cordyline fruticosa*
Corn (maize)	*Zea mays*
Irish potato	*Solanum tuberosum*
Kudzu	*Pueraria lobata*
Polynesian arrowroot	*Tacca leontopetaloides*
Queensland arrowroot	*Canna edulis*
Rice	*Oryza sativa*
Sago	*Metroxylon sagu*
Sago	*Metroxylon salomonense*
Sorghum	*Sorghum bicolor*
Sugar cane	*Saccharum officinarum*
Sweet potato	*Ipomoea batatas*
Taro	*Amorphophallus paeoniifolius*
Taro (taro tru)	*Colocasia esculenta*
Taro, Chinese	*Xanthosoma sagittifolium*
Taro, giant	*Alocasia macrorrhizos*
Taro, swamp	*Cyrtosperma chamissonis*
Wheat	*Triticum aestivum*
Yam	*Dioscorea nummularia*
Yam	*Dioscorea pentaphylla*
Yam, aerial	*Dioscorea bulbifera*
Yam, greater	*Dioscorea alata*
Yam, lesser	*Dioscorea esculenta*
Yam, white	*Dioscorea rotundata*

Vegetables

Common name	Scientific name
Aibika	*Abelmoschus manihot*
Amaranthus	*Amaranthus blitum*
Amaranthus	*Amaranthus caudatus*
Amaranthus	*Amaranthus cruentus*
Amaranthus	*Amaranthus dubius*
Amaranthus	*Amaranthus tricolor*
Bean, adzuki	*Vigna angularis*
Bean, broad	*Vicia faba*
Bean, climbing	*Phaseolus vulgaris*
Bean, common	*Phaseolus vulgaris*
Bean, dwarf	*Phaseolus vulgaris*
Bean, jack	*Canavalia ensiformis*
Bean, lablab	*Lablab purpureus*
Bean, lima	*Phaseolus lunatus*
Bean, mung (bean sprouts)	*Vigna radiata*
Bean, rice	*Vigna umbellata*
Bean, snake	*Vigna unguiculata* cv. group Sesquipedalis

Vegetables (continued)

Bean, winged	*Psophocarpus tetragonolobus*
Beetroot	*Beta vulgaris* cv. group Garden Beet
Broccoli	*Brassica oleracea* cv. group Broccoli
Brussels sprout	*Brassica oleracea* cv. group Brussels Sprouts
Cabbage, Chinese	*Brassica rapa* cv. group Chinese cabbage
Cabbage, head	*Brassica oleracea* cv. group White Headed Cabbage
Capsicum	*Capsicum annuum*
Carrot	*Daucus carota*
Castor	*Ricinus communis*
Cauliflower	*Brassica oleracea* cv. group Cauliflower
Celery	*Apium graveolens* var. *dulce*
Ceylon spinach	*Basella alba*
Chilli	*Capsicum frutescens*
Choko (chayote)	*Sechium edule*
Coral tree	*Erythrina variegata*
Cowpea	*Vigna unguiculata* cv. group Unguiculata
Cucumber	*Cucumis sativus* cv. group Slicing Cucumber
Cyanotis	*Cyanotis moluccana*
Dicliptera	*Dicliptera papuana*
Eggplant	*Solanum melongena* cv. group Common Eggplant
Ficus wassa	*Ficus wassa*
Garlic	*Allium sativum* cv. group Common Garlic
Ginger	*Zingiber officinale*
Gourd, bottle	*Lagenaria siceraria*
Gourd, wax	*Benincasa hispida* cv. group Wax Gourd
Highland *kapiak*	*Ficus dammaropsis*
Job's tears	*Coix lacryma-jobi*
Kalava	*Ormocarpum orientale*
Kangkong	*Ipomoea aquatica*
Karakap	*Solanum americanum*
Kohlrabi	*Brassica oleracea* cv. group Kohlrabi
Kumu musong	*Ficus copiosa*
Leek	*Allium ampeloprasum* cv. group Leek
Lemon grass	*Cymbopogon citratus*
Lettuce	*Lactuca sativa*
Marrow	*Cucurbita pepo*
Oenanthe (water dropwort)	*Oenanthe javanica*
Onion	*Allium cepa* cv. group Common Onion
Pak choi	*Brassica rapa* cv. group Pak Choi
Parsley	*Petroselinum crispum*
Pea	*Pisum sativum*
Peanut	*Arachis hypogaea*
Pigeon pea	*Cajanus cajan*
***Pitpit*, highland**	*Setaria palmifolia*
***Pitpit*, lowland**	*Saccharum edule*
Pumpkin	*Cucurbita moschata*
Radish, red	*Raphanus sativus* cv. group Small Radish
Rhubarb	*Rheum ×cultorum*
Rorippa	*Rorippa schlechteri*
Rungia	*Rungia klossii*
Russian comfrey	*Symphytum asperrimum*
Shallot	*Allium cepa* cv. group Aggregatum
Silverbeet	*Beta vulgaris* cv. group Spinach Beet
Soya bean	*Glycine max*
Spinach	*Spinacia oleracea*
Spring onion	*Allium cepa* cv. group Aggregatum
Tomato	*Lycopersicon esculentum*

Vegetables (continued)

Trichosanthes pulleana	*Trichosanthes pulleana*
Tulip	*Gnetum gnemon*
Turnip	*Brassica rapa* cv. group Vegetable Turnip
Valangur	*Polyscias verticillata*
Wandering Jew	*Commelina diffusa*
Watercress	*Rorippa nasturtium-aquaticum*
Yam bean	*Pachyrhizus erosus*
Zucchini	*Cucurbita pepo*

Fruit

Common name	Scientific name
Apple	*Malus* spp.
Avocado	*Persea americana*
Brazil cherry	*Eugenia uniflora*
Bukabuk	*Burckella obovata*
Bullock's heart	*Annona reticulata*
Cape gooseberry	*Physalis peruviana*
Carambola (five corner)	*Averrhoa carambola*
Cherimoya	*Annona cherimolia*
Cumquat	*Fortunella japonica*
Custard apple (sweetsop)	*Annona squamosa*
Durian	*Durio zibethinus*
Egg tree	*Garcinia xanthochymus*
Elder	*Sambucus nigra*
Golden apple	*Spondias cytherea*
Governor's plum	*Flacourtia indica*
Granadilla	*Passiflora quadrangularis*
Grapefruit	*Citrus paradisi*
Guava	*Psidium guajava*
Guava, cherry	*Psidium cattleianum*
Jackfruit	*Artocarpus heterophyllus*
Langsat	*Lansium domesticum*
Lemon	*Citrus limon*
Lime	*Citrus aurantifolia*
Longan	*Euphoria longan*
Loquat	*Eriobotrya japonica*
Malay apple	*Syzygium malaccense*
Malay apple, giant	*Syzygium megacarpa*
Mandarin	*Citrus reticulata*
Mango	*Mangifera indica*
Mango, traditional	*Mangifera minor*
Mangosteen	*Garcinia mangostana*
***Marita* pandanus**	*Pandanus conoideus*
Mon	*Dracontomelon dao*
Mulberry	*Morus nigra*
Naranjilla	*Solanum quitoense*
Nectarine	*Prunus persica* var. *nectarina*
Orange	*Citrus sinensis*
Pandanus, coastal	*Pandanus tectorius*
Parartocarpus	*Parartocarpus venenosa*
Passionfruit, banana	*Passiflora mollissima*
Passionfruit, lowland yellow	*Passiflora edulis* f. *flavicarpa*
Passionfruit, purple	*Passiflora edulis* f. *edulis*
Pawpaw (papaya)	*Carica papaya*
Peach	*Prunus persica*
Persimmon	*Diospyros kaki*
Pineapple	*Ananas comosus*
Plum, Japanese	*Prunus* spp.
Pomegranate	*Punica granatum*
Pomelo	*Citrus maxima*
Pouteria	*Pouteria maclayana*
Pulasan	*Nephelium mutabile*
Rambutan	*Nephelium lappaceum*
Raspberry, black	*Rubus lasiocarpus*
Raspberry, red	*Rubus moluccanus*
Raspberry, red	*Rubus rosifolius*

Fruit (continued)

Rockmelon (cantaloupe)	*Cucumis melo*
Rukam	*Flacourtia rukam*
Santol	*Sandoricum koetjape*
Soursop	*Annona muricata*
Star apple	*Chrysophyllum cainito*
Strawberry	*Fragaria spp.*
Suga prut (highland yellow passionfruit)	*Passiflora ligularis*
Tamarillo (tree tomato)	*Cyphomandra betacea*
Tamarind	*Tamarindus indica*
Ton (*taun*)	*Pometia pinnata*
Watermelon	*Citrullus lanatus*
Watery rose apple	*Syzygium aqueum*

Nuts

Common name	Scientific name
Breadfruit	*Artocarpus altilis*
Candle nut	*Aleurites moluccana*
Cashew	*Anacardium occidentale*
Castanopsis	*Castanopsis acuminatissima*
Dausia	*Terminalia megalocarpa*
Elaeocarpus womersleyi	*Elaeocarpus womersleyi*
Finschia	*Finschia chloroxantha*
Galip	*Canarium decumanum*
Galip	*Canarium harveyi*
Galip	*Canarium indicum*
Galip	*Canarium lamii*
Galip	*Canarium salomonense*
Karuka, planted	*Pandanus julianettii*
Karuka, wild	*Pandanus antaresensis*
Karuka, wild	*Pandanus brosimos*
Macadamia	*Macadamia integrifolia*
Macadamia	*Macadamia tetraphylla*
Okari	*Terminalia impediens*
Okari	*Terminalia kaernbachii*
Omphalea gageana	*Omphalea gageana*
Pao	*Barringtonia edulis*
Pao	*Barringtonia novae-hiberniae*
Pao	*Barringtonia procera*
Pecan	*Carya illinoensis*
Polynesian chestnut (*aila*)	*Inocarpus fagifer*
Sea almond (*talis*)	*Terminalia catappa*
Sis (*solomon*)	*Pangium edule*
Tulip	*Gnetum gnemon*

Stimulants

Common name	Scientific name
Betel nut (*buai*)	*Areca catechu*
Betel nut, highland	*Areca macrocalyx*
Betel pepper, highland	*Piper gibbilimbum*
Betel pepper, lowland	*Piper betle*
Kava	*Piper methysticum*
Marijuana	*Cannabis sativa*
Tobacco	*Nicotiana tabacum*

Cash crops

Common name	Scientific name
Annatto (bixa)	*Bixa orellana*
Balsa	*Ochroma lagopus*
Black pepper	*Piper nigrum*
Cardamom	*Elettaria cardamomum*
Chilli	*Capsicum frutescens*

Cash crops (continued)

Cinnamon	*Cinnamomum verum*
Citronella grass	*Cymbopogon nardus*
Cocoa	*Theobroma cacao*
Coffee, Arabica	*Coffea arabica*
Coffee, Robusta	*Coffea canephora* var. *robusta*
Coconut	*Cocos nucifera*
Ginger	*Zingiber officinale*
Japanese mint	*Mentha arvensis*
Jatropha	*Jatropha curcas*
Kapok	*Ceiba pentandra*
Lemon grass	*Cymbopogon citratus*
Nutmeg	*Myristica argentea*
Nutmeg	*Myristica fragrans*
Oil palm	*Elaeis guineensis*
Patchouli	*Pogostemon cablin*
Pyrethrum	*Chrysanthemum cinerariaefolium*
Rubber, assam	*Ficus elastica*
Rubber, para	*Hevea brasiliensis*
Tea	*Camellia sinensis*
Turmeric	*Curcuma domestica*
Vanilla	*Vanilla planifolia*
Vanilla	*Vanilla tahitensis*

Forest products

Common name	Scientific name
Acacia	*Acacia crassicarpa*
Acacia	*Acacia mangium*
Burckella	*Burckella* spp.
Cajuput oil	*Asteromyrtus symphyocarpa*
Calophyllum	*Calophyllum* spp.
Casuarina (*yar*)	*Casuarina oligodon*
Chinese water chestnut	*Eleocharis dulcis*
Copal gum	*Agathis* spp.
Cordyline	*Cordyline fruticosa*
Dammar	*Vatica papuana*
Dillenia	*Dillenia* spp.
Eaglewood	*Gyrinops ledermannii*
Eucalyptus (*kamarere*)	*Eucalyptus deglupta*
Eucalyptus	*Eucalyptus pellita*
Hibiscus	*Hibiscus tiliaceus*
Hoop pine	*Araucaria cunninghamii*
Ivory nut	*Metroxylon amicarum*
Kamarere	*Eucalyptus deglupta*
Klinki pine	*Araucaria hunsteinii*
Kwila	*Intsia bijuga*
Malas	*Homalium* spp.
Massoi	*Cryptocarya massoy*
Mersawa	*Anisoptera* spp.
New Guinea walnut (*mon*)	*Dracontomelon dao*
Noni	*Morinda citrifolia*
Perennial cane grass	*Coix gigantea*
Pinus	*Pinus* spp.
Planchonella	*Pouteria* spp.
Red canarium (*galip*)	*Canarium indicum*
Red cedar	*Toona ciliata*
Rosewood	*Pterocarpus indicus*
Rubber	*Hevea brasiliensis*
Sandalwood	*Santalum macgregorii*
Sisal	*Agave sisalana*
Teak	*Tectona grandis*
Terminalia	*Terminalia* spp.
Tigaso oil	*Buchanania* spp.
Tigaso oil	*Campnosperma brevipetiolata*
Ton (*taun*)	*Pometia pinnata*

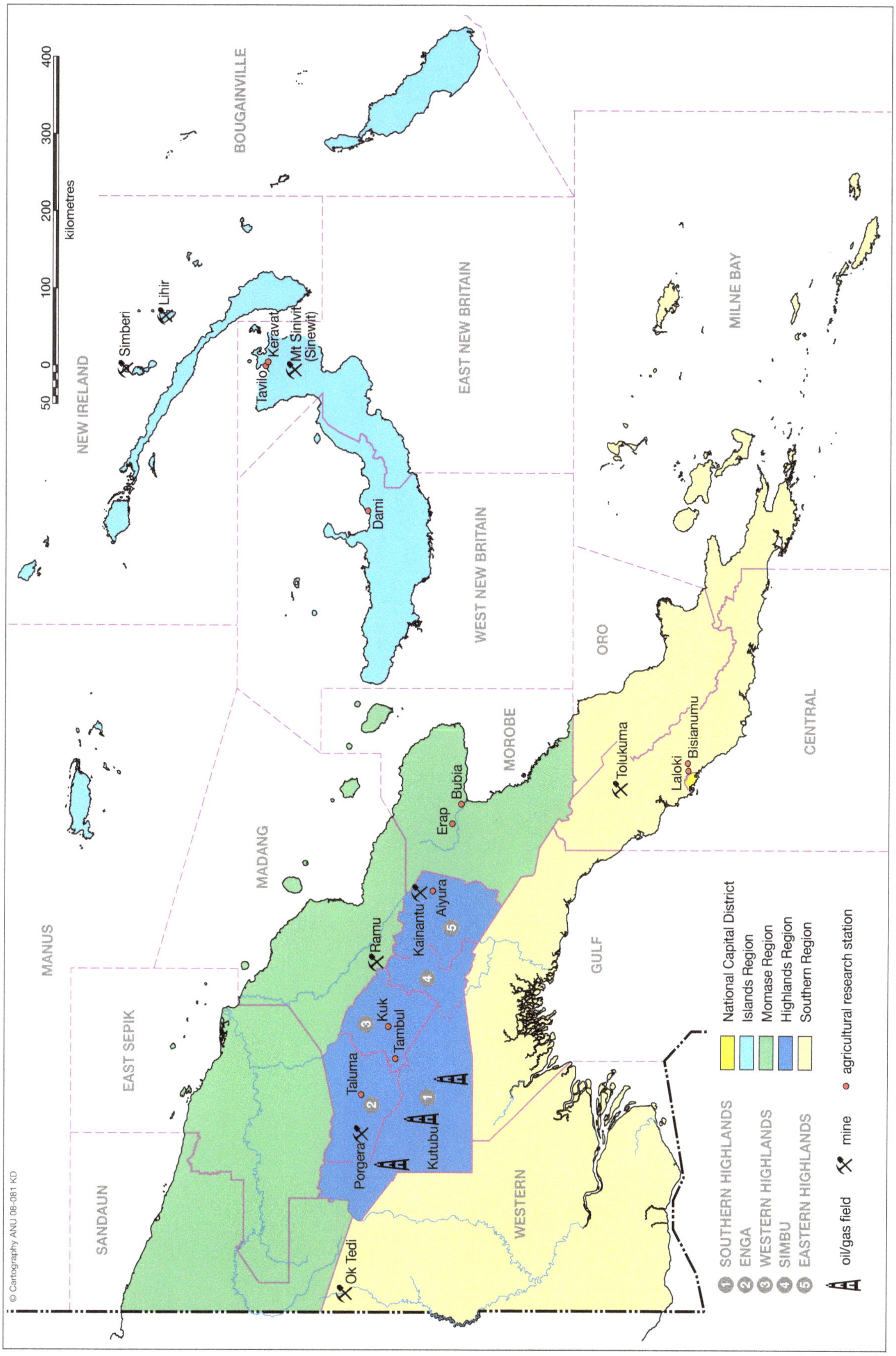

The five regions of Papua New Guinea, mine sites mentioned in the text, and agricultural research stations

ADMIRALTY ISLAN
MANUS
Bismarck Bay
LORENG
Manus Is
Fedarb
Mbuke Is
Lou Is
Pam I
Baluan Is
SANDAUN
EAST SEPIK
VANIMO
Bewani
Sissano Lagoon
Aitape
Vokeo Is
Dagua
WEWAK
Amanab
Dreikikir
Nuku
Maprik
Bainyik
Wosera
Murik Lakes
Gavien
Angoram
Sepik
Ama
Ambunti
Manam Is
Bogia
MADANG
Keram
Ramu
May
Yapsiei
Yuat
Karkar Is
ADELBERT RA.
Bagabag Is
Gogol
SCHRADER RA.
Telefomin
Tabubil
Oksapmin
Lagaip
Kaironk
Simbai
MADANG
Long Is
Rai Coast
Kopiago
Kompiam
Sirunki
Porgera
WABAG
Lai
Baiyer
Jimi
Saidor
Laiagam
Wapenamanda
Ningerum
Koroba
Kandep
Tambul
Kuk
Wahgi
Mt Wilhelm
Bundi
Gembogl
FINISTERRE RA.
Dumpu
Teptep
Wasu
Tari
MOUNT HAGEN
Kudjip
Kerowagi
Gusap
Ok Tedi
Margarima
Nebilyer
KUNDIAWA
GOROKA
Wantoat
SARAWAGET RA.
Kabwum
Kiunga
Nipa
MENDI
Gumine
Mt Elimbari
Asaro
Kainantu
Huon Peninsula
Strickland
Nomad
Nembi
Ialibu
Henganofi
Arona
Kaiapit
Fly
Great Papuan Plateau
Lake Kutubu
Lufa
Ukarumpa
Markham
Nadzab
Gobari
Oger
Situm
Karimui
Okapa
Pimaga
Erave
Mt Bosavi
Erave
Wopasali
Lamari
Imani
Patep No.2
Watut
LAE
Lake Murray
Wonenara
Wampit
Wagau
Kikori
Purari
Marawaka
Mumeng
Snake
Busama
Bosset
Bulolo
Aird Hills
Menyamya
MO
WESTERN
Kikori
Aseki
Wau
Koravaki
Kanabea
Tauri
Lakekamu
Garaina
Fly
Murua
Epo
Suki
Balimo
KEREMA
Waria
GULF
Bereina
GULF OF PAPUA
Oriomo
Yule Is
DARU
Galley Reach
Sog
Saibai Is
PORT MORESBY
1 SOUTHERN HIGHLANDS
2 ENGA
3 WESTERN HIGHLANDS
4 SIMBU
5 EASTERN HIGHLANDS
© Cartography ANU 08-081 KD

Locations mentioned in the text

Conventions, data sources and limitations

Conventions

Currency

The value of production and exports is usually expressed in PNG kina, with values in US dollars also used in some appendix tables. Australian pounds was the currency used in PNG until early 1966 and pounds have been converted to Australian dollars by multiplying by two. Australian dollars were then used until April 1975, when the kina was introduced. Values in Australian dollars prior to 1975 are expressed as kina throughout the book. Figure 4.1.1 and Table A4.1.1 display the exchange rates used in the conversions between the kina and Australian dollar and US dollar.

The symbols used for the main currencies are:

K	PNG kina
A$	Australian dollar
US$	United States dollar

The kina lost value relative to the US dollar after 1994, and particularly after 1998. Hence the value of exports from PNG in kina terms generally increased after the late 1990s (Figure 5.2.4). However, the buying power of the kina decreased at the same time, so real returns to growers are generally less than they appear in figures showing the export value for each cash crop (Part 5).

Food crop categories

A number of terms could be used to describe the most important carbohydrate foods grown in PNG. We have used the term 'staple food crop' as a category for these crops, the most important being sweet potato, sago, banana and various types of taro and yam. Strictly speaking, the term 'staple food' refers to a food that provides the most food energy in a population's diet. So a crop might be the staple food for some people, but not for others. Alternative terms we could have used and which would have been more precise in some contexts include 'starchy food', 'carbohydrate food', 'energy food', 'main food crop' or 'most important food crop.' A number of food crops grown in PNG are difficult to classify in crop categories. We have included banana, coconut, corn, Irish potato and sugar cane with staple foods. Peanut is included in the vegetable class, while breadfruit is included with edible nuts.

We use the term 'stimulant' for a group of substances that people use to alter their mood, emotion or perception. The substances covered here (tobacco, betel nut, marijuana, caffeine, alcohol and kava) are not all stimulants in a medical sense, but the term is less emotive than alternative names such as 'narcotic' or 'drug' and less technical than the more precise term 'psychoactive drug.'

The common names of food crops and plants are used as much as possible in this book. Scientific names are used where common names are not known or used, or to distinguish between similar species. For example, *galip* is the common name for a nut belonging to the genus *Canarium*, of which there are many species. In the text, the nut may be referred to by its full scientific name to make it clear which *galip* species is being discussed. A table of all plant common and scientific names is on page xiv.

Imperial measurements

Some historical data is available only in imperial measurements, such as tons or acres. We have converted all data to metric units, which have been used in PNG since 1972.

Population

Official population data from the 2000 National Census has been used throughout this book. The total population in mid 2000 was 5.2 million people. Census units, the lowest unit of the PNG census, were used to group the population into urban, rural and rural non-village classes. ('Rural non-village' locations include mining settlements, logging camps, mission stations, schools and research stations.) Of the total, 81% (4.2 million) were rural villagers; 13% (0.7 million) lived in urban areas; and 6% (0.3 million) lived in rural non-village locations (Table 1.1.1). The average number of people per household in PNG is 5.1, so to convert population figures to number of households, divide by 5.1. The mid 2008 PNG population was estimated by the Secretariat of the Pacific Community to be 6.5 million people.

Most population data used in tables and figures are for rural villagers, where we use the term 'rural population'. These figures do not include the rural non-village population. However, some analyses present data for the entire provincial or national population. Readers should note the distinction between 'rural population' and 'total population' in column headings and graph axes. In the x-axis of graphs, the term 'Central' refers to the rural village population only in Central Province, whereas the term 'Central + NCD' includes the total population of Central Province and the National Capital District (Port Moresby).

Provincial names

The names commonly used for some provinces differ from the official (and older) names. We have used the more commonly used names, that is, Oro, Sandaun and Simbu rather than Northern, West Sepik and Chimbu. We use the current province name rather than earlier province names, even when referring to historical data. Some provinces have been split over the past 60 years: they are Sepik into East Sepik and Sandaun; New Britain into East New Britain and West New Britain; Central Highlands into Southern Highlands, Western Highlands and Eastern Highlands; Eastern Highlands into Eastern Highlands and Simbu; and Western Highlands into Western Highlands and Enga.

When we commenced this book, Bougainville was one of 19 provinces. On 15 June 2005, its status changed to the Autonomous Region of Bougainville. We have used the term 'Bougainville' throughout rather than the post 2005 name.

Data sources

Data in this book come from many places and are noted under 'Sources' at the end of each section. Statements are not attributed to an author in the text, although they are sometimes attributed in footnotes.

Some data sources for tables and figures are given as organisations with or without dates. If data were obtained directly from an organisation or government department (for example, from the Department of Agriculture and Livestock, DAL), then no date is supplied. When data are from a published source, a date is given and the bibliographic details are listed at the end of the relevant section, for example, DAL (1992).

The Papua New Guinea Resource Information System (PNGRIS), developed by CSIRO in Australia and PNG Department of Primary Industry (now DAL) in the 1980s and 1990s, was widely used in Part 1. It contains information on natural resources, land use and population distribution in PNG, with data in a mappable form (or geographical information system). The database is described in a number of publications (see Section 1.15 for the references).

Another significant data source was the Mapping Agricultural Systems of Papua New Guinea Project (MASP) database, particularly for Part 3. This was developed by members of the Land Management Group at The Australian National University and is based on extensive fieldwork in the 1990s (see Section 1.15).

Much of the data are presented in about 380 tables, figures and maps. Smaller tables are contained in the body of the book. Larger and more complicated tables have been placed in an appendix. These are identified by the letter 'A' in front of the table number. For example, Table A6.7.1 is the first appendix table for Section 6.7. The appendix tables are not published in the paper version of this book, but are contained in the accompanying compact disk. Data for most figures are given in appendix tables, so that they can be re-analysed by users or extended in the future. Subject to funding availability, it is planned to make this book available on the web in a form where tables can be downloaded as Excel spreadsheets.

Data limitations

Most numerical data in this book has an error component. Where data are presented without the term 'estimated' the data are believed to be fairly accurate, that is, to be within +/– 15% of the real value. This is the case for the volume and value of most export crops and those statistics derived directly from the PNGRIS and MASP databases.

The term 'estimated' is used to indicate that the data are likely to be less accurate, often because an analysis is based on a number of assumptions. An example is estimates of production of staple food crops in 2000 (Tables 2.2.1, A2.2.1, A2.2.2).

Generally the size of the possible error is not given. Where we suspect that an estimate may have a particularly large error component, this is noted as a footnote, but we have usually chosen not to present data that seems to be particularly inaccurate.

There are two estimates of the source of foods consumed in PNG: one done as a *consumption survey* as part of the PNG Household Survey in 1996; and the second one as *production estimates* for 2006 done for this book. A comparison of the detailed results is useful as it indicates the order of errors in these two studies (see Table A2.1.1).

Data for the volume and value of export crops is likely to be fairly accurate, but errors are apparent, even with data for the same year from the same industry body. We have spent a lot of effort in attempting to resolve conflicting data, but have not always been successful. The reliability of some data appears to decrease in recent years, particularly since the late 1990s. We chose to present data on export crops from the industry bodies rather than the more commonly used data from the Bank of Papua New Guinea. Our reasoning is that data from the industry bodies is more likely to be accurate, but this appears to be not always so.

Export crops data from the Bank of PNG has the advantage of being consistent and readily available on the internet <http://www.bankpng.gov.pg/>. A disadvantage is that the categories used by the bank tend to reflect the PNG economy in past decades and have not been adjusted for changes in the economy, such as the relative decline in the importance of rubber and copra and the rise in production of other commodities. In the case of oil palm, data from the large processing companies is more accurate as the Bank of PNG data refers only to crude palm oil and does not include palm kernel oil, refined palm oil or palm kernel expellent.

Generally we have given the year when estimates were made, except for those derived mainly from PNGRIS and MASP. We have presented the most recent available data. We use older data where there is no more reliable recent data and the older data tells an important story. Data on the volume and value of export crops in Part 5 is presented up to 2006 or 2007, depending on data availability.

The estimates of cash income from the MASP database have changed since they were recorded in the early to mid 1990s and standardised in 1998 (Table 5.1.1). We have presented these estimates as there is no more recent detailed information on cash income for rural people, particularly for the informally marketed commodities such as fresh food, betel nut, firewood and fish. Despite changes in kina terms of income from agricultural sources over the past 15 years, the relative ranking of the sources of cash income has not changed greatly over this period.

Sources of information about Papua New Guinea agriculture

A large volume of information about agricultural surveys, research and many other aspects of rural development exists for PNG, and is contained in published and unpublished books, reports and papers. These documents can be identified through printed and electronic bibliographies.[1] Four major electronic bibliographies have been developed for agriculture, rural development and related topics in PNG. These databases do not give access to the papers themselves, but they indicate what literature is available and often where it can be located.

PNG National Agricultural Information System

The National Agricultural Information System (NAIS) is primarily a library catalogue of the holdings of 19 libraries of seven PNG agricultural institutions. Of the electronic systems described here, it is the only one based in PNG. The system comprises over 32 000 entries, and includes an early version of the PNG Agriculture Literature Database (see below) plus other materials found in partner libraries. Included in NAIS are over 600 agricultural titles published in PNG, from technical bulletins to extension leaflets. Many are in electronic format. The system uses Inmagic DB/TextWorks software, with all edited records conforming to international bibliographic and library standards. Subject descriptors follow the *CAB Thesaurus*. The system is coordinated from the NARI Library at Bubia.[2] The database is also accessible on the internet at <http://www.pngnais.org/>, although not all the functions of the desktop version have been implemented. The desktop version has been distributed to appropriate institutions throughout PNG.

1 For example, publications arising from conferences devoted to food production and human nutrition in PNG between 1970 and 1999 are listed by Bourke (2001).

2 In 2008 the administrator of the NAIS bibliography was Ms Irene Okpul (irene.okpul@nari.org.pg).

PNG Agriculture Literature Database

The PNG Agriculture Literature Database ('AgBib') was developed by the Land Management Group at The Australian National University. The focus of AgBib is agriculture in PNG, but it includes other topics relating to rural development. The database contains more than 16 000 entries. The original bibliography was developed by Robin Hide (see below), using EndNote software. A feature of the database is a comprehensive keyword list that is drawn from the *CAB Thesaurus*. Increasingly, where digital full text versions of papers exist, these are being linked to the related bibliographic record. Many of the references have been checked against originals and the database contains few errors. An early version of the database can be accessed on the internet at <http://database.anu.edu.au/rspas/hug/pngagbib/>. Up-to-date copies of the bibliography (about 7 MB in size) are available from the Land Management Group (lmg@anu.edu.au).

An important component of AgBib is references to unpublished reports and papers, known as 'grey literature', going back to the 1950s. Large numbers of unpublished reports produced on PNG agriculture have not been deposited in a central archive, such as the PNG National Library. The Land Management Group has copies of many of these reports, which will be archived in the Pacific Research Archive at the ANU Library. If funds become available this literature will be scanned and made available on the internet.

Hays's New Guinea bibliography

Terence Hays (thays@ric.edu) at Rhode Island College in the United States has developed and maintains an ethnographic bibliography for the island of New Guinea, that is, Indonesian Papua and mainland PNG (Momase, Highlands and Southern regions). Literature about ethnic groups is the main focus, with less information relating specifically to agriculture. About two-thirds of the entries refer to PNG and the rest to Papua. An internet version of the bibliography contains over 21 000 annotated entries and is located at <http://www.papuaweb.org/bib/hays/ng/intro.html>. A feature of this database is that all entries have been checked against originals and it is the most accurate of the electronic bibliographies reviewed here.

Hide's New Guinea bibliography

Robin Hide maintains a personal research database that has been built up over a 40-year period. It includes publications on agriculture, health, ecology and other topics focused on New Guinea (both Papua New Guinea and Papua, Indonesia). It contains some 43 000 entries, of which about 36 000 concern New Guinea (as at July 2008). The bibliography uses EndNote software. An earlier version of the bibliography formed the beginning of the PNG Agriculture Literature Database. As a personal, work-in-progress database, entries are not standardised, nor are all checked against originals. The database is not publicly available, but printed or electronic lists by topic or of specific searches are available on request (robin.hide@anu.edu.au). One element of this database – an annotated bibliography on the subject of pig husbandry in PNG – has been published (Hide 2003).

Other electronic sources

Increasingly, the internet provides a powerful means to search for information. A huge amount of information is now available on the websites of organisations and institutions. Google or other internet search engines will find these sites if sensible keywords are used. Search engines will also provide the web addresses of particular journals. A search engine designed to assist researchers, Google Scholar, will access author, title and subject listings, including citations of papers in other journals, which can often help locate a particular paper.

Bibliographic software like EndNote makes it possible to manage large numbers of references. Many journal websites allow a full reference and abstract to be downloaded in a format that EndNote recognises. A very important tool for scientific research is access to full text digital copies or images of papers from the websites of particular journals and archives. Access to digital copies is often restricted and usually a fee is charged, unless the access is made through an institution that has paid the access fee in advance.

Access to most journals is not presently available in PNG. However, in 2008, the National Research Institute, National Agricultural Research Institute, Office of Higher Education and the six PNG universities established the PNG Academic and Research Network (PNGARNet). PNGARNet uses satellite communications to overcome problems created by the geographical isolation of these institutions from each other by providing high speed internet connections. If funding becomes available, the members of PNGARNet will provide access to their individual library catalogues and full text journal databases.

Another source of references to information on PNG (but not the information itself) is web-based library catalogues.[3] Many library catalogues can be accessed from outside the institution that supports them. The Australian National Library is a good example <http://catalogue.nla.gov.au/>. Most Australian university library catalogues can be accessed by external users.

How to use this book

Some readers may choose to read this book from the start to the end, while others will refer to a particular section that is of interest to them. However, information on a particular topic may be found in a number of places. The book contains a comprehensive index and we encourage readers to use the index to find all the references to information on topics of interest.

[3] For an overview of libraries in PNG, see a paper by John Evans titled *Libraries in Papua New Guinea*. An early version of this paper can be accessed on the web at <http://www.pngbuai.com/000general/libraries/library-development/png-libraries/Encx-libraries-png-1.html#national-archives>.

Sources

Bourke, R.M. (2001). Previous conferences devoted to food production or human nutrition in PNG (1970–99). In Bourke, R.M., Allen, M.G. and Salisbury, J.G. (eds). *Food Security for Papua New Guinea. Proceedings of the Papua New Guinea Food and Nutrition 2000 Conference*. ACIAR Proceedings No. 99. Australian Centre for International Agricultural Research, Canberra. pp. 889–892.

Harwood, T. and Vlassak, V. (2004). *Papua New Guinea Agriculture Literature Database Guidelines*. Land Management Group, Department of Human Geography, Research School of Pacific and Asian Studies, The Australian National University, Canberra.

Hide, R. (2003). *Pig Husbandry in New Guinea: A Literature Review and Bibliography*. ACIAR Monograph No. 108. Australian Centre for International Agricultural Research, Canberra.

Twenty myths about Papua New Guinea agriculture

R. Michael Bourke and Bryant Allen

A myth is a popularly held belief that has no basis in fact. A great deal of misinformation exists about agriculture in PNG. Some of it is repeated often in official reports, newspaper articles, letters to newspapers and in public statements by prominent people and thus becomes a myth. One of the aims of this book is to present facts about PNG agriculture, with the supporting evidence, so that those who debate policy and make recommendations for change can base their arguments on established truths, instead of myth.

Myth No. 1 Food production is not keeping pace with population growth.

In almost all places in PNG, villagers have increased production as population has grown so that the food supply has been maintained. The rate of increase of locally grown foods has varied over time. For example, in the period 1998 to 2005, production of staple foods for subsistence and sale increased faster than population growth and imports of rice and wheat per person declined (Section 2.7). In a limited number of locations locally produced food has not kept pace with population growth and people are either short of food or sell other produce to buy food (see Box 5.3).

Myth No. 2 PNG is a food-deficit country.

This myth is based on an FAO definition that labels a country as 'food-deficit' if imports of grain exceed exports of grain. PNG does not export grain. However, an estimated 83% of food energy and 76% of protein consumed in PNG comes from locally grown foods (Figure 2.1.1). PNG exports other foods including palm oil, copra oil, coffee, cocoa and tea. The FAO definition is not questioned in PNG but is repeated inside PNG and at international conferences, possibly in order to try to attract aid funding.

Myth No. 3 Papua New Guineans live mainly on imported rice.

This myth is believed by many urban people for whom imported rice is their most important food. But it is not true for most people. Imported rice provides an estimated 9% of food energy in PNG (and wheat-based foods a further 5%) (Figure 2.1.2). Locally grown staples provide an estimated 68% of food energy, with sweet potato by far the most important of those crops (Figure 2.2.2). Papua New Guineans consume about 30 kg of imported rice per person per year. This can be compared with their annual consumption of over 500 kg of root crops, banana and sago.

Myth No. 4 Imports of rice are increasing rapidly.

This myth derives from the 1960s and 1970s, when rice and wheat imports were increasing every year and projections of the 1970s trend caused alarm. But per person consumption of rice slowed greatly in the 1980s and actually fell from 34 kg/person/year in the mid 1990s to 27 kg/person/year in 2001–2005. The consumption rate has increased slightly from 2006 as the economy has recovered and rural people have more money to spend on 'luxury' foods.

Myth No. 5 The Australian Administration did not promote rice production in PNG and Australians are attempting to stop local production to protect the Australian rice industry.

From the 1950s to the 1970s the Australian Administration invested significantly in attempts to produce rice, with very limited success (Section 2.5). Increased production of rice, based on the idea that rice imports should be replaced by locally grown rice, has been promoted by many individuals and organisations from the 1950s to the present. Despite considerable effort and financial investment, very little rice is grown in PNG, with annual production a minute fraction of the production of root crops, banana and sago (Table 2.2.1) and a tiny proportion of the amount of imported rice. Rice has not been widely adopted as a cash or subsistence crop in PNG for a number of reasons. The most important is that growing rice for sale results in poor returns to labour relative to other possible activities, such as growing coffee, cocoa or sweet potato (Table 5.20.1). As long as these conditions prevail, there is little potential for expanded rice production.

Myth No. 6 During the 1997–98 food shortages, Australia saved many Papua New Guineans from starving to death, with an emergency famine relief program.

In 1997–98 a countrywide drought and repeated frosts in the highlands completely disrupted some food production systems and severely reduced others (Section 1.6). In 1997, rice imports increased by 66 000 tonnes or about 40% more than the previous year. Most of the additional rice (75%) was sold through retail outlets. This rice was purchased by rural villagers and their urban-based relatives with cash earned from wages, from savings, or from selling export cash crops, fresh food and pork meat. The remainder was purchased by the PNG Government (8%), the Australian Government (4%), and other donors. Rice and other food was delivered by the Australian Defence Force to people in remote areas. If this had not been done, many of these people would probably have died. But the majority of rural Papua New Guineans and their urban-based *wantoks* saved themselves.

Myth No. 7 Imported meat, particularly lamb flaps from Australia and New Zealand, is increasing rapidly in volume.

Consumption of imported meat increased in PNG until 1994. From 1994 the quantity of imported meat eaten per person has fallen by almost two-thirds (Figure 2.9.1). When the kina fell in value relative to other currencies, imported foods became more expensive in PNG and people chose to reduce the amount of meat eaten. Consumption of imported mackerel fish has also decreased (Figure 2.10.1). This has been replaced to some degree by increased consumption of tuna, which is caught in PNG waters and canned onshore.

Myth No. 8 Lamb flaps are an unhealthy food.

Lamb flaps have a high fat content. Consumption of too much animal fat, including lamb flaps, is not good for people's health. However, many rural people eat very little animal fat (and vegetable oil) and consumption of a small amount of fatty food, such as lamb flaps, from time-to-time, is good for people's health. Furthermore, most rural people do hard physical work every day and so use the fat from lamb flaps as energy. Most rural diets are also very low in protein and lamb flaps provide protein. Most urban people, on the other hand, do little physical exercise. They would be healthier if they ate less fatty foods, including lamb flaps. So the idea that consumption of lamb flaps is unhealthy is basically true for most urban people, but is not true for most rural villagers who live an active lifestyle.

Myth No. 9 PNG agriculture has not changed for thousands of years. The practices and crops that are used today are traditional and unchanging.

Agriculture in PNG has a long and diverse history that is characterised by a high degree of innovation and openness to change on the part of Papua New Guineans. New crops have been adopted and old ones discarded or their importance has been reduced (Figures 3.1.2, 5.2.3). As well, many new techniques have been adopted or invented, such as composting, planting tree in fallows and crop rotations, particularly as people have intensified land use in response to population increase (Sections 3.6, 3.7). The introduction of sweet potato into the highlands about 300 years ago resulted in many changes in production techniques and in highland societies. The rate of change has increased in the past 140 years (see History of agriculture).

Myth No. 10 PNG has an abundance of high-quality land for agriculture and any tropical crop will grow well anywhere in PNG.

Only a quarter of the PNG landmass is used for agriculture (Table 1.2.1). The rest of the landmass is unsuitable for agricultural production because it is too steep, too high in altitude (too cold), rainfall is very high, or the land is flooded every year (Sections 1.5, 1.7, 1.9, 1.11). About 63% of the land used for agriculture in PNG is on mountains and hills (Section 1.10). Only 7% of the land area is classed as high or very high quality for agricultural production, with a further 20% of moderate quality (Section 1.12).

Myth No. 11 With the exception of oil palm, production of export cash crops is static (sometimes expressed as: production is the same now as it was in 1975 at Independence).

Production of most export cash crops, except copra, has increased over the past 30 years (Table 5.2.2). Production *by plantations* of coffee, cocoa, copra, rubber and tea has declined and this has reduced the overall growth rate. However, production of most export cash crops *by villagers* has increased (Table 5.2.1). From 2002 to 2007 the renewable resources sector (agriculture, forestry and fisheries) grew at 2.9% per year, which is a little faster than the population growth rate.

Myth No. 12 Women do most of the work in producing food in PNG.

Agricultural work is 'gendered' in PNG, that is, there are tasks which are mostly done by women and those which are mostly done by men. But both women and men contribute labour to food and cash crop production, usually as husband and wife. As well, some men do tasks that are considered mainly women's work and vice versa (Section 6.2). Many more women than men sell fresh food in markets, while more men than women sell export cash crops. For this reason men tend to earn more money from agriculture than women.

Myth No. 13 Villagers have a lot of spare time and it does not matter to them how much labour is needed to produce a certain crop.

One of the most important determinants of whether people will adopt a new crop or practice is the amount of food or money that they get in return for the amount of work they have to do to produce the crop. This applies to both cash crops and subsistence crops (Section 5.20).

Myth No. 14 Agricultural production is seriously constrained by customary land tenure arrangements.

Virtually all food crops, betel nut, vanilla and most coffee, cocoa, copra and rubber in PNG is grown on customary land. It is difficult but not impossible to access large areas of land for plantations. Since the mid 1990s, all increases in agricultural production for both smallholders and plantations have been on customary land. Internal migration is significant and many people can access land for agriculture where they settle (see Section 1.4). People are moving from poor agricultural environments to better ones, where the population density is greater (Figure 1.12.4). Many settlement schemes, where settlers have registered title to their land, have been economically unsuccessful (Section 6.7). Similarly, plantation production of all export cash crops, except oil palm, has declined for the past 30 years (Table 5.2.1). There are a number of reasons for this, but having a registered land title did not solve the problems of the plantation sector.

Myth No. 15 There are few roads in PNG and this reduces agricultural production.

More than half of the total population live within 5 km of a national road and a further 10% live within 10 km of a national road. Considerably more also live near a provincial or district road (Section 6.9). When roads were built in PNG, they usually went through the most densely populated places. This myth is fuelled by the fact that Port Moresby is not connected by road to any province, other than Central and Gulf. Unfortunately it is now true that many rural roads and bridges have not been properly maintained and many are impassable in wet weather. This is a significant limitation on agricultural production.

Myth No. 16 There is little information about PNG agriculture with which to develop sound policy, or for planning.

A large amount of information exists on the environment and agriculture in PNG, significantly more than for most other developing countries. The purpose of this book is to bring much of that information together in one place and to indicate where more can be found. Much more remains to be learned, and there are gaps in the knowledge of important topics, such as how many hours of labour are needed to grow particular crops, or which crops are best suited for PNG's different environments. But much is already known about PNG agriculture. (Also see Sources of information about PNG agriculture, page xxv.)

Myth No. 17 There is significant potential to export fresh food to New Zealand, Australia and South-East Asia.

There is very little potential to export fresh food from PNG because of quarantine issues (including a serious fruit fly problem), poor presentation of food, expensive and unreliable air and sea transport and lack of price competitiveness (Section 5.3). Limited possibilities for certain niche markets exist, but many obstacles remain. In contrast, there is significant unrealised potential for expanded sale of fresh food within PNG. Certain indigenous edible nuts, such as *galip*, *karuka* and *okari*, have considerable potential as export crops.

Myth No. 18 Global climate change is now causing significant problems for many people on very small islands.

Sea level rise, temperature increases, higher rainfall and possibly a greater incidence of extreme climatic events have the potential to cause significant problems for people in many locations in PNG, including those living on very small islands (Section 1.8). Some crops are now bearing at higher altitudes in the highlands because of higher temperatures. But overall the impact of climate change has been relatively small so far. Some problems attributed to rising sea levels are caused by overpopulation (Carteret Islands in Bougainville Province) or sinking land associated with geological activity (Duke of York Islands in East New Britain Province).

Myth No. 19 There is no poverty in rural PNG because there is plenty of food to eat.

There is significant poverty in rural PNG where one million people live in severe poverty (Section 6.10). In PNG poverty is heavily influenced by where people live. 'Poor places' are overwhelmingly rural (94% of poor people live in rural areas). In these locations, carbohydrate food is generally sufficient, but protein, fats and oils are not. Cash incomes are very low so people cannot buy foods that could increase protein in their diets. Health and education services are poor. As a result, life expectancy is short; many infants die before they are one year old; and all measures of health, education and life outcomes are among the worst in the Asia–Pacific region.

Myth No. 20 Poor governance of agricultural institutions does not matter because rural people grow their own food and look after themselves.

A number of the bodies that govern the production, purchasing, processing and marketing of PNG's cash crops have seriously disadvantaged producers by interference in their management by politicians, the appointment of people who do not have the skills to be effective managers and by some very bad policy making (Section 6.4). The best outcomes for PNG producers have occurred when government has stayed out of marketing and exporting, and has regulated in favour of village producers. An example of poor governance affecting marketing is the collapse of chilli exports in 1982, despite a large increase in prices. This was caused by marketing problems with provincial government buying systems (Figure 5.15.1). An example of how good governance has a positive influence on exports is the increase in exports of tuna and increased revenue to PNG in the late 1990s, following the establishment of the National Fisheries Authority and that body's proper regulation of PNG fisheries (Figure 5.9.2).

Introduction

R. Michael Bourke and Bryant Allen

Agriculture is the most important activity carried out by the vast majority of Papua New Guineans. For most people, agriculture fills their lives, physically, culturally, economically, socially and nutritionally. Yet agriculture is the most undervalued and misunderstood part of PNG life (see Twenty myths about PNG agriculture, page 1). The reasons for this are partly because mineral and oil exports make PNG comparatively wealthy for a developing country; partly because agriculture is practised in the countryside, away from towns, and is therefore largely 'invisible' to urban people and international visitors; and partly because agriculture is viewed as not being 'modern'.

However, as this book shows, agriculture feeds most Papua New Guineans, houses them, provides a significant amount of food to townspeople and earns a significant amount of foreign exchange. By doing these things, agriculture provides PNG with a social, economic and political stability which has meant that, despite sometimes poor economic management, the country has been able to weather a number of economic and political crises. Furthermore, agriculture will continue in this role, long after all the minerals and oil have been dug up and exported.

The primary purpose of this book is to demonstrate how important agriculture is to PNG and to provide the knowledge that will enable policy makers to make sensible plans for it, within the PNG economy. For a developing country, PNG is rich in information for planning, but most of it is not easily accessible to those who need it. This book was initially conceived to bring this 'lost' information on PNG agriculture together in a single publication where it would be easy to access. It began as a compendium of statistics but has evolved into a comprehensive book of 72 sections on PNG agriculture and related topics. The number of sections alone demonstrates how agriculture pervades almost every aspect of life in PNG.

Sections include information about population; land use; climate and other aspects of the physical environment in which agriculture occurs; overviews of subsistence food production, with detailed descriptions of production techniques and the most important food crops; descriptions of cash crop production for both domestic and export markets; agriculture in the broader economy; and a series of papers on agricultural development, policies and governance. An overview of 50 000 years of agricultural history in PNG follows this introduction.

An outstanding aspect of this book is the emphasis it places on village agriculture, the most important part of PNG agriculture. It provides details that have not been previously available on subsistence food production systems; on the informal sector that provides most of the marketed fresh food, betel nut, firewood, fish and other products; and on all of the major and minor cash crops, with new analyses and datasets, many of which contain data runs over 40 years long.

The book has a dual focus: food production and cash income. Almost every section relates to how people manage the environment, the production and the marketing of the commodities they produce. The book is dedicated to the village women, men and children who are the skilled agriculturalists of PNG and who produce all of the fresh food in the country as well as most of the export cash crops and the domestic animals.

Agriculture in Papua New Guinea

Subsistence food production is the most important part of PNG agriculture. It provides most of the food consumed in the country – an estimated 83% of food energy and 76% of protein (Figure 2.1.1). Its importance only becomes apparent when a rare partial failure of food production occurs, as happened in the very severe drought and frosts of 1997 (Section 1.6). People's sustenance also comes from hundreds of food crops, husbandry of domestic animals, fish and locally produced sugar (Sections 2.6, 5.9, 5.10). This is supplemented by imported rice, wheat-based foods, some meat and some fish (Section 2.1).

Cash income is provided by sales of Arabica coffee, fresh food, cocoa, betel nut, copra, oil palm, firewood, tobacco, fish and many minor products including vanilla, rubber, balsa and tea (Section 5.1).

There is a long history of agriculture in what is now PNG, with the first settlers arriving about 50 000 years ago. Agriculture began in PNG at about the same time as it appeared in the Middle East and in China, about 10 000 years ago (see page 10). Some food plants were domesticated in New Guinea and others were introduced from Asia and other Pacific islands. In the 1700s, new crops began to reach PNG from the Americas and one, sweet potato, has become PNG's most important crop. Many new food and potential cash crops were introduced during the colonial period, from the early 1870s onwards. A feature of the long history of agriculture in PNG is the careful adoption of some new crops and the rejection of others. It is also characterised by innovations in the management of land including draining, fencing, tree planting, composting, soil erosion control and mounding. Many of these changes have been associated with the intensification of land use, in which more food is produced from the same area of land (Sections 3.6 to 3.12).

Influences of global trends in 2007–2008

Evidence is building that fundamental worldwide changes are occurring that go beyond global climate change, and which will pose multiple challenges for agriculture everywhere. Land once used to grow food is growing biofuels; the demand for many products has increased as the economies of China and other rapidly developing nations grow; agricultural land is being lost to urbanisation and industrialisation; fertilisers and herbicides are becoming more expensive; surface water is being reduced in many locations; soil nutrients are being lost; pest and disease problems are increasing; a neglect of agricultural research is reducing technological solutions; marine harvests are declining; and prolonged droughts exist in a number of countries, possibly associated with global climate change. As a result, the prices of rice, wheat, corn (maize), vegetable oil, dairy products, meat and fish have all increased markedly in 2007–2008.

These global changes pose challenges for PNG agriculture producers and consumers, including much higher prices for imported foods. At present the PNG economy is benefiting from these changes because prices of agricultural commodities that are linked to crude oil have risen sharply in recent years, including palm oil, copra oil and natural rubber. The use of fertilisers and herbicides in PNG agriculture continues to be negligible, with most villagers using little or no inputs into production other than their land and their labour (Section 5.19).

The increase in exports of marine products, mainly tuna (Figure 5.9.2), round logs (Figure 5.8.3) and palm oil from PNG (Figure 5.7.4) is the outcome of higher global demand. The changing global economy is reflected in the greater proportion of logs and balsa from PNG exported to China and other emerging economies in Asia (Figures 5.8.4, 5.13.2).

As this book goes to press in late 2008, global financial markets are in turmoil. This is likely to have far-reaching consequences for the global economy, but much remains unknown. Papua New Guinea rural producers will be affected by these changes, but will be largely sheltered by subsistence food production and their limited dependence on external inputs.

Strengths of PNG agriculture

PNG agriculture has many strengths. Firstly, by providing most of the food consumed within PNG, it imparts great stability to the nation. Agriculture is also the main source of cash income for a great majority of people. People use money to better their lives, including supplementing protein-poor diets, building better houses, educating their children and accessing medical services. Most rural people are employed as producers, processors or intermediate traders in agriculture. As well, there are vigorous informal businesses where people grow, transport, trade and retail fresh food, betel nut, firewood, fish and other commodities (Sections 5.1, 5.17). Agriculture also provides 17% of PNG's exports by value (Figure 5.2.1).

Agriculture is a major contributor to development within PNG. A vibrant cash crop component exists in both the formal and informal sectors, in almost all of the more-developed locations. In contrast, the least-developed parts of PNG, where most people are poor, are characterised by subsistence food production only.

A number of environmental factors severely constrain agriculture in PNG, including excessively high rainfall, steep slopes, inundated land and extensive cloud cover. Villagers have a detailed understanding of the relationship between crops and the environment (Section 1.13). More important than a good physical environment, however, is the energy, drive and adaptability of the people in more than one million rural households. They respond to new opportunities, including higher prices and better returns to their labour (Section 5.20). They form the core of what is a largely unrecognised significant private sector group.

Other parts of the private sector consist of people involved in trading and processing commodities for the domestic and export markets. Thousands of people are involved in trading and transporting betel nut alone, as it is moved from lowland locations to the highlands, where it does not grow, and to urban centres and mines. A few small companies participate in the formal economy, trading and processing produce. A number of large firms grow, buy, process and export oil palm, sugar cane, coffee, cocoa, copra, rubber and tea.

Villagers are often portrayed as unresponsive to price or market messages. The devaluation of the PNG currency in the late 1990s (Figure 4.1.1) is a clear demonstration that this is false. Large increases in the prices for many goods as measured by the consumer price index occurred in the late 1990s (Figure 4.2.1), including imported rice (Figure 4.3.3) which, in turn, led to an increase in the price of marketed sweet potato (Figure 4.3.1). This was also a time of low returns for coffee in the highlands (Figure 5.4.3). Many highlanders responded to the increased price for sweet potato and the reduced price of coffee by growing more sweet potato for sale in urban markets. The increase in production resulted in lower prices for consumers, so that sweet potato became more competitive compared with imported rice, leading to increased consumption of sweet potato. A similar response is occurring in 2008 following rapid increases in the price of rice and other imported foods.

Another strength of PNG agriculture is the customary land tenure system (Section 6.1). Individuals and companies who wish to access large areas of land for agricultural development can be frustrated by customary tenure, but the system is sufficiently flexible to accommodate increasing population and internal migration (Sections 1.1, 1.3, 1.4). It has been argued that economic development will not occur unless all land is privatised and registered to individuals, but individual titles to land on settlement schemes has often resulted in poor economic outcomes (Section 6.7). Nor has the holding of a title helped the plantation sector, where production of all export cash crops except oil palm has declined over the past 30 years (Table 5.2.1; Figures 5.4.4, 5.5.2, 5.7.5, 5.11.3). PNG does not have the administrative capacity to survey customary land, identify customary owners, settle the myriad disputes that would result, and issue titles, let alone record the thousands of changes of ownership that will occur. Only where uncertainties in tenure are causing real problems, such as in the peri-urban areas of the larger towns, should attempts be made to interfere with customary tenure.

Marketing systems for oil palm, coffee, cocoa and copra are efficient and return to growers a reasonable share of the international market price (Section 5.21). Marketing systems are less efficient for vanilla and most domestically marketed produce, particularly fresh food (Sections 5.3, 5.14).

Challenges and constraints

This book identifies a number of significant challenges and constraints which presently limit the potential of agriculture in PNG. If the sector is to achieve its full potential, these will need to be addressed by the PNG Government and private sector, supported by international donors. The major constraints are:

- Pressure on land associated with rapid population growth, particularly in parts of the highlands and on small islands (Section 1.3). This is affecting soil fertility and the ability to grow food.
- Poorly maintained transport infrastructure, particularly roads and bridges (Section 6.9).
- Limited new technology being generated from research. Some commodities, particularly palm oil and sugar cane, are supported by excellent research, but most commodities are not adequately supported, including subsistence and marketed fresh food. Research is frequently focused on a crop, rather than the agricultural system of which it is a part.
- Research capacity directed at commodities that have little or no chance of expanding or being adopted, including rice, wheat and pulses.
- Very limited effective outreach and agriculture extension capacity. Despite the availability of a moderately large body of information which would advantage producers, little of this is effectively communicated to them.
- Climate change as a significant long-term challenge, particularly rising sea levels, greater rainfall, increasing temperatures and possibly greater frequency of extreme climatic events (Section 1.8).
- Mismanagement of the national economy, poor performance of institutions involved in the governance and administration of agriculture, and poor policy making. This includes policies that result in distortions of the terms of trade and large movements in the currency exchange rate (Sections 4.1, 6.3, 6.4).
- Insufficient involvement of women in some aspects of agriculture, particularly trading (Section 6.2).
- An HIV/AIDS epidemic, which will severely affect agricultural labour supply in some places (Section 1.1).
- Inadequate security for people and property.
- Insufficient access to credit for intermediate traders.
- Inadequate attention to quality control for some commodities, particularly coffee, fresh food and vanilla.
- Poor communication among growers, middlemen, processors and retailers in the production and marketing chain.
- Insufficient attention given to marketing and promotion of PNG produce.
- Severe poverty for about a sixth of the population, which limits their ability to participate in the formal economy (Section 6.10).

Future focus for research and development

The greatest returns to investment in PNG agriculture will come from research and development in the following components of PNG agriculture:

- Subsistence food production, particularly sweet potato; agricultural system management; soil fertility maintenance; and pest and disease management (Part 3).
- Domestic marketing of food, particularly root crops and certain fruit, including mandarin, mango, mangosteen and rambutan (Section 3.3).

- Major export tree crops of coffee, cocoa, copra and oil palm (Sections 5.4, 5.5, 5.6, 5.7).
- Minor export crops of rubber, balsa and certain spices (Sections 5.11, 5.13, 5.14, 5.15).
- Fuelwood and hardwood timber species, grown in plantations or in village plots (Section 5.8).
- Edible nuts for domestic and export markets, notably *galip*, *karuka*, *okari*, *pao* and sea almond (Section 3.4).
- Domesticated animals, particularly pigs, chickens, cattle and goats (Section 2.6).

Currently there is investment or interest in a number of new uses for cash crops. For export these include biofuel from cassava, jatropha, castor, corn and coconut; bamboo for edible shoots and timber; sago starch; sandalwood; kava; cashew nuts; and organically certified essential oils from spice plants. Village plots of timber for fuelwood, construction, pulp and hardwood; peanuts; *galip* nuts; noni; and prawn farming are being investigated for both the domestic market and for export. Cut flowers could be grown for the domestic market.

Many of these have good potential. Success will depend on a number of factors, particularly markets, prices and returns to labour or capital investment as well as the removal of at least some of the constraints listed above.

There are proposals for the development of a number of other agricultural commodities, including fresh food for export; processing food for the domestic market; and local production of rice, wheat and pulses to replace imports. We believe that poor returns and other limiting factors mean that these proposals have little or no realistic chance of success. This assessment is based not only on returns to labour and capital investment, but also on a long history of past failures.

Agriculture in PNG has a long and dynamic history in which the outstanding feature has been the willingness of villagers to adopt new crops and to change technologies. Changes in staple foods (Figure 3.1.2) and export cash crops (Figure 5.2.3) over time demonstrate this willingness. Many exciting new possibilities exist for further development of food and cash crops in PNG, but it is important to learn from the past and trust the judgements of village producers. Their willingness to change and develop in the past must be placed alongside their refusal to adopt crops and ideas for which they can see no future. Much of the information in this book was gathered from villagers in every district in PNG. Their knowledge and their needs must be acknowledged if PNG agriculture is to continue to grow.

History of agriculture in Papua New Guinea

R. Michael Bourke

Introduction

The history of agriculture in PNG is about 10 000 years old. This history is reviewed here in the context of 50 000 years of human occupation of the Australia – New Guinea region.[1] More is known about what has happened nearer to the present, especially since 1870, than about the distant past. Much of the early history (prehistory) of PNG was unknown until about 50 years ago, but since 1959 there has been a lot of research on the prehistory of PNG, with a major focus on agriculture. However, this is a rapidly evolving field of study and our understanding of the history of agriculture in PNG is still incomplete. The information that is summarised here will be expanded and modified by future research.

Historical evidence is reviewed in a number of periods: the arrival of humans in New Guinea some 50 000 years ago; the beginnings of agriculture about 10 000 years ago; the appearance of Austronesian-speaking people from island South-East Asia about 3500 years ago, bringing with them more domesticated crops and animals;[2] the introduction of sweet potato about 300 years ago; permanent settlement by Europeans and other outsiders, with many introductions of plants and animals after 1870; and the period of rapid social and economic change that commenced about 70 years ago in 1940.

The peopling of New Guinea

When the first humans came to New Guinea about 50 000 years ago the climate was very different from now. Worldwide, temperatures were lower, the polar ice caps were larger, glaciers were more common, and sea levels were lower. As a result, the South-East Asia mainland extended as far east as Bali and Borneo to form a landmass that is known as Sunda. The Asian mainland (Sunda) and New Guinea were always separated by ocean but, at that time, New Guinea was not an island, but formed the northern part of a large continent that also included Australia and Tasmania, known as Sahul (Figure 1). The Bismarck Archipelago and the Solomon Islands chain have always been separated from the Sahul continent by ocean.

The world climate started to warm from about 18 000 years ago. The sea level began to rise from the melting of ice caps and glaciers and the tree line became higher. The extensive low-lying plains between New Guinea and Australia were flooded. By 10 000 years ago, only a narrow strip of land linked southern New Guinea with the Australian mainland. Around 8500 years ago this land bridge was broken

---

1 Prehistorians do not agree how long humans have occupied the Sahul continent (Australia, New Guinea and Tasmania). The figure of 50 000 years used here is a compromise between the shorter time period of about 45 000 years argued by some scholars and the longer one of 50 000–60 000 years argued by others.

2 See box on page 11 for a definition of domestication.

when Torres Strait became flooded and the northern part of the great Sahul continent became the island of New Guinea, with a coastline similar to the present.

The first people to settle the Sahul continent are likely to have come in small groups. They would have made scattered landings on the coastline following earlier movements from the Asian mainland via the eastern islands of the Indonesian archipelago. Following the initial colonisation, human settlement spread to different parts of what is now PNG.

People probably reached the islands of New Britain and New Ireland by 40 000 years ago, soon after the initial colonisation of the Sahul mainland. By 28 000 years ago there were people on what is now Buka Island, at that time the northern end of a single island that included most of the Solomon Islands. The trip from New Ireland to Buka required some time at sea without view of the target land. Manus was settled by at least 20 000 years ago. Colonisation of Manus involved an open sea crossing of more than 200 km, of which 75 km would have been out of sight of land. Human settlement in the Pacific islands extended as far as the end of the Solomon Islands until about 3500 years ago.

Definitions of terms

Archaeology. The scientific study of a prehistoric culture by excavation and description of its remains.

Bismarck Archipelago. The islands of Manus, New Ireland and New Britain and smaller nearby islands, north-east of mainland PNG.

Domestication. The process whereby people transform a wild plant or animal population into one with more desirable characteristics, usually with an edible product such as a grain, tuber, fruit or nut (in the case of plants). This is done by selection and propagation of plants or animals with the desired characteristic.

Glacier. A river of ice. Small glaciers still exist at high altitudes in west New Guinea (Indonesian Papua), although these are disappearing as the climate warms (see Section 1.8). There were a number of glaciers in the mountains of east New Guinea up to about 18 000 years ago.

New Guinea. The second largest island in the world (after Greenland), lying just south of the equator. It is split into two national units. The eastern half is part of Papua New Guinea, an independent nation, while the western half (west of 141° E longitude) is the Indonesian province of Papua, formerly known as Irian Jaya. The term New Guinea is used here to refer to the island, not to a political unit.

Papua. This is a confusing term as it has a number of meanings. Papuan languages are a group of related languages spoken mainly on the island of New Guinea, but also by some groups in New Britain, New Ireland, the Solomon chain and the Timor and Halmahera areas of east Indonesia. Papua is the current name of the Indonesian province that occupies the western half of the island of New Guinea. It is also the name of the former Australian colony, now known as the Southern Region of PNG; and thus it has been incorporated into the name for the nation of Papua New Guinea.

Prehistory. The history of humans in the period before events were recorded in documents, known mainly through archaeological research.

Solomon Islands. A chain of islands lying south-east of New Britain, extending from Buka to San Cristobal. The two larger north-west islands (Buka and Bougainville) lie in PNG; the others in the political state of Solomon Islands. The term is used here in a geographic rather than political sense.

Tree line. The distance above sea level, or altitude, above which trees do not grow because the temperature is too low. In PNG at present the tree line is around 3800 m above sea level.

The earliest indications of human activity in the mountains of New Guinea are thought to be 35 000 years old and are evidence of disturbance of the vegetation by burning. This may have been caused by hunting and exploitation of seasonal foods, especially pandanus nuts, rather than long-term occupation. From about 18 000 years ago, as the climate became warmer, the vegetation in the highlands changed and there was greater use of the highland valleys by people.

Around 3500 years ago a group of people came to the New Guinea area. They were pottery-making agriculturalists and possessed what archaeologists call the 'Lapita culture', named after a style of pottery that they made. Over the next 500 years 'Lapita' people moved beyond the limit of previous settlement at the end of the Solomon Islands and reached New Caledonia, Fiji, Tonga and Samoa. They spoke languages known as Austronesian, which may have originated in Taiwan. These languages are now found

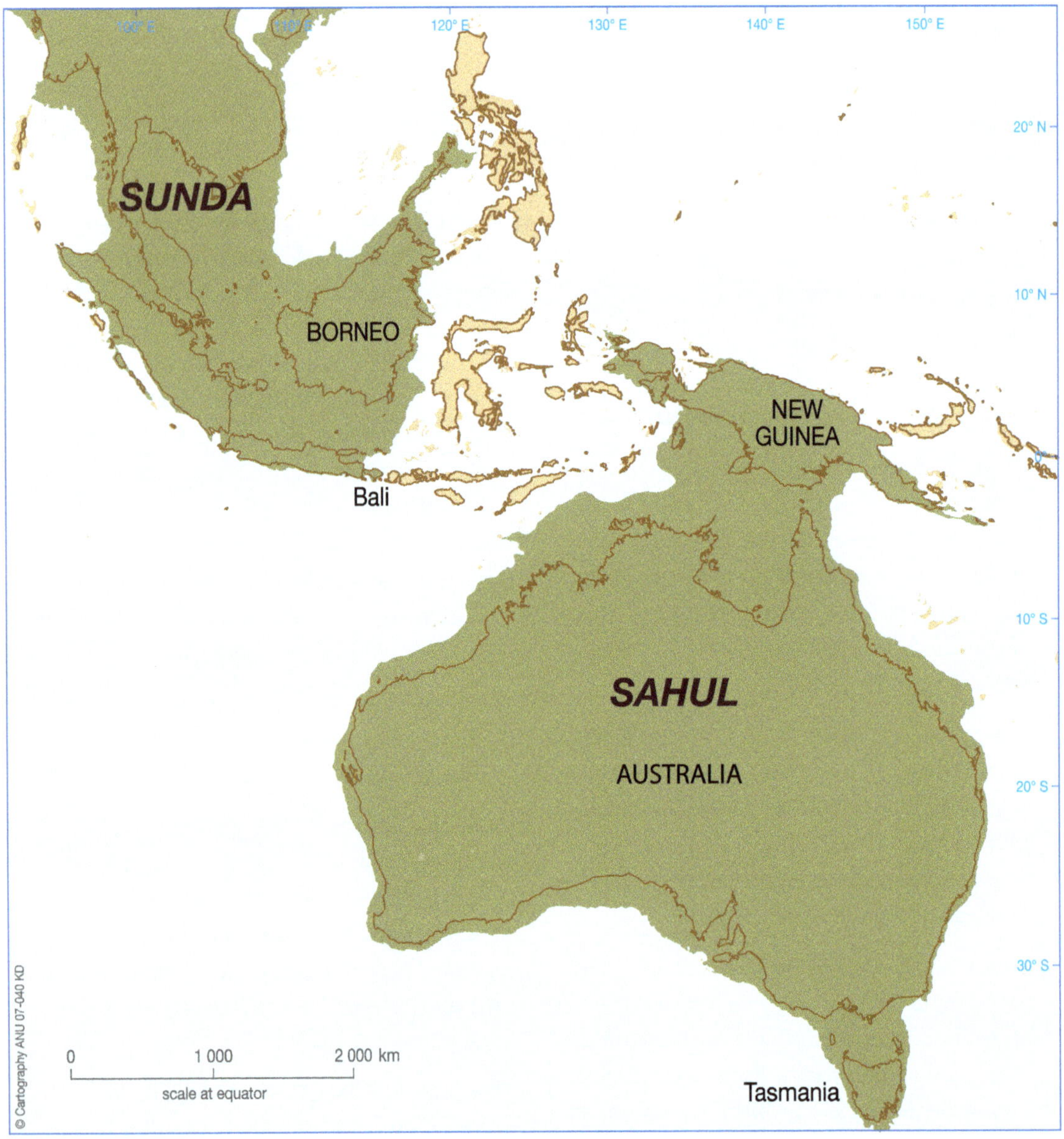

Figure 1 The Sunda and Sahul landmasses at about 50 000 years ago when people first came to Sahul.
Note: Some modern islands were connected to the two large landmasses and others, such as New Britain, New Ireland and the Solomon chain, were always separate. Source: Cartographic Services, ANU.

over a very large area in the Pacific, Indonesia, parts of mainland South-East Asia, and Madagascar in the western Indian Ocean. In PNG they are much better represented in the Bismarck Archipelago and other islands than on the New Guinea mainland, where they are scattered around the coast, particularly in the north and east. Austronesian speakers later spread throughout the Pacific as far as Easter Island, New Zealand and Hawaii. Some of their Polynesian descendents came back to settle on small islands in the New Guinea region within the last 1000 years.

The history of the New Guinea region is made up of many movements of people and Papua New Guineans are heirs to a long and varied genetic, linguistic and cultural history. Most details of the early settlement history are unknown and may never be known. The early colonisation of Sahul would have been made up of small independent movements from different starting points to different places on the coast. Similar movements would have continued after initial colonisation, as illustrated by the transport of plants, animals and raw materials described below.

Subsistence for the first settlers (50 000 to 10 000 years ago)

The first settlers in New Guinea and nearby islands obtained their food from hunting, fishing and gathering. The animals hunted included giant marsupials, now extinct and possibly hunted out of existence by the migrants. The people would have exploited local plants for food, including sago. It is probable that sago was domesticated by the selection of plants with a high content of starch in the trunk. Human populations were probably very small.

Stone tools, dated to some 40 000 years ago, have been found on terraces on the north side of the Huon Peninsula of Morobe Province. These tools were possibly used to thin, trim and ringbark trees to assist the growth of desirable plants that provided food or to obtain starchy food from sago or cycad trees.

It is likely that very early people started to use trees that had, for example, larger edible nuts, and to cut down trees that had smaller nuts. If this was done over a long period, the best-yielding trees will now dominate the forests where people are living. *Galip* nut (*Canarium* species) provides evidence for this practice. Seed remains of *galip* nut have been dated as early as 17 000 years ago in archaeological excavations in the middle Sepik area. They have been dated at 15 500 years ago on Manus, at 11 500 years ago on Buka and 9000 years ago on New Ireland. It seems that *galip* was domesticated by people on the north coast of New Guinea and then introduced to the Bismarck Archipelago and Solomon Islands.

Edible nuts of one pandanus species (*Pandanus antaresensis*) have probably been used for about 30 000 years, and the high-altitude pandanus nut (*P. brosimos*) was possibly first used about 10 000 years ago. The form of nut pandanus that is common at 1800–2600 m altitude in the highlands now (*P. julianettii*) appears to have been domesticated from *P. brosimos*, possibly about 2000 years ago (see Section 3.4). It is likely that people were domesticating other plants at this time, even before they used agriculture as we now know it. Such plants could have included *marita* pandanus and other nut- and fruit-bearing plants. People may also have been exploiting wild taro plants a long time before the beginning of agriculture: taro starch has been found on stone tools from Buka Island that were used as long as 28 000 years ago.

We also know that people were trading obsidian (a black, glass-like stone formed in volcanoes that was used to make sharp cutting tools) a long time ago. Obsidian from Talasea on the north coast of New Britain first appeared in New Ireland about 23 000 years ago and has been found as far away as Borneo. We also know that people were moving wild animals such as the cuscus and bandicoot from the New Guinea mainland to the islands, with the first movement as early as 23 000 years ago. Presumably, wild animals were transported so they could be hunted for food. The cassowary is the only indigenous animal that has undergone some degree of domestication. People hunt it and rear captured chicks. It is not known for how long people have done this.

The beginning of agriculture (10 000 years ago)

By 10 000 years ago the climate had warmed to modern temperature levels. It seems that people started practising agriculture, at least in the New Guinea area, from about 10 000 years ago. Certainly, from 7000 years ago, the evidence for agriculture is very clear. It is also likely that agriculture was invented in the New Guinea highlands at about the same time as it appeared in other parts of the world,[3] and that the development of agriculture in New Guinea was independent of what happened elsewhere. The evidence comes from a site called Kuk in the upper Wahgi Valley in Western Highlands Province. Extensive research at Kuk over a 30-year period suggests that:

- Plants were being exploited and some cultivation was occurring about 10 000 years ago. Archaeological research has found features that indicate planting, digging and staking of plants, and possibly localised swamp drainage. Taro starch found on stone tools excavated at Kuk that are about 10 000 years old suggests that taro was being planted at Kuk at this time.
- A network of small island beds and associated basins had been constructed by 7000 years ago, so that water-tolerant plants could be cultivated in the basins and those requiring drier conditions could be planted on the island beds.
- Banana was intensively cultivated from 7000 years ago.
- From 4500 to 5000 years ago, swamp gardens were drained by straight line ditches dug at right angles to each other that drained into large channels.

3 Apart from PNG, other centres of early agriculture were the Fertile Crescent of the Middle East and the Yangtze and Yellow river basins of central China, with dates of 11 000 and 9 000 years ago respectively.

Arrival of the Austronesians (3500 years ago)

The arrival of the Austronesians is associated with the appearance in the Bismarck Archipelago of the distinctive Lapita pottery and the first domesticated animals – the pig, chicken and dog. The newcomers were agriculturalists and brought many of their crops with them. Some of these were of the same species of plant that people had domesticated in the New Guinea region. Indeed, some of these crops may originally have been domesticated in New Guinea and carried back into South-East Asia.

European exploration and transfer of crops (late 1400s to 1870)

In the late 1400s, explorers, missionaries and traders from Spain and Portugal, and later the Netherlands, France and England, moved and settled around the globe. They took plants and domestic animals from one region and introduced them to other regions. Many of these plants became important economically in the new places. The production of important food and cash crops is now often greater in locations distant from where the crops were initially domesticated. For example, the most important palm oil-producing area is now South-East Asia (Malaysia, Indonesia and Thailand), but oil palm was domesticated in West Africa. Similarly, wheat was domesticated in the Middle East, but the main wheat-exporting nations are now the United States, Canada, Australia, France and Argentina. Sweet potato was domesticated by people in the American tropics, but today the major sweet potato-producing country is China. Sweet potato is now the most important staple food for people in the western Pacific (Solomon Islands, PNG and Indonesian Papua) and production has expanded greatly in that region in the last 60 years.

This major transfer of plant materials around the world by European explorers and colonists, which has had such a large impact on global agricultural production, occurred in the western Pacific later than elsewhere. Europeans and other travellers made sporadic contact with Papua New Guineans from the early to mid 1500s, but there is no evidence that the early European explorers made any plant introductions into PNG at that time.

A small number of species from the Americas were introduced by Europeans into Indonesia and spread from there to PNG before 1870 (Table 1). The most important of these was sweet potato (see page 17). Another crop of American origin that became important in PNG is tobacco. Tobacco was introduced by Europeans to the Moluccas in eastern Indonesia before 1600, from where it spread to New Guinea. It is likely to have come into PNG at a number of locations. One of these is the Trans Fly area in the south of Western Province, where Moluccan traders probably introduced tobacco when they came to this area seeking dammar between 1645 and 1790.[4] The first written record for tobacco in New Guinea is by the Dutch explorer Schouten in 1616, who saw it on Arimoa Island in north-western Papua (Indonesia). Tobacco diffused through New Guinea over several centuries, but it had not reached south-east New Guinea (Oro, Central and Milne Bay provinces of PNG) by the time of first sustained contact with foreigners from the 1870s.

Lima bean was also probably transported from east Indonesia to New Guinea some time between 1700 and 1870. It had been introduced to east Indonesia prior to 1650. Many villagers in the highlands believe that lima bean was used by distant ancestors, but others say it is an introduction (Table 1).

Cassava is another food crop that was probably introduced into parts of mainland PNG from west New Guinea some time after 1800. People in the western part of PNG, in particular Western and Sandaun provinces, consider it to be a 'traditional' crop. It seems that cassava was introduced directly to some islands by European sailors around the same time.

Bixa, a plant used as a bright red dye and body paint, also reached PNG before 1870. Some villagers consider bixa to be a traditional plant, but others do not. It is likely that bixa seed was also spread from Indonesia, where it had been introduced by Europeans from the Americas (Table 1).

Table 1 Proportion of villagers who consider five crops of American origin to be post-European introductions[a]

Crop	Number of locations surveyed	Post-European introduction (%)	Pre-European introduction (%)
Bixa	32	31	69
Cassava	65	63	37
Lima bean	25	20	80
Sweet potato	52	54	46
Tobacco	52	17	83

[a] These crops were introduced to PNG between 1600 and 1870.

Sources: Extracted from published reports and author surveys in various locations in the PNG lowlands and highlands.

Plants used for agriculture until 300 years ago

Prior to permanent settlement by foreigners in the 1870s, more than 170 plant species were used by Papua New Guineans for food. As well, hundreds of other species provided materials for shade, firewood, medicine, tools, weapons, house and fence construction, decoration, rope, string, food wrappings, bark cloth, dress, personal adornment, canoe and raft construction, and ritual and magic purposes. The most important staple (carbohydrate) foods were taro, banana, sago and yam (Figure 3.1.2).

Many plant species that today provide carbohydrate food, vegetables, fruit and nuts were either domesticated in the New Guinea area or were introduced into PNG thousands of years ago (Table 2). Other species were domesticated in Asia or elsewhere in the Pacific and then introduced into PNG somewhere between several hundred to several thousand years ago (Table 3). Many other minor foods in PNG are likely to have been important in the past, but have been displaced by more recent introductions.

4 Dammar is a resin that comes from a number of trees, including *Vatica papuana*, which grows in south-west PNG (Swadling et al. 1996:157–65). It was used for lighting, as well as for coating and sealing pottery.

Most of the species listed in Table 3 were domesticated in Asia, but two came from the Pacific. The first is _pao_ nut (_Barringtonia procera_), which was probably domesticated in the Solomon Islands and introduced into New Ireland and the Admiralty group relatively recently, perhaps less than 1000 years ago. It was taken by migrants from southern New Ireland to the Gazelle Peninsula of New Britain about 400 years ago. It has spread to mainland New Guinea and elsewhere in New Britain over the past 50 years.[5] Kava was probably also domesticated in the Pacific, in Vanuatu, and introduced before 1870 into a limited number of locations in PNG including the Madang area, some islands off Manus Island, and parts of Western Province (see Section 3.5).

[5] Related species (_B. novae-hiberniae_ and _B. edulis_) with edible nuts are found on New Guinea as well as the Bismarck Archipelago, Solomon Islands and Vanuatu.

Table 2 Crops domesticated in the New Guinea area, or very ancient introductions

Staple (carbohydrate) foods	*Oenanthe	**Nuts**
*Banana	*_Pitpit_, highland	*Breadfruit
Coconut	*_Pitpit_, lowland	Candle nut
*Cordyline	*Rorippa	Castanopsis
Kudzu (Pueraria)	*Rungia	_Dausia_
Polynesian arrowroot	_Trichosanthes pulleana_	*_Elaeocarpus womersleyi_
*Sago	_Tulip_	*Finschia
*Sugar cane	Wandering Jew	*_Galip_ (_Canarium decumanum_)
*Taro (_Colocasia_)	**Fruit**	*_Galip_ (_Canarium indicum_)
Taro (_Alocasia_)	*_Bukabuk_	*_Galip_ (_Canarium lamii_)
Taro, swamp	Coastal pandanus	*_Karuka_, planted (_Pandanus julianettii_)
*Yam, greater	Golden apple	*_Karuka_, wild (_Pandanus antaresensis_)
Yam, aerial	Mango (_Mangifera minor_)	*_Karuka_, wild (_Pandanus brosimos_)
Yam (_Dioscorea nummularia_)	*_Marita_ pandanus	*_Okari_ (_Terminalia impediens_)
Yam (_Dioscorea pentaphylla_)	_Mon_	*_Okari_ (_Terminalia kaernbachii_)
Vegetables	*_Parartocarpus venenosa_	*_Omphalea gageana_
*_Dicliptera papuana_	*_Pouteria maclayana_	Polynesian chestnut (_aila_)
*_Ficus wassa_	Raspberry, red (_Rubus moluccanus_)	Sea almond (_talis_)
*Highland _kapiak_	Raspberry, red (_Rubus rosifolius_)	_Sis (solomon)_
Job's tears	*_Ton_	**Stimulants**
*_Kumu musong_		Betel nut, highland
		Betel pepper, highland

Note: Species with an asterisk (*) are likely to have been domesticated by people in the New Guinea area. The other species in this table may have been domesticated in the New Guinea area, but the evidence is less clear.

The adoption of sweet potato in the highlands (about 300 years ago)

Sweet potato was taken from its American homeland by Polynesians who introduced it into many Pacific islands and New Zealand about 1000 years ago. However, it came to PNG from Indonesia. Sweet potato was taken back to Europe from the West Indies after the first voyage in 1492 by Christopher Columbus, an Italian navigator and maritime explorer who crossed the Atlantic Ocean under Spanish sponsorship. Portuguese explorers then took sweet potato to Africa, India and their colony in the Moluccas in eastern Indonesia. From there it was traded by local people into New Guinea. Oral history research from the Tari basin in Southern Highlands Province and archaeological research at Kuk in Western Highlands Province indicates that sweet potato was adopted in the highlands some decades after the major volcanic eruption of Long Island off the north coast of New Guinea. This blanketed the highlands in ash which leaves a record in the soil as well as in oral history (related as the 'Time of Darkness'). The eruption has been dated to 1665. Further oral history research shows sweet potato was traded into the Lagaip Valley of northern Enga from the Sepik area. It may have been adopted into the highlands of Indonesian Papua somewhat earlier than 1700.

Table 3 Crops introduced into PNG from Asia or the Pacific several hundred to several thousand years ago

Staple (carbohydrate) foods
Yam, lesser
Vegetables
Aibika
Amaranthus tricolor
Bean, lablab
Bean, winged
Castor
Coral tree
Cucumber
Ginger
Gourd, bottle
Gourd, wax
Lemon grass
Fruit
Malay apple
Rukam
Nuts
Pao (Barringtonia procera)
Stimulants
Betel nut
Betel pepper, lowland
Kava

Note: Kava was probably domesticated in Vanuatu. *Pao* was probably domesticated in Solomon Islands. The other crops in this list most likely came to PNG from Asia.

Prior to the adoption of sweet potato in the New Guinea highlands, people depended on taro as their main food, supplemented by banana and yam (*Dioscorea alata*). The adoption of sweet potato brought major changes in highland societies. First, sweet potato makes good pig fodder, and can be fed to pigs raw, whereas taro must be cooked.[6] The adoption of sweet potato gave an advantage: people could produce more pigs, thus becoming wealthier than their neighbours. Also, their diet possibly improved. Second, sweet potato will grow at higher altitudes than taro. The adoption of sweet potato meant that people could occupy higher altitude land on a permanent basis. Settlements spread from around 2200 m up to 2800 m above sea level. Third, the adoption of the new crop resulted in significant changes in the social organisation. Some of these changes are known from oral history research in Enga Province.

By the time that Europeans penetrated the highlands of PNG in the 1920s and 1930s, sweet potato was the main food for almost all highlanders. There were some exceptions. West of the Strickland River, in the Oksapmin and Telefomin areas, taro remained the

6 Taro contains crystals of oxalic acid that cause severe irritation to the mouth and throat of humans and pigs. Cooking destroys these crystals and makes taro edible.

most important food. Several groups of people living in the Lamari and Imani valleys south of Kainantu still depended on a mix of taro, yam and sweet potato. People in these two valleys changed to a diet based on sweet potato after about 1980. At the time of contact with outsiders, sweet potato was present at a number of locations along the north coast of New Guinea as far east as the Huon Peninsula. It was not present in the Bismarck Archipelago and Solomon Islands until the early to late 1800s, where it was introduced by European traders and settlers.

Some people in the PNG highlands, for example in the Tari basin, have stories about the time when they did not have sweet potato and taro was their staple food. But most highlanders think that their ancestors have always lived on sweet potato. In fact, it has been the most important food in the highlands only for about 10–12 generations, or 300 years. The changes brought about by its adoption are still occurring today, for example, in the continuing intensification of land use (see Section 3.6).

Settlement by foreigners and introduction of many new crops (since 1870)

The settlement in PNG by foreigners first occurred in the early 1870s. Europeans, Asians and other Pacific islanders settled in many coastal locations,[7] bringing with them new plant species. The Russian scientist Nikolai Miklouho-Maclay lived on the Rai Coast near Madang for several years from 1871. In the same year, the London Missionary Society (LMS) placed teachers on three islands in the Torres Strait near PNG. In the following year (1872), the LMS established stations west of Port Moresby and on the coast of Western Province. Also in 1872, a trading post was established on the Duke of York Islands between New Britain and New Ireland. In 1873, the German firm Godeffroy set up trading stations in Blanche Bay on New Britain. Methodist missionaries established a station on the Duke of York Islands in 1875.

Records of crop introductions are limited, although some early accounts exist. French Marist missionaries introduced a number of food crops to Woodlark Island in Milne Bay Province in 1847, including beans, pumpkin, corn (maize) and watermelon. Miklouho-Maclay introduced pumpkin, watermelon, corn and pawpaw (papaya) to the Rai Coast in 1871 with seed brought from Tahiti. He noted that the pawpaw, watermelon and corn 'became the favourites and were soon introduced in the plantations [gardens] and the villages on the coast'. In 1873 he introduced mangosteen, durian, orange, lemon, coffee and other species with seed from east Indonesia.[8] Methodist missionaries introduced a number of food crops to the Duke of York Islands in 1875, including orange, lemon, lime, custard apple, guava and new varieties of banana. Many of the early introductions in PNG were made by Pacific island missionaries, as well as by Europeans and Asians.

When many of these introductions occurred is not known, but it is known which species have been introduced since 1870. These include hundreds of potential cash crops, as well as fodder plants (grasses and legumes), shade crops, decorative plants and weeds.[9] Many food crops were introduced into PNG

7 Prior to 1870 there were a few scattered and generally short-lived settlements in the New Guinea region. From 1793 to 1794 there was a British settlement at Restoration Bay near Manokwari in west New Guinea. The Dutch made a settlement on the south coast of west New Guinea at Triton Bay in 1828–36. Both locations are a long way from PNG, so it is unlikely that crops introduced by the British or Dutch settlers were transported to PNG, but it is possible. A mission station was established at Guasopa on Woodlark Island in Milne Bay Province by French (and later Italian) Marist Catholic missionaries between 1847 and 1855. The Marists established a settlement on Umboi Island between New Britain and New Guinea in 1848–49. These settlements had limited and local impacts.

8 Villagers at Bongu village and nearby locations still use their version of Russian names for some of the plants and items introduced by Miklouho-Maclay, including watermelon, corn, pumpkin, cucumber, knife and axe. The term for corn (*gugurus*), derived from the Russian word for maize (*kukuruz*), is used elsewhere in coastal Madang Province.

9 For example, the Experiment Station at Rabaul had planting material of 115 species available for distribution in 1926. Most were introduced species and included fruits, vegetables, other food crops, cover crops, fodder grasses and actual or potential cash

after 1870 (Table 4). The Department of Agriculture, Stock and Fisheries introduced more than 2200 varieties of 90 food crop species between 1950 and 1975. Many of these species were already in PNG by then; further introductions were made to identify varieties with superior qualities.[10]

Foreign settlement, particularly the introduction of new crops and cash cropping, resulted in an important new era in PNG agriculture. Some of the new crops were adopted by villagers and had a significant impact on village agriculture. For example, the anthropologist Malinowski, who conducted fieldwork in the Trobriand Islands in Milne Bay from 1915 to 1918, noted that, since the adoption of sweet potato (in the 1890s) and the availability of imported rice, there had not been a famine. Sweet potato, and later cassava, were widely adopted throughout the islands of Milne Bay Province and greatly improved food security. At a village in the Aiyura basin in Eastern Highlands Province in 1980, villagers grew 87 species of food and cash crops. Almost 60% of these (51 species) had been introduced and adopted during the previous 50 years, including some grown in significant quantities such as peanut, coffee, common bean, Chinese taro, corn and pak choi.

Some introduced crops moved inland ahead of European colonisation. Corn was widely grown in Eastern Highlands, Simbu and Western Highlands provinces when the first European explorers and missionaries visited those areas in the 1930s. It was spread by villagers after being introduced to the Madang area by Miklouho-Maclay. The naturalist MacGillivray further distributed corn in Milne Bay in 1849. Corn was present on some other islands when first visited by European explorers in the 1870s. Similarly, some people in the highlands were growing common bean by the 1930s. It also seems to have been introduced into some coastal locations in the 1870s or 1880s and to have spread into the highlands over the next 40–50 years. Another crop that was adopted quickly in the highlands was pumpkin. Villagers in some locations in the Sepik River area believe *kangkong* to be a traditional crop, which suggests that it was possibly introduced to that area some decades before 1870.

Foreign cash cropping (1880s onwards)

Individual foreigners and overseas companies have been involved in PNG agriculture since the 1880s. This history is partially covered in a number of publications and is reviewed only briefly here. Some of the first foreign settlers in PNG came to trade for coconut to make copra. Villagers in many coastal locations in the islands and on the New Guinea mainland responded by planting significantly more palms. Copra was the most important cash crop in PNG from the 1880s to the early 1970s. Foreigners in both New Guinea and Papua produced and exported a wide range of cash crops from the 1880s onwards, generally in small quantities. These included tobacco, cotton, kapok, rubber, cocoa, sisal and coffee. Cattle and rice were grown for the local market. Other agricultural exports collected from natural stands of trees or the sea included ivory nut (the seed of a palm related to sago), sandalwood, bêche de mer (sea cucumber), trochus shell and pearl shell.

Prior to 1940 experimental plantings were made of a wide range of other crops including sugar cane, corn, castor, tea, cinchona (for quinine), teak, oil palm, vanilla, ginger and peanut, with a view to developing export industries. During the 1950s copra remained the most important export crop, supplemented by rubber, cocoa and coffee. Coffee and cocoa increased in significance in the 1960s and 1970s (Figure 5.2.3). Oil palm, tea, tobacco, corn, sorghum and peanut were also grown by individual foreign settlers and large plantation companies. Village cash cropping increased in significance from the 1950s and the plantation sector declined after 1980. Smallholders now dominate production of all cash crops except oil palm (Table 5.2.1).

crops. The cash crops included those producing fibres, oils, spices, rubber, West African oil palm, bixa, coffee and cocoa. There were a number of varieties of some species (Hopkins 1926).

10 Rice (627 varieties) and wheat (272 varieties) accounted for 40% of the varieties of food plants introduced between 1950 and 1975 (Charles 1976). This was part of a significant but unsuccessful research effort by the Department of Agriculture, Stock and Fisheries to make PNG self-sufficient in rice production (see Section 2.5).

Table 4 Some of the food crops introduced into PNG after 1870

Staple (carbohydrate) foods
Corn (maize)
Irish potato
Queensland arrowroot
Rice
Taro, Chinese
Wheat
Vegetables
Amaranthus blitum
Amaranthus caudatus
Amaranthus cruentus
Bean, broad
Bean, common
Bean, snake
Beetroot
Broccoli
Cabbage, Chinese
Cabbage, head
Capsicum
Carrot
Cauliflower
Celery
Chilli
Choko
Eggplant
Garlic
Kangkong
Kohlrabi
Leek
Lettuce
Onion
Pak choi
Parsley
Pea
Peanut
Pumpkin
Radish
Rhubarb
Shallot
Silverbeet
Soya bean
Spring onion
Tomato
Turnip
Watercress
Yam bean
Zucchini
Fruit
Apple
Avocado
Brazil cherry
Bullock's heart
Cape gooseberry
Carambola
Cherimoya
Cumquat
Custard apple
Durian
Elder
Governor's plum
Granadilla
Grapefruit
Guava
Guava, cherry
Jackfruit
Langsat
Lemon
Lime
Loquat
Mandarin
Mango (*Mangifera indica*)
Mangosteen
Mulberry
Naranjilla
Nectarine
Orange
Passionfruit, banana
Passionfruit, lowland yellow
Passionfruit, purple
Pawpaw
Peach
Persimmon
Pineapple
Plum, Japanese
Pomegranate
Pomelo
Pulasan
Rambutan
Raspberry, black
Rockmelon
Santol
Soursop
Star apple
Strawberry
Suga prut (highland yellow passionfruit)
Tamarillo (tree tomato)
Tamarind
Watermelon
Watery rose apple
Nuts
Cashew
Macadamia integrifolia
Macadamia tetraphylla
Pecan

Changes in village agriculture since 1940

After 1940 the rate of change in PNG agriculture increased greatly. Factors driving these changes were:

- Population increase and pressure on land. The population of PNG rose from 2.2 million in 1966 to 5.2 million in 2000 (an increase of 138%) (Table 1.1.3).
- Alienation of land in some locations, resulting in increased land pressure. For example, in the Cape Hoskins to Talasea area of New Britain, land was alienated for growing oil palm (see Sections 5.7 and 6.1).
- Cash cropping by smallholders. Crops include cocoa, coffee and coconut (see Part 5).
- Plant diseases, especially taro blight, but also a root rot in Chinese taro.

The first major change since 1940 was the replacement of taro by sweet potato as the staple food in Bougainville and the rest of the Solomon Islands. This occurred because of the devastating impact of taro blight, which was introduced there in the early 1940s. Change did not occur evenly in all parts of the country. On the Gazelle Peninsula of East New Britain Province, for example, major changes in the food crops grown, production techniques used and the adoption of cocoa as a cash crop took place between 1945 and 1965, while in adjacent West New Britain Province the widespread planting of oil palm and changes in the main food crops grown did not take place until after 1970.

The responses people made to the new social and economic conditions include:

- Adoption of new crop species. Production of sweet potato and cassava in particular has expanded greatly since 1940 (Figures 2.2.1, 2.2.3). Other crops that have been widely adopted and become relatively important foods include corn, peanut and Chinese taro. Irish potato has become important above 2000 m altitude.
- Adoption of more productive varieties of some crops, including sweet potato and banana.
- More intensive land use characterised, for example, by shorter fallow periods and longer cropping periods (see Sections 3.6 and 3.8).
- Development or adoption of techniques to maintain soil fertility. These include managing fallow species composition by planting trees (especially casuarina in the highlands); crop rotations, especially of sweet potato and peanut; and green manuring (composting) (see Sections 3.7 to 3.12).
- Development of new agricultural systems by the integration of export cash crops into food crop systems. These systems include coffee–casuarina–food crops in the highlands and cocoa–food crops–leucaena–banana on the Gazelle Peninsula.

Agricultural techniques

Villagers use a variety of agricultural techniques in the cultivation of food and cash crops (see Sections 3.7 to 3.12). These techniques are used in different combinations, depending on climate, soil type, fallow vegetation and pressure on land. Archaeological research provides evidence as to how long some of these techniques have been practised.

- **Stone tools** were used as early as 40 000 years ago. Stone tools were used for clearing trees until the late 1800s, when they were replaced by introduced steel tools. In some highlands locations, stone tools were used until about 1950. In some places stone tools are still used to extract sago starch.
- **Burning** has a very long history in PNG and has been used for clearing land for at least 30 000 years (see Section 3.8).
- **Drainage** of agricultural land is widespread in PNG (see Section 3.12). Field drains have been dug in the highlands to remove excess water from food gardens for 4500–5000 years.

- **Mounding** is a widespread technique in PNG (see Section 3.11) and has been observed from 7000 years ago at Kuk.
- **Fences** are commonly built to exclude domestic and wild pigs from food gardens (see Section 3.12). This technique has not been dated, but presumably it was only adopted after pigs were introduced into PNG about 3500 years ago.[11]
- **Green manuring (composting)** was widespread in large areas of Enga, Southern Highlands and Western Highlands provinces when Europeans first visited the region in the 1930s (see Section 3.11). It is possible that the technique was invented in this form about 150 years ago.[12]
- **Planting trees** in fallow land to improve soil fertility is a technique used in a limited number of locations (see Section 3.10). This technique has increased in importance since the 1920s, but it is not known when it first developed. An examination of swamp deposits in a number of highlands locations has shown that at around 1200 years ago there was a marked increase in the numbers of casuarina pollen grains compared to older levels. This has been interpreted to mean that people began deliberately planting casuarina trees around this time, perhaps to provide timber as natural stands were depleted.

11 Pigs were introduced into the Bismarck Archipelago about 3500 years ago. The earliest record for pigs on the New Guinea mainland is 2000 years ago and the first record for the highlands is 1000 years ago.

12 The practice of placing organic matter in large mounds or beds to form compost was still spreading in recent decades (early 1960s to late 1980s) on the edge of the 'composting zone'. This suggests that adoption of the technique was as recent as the nineteenth century. Given the initial boost that sweet potato would have given to food production, there would not have been a need to adopt such a technique immediately after adoption of sweet potato. Development of the large composted mounds in Enga probably occurred in central or western Enga some years after the initial introduction of sweet potato (Wiessner and Tumu 1998:115).

Summary

People were in what is now the island of New Guinea about 50 000 years ago and in the Bismarck Archipelago and the Solomon Islands around 40 000 and 30 000 years ago respectively. We do not know who the earliest occupants of New Guinea were, nor their relationship to modern populations. The final period of prehistoric settlement is associated with the arrival of Austronesian speakers from island South-East Asia. They entered the Bismarck Archipelago and the Solomon Islands over the period 3500 to 3000 years ago and had a marked influence on the subsequent history of those regions. Austronesian impact on the New Guinea mainland was later, more uneven, largely restricted to the coast and intensive only in particular places. Settlement by Europeans, Asians and other Pacific islanders from 1870 onwards caused major changes in agricultural production.

The very early New Guineans depended on hunting and gathering, but it is probable that they began to manage nut-producing trees to encourage better production. There is evidence from starch grains on stone tools for use of taro 28 000 years ago on Buka Island. It is likely that early migrants domesticated sago, *galip* nut and other native species. Agriculture was invented in New Guinea independently of developments elsewhere in the world but at a similar time to its beginnings in the Middle East and central China.

It is now generally accepted that many of the PNG food crops that were important before 1870 were domesticated in New Guinea or nearby areas, including the Bismarck Archipelago, as well as in Asia. Important foods probably domesticated in the New Guinea area are taro, some yam species, banana, breadfruit, sago, many plants used as green vegetables, and some fruits and nuts. Some species were domesticated in both the New Guinea area and Asia independently. In contrast to the important food crops, the most important domestic animal species – the pig, chicken and dog – were introduced into New Guinea after being domesticated elsewhere.

The indigenous PNG cassowary has been hunted and captured chicks reared in captivity, so it could be said to be partially domesticated.

There is a long but poorly known history of adoption and domestication of new crop species. It is likely that a number of new species were introduced by Austronesian-speaking migrants from about 3500 years ago. The overall direction has been replacement of crops by others that have higher productivity, can cope better with environmental or disease problems, and have superior eating properties. For example, it is likely that people once ate wild yam with poor eating properties. Some of these species were domesticated and the quality of the tubers improved. There are indications that some species of yam, such as *Dioscorea pentaphylla* and *D. nummularia*, are very ancient crops in PNG, perhaps introduced from elsewhere but most likely domesticated in the New Guinea area. Tubers of *D. pentaphylla*, for example, have inferior eating qualities and yields appear to be poor, yet people still grow the occasional plant, probably for its cultural value ('*bilong tumbuna*') rather than for food. It is likely that superior varieties of the greater yam (*D. alata*) were developed and that greater yam became the most important species of yam before the introduction of lesser yam (*D. esculenta*).

Lesser yam is likely to have been a later introduction into PNG.[13] It is agronomically superior to other yam species in PNG, including greater yam, has fewer disease problems, a greater yield per plant and tubers that are more easily prepared for cooking than those of most yam species. More people grow greater yam than lesser yam, although greater yam is not usually an important food (Tables 3.1.1, 2.2.1). *Dioscorea esculenta* is less important for ritual purposes than the other yam species, particularly greater yam. All of this suggests that lesser yam has been adopted because of its ability to provide food energy, while the other species have been retained for different reasons.

The many minor species of green vegetables used in PNG may also be what the ethnobotanist Jacques Barrau calls 'witnesses of the past'. That is, their continuing use tells us that they were once important foods, but have been superseded by superior species. The displacement of older species by ones with superior characteristics has continued into recent times. For example, sweet potato replaced taro as the main food in the highlands about 300 years ago. In the lowlands, food crops of American origin, particularly sweet potato and to a lesser extent cassava and Chinese taro, have displaced the older Asia–Pacific crops of taro, yam, banana and sago (Figure 3.1.2).

The adoption of sweet potato in the New Guinea highlands led to significant changes in the social and economic conditions, as well as allowing settlement at higher altitudes. Sweet potato and tobacco were the first of a wave of new crop introductions. They reached PNG between 300 and 400 years ago. Since permanent settlement by foreigners in the 1870s, a large number of food crops, cash crops, other plant species and domestic animals have been introduced. The new food species, the growing importance of export commodity markets and other social and economic changes have resulted in many important changes to agricultural systems since the 1940s.

Agriculture has had a long and successful history in PNG. The history of agriculture has been one of continuous change and evolution, with the rate of change increasing towards the present. The outcome has been agricultural systems which, despite significant social and economic change in other sectors of the economy, still feed more than 80% of the PNG population. The adoption of new crops and the invention of new techniques to cope with changing environmental conditions, increasing population pressure and social change has been carried out with great skill and imagination. People will have to continue to adapt to different circumstances, including to the HIV/AIDS epidemic, climate change and new economic conditions. There is every reason to believe that they will continue to do so as they have done for the past 50 000 years.

13 Lesser yam may have been introduced 1000–3000 years ago, but this is a crude estimate. Austronesian migrants possibly introduced it 3500 years ago. However, linguists have not been able to reconstruct a word for this species in Proto-Oceanic, the language spoken by the Austronesian migrants (Malcolm Ross, pers. comm.). This suggests that lesser yam may have arrived in PNG some time after 3500 years ago.

Sources

Affleck, D. (1983). Notes and documents: Manuscript XVIII: Information on customs and practices of the people of Woodlark Island. A translation of 'Ragguagli sugli usi e costumi del popolo Woodlarkese' by The Reverend Father Carlo Salerio, P.I.M.E., with notes by David Lithgow. *The Journal of Pacific History* 18(1):57–72.

Allen, J., Golson, J. and Jones, R. (eds) (1977). *Sunda and Sahul: Prehistoric Studies in Southeast Asia, Melanesia and Australia*. Academic Press, London.

Allen, M.G. (2005). The evidence for sweet potato in Island Melanesia. In Ballard, C., Brown, P., Bourke, R.M. and Harwood, T. (eds). *The Sweet Potato in Oceania: A Reappraisal*. Ethnology Monographs 19 / Oceania Monograph 56. Oceania Publications, University of Sydney and Ethnology, Department of Anthropology, University of Pittsburgh, Sydney and Pittsburgh. pp. 99–108.

Ballard, C. (1995). The death of a great land: ritual, history and subsistence revolution in the Southern Highlands of Papua New Guinea. PhD thesis. The Australian National University, Canberra.

Ballard, C. (2005). Still good to think with: the sweet potato in Oceania. In Ballard, C., Brown, P., Bourke, R.M. and Harwood, T. (eds). *The Sweet Potato in Oceania: A Reappraisal*. Ethnology Monographs 19 / Oceania Monograph 56. Oceania Publications, University of Sydney and Ethnology, Department of Anthropology, University of Pittsburgh, Sydney and Pittsburgh. pp. 1–13.

Barrau, J. (1965). Witnesses of the past: notes on some food plants of Oceania. *Ethnology* 4(3):282–294.

Bayliss-Smith, T. (1988). Prehistoric agriculture in the New Guinea Highlands: problems in defining the altitudinal limits to growth. In Bintliff, J.L., Davidson, D.A. and Grant, E.G. (eds). *Conceptual Issues in Environmental Archaeology*. Edinburgh University Press, Edinburgh. pp. 153–160.

Bayliss-Smith, T., Golson, J., Hughes, P., Blong, R. and Ambrose, W. (2005). Archaeological evidence for the Ipomoean Revolution at Kuk swamp, upper Wahgi Valley, Papua New Guinea. In Ballard, C., Brown, P., Bourke, R.M. and Harwood, T. (eds). *The Sweet Potato in Oceania: A Reappraisal*. Ethnology Monographs 19 / Oceania Monograph 56. Oceania Publications, University of Sydney and Ethnology, Department of Anthropology, University of Pittsburgh, Sydney and Pittsburgh. pp. 109–120.

Bellwood, P. (2005). *First Farmers: The Origins of Agricultural Societies*. Blackwell Publishing, Oxford.

Bourke, R.M. (1992). Fifty years of agricultural change in a New Guinea highland village. In Levett, M.P., Earland, J. and Heywood, P. (eds). *Proceedings of the First Papua New Guinea Food and Nutrition Conference: Changes in Food and Nutrition in the Last Three Decades*. University of Papua New Guinea Press and Department of Agriculture and Livestock, Port Moresby. pp. 26–53.

Bourke, R.M. (2001). Intensification of agricultural systems in Papua New Guinea. *Agricultural Transformation and Intensification. Asia Pacific Viewpoint Special Issue* 42(2/3):219–235.

Bourke, R.M. (2005). The continuing Ipomoean Revolution in Papua New Guinea. In Ballard, C., Brown, P., Bourke, R.M. and Harwood, T. (eds). *The Sweet Potato in Oceania: A Reappraisal*. Ethnology Monographs 19 / Oceania Monograph 56. Oceania Publications, University of Sydney and Ethnology, Department of Anthropology, University of Pittsburgh, Sydney and Pittsburgh. pp. 171–179.

Bourke, R.M. (2007). Managing the species composition of fallows in Papua New Guinea by planting trees. In Cairns, M.F. (ed). *Voices from the Forest: Integrating Indigenous Knowledge into Sustainable Upland Farming*. Resources for the Future Press, Washington, DC. pp. 379–388.

Brookfield, H.C. and White, J.P. (1968). Revolution or evolution in the prehistory of the New Guinea highlands: a seminar report. *Ethnology* 7(1):43–52.

Charles, A.E. (1976). DASF food plant introduction work. In Wilson, K. and Bourke, R.M. (eds). *1975 Papua New Guinea Food Crops Conference Proceedings*. Department of Primary Industry, Port Moresby. pp. 379–380.

De Langhe, E. and de Maret, P. (1999). Tracking the banana: its significance in early agriculture. In Gosden, C. and Hather, J. (eds). *The Prehistory of Food: Appetites for Change*. Routledge, London. pp. 377–396.

Denham, T. (2005). Agricultural origins and the emergence of rectilinear ditch networks in the highlands of New Guinea. In Pawley, A., Attenborough, R., Golson, J. and Hide, R. (eds). *Papuan Pasts: Cultural, Linguistic and Biological Histories of Papuan-speaking Peoples*. Pacific Linguistics 572. Pacific Linguistics, Research School of Pacific and Asian Studies, The Australian National University, Canberra. pp. 329–361.

Denham, T.P., Haberle, S.G., Lentfer, C., Fullagar, R., Field, J., Therin, M., Porch, N. and Winsborough, B. (2003). Origins of agriculture at Kuk swamp in the highlands of New Guinea. *Science* 301:189–193.

Denham, T.P., Golson, J. and Hughes, P.J. (2004). Reading early agriculture at Kuk swamp, Wahgi Valley, Papua New Guinea: the archaeological features (Phases 1–3). *Proceedings of the Prehistoric Society* 70:259–297.

Denham, T., Haberle, S. and Lentfer, C. (2004). New evidence and revised interpretations of early agriculture in highland New Guinea. *Antiquity* 78(302):839–857.

Denoon, D. and Snowden, C. (eds) (1981). *A History of Agriculture in Papua New Guinea: A Time to Plant and a Time to Uproot*. Institute of Papua New Guinea Studies, Port Moresby.

French, B.R. (1986). *Food Plants of Papua New Guinea: A Compendium*. Sheffield, Tasmania.

Golson, J. (1977). No room at the top: agricultural intensification in the New Guinea highlands. In Allen, J., Golson, J. and Jones, R. (eds). *Sunda and Sahul: Prehistoric Studies in Southeast Asia, Melanesia and Australia*. Academic Press, London. pp. 601–638.

Golson, J. (1991). The New Guinea highlands on the eve of agriculture. *Indo-Pacific Prehistory Association Bulletin* 11:82–91.

Golson, J. (2002). Gourds in New Guinea, Asia and the Pacific. In Bedford, S., Sand, C. and Burley, D. (eds). *Fifty Years in the Field. Essays in Honour and Celebration of Richard Shutler Jr's Archaeological Career*. New Zealand Archaeological Association Monograph 25. New Zealand Archaeological Association, Auckland. pp. 69–78.

Golson, J. (2005). Introduction to the chapters on archaeology and ethnology. In Pawley, A., Attenborough, R., Golson, J. and Hide, R. (eds). *Papuan Pasts: Cultural, Linguistic and Biological Histories of Papuan-speaking Peoples*. Pacific Linguistics 572. Pacific Linguistics, Research School of Pacific and Asian Studies, The Australian National University, Canberra. pp. 221–233.

Groube, L., Chappell, J., Muke, J. and Price, D. (1986). A 40,000 year-old human occupation site at Huon Peninsula, Papua New Guinea. *Nature* 324:453–455.

Haberle, S. (2007). Prehistoric human impact on rainforest biodiversity in highland New Guinea. *Philosophical Transactions of the Royal Society B* 362:219–228.

Hays, T.E. (1991). 'No Tobacco, No Hallelujah': missions and the early history of tobacco in eastern Papua. *Pacific Studies* 14(4):91–112.

Hope, G. and Golson, J. (1995). Late Quaternary change in the mountains of New Guinea. *Transitions: Pleistocene to Holocene in Australia and Papua New Guinea. Antiquity* 69(265):818–830.

Hope, G.S. and Haberle, S.G. (2005). The history of the human landscapes of New Guinea. In Pawley, A., Attenborough, R., Golson, J. and Hide, R. (eds). *Papuan Pasts: Cultural, Linguistic and Biological Histories of Papuan-speaking Peoples*. Pacific Linguistics 572. Pacific Linguistics, Research School of Pacific and Asian Studies, The Australian National University, Canberra. pp. 541–554.

Hope, J.H. and Hope, G.S. (1976). Palaeoenvironments for man in New Guinea. In Kirk, R.L. and Thorne, A.G. (eds). *The Origin of the Australians*. Human Biology Series No. 6. Australian Institute of Aboriginal Studies, Canberra. pp. 29–54.

Hopkins, R.A. (1926). Seed, plants, etc., available at Experiment Station, Rabaul. Leaflet No. 51. Department of Agriculture, Territory of New Guinea, Rabaul.

Lebot, V. (1999). Biomolecular evidence for plant domestication in Sahul. *Genetic Resources and Crop Evolution* 46(6):619–628.

Lilley, I. (1992). Papua New Guinea's human past: the evidence of archaeology. In Attenborough, R.D. and Alpers, M.P. (eds). *Human Biology in Papua New Guinea: The Small Cosmos*. Clarendon Press, Oxford. pp. 150–171.

Lilley, I. (2006). Archaeology in Oceania: themes and issues. In Lilley, I. (ed). *Archaeology of Oceania: Australia and the Pacific Islands*. Blackwell Studies in Global Archaeology No. 8. Blackwell Publishing, Malden, Massachusetts. pp. 1–28.

Loy, T.H., Spriggs, M. and Wickler, S. (1992). Direct evidence for human use of plants 28,000 years ago: starch residues on stone artefacts from the northern Solomon Islands. *Antiquity* 66(252):898–912.

McKillop, R. and Firth, S.G. (1981). Foreign intrusion: the first fifty years. In Denoon, D. and Snowden, C. (eds). *A History of Agriculture in Papua New Guinea: A Time to Plant and a Time to Uproot*. Institute of Papua New Guinea Studies, Port Moresby. pp. 85–103.

Miklouho-Maclay, N. de (1885). List of plants in use by the natives of the Maclay-Coast, New Guinea. *Proceedings of the Linnean Society of New South Wales* 10(3):346–358.

Pawley, A. (2005). Introduction to the chapters on historical linguistics. In Pawley, A., Attenborough, R., Golson, J. and Hide, R. (eds). *Papuan Pasts: Cultural, Linguistic and Biological Histories of Papuan-speaking Peoples*. Pacific Linguistics 572. Pacific Linguistics, Research School of Pacific and Asian Studies, The Australian National University, Canberra. pp. 1–14.

Pawley, A. (2007). Recent research on the historical relationships of the Papuan languages, or, what does linguistics say about the prehistory of Melanesia? In Friedlaender, J.S. (ed). *Genes, Language, and Culture History in the Southwest Pacific*. Oxford University Press, New York. pp. 36–58.

Specht, J. (2005). Revisiting the Bismarcks: some alternative views. In Pawley, A., Attenborough, R., Golson, J. and Hide, R. (eds). *Papuan Pasts: Cultural, Linguistic and Biological Histories of Papuan-speaking Peoples*. Pacific Linguistics 572. Pacific Linguistics, Research School of Pacific and Asian Studies, The Australian National University, Canberra. pp. 235–288.

Swadling, P., Wagner, R. and Laba, B. (1996). *Plumes From Paradise: Trade Cycles in Outer Southeast Asia and Their Impact on New Guinea and Nearby Islands Until 1920*. Papua New Guinea National Museum in association with Robert Brown and Associates (Qld), Boroko, Papua New Guinea.

Swadling, P. and Hide, R. (2005). Changing landscape and social interaction: looking at agricultural history from a Sepik–Ramu perspective. In Pawley, A., Attenborough, R., Golson, J. and Hide, R. (eds). *Papuan Pasts: Cultural, Linguistic and Biological Histories of Papuan-speaking Peoples*. Pacific Linguistics 572. Pacific Linguistics, Research School of Pacific and Asian Studies, The Australian National University, Canberra. pp. 289–327.

Wiessner, P. (2005). Social, symbolic, and ritual roles of the sweet potato in Enga, from its introduction until first contact. In Ballard, C., Brown, P., Bourke, R.M. and Harwood, T. (eds). *The Sweet Potato in Oceania: A Reappraisal*. Ethnology Monographs 19 / Oceania Monograph 56. Oceania Publications, University of Sydney and Ethnology, Department of Anthropology, University of Pittsburgh, Sydney and Pittsburgh. pp. 121–130.

Wiessner, P. and Tumu, A. (1998). *Historical Vines: Enga Networks of Exchange, Ritual, and Warfare in Papua New Guinea*. Crawford House Publishing, Bathurst.

Yen, D.E. (1991). Domestication: the lessons from New Guinea. In Pawley, A. (ed). *Man and a Half: Essays in Pacific Anthropology and Ethnobiology in Honour of Ralph Bulmer*. The Polynesian Society, Auckland. pp. 558–569.

Yen, D.E. (1995). The development of Sahul agriculture with Australia as bystander. *Transitions: Pleistocene to Holocene in Australia and Papua New Guinea. Antiquity* 69(265):831–847.

PART 1

People, Land and Environment

Bryant Allen and R. Michael Bourke

1.1 Total population

The numbers of people in a country and the rate at which they are increasing is a critical issue in any discussion of agriculture and the production of food.

The 2000 National Census gives the total population of Papua New Guinea as 5.2 million. Around 81%, or about 4.2 million of these people live in rural villages (Table 1.1.1). Around 5% live in the National Capital District city of Port Moresby, 8% in other urban areas and 6% in small stations, missions, schools, logging camps and mines, known as 'rural non-village' locations. Only Rwanda, Bhutan, Nepal and Uganda have a greater proportion of people living in rural areas than PNG.

Population growth

Population growth rates are conventionally determined on the basis of the difference between birth rates and death rates. However, almost all births and deaths in PNG are not registered, so estimates of population growth rates must be calculated from census totals. If censuses are flawed for any reason, estimates of growth will also be inaccurate.

The total population of PNG has been increasing at between 2.1% and 3.2% per year since 1966 (Table 1.1.2), with an average rate of growth over this period of about 2.5% per year. At this rate of increase, the total population will double around every 30 years. Every year approximately 105 000 people are added to the population. These people must be fed, clothed, housed, educated and provided with access to health care. If the 2000 population continues to increase at 2.5% per year, it will reach 10.7 million by 2030 (Figure 1.1.1).

Population growth rates between 1966 and 2000 have been highest in the Islands Region at 2.7% per year, with the Highlands and Momase regions growing at around 2.5% per year (Table 1.1.2, Figure 1.1.2). Growth rates in the Southern Region have fluctuated between 1966 and 2000, but overall have been the lowest.

Between the 1990 and 2000 censuses, growth rates in the Highlands and Momase regions appear to have increased significantly to 3.6% per year and 3.3% per year respectively, while in the Islands and Southern regions growth rates have fallen, with almost no growth in the Southern Region. The volcanic eruption in September 1994 at Rabaul, the civil war on Bougainville (1989–1997), and continued growth in the National Capital District will have influenced this pattern, but under-enumeration in the 1990 census has probably inflated the apparent rates of increase in the Highlands Region, as has probable overestimates in the population of Southern Highlands Province in the 2000 census. The 1990 to 2000 increase in the Southern Highlands population is demographically impossible, particularly in light of the known out-migration which is occurring from this province (Figure 1.4.3). The increases are restricted mainly to the central part of the province and suggest errors in the 2000 census data.

Table 1.1.1 Rural, rural non-village, and urban populations of PNG, 2000

Province	Rural	%	Rural non-village	%	Urban	%	Total
Western	107,837	70	12,445	8	33,022	22	153,304
Gulf	92,265	86	3,620	3	11,013	10	106,898
Central	157,058	85	21,165	12	5,760	3	183,983
National Capital District	0	0	0	0	254,158	100	254,158
Milne Bay	188,334	90	9,327	4	12,751	6	210,412
Oro	106,288	80	15,406	12	11,371	9	133,065
Southern Highlands	526,398	96	8,813	2	11,054	2	546,265
Enga	283,498	96	4,014	1	7,519	3	295,031
Western Highlands	371,014	84	39,094	9	29,917	7	440,025
Simbu	242,748	93	7,201	3	9,754	4	259,703
Eastern Highlands	393,418	91	13,243	3	26,311	6	432,972
Morobe	356,100	77	46,869	10	57,743	13	460,712
Lae City	0	0	0	0	78,692	100	78,692
Madang	308,135	84	18,626	5	38,345	11	365,106
East Sepik	303,706	88	7,492	2	31,983	9	343,181
Sandaun	166,919	90	4,508	2	14,314	8	185,741
Manus	34,899	80	1,276	3	7,212	17	43,387
New Ireland	103,259	87	4,346	4	10,745	9	118,350
East New Britain	174,230	79	35,613	16	10,290	5	220,133
West New Britain	109,299	59	54,969	30	20,240	11	184,508
Bougainville	167,156	95	3,897	2	4,107	2	175,160
Papua New Guinea	4,192,561	81	311,924	6	686,301	13	5,190,786

Note: The PNG 2000 National Census records for census units have 2-digit or 3-digit codes for Province (2), District (2), Local Level Government Area (2), Local Level Government Area Ward (2) and Census Unit (3). These are combined to form a unique 11-digit geocode for every census unit in PNG.

This table was created by grouping and summing the populations for all census units for which the ward number is between 80 and 88 as *urban census units*, and all census units for which the census unit number is between 400 and 599 as *rural non-village census units*. National Capital District (Province=04) and Lae City (Province=12, District=03 and 05) were extracted separately. *Rural census units* are all the rest.

Source: NSO (2002).

Table 1.1.2 Population growth rates[a] by region, 1966–2000 (%)

Region	1966–1971	1971–1980	1980–1990	1990–2000	1966–2000
Southern	3.35	1.81	2.72	0.21	1.84
Highlands	2.07	1.95	2.05	3.65	2.49
Momase	2.35	2.30	1.82	3.36	2.48
Islands	3.59	2.40	2.90	2.30	2.69
National Capital District	–	–	4.62	2.57	3.60[b]
Papua New Guinea	2.61	2.09	2.25	3.24	2.55

[a] Growth rates are calculated as follows: $LogN(P_n) - LogN(P_0)$/intercensal period × 100. The intercensal periods used are 1966–1971, 5 years; 1971–1980, 9.1 years; 1980–1990, 9.92 years; 1990–2000, 9.92 years; 1966–2000, 33.94 years.

[b] 1980–2000 intercensal period is 19.84 years.

Sources: 1966–1990: DNPM (1999); 2000: NSO (2002).

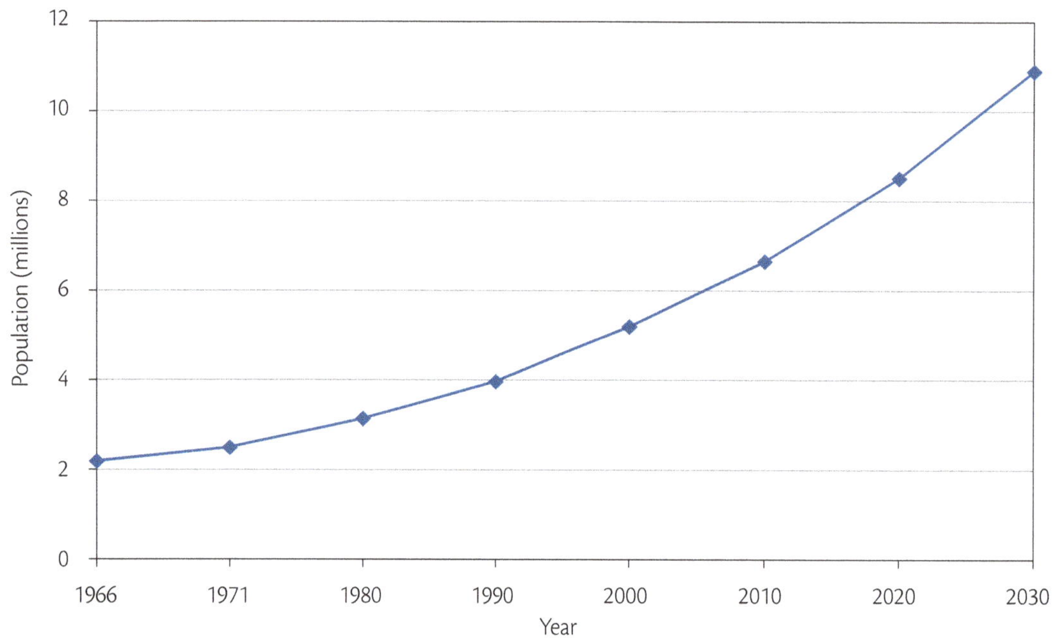

Figure 1.1.1 Population growth 1966–2000 and projected growth 2000–2030 at a constant 2.5% per year.
Sources: 1966–1990: DNPM (1999); 2000: NSO (2002).

The National Statistical Office Population Projections Task Force predicted in 1999 that the Highlands Region population will reach 3.9 million and the Momase Region population 2.9 million by around 2030.

Population distribution

The total population of PNG is not spread evenly throughout PNG. Half of the population lives in six provinces: Southern Highlands, Morobe, Western Highlands, Eastern Highlands, Madang and East Sepik. Another six provinces contain only 14% of the total population: Western, Bougainville, Oro, New Ireland, Gulf and Manus.

In 2000, 38% of the population lived in the five provinces that make up the Highlands Region, 28% lived in the Momase Region, 15% in the Southern Region, 14% in the Islands Region and 5% in the National Capital District. These proportions have remained reasonably stable over the 34-year period 1966 to 2000 (Table 1.1.3, Figure 1.1.3).

Of the total population, 686 301 (13%) are located in urban areas, notably Port Moresby (254 000), Lae (78 700), Mount Hagen (28 500), Madang (27 900), Wewak (20 250) and a number of much smaller towns and administrative centres. There are 73 urban areas, 40 of which have populations of more than 1000 people. A further 311 924 (6%) live in settlements that are classified as 'rural non-village' census units. These are boarding schools, mission stations, sawmills, logging camps and similar settlements located in rural areas, but which are not villages. The rural village population is 4 192 561 (81%).

The effect of HIV/AIDS

PNG is in the early stages of a serious epidemic of HIV/AIDS (Figure 1.1.4). The long-term influence of this epidemic on the distribution of the population is difficult to predict. The impacts on agriculture are also speculative. Most males diagnosed with HIV are aged 25–34 years old and most females are 20–34 years old.

The experience of HIV/AIDS in rural East Africa suggests that the loss of persons in the 25–45 year age group from rural communities will have a significant effect on food and cash crop production, and on the wellbeing of children. HIV/AIDS will possibly have the greatest impact in places that already have a chronic shortage of food (energy and protein), an existing labour shortage (from migration, for example), or an inability to substitute less labour-demanding crops (such as cassava) for present staple crops.

To date, by far the greatest number of infections has been diagnosed in Port Moresby. However, given the propensity of people in PNG to move between the cities and their home villages, it is also important to know the origins of persons infected because such figures may indicate which areas are at greatest risk

Table 1.1.3 Total population by region, 1966–2000

Region	1966	%	1971	%	1980	%	1990	%	2000	%
Southern[a]	422,233	19	499,273	20	588,700	19	771,193	19	787,662	15
Highlands	846,520	39	938,780	38	1,121,258	36	1,373,673	35	1,973,996	38
Momase	618,651	28	695,857	28	857,773	27	1,027,600	26	1,433,432	28
Islands	297,578	14	356,037	14	442,996	14	590,488	15	741,538	14
National Capital District	–		–		123,624	4	195,570	5	254,158	5
Papua New Guinea	2,184,982	100	2,489,947	100	3,134,351	100	3,958,524	100	5,190,786	100

[a] The total population of Southern Region in 1966 and 1971 includes the population of the present National Capital District, which was not established until Independence in 1975.

Sources: 1966–1990: DNPM (1999); 2000: NSO (2002).

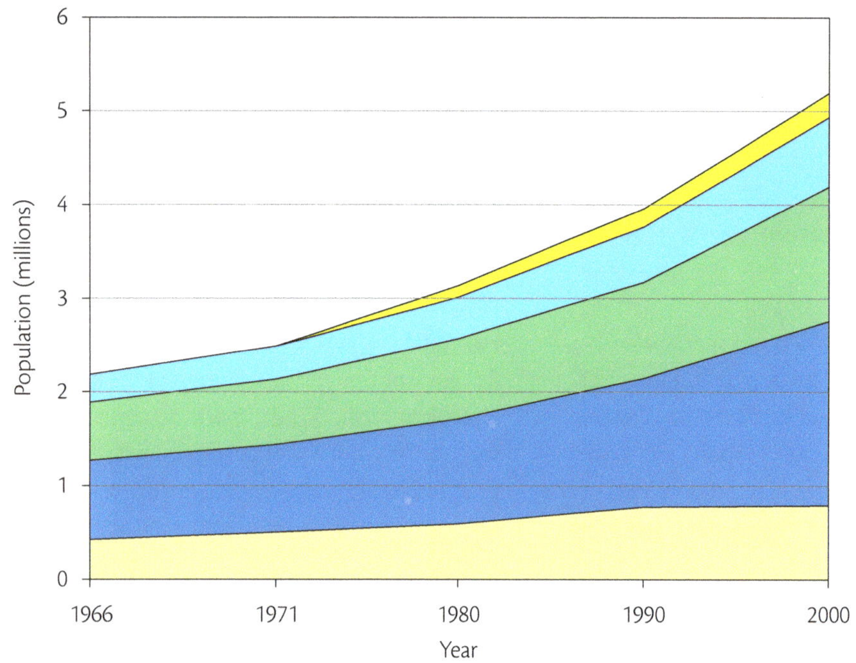

Figure 1.1.2 Population growth by region, 1966–2000. Sources: 1966–1990: DNPM (1999); 2000: NSO (2002).

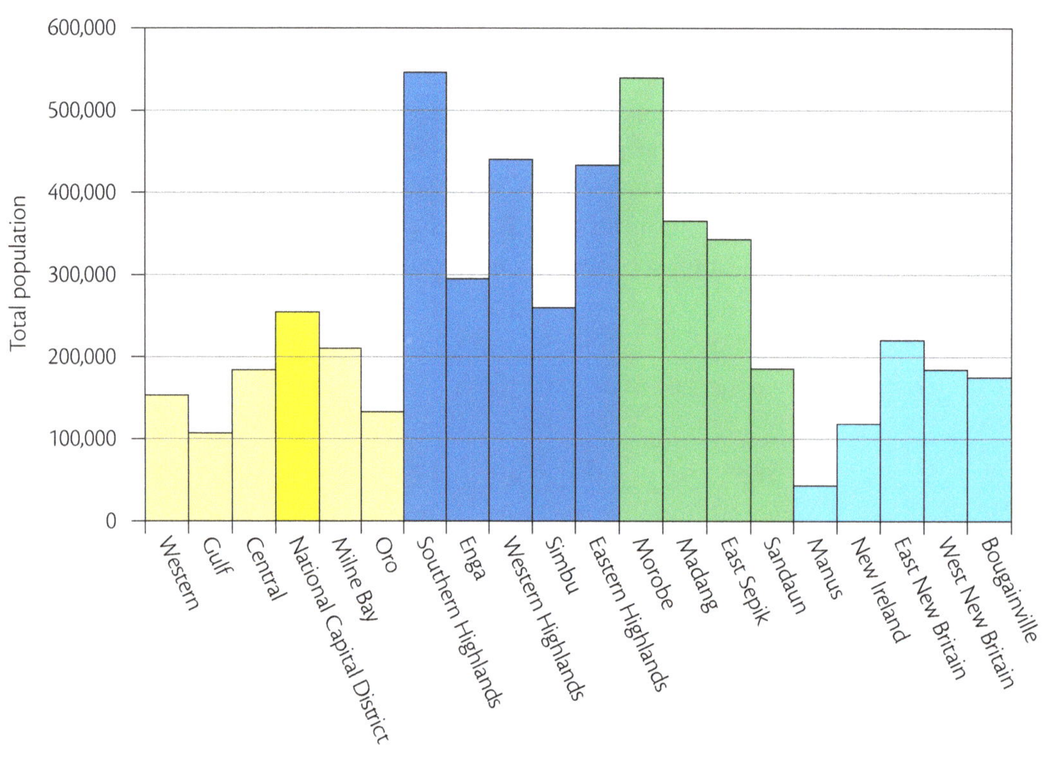

Figure 1.1.3 Total population by province, 2000. Source: NSO (2002).

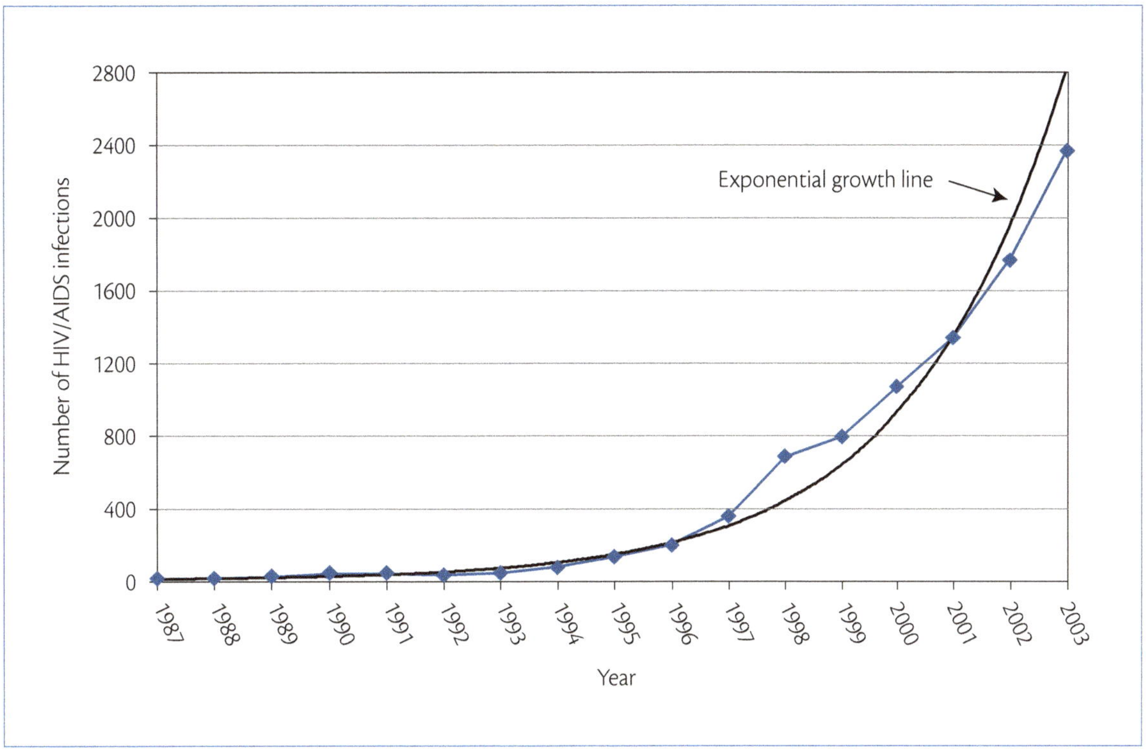

Figure 1.1.4 Number of reported HIV/AIDS infections in PNG, 1987–2003. Source: NACS (2003).

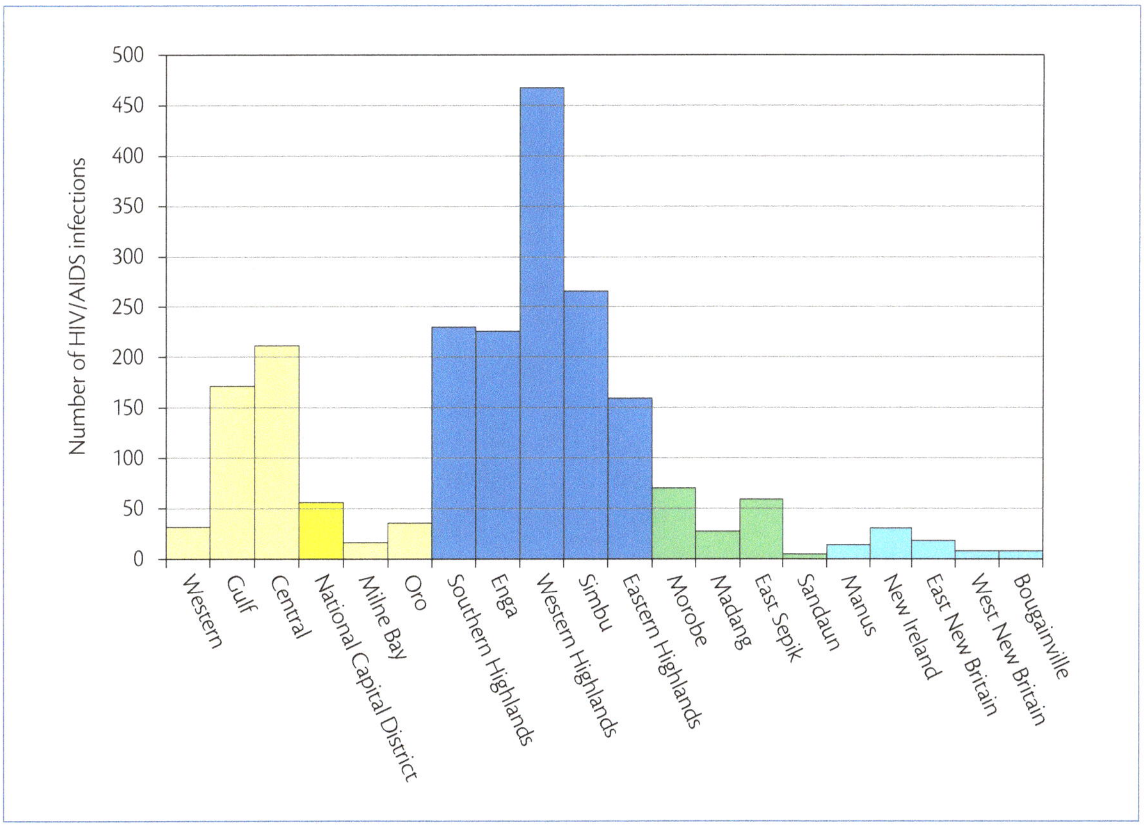

Figure 1.1.5 Number of detected HIV/AIDS infections by province of origin, 1987–2003. Source: NACS (2003).

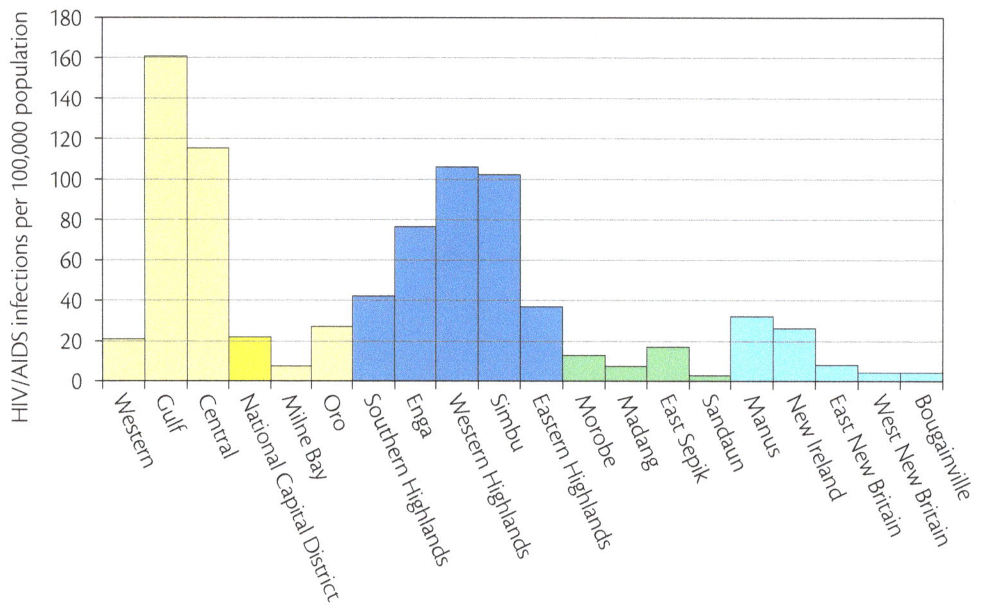

Figure 1.1.6 Rate of confirmed HIV/AIDS infections by province, 1987–2003. Source: NACS (2003).

of local epidemics. Available data, which do not represent all infections, indicate that in 2003 the largest number of infections was in people resident in Central, Gulf and the five highlands provinces (Figure 1.1.5). On the other hand, crude estimates of the number of people infected per total provincial population suggest that infection rates in people from the Southern and Highlands regions (between 55 and 58 per 100 000) are up to five times higher than infection rates in people from the Momase and Islands regions (10 to 11 per 100 000) (Figure 1.1.6). This difference is probably associated with the relative ease of access to Port Moresby, Lae and the Highlands Highway, and possibly also to the relative frequency of testing for the infection in different parts of PNG. In mid 2004 it was reported that around one in every 50 (or 2000 per 100 000) pregnant women admitted to Lae Hospital were HIV positive. Pregnant women were being tested in Lae as part of their antenatal examination. Because most pregnant women are married, and married women are not usually seen as being at 'high risk' of infection, this statistic suggests that the epidemic has moved into the general population and is not restricted to groups who practice risky personal behaviour, such as frequent promiscuous unprotected sexual activity or drug injecting.

Sources

Allen, B. (1997). HIV/AIDS in rural Melanesia and South-East Asia: divination or description. In Linge, G. and Porter, D. (eds). *No Place for Borders: The HIV/AIDS Epidemic and Development in Asia and the Pacific*. Allen & Unwin, St Leonards, NSW. pp. 114–123.

Barnett, T. and Blaikie, P. (1992). *AIDS in Africa: Its Present and Future Impact*. Belhaven Press, London.

DNPM (Department of National Planning and Monitoring) (1999). *Papua New Guinea National Population Policy 2000–2010*. Department of National Planning and Monitoring, Port Moresby.

NACS (National AIDS Council Secretariat and Department of Health) (2003). HIV/AIDS quarterly report December 2003. National AIDS Council Secretariat and Department of Health, Boroko, Papua New Guinea.

NSO (National Statistical Office of Papua New Guinea) (2002). Papua New Guinea 2000 Census: Final Figures. National Statistical Office of Papua New Guinea, Port Moresby.

1.2 Land use

The total land area of PNG is 459 854 km^2. In land use studies of PNG, two different measures of 'land use intensity' have been developed. The first, used in the Papua New Guinea Resource Information System (PNGRIS), examines the proportion of the total land area that is in use at a particular time. The second, applied in the Mapping Agricultural Systems of Papua New Guinea Project (MASP), looks at how often land is used through time (see Section 1.15 for detailed descriptions of these databases).

Understanding land use intensity in PNGRIS

PNGRIS, developed by the Commonwealth Scientific and Industrial Research Organisation (CSIRO) Division of Water and Land Resources, is based on the interpretation of air photographs. CSIRO approached the problem of how to measure land use intensity in shifting cultivation systems over the whole of PNG by first making a distinction between 'used land' and 'unused land' (Figure 1.2.1, Table 1.2.1).

Unused land is all land that is not classed as being used and is usually forest that has never been used for agriculture, but may include swamps. This land may be used for hunting and gathering.

Used land is divided into 'cultivated land' and 'uncultivated land'. Cultivated land includes 'land in current use', that is, land which is currently producing food or cash crops, and 'land in fallow', which is not currently producing, but which has been productive in the past and will be used again to produce food (see Section 3.8). **Everywhere else in this book we use the term 'land used for agriculture' to refer to the CSIRO class 'cultivated land'.** Much fallow land is covered in secondary forest at various stages of regrowth and appears to observers unfamiliar with shifting cultivation systems to be not in use. The inclusion of fallow land as used land is justified because fallows are an integral part of a rotational production system. Tree crops, including coffee, cocoa (and the trees used to shade them) and fruit and nut trees, form an integral part of this complex mix of productive and fallow used land, but are moved about the landscape less than food crop-producing areas.

Used land subclassed as uncultivated land in PNGRIS is land that is not part of a rotational production system, but is nevertheless irregularly interfered with by people. Thus grasslands (including subalpine, alpine and savanna grasslands) are classed as used land in PNGRIS. Grasslands are included because people burn them, and sago is included because some sago is managed and planted by people.

PNGRIS also employs the concept of 'anthropogenous' vegetation, a term first used by the American geographer Carl Sauer in the 1950s in his studies of historical changes in landscapes brought about by people.[1] The term is not clearly defined in PNGRIS but is applied to both cultivated and uncultivated

[1] Sauer (1952) used the term 'anthropogenic' vegetation.

land. When applied to cultivated land it refers to the natural vegetation growing on fallow land, and is used as an alternative measure of land use intensity. When applied to uncultivated land it indicates that part or all of an area could have been interfered with by people. Anthropogenous vegetation is not applied to sago stands or alpine grassland.

Land use intensity measures are developed only for cultivated land. They are based on a subjective interpretation (that is, the measure depends on the person doing the interpretation and not on a fixed and defined measure that can be used by anyone) of patterns on air photographs of the cultivated land that has anthropogenous vegetation growing on it and the area of land that is in current use. Land use is highest where an estimated 75% of cultivated land has vegetation that is influenced by human activity and over 10% is estimated to be in current use (Figure 1.2.1, Table 1.2.1).

While this conceptual structure can be difficult to understand and can lead to misinterpretation, in practice, with careful use of the PNGRIS land use classes, it is easy to distinguish between cultivated land and all other land, and to examine patterns of land use intensity on cultivated land.

Table 1.2.1 Main land uses and land use intensity in PNGRIS

Land use	**Total land area**		**Anthropogenous (%)**	**Current use (%)**
	(km²)	**(%)**		
Used land				
Cultivated land[a]				
Very high with tree crops	2,881	0.6	≥75	>20
Very high	1,711	0.4	≥75	10–20
High	6,237	1.4	>50	5–10
Moderate	15,291	3.3	20–50	1–5
Low	34,115	7.4	20–50	<1
Extremely low and very low	57,623	12.5	<20	<1
Total	117,858	25.6		
Uncultivated land				
Grassland	9,482	2.1	up to 100	0
Sago stands	8,189	1.8	0	0
Subalpine grassland	1,232	0.3	up to 100	0
Alpine grassland	1,033	0.2	0	0
Savanna woodland	2,529	0.5	up to 100	0
Total	22,465	4.9		
Unused land				
Forest	319,531	69.5		
Total land area	459,854	100.0		

[a] Everywhere else in this book we use the term 'land used for agriculture' to refer to the CSIRO class 'cultivated land'.

Sources: Bellamy (1986:116); PNGRIS.

An analysis of PNGRIS shows that in 1975:

- 319 531 km^2, or about 70% of the total land area of PNG was unused (Table 1.2.1, Figure 1.2.2).
- Of the 140 323 km^2 of used land, 22 465 km^2 (5% of the total land area and 16% of used land) was uncultivated and was grassland, subalpine or alpine grassland, savanna woodland, or sago stands.
- A further 117 858 km^2 (25% of the total land area and 84% of the used land), was cultivated. This is land that is in current use *and* land under fallow. Some of the fallow land is covered in secondary forest.

The PNGRIS land use map was updated in 1996 when it was estimated that the area of 'significant land use' had expanded by about 15% between 1975 and 1996, from 60 235 km^2 to 69 183 km^2. These figures include land uses such as mining, oil palm estates and reforestation.

The greatest area of used land was in Madang Province, which accounted for nearly 14% of all used land in PNG. Morobe Province accounted for 10% of the national total (Table 1.2.2). In four provinces, the area of used land as a proportion of the total provincial area was 50% or greater: Madang (56%), Bougainville (55%), Western Highlands (50%) and Eastern Highlands (50%). Western Province had the smallest proportion of used land (8%). The very high figure for Manus Province (83%) reflects an error in the 1975 air photo interpretation of the area of anthropogenous vegetation on Manus Island.

Almost half of the cultivated land is used at extremely low and very low intensities (Figure 1.2.3, Table A1.2.1). On this land, less than 20% of the vegetation is anthropogenous and less than 1% is in current use. A further 29% of used land is used at low intensity. At the other extreme, only about 4% is used at very high intensity, where more than 75% of vegetation is anthropogenous and more than 10% of land is in current use.

The highlands provinces have the greatest proportions of cultivated land used at high intensities; more than one-third of cultivated land in both Enga and Western Highlands is used at high intensity (Figure 1.2.4). Simbu Province has a significant area (43%) of very

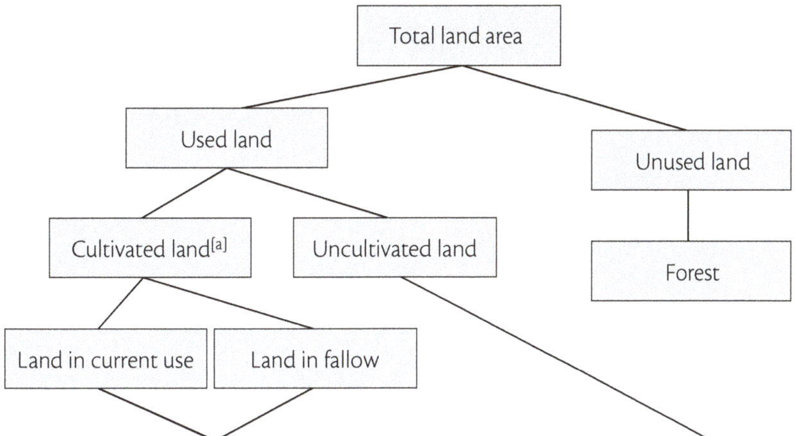

Code	Class	Anthropogenous (%)	Current use (%)
0	Very high with tree crops	»75	>20
1	Very high	»75	10–20
2	High	>50	5–10
3	Moderate	20–50	1–5
4	Low	20–50	<1
5	Very low	10–20	<1
6	Extremely low	<10	<1

Code	Class	Anthropogenous (%)	Current use (%)
7	Grassland	Up to 100	0
8	Sago stands significant	0	0
9	Subalpine grassland	Up to 100	0
10	Alpine grassland	0	0
11	Savanna woodland	Up to 100	0

[a] Everywhere else in this book we use the term 'land used for agriculture' to refer to the CSIRO class 'cultivated land'.

Figure 1.2.1 PNGRIS classification of land use and land use intensity. Source: PNGRIS.

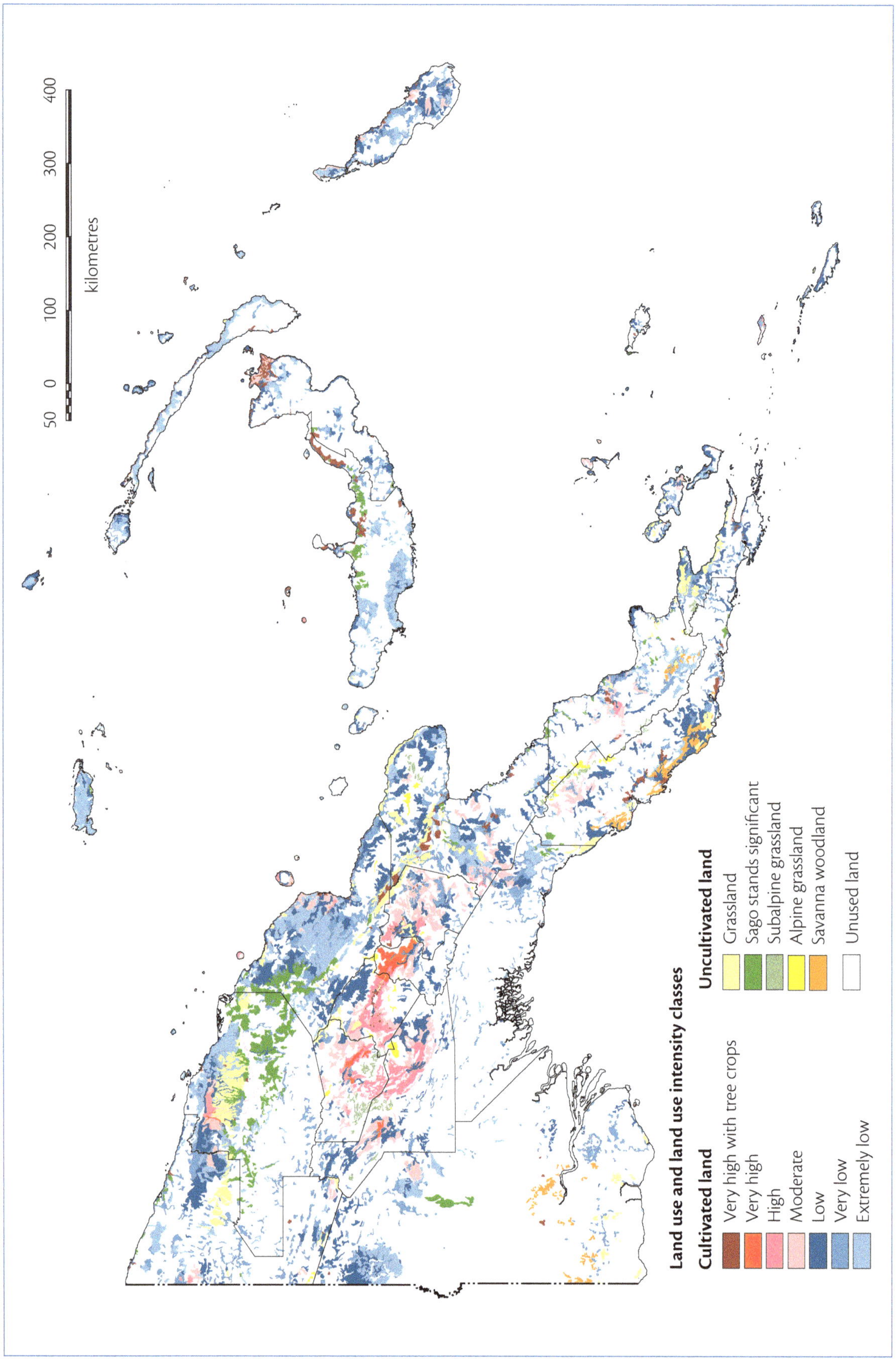

Figure 1.2.2 Land use and land use intensity from PNGRIS, 1975. Source: PNGRIS.

high intensity land use. East New Britain, a lowlands province, stands out as the only province in PNG where a significant area of land is used at very high intensity with tree crops.

The area of 'village only' land use increased by around 11% between 1975 and 1996 (Table 1.2.3). The expansion in the area of used land is much smaller proportionately than the increase in population over the same period. Rather than expanding agriculture into land not presently in use, people in PNG are using land more intensively.

Understanding land use intensity in MASP

Land use intensity in the Mapping Agricultural Systems of Papua New Guinea Project (MASP) database is based on Ruthenberg's measure, known as R, which is a measure of how often land is used. It is particularly useful when applied to shifting cultivation systems because it compares the time that land is in cultivation with the time that it is in

Table 1.2.2 Total land area and cultivated land area by province, 1996

Province	Total land area		Cultivated land area[a]		Cultivated land as a proportion of total land area (%)
	(km²)	(%)	(km²)	(%)	
Western	97,065	21.1	7,931	6.7	8.2
Gulf	33,847	7.4	3,801	3.2	11.2
Central	29,954	6.5	6,407	5.4	21.4
Milne Bay	14,125	3.1	5,691	4.8	40.3
Oro	22,510	4.9	4,258	3.6	18.9
Southern Highlands	25,698	5.6	7,112	6.0	27.7
Enga	11,839	2.6	3,749	3.2	31.7
Western Highlands	8,897	1.9	4,482	3.8	50.4
Simbu	6,022	1.3	2,515	2.1	41.8
Eastern Highlands	11,006	2.4	5,539	4.7	50.3
Morobe	33,525	7.3	12,245	10.4	36.5
Madang	28,732	6.2	16,046	13.6	55.8
East Sepik	43,720	9.5	8,992	7.6	20.6
Sandaun	36,010	7.8	8,211	7.0	22.8
Manus	2,098	0.5	1,747	1.5	83.3
New Ireland	9,615	2.1	4,531	3.8	47.1
East New Britain	15,109	3.3	3,772	3.2	25.0
West New Britain	20,753	4.5	5,676	4.8	27.4
Bougainville	9,329	2.0	5,153	4.4	55.2
Papua New Guinea	459,854		117,858		25.6

[a] Everywhere else in this book we use the term 'land used for agriculture' to refer to the CSIRO class 'cultivated land'.

Source: PNGRIS.

fallow. R is the ratio between the length of active cultivation to the length of the total cultivation cycle (active cultivation and fallow), in years, multiplied by 100. The highest value that R can reach is 100, which represents permanent use. In MASP, land use intensity is measured by converting classes of cultivation period and classes of fallow period into an R-value.

The agricultural systems in MASP are located only within the 117 858 km² of land classified in PNGRIS as used and cultivated, that is, 25% of the total land area of PNG (Figure 1.2.5).

If R is used as the measure of land use intensity, around 71% of the cultivated land in PNG is used at very low intensity (R-value less than 10) and a further 20% at low intensity (Figure 1.2.6, Table A1.2.2). At the other extreme, only 2% is used at high and very high intensity.

Provinces with the largest areas of land used at very low and low intensity are Western, East Sepik, Madang, Sandaun and Gulf (Figure 1.2.7). Large areas used at high and very high land use intensity occur in Enga, Southern Highlands, East New Britain, Eastern Highlands and Western Highlands provinces. Simbu Province, which has large areas of very high intensity land use when using the PNGRIS measure, does not stand out in the MASP data.

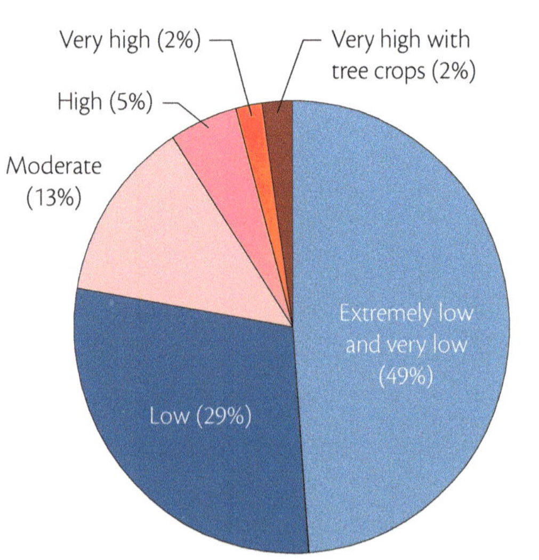

Figure 1.2.3 Proportion of cultivated land by PNGRIS land use intensity class, 1975 (km²). Source: PNGRIS.

Table 1.2.3 Change in village land use, 1975–1996

Province	Increase in village land use 1975–1996	
	(km²)	(%)
Western[a]	709	26
Gulf	141	16
Central	440	9
Milne Bay	134	3
Oro[b]	417	20
Southern Highlands	329	5
Enga	168	5
Western Highlands	312	7
Simbu	86	4
Eastern Highlands	49	1
Morobe	667	11
Madang	508	8
East Sepik	304	11
Sandaun	531	18
Manus[c]	463	228
New Ireland	121	8
East New Britain	451	21
West New Britain[b]	594	64
Bougainville	n.a.	n.a.
Papua New Guinea	6424	11

[a] Refugee resettlement areas and Ok Tedi mine.

[b] Oil palm land settlement schemes occur in these provinces.

[c] Manus figures are an error caused by a misinterpretation of the area of anthropogenous vegetation in 1975.

n.a. = not available

Source: McAlpine et al. (2001:279).

About 2.9 million people, or 70% of the total rural population, live in areas of very low or low land use intensity. Just over 12% of the rural population (513 000 people) live in areas of high or very high land use intensity (Table A1.2.3). Provinces with the largest numbers of people living in areas of high or very high land use intensity are Enga (with over 190 000 people living in very high land use intensity areas), Southern Highlands, East New Britain, Eastern Highlands and Western Highlands (Figure 1.2.8).

Land use intensity measures in PNGRIS and MASP compared

The two land use intensity measures might be expected to produce similar results. However, they have been created using different methods and they measure different things.

Land use intensity in PNGRIS is measured as the proportion of the *total* land area under different uses in the 1970s. It is derived from an interpretation of air photographs of the environmental outcomes of cultivation. The quality of an environment will influence the outcome, because better environments will recover faster from cultivation than poorer environments.

Land use intensity in MASP, on the other hand, is based on estimates of the time that land is under cultivation compared to the time it is in fallow. These estimates are based on interviews with villagers and field observations of land use practices throughout PNG between 1990 and 1995.

Rather than making a direct comparison of the two measures, a more useful analysis is to examine the PNGRIS measure of land use intensity (derived from vegetation analysis) against the MASP R-values (derived from a measure of how often land is used) in order to see where similar R-values result in different vegetation patterns, and then to see if these differences can be explained.

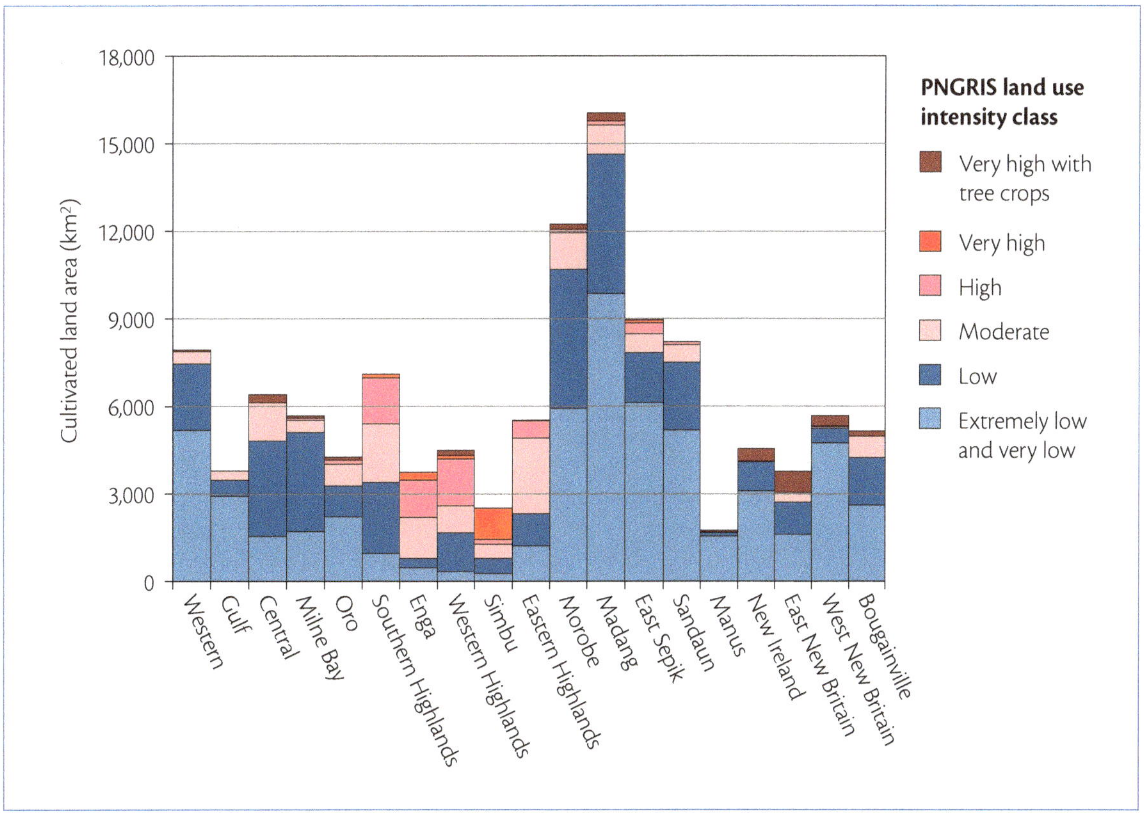

Figure 1.2.4 Cultivated land area by PNGRIS land use intensity class and province, 1975. Source: PNGRIS.

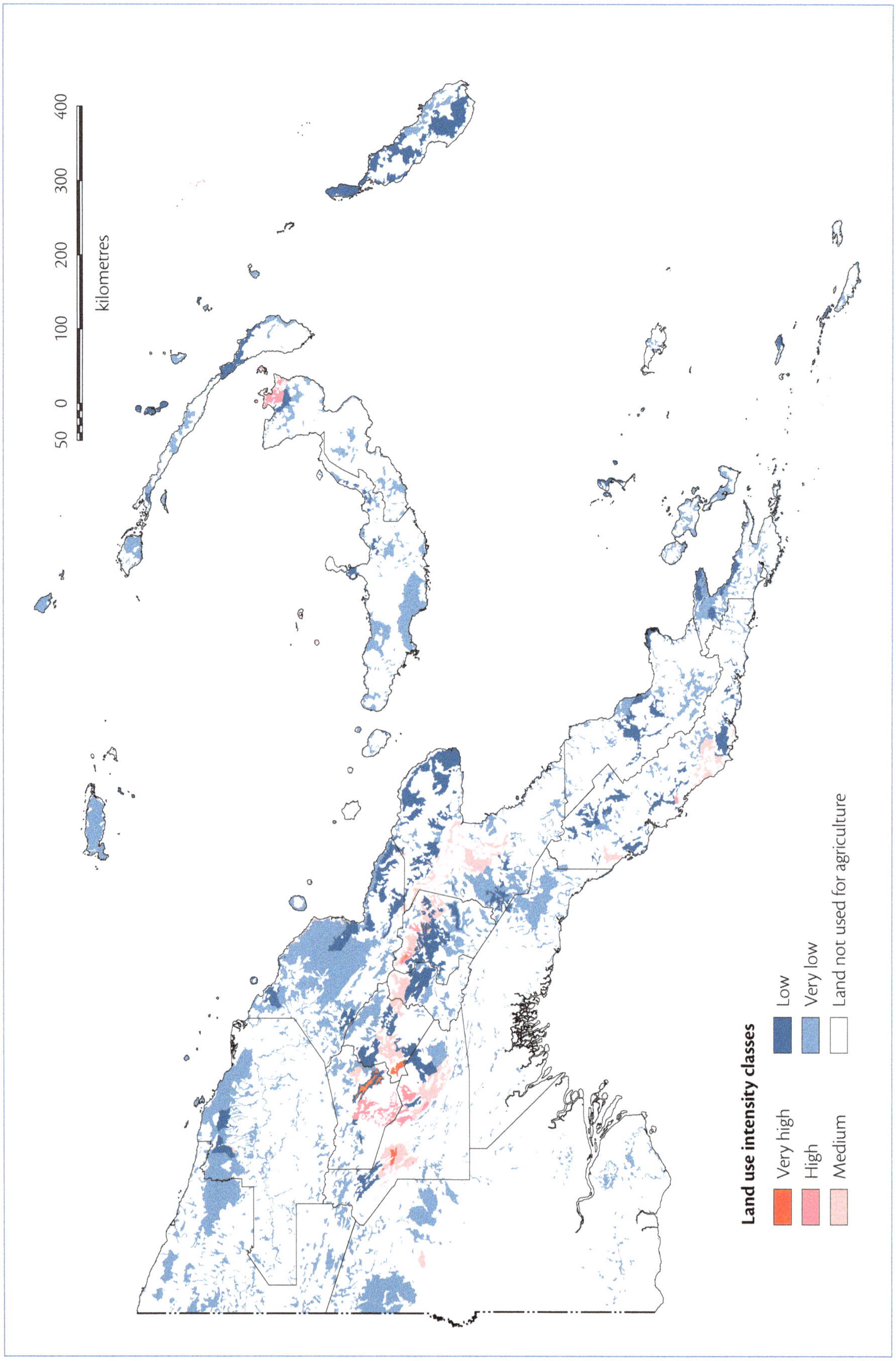

Figure 1.2.5 Land use intensity (R-value) from MASP, 1995. Source: MASP.

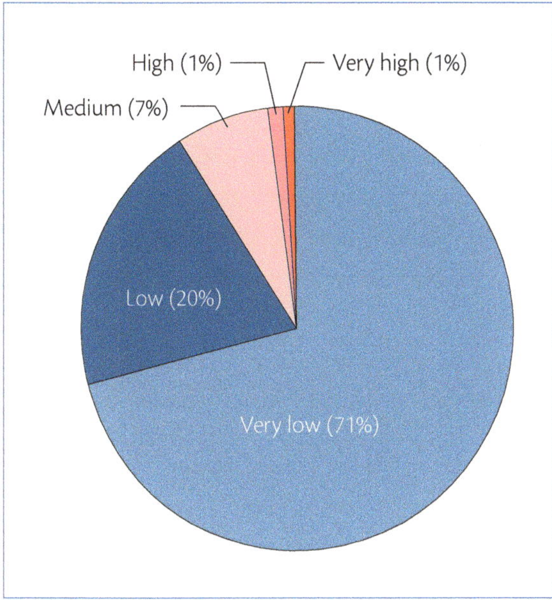

Figure 1.2.6 Proportion of cultivated land by MASP land use intensity class (R-value), 1995 (km²). Source: MASP.

Land use intensity and population density

Logically, a positive association should be expected between population density and land use intensity. Population density is, on average, higher where land use intensity is higher (Figures 1.2.9, 1.2.10, Table A1.2.4), but a lot of variation occurs. Reasons for the high levels of variation between population density and land use intensity may relate to the mapping units used; in this case agricultural systems, which were defined on attributes not associated with population density or land use intensity. It may also relate to a slowness, or an inability, in some systems, of people to respond to increases in population by intensifying their agricultural systems. The R-value is a measure of land use intensity on land used for food production only, and the ability to earn cash incomes from the land may be another reason why there is not a better match between population density and land use intensity, when R-value is used as a measure of land use intensity.

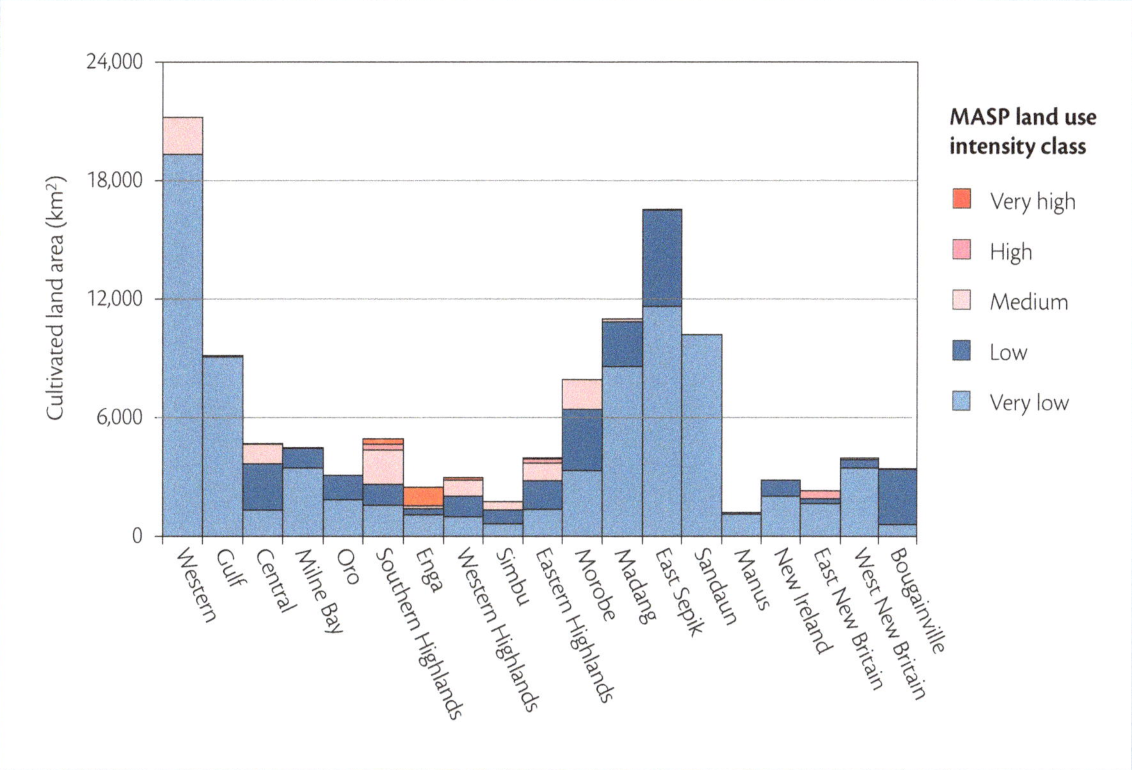

Figure 1.2.7 Cultivated land area by MASP land use intensity class (R-value) and province, 1995. Source: MASP.

FAO statistics on land use in PNG

Significant differences exist between statistics presented by the Food and Agriculture Organization of the United Nations (FAO) on land use in PNG and those that can be derived from PNGRIS. The FAO statistics require comment because they are easily available from the FAO website and because they are frequently quoted in PNG Government documents and consultants' reports.

In 1999, the FAO figures estimated 'agricultural land' in PNG to be 760 km^2 and 'arable land' as 60 km^2, compared with the PNGRIS estimate of 117 858 km^2 'cultivated land'. This is a difference of several orders of magnitude.

Some of the difference arises because the FAO figures refer to 'agricultural land' and the PNGRIS figures refer to 'cultivated land'. FAO defines 'agricultural land' as the sum of the areas of 'arable land', 'permanent crops' and 'permanent pasture'. 'Arable land' is defined by the FAO as 'land under temporary crops, temporary meadows for mowing or pasture, land under market and kitchen gardens, and land temporarily under fallow (less than five years)'. The FAO category of 'agricultural land' excludes the large amount of land in PNG that is under fallow for periods of more than five years. In contrast, the PNGRIS class of 'cultivated land' includes all land in fallow.

If the PNG population is brought into the analysis, serious problems that go beyond definitions are revealed in the FAO figures. For example, if around 0.08 hectares of land is cultivated for food production for every person in PNG (a reasonable estimate), and with a rural village population of 4.2 million in 2000, it follows that there had to have been at least 336 000 ha, or 3360 km^2, of 'arable land' in PNG in 2000,[2] not 60 km^2, or even 760 km^2.

2 4.2 million people × 0.08 ha = 336 000 ha or 3360 km^2 (there are 100 ha in 1 km^2).

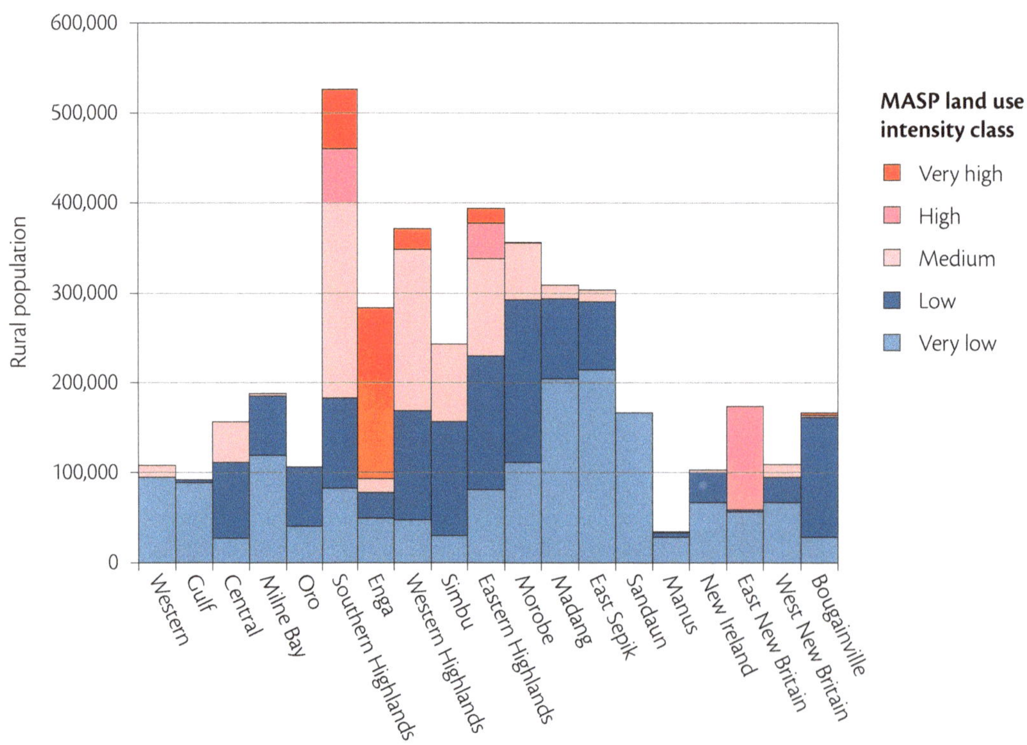

Figure 1.2.8 Number of rural people living on cultivated land by MASP land use intensity class (R-value) and province.
Sources: NSO (2002); MASP.

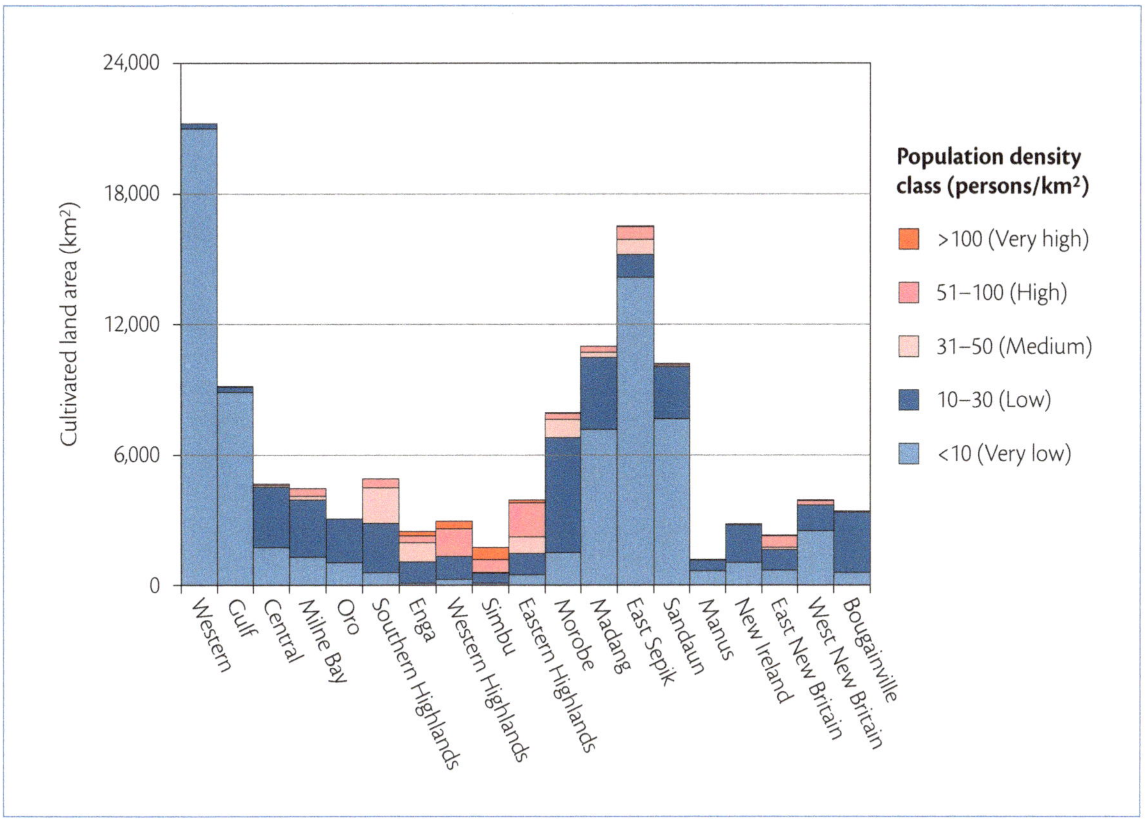

Figure 1.2.9 Cultivated land area by population density class and province, 1995. Sources: NSO (2002); MASP.

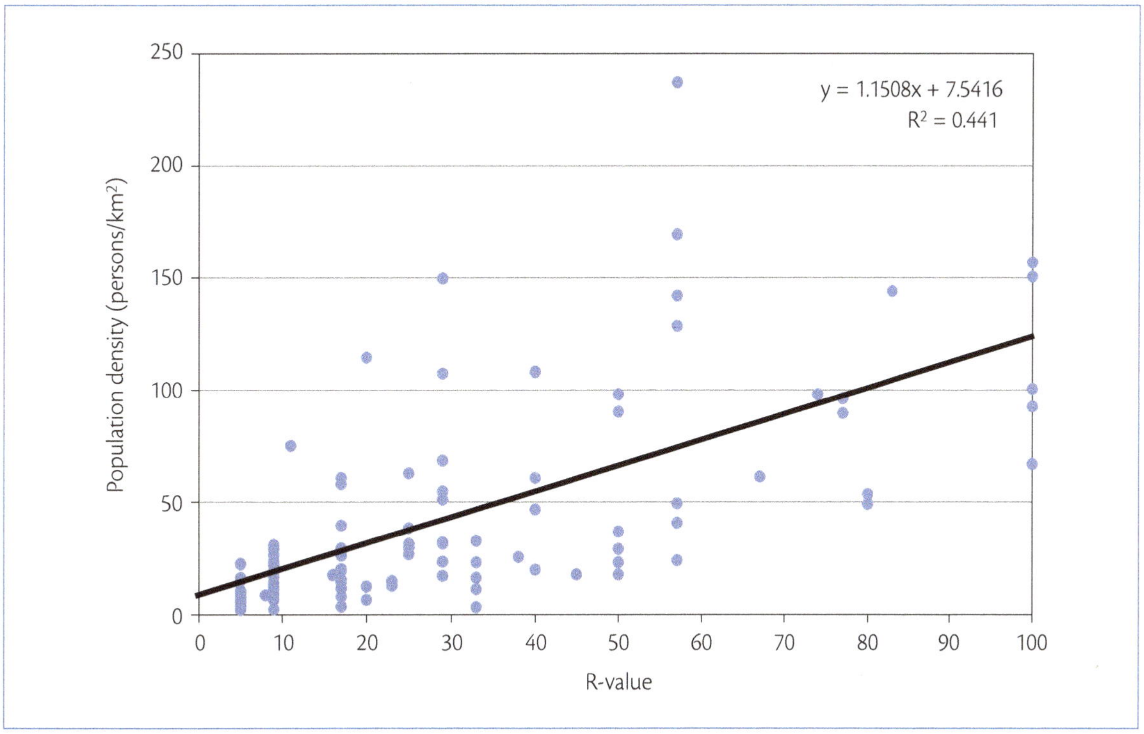

Figure 1.2.10 Association between MASP land use intensity (R-value) and population density, 1995. Sources: NSO (2002); MASP.

Furthermore, if the area of 'agricultural land' in PNG was only 760 km^2, as the FAO figures state, and the rural village population was 4.2 million in 2000, then the average population density on 'agricultural land' in all PNG would have been around 5526 persons per square kilometre.[3] This population density is not reached anywhere in PNG, nor in many rural places in the world.

Sources

Allen, B.J. (2001). Boserup and Brookfield and the association between population density and agricultural intensity in Papua New Guinea. *Agricultural Transformation and Intensification. Asia Pacific Viewpoint Special Issue* 42(2/3):237–254.

Bellamy, J.A. (1986). *Papua New Guinea Inventory of Natural Resources, Population Distribution and Land Use Handbook.* Natural Resources Series No. 6. Division of Water and Land Resources, Commonwealth Scientific and Industrial Research Organisation, Canberra.

Bellamy, J.A. and McAlpine, J.R. (1995). *Papua New Guinea Inventory of Natural Resources, Population Distribution and Land Use Handbook.* Second edition. PNGRIS Publication No. 6. Australian Agency for International Development, Canberra.

Bourke, R.M. (2001). Intensification of agricultural systems in Papua New Guinea. *Agricultural Transformation and Intensification. Asia Pacific Viewpoint Special Issue* 42(2/3):219–235.

Bourke, R.M., Allen, B.J., Hobsbawn, P. and Conway, J. (1998). *Papua New Guinea: Text Summaries.* Agricultural Systems of Papua New Guinea Working Paper No. 1. Two volumes. Department of Human Geography, Research School of Pacific and Asian Studies, The Australian National University, Canberra.

Food and Agriculture Organization of the United Nations website. PNG land use information can be found at <http://faostat.fao.org/>.

McAlpine, J.R. and Freyne, D.F. (2001). Land use change and intensification in Papua New Guinea 1975–1996. *Agricultural Transformation and Intensification. Asia Pacific Viewpoint Special Issue* 42(2/3):209–218.

McAlpine, J.R., Freyne, D.F. and Keig, G. (2001). Land use and rural population change in PNG, 1975–96. In Bourke, R.M., Allen, M.G. and Salisbury, J.G. (eds). *Food Security for Papua New Guinea. Proceedings of the Papua New Guinea Food and Nutrition 2000 Conference.* ACIAR Proceedings No. 99. Australian Centre for International Agricultural Research, Canberra. pp. 277–283.

NSO (National Statistical Office of Papua New Guinea) (2002). Papua New Guinea 2000 Census: Final Figures. National Statistical Office of Papua New Guinea, Port Moresby.

Ruthenberg, H. (1980). *Farming Systems in the Tropics.* Third edition. Clarendon Press, Oxford.

Sauer, C.O. (1952). *Agricultural Origins and Dispersals.* Bowman Memorial Lectures Series 2. American Geographical Society, New York.

Saunders, J.C. (1993). *Agricultural Land Use of Papua New Guinea: Explanatory Notes to Map.* PNGRIS Publication No. 1. Australian International Development Assistance Bureau, Canberra.

[3] 4.2 million people ÷ 760 km^2 = 5526 persons/km^2.

1.3 Population density

Population density is a measure of the numbers of people living within a defined area of land and is usually expressed as 'persons per square kilometre' (persons/km^2).[1]

Population density is an average figure. A population density of one person per km^2 does not mean that each person lives at least 1000 metres away from every other person. Rather, it may mean for example, that 100 people live close together in a small village, but occupy a defined area of land that measures 100 km^2.

It is important to know what land area is involved for each population density calculation. For example, the total land area of PNG is 459 854 km^2, and if there were 5.2 million people in PNG in the year 2000, the population density on the *total land area* of PNG would be 11 persons/km^2. But if the area of land used for agriculture in PNG was 117 858 km^2, then the population density on *land used for agriculture* would be 44 persons/km^2.

This section presents a general overview of population density. The associations between population density and factors such as land use intensity, altitude, or land quality are presented in other sections. The figures presented in this section are population densities within PNG agricultural systems (see Part 3) on *land used for agriculture* only.

1 A square kilometre (km^2) is a square 1000 metres long by 1000 metres wide. There are 100 hectares in one square kilometre.

Distribution of population density

The most striking thing about population density in PNG is that around 66% of the total land area is *not* occupied by people (Figure 1.3.1, Table A1.3.1), although some of this unoccupied land is used for hunting and collecting wild foods. A further 30% of the total land area is occupied at low population densities of less than 31 persons/km^2. Only 4% of the total land area is occupied at densities greater than 30 persons/km^2. The highest population densities in PNG occur in the highlands; in the Maprik area of the East Sepik; on the Gazelle Peninsula in East New Britain; and on many small offshore islands in a number of provinces. However, areas of high population density and very low population density commonly occur within the same province.

The provinces with the largest areas of land occupied at very low densities (<10 persons/km^2) are Western, East Sepik, Gulf, Sandaun and Madang.

The most densely occupied province is Simbu, where 13% of the total land area is occupied at over 100 persons/km^2 and a further 13% is occupied at densities of 51–100 persons/km^2. Other provinces with relatively large areas of land occupied at densities of more than 50 persons/km^2 are Eastern Highlands and Western Highlands. Lowlands provinces with significant areas of high population density are East Sepik, East New Britain and Milne Bay.

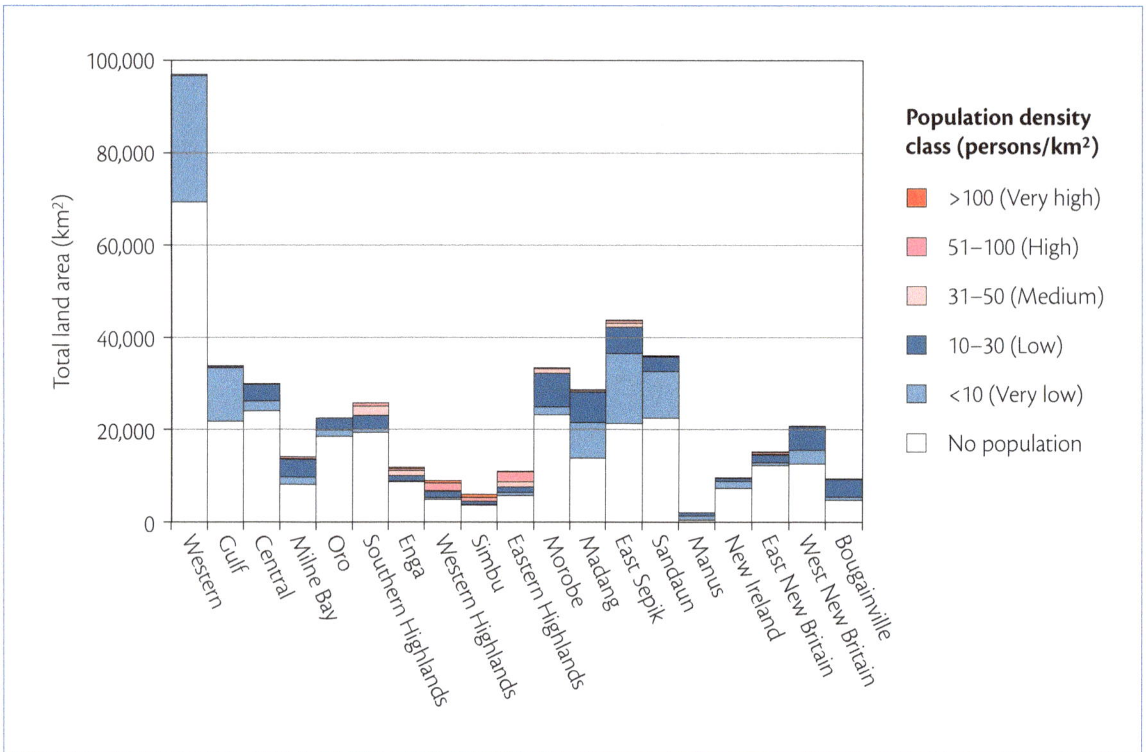

Figure 1.3.1 Total land area by population density class and province. Sources: McAlpine and Quigley (c. 1995); NSO (2002); MASP.

Importantly, however, the highest population densities in PNG occur on small offshore islands in Milne Bay, Morobe, Manus, Bougainville and other provinces (Table 1.3.1). Because they are small and isolated, these very high population density places are often overlooked. The highest population density in PNG is in the Carteret Islands north-east of Buka Island in Bougainville Province. In mid 2000, the resident population density was more than 1200 persons/km^2, and it has increased since then.

Population distribution and population density

Around 12% (480 000 people) of PNG's rural population live at high and very high population densities (>50 persons/km^2) (Table A1.3.2). The largest numbers of people living at these densities are in Enga Province (190 000 people; 67% of the provincial population), East New Britain Province (66%), Southern Highlands Province (24%) and Eastern Highlands Province (14%). Another 766 000 people (18% of the total rural population) live at medium densities (31–50 persons/km^2).

It needs to be remembered, however, that most people in PNG (2.9 million people or 70% of the population) live in areas where population densities are less than 31 persons/km^2. In three provinces – Gulf, Oro and Sandaun – the entire populations live at low and very low densities (Figure 1.3.2). In six other provinces – Milne Bay, Manus, Bougainville, New Ireland, East Sepik and Madang – more than 90% of the populations live at densities of less than 31 persons/km^2.

Table 1.3.1 Resident population in 2000, total area and population density of selected islands

Island	Population	Area[a] (km^2)	Population density ($persons/km^2$)
Milne Bay Province			
Iwa Island	766	3.0	255
Koyagaugau/Ole islands[b]	187	1.4	133
Kwaraiwa Island[b]	358	1.9	190
Naluwaluwali Island[b]	282	2.8	100
Wari Island[b]	707	2.2	321
Morobe Province			
Malai Island	503	1.0	503
Tuam Island	333	1.0	333
Manus Province			
M'Buke Islands	406	3.0	135
West New Britain Province			
Garove Island	3,617	52.0	70
Mundua Islands	1,315	8.0	164
Unea (Bali) Island	8,802	33.0	267
Bougainville Province			
Bougainville Island[c]	130,371	8688.0	15
Buka Island	33,809	586.0	58
Carteret Islands	979	0.8	1224
Matsungan Island	497	0.8	621
Mortlock Islands	443	1.1	403
Nissan Island	4,824	30.1	160
Nuguria Islands	502	6.1	82
Petats Island	1,145	1.9	603
Pinipel Island	901	8.6	105
Pororan/Hetau islands[d]	1,225	2.0	613
Tasman Islands	464	3.1	150

[a] The areas given are the *total* land areas. This includes land that has been used for agriculture and land that is unsuitable for agriculture because it is swampy or too steep or too high.

[b] Part or all of these populations have access to garden land on nearby (mostly very small) islands. The populations for Koyagaugau and Ole islands have been combined, as many Ole people have gardens on Koyagaugau.

[c] The population and area figures for Bougainville Island include a number of small islands near the main island where villagers cultivate land on the mainland.

[d] The figures for Pororan and Hetau islands have been combined, as Hetau people have no garden land on their island and make some gardens on Pororan Island.

Sources: Milne Bay Province: Foale (2005); Bougainville Province: Bourke and Betitis (2003); other provinces: compiled by the author.

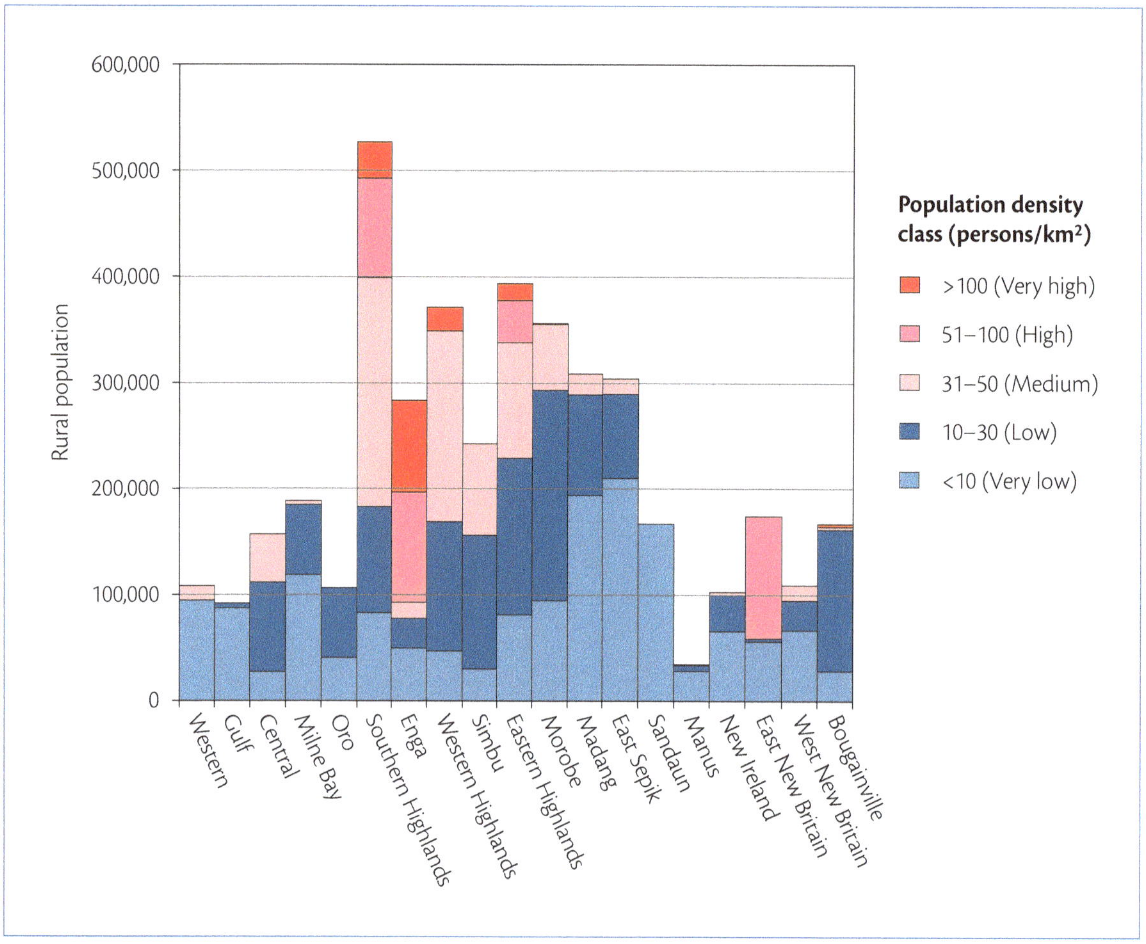

Figure 1.3.2 Rural population by population density class and province. Sources: NSO (2002); MASP.

Sources

Bellamy, J.A. and McAlpine, J.R. (1995). *Papua New Guinea Inventory of Natural Resources, Population Distribution and Land Use Handbook*. Second edition. PNGRIS Publication No. 6. Australian Agency for International Development, Canberra.

Bourke, R.M., Allen, B.J., Hobsbawn, P. and Conway, J. (1998). *Papua New Guinea: Text Summaries*. Agricultural Systems of Papua New Guinea Working Paper No. 1. Two volumes. Department of Human Geography, Research School of Pacific and Asian Studies, The Australian National University, Canberra.

Bourke, R.M. and Betitis, T. (2003). Sustainability of agriculture in Bougainville Province, Papua New Guinea. Land Management Group, Department of Human Geography, Research School of Pacific and Asian Studies, The Australian National University, Canberra.

Foale, S. (2005). *Sharks, sea slugs and skirmishes: managing marine and agricultural resources on small, overpopulated islands in Milne Bay, PNG*. Resource Management in Asia-Pacific Working Paper No. 64. Resource Management in Asia-Pacific Program, Research School of Pacific and Asian Studies, The Australian National University, Canberra.

McAlpine, J. and Quigley, J. (c. 1995). *Natural Resources, Land Use and Population Distribution of Papua New Guinea: Summary Statistics from PNGRIS*. PNGRIS Report No. 7. Australian Agency for International Development, Canberra.

NSO (National Statistical Office of Papua New Guinea) (2002). Papua New Guinea 2000 Census: Final Figures. National Statistical Office of Papua New Guinea, Port Moresby.

1.4 Internal migration

Significant numbers of people are moving temporarily and permanently within PNG.

From around 1890 until World War II, colonial governments administered indentured labour schemes that brought men from undeveloped areas with surplus labour to plantations and mines that required labour. At the end of a contract indentured labourers were returned to their homes.

During World War II, large numbers of young men were recruited as carriers and labourers by both Japanese and Allied forces. After 1946, the Highland Labour Scheme brought young men from the highlands to work in coastal locations for two years at a time. In the 1970s, the government sponsored the movement of selected rural families from mainly East Sepik, East New Britain and Simbu provinces to the north coast of West New Britain Province to produce oil palm on land settlement schemes (see Sections 5.7 and 6.7). Attempts to establish land settlement schemes based on rubber production at Gavien in East Sepik Province and Cape Rodney in Central Province have not resulted in long-term rubber production, although the settlers have remained (see Section 5.11).

Migration today is overwhelmingly informal. Young men still move more than young women, but increasingly whole families are moving.

A measure of internal migration is 'net migration', which is the difference between the movements into a province and the movements out of it. In 1980, this measure showed that people were moving out of Gulf, Simbu, Manus, Central, Southern Highlands, East Sepik, Enga and Milne Bay provinces, and were moving into West New Britain, Bougainville and Western Highlands provinces (Figure 1.4.1). The patterns of movement established in 1980 were largely maintained into 1990 and 2000 (Figures 1.4.2, 1.4.3).

The most outstanding differences between 1980, 1990 and 2000 are the movement of people out of Bougainville as a result of the civil war (1989–1997); into Western Province between 1980 and 1990 after the opening of the Ok Tedi mine; and out of East New Britain between 1990 and 2000 following the 1994 volcanic eruptions (Table 1.4.1). The proportion of families resident in National Capital District, but born elsewhere, increased considerably between 1990 and 2000.

These movements are associated with the ability to earn cash incomes in particular provinces, with most migrants moving from provinces where incomes are lowest, to provinces where incomes are highest (Figure 1.4.4).

These broad patterns at the province level were confirmed by a finer-scale study of increases and decreases in Census Division populations between the 1980 and 1990 censuses (excluding Bougainville) (Keig 2001). Keig's study revealed that four main types of internal migration are occurring within PNG:

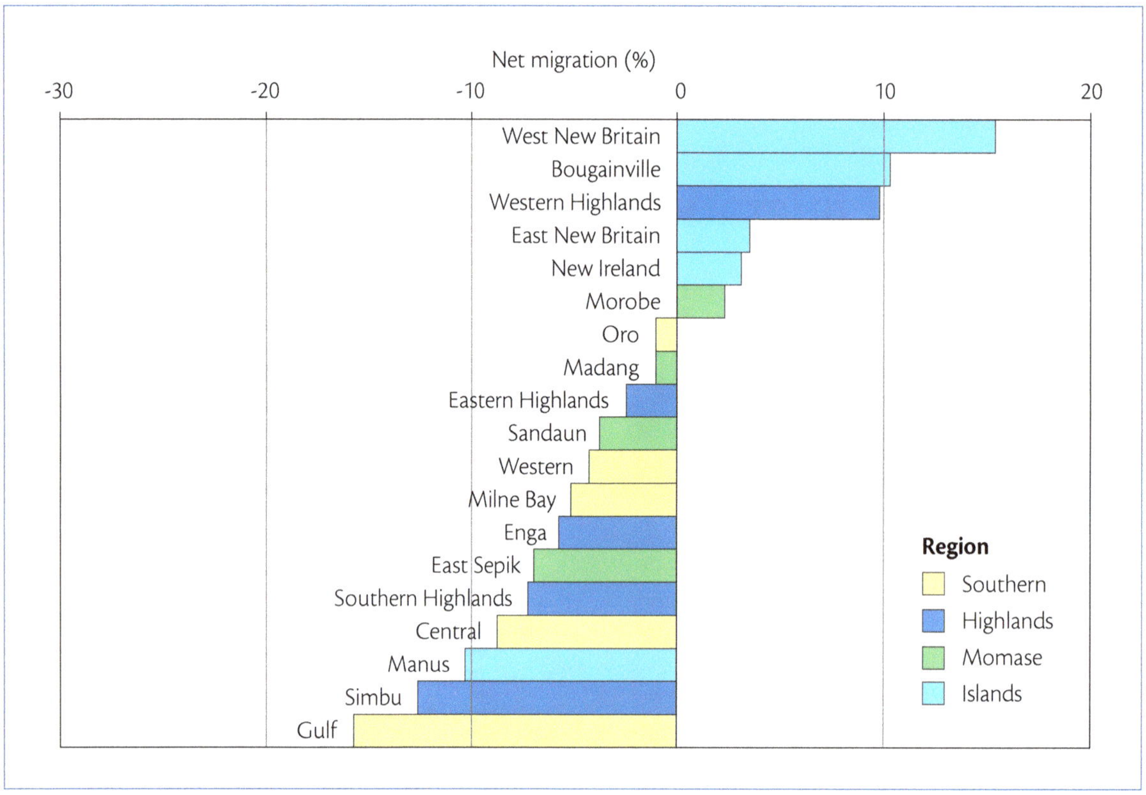

Figure 1.4.1 Net migration by province, 1980. Source: Goodman et al. (1985).

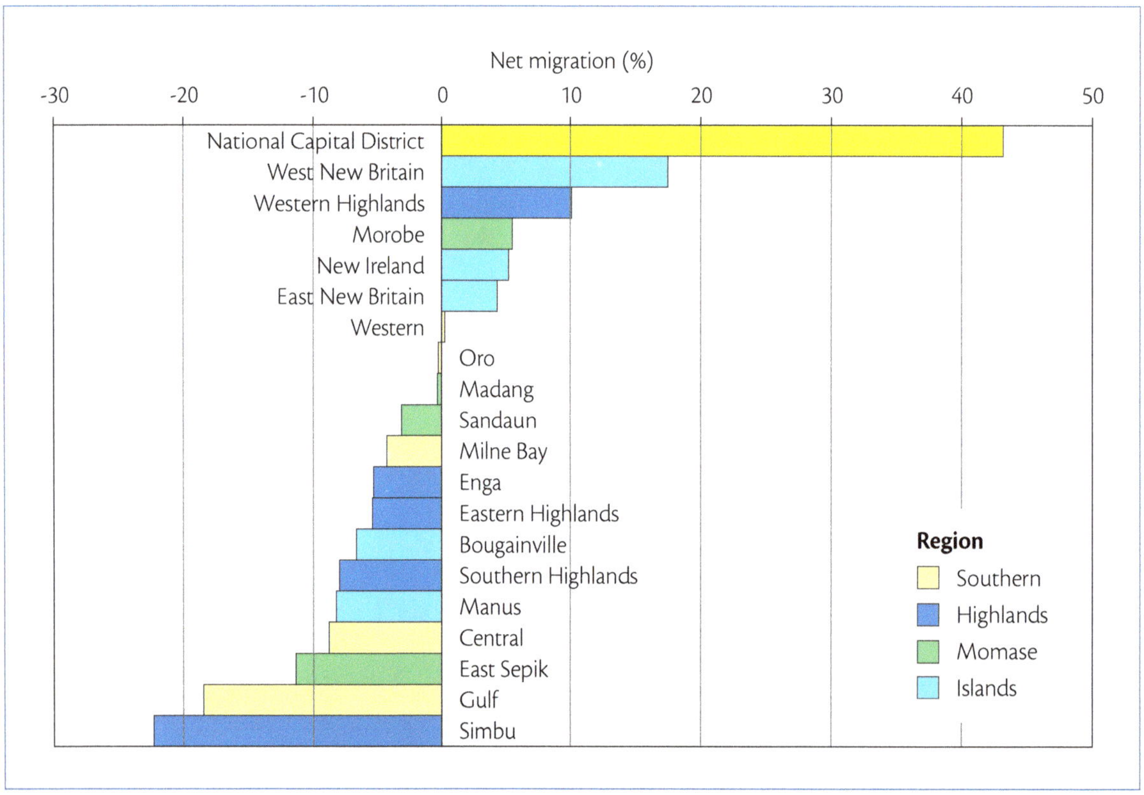

Figure 1.4.2 Net migration by province, 1990. Source: DNPM (1999).

- **Rural-to-urban** – from rural areas to the largest cities of Port Moresby, Lae, Madang, Mount Hagen and Goroka. Large informal 'settlements' have developed within these cities.
- **Rural-to-periurban** – from rural areas to the areas surrounding, but not within, main centres, such as Port Moresby, Lae, Goroka, Mount Hagen and Kokopo. Significant growth has also occurred in areas close to the smaller towns of Kimbe, Bulolo, Popondetta, Kavieng, Alotau, Mendi, Tari and Kainantu. Periurban 'settlements' are occupied at a lower density than urban settlements and a considerable amount of subsistence food is produced from them.
- **Rural-to-rural** – from rural areas in Simbu, East Sepik, Southern Highlands, Eastern Highlands, Enga, Gulf, Central and Bougainville provinces to rural areas in West New Britain, Western Highlands, Morobe, East New Britain and New Ireland provinces. Significant growth has occurred on a number of small offshore islands, such as Losuia and Misima in Milne Bay Province and the Feni Islands in New Ireland Province. However, this growth is primarily a result of natural population increase (see Section 1.1), rather than net in-migration. Complex informal extra-legal tenurial arrangements are thought to exist between landowners and migrants. These range from sharefarming, to labour in exchange for rent, to cash for rent, to the purchase of land.

Significant internal migration is also occurring within provinces. Generally people move from locations with poor environments and poor road access to those more favourable for agricultural production or close to a road. These movements are poorly documented as they are not recorded in national census data.

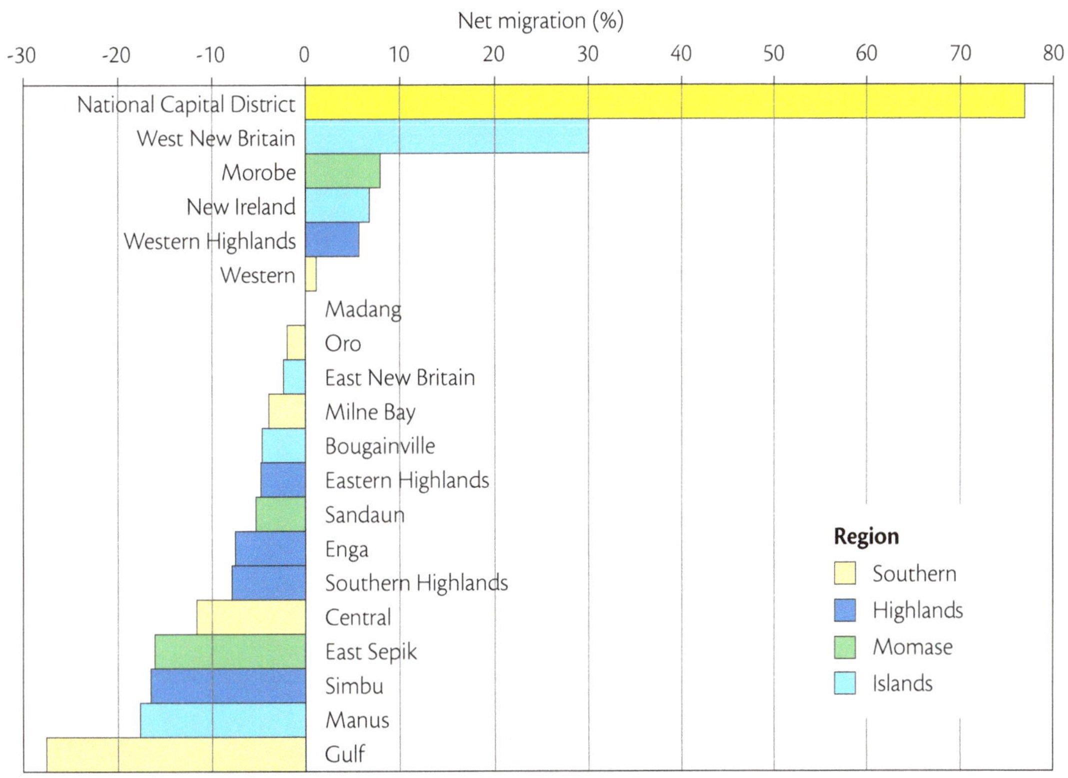

Figure 1.4.3 Net household migration by province, 2000. **Note:** This figure differs from Figures 1.4.1 and 1.4.2 in that it is based on the place of birth and the residence of household heads at the 2000 census, and not of individuals. The proportions are of the total number of households, not the total population.
Source: Calculated by the author from the PNG 2000 National Census household tables.

Table 1.4.1 Net inter-provincial migration (%)

Province	1980	1990	2000	Province	1980	1990	2000
Western	−4.2	0.23	1.18	Eastern Highlands	−2.4	−5.50	−4.78
Gulf	−15.8	−18.42	−27.56	Morobe	2.3	5.55	7.97
Central	−8.7	−8.82	−11.66	Madang	−1.0	−0.33	0.03
National Capital District	n.a.	43.20	76.91	East Sepik	−6.9	−11.36	−16.10
Milne Bay	−5.1	−4.33	−3.92	Sandaun	−3.7	−3.13	−5.30
Oro	−1.0	−0.28	−1.94	Manus	−10.3	−8.27	−17.64
Southern Highlands	−7.2	−8.01	−7.88	New Ireland	3.1	5.24	6.81
Enga	−5.7	−5.42	−7.54	East New Britain	3.5	4.34	−2.35
Western Highlands	9.8	10.11	5.73	West New Britain	15.4	17.53	30.06
Simbu	−12.6	−22.29	−16.52	Bougainville	10.3	−6.74	−4.65

Note: Net inter-provincial migration is the difference between the number of people who were born in a province and were resident in another province at the time of the census, and the number who were resident in a province at the census, but were born in another province, as a proportion of the total provincial population. The 2000 data are based on household heads, not on the total population.

Sources: 1980: calculated by Christine McMurray and published in Goodman et al. (1985:79); 1990: DNPM (1999); 2000: calculated by the author from the PNG 2000 National Census household tables.

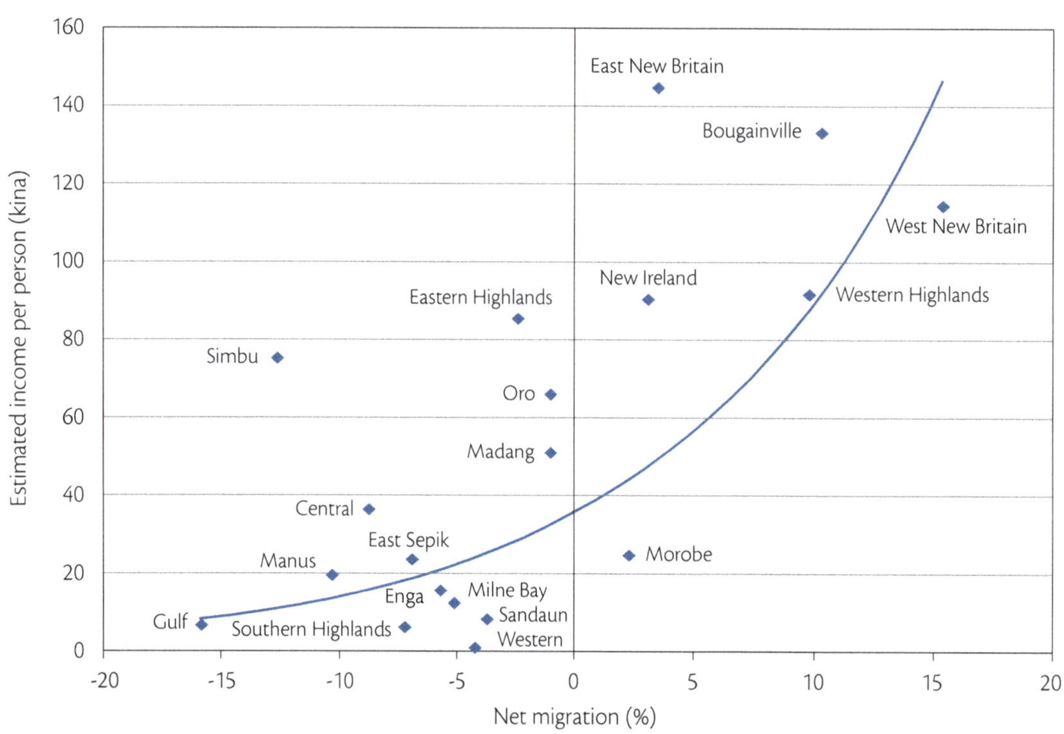

Figure 1.4.4 Association between net migration, 1980, and estimated cash income per person, 1985.
Source: Goodman et al. (1985).

In some locations people move temporarily to lower or higher altitudes for some weeks every year to make sago or to harvest pandanus nuts, sources of food which do not grow at the altitude of their settlements (see Section 1.13).

- **Rural-to-resource projects** – from rural areas to the vicinity of major mining and oil projects, such as to the Porgera gold mine (Enga Province) and to the Ok Tedi copper mine (Western Province).

Destination areas are in general around, as well as in, major towns or large mining projects; around successful land settlement schemes (see Section 6.7); on good quality land (see Section 1.12); and in places with good access to roads, markets and services (see Section 1.14).

Source areas are mainly locations with poor access to services, low quality land, few employment opportunities, or 'troubled' areas (areas with customary land tenure issues, sorcery, or criminal activity, for example). It is possible that difficult access resulted in a poor census in 1990 that has exaggerated the loss of people from such areas. Nevertheless, the trends are very clear.

It can be concluded that in many parts of PNG people are moving from areas they perceive as disadvantaged to areas they perceive as advantaged. That is, from poor quality land with poor access to markets and services, to higher quality land with better access to markets and services and with increased chances to engage in the cash economy.

Sources

DNPM (Department of National Planning and Monitoring) (1999). *Papua New Guinea National Population Policy 2000–2010*. Department of National Planning and Monitoring, Port Moresby.

Goodman, R., Lepani, C. and Morawetz, D. (1985). *The Economy of Papua New Guinea: An Independent Review*. Development Studies Centre, The Australian National University, Canberra.

Keig, G. (2001). Rural population growth in Papua New Guinea between 1980 and 1990. *Agricultural Transformation and Intensification. Asia Pacific Viewpoint Special Issue* 42(2/3):255–268.

NSO (National Statistical Office of Papua New Guinea) (2002). Papua New Guinea 2000 Census: Final Figures. National Statistical Office of Papua New Guinea, Port Moresby.

1.5 Rainfall

Rainfall is a direct influence on plant growth and must be taken into account in any discussion of agriculture. PNG is one of the wettest countries on earth; much of PNG regularly receives 2000–4000 mm of rain per year, and a few areas of the country receive more than 7000 mm of rain every year. Conversely, in other places annual rainfall is below 1500 mm (for example, Port Moresby). In some places there is no discernible seasonal pattern to rainfall, while in others rainfall is strongly seasonal. These rainfall patterns are the most important determinant of soil water availability.

Annual rainfall

Two digital maps of annual rainfall exist for PNG. The first is derived from PNGRIS in which the annual rainfall for each resource mapping unit (RMU) is mapped (Figure 1.5.1). (See Section 1.15 for an explanation of PNGRIS and a definition of an RMU.) The second was created by the Centre for Resource and Environmental Science (CRES) at The Australian National University. It is created by applying a mathematical procedure to the rainfall data for individual stations that creates an interpolated and smoothed rainfall 'surface' for the whole of PNG (Figure 1.5.2). Both techniques have problems: the PNGRIS map has sudden changes in rainfall along provincial borders and assumes the same rainfall across the whole of an RMU; the CRES map appears to create spurious values when, for example, only one or two rainfall stations are located along a coastline and there are no points inland, or out to sea, for the procedure to work with. Nevertheless, both maps provide a similar picture of the distribution of annual rainfall in PNG.

These maps of the national pattern of annual rainfall (Figures 1.5.1, 1.5.2) show that very high rainfall occurs on three sides of the main highlands valleys: to the west and along both the north and south sides of the main range. High rainfall is also received on the south coast of New Britain and on south Bougainville Island. Very high annual rainfall – over 7000 mm per year – is received from Ok Tedi in the far north-west of Western Province, south-east into northern Gulf Province. Within local areas, rainfall increases with altitude. But over the whole country, altitude is not associated with higher annual rainfall.

Areas where annual rainfall is below 2000 mm occur in the northern part of East Sepik Province, the Markham Valley in Morobe Province, part of the adjacent Ramu Valley in Madang Province, the northern part of Eastern Highlands Province, the southern third of Western Province, the coastline of Central Province, and the Cape Vogel – Rabaraba area of Milne Bay Province.

The ideal annual rainfall for many tropical crops is 1500 mm to 3000 mm (see Section 1.13). Most of the PNG rural population live in places where annual rainfall is in the range 1800–3500 mm. Localities where the annual rainfall is more than 4000 mm tend to be too wet and have too much cloud cover for good agricultural production (see Sections 1.7 and 1.13). Population densities are lower in wetter areas.

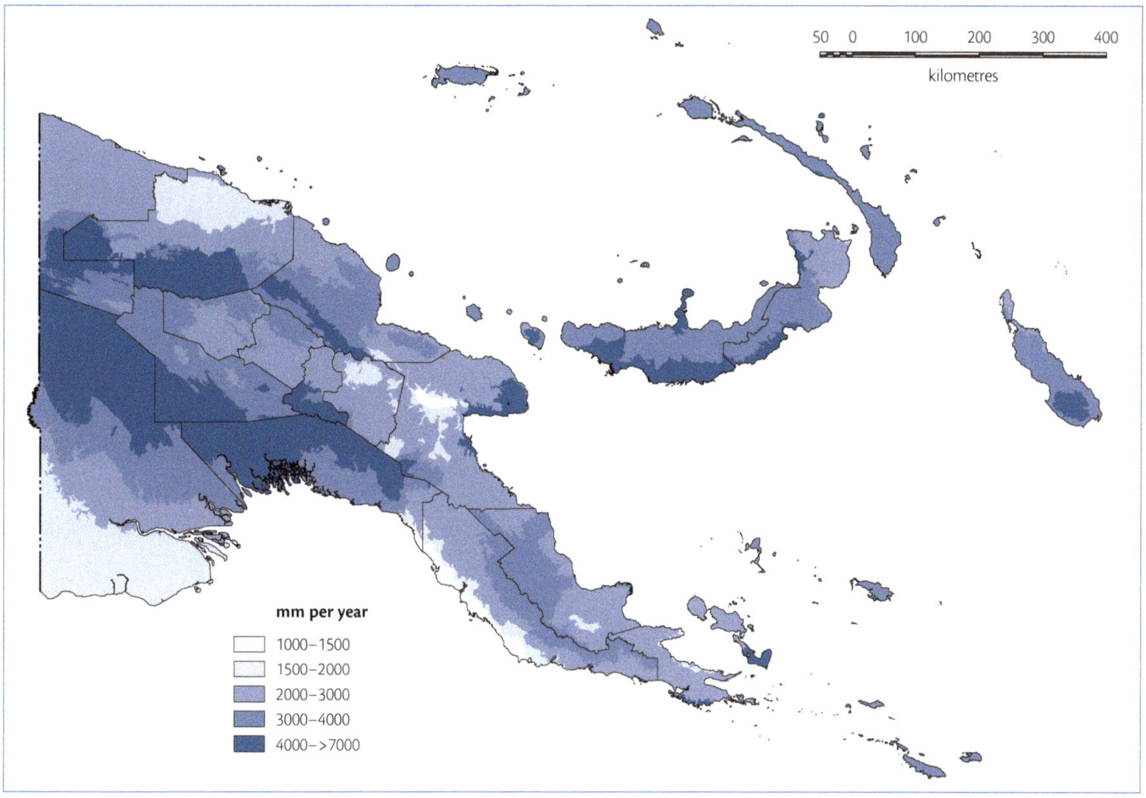

Figure 1.5.1 Annual rainfall from PNGRIS. Source: PNGRIS.

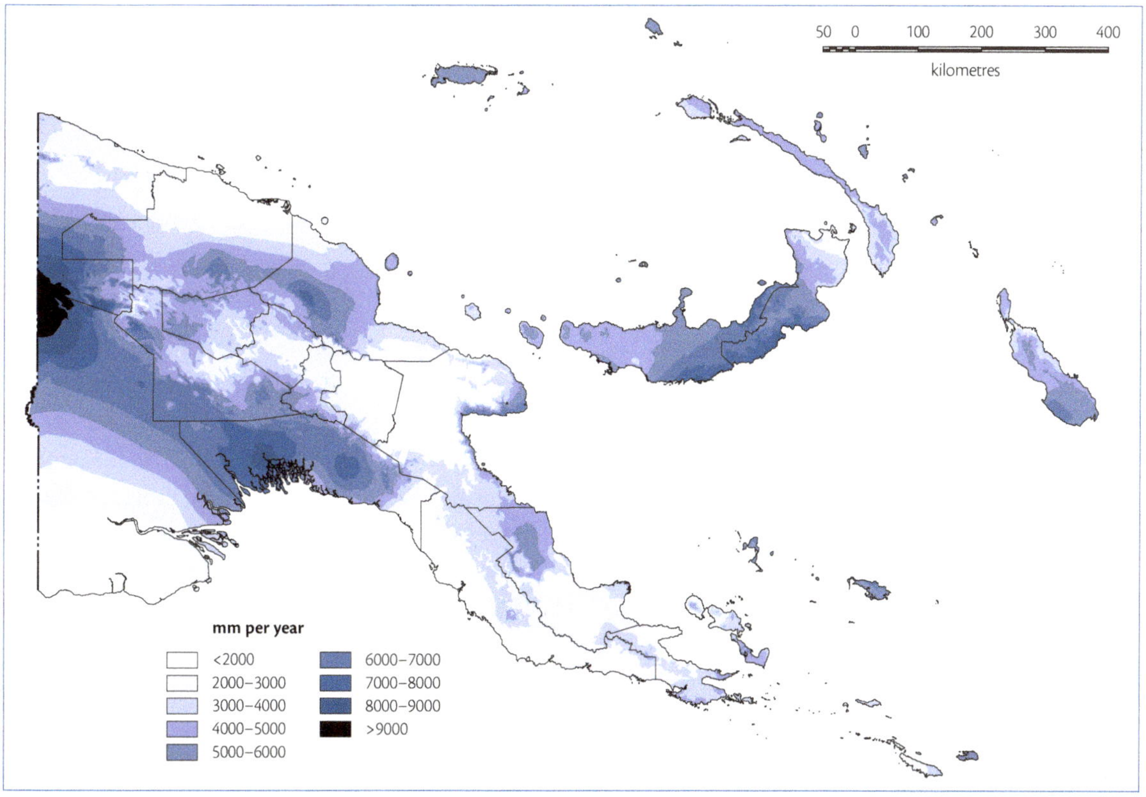

Figure 1.5.2 Annual rainfall from CRES. Source: Centre for Resource and Environmental Science, ANU.

Rainfall seasonality

Most rain falls between January and April in many parts of PNG, with the least falling between May and August. In some parts of the country this pattern is reversed and more rain is received between May and August (Gulf Province, the Huon Gulf around Lae and Finschhafen, the southern part of mainland Milne Bay Province, the southern coast of New Britain and the south of Bougainville Island).

Seasonal rainfall differences can be measured as the relative difference in rainfall between the dry season and the wet season. The most seasonal parts of PNG are the southern half of Western Province, inland and coastal south-east Central Province, the north coast of mainland Milne Bay Province, most of Eastern Highlands Province, and the Markham Valley and north coast of the Huon Peninsula in Morobe Province (Figure 1.5.3).

In other places, rain is received all year round and has no seasonal pattern. There is no seasonal difference in rainfall in the northern part of Western Province, much of Southern Highlands Province, the southern parts of Sandaun and East Sepik provinces, Manus Province, and some of the islands in Milne Bay Province. There are no parts of PNG that are dry all year round.

Rainfall variability and drought

PNG possesses highly reliable annual rainfall that does not vary greatly from year to year for most of the country. Year to year variability is lowest in the highlands.

High rainfall variability occurs in only a few areas: in western Gulf Province, in southern Western Province around the mouth of the Fly River, on East Cape and the islands of Milne Bay Province, and on the south coast of New Britain.

From time to time PNG experiences periods of uncharacteristically low rainfall that are associated with the El Niño Southern Oscillation (ENSO) phenomenon. These events can seriously disrupt food production in PNG (see Section 1.6).

Soil water surpluses and deficits

The amount of water in the soil available to plants is critical to agriculture. Soil water is measured as 'water balance'. Different types of soils can hold differing amounts of water and water can be absorbed or lost from different soils at different rates. The amount of water a soil can hold is known as the 'field capacity' of a soil. The balance of water in the soil (measured in millimetres per day) is the difference between the amount of water entering the soil as rain and the amount of water lost from the soil as evaporation (from the surface) or transpiration (lost from the leaves of plants), or drained downwards through the soil beyond the reach of agricultural plants (Figure 1.5.4). When a soil is capable of absorbing more water (that is, the soil is below field capacity), but no water is supplied by rainfall or irrigation, the water balance is said to be in 'deficit'. When water begins to run off the soil surface and the soil can absorb no more water, it is said to be 'saturated' and the soil water balance is said to be in 'surplus'. Although different soils absorb and lose water at different rates, the most important determinant of soil water surpluses and deficits is rainfall.

Five patterns of soil water balance can be observed in PNG (Figure 1.5.5):

- Regular, seasonal, severe soil water deficits. These occur in the southern part of Western Province and the coast of Central Province, east and west of Port Moresby.
- Irregular, moderate soil water deficits. These occur in the northern part of East Sepik Province, north-west coastal Madang Province, the northern part of Eastern Highlands Province, the Markham Valley in Morobe Province, the Ramu Valley in Madang Province, the central part of Western Province, inland coastal Central Province, parts of Milne Bay Province and the north-east lowlands of the Gazelle Peninsula in East New Britain Province.

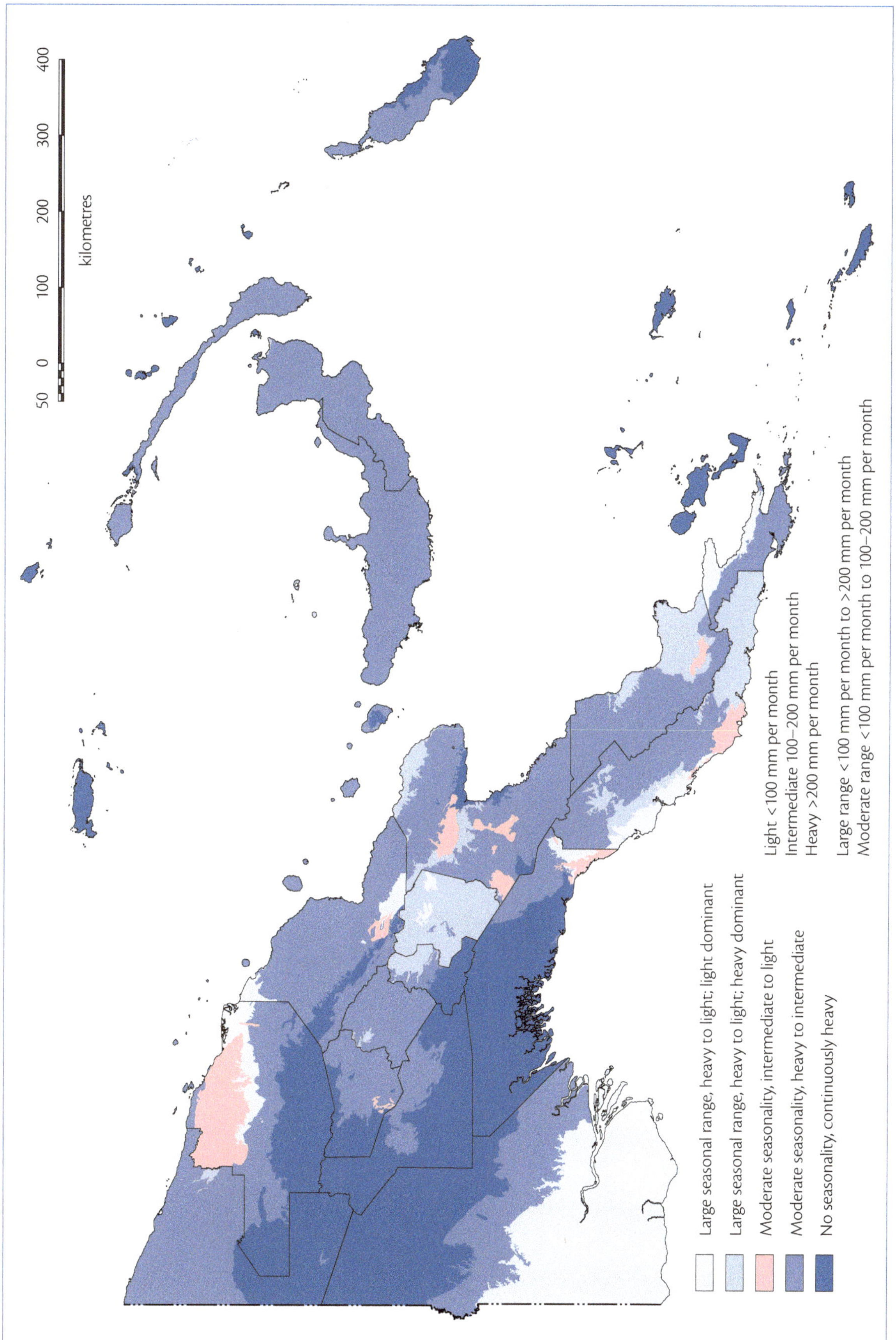

Figure 1.5.3 Rainfall seasonality. Source: PNGRIS.

- Infrequent, slight soil water deficits. This is the pattern over much of the New Guinea mainland lowlands (below 1200 m altitude), on much of the Gazelle Peninsula of East New Britain Province, and on New Ireland Province.
- Rare deficits with moderate soil water surpluses. This is the pattern on the New Guinea mainland at middle altitudes (1200–1500 m) and over much of Manus, New Britain and Bougainville islands.
- Rare deficits with large soil water surpluses. This is the pattern in the highlands of the western mainland, stretching down the southern side of the highlands into Gulf Province, on the eastern end of the Huon Peninsula and along the south coast of New Britain.

These patterns of soil water balance have a strong influence on agricultural systems (see Part 3). Where rainfall is high and regular and soils are usually saturated, digging drains to remove water from the soil and planting certain crops in mounds to raise their roots above the saturated soil is critical for successful agricultural production.

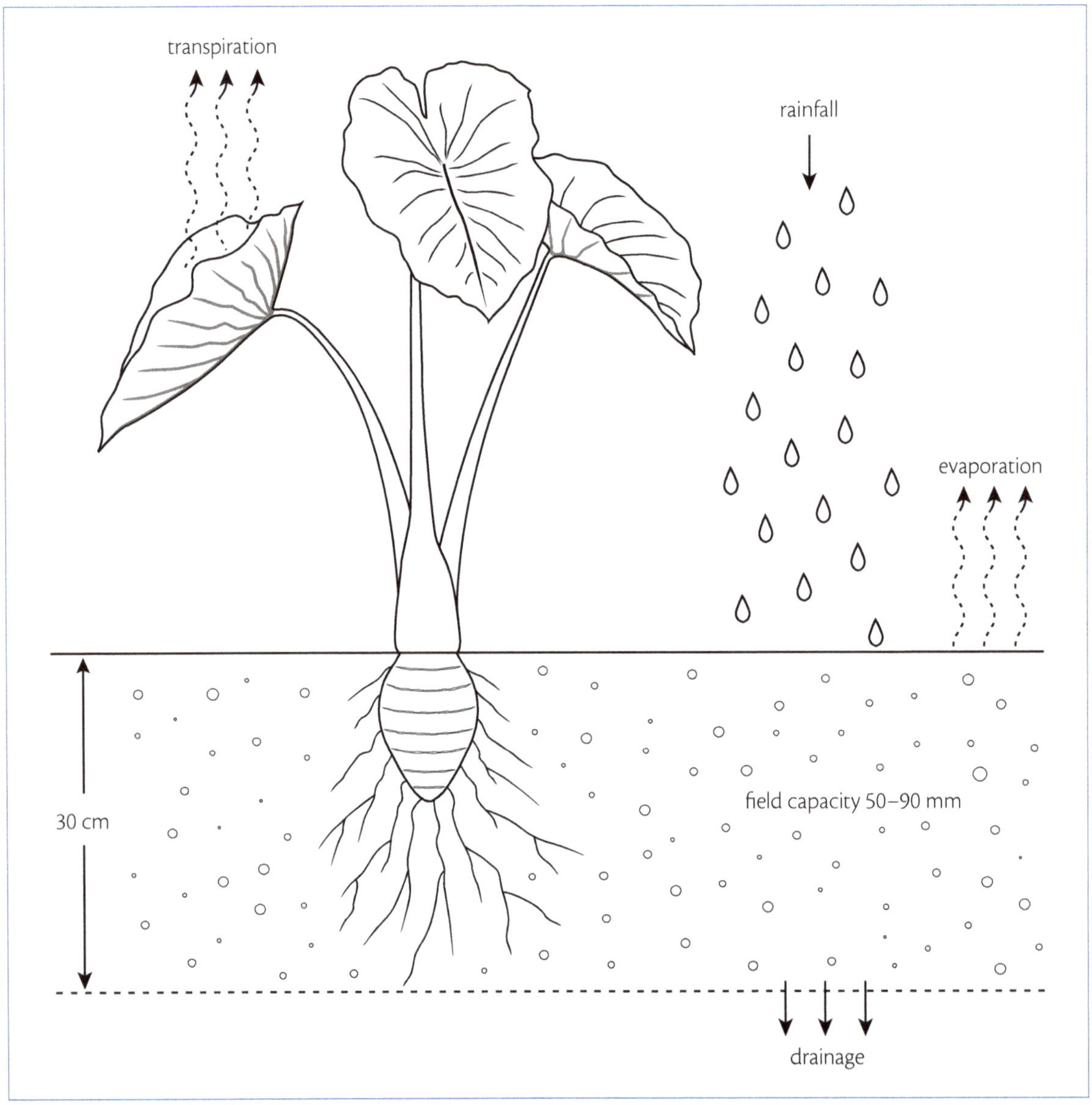

Figure 1.5.4 Water cycle and soil field capacity. Drawing by Catherine Eadie.

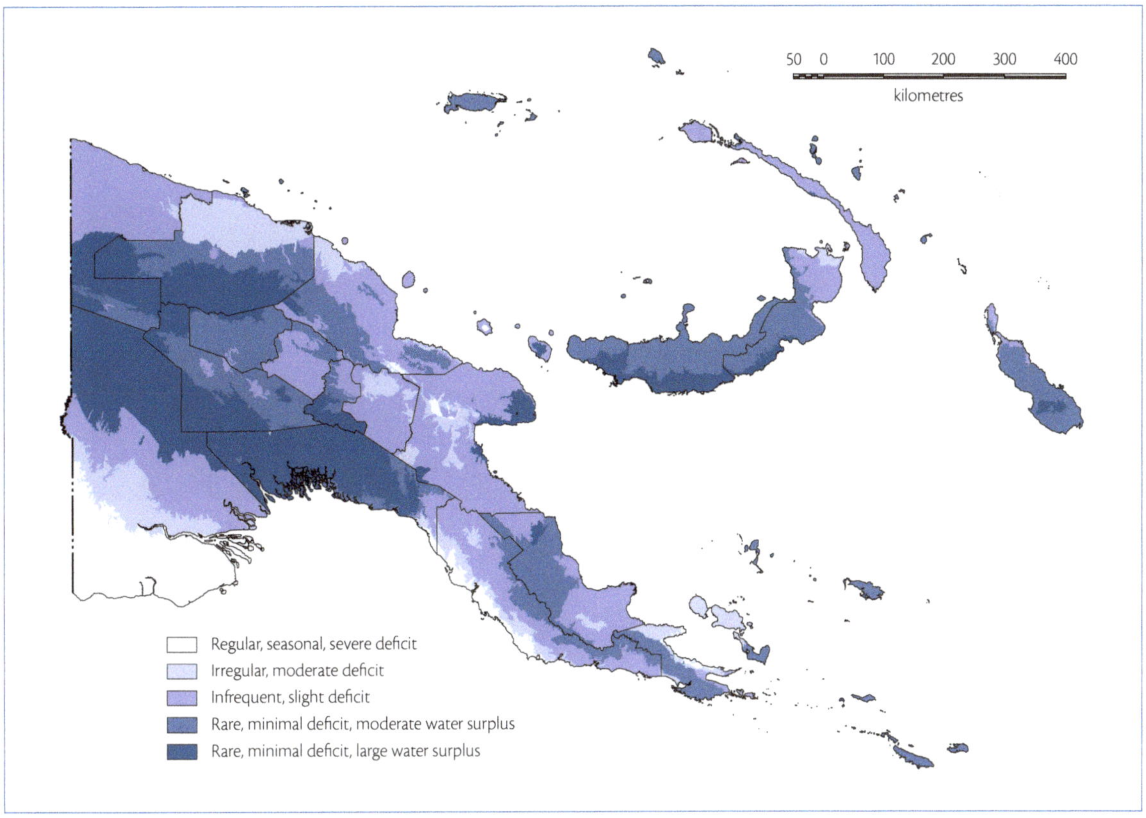

Figure 1.5.5 Soil water deficit and surplus. Source: PNGRIS.

Where regular seasonal soil water deficits occur, a number of techniques are used to overcome the lack of soil water. The most common technique is to plant a mix of different crop species. For example, taro, yam, banana and sweet potato are usually planted together at the start of the wetter months (October–November in places where rainfall seasonality is experienced). This makes food available throughout the year because of the different maturation times of these crops (Figure 1.13.1). Irrigation in PNG was not common and is now only a significant practice in a few villages in the Rabaraba area of mainland Milne Bay Province. It was practised more widely in the past, mostly with taro. The last remnants of such systems are still practised in a few locations such as the Lamari Valley of Eastern Highlands Province and the Kabwum area of the Huon Peninsula.

Sources

Bellamy, J.A. and McAlpine, J.R. (1995). *Papua New Guinea Inventory of Natural Resources, Population Distribution and Land Use Handbook*. Second edition. PNGRIS Publication No. 6. Australian Agency for International Development, Canberra.

McAlpine, J.R., Keig, G. and Short, K. (1975). *Climatic tables for Papua New Guinea*. Division of Land Use Research Technical Paper No. 37. Commonwealth Scientific and Industrial Research Organisation, Melbourne.

McAlpine, J.R., Keig, G. and Falls, R. (1983). *Climate of Papua New Guinea*. Commonwealth Scientific and Industrial Research Organisation in association with Australian National University Press, Canberra.

Radcliffe, D.J. (1984). *The land resources of Upper Mendi*. AFTSEMU Technical Report No. 8. Agricultural Field Trials, Studies, Extension and Monitoring Unit, Southern Highlands Rural Development Project, Mendi.

1.6 El Niño Southern Oscillation (ENSO) and food supply

The El Niño Southern Oscillation (ENSO) phenomenon is a very important influence on high and low rainfall extremes in PNG. The Southern Oscillation is a global phenomenon in which the temperature of the sea, the air pressure over the sea, and the circulation of air across the oceans move in unison, from one extreme to another (that is, they oscillate).

An El Niño event occurs when the Southern Oscillation moves towards one of its extremes. Extreme El Niño events are associated both with periods of very high rainfall and very low rainfall, usually in a sequence of a period of higher than normal rainfall at the beginning of an ENSO event, followed by a period of lower than normal rainfall and finally a period of higher than normal rainfall as the ENSO event fades away. Excessively wet periods can seriously disrupt food supply but, in a country where rainfall is usually high, the most spectacular disruptions are caused by prolonged droughts.

As well as causing rainfall extremes over large areas of PNG, El Niño events also cause clear skies over what is normally one of the cloudiest countries in the world. Clear skies at night and lower than normal relative humidity allow the heat received from the sun during the day to radiate into the sky at night. Under these conditions, night-time temperatures can fall to below freezing (0 °C) at locations above 2200 m altitude. Succulent plants are damaged by low temperatures and food supply may be disrupted.

The El Niño Southern Oscillation (ENSO) phenomenon

Much of the time, the ocean in the western Pacific and around PNG is warmer than the waters in the eastern Pacific. Warm sea surface temperatures cause low-pressure air to rise over the western Pacific from where it moves east across the Pacific, and descends over the eastern Pacific. This is known as the Walker circulation (Figure 1.6.1).

Under conditions of warm seas and low air pressure, large amounts of water from the sea can be absorbed into the air as water vapour. When this warm, wet air moves over land and rises up over mountain ranges, it cools, and the water vapour in the air condenses and falls as rain. During an El Niño event however, the ocean around PNG becomes cooler than the ocean in the eastern Pacific. The cooler seas cause the air above it to cool and to descend, creating higher air pressure on the surface. As a result, the Walker circulation reverses, and cooler, drier, high-pressure air descends over PNG (Figure 1.6.1). Cooler, drier, higher-pressure air can absorb less water from the ocean. As a result, less rain falls over PNG. In addition, cloud is reduced and skies are clearer, especially at night.

It is possible to study the behaviour of the Southern Oscillation across the Pacific for the past 130 years. Since 1876, the air pressure at sea level has been measured at Tahiti in the eastern Pacific and at Darwin in the western Pacific. Using these two sets of records, the difference between the pressure at Tahiti and at Darwin has been standardised into what is known as the Southern Oscillation Index (SOI).

The SOI is usually a positive number because it is calculated in a way that represents the usual situation in which the air pressure at Tahiti is higher than the pressure at Darwin. During an ENSO event, the SOI becomes negative because the pressure at Darwin becomes higher than the pressure at Tahiti (Figure 1.6.2 shows the SOI during the 1997–98 El Niño event).

Although the SOI is now only one of a number of measures of ENSO, the SOI record allows the past 130 years of ENSO events to be identified and their impacts on PNG examined. Between 1876 and 2005 there were 25 years when the SOI was less than −10 in April and October. In 11 of these 25 years, official reports, newspapers and oral histories describe the occurrence of some or all of the following: widespread droughts, repeated frosts in the highlands, severe food shortages, forest fires, and migrations (of people from areas with inadequate food).

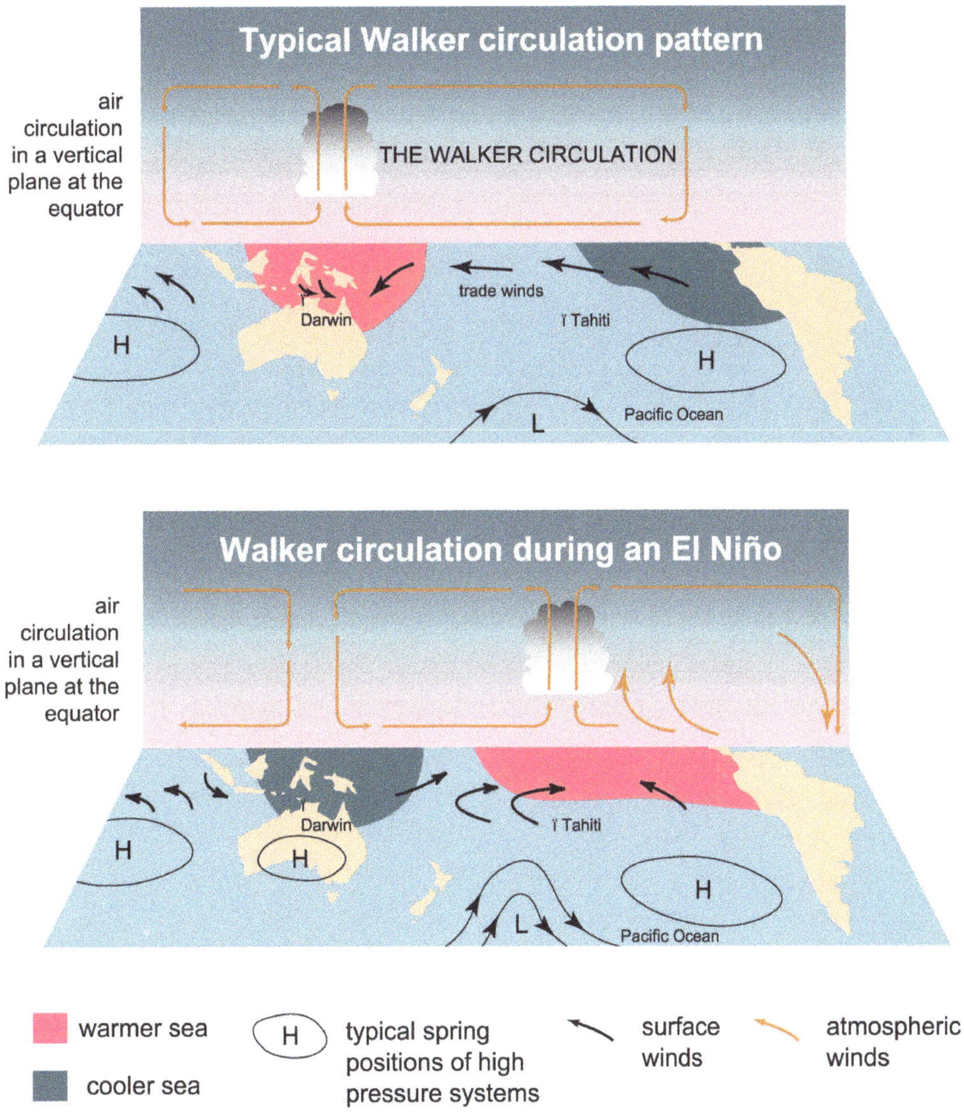

Figure 1.6.1 The circulation of air over the Pacific Ocean and the Southern Oscillation.
Source: Australian Government Bureau of Meteorology website <http://www.bom.gov.au/climate/glossary/elnino/elnino.shtml>. © Commonwealth of Australia, reproduced by permission.

Of these 11 years, there are four years in which all these impacts are reported to be very serious. These years are 1902, 1914, 1941 and 1997.[1] Importantly, not all large ENSO events have severe impacts in PNG. This makes it difficult to predict which particular ENSO event will have a severe impact, but if the frequency of severe impacts since 1876 does not change (with climate change for example), an ENSO event can be expected to have a severe impact on PNG about once every 30 years. This is an average figure and it cannot be used to predict a particular event.

ENSO, food production and famine relief

A strong ENSO event will reduce rainfall significantly in normally wet areas and prolong dry seasons in areas that are seasonally dry, severely reducing food production.[2] Water sources that are used to make sago can dry up and although the larger rivers will continue to flow, the reduction in flow allows salt water from the ocean to invade estuaries, preventing sago production near the coast. At high altitudes, above 2200 m, repeated frosts will completely disrupt food production. In the period preceding the low rainfall, there is often a period of exceptionally high rainfall, which can reduce food production in some systems. In sweet potato systems, the reduction in food supply will occur around 4–5 months after the heavy rainfall.

Most people in PNG have strategies that enable food supply to be maintained during short periods of disruption, including the cultivation of a large number of edible plant species and a knowledge of uncultivated (or wild) edible species that can be collected. However, a severe ENSO event will seriously disrupt the supply of food to large numbers of people over much of PNG for a number of months. Oral histories and colonial reports describe unknown but significant numbers of people dying in the large ENSO events of 1902, 1914 and 1941.

Since at least 1914, colonial administrations have attempted to prevent deaths from starvation during severe ENSO events by the provision of food relief, initially in an unsystematic manner. The first large-scale, systematic provision of food occurred in 1972 following repeated frosts in the PNG highlands. The then Australian colonial administration transported rice, tinned fish and cooking oil into the highlands to prevent what they thought would be a catastrophic movement of starving people out of the frost-affected areas. However, ethnographic research conducted during and after the event showed that migration was a traditional means of coping with frost damage and the migrations were orderly and to places occupied by relatives. Even so, many old people and children are known to have died on the journey through the mountains to the lower valleys.

Following Independence in 1975, the provision of food by government agencies to people said to be starving happened more frequently, particularly in election years, when local political representatives tried to get food delivered to people within their electorates, often on dubious grounds. By the 1990s, requests for food relief by political representatives had come to be viewed with suspicion and cynicism by administrators.

The 1997–98 ENSO event had severe impacts in PNG. By October 1997 it was estimated 150 000 people were eating wild foods and in December this estimate, based on a second nationwide field assessment, had risen to 260 000. A further 980 000 people were assessed to be eating poor quality garden food, in reduced quantities. Many people were forced to walk for hours to collect drinking water of questionable quality. Death rates in some isolated places increased, many schools closed, and many health centres were not staffed and had no medical supplies anyway. In some centres town water supplies were threatened. The hydro-electrical station at Sirinumu, inland of Port Moresby, was forced to stop generating (in order to conserve drinking water in the dam), causing serious power supply disruptions in Port Moresby city. The Ok Tedi mine in Western Province

1 It is possible that 1884 was also a year of severe impacts. The longer ago the event, the less likely it will have been reported. For example, written reports from the Eastern Highlands begin in the late 1930s but they do not begin in the Southern Highlands until the 1950s.

2 South Bougainville Island has a different pattern, and the rainfall tends to be much higher in an ENSO event.

closed for seven months because the Fly River became un-navigable and the Porgera mine in Enga Province closed for six weeks through lack of water for processing operations. The slump in mineral exports resulted in a severe loss of foreign currency to the PNG economy.

AusAID and the Australian Defence Force instigated a relief program that provided food to more than 100 000 people in areas accessible only by air. The PNG national government purchased around 23 500 tonnes of rice for relief in 1997–98 (compared to around 9400 tonnes purchased by AusAID and 7000 tonnes purchased by individual provincial governments), but much of this food was delivered after the most critical period in December 1997 had passed.

Rural people adopted a number of different strategies to survive in the drought:

- They ate 'famine foods'; either foods that are not eaten often, or that are only eaten in times of hunger.
- They raised small amounts of cash by killing and selling pigs, cooking and selling pork and vegetables, buying packets of cigarettes and selling them individually, and by selling artefacts. With the cash earned they purchased imported rice and flour.
- They moved to areas where food was available. It was estimated that in Enga Province up to 75% of people moved out of the high-altitude Kandep and Marient basins during 1997 and walked over mountain passes into the Tsak and Lai valleys. In 1997 however, the drought was so widespread and severe that food in these areas was also critically short. Many people moved further to stay with relatives in towns. For example, an estimated 20–25% of the population from villages at Elimbari in Simbu Province migrated to Goroka, Lae and Port Moresby.
- People employed in urban areas or at mines either sent money to their rural relatives or purchased rice and sent it to them.

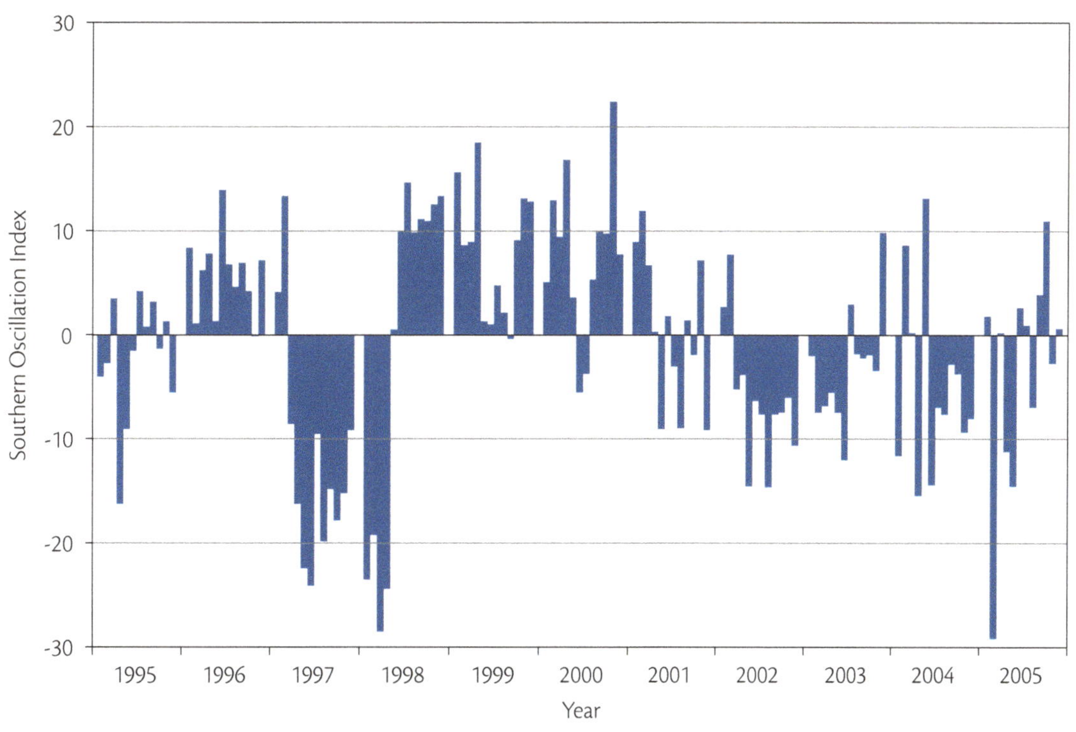

Figure 1.6.2 The Southern Oscillation Index, 1995–2005. Source: Australian Government Bureau of Meteorology website. Data available at <ftp://ftp.bom.gov.au/anon/home/ncc/www/sco/soi/soiplaintext.html>.

Rice imports into PNG during 1997–98 increased by 66 000 tonnes (38%) (see Figure 2.7.1, Table A2.7.1). More than 80% (54 000 tonnes) of this additional rice was sold through retail outlets. It was purchased by people who used their savings or raised small amounts of cash in local markets, and by wage and salary earners who purchased rice for their rural relatives.

Lessons from the ENSO record in PNG

The following are the most important lessons to be learned from the record of the impact of ENSO events in PNG:

- Minor ENSO events occur around every five to six years and may cause local frosts at high altitudes and minor food shortages in the highlands from high rainfall.
- Around every 12 years an ENSO event will have a significant impact on PNG food production because of drought and high rainfall, mainly in the highlands. Widespread frosts may occur. Urban water supplies and electricity generation may be affected.
- Around every 30 years an ENSO event will have a very significant effect on PNG food production all over the country. Sago production will be also affected. Repeated and widespread frosts will occur, completely disrupting food production at higher altitudes. Large bushfires may occur. Urban water supplies and electricity generation will be affected. In isolated areas, the death rate will increase.
- It is not yet possible to predict, with any level of confidence, what severity of impact any particular ENSO event will have on PNG.
- The impact of the 1997 event was worst further away from the equator and in the poorest and most isolated parts of PNG. People with access to earnings from cash cropping or wage labour were generally able to look after themselves, with help from their relatives, by purchasing imported food through retail outlets. Since the 1997 event, their ability to do this may have been significantly reduced by a decline in the value of the kina and a consequent increase in the cost of imported foods.
- When an ENSO event of the severity of 1997 occurs, there is not much people can do to maintain food production. It is not realistic to expect people to protect field crops against repeated severe frosts for months at a time, by covering crops or by setting smoky fires at night.
- Conversely, there *is* much that can be done to assist people to recover after the drought is over. In 1997 impressively large areas of food crops were planted as soon as adequate rain began to fall.
- Overall, food security in PNG is threatened more often by too much water than by too little water. But the food supply problems caused by excessive rainfall are insidious, delayed, difficult to identify and do not affect the whole country at the same time. Food shortages caused by drought are immediate, spectacular and widespread. Contingency planning should take into account threats to food security from all environmental causes, not just from drought.
- People in areas that have a regular dry season use agricultural systems that are adapted to a lack of water for part of the year. They are adversely affected only by very severe ENSO events. People in areas that do not have a regular dry season use agricultural systems that are adapted to deal with excessive water, thus they can be severely impacted by drought.
- Long-term PNG food security strategies, including drought contingency plans, must include the use of imported foods to feed a significant proportion of the population for a short time from time-to-time. But the government need not be directly involved in either the importation or the distribution of this food.

Sources

Allen, B.J. and Bourke, R.M. (2001). The 1997 drought and frost in PNG: overview and policy implications. In Bourke, R.M., Allen, M.G. and Salisbury, J.G. (eds). *Food Security for Papua New Guinea. Proceedings of the Papua New Guinea Food and Nutrition 2000 Conference.* ACIAR Proceedings No. 99. Australian Centre for International Agricultural Research, Canberra. pp. 155–163.

Allen, B., Brookfield, H. and Byron, Y. (eds) (1989). Frost and Drought in the Highlands of Papua New Guinea: A Special Collection of Papers. *Mountain Research and Development*. 9(3):199–334.

Australian Government Bureau of Meteorology website <http://www.bom.gov.au/climate/glossary/elnino/elnino.shtml>.

Bang, S.K. and Sitango, K. (2002). Indigenous drought coping strategies and risk management in Papua New Guinea. In Yokoyama, S. and Concepcion, R.N. (eds). *Coping against El Niño for Stabilizing Rainfed Agriculture: Lessons from Asia and the Pacific. Proceedings of a Joint Workshop held in Cebu, the Philippines*. CGPRT Centre Monograph No. 43. CGPRT Centre, Bogor, Indonesia. pp. 147–163.

Bang, S.K., Poloma, S. and Allen, B. (2003). *Stabilization of Upland Agriculture Under El Niño-Induced Climatic Risk: Impact Assessment and Mitigation Measures in Papua New Guinea*. CGPRT Centre Working Paper No. 73. CGPRT Centre, Bogor, Indonesia.

Barter, P. (2001). Responses to the 1997–98 drought in PNG. In Bourke, R.M., Allen, M.G. and Salisbury, J.G. (eds). *Food Security for Papua New Guinea. Proceedings of the Papua New Guinea Food and Nutrition 2000 Conference.* ACIAR Proceedings No. 99. Australian Centre for International Agricultural Research, Canberra. pp. 260–264.

Bourke, R.M. (2000). Impact of the 1997 drought and frosts in Papua New Guinea. In Grove, R.H. and Chappell, J. (eds). *El Niño – History and Crisis: Studies from the Asia-Pacific Region*. The White Horse Press, Cambridge. pp. 149–170.

Bourke, R.M. (2005). How Papua New Guinea villagers survived the 1997 drought and frosts. *Development Bulletin* (67):27–29.

Kapal, D., Bang, S., Askin, D. and Allen, B. (eds) (2003). *Drought Response: On-Farm Coping Strategies*. NARI Information Bulletin No. 6. National Agricultural Research Institute, Lae.

Sudradjat, A. (2001). Australia's response to the 1997 PNG drought. In Bourke, R.M., Allen, M.G. and Salisbury, J.G. (eds). *Food Security for Papua New Guinea. Proceedings of the Papua New Guinea Food and Nutrition 2000 Conference.* ACIAR Proceedings No. 99. Australian Centre for International Agricultural Research, Canberra. pp. 218–221.

Whitecross, N. and Franklin, P. (2001). The role of rice in the 1997 PNG drought. In Bourke, R.M., Allen, M.G. and Salisbury, J.G. (eds). *Food Security for Papua New Guinea. Proceedings of the Papua New Guinea Food and Nutrition 2000 Conference.* ACIAR Proceedings No. 99. Australian Centre for International Agricultural Research, Canberra. pp. 255–259.

1.7 Temperature, cloudiness and sunshine

Temperature and light are critical determinants of plant growth (see Section 1.13). In PNG, a country of high mountains, altitude is a major influence on maximum and minimum temperatures and on cloud cover. Cloud cover influences the amount of sunshine available to plants.

Temperature

Altitude has the greatest influence on temperature in PNG. Above 500 m, temperature falls at a regular rate of 0.5 °C for every 100 m increase in altitude, or 5 °C for every 1000 m. This decline in temperature with altitude is known as the 'lapse rate'. The lapse rate causes differences in maximum and minimum temperatures of up to 16 °C over the range of occupied land in PNG, from sea level to 2800 m (Table 1.7.1, Figure 1.7.1).

The average temperature can be predicted for any altitude in PNG. The following formulas can be used to calculate the maximum, minimum and mean temperatures for locations away from the coast (these formulas do not apply to coastal locations). The actual average temperatures may vary a little from those calculated.

$$Y_{max.} = 32.67 - 0.0052\,x$$
$$Y_{ann.} = 27.32 - 0.0052\,x$$
$$Y_{min.} = 22.08 - 0.0052\,x$$

where: $Y_{max.}$, $Y_{ann.}$ and $Y_{min.}$ are mean maximum, mean annual and mean minimum air temperatures (°C) respectively; and x is the altitude in metres

Temperature is also influenced by latitude, or distance from the equator. In PNG, the further away from the equator a place is located, the greater the range in temperatures during the year. In places where the rainfall pattern is reversed (that is, most rain falls between May and September, which is the Southern Hemisphere winter – see Section 1.5), the annual differences in temperature are greater than at locations which receive more rain in January to April or evenly throughout the year (see Table 1.13.1).

Cloudiness and sunshine

Cloudiness influences the amount of sunshine that reaches plants, and the amount of sunshine received by plants influences their growth and productivity. PNG is a very cloudy country (Figure 1.7.2). Between 2 pm and 4 pm from December to February and from June to August, the skies over most of the

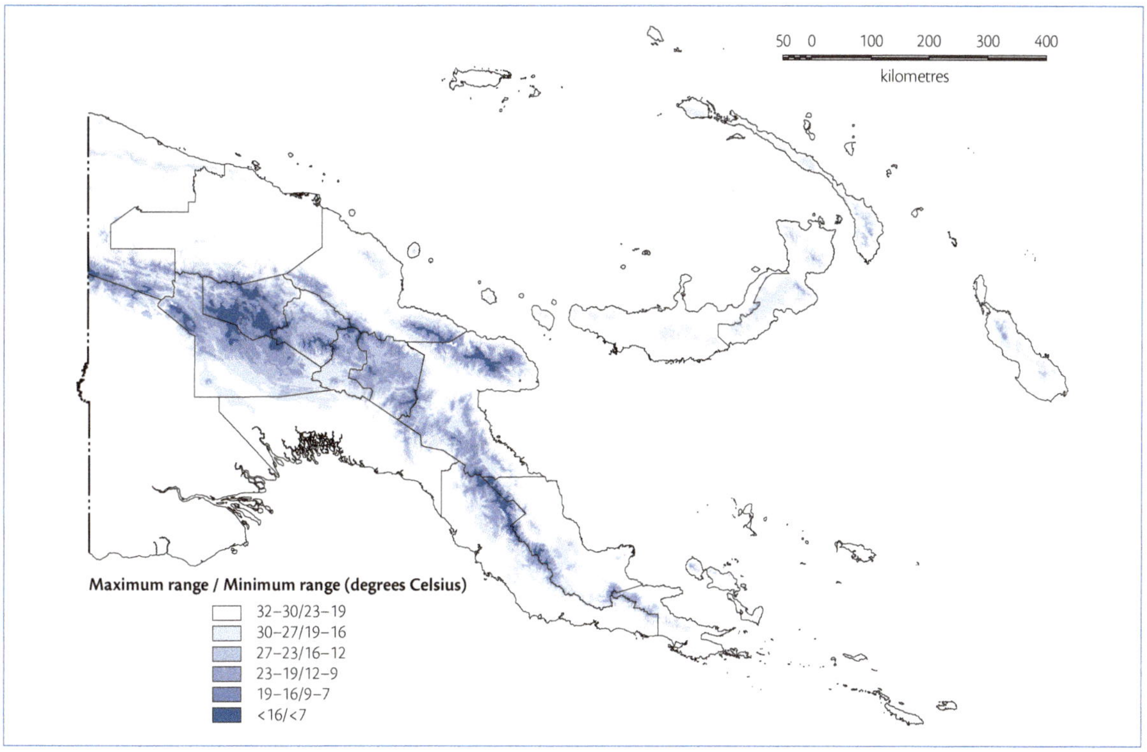

Figure 1.7.1 Generalised temperature zones in PNG in January, based on lapse rates and altitude (°C) (July temperatures are less than 3 °C lower). Sources: McAlpine et al. (1983:95); PNGRIS.

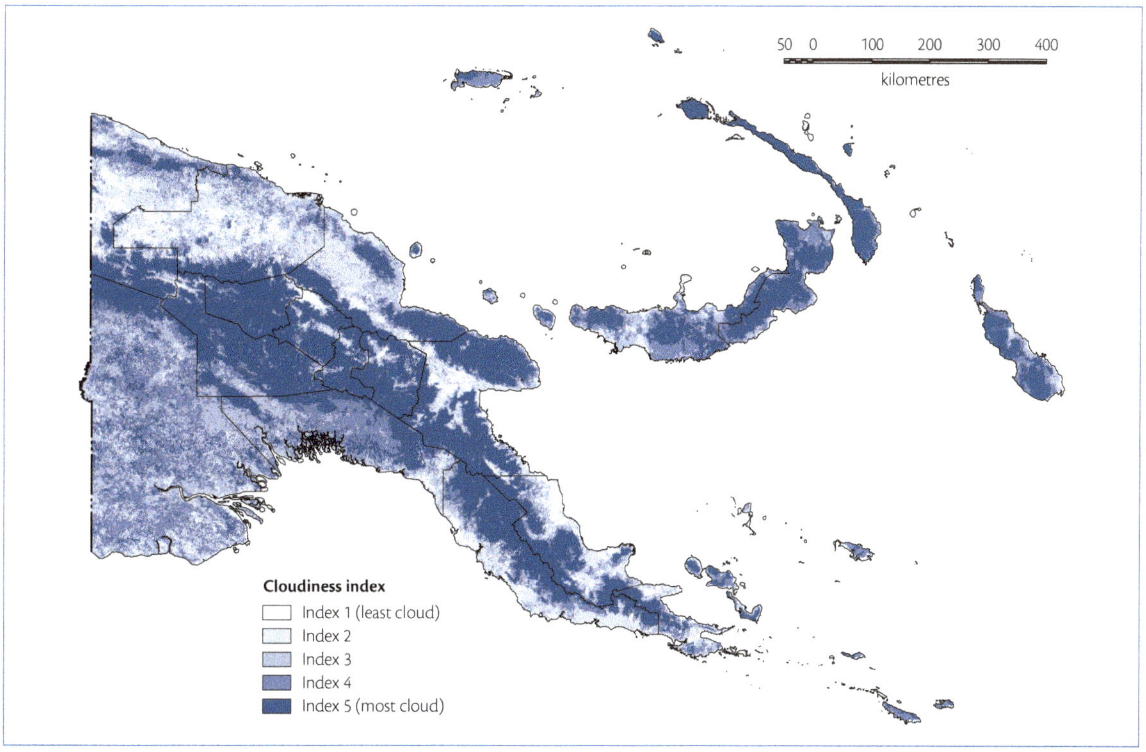

Figure 1.7.2 Cloud cover over PNG, 1997–1998. Source: Developed from a composite set of Advanced Very High Resolution Radiometer (AVHRR) satellite images for twelve months in 1997 and 1998, provided by the CSIRO Division of Land and Water.

Table 1.7.1 Altitude classes and associated maximum and minimum temperatures

Altitude class (metres above sea level)	Maximum temperature (°C)	Minimum temperature (°C)
0–600	32–30	23–19
600–1200	30–27	19–16
1200–1800	27–23	16–12
1800–2400	23–19	12–9
2400–2800	19–16	9–7
>2800	<16	<7

Source: Bellamy and McAlpine (1995:89).

country are up to 63% cloud-covered (five oktas).[1] Mountain areas have consistently heavier cloud cover. In areas of seasonal rainfall, cloud cover is greatest during the wetter season.

In PNG a relationship exists between rainfall, cloudiness and sunshine. The greatest amount of sunshine is received in places with the lowest rainfall. The highlands have less sunshine than other parts of the country. For example, at Goroka in Eastern Highlands Province, sunlight ranges from four hours per day to six hours per day over the year. Nevertheless, most places in PNG are, on average, likely to receive some sunshine every day.

Daylength

The time a plant is exposed to light in a day is called the photoperiod. It is dependent on daylength. There are only small differences in daylength during the year in most of PNG, because the country is aligned east to west just south of the equator. Daylength is influenced by distance from the equator. At Lorengau (latitude 2° S) on Manus Island, the difference between the longest day (in December) and the shortest day (in June) is only 14 minutes. At Port Moresby (latitude 9° 29' S), the difference is 66 minutes. Differences are greater still in the southern part of Milne Bay Province.

Sources

Bellamy, J.A. and McAlpine, J.R. (1995). *Papua New Guinea Inventory of Natural Resources, Population Distribution and Land Use Handbook*. Second edition. PNGRIS Publication No. 6. Australian Agency for International Development, Canberra.

McAlpine, J.R., Keig, G. and Falls, R. (1983). *Climate of Papua New Guinea*. Commonwealth Scientific and Industrial Research Organisation in association with Australian National University Press, Canberra.

McVicar, T.R. and Bierwirth, P.N. (2001). Rapidly assessing the 1997 drought in Papua New Guinea using composite AVHRR imagery. *International Journal of Remote Sensing* 22(11):2109–2128.

1 Cloud cover is measure in oktas. One okta is one-eighth of the sky. If the sky is completely covered in cloud, the cloud cover is given as eight oktas, or 8/8. If the sky is half covered, the cloud cover is given as four oktas, or 4/8.

1.8 Climate change

The global climate is changing, almost certainly as a result of human activity. The most commonly reported aspect of climate change is an increase in average temperature, but this is associated with many other changes. This section provides a brief overview of the causes of global climate change; what changes have occurred globally and in PNG so far; and some implications for agricultural production in PNG.

Why the global climate is changing

The earth is a natural greenhouse. Energy from the sun is absorbed by the earth and radiated back into space. But not all of the energy from the sun is lost back into space. The atmosphere, which is made up of a number of different gases, is warmed and so traps some of the heat from the sun. As long as the composition of the atmosphere remains unchanged, the amount of heat gained from the sun and lost again into space remains constant and the temperature of the planet remains relatively stable (Figure 1.8.1). The trapping of heat in the atmosphere is known as the 'greenhouse effect'.

However, it is now clear that the composition of the gases in the atmosphere is changing. The proportions of carbon dioxide, methane and nitrous oxide are increasing. The amount of carbon dioxide in the atmosphere is at a record level. For example, the concentration of carbon dioxide in the atmosphere over the past 400 000 years has been in the range 180–280 parts per million (ppm). It has risen from 270 ppm in around 1850 to 380 ppm in 2005 (Figure 1.8.2). The Intergovernmental Panel on Climate Change (IPCC) reports that the concentration of greenhouse gases is at the highest level for at least 650 000 years. The addition of these gases to the atmosphere is causing it to become warmer. This is because there is a close relationship between the amount of carbon dioxide, methane and nitrous oxide in the atmosphere and the amount of heat the atmosphere can hold and hence the air temperature at the surface of the earth (Figure 1.8.2).

The reason for the changes in the atmosphere is assumed to be human activity. Carbon dioxide is released when oil, petrol, gas and coal (the fossil fuels) are burned. Many sources of energy on the earth involve the burning of these products. Carbon dioxide is also released when forests are burned; when permafrost in the sub-Arctic region melts; and when organic matter is lost from soil as a result of agricultural activity. Methane comes from a number of sources, including wet rice cultivation and digestion of food by sheep and cattle. Nitrous oxide comes from industrial production and losses of nitrogen from fertiliser applied to crops.

Some of the heat in the atmosphere is absorbed by the oceans. So as the atmosphere warms because of the changes in its composition, the oceans also warm and expand (thermal expansion). A warmer atmosphere also causes the ice at poles and in glaciers to melt. The melted ice adds water to the oceans. Thermal expansion and melting ice result in higher sea levels. The rise in atmospheric and oceanic temperatures will cause changes in global wind patterns. These three things, warmer air, warmer seas and changes in wind patterns, may cause changes in rainfall and snowfall. Annual rainfall could increase or decrease or the time of the year when rain is received might change. These changes could also result in a change in the number and severity of extreme climatic events, including more cyclones and episodes of very high rainfall or drought.

Most of our understanding of the climate is based on analyses of records of the climate in the past. Prediction of climate in the future is based on mathematical models developed from these analyses of climatic records. The accuracy of these models is not yet known and different models give somewhat different predictions, but all models predict more warming and significant changes to the climate.

Global climatic change so far and predictions for the future[1]

Firstly, a short overview of changes in global climate up to now and IPCC predictions for future global change is given. The next section focuses on what we know from PNG.

Global temperature

Global mean temperatures have risen 0.8 °C since the late 1800s. However, the trend is not linear and rates of increase in surface temperature are greater after the mid 1970s. From 1979 to 2005, the linear trend was 0.16–0.18 °C per decade, giving total warming of 0.46 °C since 1979. The year 2005 was one of the warmest two years on record.

Changes in extremes of temperature are also consistent with warming of the climate. There has been a widespread reduction in the number of frost days in mid-latitude regions, an increase in the number of warm extremes, and a reduction in the number of daily cold extremes. The most marked changes are for cold nights, which have become rarer over the period 1951–2003 for most of the land regions studied. Warm nights have become more frequent across most of the same regions. Diurnal temperature range (the difference in temperature between day and night) decreased by an average of 0.07 °C per decade over the period 1950–2004.

A summary of all expert assessment for the future is that warming is likely to be in the range 2.0–4.5 °C, with a likely value of about 3 °C by 2100 compared with the period 1980–2000. A temperature rise of 2.0–4.5 °C may not sound very much. But even at the coldest period during the most recent ice age, about 15 000 years ago, the average temperature of the planet was only about 5 °C colder than the present climate. So the increase in temperature in the next 100 years could be as much as the difference between the average global temperature some 15 000 years ago and the present. The average temperature in New Guinea was also about 5 °C colder than the present climate about 18 000 years ago.

Global rainfall

Over the past 30 years (since 1976) there has been a downward trend in rainfall in the tropics between 10° south and 10° north, but this is a very uneven trend. It seems that there has been an increase in the number of heavy rainfall events in many regions of the globe, even in those places where a reduction in total rainfall has occurred.

Droughts have become widespread in various parts of the world since the 1970s. In Australia and Europe, direct links to global warming have been inferred through the extreme nature of high temperatures and heat waves accompanying recent droughts. More generally, decreased rainfall and higher temperatures that increase evaporation and transpiration are important factors that have contributed to more regions experiencing drought (see Figure 1.5.4 for an illustration of these terms).

---

1 Information in this section is taken from the Intergovernmental Panel on Climate Change Draft Fourth Assessment Report. This is due for official release in 2007 after peer review and was made available in May 2006 on a US government website. The IPCC has operated since 1988 and reports about every five years. The working groups are composed of experts in climate and related fields, and their papers are subject to rigorous peer review.

Global warming models predict that rainfall will generally increase in the tropics and over the tropical Pacific in particular, with general decreases in the subtropics, and increases at high latitudes. The intensity of rainfall events is also predicted to increase, particularly in tropical and high-latitude areas that experience increases in mean rainfall. Annual rainfall is likely to increase slightly in the southern Pacific; however, this is a regional prediction and the outcome could be quite different at different locations in PNG.

Global sea level rise

The mean sea level rose in the second half of the 20th century at 1.8 mm per year (Figure 1.8.3). Projections of global average sea level rise for the 21st century due to thermal expansion and melting ice are in the range 130–380 mm by 2100. Thermal expansion will make the greatest contribution to sea level rise.

Most climatic models predict that sea surface temperatures will be warmer in the central and eastern Pacific near the equator, compared with the western Pacific, that is, an El Niño-like response (see Figure 1.6.1). This suggests that PNG and other western Pacific locations will experience more droughts, but there are considerable uncertainties with these predictions.

Climate change in PNG

The data on climate in PNG presented in this book (Sections 1.5, 1.6, 1.10 and 1.11) is largely based on climatic data recorded prior to 1970. Less information has been recorded over the past 30 years, but enough information exists to conclude that the climate is changing in PNG as it is elsewhere in the world. This can be inferred from direct measurements of

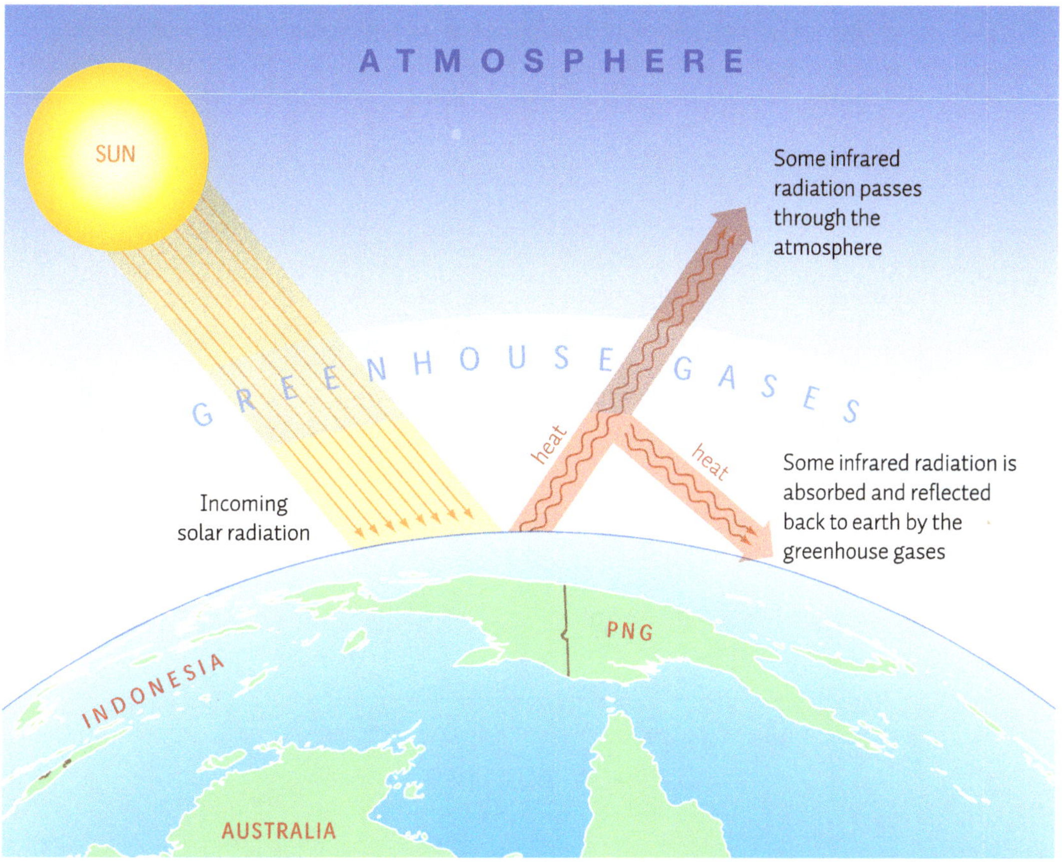

Figure 1.8.1 The greenhouse effect. Source: Cartographic Services, ANU.

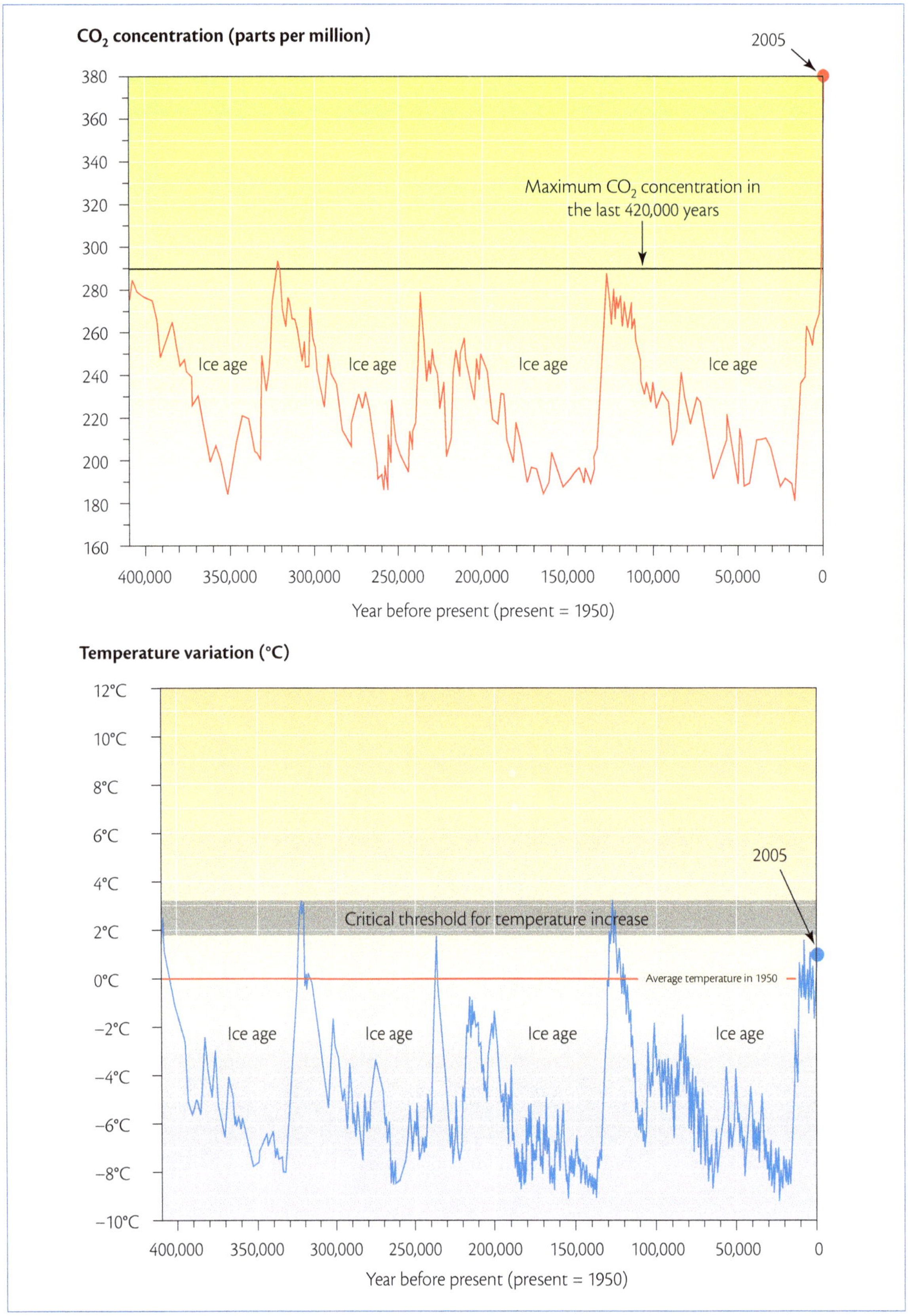

Figure 1.8.2 Record of carbon dioxide concentrations and average global temperature over the past 400 000 years.
Sources: UNEP/GRID-Arendal (2006) and <http://monde-diplomatique.fr/cartes> from data in Petit et al. (1999). Graphic design: Philippe Rekacewicz.

temperature and rainfall, observations of crop growth made by villagers and scientists, and extrapolation from nearby locations, including Indonesia and Australia.

Temperature in PNG

Studies of temperature change in PNG have found that temperatures increased during the 20th century, with most of the increase occurring from about the mid 1970s. An important finding from an agricultural perspective was that the rate of temperature increase was greater for minimum than for maximum temperatures.

An analysis of temperature change from nine coastal locations found that minimum, mean and maximum air temperatures had increased by an average of 0.2 °C per decade. The trends from three of these sites – at Port Moresby, Madang and Kavieng – are illustrated in Figures 1.8.4 and 1.8.5. The rate of temperature increase and the faster rise since the mid 1970s in PNG are similar to the global trends described earlier. Temperatures tended to be lowest in El Niño years and higher in La Niña years that followed. Only limited long-term temperature data runs are available for highland stations in PNG, but they show similar trends to the lowland stations. For example, at Aiyura in Eastern Highlands Province, maximum temperature increased by 0.75 °C (0.3 °C per decade) over the 25 years 1977–2001 (although the daily minimum did not increase over this period).

Another clear indication that temperatures are increasing in the highlands comes from observations on the upper and lower altitudinal limits of crops (Figure 1.13.3). For example, between 1980 and 1984, the author observed coconuts bearing up to an average altitude of 1000 m. Coconuts grew up to 1700–1800 m in the highlands, but did not bear nuts at those altitudes. Occasionally, they had very small nuts at altitudes as high as 1310 m. By the year 2000, coconuts were bearing very small nuts at up to 1450 m altitude in at least three highland valleys. This suggests that temperatures in the highlands have increased by about 0.7 °C over 20 years, which is consistent with the changes in maximum temperature recorded at Aiyura.

Highland villagers also report that crops have been bearing at higher altitudes since around the mid 1990s. As well as coconut, they say that betel nut, mango and breadfruit are now bearing where the trees previously grew but did not bear. Such changes have been documented at a village at about

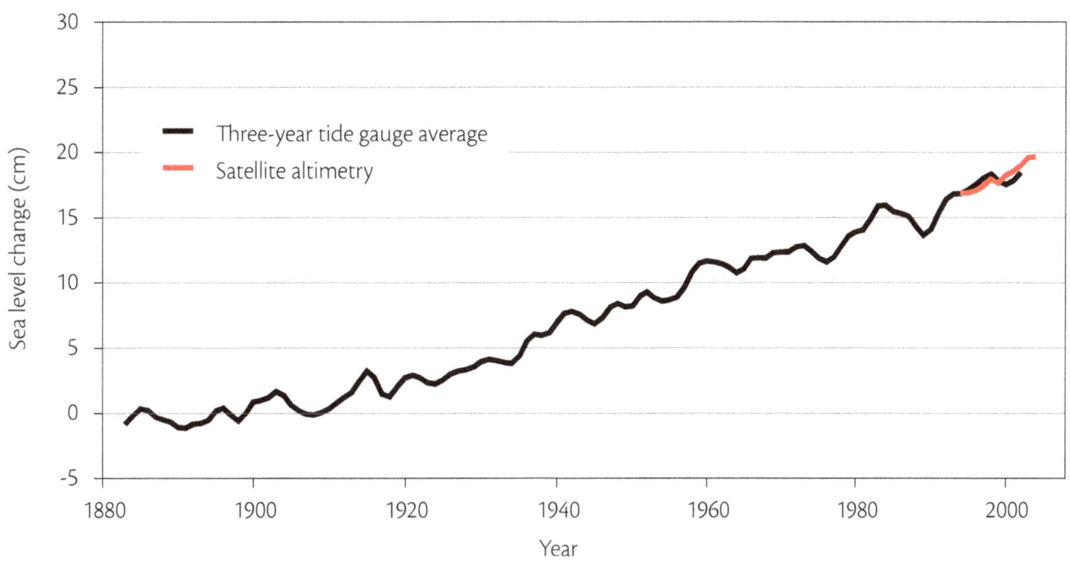

Figure 1.8.3 Global mean sea level rise, 1883–2002. **Note:** The tide gauge data spans 1883–2002 and the satellite altimetry data spans 1993–2004. Source: Robert A. Rohde / <http://www.globalwarmingart.com/>.

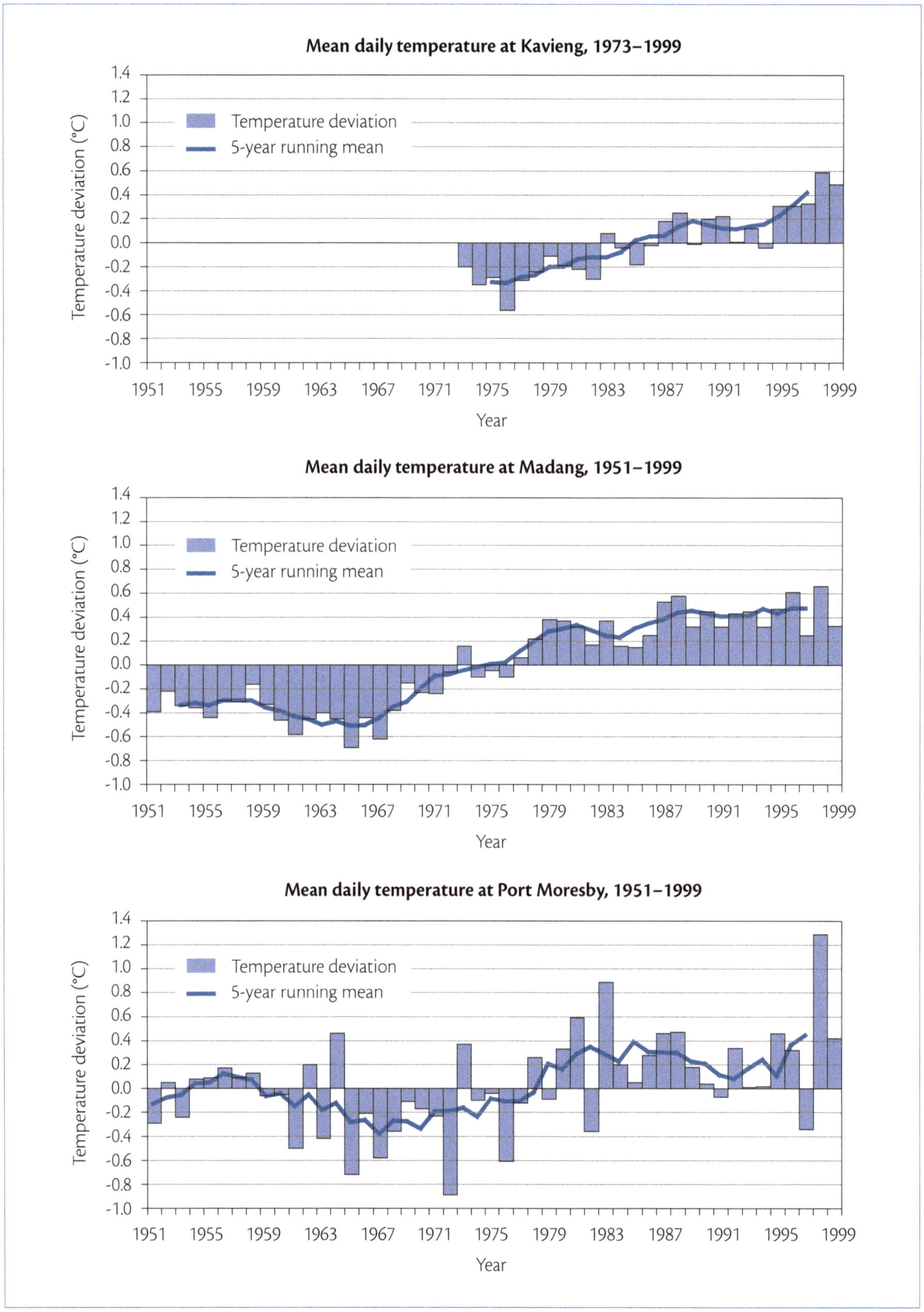

Figure 1.8.4 Temperature increases at Kavieng (1973–1999), Madang (1951–1999) and Port Moresby (1951–1999).
Note: Temperature deviation is the difference between the mean temperature for any one year from the long-term average for the entire period. Temperature data for Kavieng for the years 1951–1972 was not available.
Source: Bourke et al. (2002).

1900 m altitude near Teptep on the Huon Peninsula. Villagers there believe that the climate has warmed in the decade prior to 1988. It had become possible to grow *marita* pandanus and a cultivated palm by the late 1980s. Previously, these crops could only be grown at lower altitudes. In addition, a bird species that was previously found only at lower altitudes had moved into higher altitude locations.

Rainfall in PNG

There is less evidence for changes in rainfall patterns in PNG over the past 30 years. The majority of rural rainfall recording stations in PNG ceased recording around 1980, leading to a loss of information about rainfall. An analysis of monthly rainfall patterns at Goroka in Eastern Highlands Province from 1946 to 2002 found that there had been a shift to longer, but less pronounced, rainy seasons. Throughout the lowlands and highlands, villagers report similar changes in rainfall patterns. People say that seasonal rainfall patterns (see Section 1.5) have changed or are less predictable. They generally do not report an overall increase or decrease in rainfall. But because rainfall is already high at most locations in PNG, it is unlikely that villagers would notice an increase or decrease unless it was very large.

Sea level rise in PNG

There is much anecdotal evidence for rises in sea level in PNG. For example, a study of agriculture in Bougainville Province reported that villagers in many islands believed that the sea was eroding the coastline and that these changes had commenced some 30–40 years ago. In the Mortlock Islands, villagers were concerned about poor growth of their staple food, swamp taro, in some of the taro pits on the islands. They believed that sea water was invading the underground freshwater lens in which the swamp taro was growing and was responsible for the poor growth in some plots. A similar report was made by villagers on Ontong Java, north of Malaita Island in Solomon Islands. Other reports of encroachment by

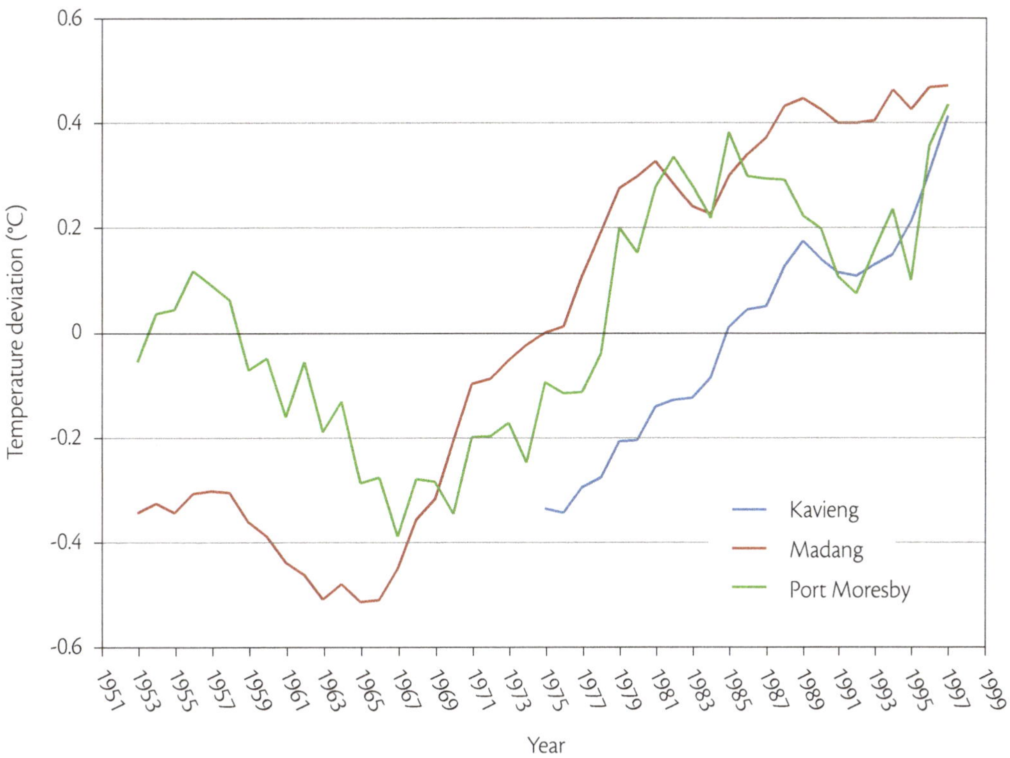

Figure 1.8.5 Temperature increases at Kavieng (1973–1999), Madang (1951–1999) and Port Moresby (1951–1999) (5-year running means). Source: Bourke et al. (2002).

the sea in recent years come from the Murik Lakes in East Sepik Province, along the Wewak waterfront, some small islands in Manus Province, and along the coast in Gulf and Western provinces.

However, caution is advisable in attributing the changes in shorelines to sea level rise. The Duke of York Islands off the north-eastern coast of East New Britain Province are reported to be swamped by rising sea levels. In fact, these islands are sinking because of tectonic (volcano-related) activity. Tectonic movements can cause land to rise as well as fall and are the reason why land at nearby Rabaul is rising. Elsewhere, sea level rise is blamed for other causes of failure of food supply. People in the Carteret Islands in Bougainville Province suffer from chronic food shortages because of a very high population density (see Table 1.3.1) and shortage of land for food gardens. In late 2005, plans were made to move the islanders to Bougainville Island (they had previously been relocated in the early 1980s, but returned to their islands some years later). Newspaper reports described the Carteret islanders as 'some of the world's first climate change refugees'. In fact, their problems are caused by the highest population density anywhere in PNG and having a mainly subsistence economy, rather than by rising sea levels.

In summary, there is much evidence from direct recordings, the bearing pattern of certain plants and villagers' observations that average temperatures have increased in the lowlands and highlands of PNG, particularly since the mid 1970s. There are indications that the distribution of rainfall within the year has changed, if not the total rainfall. There are also indications that rises in sea level are having an impact on shorelines and growth of crops on atolls.

Implications for agriculture in PNG

Climatic change is already influencing agriculture in PNG but, in most cases, it remains difficult to predict what outcomes will be. This is because we do not know enough about probable changes in patterns of rainfall and rainfall extremes. Furthermore, agricultural responses to climate change will be complex, and will depend on how people respond, as well as the responses of plants. For example, an increase in temperature might make coffee rust more severe in the highlands, but this impact could be reduced if the rainfall and humidity was reduced. Conversely, the impact could be greater if the rainfall increased as well as the temperature. This impact would be reduced if villagers adopted improved rust control techniques or rust resistant coffee varieties.

Temperature and agriculture

Increases in minimum and maximum temperature are already having a small influence on agricultural production and will have a greater influence in the future. Some tree crops are bearing at higher altitude in the highlands. Given the direct relationship between altitude and temperature in PNG (see Section 1.7), it is likely that many crops will be able to be grown at marginally higher altitudes in the future and so the areas where they can be grown will expand. However, the lower altitudinal limit of some crops, such as Irish potato, Arabica coffee and *karuka* (*Pandanus julianettii* and *P. brosimos*), will increase because of increasing temperatures (Figure 1.13.3).

Sweet potato is the most important food crop in PNG and provides almost two-thirds by weight of the staple food crops (see Figures 2.2.1 and 2.2.3). It is an important food in many lowland locations as well as throughout the highlands. Tuber formation in sweet potato is significantly reduced at temperatures above 34 °C. Maximum temperatures in the lowlands are now around 32 °C, so an increase of 2.0–4.5 °C within a hundred years could reduce sweet production in lowland locations, perhaps within one or two generations.

Diurnal temperature ranges also have an important influence on productivity (see Section 1.13). In PNG, the diurnal temperature range is greater in the highlands than in the lowlands, and is one reason why crop yields are higher in the highlands. The IPCC reports that, for most of the planet, diurnal range decreased over the period 1950–2004. If this is also occurring in PNG, it will tend to reduce crop yields to an unknown extent. It is likely that overall temperature increases will marginally reduce productivity in the lowlands and in the main highland valleys, but will marginally increase productivity at locations above 2000 m.

Increases in temperature may also change the incidence of some diseases, especially those influenced by rainfall and humidity. Taro blight, caused by the fungus *Phytophthora colocasiae*, is less severe in PNG at a few hundred metres above sea level than at sea level, and is rarely found above an altitude of 1300 m. The fungus is sensitive to temperature and a small rise in temperature could mean that the fungus would reduce taro yield at higher altitudes than occurs now. Similarly, coffee rust is present in the main highland valleys at 1600–1800 m, but does little damage there. It is a problem for coffee production at lower altitudes, below about 1200 m. So again, a rise in temperature is likely to increase the altitude at which coffee rust has a severe impact on coffee production.

Changes associated with El Niño Southern Oscillation (ENSO) events are not known. If ENSO events occur more often as models predict, more frosts will result at high-altitude locations (above 2200 m) (see Section 1.6). This will have an adverse impact on agricultural production.

Rainfall and agriculture

The IPCC prediction is for higher rainfall in the South Pacific, including PNG. Rainfall patterns in PNG are complex (see Section 1.5) and it is likely that any changes in rainfall patterns will also be complex and so difficult to predict.

With a few exceptions, most places in PNG receive high annual rainfall. In the lower-rainfall areas, an increase in rainfall and reduced seasonality of distribution would be beneficial for agriculture. However, for most of PNG, an increase in total rainfall and a less seasonal distribution would have a negative impact on agriculture. The most vulnerable locations are those where rainfall is already over 3500 mm per year.

If ENSO events become more common, more droughts could result as well as more episodes of very high rainfall. However, the likely changes in the pattern of ENSO events are not well understood. Another outcome of increased rainfall would be greater cloudiness and less bright sunshine. If that was to occur, it would probably reduce crop productivity, especially where cloud cover is already very high (see Section 1.7).

Sea level rise and agriculture

There are indications that rising sea levels are already having a minor negative impact on very small islands and other coastal locations because of coastal erosion. Rising sea water is also possibly contaminating the freshwater lens on atolls where swamp taro is grown. Rising sea levels do not necessarily cause very small islands to be covered by sea water: coral reefs respond to rising sea level by growing upwards, and the extra coral sand may cause the islands to rise. Nevertheless, it is quite possible that a rapid rise in sea levels will cause major problems for villagers living on very small islands and in other low-lying coastal locations, such as in the Gulf of Papua.

In the Pacific, attention has focused on small island states such as Tuvalu. However, there are about 100 000 people in PNG living on what have been defined as 'Small Islands in Peril'. These are about 140 islands smaller than 100 km^2 in size and with population densities greater than 100 persons/km^2. It is these people who are likely to suffer the most severe consequences of rising sea levels.

Other implications

There are implications from climate change in PNG that are not considered here. These include human health, especially malaria incidence in the highlands, water supply, fish availability, and health of coral reefs and other marine ecosystems.

Climate change is going to alter the global economy in coming decades and some of these changes are likely to impact, positively and negatively, on PNG. Such changes could include a reduction in availability of fossil fuels; a greater demand for biofuels from oil crops such as oil palm and coconut; increased demand for hydro-electricity (which is potentially abundant in PNG) for industrial uses; and carbon trading whereby companies pay for trees to be planted on a large scale to absorb carbon dioxide. Some of these changes present economic opportunities that could be beneficial for PNG.

It is known that a higher level of carbon dioxide in the atmosphere results in increased photosynthesis and decreased evaporation and transpiration. This is a positive for crop production. Given that atmospheric carbon dioxide is at record levels and is forecast to keep increasing, plant productivity is likely to increase. However, possible benefits are likely to be offset by decreases in productivity caused by higher temperatures and altered rainfall patterns.

Sources

Bourke, R.M. (1989). Altitudinal limits of 230 economic crop species in Papua New Guinea. Unpublished paper. Department of Human Geography, Research School of Pacific and Asian Studies, The Australian National University, Canberra.

Bourke, R.M., Humphreys, G.S. and Hart, M. (2002). Warming in Papua New Guinea: some implications for food productivity. Unpublished paper. Department of Human Geography, Research School of Pacific and Asian Studies, The Australian National University, Canberra.

Bourke, R.M. and Betitis, T. (2003). Sustainability of agriculture in Bougainville Province, Papua New Guinea. Land Management Group, Department of Human Geography, Research School of Pacific and Asian Studies, The Australian National University, Canberra.

Brown, J.S., Kenny, M.K., Whan, J.H. and Merriman, P.R. (1995). The effect of temperature on the development of epidemics of coffee leaf rust in Papua New Guinea. *Crop Protection* 14(8):671–676.

Cabanes, C., Cazenave, A. and Le Provost, C. (2001). Sea level rise during past 40 years determined from satellite and in situ observations. *Science* 294(5543):840–842.

Flannery, T. (2005). *The Weather Makers: The History and Future Impact of Climate Change.* The Text Publishing Company, Melbourne.

Haberle, S.G., Hope, G.S. and van der Kaars, S. (2001). Biomass burning in Indonesia and Papua New Guinea: natural and human induced fire events in the fossil record. *Palaeogeography, Palaeoclimatology, Palaeoecology* 171(3–4):259–268.

Hulm, P. (1989). *A Climate of Crisis: Global Warming and the Island South Pacific.* Association of South Pacific Environmental Institutions, Port Moresby.

Inape, K. and Humphrey, B. (2001). Potential impact of global climatic change on smallholder farmers in PNG. In Bourke, R.M., Allen, M.G. and Salisbury, J.G. (eds). *Food Security for Papua New Guinea. Proceedings of the Papua New Guinea Food and Nutrition 2000 Conference.* ACIAR Proceedings No. 99. Australian Centre for International Agricultural Research, Canberra. pp. 73–78.

IPCC (Intergovernmental Panel on Climate Change) (2006). Working Group I Contribution to the Fourth Assessment Report of the IPCC. Climate Change 2007: The Physical Science Basis. Accessed 10 May 2006 at <http://www.climatescience.gov/Library/ipcc/wg14ar-review.htm>.

Kocher Schmid, C. (1991). *Of People and Plants: A Botanical Ethnography of Nokopo Village, Madang and Morobe Provinces, Papua New Guinea.* Basler Beiträge zur Ethnologie, Band 33. Ethnologisches Seminar der Universität und Museum für Völkerkunde, Basel.

Mueller, I., Namuigi, P., Kundi, J., Ivivi, R., Tandrapah, T., Bjorge, S. and Reeder, J.C. (2005). Epidemic malaria in the highlands of Papua New Guinea. *American Journal of Tropical Medicine and Hygiene* 72(5):554–560.

Petit, J.R., Jouzel, J., Raynaud, D., Barkov, N.I., Barnola, J.-M., Basile, I., Bender, M., Chappellaz, J., Davis, M., Delaygue, G., Delmotte, M., Kotlyakov, V.M., Legrand, M., Lipenkov, V.Y., Lorius, C., Pépin, L., Ritz, C., Saltzman, E. and Stievenard, M. (1999). Climate and atmospheric history of the past 420,000 years from the Vostok ice core, Antarctica. *Nature* 399(6735):429–436.

Rosenberg, N.J. and Scott, M.J. (1994). Implications of policies to prevent climate change for future food security. *Global Environmental Change* 4(1):49–62.

Schneider, S.H. (1996). The future of climate: potential for interaction and surprises. In Downing, T.E. (ed). *Climate Change and World Food Security.* NATO ASI Series. Series I: Global Environmental Change, Volume 37. Springer, Berlin. pp. 77–113.

UNEP/GRID-Arendal (ed.) (2006). *Planet in Peril: An Atlas of Current Threats to People and the Environment.* UNEP/GRID-Arendal and Le Monde diplomatique, Norway and France.

1.9 Soils

There are five major soil-forming factors: climate, vegetation, slope, parent material and time. Changes in the combination of these factors can result in soils varying over relatively short distances. Soils are also significantly influenced by human activities, for example, by tillage, composting, mounding or erosion caused by cultivation on steep slopes. As a result, it is difficult to generalise about soils and their properties in PNG.

PNG soils are relatively young, particularly in areas where sedimentation is occurring or volcanoes are depositing ash. These young soils have a fair potential for agricultural production because they can provide adequate amounts of water and nutrients to plants.

Soil classification in PNG

Knowledge on the distribution of soils and their abilities to support plant growth is essential for managing agricultural production. In PNG, most villagers have names for their soils and they are knowledgeable about soil properties. In textbooks and publications on soils in PNG, the Soil Taxonomy classification system of the United States Department of Agriculture (USDA) is used. USDA Soil Taxonomy uses measurable properties of a soil, including soil depth; colour; texture (e.g. sand, sandy loam, loam, clay); structure (e.g. blocky, granular, crumb, columnar); consistency (e.g. sticky, porous); soil water balance (see Section 1.5); and soil chemical properties such as pH (a measure of acidity) and the ability to retain nutrients. These properties are used to group soils into classes. A full classification under the USDA Soil Taxonomy requires chemical and physical analyses.

PNG uses the USDA Soil Taxonomy because it is closely linked to the development of the Papua New Guinea Resource Information System (PNGRIS; for further information see Section 1.15). An internationally recognised soil classification system has advantages when exchanging information; however, some problems occur when USDA Soil Taxonomy is used to classify PNG soils. USDA Soil Taxonomy relies heavily on laboratory analysis, which is expensive. Many of the previous analyses of PNG soils do not have sufficient analytical data to allow a full USDA Soil Taxonomy classification. In addition, soil moisture and soil temperature information is needed in USDA Soil Taxonomy, but only limited data are available for PNG soils.

Information on the soils of PNG is available in PNGRIS. The information in PNGRIS covers the whole of PNG. It was created by using information from soil surveys in particular places and predicting the soil types that will occur in the PNGRIS resource mapping units (RMUs) by using the associations that occurred in the soil surveys between parent material, rainfall, inundation and slope. (See Section 1.15 for a definition of an RMU.) PNGRIS lists three possible soil types in each RMU, depending on the relative area of each soil. PNGRIS cannot be used to produce detailed soil maps and should not be used above a scale of 1:500 000. The map in this section is based on only the first-listed soil in PNGRIS RMUs[1] (Figure 1.9.1).

In USDA Soil Taxonomy, the highest level is soil order. The following soil orders occur in PNG:

- **Entisols** – very young soils, with little or no profile. These soils occur mainly on recent alluvium or on steep slopes where soil erosion takes place.
- **Histosols** – soils that contain very high levels of organic matter (peat soils). These soils are mostly dark brown to black in colour, and occur in swampy areas. They are saturated with water for much of the year.
- **Inceptisols** – moderately weathered soils, with no strongly contrasting horizons. They include soils derived from volcanic ash.
- **Andisols** – volcanic ash soils. These soils are very important in PNG. They do not appear in PNGRIS, where they are classed with Inceptisols. In Tables A1.9.1 and A1.9.2 and Figures 1.9.2 and 1.9.3 they are included with the Inceptisols. However, in Figure 1.9.1 a class called Andisols has been created by mapping from PNGRIS soils that are derived from lightly weathered volcanic ash.
- **Vertisols** – soils with high clay content that are sticky when wet and very hard when dry. These soils swell when wet and crack when dry and are generally of high fertility. They are not common in PNG.
- **Mollisols** – soils in which there is accumulation and decomposition of organic matter. These soils generally have a high base (e.g. calcium, magnesium) content.
- **Alfisols** – moderately weathered soils that have an argillic horizon (a soil layer with higher clay content due to movement of clay from the top to lower layers). These soils are usually highly fertile.
- **Ultisols** – strongly weathered and acid soils with an argillic horizon. These soils have a low base saturation and are usually relatively infertile.
- **Oxisols** – very strongly weathered soils, with low fertility. These soils occur on old land surfaces and are not common in PNG.

The distribution of soils in PNG

The most common soils in PNG are Inceptisols (with Andisols), which are found over almost half of the total land area (Figures 1.9.1, 1.9.2, Table A1.9.1). Inceptisols cover more than 80% of the land area of Western Highlands, Simbu, Eastern Highlands and West New Britain provinces. Inceptisols are least common in Western, East Sepik and Gulf provinces.

The next most common soils are Entisols. These young soils cover more than a quarter of the total land area, reflecting the high geological activity that occurs in PNG. Entisols are common soils in East Sepik, Gulf, Morobe, Central, Western, Oro, Sandaun and Madang provinces.

Ultisols (strongly weathered soils) cover approximately 14% of the land area of PNG, and occur mainly in Western Province, where they occupy more than half of that province. Alfisols are common in New Ireland where they cover over a quarter of the land area. Mollisols are most prevalent in East New Britain, where they occur on a third of the land area.

---

1 In PNGRIS, soils information is mapped into mapping units that are defined by attributes other than soils (landform, rock type, altitude, relief, rainfall, inundation, province). This means that more than one soil can occur in an RMU. PNGRIS deals with this by allowing each RMU to have up to three soils. Soils that cover less than 20% of an RMU are not listed. If only one soil is listed it covers around 80% of the RMU area. If two soils are listed, the first covers 40%–60% of the RMU and the second 20%–40%. If three soils are listed, the first covers 30%–50% of the RMU area and the second and third 20%–40%. The information presented in this section is based on *only* the first soil listed. Thus the figures presented in Tables A1.9.1 and A1.9.2 are approximations.

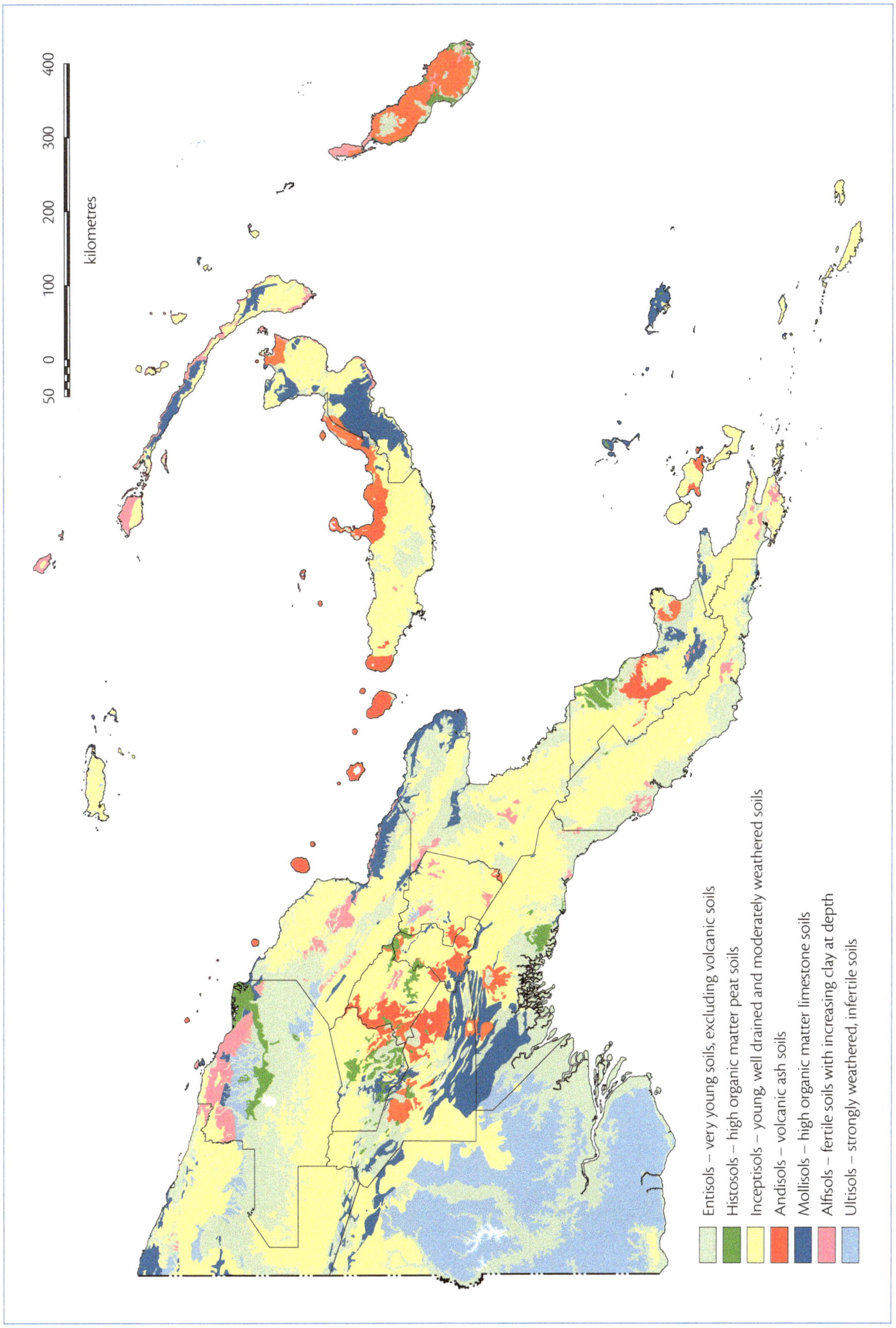

Figure 1.9.1 Distribution of USDA Soil Taxonomy soil orders. **Note:** The areas covered by Vertisols and Oxisols are too small to be shown on the map (see Table A1.9.1). Source: PNGRIS.

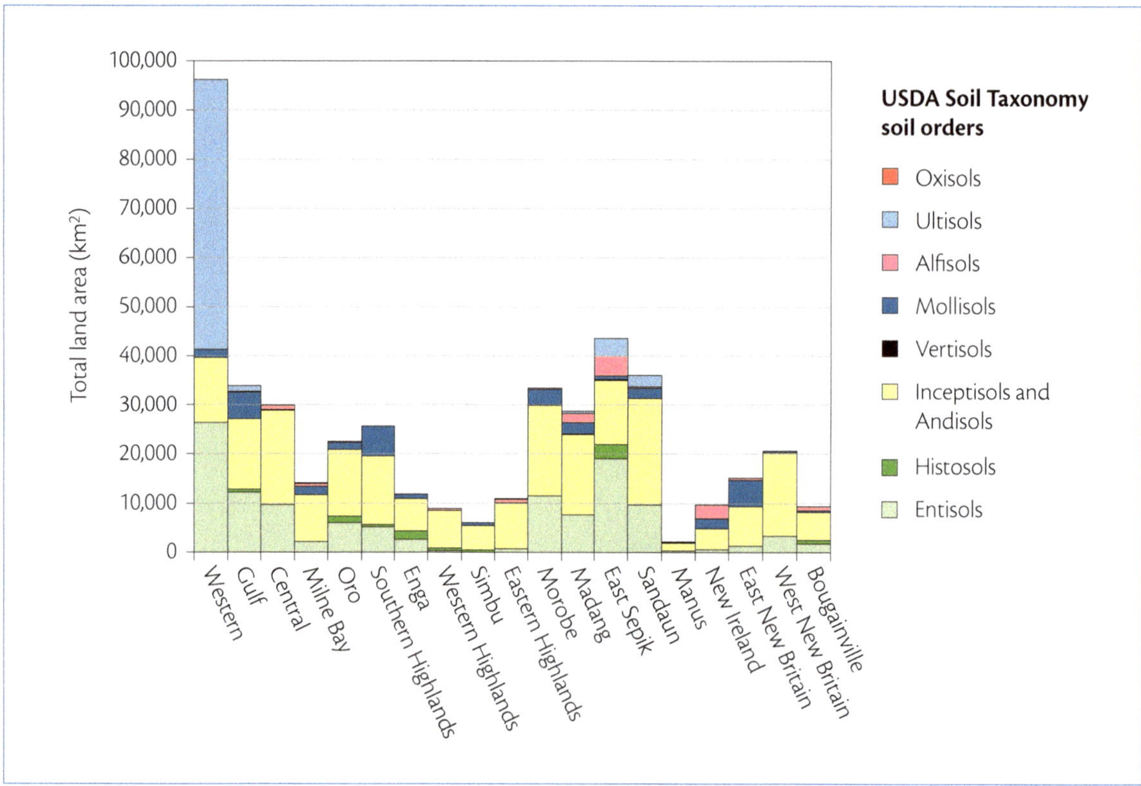

Figure 1.9.2 Distribution of USDA Soil Taxonomy soil orders by land area and province. Source: PNGRIS.

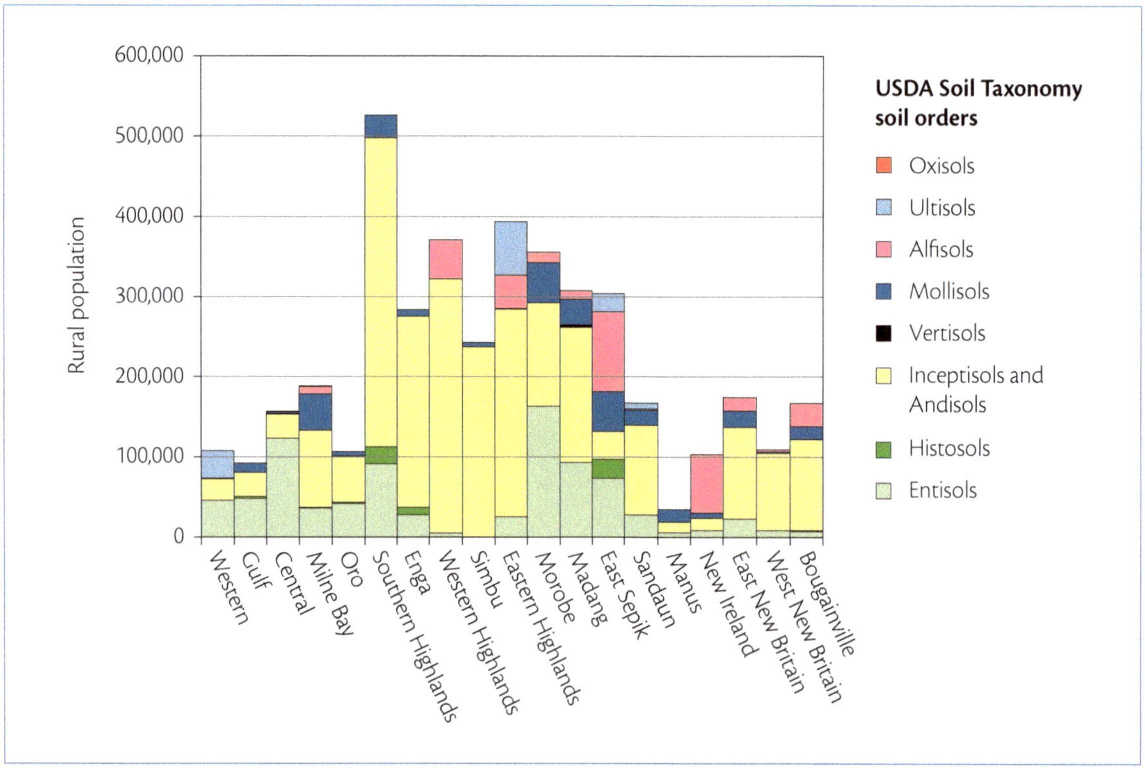

Figure 1.9.3 Distribution of USDA Soil Taxonomy soil orders by rural population and province.
Sources: NSO (2002); PNGRIS.

People and soils

This section discusses the numbers of people who live on the soil orders described above. As the distribution of soils is complex, the information presented here can be used only as a general indication of the importance of particular soils to the wellbeing of people in PNG. Landform, climate and other factors also help to determine the distribution of people (see Sections 1.10, 1.11 and 1.12).

Almost 2.5 million people (59%) of PNG's rural population live on Inceptisols (Figure 1.9.3, Table A1.9.2). Inceptisols (with Andisols) support more than 80% of the populations of Simbu, West New Britain, Western Highlands and Enga provinces. Entisols support 20% of the population and are most important in Central Province, where 78% of the provincial population use these soils. Alfisols and Mollisols together support a further 16% of the rural population. Alfisols are important in New Ireland Province and Mollisols are important in Manus Province. Around 32% of the population in Western Province, and 17% of the population in Eastern Highlands Province, cultivate Ultisols. Although Ultisols are usually relatively infertile, those in Eastern Highlands are well-drained, humus-rich, reddish Ultisols that occur mostly on old stable surfaces. These Ultisols have few available minerals but are able to support a large population (66 500 people on 226 km^2) because of their structure and the practice of intensive tillage.

Soil nutrients

The availability of nutrients for plants depends on several factors. Low levels of nutrients in the soil may be caused by low amounts of nutrients in the parent material from which the soil is derived. The nutrients may also become chemically fixed in the soil and so not available to plants. Nutrient imbalances (for example, high calcium, low potassium) in the soil may have a similar effect. High rainfall can leach nutrients from the soil. Low nutrient levels may also result from cultivation, when agricultural crops remove nutrients which are then not replaced by humans or by natural processes.

Soil nutrients that plants require in relatively large amounts are nitrogen (N), phosphorus (P), potassium (K), calcium (Ca), magnesium (Mg) and sulfur (S). A number of other elements required in very small amounts by plants are called micronutrients, and include boron, zinc, manganese, iron and copper. Nutrient deficiencies affect crop production in PNG. This problem is likely to worsen in the future as population increases and land is used more intensively.

Soil nitrogen availability is determined in part by the length and type of fallow (see Section 3.8), the introduction of organic matter (plant materials) into the soil (see Sections 3.10 and 3.11), and climate (rainfall and temperature) (see Sections 1.5 and 1.7). Most nitrogen in PNG soils is derived from organic matter. Soil nitrogen tends to be higher in highlands soils, where temperatures are cooler and organic matter decomposes more slowly.

In PNG soils the availability of phosphorus is also dependent mostly on the organic matter content. A small part comes from the weathering of parent material or secondary minerals. Phosphorus is usually found in combination with calcium, magnesium, iron and aluminium. Although relatively large amounts of total phosphorus may be present in the soil, little may be available to plants because it combines with other elements (iron and aluminium) and becomes insoluble, or 'fixed' in the soil. Phosphorus fixation is widespread in the volcanic ash soils that support large numbers of people in the highlands, and in Oro and Bougainville provinces (Table A1.9.2). Phosphorus fixation can also be severe in Ultisols and Oxisols.

The availability of soil potassium is related to rock type and the mineralogy and the stage of weathering of the soil. Potassium-deficient soils are usually highly weathered and leached with limited amounts of mineral reserves. Volcanic ash soils usually have high levels of potassium. Soils that develop on limestone and that have high levels of calcium and magnesium may have a potassium deficiency because the calcium holds the potassium in the soil and makes it unavailable to plants. This is common, for example, in New Ireland Province.

Research on nutrient deficiencies of agricultural crops started in the 1950s, but little active research is being carried out now on soil nutrient deficiencies or on soil nutrient management strategies in PNG. Soil nutrient problems exist in parts of the country. Further intensification of land use will affect soil fertility, and nutrient deficiencies are therefore likely to increase, particularly in food crops where inorganic fertilisers are not being used.[2] There is a need to monitor the development of nutrient deficiencies as well as to properly identify them through trials and soil and foliar (leaf) analyses.

Sources

Bellamy, J.A. (1986). *Papua New Guinea Inventory of Natural Resources, Population Distribution and Land Use Handbook*. Natural Resources Series No. 6. Division of Water and Land Resources, Commonwealth Scientific and Industrial Research Organisation, Canberra.

Bleeker, P. (1983). *Soils of Papua New Guinea*. Commonwealth Scientific and Industrial Research Organisation and Australian National University Press, Canberra.

Bourke, R.M. (1983). Crop micronutrient deficiencies in Papua New Guinea. Technical Report 83/3. Department of Primary Industry, Kainantu.

Dearden, P.N., Freyne, D.F. and Humphreys, G.S. (1986). Soil and land resource surveys in Papua New Guinea. *Soil Survey and Land Evaluation* 6(2):43–50.

Hartemink, A.E. and Bourke, R.M. (2001). Nutrient deficiencies in export tree and food crops: literature review and field observations. In Bourke, R.M., Allen, M.G. and Salisbury, J.G. (eds). *Food Security for Papua New Guinea. Proceedings of the Papua New Guinea Food and Nutrition 2000 Conference.* ACIAR Proceedings No. 99. Australian Centre for International Agricultural Research, Canberra. pp. 316–327.

Humphreys, G.S. (1991). Soil maps of Papua New Guinea: a review. *Science in New Guinea* 17(2):77–102.

NSO (National Statistical Office of Papua New Guinea) (2002). Papua New Guinea 2000 Census: Final Figures. National Statistical Office of Papua New Guinea, Port Moresby.

Wood, A.W. (1982). The soils of New Guinea. In Gressitt, J.L. (ed). *Biogeography and Ecology of New Guinea. Monographiae Biologicae*. Volume 42. Dr W. Junk Publishers, The Hague. pp. 73–83.

2 Organic fertilisers include animal manure, compost and mulch made from leaves (see Section 3.11). Inorganic fertilisers are usually manufactured, for example, urea, potassium chloride or mixed nitrogen, phosphate and potash fertiliser. Most organic fertiliser is not purchased in PNG, whereas most inorganic fertiliser is purchased. Most of the inorganic fertiliser used in PNG is applied to export tree crops or to commercial crops grown for the domestic market, such as sugar cane (see Section 5.19).

1.10 Landforms and altitude

Landforms

Landforms are the structural features of the landscape. Landforms have a direct relationship to relief and slope and, in turn, determine rates of erosion, runoff and whether land is flooded. Landforms influence agricultural land use.

The PNG landscape can be divided into five basic landforms (Figure 1.10.1):

- Mountains and hills (not of volcanic origin).
- Landforms of volcanic origin (including volcanic foot slopes and plains).
- Plains and plateaus.
- Floodplains.
- Raised coral reefs and littoral areas (beach ridges, tidal flats, mangrove swamps and other coastal features).

The following summary points can be made about these landforms in PNG:

- Around half (52%) of the total land area of PNG is mountains and hills; almost 19% is plains or plateaus; 18% is floodplains; and a considerably smaller proportion is volcanic landforms or raised coral reefs and littoral areas (Table 1.10.1, Table A1.10.1).
- The provinces with the greatest proportion of total land area comprising mountains and hills are Enga (91%), Eastern Highlands (90%), East New Britain (83%), Simbu (79%), Central (78%) and Morobe (77%).
- Almost two-thirds (63%) of the land used for agriculture in PNG (see Section 1.2) is on mountains and hills; 12% is on volcanic landforms; 11% is on plains and plateaus; and 9% is on floodplains (Table 1.10.1, Figure 1.10.2, Table A1.10.2).
- The provinces with the greatest proportion of land used for agriculture on mountains and hills are Eastern Highlands (91% of land used for agriculture), Enga (90%), Simbu (86%), Madang (76%), Sandaun (76%), Morobe (76%) and Gulf (75%).
- Volcanic landforms used for agriculture are most important in Bougainville Province, where 61% of land used for agriculture is of volcanic origin. This is also a significant landform for agriculture in Oro (46%) and Southern Highlands provinces (32%).
- Although 59% of the land used for agriculture on plains and plateaus is in Western Province, much of this land is used at very low intensity (see Section 1.2).
- Provinces where a high proportion of land used for agriculture is floodplain are East Sepik (22%), Central (22%), Oro (16%), Gulf (16%) and Sandaun (14%).

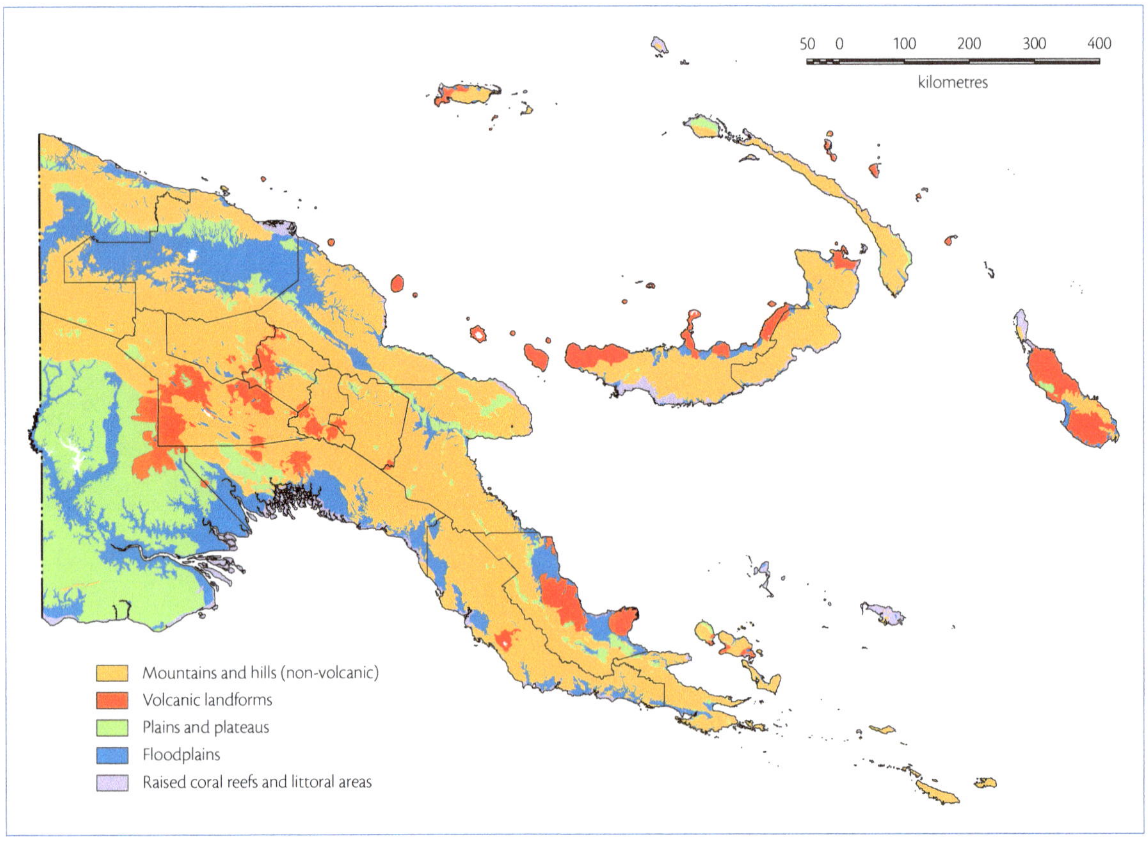

Figure 1.10.1 The five basic landforms of the PNG landscape. Sources: McAlpine and Quigley (c. 1995); PNGRIS.

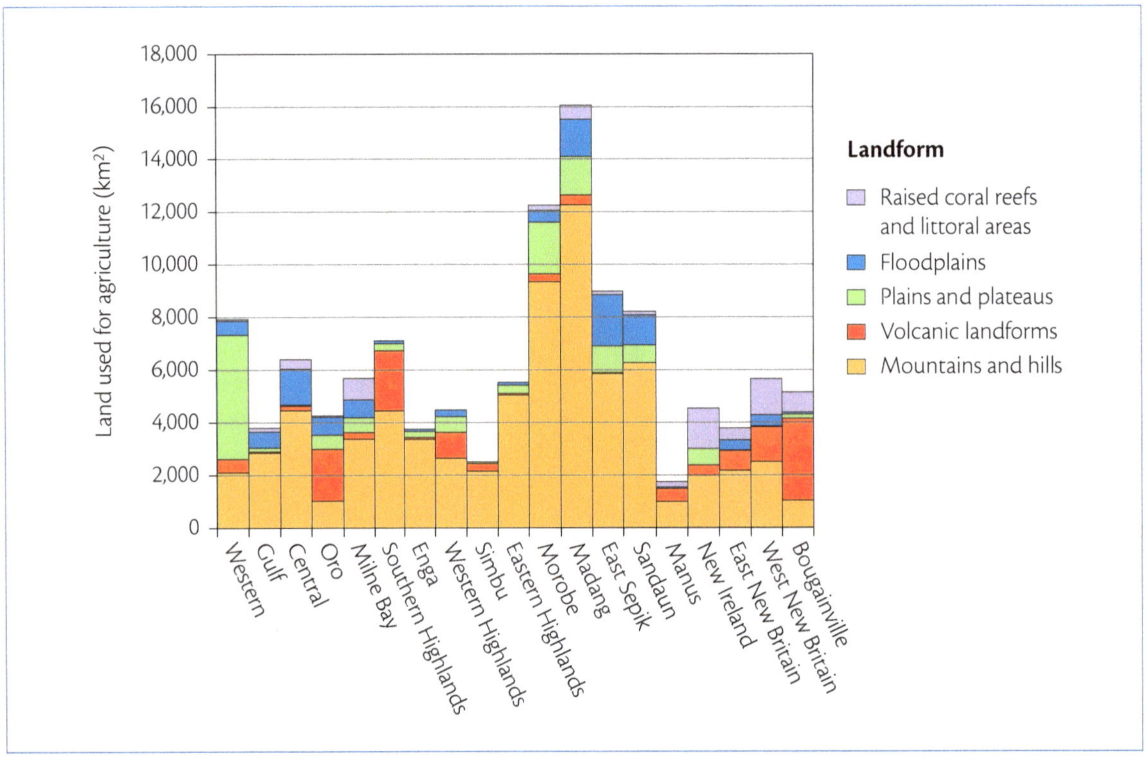

Figure 1.10.2 Land area used for agriculture by landform and province. Sources: McAlpine and Quigley (c. 1995); PNGRIS.

- Coastal landforms used for agriculture are most important in the Islands Region and in Milne Bay Province.
- Approximately half of the population of PNG live on mountains and hills; 17% live on volcanic landforms; 13% on plains and plateaus; and 9% on floodplains. Raised coral reefs and littoral areas – the landform that makes up the smallest area of PNG (only 4%) – support 11% of the population (Table 1.10.1, Figure 1.10.3, Table A1.10.3).
- PNG landforms with the highest population densities on land used for agriculture are raised coral reefs and littoral areas (70 persons/km^2) (Table 1.10.1; see also Table 1.3.1). Population densities on volcanic landforms average 50 persons/km^2 and on plains and plateaus are around 40 persons/km^2. Population densities are lowest on mountains and hills.

Altitude

In PNG, people live and practise agriculture from sea level to around 2800 m above sea level. Above 2800 m, people maintain *Pandanus* orchards and hunt, but they do not cultivate land or maintain settlements. Because PNG is located close to the equator, there is little variation in temperature in most places from month to month during the year (see Table 1.13.1). However, average temperature declines with increasing height above sea level at a rate of 5 °C for every 1000 m of altitude (see Section 1.7), so altitude is an alternative measure for temperature in PNG.

Temperature is a critical determinant of plant growth (see Section 1.13). It also determines the survival range of some insect pests, for example, the *Anopheles* mosquito that transmits malaria. There is evidence that average temperatures are slowly increasing in PNG as a result of climate change (see Section 1.8).

Six altitude classes from PNGRIS, and the maximum and minimum temperatures associated with them, are used here:

- Lowlands (sea level to 600 m), 32–30 °C to 23–19 °C.
- Intermediate (600–1200 m), 30–27 °C to 19–16 °C.
- Highlands (1200–1800 m), 27–23 °C to 16–12 °C.
- High altitude (1800–2400 m), 23–19 °C to 12–9 °C.
- Very high altitude (2400–2800 m), 19–16 °C to 9–7 °C.
- Uninhabited (>2800 m), <16 °C to <7 °C.

These altitude/temperature classes are used because of the crops that grow in them (see Figure 1.13.3). Some crops only grow well up to around 600 m, for example, Polynesian chestnut, *pao* nut and *kangkong*, so 600 m is a convenient place to separate 'lowlands' from the 'intermediate' class. Many tree crops in PNG, such as coconut, betel nut, *tulip* and breadfruit, grow to about 1000–1200 m above sea level, so 1200 m is used to separate the intermediate class from the 'highlands'. Many food and cash crops do not grow well above 1800–2000 m, so 1800 m is a convenient break between the highlands and 'high altitude' classes. Above 2800 m there is no permanent settlement or agriculture. This class is 'uninhabited'.

The following summarise the main points about altitude and agriculture, and altitude and population in PNG:

- Two-thirds of PNG's total land area lies below 600 m altitude and falls into the lowlands environment. The intermediate altitude class is 15% of the total land area and the highlands altitude class 9%. Less than 10% of the total land area is above 1800 m and only 2.5% is above 2800 m (Table 1.10.2, Figure 1.10.4, Table A1.10.4). Above 2200 m, temperatures may fall below freezing on cloud-free nights (see Sections 1.6 and 1.7). The highest point in PNG is Mt Wilhelm (4509 m), on the border of Western Highlands, Simbu and Madang provinces.
- The greatest proportion of land used for agriculture lies between sea level and 600 m (62%) (Figure 1.10.5, Table A1.10.5). The lowest proportion of land used for agriculture is in the very high altitude class, 2400–2800 m (1.4%).

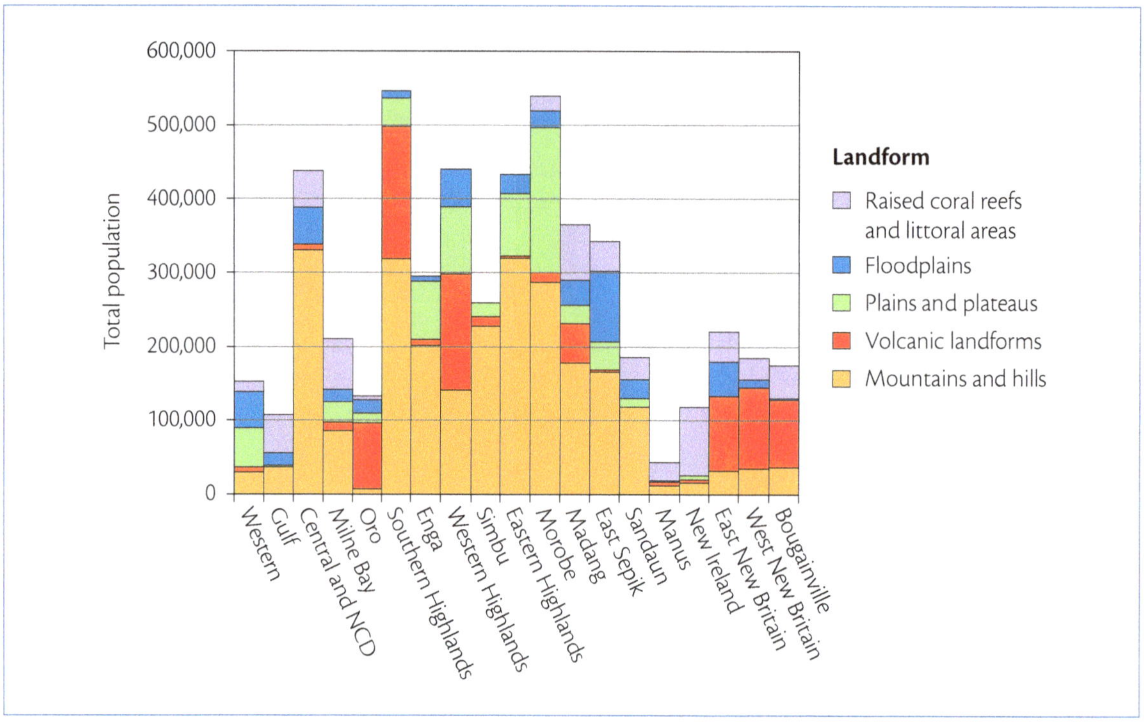

Figure 1.10.3 Total population (including urban and rural non-village populations) by landform and province. Sources: McAlpine and Quigley (c. 1995); NSO (2002); PNGRIS.

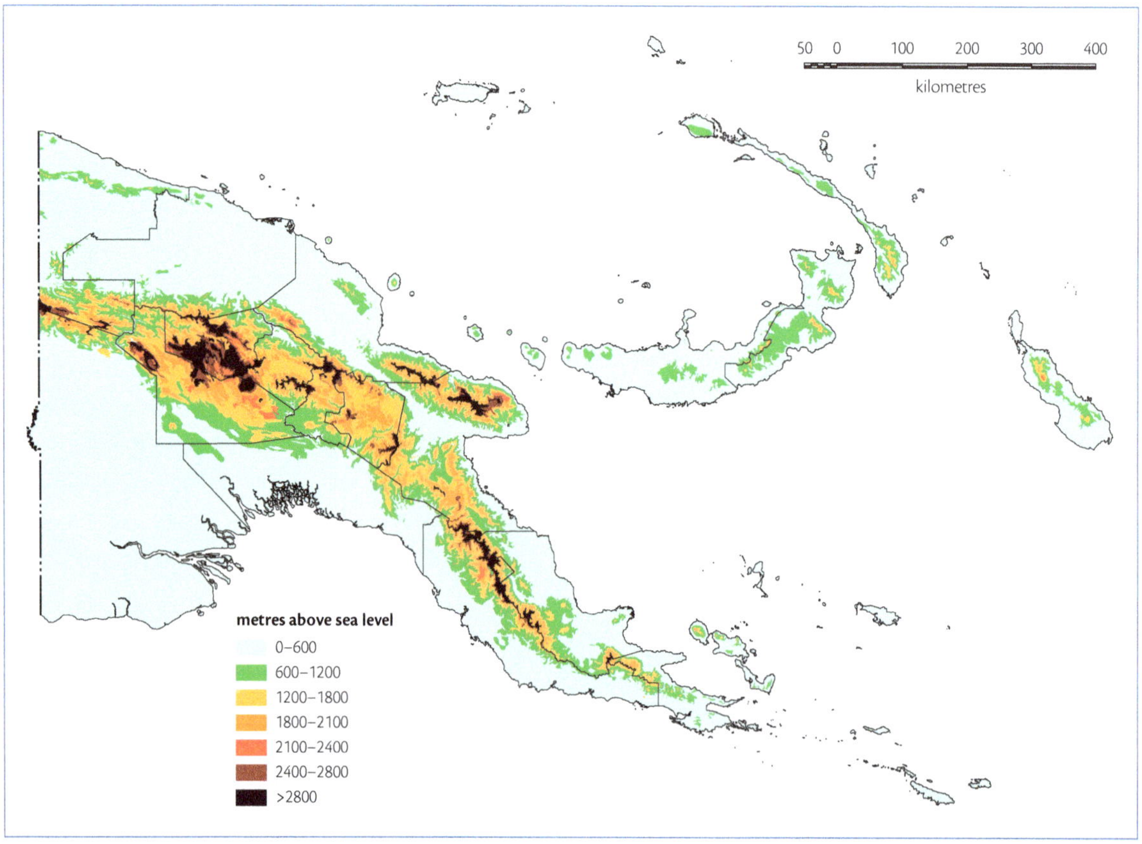

Figure 1.10.4 Altitude classes of the PNG landscape. Sources: McAlpine and Quigley (c. 1995); PNGRIS.

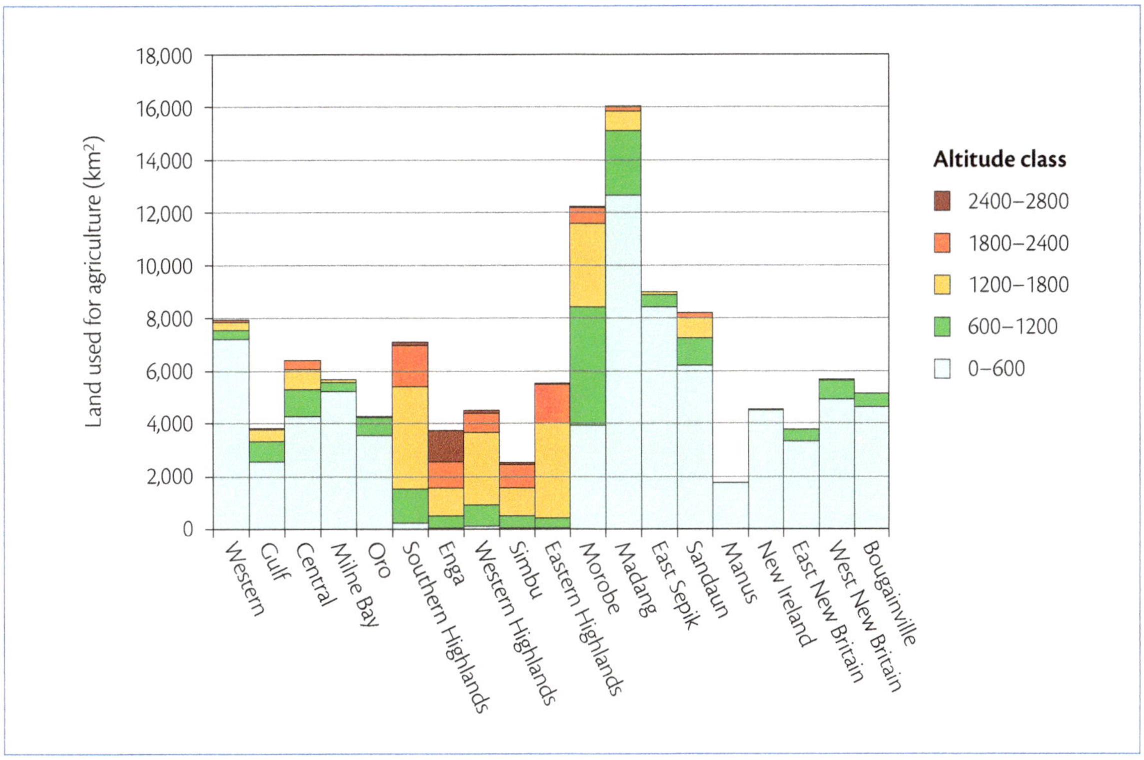

Figure 1.10.5 Land area used for agriculture by altitude class and province.
Sources: McAlpine and Quigley (c. 1995); PNGRIS.

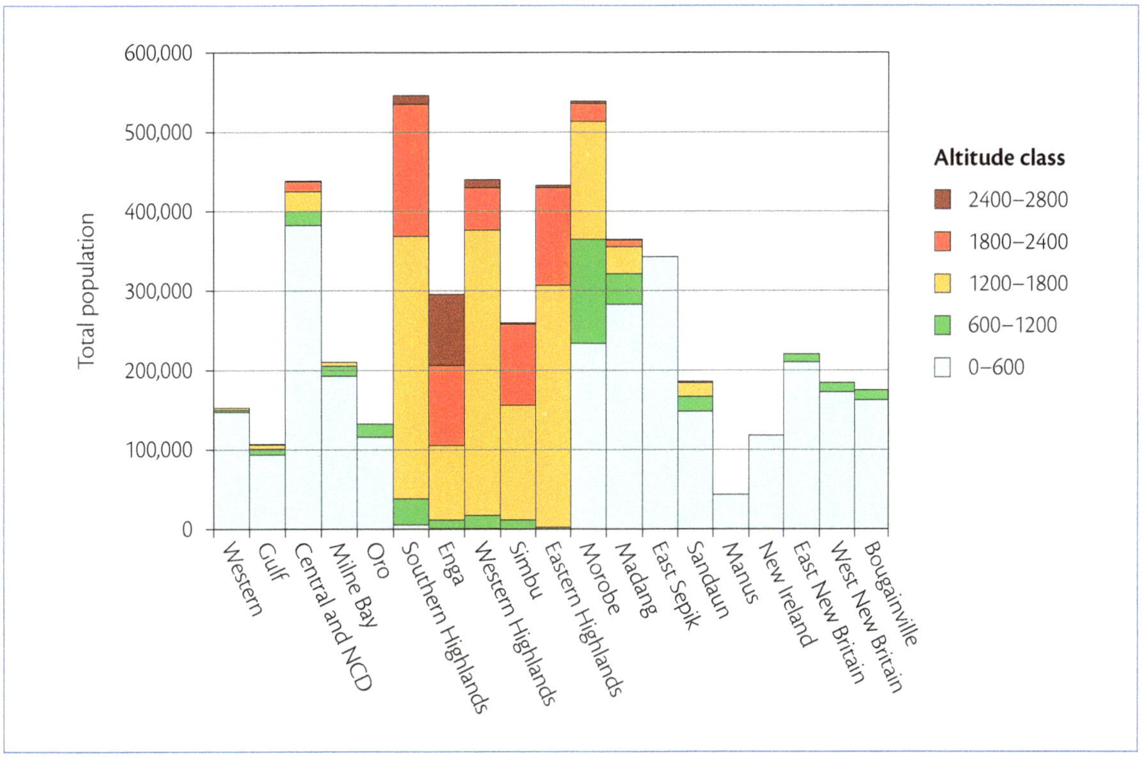

Figure 1.10.6 Total population (including urban and rural non-village populations) by altitude class and province.
Sources: McAlpine and Quigley (c. 1995); NSO (2002); PNGRIS.

- Only in the Highlands Region is agriculture significant above 2400 m, with Enga Province having the highest proportion of land used for agriculture in the 2400–2800 m altitude class (32%).
- In the 1200–1800 m altitude class, the proportion of land used for agriculture, at 43%, is greater than for any other altitude class (Table 1.10.2). This class includes all of the main highland valleys.
- Half of all people in PNG live below 600 m and 42% live above 1200 m (Table 1.10.2).
- There are two altitude classes where fewer people live. These are 600–1200 m and 2400–2800 m. These areas are generally characterised by steep land, very high rainfall and high levels of cloud cover.
- People live above 2400 m in eight provinces, but only in Enga Province do a significant number (89 000) live in the very high altitude class (Figure 1.10.6, Table A1.10.6). Eastern Highlands is the only province in PNG in which no people are recorded living below 600 m.
- A disproportionate number of people, in relation to land area occupied, live between 1200 m and 1800 m (where about 28% of the total population live on 9% of the total land area) and between 1800 m and 2800 m (where about 14% of the total population occupy 7% of the total land area) (Table 1.10.2).
- Population densities between 1200 m and 2800 m are two to four times higher than in the lowlands and intermediate altitude classes.

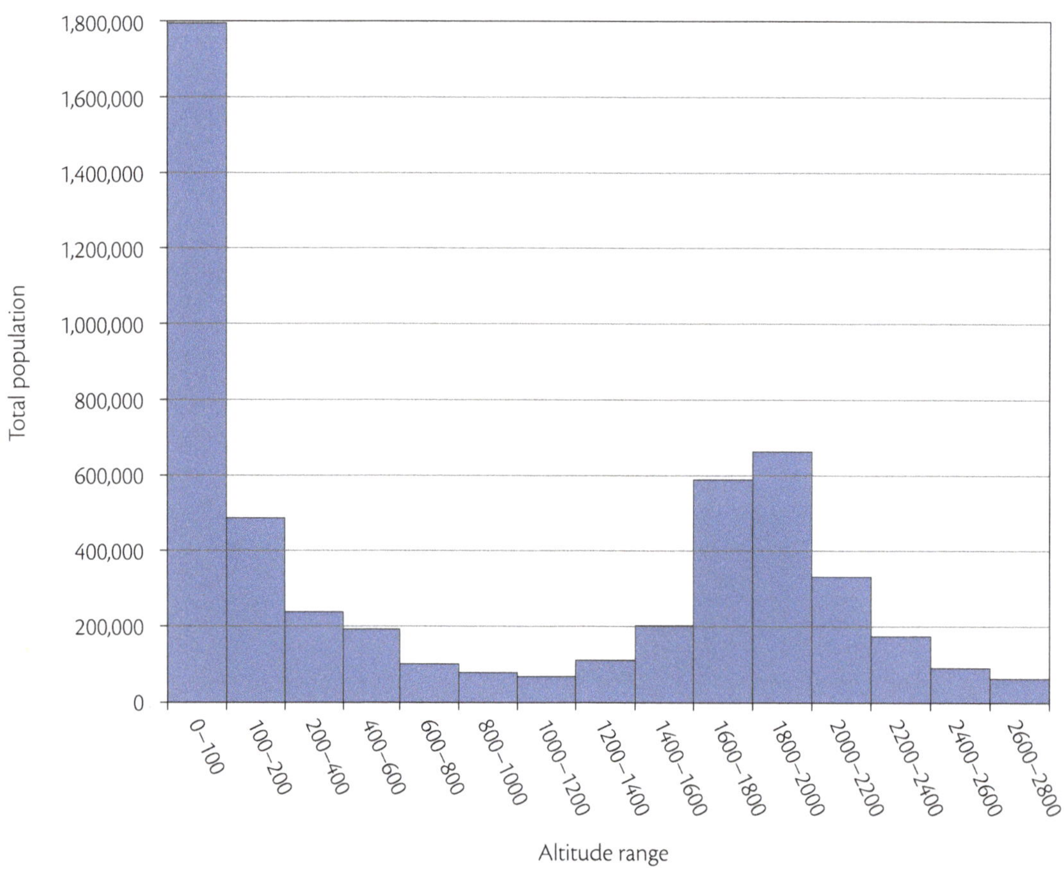

Figure 1.10.7 Total population (including urban and rural non-village populations) by altitude range.
Sources: NSO (2002); digital 1:250 000 topographic map; PNGRIS.

Table 1.10.1 Total land area, land used for agriculture, rural population, and rural population density, by landform

Landform	Total land area		Land used for agriculture		Rural population	Proportion of rural population (%)	Land used for agriculture as a proportion of total land area (%)	Rural population density on land used for agriculture (persons/km²)
	(km²)	(%)	(km²)	(%)				
Mountains and hills	238,028	51.8	73,862	62.7	2,081,808	49.7	31.0	28
Volcanic landforms	36,734	8.0	13,542	11.5	693,711	16.5	36.9	51
Plains and plateaus	85,933	18.7	13,399	11.4	553,461	13.2	15.6	41
Floodplains	81,556	17.7	10,277	8.7	391,000	9.3	12.6	38
Raised coral reefs and littoral areas	17,603	3.8	6,778	5.8	472,580	11.3	38.5	70
Papua New Guinea	459,854	100.0	117,858	100.0	4,192,561	100.0	25.6	36

Sources: McAlpine and Quigley (c. 1995); NSO (2002); PNGRIS.

Table 1.10.2 Total land area, land used for agriculture, total population (including urban and rural non-village populations), and total population density, by altitude class

Altitude class (metres above sea level)	Total land area		Land used for agriculture		Total population	Proportion of total population (%)	Land used for agriculture as a proportion of total land area (%)	Total population density on land used for agriculture (persons/km²)
	(km²)	(%)	(km²)	(%)				
0–600	303,844	66.1	73,531	62.4	2,654,521	51.1	24.2	36
600–1200	69,505	15.1	16,766	14.2	354,262	6.8	24.1	21
1200–1800	43,416	9.4	18,844	16.0	1,471,042	28.3	43.4	78
1800–2400	25,359	5.5	7,126	6.0	590,928	11.4	28.1	83
2400–2800	6,275	1.4	1,591	1.4	120,033	2.3	25.4	66
>2800	11,455	2.5	0	0.0	0	0.0	0.0	0
Papua New Guinea	459,854	100.0	117,858	100.0	5,190,786	100.0	25.6	44

Sources: McAlpine and Quigley (c. 1995); NSO (2002); PNGRIS.

The altitude data above was extracted from PNGRIS (see Section 1.15) where resource mapping units (RMUs) are allocated an altitude class and the census units (CUs) located in all the RMUs are summed by altitude class. Another means of deriving information about population and altitude is to use geographic information system software to join a digital map of CUs to a map of contour lines and to give the CUs the altitude of the nearest contour line.

An analysis of population distribution by this second method is illustrated in Figure 1.10.7 and Table A1.10.7. It shows the largest concentration of people, both rural and urban, is from sea level to 100 m altitude (35% of the total population). The other altitude zone where there is a high concentration of people is in the valleys and basins of the highlands over the altitude range of 1600–2000 m (24%). Note that these two techniques give a somewhat different picture of the distribution of the population by altitude class, especially in the highlands. However, the overall trends are the same using both methods.

Sources

Bellamy, J.A. and McAlpine, J.R. (1995). *Papua New Guinea Inventory of Natural Resources, Population Distribution and Land Use Handbook.* 2nd edition. PNGRIS Publication No. 6. Australian Agency for International Development, Canberra.

Löffler, E. (ed.) (1977). *Geomorphology of Papua New Guinea.* Commonwealth Scientific and Industrial Research Organisation and Australian National University Press, Canberra.

McAlpine, J. and Quigley, J. (c. 1995). *Natural Resources, Land Use and Population Distribution of Papua New Guinea: Summary Statistics from PNGRIS.* PNGRIS Report No. 7. Australian Agency for International Development, Canberra.

NSO (National Statistical Office of Papua New Guinea) (2002). Papua New Guinea 2000 Census: Final Figures. National Statistical Office of Papua New Guinea, Port Moresby.

1.11 Agricultural environments

Agriculture involves the production, from plants and animals, of food and other products useful to people. Environmental conditions such as temperature, rainfall, rainfall seasonality, flooding, soil fertility, soil structure and slope all influence agricultural production. Different combinations of environmental conditions will favour or constrain production from particular plants and animals.

An understanding of environments and the opportunities and limitations they present to agriculture is important. If we know what combinations of environmental conditions best suit particularly successful economic plants or agricultural practices, we can look for other places that might be suitable for the introduction of those plants or practices. Knowledge of environmental conditions can be used to assess where investment in agricultural research will give the greatest returns or will benefit the largest possible number of people.

A powerful way to examine agricultural environments is to produce a map of them (for example, Figure 1.11.1). What combinations of environmental conditions are included or excluded when a map of agricultural environments is drawn depends on the purpose for creating the map. The same basic information about the environment can be combined in many different ways for different purposes.

The construction of an agricultural environment map also depends on the availability and reliability of information about individual environmental conditions. For example, rainfall data may be of a high quality, but information about soils poor. In such a case, it would be better not to use that soils information, or to reduce its influence, in the creation of a map of agricultural environments. Temperature, rainfall and inundation are environmental attributes that are reasonably consistent over large areas. Characteristics like soil fertility, soil structure and slope often vary greatly over shorter distances and so are more difficult to use.

An agricultural environment map of PNG

In this section, three environmental conditions important for PNG agriculture are combined to create a map of agricultural environments (Figure 1.11.1). This type of map is particularly suitable for research planning. The conditions are:

- Altitude (as an alternative for temperature).
- Soil water deficit (the likelihood of drought, based on a soil water balance from 54 PNG climate stations for a 15-year period).
- Inundation (the degree of flooding).

Information about these conditions is taken from the Papua New Guinea Resource Information System (PNGRIS) (see Section 1.15).

Altitude/temperature

Temperature has a significant influence on crop production. In PNG, temperature is most influenced by altitude (see Sections 1.7 and 1.10).

For the purpose of analysis here, the six altitude classes described in Section 1.10 have been reduced to the following four classes:

- Lowlands (sea level to 600 m).
- Intermediate (600–1200 m).
- Highlands (1200–1800 m).
- High altitude (1800–2800 m).

Soil water deficit/rainfall deficit

A lack of water in the soil can limit production of most common agricultural food plants. In PNG, too much water in the soil frequently limits agricultural production rather than not enough water. However, it is important to note that soils do dry out for a period of weeks in PNG. Dry periods occur in some places every year and in most places on average every 10–15 years. A measure of when low soil moisture conditions limit plant growth is 'soil water deficit' (see Section 1.5).

The five patterns of soil water balance described in Section 1.5 have been collapsed into three classes for analysis here:

- Strong soil water deficit (a regular and severe soil water deficit – plant growth is limited by soil moisture conditions for 5–7 months every year).
- Moderate soil water deficit (infrequent deficit – plant growth is limited by soil moisture conditions for 2–4 months every year).
- No soil water deficit (places where sufficient rain falls in all months of the year in most years – plant growth is not limited by soil moisture conditions or is limited for only one month every year).

Note that these classes of soil water deficit (5–7 months, 2–4 months and 0–1 dry month) are based on average rainfall, and actual soil moisture levels vary from year to year. Soil water deficits also vary with the type of soil, although this is less important than rainfall. The consequences of soil water deficits vary from crop to crop but, in general, they determine when crops grow during the year: the three classes correspond to growing periods of 5–7 months, 8–10 months and 11–12 months respectively.

Nearly half of the country experiences moderate soil water deficit. However, 43% of the land has no significant deficit and only 8% has a strong deficit (Figures 1.11.2, 1.5.5). Strong water deficit occurs only in the lowlands. In the other three inhabited altitude classes, only 'moderate soil water deficit' or 'no soil water deficit' occur (Table 1.11.1).

Inundation (flooding)

Inundation occurs when the soil surface is covered with water, or flooded. Flooding is a severe limitation to most forms of agriculture. Inundation is a reliable indicator of where most plants are unlikely to grow well. A notable exception is sago; inundated locations are often places where sago is an important food.

The nine classes of inundation described in PNGRIS are here collapsed into two classes:

- No inundation (PNGRIS Class 0: land is never flooded).
- Inundation (PNGRIS Classes 1–8: land is flooded briefly, seasonally, or permanently flooded, as in a swamp).

In PNG, 33% of the total land area is inundated in some way, which leaves 67% that is not inundated. Almost all inundated land is in the lowlands altitude class. A number of very small areas of inundation occur in the highlands altitude class, but they have been excluded in order to simplify this analysis.

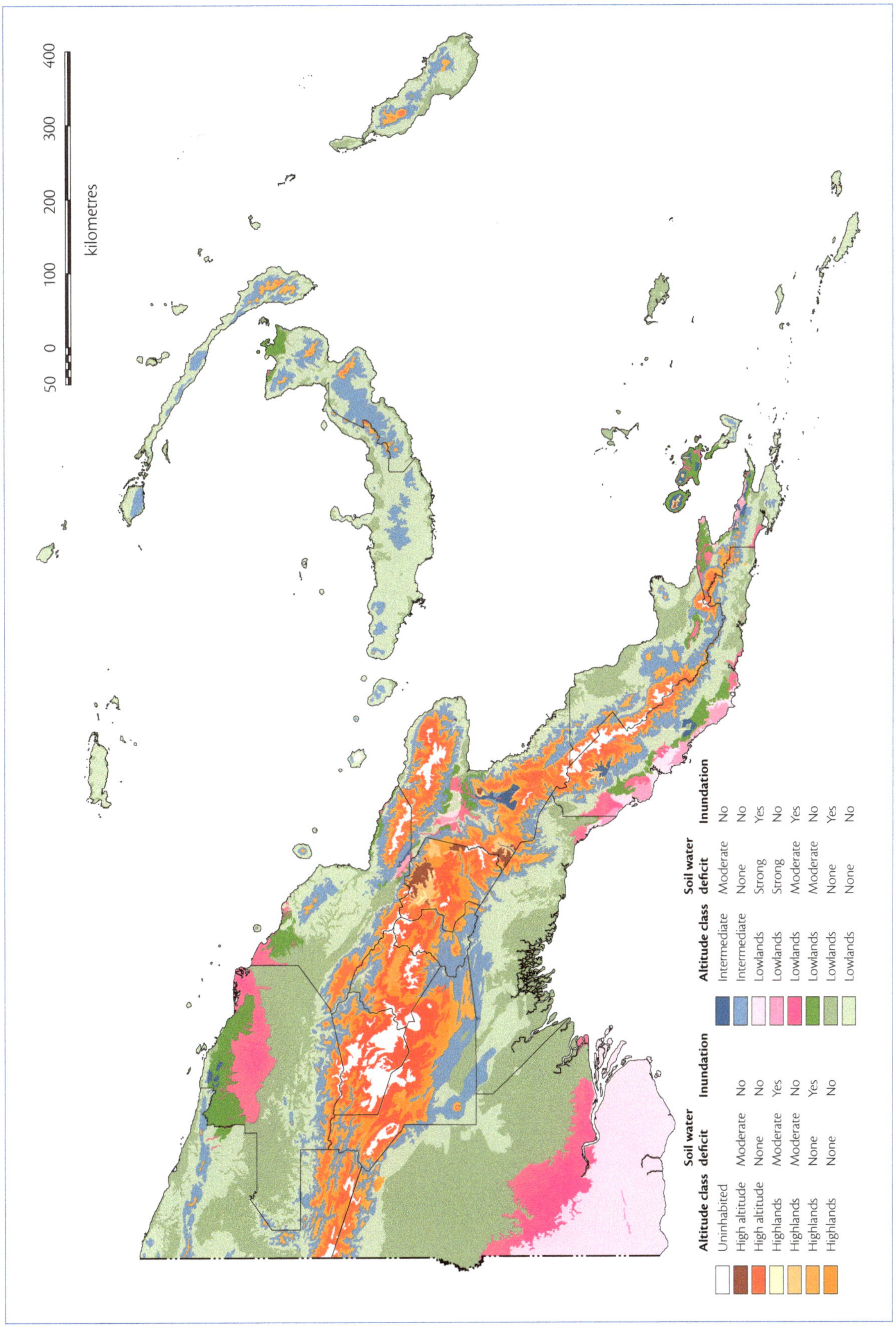

Figure 1.11.1 Agricultural environments based on altitude (temperature), soil water deficit (rainfall deficit) and inundation classes. Source: PNGRIS.

The three conditions combined into agricultural environments

When all possible combinations of the environmental conditions of altitude/temperature, soil water deficit and inundation described in this section are combined and mapped, the outcome is 14 agricultural environments, plus the very high altitude environment that is presently uninhabited and not used for agriculture. Within each environment, there are big differences in soil fertility and slope that cannot be shown at the scale used here. However, these 14 broad classes can be used to identify where a particular crop might grow best or where an agricultural practice that has proved successful elsewhere might be suitable. An example is the practice of planting casuarina trees in fallow gardens. Within each class, further work in the field can identify local land that is not too steep and is reasonably fertile.

The distribution of PNG's population by agricultural environment is of interest for a number of reasons. It provides an indication of the numbers of people who would benefit from research or other activities within a given environment. It also gives insights into what sort of environments have been favoured by Papua New Guinean agriculturalists over long periods of time.

Almost half of all rural Papua New Guineans live in lowlands environments. A further 27% of the rural population live in highlands environments and 17% in high altitude environments, leaving only 7% in intermediate altitude environments (Table 1.11.1).

Of those who live in lowlands environments, around 70% favour environments which are not inundated and which do not have a strong soil water deficit. Across the country, inundation is associated with low population densities. In highlands environments population densities are higher where there is a moderate soil water deficit and no flooding.

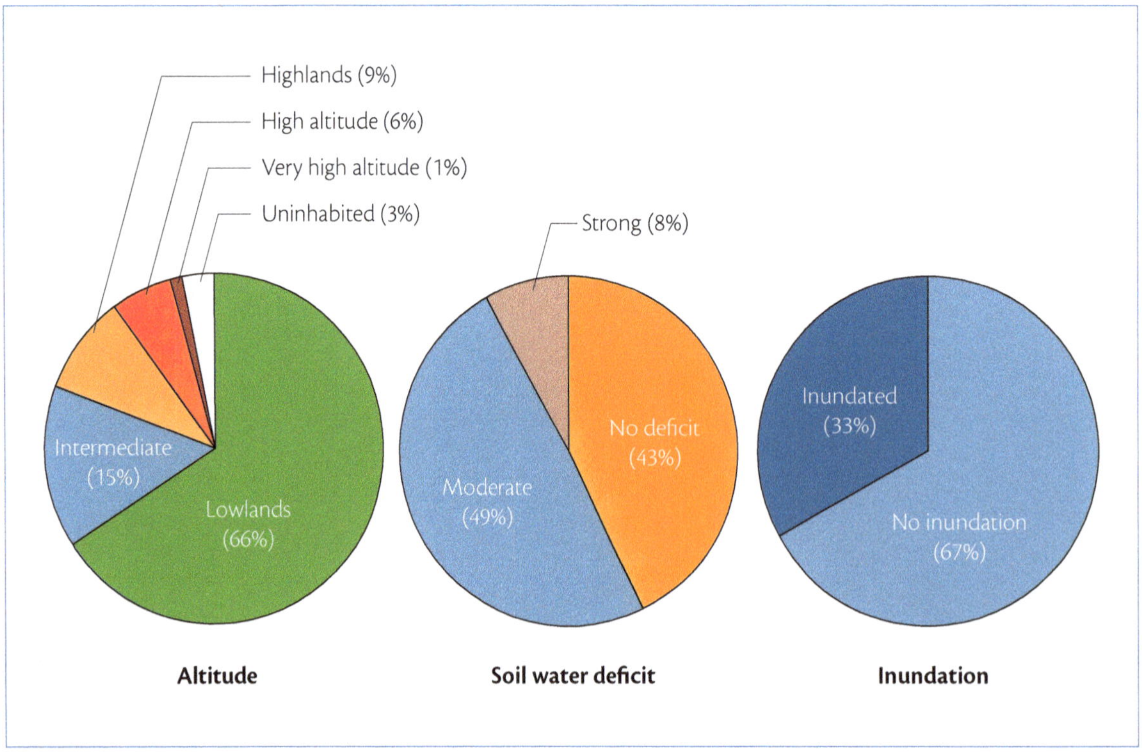

Figure 1.11.2 Proportion of the total land area by altitude (temperature), soil water deficit (rainfall deficit) and inundation classes. Source: PNGRIS.

Table 1.11.1 Total land area, rural population and rural population density, by agricultural environment

Agricultural environment	Total land area		Rural population	Proportion of rural population (%)	Proportion of population in altitude class (%)	Rural population density on total land area[a] (persons/km^2)
	(km^2)	(%)				
Lowlands–Strong soil water deficit–Inundation	31,200	6.8	36,675	0.9	1.8	1.2
Lowlands–Strong soil water deficit–No inundation	3,828	0.8	126,466	3.0	6.2	33.0
Lowlands–Moderate soil water deficit–Inundation	61,744	13.4	218,048	5.2	10.6	3.5
Lowlands–Moderate soil water deficit–No inundation	67,066	14.6	893,196	21.3	43.5	13.3
Lowlands–No soil water deficit–Inundation	56,118	12.2	256,559	6.1	12.5	4.6
Lowlands–No soil water deficit–No inundation	83,889	18.2	523,523	12.5	25.5	6.2
Lowlands altitude class (sea level to 600 m) total	303,844	66.1	2,054,466	49.0	100.0	6.8
Intermediate–Moderate soil water deficit–No inundation	25,821	5.6	137,770	3.3	49.6	5.3
Intermediate–No soil water deficit–No inundation	43,684	9.5	140,122	3.3	50.4	3.2
Intermediate altitude class (600–1200 m) total	69,505	15.1	277,893	6.6	100.0	4.0
Highlands–Moderate soil water deficit–Inundation	741	0.2	72,453	1.7	6.4	97.8
Highlands–Moderate soil water deficit–No inundation	19,398	4.2	702,453	16.8	62.3	36.2
Highlands–No soil water deficit–Inundation	114	0.0	544	0.0	0.0	4.8
Highlands–No soil water deficit–No inundation	23,163	5.0	352,951	8.4	31.3	15.2
Highlands altitude class (1200–1800 m) total	43,416	9.4	1,128,400	26.9	100.0	26.0
High altitude–Moderate deficit–No inundation	14,091	3.1	303,956	7.2	41.5	21.6
High altitude–No soil water deficit–No inundation	17,543	3.8	427,845	10.2	58.5	24.4
High altitude class (1800–2800 m) total	31,634	6.9	731,802	17.5	100.0	23.1
Uninhabited altitude class–Moderate deficit–No inundation	11,455	2.5	0	0.0	0.0	0
Uninhabited altitude class (>2800 m) total	11,455	2.5	0	0.0	0.0	0.0
Papua New Guinea	459,854	100.0	4,192,561	100.0		9.1

[a] This is rural population density on the *total* land area. Do not confuse it with the total population density on land used for agriculture – see Table 1.10.2 and Sections 1.2 and 1.3.

Sources: NSO (2002); PNGRIS.

Sources

Bellamy, J.A. and McAlpine, J.R. (1995). *Papua New Guinea Inventory of Natural Resources, Population Distribution and Land Use Handbook.* Second edition. PNGRIS Publication No. 6. Australian Agency for International Development, Canberra.

Bourke, R.M. (1981). Agriculture in Papua New Guinea. *Alafua Agricultural Bulletin* 6(3):77–82.

Hanson, L., Allen, B.J. and Bourke, R.M. (2001). Mapping land resource vulnerability in the highlands of PNG. In Bourke, R.M., Allen, M.G. and Salisbury, J.G. (eds). *Food Security for Papua New Guinea. Proceedings of the Papua New Guinea Food and Nutrition 2000 Conference.* ACIAR Proceedings No. 99. Australian Centre for International Agricultural Research, Canberra. pp. 284–290.

Hanson, L.W., Allen, B.J., Bourke, R.M. and McCarthy, T.J. (2001). *Papua New Guinea Rural Development Handbook.* Land Management Group, Department of Human Geography, Research School of Pacific and Asian Studies, The Australian National University, Canberra.

NSO (National Statistical Office of Papua New Guinea) (2002). Papua New Guinea 2000 Census: Final Figures. National Statistical Office of Papua New Guinea, Port Moresby.

1.12 Land quality

Measuring land quality

Land quality can be assessed in a number of ways. It is measured here as 'resource potential', or the ability of the land to grow particular crops at their optimum production over a long period of time. Resource potential is determined by identifying conditions that will prevent optimum plant growth. These are known as 'constraints' and include, for example, infertile soils or soils that will lose fertility quickly, too much or too little water in the soil, steep slopes that will erode if cultivated, and low temperatures. Each constraint is given a score depending on how critical it is to plant growth over the long term, relative to all the other constraints. The constraints are added together and areas of land are examined to see how well they score against all the constraints.

People commonly overcome some constraints by modifying the environment. So swampy land is drained to lower soil water levels (see Sections 1.5, 1.11 and 3.12), physical barriers are constructed across slopes to slow or prevent soil erosion (see Section 3.9), and green manures are used to maintain soil fertility over many cropping cycles (see Section 3.11). The type of modification that is used often depends on the crop species being grown. For example, taro is constrained less by high soil water levels than sweet potato, and so drainage is less important if taro is to be grown than if sweet potato is the chosen crop.

Methods of classifying land according to its quality depend on the accuracy of the information available. The resource mapping units (RMUs) of PNGRIS provide a convenient unit of analysis and contain information on natural resources that is reasonably accurate at provincial and national scales (see Section 1.15). Resource potential, or land quality, is usually estimated relative to a particular crop or land use. The constraints to optimum growth are estimated for particular species and the outcome applies only to that species, or to other similar plants.

The assessment of land quality presented here (see Figure 1.12.1) is based on the potential of land in PNG to grow sweet potato. It was carried out using the following information:

- Altitude (as an alternative for temperature) (from PNGRIS).
- Inundation or flooding (from PNGRIS).
- Slope (from PNGRIS).
- Soil type (from PNGRIS).
- Annual rainfall (from the ANU's Centre for Resource and Environmental Studies rainfall surface for PNG; see Figure 1.5.2).
- Light (measured as cloud cover by the Australian Geological Survey from the United States' National Oceanic and Atmospheric Administration satellite imagery data).

Sweet potato was chosen because it is grown in almost all inhabited parts of PNG, it is the staple food for more than 60% of the population and a lot is known

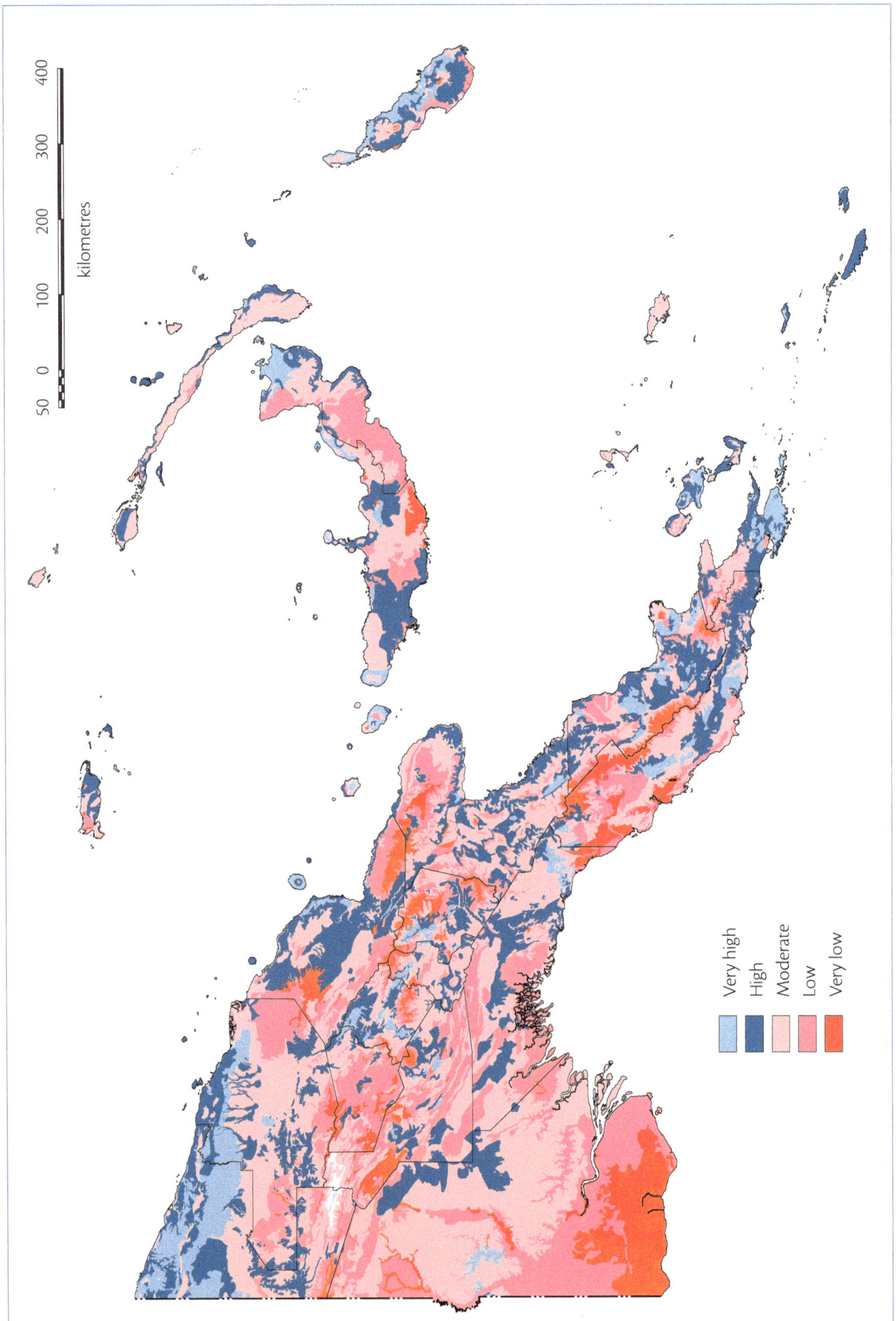

Figure 1.12.1 Land quality. Source: PNGRIS.

about constraints to its optimum growth. In the analysis presented here, modifications to the environment, such as draining or composting, are *not* taken into account. For an example of an analysis where environmental modifications are taken into account, see Hanson, Bourke, Allen and McCarthy (2001).

Land quality in PNG

Using the above measures to determine the potential of land to grow sweet potato, the following points can be made about land quality in PNG:

- It is predominantly of low quality. More than 70% of the total land area is of low or very low quality (it must be remembered, however, that only 25% of the total land area of PNG is used for agriculture (see Section 1.2), and much of the poorest quality land is not occupied by people); 20% of the total land area is of moderate potential and only 7% is of high or very high potential (Table 1.12.1, Figure 1.12.2, Table A1.12.1).
- Most of the people in PNG produce food from land of moderate to low quality. Up to 80% of the population occupies land of moderate or lower potential. Only 20% occupy high or very high quality land (Table 1.12.1).
- High quality land is associated with volcanic landforms or with land covered with volcanic tephra (fine material that has fallen on the land following a volcanic eruption). The largest area of high quality land that is presently unoccupied surrounds active and dangerous volcanoes that are known to have erupted and killed people in the relatively recent past.
- Of the estimated 983 000 people who occupy high and very high quality land, almost 30% live in either East New Britain Province or Western Highlands Province. A further 12% live in East Sepik Province and 10% live in Sandaun Province (Figure 1.12.3, Table A1.12.2).
- Of the estimated 2.8 million people who occupy very low or low quality land, 24% live in Morobe Province or Southern Highlands Province and a further 18% in Eastern Highlands or Enga provinces.
- People are not evenly distributed over poor and good quality land, but are concentrated on better quality land. Average population densities on high and very high quality land are between four and eight times those on poor quality land (Table 1.12.1, Figure 1.12.4).

Over the long term people can produce food from poor quality land either because they use it at low intensities (land may be used only once every 20 to 30 years) (see Section 1.2) or because they use special techniques to overcome constraints to crop growth, such as draining a swamp or using green manure to maintain soil fertility.

Table 1.12.1 Total land area, rural population and rural population density, by land quality (the potential of the land to grow sweet potato)

Land quality	Total land area (km²)	Total land area (%)	Rural population	Proportion of rural population (%)	Rural population density (persons/km²)
Very low	85,270	18.5	456,989	10.9	5
Low	251,563	54.7	1,773,453	42.3	7
Moderate	92,121	20.0	1,123,606	26.8	12
High	20,532	4.5	419,256	10.0	20
Very high	10,368	2.3	419,256	10.0	40
Total	459,854	100.0	4,192,561	100.0	9

Sources: NSO (2002); PNGRIS; land quality calculated by Luke Hanson.

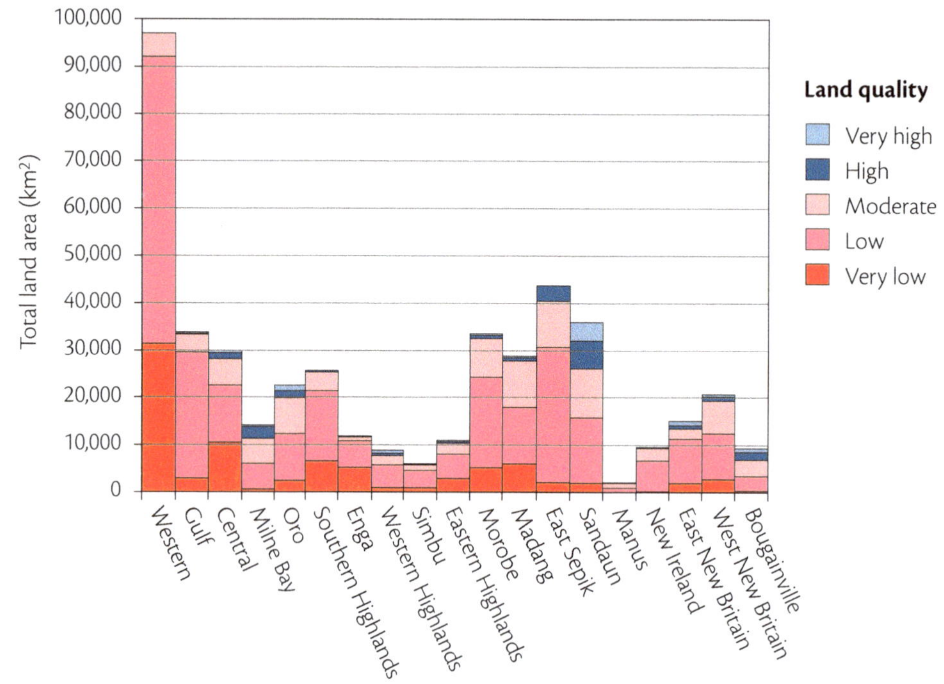

Figure 1.12.2 Total land area by land quality and province. Source: PNGRIS.

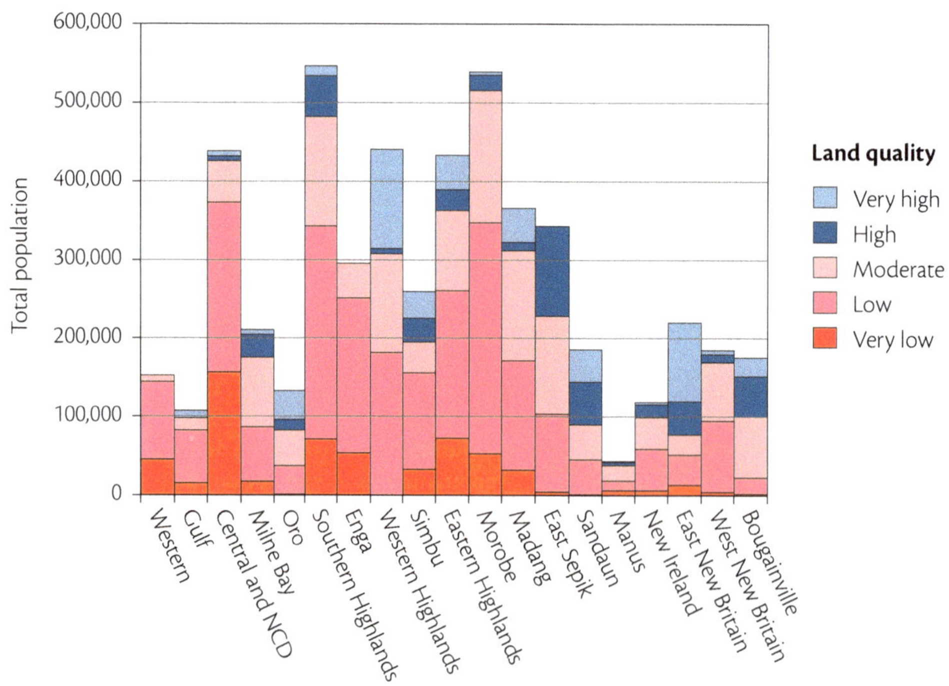

Figure 1.12.3 Total population (including urban and rural non-village populations) by land quality and province. Sources: NSO (2002); PNGRIS.

The results of land quality analyses always require some qualification. The results of the analysis here are no different. This analysis of land quality classifies high altitude areas as low quality, for reasons associated with the boundaries of PNGRIS RMUs and the altitude classes used in PNGRIS. Hence highlands provinces are almost certainly over-represented in the classes of lower land quality. Although sweet potato is known to produce relatively slowly at higher altitudes (6–8 months to main harvest compared to 3–5 months at lower altitudes), high altitude land quality is not as poor as this analysis suggests. This analysis also produces other results that are difficult to interpret, such as the large areas of high quality land in Sandaun Province. These outcomes may result from errors in PNGRIS data for this province.

These results emphasise the need to interpret the findings of land quality analyses in the light of other knowledge. Findings should never be accepted unquestioningly. That said, the broad overall pattern of land quality in PNG presented here will not change greatly if another crop is substituted for sweet potato, or other adjustments are made. The greatest changes will occur if tree crops, banana or sago are used in the analysis. An analysis of the suitability of land for producing cocoa and coconuts was done by Hanson et al. (1998).

Sources

Bourke, R.M. (2005). Sweet potato in Papua New Guinea: the plant and people. In Ballard, C., Brown, P., Bourke, R.M. and Harwood, T. (eds). *The Sweet Potato in Oceania: A Reappraisal. Ethnology Monographs 19 / Oceania Monograph 56*. Oceania Publications, University of Sydney and Ethnology, Department of Anthropology, University of Pittsburgh, Sydney and Pittsburgh. pp. 15–24.

Hanson, L.W., Bourke, R.M. and Yinil, D.S. (1998). *Cocoa and Coconut Growing Environments in Papua New Guinea. A Guide for Research and Extension Activities*. Australian Agency for International Development, Canberra.

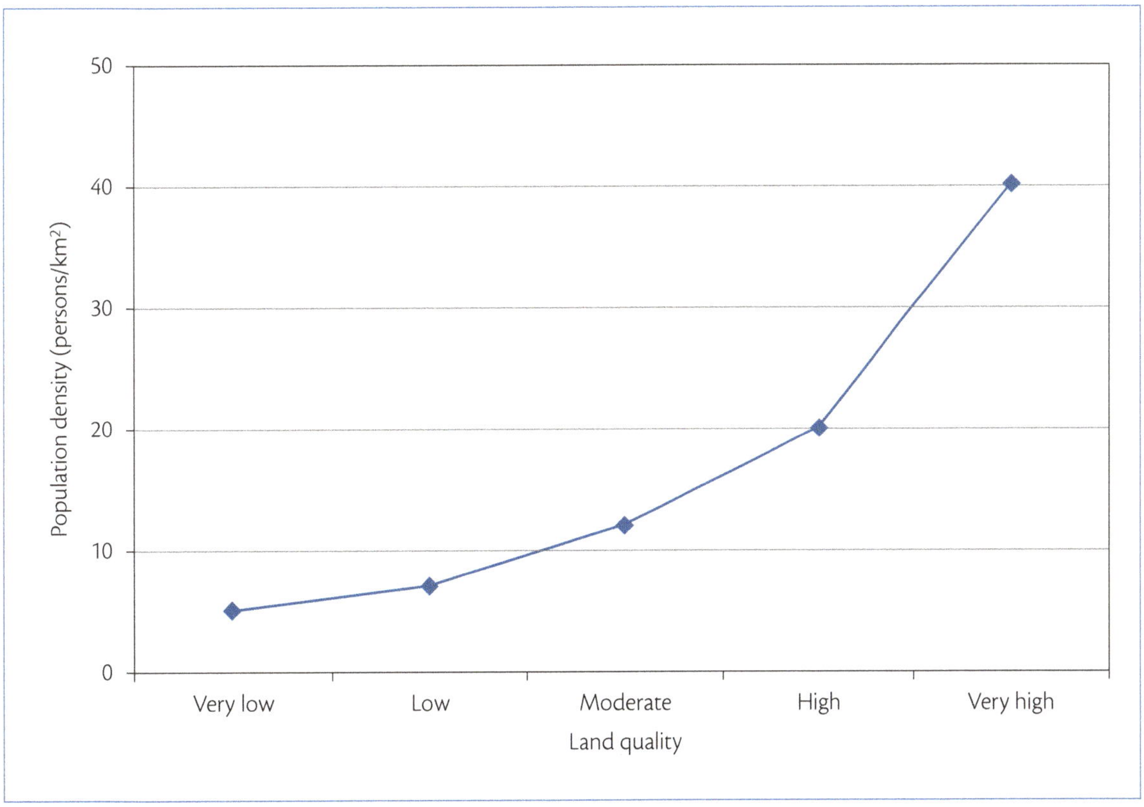

Figure 1.12.4 Association between land quality and population density. Sources: NSO (2002); PNGRIS.

Hanson, L., Allen, B.J. and Bourke, R.M. (2001). Mapping land resource vulnerability in the highlands of PNG. In Bourke, R.M., Allen, M.G. and Salisbury, J.G. (eds). *Food Security for Papua New Guinea. Proceedings of the Papua New Guinea Food and Nutrition 2000 Conference*. ACIAR Proceedings No. 99. Australian Centre for International Agricultural Research, Canberra. pp. 284–290.

Hanson, L.W., Bourke, R.M., Allen, B.J. and McCarthy, T.J. (2001). Mapping land resource potential and agricultural pressure in Papua New Guinea: an outline of new methods to assist rural planning. ACIAR Technical Reports No. 50. Australian Centre for International Agricultural Research, Canberra.

NSO (National Statistical Office of Papua New Guinea) (2002). Papua New Guinea 2000 Census: Final Figures. National Statistical Office of Papua New Guinea, Port Moresby.

Rossiter, D.G. (1990). ALES: a framework for land evaluation using a microcomputer. *Soil Use and Management* 6(1):7–20.

Trangmar, B.B., Giltrap, D.J., Burgham, S.J. and Savage, T.J. (1995). *Land Suitability Assessments for Selected Crops in Papua New Guinea*. PNGRIS Publication No. 8. Australian Agency for International Development, Canberra.

Venema, J.H. and Daink, F. (1992). Papua New Guinea Land Evaluation Systems (PNGLES). AG: TCP/PNG/ 0152 Field Document 1. Department of Agriculture and Livestock and Food and Agriculture Organization of the United Nations, Port Moresby and Rome.

1.13 Crops, people and the environment

Planted crops are the basis for agriculture in PNG. The natural physical environment influences which crops can be grown in a particular location, how well a crop performs, and other matters such as when flowering occurs, the time from flowering to fruit maturity, and fruit flavour. People can influence the relationship between the physical environment and a crop to some degree, either by selecting the environment for planting a crop or by modifying the environment. An example of people selecting the environment would be planting a crop that requires high soil fertility in a part of a food garden where the soil is deeper and more fertile. Examples of people modifying the environment include transferring organic material into a garden (composting) or digging drains to reduce the amount of water in the soil.

This section examines how aspects of the physical environment affect plant growth and yield and how people influence this in PNG. The information is organised by main environmental factors. Sometimes more than one environmental factor is important in determining how well crops grow, and sometimes people's practices can have multiple effects.

An example of a practice having multiple benefits is the use of mixed species plantings in food gardens, a technique widely used in PNG. This is a very efficient practice and generally results in higher total yields and reduced labour inputs for the same area, compared with planting a single species (monocropping). The benefits from mixed species cropping include more efficient use of available sunlight, with shade-tolerant crops planted under sun-loving species. An example is Chinese taro planted under banana.

Another benefit of mixed species planting is the more efficient use of soil nutrients and soil moisture, with quick-growing species using some nutrients before they are needed by slower-growing species, and before they can be leached from the soil. Different crops also have different needs for soil nutrients, so planting a mix of species can mean that the species are not competing for the same nutrients. For example, green vegetables have a high demand for nitrogen, whereas root crops and banana have a high demand for potassium, a nutrient that is less important for most leafy greens. Roots of the different species also use nutrients from different soil depths. As well as the total crop yield being greater from mixed plantings, the labour inputs are often less because of a reduced need for weeding.

Aspects of the physical environment that have the greatest influence on plant growth are rainfall, cloud cover, temperature, daylength, inundation (flooding) and soil fertility. Some of these factors are in turn influenced by others, such as altitude, latitude, slope, landform and underlying rock type. The various attributes outlined here are described in more detail in sections elsewhere: 1.5 (rainfall), 1.7 (temperature, cloudiness and sunshine), 1.9 (soils) and 1.10 (landforms and altitude). Combinations of these factors are discussed in Sections 1.11 (agricultural environments) and 1.12 (land quality).

Rainfall

A number of aspects of rainfall influence plant production, including total annual rainfall, its seasonal distribution, variability between years, extremes (drought or periods of particularly high rainfall) and intensity (millimetres per hour) (see Section 1.5). The optimum annual rainfall for many crops in PNG is 1500 to 3000 mm per year (125 to 250 mm per month). For many crops, a period during the year of 1–3 months where the monthly rainfall is somewhat less (50 to 100 mm) is also a benefit. Crops tend not to grow so well in locations where the total annual rainfall is greater than 4000 mm. Some crops are more tolerant of high rainfall than others. For example, oil palm produces well in locations with an annual rainfall of up to 3500 mm, whereas cocoa does best where the annual rainfall is between 1800 mm and 2600 mm. Sweet potato is vulnerable to high levels of soil moisture, whereas taro is much more tolerant of wetter soil conditions. Despite this, sweet potato is grown in locations where the rainfall is high to very high and people use mounds or drains to reduce soil water levels.

Rainfall seasonality influences crop growth in a number of ways. Some species require a drier period in the year for flowering and fruiting, although many of the crops grown in PNG produce even without a drier period each year. Mango is a good example of a food crop that requires a period of drier weather to induce flowering. Higher rainfall and humidity during the period that mango flowers are forming can allow fungal growth that causes flowers to abort, thereby reducing fruit yield. *Karuka* nut pandanus is another crop where flowering and fruiting is induced by drier weather. In places that experience a regular drier period each year, such as Eastern Highlands Province, fruiting is more regular. However, in the western part of the highlands where the annual rainfall is less seasonal, production of fruit is less regular. The best yields of *karuka* nut follow mild droughts.

In locations where there is regular variation in the seasonal distribution of rainfall, villagers adapted their agricultural systems to avoid interruptions to food supply. Where rainfall is well distributed throughout the year, taro was generally the main staple food in the lowlands. In more seasonal environments, people grew more banana and yam. Where the rainfall distribution was markedly seasonal, the agricultural system was based on a combination of taro, yam and banana, sometimes with sago being used for part of the year. Using a mix of crops with varying times to maturity spread the supply of food more evenly throughout the year. In some strongly seasonal locations, especially on small islands and in Milne Bay Province, foods from trees including breadfruit and Polynesian chestnut

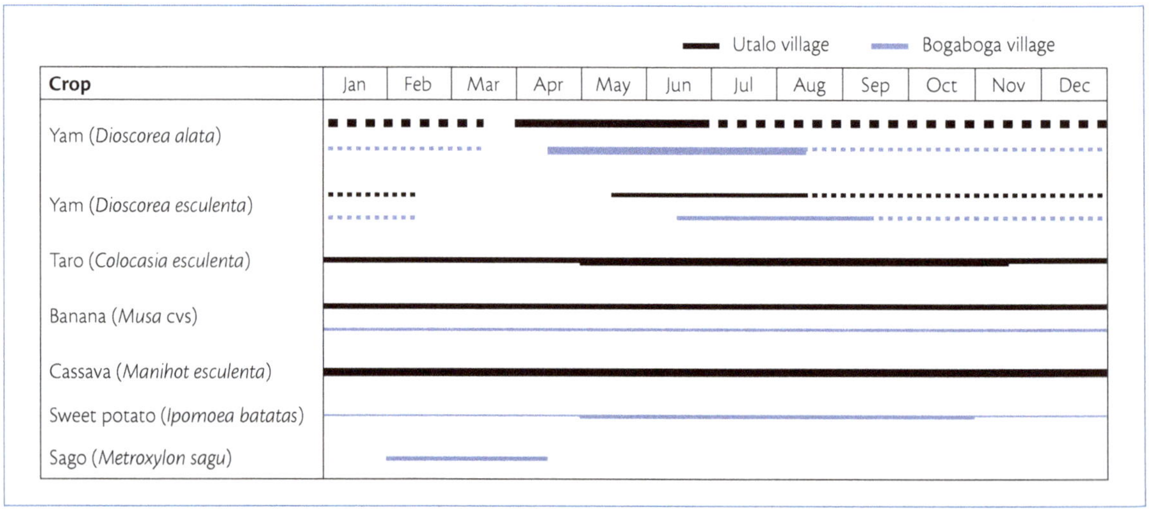

Figure 1.13.1 Food availability at Utalo village, Goodenough Island, and Bogaboga village, Cape Vogel area, Milne Bay Province. **Note:** A solid line indicates availability directly from food gardens. A dashed line indicates availability from storage. The thickness of the bands indicates relative abundance. Source: Mogina (2002:159).

(*aila*) were important when the supply of garden food was low. The availability of foods at two locations in Milne Bay Province is illustrated in Figure 1.13.1. Over the past 130 years, villagers have incorporated sweet potato, cassava, corn and other introduced foods into their agricultural systems. This has allowed greater flexibility in the food production systems and spread the supply of food more evenly throughout the year.

Seasonality of rainfall influences the timing of crop planting. In some locations in PNG, the main food gardens are planted prior to the wetter part of the year, often in about September–October. In the highlands, the mixed vegetable gardens are commonly planted at this period, which results in a greater supply of some foods in the months that follow. Figure 1.13.2 shows when amaranthus, common bean, corn and cucumber are most abundant in a number of highland markets.

A period of lower rainfall is often desirable to break a cycle of disease build-up, and is particularly important in PNG where there is little seasonal change in temperature. Taro blight, a fungal disease, is worst in locations where there is no annual drier period. A number of crops, including watermelon, cucumber and rockmelon, only bear well when planted during drier months, almost certainly because pests and diseases are worse when rainfall is higher.

Drought is an extreme form of rainfall variability. It is difficult for villagers to devise an agricultural system to accommodate an event such as a major drought that occurs infrequently and in an unpredictable manner (see Section 1.6). People have many responses to surviving a drought, including eating plants and parts of plants that are not normally eaten, such as very small tubers, or eating uncultivated ('wild') plants, including ferns, wild yams and *kudzu* tubers. Another strategy is to migrate to locations where the food supply is better. In the modern context, people often buy locally grown or imported food using cash.

People make a number of modifications to the environment to influence soil moisture so that it does not interfere with plant growth. In most of PNG, the main concern is removal of excess water, and the construction of drains is a common technique, especially in sweet potato gardens in the highlands. Crops, especially sweet potato and yam, may also be planted on mounds to allow excess moisture to drain away from the root zone. In the past, where a lack of water for part of a year was the problem, irrigation of taro and other crops was practised, but irrigation is now restricted to a limited number of places, the most important being the Rabaraba area of Milne Bay Province. In Eastern Highlands and Morobe provinces, taro was irrigated using bamboo pipes. At Rabaraba, streams at the top of large alluvial fans are diverted into ditches that allow water to be distributed to areas under cultivation, while other areas are left in fallow, uncultivated and unirrigated (see Section 3.12).

Cloud cover

High levels of cloudiness tend to be associated with high rainfall and less direct sunlight (see Section 1.7). The combination of excessive soil moisture and less direct sunlight results in poorer plant growth for many crops. This is one reason why yields of grain crops, such as rice and corn, are not particularly high in PNG. The highest yields of rice occur where there are a high number of hours of bright sunshine in the day and water supply is plentiful. These conditions occur, for example, in the Murrumbidgee Irrigation Area of New South Wales, Australia.

Yields of sweet potato and other crops tend to be lower on the southern sides of the main mountain ranges, for example, in Southern Highlands Province and mountainous parts of Gulf Province. This is because of both excessively high rainfall and high levels of cloudiness. In mountainous locations where clouds form early in the day and reduce sunlight, human settlement and agriculture is generally absent.

Temperature

Altitude has the greatest influence on temperature in PNG (see Sections 1.7 and 1.10). Above 500 m, there is a regular decline in temperature with increasing altitude. Other factors that have an influence on temperature are geomorphology (landforms; see Section 1.10) and latitude (distance from the equator). Seasonal differences in temperature are very small at locations near the equator, for

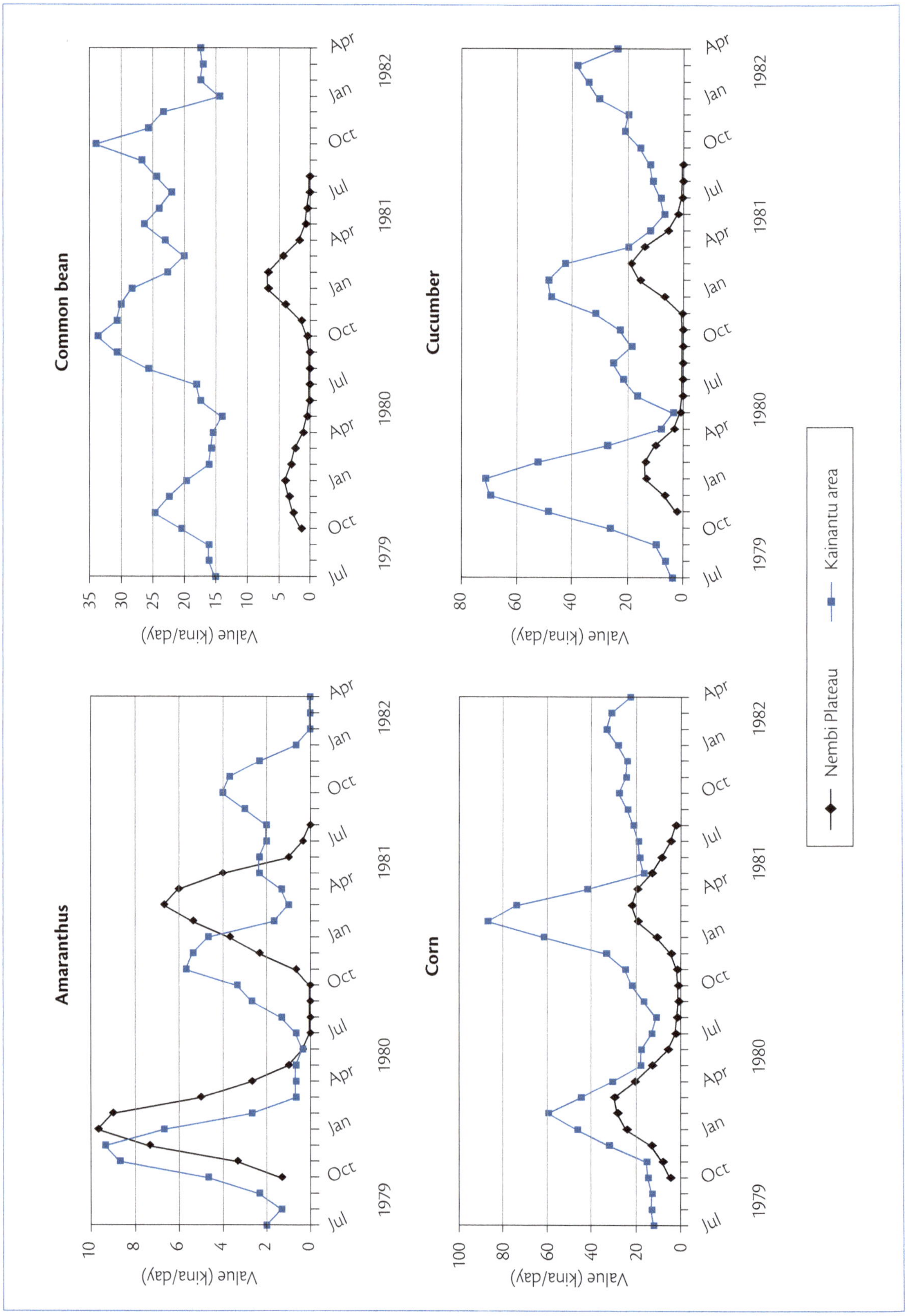

Figure 1.13.2 Availability of amaranthus (*Amaranthus* spp.), common bean (*Phaseolus vulgaris*), corn (*Zea mays*) and cucumber (*Cucumis sativus*) over a 3-year period in local food markets in the Kainantu area, Eastern Highlands Province, and Nembi Plateau, Southern Highlands Province. Data presented as a 3-month running mean. Source: Bourke et al. (2004).

example, on Manus Island. The differences increase at locations further south, particularly in Oro, Milne Bay and Central provinces and the southern part of Western Province (Table 1.13.1). Seasonal differences are also greater where the wettest part of the year coincides with the Southern Hemisphere winter. These conditions occur, for example, in Lae, the Alotau area and the south coast of New Britain. However, even at these locations, the difference in the mean maximum temperature between the coldest part of the year (about July) and the warmest (about February) is only around 4 °C.

Another feature of temperature is the diurnal range, that is, the difference between daytime maximum and night-time minimum temperatures. There are small differences in the diurnal ranges in different locations in PNG, but there is little variation from day to day or even from season to season at any given location. At locations near the ocean the difference between day and night-time temperature is typically about 8 °C. This difference can be less on very small islands where the volume of land is small. In inland locations, for example in the highlands, the diurnal range is typically about 10 °C.

Temperature influences crop growth in a number of ways:

- Minimum or maximum temperatures determine where crops will grow and where they will produce. (These are not always the same. For example, coconut palms grow in the highlands up to 1700–1800 m, but usually bear nuts only up to about 1000 m.)[1]
- Crops generally take longer to mature in a cooler climate. In PNG, this means that crops take longer to mature with increasing altitude.
- One type of temperature extreme, frost, may cause considerable damage to crops. Most crops that grow in PNG are not frost tolerant.

Because altitude has such a strong influence on temperature in PNG, it has a strong influence on where crops will grow and produce. It is possible to define the altitudinal range of crops in PNG. This is illustrated in Figure 1.13.3 and Table A1.13.1 where the usual and extreme altitudinal ranges of 22 crops are shown. The upper or lower altitudinal limits for each species vary somewhat between locations but, for most crops, the upper or lower limit is within ±100 m of the average altitudinal range.

Villagers, especially in the highlands, are very conscious of the altitudinal limits of individual crops. Clan land commonly covers a range of altitudes, so that people can exploit temperature differences to grow a range of crops. For example, a village might be located at 1700 m altitude, with clan land covering the range 1500–2000 m. Crops that require warmer temperatures, such as *marita* pandanus and pawpaw, will be grown only on the lower land, whereas crops that require cooler conditions, such as *karuka* nut pandanus, Irish potato or cabbage, will be located on higher land. Where village land straddles an altitudinal range where a number of crops stop or start to grow, for example at about 1000–1200 m or about 1800–2000 m, it is common for people to speak of 'hot' and 'cold' locations.

For many crops, lower temperatures slow growth and result in a longer growing period. Corn, for example, requires about 90 days to mature below 600 m and 110–120 days at 1600 m altitude. Similarly, sweet potato matures in 3–5 months in the lowlands, but requires 5–8 months at 1600 m and 8–12 months at 2300 m. The longer time to maturity is not necessarily a problem, but is generally a positive condition in most of PNG. This is because, at a given altitude, where soil moisture and nutrients are not limiting growth, the other main environmental factor limiting growth will be lack of bright sunshine. A longer time to maturity allows the plants to accumulate carbohydrate (critical to human nutrition) for longer and the result is often a higher yield. For example, sweet potato yields in the highlands are typically 20–30 tonnes per hectare (t/ha) where the soil is fertile and well drained and the rainfall is adequate. In the lowlands, comparable yields are 15–20 t/ha, although this is achieved in a shorter time.

---

1 The figure of 1000 m is the average altitude of 20 locations where coconut usually bears nuts in PNG. Coconut palms occasionally bear some tiny nuts at altitudes as high as 1310 m, but this is exceptional. The altitudinal limit data quoted here was recorded in 1980–1984. Temperatures are rising in PNG (see Section 1.8), as they are worldwide, and some crops are now bearing at a higher altitude.

Temperature also influences flowering and fruiting of some crops. For example, fruit production of pawpaw, guava and carambola (five corner) is not seasonal in the lowlands, but is seasonal in the highlands near the upper altitudinal limit of these crops. Cooler temperatures have an influence on the period that *marita* pandanus fruit is available. In the lowlands, fruit is available all year. With increasing altitude, the period when *marita* is available is reduced. Near its upper altitudinal limit (1600–1700 m), it only fruits for about four months each year (Figure 1.13.4). For some crops however, such as avocado, the fruiting period is similar across a wide range of altitudes. For these crops, we can conclude that temperature does not influence flowering and fruiting.

A number of crops bear more under cooler conditions. Irish potato is one example; it will only bear tubers when the night temperature is less than 18 °C. This occurs at about 800 m altitude in most of PNG, but at slightly lower altitudes in the south of the country. Although tuberisation (the production of potato tubers) occurs at 800 m in PNG, yields are low at this altitude. Irish potato yields best in the main highland valleys (1600–1800 m) and especially at high-altitude locations (above 2000 m). The relationship between temperature and yield of pyrethrum flowers is particularly marked. The yield of pyrethrum flowers increases rapidly with increases in altitude between 1800 m and 2700 m (Figure 5.16.1).

Even though in much of PNG temperature changes may be small from day to day or even from month to month, these small changes can be enough to induce flowering in some plants. This is the case for pineapple, where lower night-time temperatures induce flowering and fruiting. It can be difficult to separate the influence of changes in minimum temperature from changes in daylength (see below) because both occur to a greater degree at locations further from the equator, particularly in the Southern Region. However, where flowering and fruiting occur at about the same time each year, it is more likely that changes in daylength are more important than changes in temperature. Where flowing and fruiting vary a lot from year to year, as they do with pineapple, it is more likely that minimum temperature is the more important factor.

Daylength

Change in daylength has a major influence on when plants flower. The length of the day varies during the course of a year in a predictable manner. Near the equator, for example in Manus and northern New Ireland, daylength does not vary greatly throughout the year. At locations further south, in Milne Bay, Oro and Central provinces and the southern part of Western Province, there is a greater difference in daylength between the period of shortest days (June) to that with the longest (December) (see Section 1.7). These differences are still small compared with places in the temperate zone, for example, in southern Australia.

Table 1.13.1 Average seasonal temperature differences between a low-latitude and a high-latitude location

Location	Latitude, south		Temperature (°C)		
			February	July	Difference
Momote, Manus Island[a]	2°03′	Mean minimum	24.2	24.5	–0.3
		Mean maximum	30.0	29.5	0.5
Samarai, Milne Bay Province[b]	10°37′	Mean minimum	24.7	22.8	1.9
		Mean maximum	31.6	27.0	4.6

[a] Data collated from 20 years of records between 1949 and 1970.

[b] Data collated from 44 years of records between 1891 and 1970.

Source: McAlpine et al. (1975).

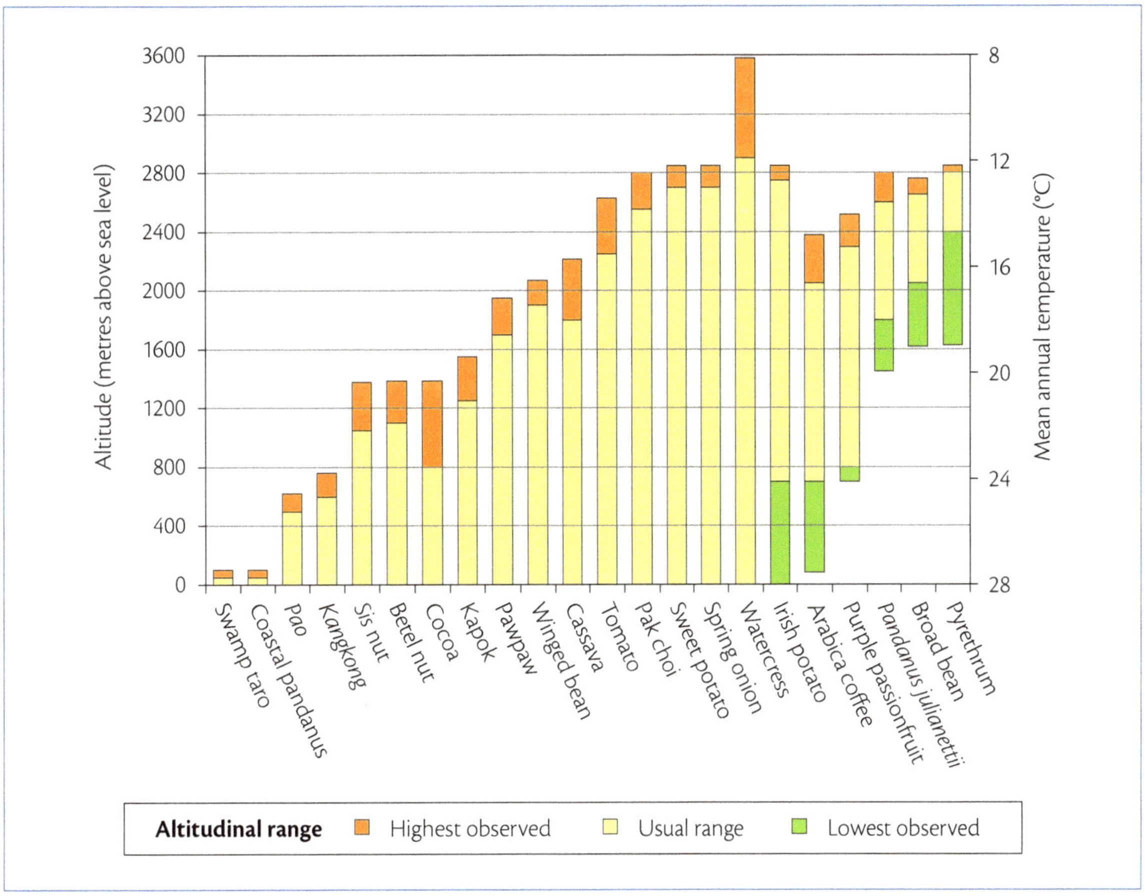

Figure 1.13.3 The usual and extreme altitudinal ranges of 22 crops in PNG. Source: Bourke (1989).

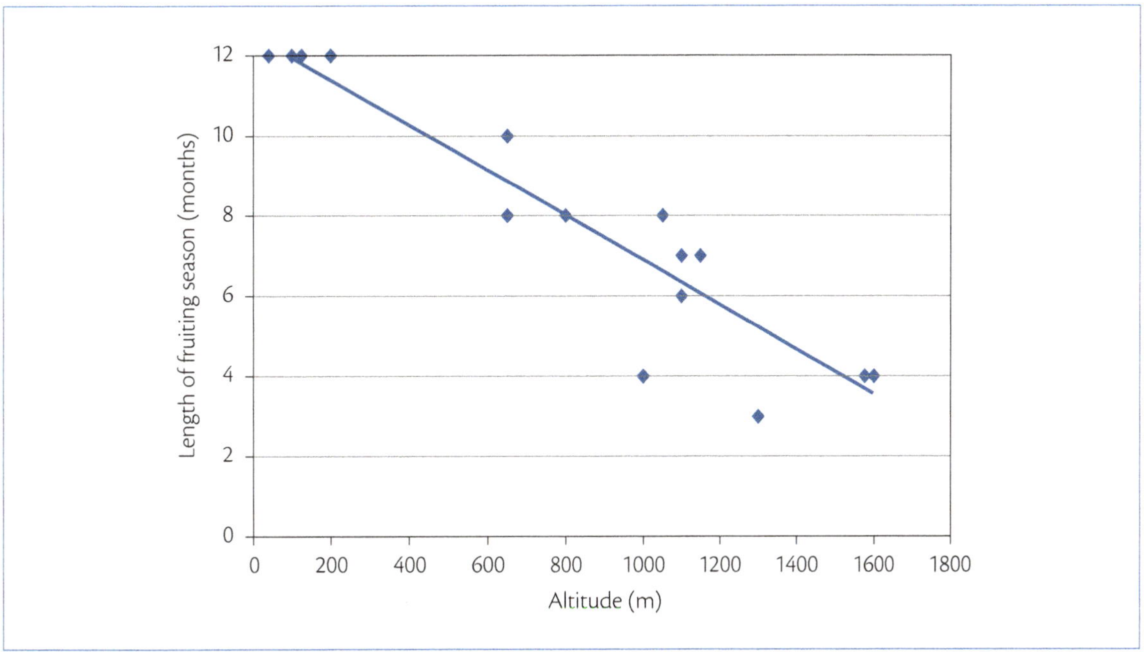

Figure 1.13.4 The length of the fruiting season of *marita* pandanus (*Pandanus conoideus*) versus altitude in PNG. Source: Bourke et al. (2004).

Where a plant species flowers or fruits at about the same time every year, we can infer that it is changes in daylength that induce flowering. This is the case for a number of crops in PNG, including *sis* nut, *marita* pandanus, purple passionfruit and *okari* nut (*Terminalia kaernbachii*). Some species, including breadfruit and Polynesian chestnut (*aila*), appear to produce in a regular seasonal manner from about 8° south of the equator and further south, but fruit in an irregular manner at locations nearer the equator. It is likely that changes in daylength are too small near the equator to induce flowering in a regular way.

Inundation (flooding)

Some locations in PNG are flooded for part of the year (see Section 1.11). This has an important influence on agricultural production in those locations, as many crops cannot grow under flooded conditions. One notable exception is sago palm, which thrives in flooded locations. Sago is a very important food in many places in East Sepik, Sandaun, Western and Gulf provinces. It is notable that sweet potato is a minor crop in those locations, as it only produces where the soil is well drained. Other crops that can tolerate inundation for short or long periods include oil palm, swamp taro, *kangkong*, taro and oenanthe.

A striking feature of PNG agriculture, especially in the highlands, is the use of steep, sloping land for agricultural production. The absence of mechanical or animal-drawn cultivation and the generally low rates of erosion allow the use of steep slopes for agriculture in the highlands. But the most important reason for use of steep land is the desire of villagers to plant sweet potato on well-drained sites. Land that floods seasonally, such as alluvial plains in the wetter season, is often used intensively when it is not flooded and abandoned when it is. In many places along the levee banks of the Sepik and Ramu rivers, villagers plant food gardens as flood levels are falling and harvest before waters rise again around six months later. Villagers favour crops that mature faster, for example, *Dioscorea alata* yam rather than *D. esculenta* yam, because *D. alata* produces tubers faster (see Section 3.12).

Soil fertility

Soil is vitally important for crop growth. It provides support, nutrients, water and aeration for plants. Soil fertility depends on rainfall, the nature of the underlying rock and natural erosion on steep slopes (see Sections 1.9 and 3.7). Soil fertility is commonly reduced by more intensive cropping. As a general rule, soil fertility is reduced faster by intensive cropping in the lowlands than it is in the highlands. This is because higher temperatures in the lowlands cause natural processes to occur at a faster rate than in the cooler highlands environments. Rainfall in the highlands tends to be less intensive than in the lowlands. This means that leaching of elements from soils exposed by cultivation occurs more slowly in the highlands.

Two aspects of soil fertility, the structure and nutrient content, have an important influence on crop growth and yield. For example, under high nitrogen conditions plants tend to grow larger, take longer to mature and yield more. There are exceptions to this, for example, if there is too much nitrogen in the soil, sweet potato will produce a lot of top growth, but the tuber yield may be reduced. Some crops, such as taro, corn, tobacco and common bean, are very sensitive to low levels of plant nutrients. In contrast, some crops, for example cassava, highland *pitpit* and some types of banana (such as Yava, Kalapua or Tukuru), bear reasonably well even when the levels of nutrients are low. The soil structure is more important for some crops than for others. Sweet potato grows best in soils that have a good structure, that is, one that is crumbly and allows water, air and roots to penetrate. If such soils are not available, because they are a heavy clay for example, villagers will improve the structure by tillage (see Section 3.11).

People can influence soil fertility in a number of ways. The first is selecting sites for planting food or cash crops. The second is by deciding when to move from the fallow phase to the garden phase. The third is by modifying the environment in which plants grow.

Villagers select sites for gardens and cash crops on a number of criteria. Sites on flatter land are more likely to be more fertile, but perhaps have poorer drainage than those on sloping land. In the highlands, mixed vegetable food gardens are typically planted on flatter and more fertile land, while sweet potato is commonly planted on better-drained sites, which are often on a gentle to steep slope. These factors are influenced by the nature of the soil, for example, some sites on flat land are well drained. The fallow period is an important determinant of soil fertility (see Section 3.8). Villagers judge how well soil fertility has been restored after the fallow phase by the growth of natural vegetation on the site and not by the period of time the land has been in fallow, although the two factors are related.

Villagers in PNG use a range of methods to improve soil fertility. These include tilling the soil, which is common where the fallow vegetation is grass (Section 3.11); transferring organic matter as green manure (termed 'compost' in PNG) to the soil surface or into a large mound (Section 3.11); planting certain tree species, especially casuarina in the highlands, into food gardens to hasten soil fertility restoration (Section 3.10); and planting a leguminous food, such as peanut, in a rotation with a root crop, especially sweet potato (Section 3.10). People also reduce soil erosion by erecting soil retention barriers or, less commonly, building small terraces to plant crops (Section 3.9). Within a garden plot, people commonly plant those crops that have the highest requirement for fertile soil in sites where fallow vegetation has been burnt and ash has accumulated. In a few locations in the highlands, people burn twigs and leaves of casuarina trees in a pile and plant the crops that demand high soil fertility in those sites, particularly some of the leafy green vegetables.

Sources

Bourke, R.M. (1989). Altitudinal limits of 230 economic crop species in Papua New Guinea. Unpublished paper. Department of Human Geography, Research School of Pacific and Asian Studies, The Australian National University, Canberra.

Bourke, R.M. (2005). Sweet potato in Papua New Guinea: the plant and people. In Ballard, C., Brown, P., Bourke, R.M. and Harwood, T. (eds). *The Sweet Potato in Oceania: A Reappraisal.* Ethnology Monographs 19 / Oceania Monograph 56. Oceania Publications, University of Sydney and Ethnology, Department of Anthropology, University of Pittsburgh, Sydney and Pittsburgh. pp. 15–24.

Bourke, R.M., Camarotto, C., D'Souza, E.J., Nema, K., Tarepe, T.N. and Woodhouse, S. (2004). *Production Patterns of 180 Economic Crops in Papua New Guinea.* Coombs Academic Publishing, Canberra.

Bourke, R.M., Allen, M.G. and Salisbury, J.G. (eds) (2001). *Food Security for Papua New Guinea. Proceedings of the Papua New Guinea Food and Nutrition 2000 Conference.* ACIAR Proceedings No. 99. Australian Centre for International Agricultural Research, Canberra. (There are 17 papers in the section 'Food shortages and the 1997 drought and frosts', pp. 153–274, that include discussion of the impact of the drought and frosts on crop performance).

Hanson, L.W., Bourke, R.M. and Yinil, D.S. (1998). *Cocoa and Coconut Growing Environments in Papua New Guinea. A Guide for Research and Extension Activities.* Australian Agency for International Development, Canberra.

McAlpine, J.R., Keig, G. and Short, K. (1975). *Climatic tables for Papua New Guinea.* Division of Land Use Research Technical Paper No. 37. Commonwealth Scientific and Industrial Research Organisation, Melbourne.

Mogina, J. (2002). Changing knowledge of plants in transitional societies at Milne Bay, Papua New Guinea. PhD thesis. The Australian National University, Canberra.

Smith, J.M.B. (1977). Man's impact upon some New Guinea mountain ecosystems. In Bayliss-Smith, T.P. and Feacham, R.G. (eds). *Subsistence and Survival: Rural Ecology in the Pacific.* Academic Press, London. pp. 185–214.

1.14 Access to markets and services

A measure of how easy it is to travel from a village in PNG to a service centre or market is defined here as 'accessibility'. It is measured by how long it takes to travel from a village to the nearest service centre or market, and by the level of services available at the service centre.

Before the establishment of a colonial administration in PNG, rural communities grew their own food, built their own houses, manufactured their own clothing, educated their children, used local medicinal plants and magic to treat illnesses and injuries, defended their territory and administered their 'laws'. Although groups of people had trade, ritual and marriage relationships with nearby groups, many individuals did not travel more than a few kilometres from where they were born for the whole of their lives. Their village was the centre of their world.

The colonial state established administrative centres that were places where education and health services and forces of law and order were located. Other places of wealth and employment, plantations and mines, also appeared. The colonial administration made it possible to travel longer distances in safety and to buy and sell commodities, including labour.

When people, commodities and services began to move longer distances in PNG, some places were advantaged over other places. It was easier for some people to receive education and health services, or to sell commodities or labour, because they were closer, in terms of distance, time or effort, to places where the services were offered, or where the commodities or labour were in demand. For them, these places were more accessible.

Because some people had better access to markets and services, in a relatively short span of time they became better educated, healthier and wealthier than other people. Where these differences in *access* to markets or services remain in place, differences in levels of health, education and cash incomes can be expected.

Accessibility classes

Access can be measured in a number of ways, but here accessibility is divided into the following classes:

- Very poor access – more than one day's travel to reach any level of service centre.
- Poor access – between 4 and 8 hours travel to reach a minor service centre.
- Moderate access – between 4 and 8 hours travel to reach a major service centre.
- Good access – between 1 and 4 hours travel to a major service centre.
- Very good access – less than one hour's travel to a major regional centre.

This classification of accessibility is based on the personal experiences of the authors in every district of PNG and takes into consideration terrain, road coverage, road quality, the presence of public road transport and shipping in the late 1990s and the services offered at various centres. Travel is defined as surface travel by a person on foot, in a vehicle or in a boat. Air travel is excluded because most people cannot afford it on a regular basis.

The unit for which accessibility is estimated is the MASP Agricultural System (for more information on MASP see Section 1.15). Accessibility is measured from the centre of the system. Because agricultural systems were defined using attributes that did not include accessibility, some minor anomalies exist. In these cases, which occur close to Lae, Tabubil and Wewak, although it looks on the map as though accessibility to these main centres is poor, it is in fact moderate to very good (Figure 1.14.1).

The estimate of accessibility is based on time to travel and not on the different costs of fuel. Per hour motorboat travel is considerably more expensive than motor vehicle travel.

Accessibility and population

Many people in PNG have good or very good access to service centres. An estimated 46% of the rural population live within four hours travel to a major service centre. A further 38% live within eight hours travel to a major centre (Figure 1.14.2). This situation is the outcome of the colonial administration establishing service centres and building roads in the most populated places.

The greatest numbers of people who have very good access to services live in East New Britain Province and Western Highlands Province. In each of Eastern Highlands, Southern Highlands, Western Highlands and Simbu provinces, more than 200 000 people have good accessibility to a major service centre (Figure 1.14.3, Table A1.14.1). A further 148 000 people in each of East Sepik and Enga provinces live within four hours travel of a major service centre.

Around 16% of the PNG population has poor or very poor access to services. These people live further than four hours travel from any service.

The greatest number of people with very poor access to services live in Sandaun Province, where almost 37 000 people (22% of the provincial population) live more than a day's travel from a minor service centre. More than 20 000 people in each of Southern Highlands and East New Britain provinces are similarly isolated.

Larger numbers of people have poor access to services. In Madang Province 106 000 people live more than four hours from a minor service centre. In each of Southern Highlands, Milne Bay, Western Highlands and Morobe provinces, over 40 000 people have poor access to services.

Summary of access to markets and services

A number of points can be made about access to markets and services in PNG:

- Slightly less than half of the total population of PNG have good or very good access to services.
- Around 16% have poor or very poor access to services.
- In the same province, large numbers of people in one part of the province can have very poor or poor accessibility, and in another part of the province large numbers can have good or very good accessibility. East New Britain Province is an example of this.
- Areas of very poor access to services are located in the Sepik Valley and the highlands of Sandaun Province, along the northern edge of the highlands, in inland Gulf and Central provinces, in south-east Oro Province and in inland East New Britain Province.
- Although the populations of small islands often have moderate access to services because boat travel takes them directly to major service centres, the cost of outboard motor fuel is a severe constraint on their ability to travel. Small boat travel is also dangerous during parts of the year because of weather conditions.

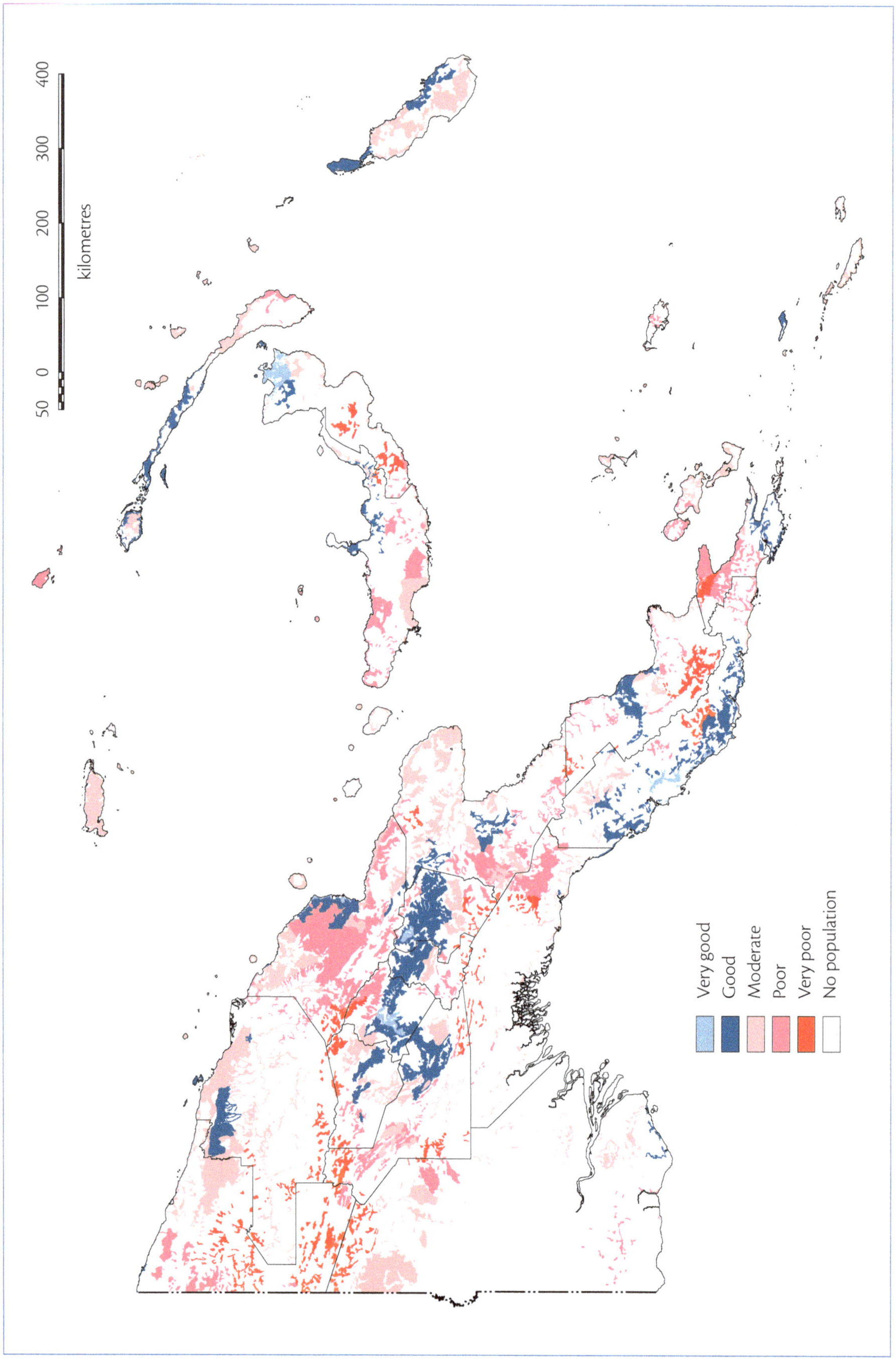

Figure 1.14.1 Access to markets and services. Source: MASP.

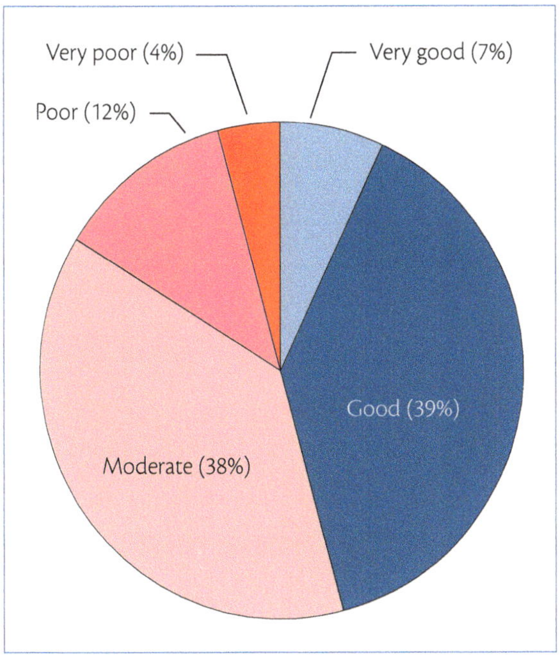

Figure 1.14.2 Access to markets and services by rural population. Sources: NSO (2002); MASP.

- Poor accessibility can be shown to be associated with low incomes and poor levels of health and education.
- Accessibility to some areas has deteriorated over the last 30 years, with the lack of road maintenance, the failure of bridges, and increases in the price of imported fuel for vehicles and boats caused by falls in the international value of the PNG currency.
- Increased fuel prices have caused significant shrinkage in commercial air service networks in PNG since around 2000. This has had the effect of isolating small service centres from the rest of the country, even though accessibility to the centre from surrounding villages may not have changed.

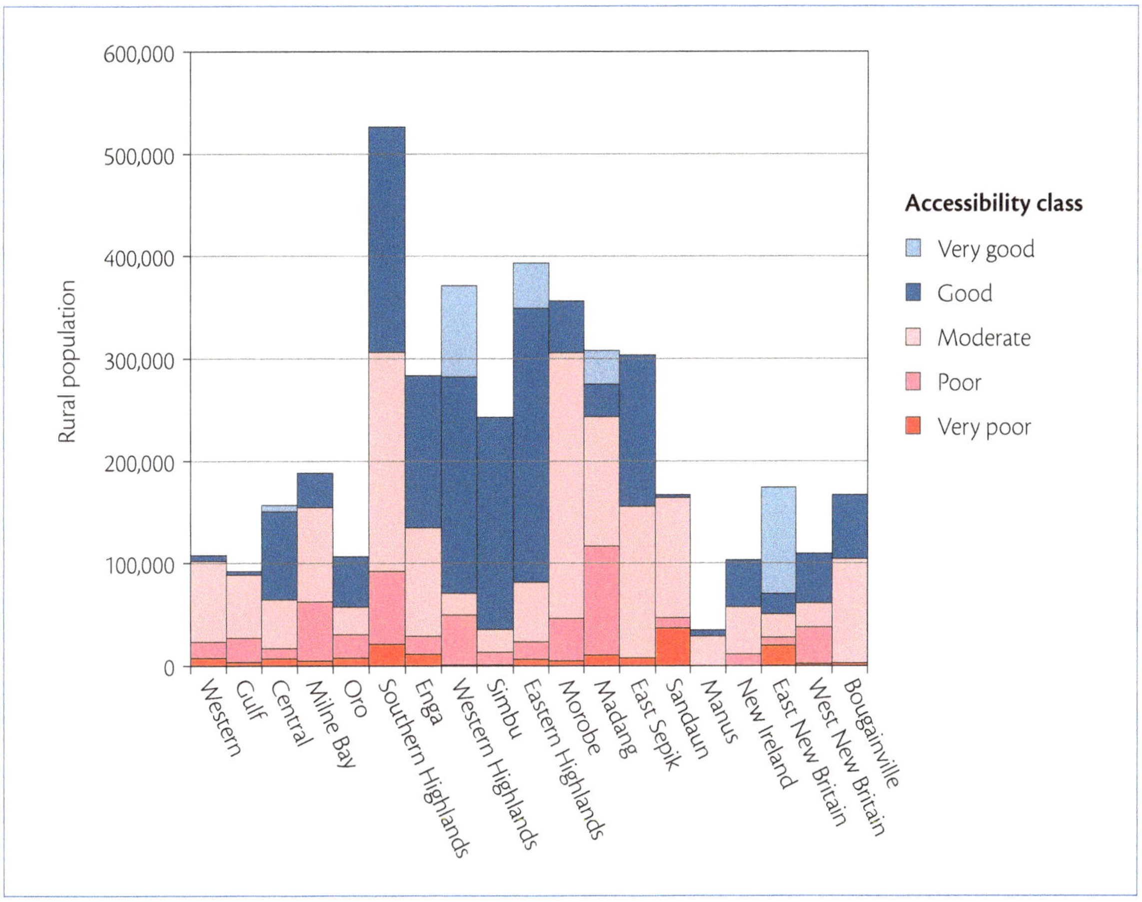

Figure 1.14.3 Rural population by accessibility class and province. Sources: NSO (2002); MASP.

Sources

Bourke, R.M., Allen, B.J., Hobsbawn, P. and Conway, J. (1998). *Papua New Guinea: Text Summaries.* Agricultural Systems of Papua New Guinea Working Paper No. 1. Two volumes. Department of Human Geography, Research School of Pacific and Asian Studies, The Australian National University, Canberra.

NSO (National Statistical Office of Papua New Guinea) (2002). Papua New Guinea 2000 Census: Final Figures. National Statistical Office of Papua New Guinea, Port Moresby.

1.15 Geographical information systems

PNG is rich in information about natural resources, forests and agriculture. Much of this information is contained in a number of computer-managed databases, some of which are accessible with geographical information systems (GISs) software. PNG has four major GISs relevant to agriculture:

- The PNG Resource Information System (PNGRIS).
- The Mapping Agricultural Systems of PNG Project database (MASP).
- The PNG Land Use maps 1975 and 1996.
- The Forest Resource Information System (FIMS).

A number of sub-national GISs have been developed using methods similar to PNGRIS, including the Madang Resource Information System and the Upper Ramu Resource Information System developed by Lincoln University in New Zealand; and the West New Britain Provincial GIS developed by the Kandrian Gloucester Integrated Development Project. Only PNGRIS and MASP are described here. See the sources at the end of this section for information about other spatial databases in PNG.

Papua New Guinea Resource Information System (PNGRIS)

The Papua New Guinea Resource Information System contains information on natural resources, land use and population distribution. PNGRIS is based on air photo interpretation of the *Skaipiksa* series.[1] Those air photographs were used to extrapolate the detailed information created by the Commonwealth Scientific and Industrial Research Organisation (CSIRO) land system studies to the whole of PNG. This information was mapped onto the 1:500 000 scale Tactical Pilotage Chart (TPC). The TPC map scale is suited to national-level and provincial-level planning, but has limitations for planning at the district level and below.

PNGRIS data is organised into mapping units known as resource mapping units or RMUs. The RMU is a relatively complex area of land characterised by a set of natural resource attributes that are unique. RMUs are defined by combinations of the following attributes:

- Landform (see Section 1.10).
- Rock type (geology).

1 A series of 1:105 000 scale, black and white vertical air photographs covering all of PNG, taken between 1972 and 1975 by the Royal Australian Air Force.

Box 1.15

The most recent version of PNGRIS was developed by a collaborative project between CSIRO and DAL, funded by AusAID. It uses *MapInfo* (version 4.5) and manages the data using *FoxPro* database management software. A *FoxPro* routine allows users to make queries that pass the outcomes to *MapInfo*, which produces a map. However, this version of PNGRIS has been overtaken by newer, more powerful versions of GIS software. The most recent versions of *MapInfo* will not run with the PNGRIS *FoxPro* software. Most users now access the PNGRIS map and data files directly using *MapInfo* or *ArcView* GIS software packages. The *MapInfo* files can be converted to *ArcView* files using conversion tools available in both packages, but the converted files should be carefully checked for missing records. In addition, in order to conduct complex queries or to produce tables, the dBase file format of the PNGRIS data files enables them to be loaded into most modern database management software packages, including Microsoft *Access* and Microsoft *Excel*. Summary tables from the defining attributes of PNGRIS are presented in McAlpine and Quigley (c. 1995).

- Altitude (as an alternative for temperature – see Sections 1.7 and 1.10).
- Relief (the difference in altitude between the hill tops and the valley bottoms).
- Mean annual rainfall (see Section 1.5).
- Inundation (flooding).
- Province.

The basic defining attribute of an RMU is landform, which is divided by rock type and altitude. The basic RMU is then subdivided by relief, inundation and rainfall. All RMUs are distinguished by province. Thus RMUs with identical attributes are given a different identifier when they cross a province boundary. A total of 4566 unique RMUs are identified for the whole of PNG, including those that are identical on either side of a provincial boundary (Figure 1.15.1, Table 1.15.1).

Population data from the 1980 and 1990 population censuses are listed by RMU in PNGRIS with census unit, census division, district and province tags attached. The authors of this book have added year 2000 census data to PNGRIS. Many of the tables and figures presented in Part 1 of this book are derived from PNGRIS.

PNGRIS also contains information on vegetation, soils, rural population, land use and land use intensity and possible physical constraints to agriculture. However, it is important to understand that the RMUs are not defined on the basis of these attributes. Rather, the additional information is mapped into the RMUs defined on the basis of the seven attributes listed above. Furthermore, information about the spatial distribution of these additional attributes within an RMU is sometimes presented in a complex way. For example, because more than one soil can occur in an RMU, PNGRIS deals with this by allowing each RMU to have up to three soils. Soils that cover less than 20% of an RMU are not listed. If only one soil is listed it covers around 80% of the RMU area. If two soils are listed, the first covers 40–60% of the RMU and the second 20–40%. If three soils are listed, the first covers 30–50% of the RMU area and the second and third 20–40%. Vegetation is presented in a similar way. It is not uncommon for users of PNGRIS to misunderstand this approach.

Because some data in PNGRIS is dependent on other data, care must be taken when analysing the relationships between attributes. For example, the estimate of inundation relies heavily on the type of vegetation, and altitude is an alternative for temperature. Information about soils is created from a combination of field observations, extrapolation from these observations to RMUs with similar environmental characteristics, and field experience.

Thus it would be a serious mistake to use PNGRIS to investigate relationships between soil type and slope, or vegetation and flooding.

The information in PNGRIS is coded. That is, a number is used to represent a class of attribute. The codes are contained in Appendix III in Bellamy (1986) (reprinted in 1995).

The Mapping Agricultural Systems of Papua New Guinea Project (MASP)

The Mapping Agricultural Systems of Papua New Guinea Project identified, described and mapped 'agricultural systems' for the whole of PNG. The project was carried out by the Land Management Group in the Department of Human Geography, Research School of Pacific and Asian Studies, The Australian National University.

The primary objective of MASP was to identify and describe agricultural activities in a way that would allow them to be assessed against the natural resource attributes of PNGRIS, in order to examine the question of agricultural sustainability under conditions of rapid demographic and socio-economic change. MASP uses the same 1:500 000 scale Tactical Pilotage Chart as PNGRIS and was designed to be compatible with PNGRIS. However, a decision was made to map agricultural systems independently of PNGRIS RMU boundaries so that

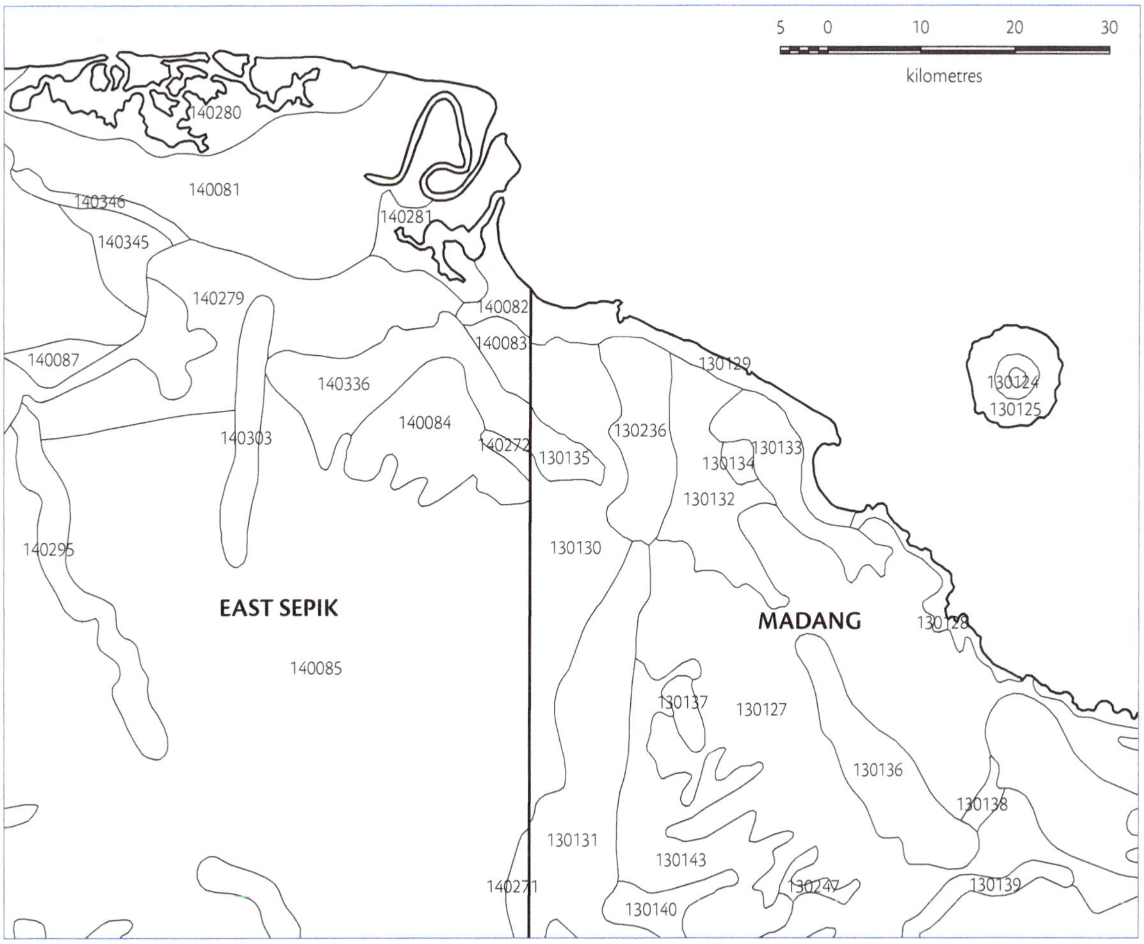

Figure 1.15.1 The PNGRIS GIS map showing resource mapping unit (RMU) numbers for part of East Sepik and Madang provinces. **Note:** To avoid cluttering the map, not all province RMU numbers are shown. Source: PNGRIS.

an analysis of the associations between agriculture and natural resources would be based on two independent databases.[2]

The mapping unit of MASP is the agricultural system (AGSYS). Agricultural systems are identified as areas in which the combinations of six agricultural activities are unique. Field observations and interviews with villagers during traverses across every district in PNG between 1990 and 1995, complemented by published and unpublished literature, supplied the information used to identify agricultural systems.[3]

2 The MASP GIS is available in both *MapInfo* and *ArcView* formats. The data files are held in the same dBase file format as PNGRIS.

3 The exception is Bougainville Province, which was not visited at that time because of the civil war (1989–1997). Identification of agricultural systems in Bougainville Province was based on interviews conducted in 1996, updated by fieldwork in 2002.

The attributes used to define an AGSYS are:

- Fallow vegetation – the type of vegetation cleared from garden sites at the beginning of planting (see Section 3.8).
- The number of times land is planted before it is fallowed.
- The period of time that land is fallowed.
- The most important staple food crops (see Section 3.1).
- Techniques used to maintain soil fertility (other than a long fallow) (see Sections 3.7–3.12).
- Segregation of crops within or between garden sites.
- Province.

Table 1.15.1 Part of the PNGRIS database showing records for resource mapping units (RMUs) for Madang and East Sepik provinces

A	B	C	D	E	F	G	H	I	J	K	L	M	N	O	P	Q	R	S	T	U	V	W	X	Y
130269	13	MAD	269	5	5	48	145	17	45	51	23	0	6	5	5	3	6	3	4	0	0	1	0	0
130270	13	MAD	270	2	4	57	145	38	27	51	13	0	6	5	4	2	4	3	3	0	0	1	95	0
130271	13	MAD	271	2	5	38	145	50	74	51	15	0	5	0	4	2	3	3	3	0	0	1	0	0
140001	14	ESK	1	2	3	11	143	16	4	6	15	0	1	0	1	1	4	3	4	1	1	95	0	0
140002	14	ESK	2	2	3	13	143	18	11	6	15	0	1	0	1	1	4	3	4	1	1	95	0	0
140003	14	ESK	3	2	3	19	143	34	46	51	31	13	5	6	5	1	4	3	4	0	0	1	95	0

Key

A	Province RMU number (unique in PNG)	**J**	Area of RMU (km^2)	**S**	Rainfall seasonality code
B	Census province code	**K**	Landform code	**T**	Rainfall deficit code
C	Province	**L**	Lithology code (rock type) 1	**U**	Inundation code
D	RMU number (unique in province)	**M**	Lithology code 2	**V**	Inundation extent code
E	1990 census district code	**N**	Slope code 1	**W**	Vegetation code 1
F	Latitude of RMU centre (degrees)	**O**	Slope code 2	**X**	Vegetation code 2
G	Latitude of RMU centre (minutes)	**P**	Relief code	**Y**	Vegetation code 3
H	Longitude of RMU centre (degrees)	**Q**	Altitude code		
I	Longitude of RMU centre (minutes)	**R**	Annual rainfall code		

Source: PNGRIS.

Agricultural systems are identified only for those parts of PNG that are classed as 'cultivated' in PNGRIS (see Section 1.2). Many PNG agricultural systems exploit microenvironments and the outcome is a complex spatial pattern of agricultural activities, often on a scale too small to be mapped at 1:500 000. This problem is dealt with by the introduction of subsystems. The subsystem boundaries are not mapped, but information is presented for all subsystems within an AGSYS, and an estimate is provided of the area occupied by each subsystem. Text notes described the relationship between the environment and the subsystem locations.

Like PNGRIS, MASP contains additional information, mapped into AGSYSs, that is not used to define the boundaries of the agricultural system. Another 102 attributes are mapped into AGSYSs.

Excluding systems that are identical on both sides of a provincial boundary, a total of 287 unique AGSYSs were identified for PNG (Figure 1.15.2, Table 1.15.2). The information for every province, including the codes used in the dBase files and GIS, has also been published in a series of Working Papers.[4] The MASP database was used to generate most of the tables and figures presented in Part 3 of this book.

4 There are 22 papers in the series: a two-volume summary (Working Paper No. 1), a paper for each province (19 papers), a technical information paper and a separate bibliography. The Working Papers have been distributed widely within PNG and are available in book form and as *Acrobat* PDF files.

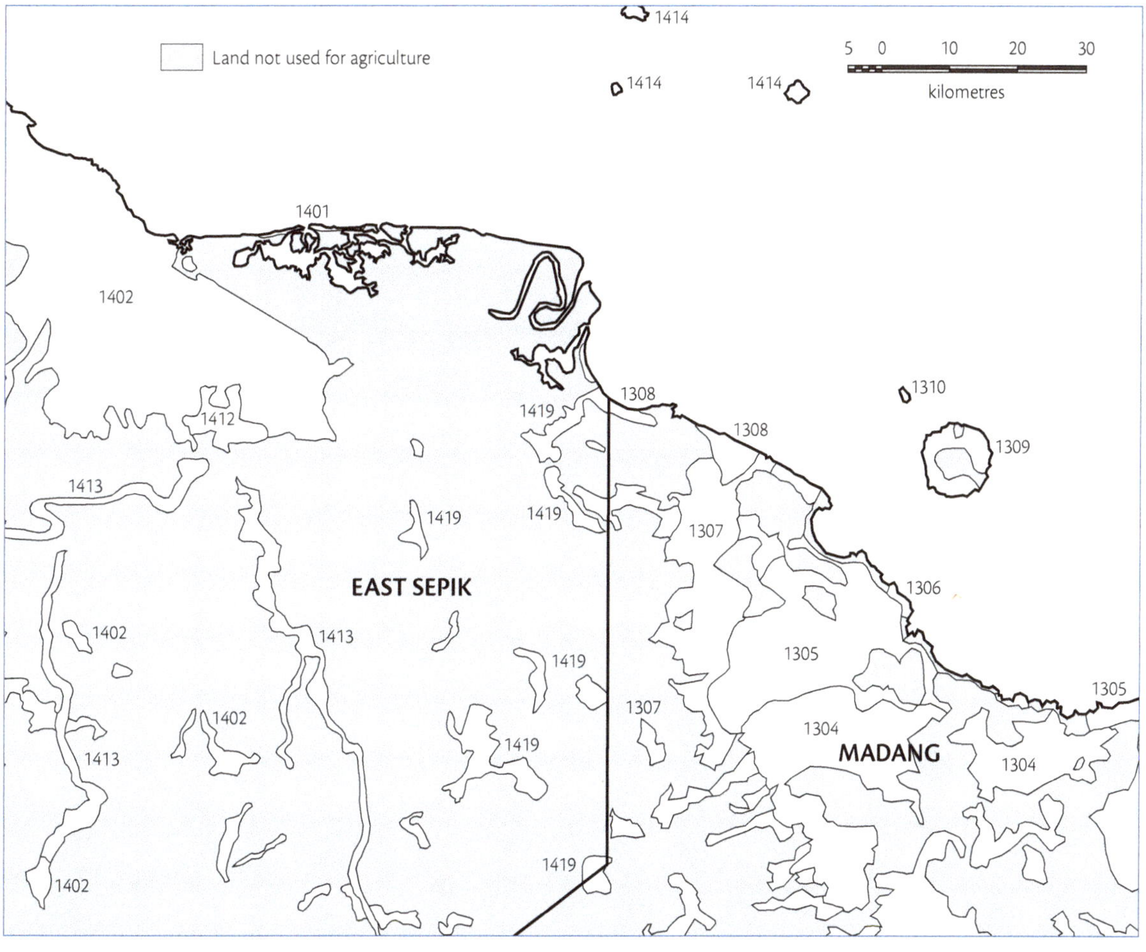

Figure 1.15.2 The MASP GIS map showing agricultural system (AGSYS) numbers for part of East Sepik and Madang provinces. **Note:** The shaded areas represent land that is not used for agriculture (see Section 1.2) and is excluded from MASP. Not all agricultural system numbers are shown. Source: MASP.

Table 1.15.2 Part of the MASP database showing records for agricultural systems (AGSYSs) for Madang and East Sepik provinces

A	B	C	D	E	F	G	H	I	J	K
1324	1	2	20	150	2	4	0	2	2	02-05
1325	1	2	40	100	3	4	0	2	2	00
1326	1	2–5	150	650	5	5	0	2	1	05–13
1327	1	1	2000	2600	2	2	2	2	4	11
1328	1	1–2	300	1600	2	5	1	2	2	11
1329	2	1	1600	2000	3	5	2	3	3	11
1330	1	5	800	1800	3	5	0	3	2	11
1331	2	5	200	450	1	1	0	2	2	02–11
1332	1	5	600	2200	5	4	0	2	1	11
1401	1	1	0	10	1	0	0	0	0	00
1402	1	1–2–3–4	0	800	5	5	0	3	1	09

Key

A	Agricultural system identifier based on province code (e.g. 13 for Madang) and the AGSYS number in that province
B	Number of subsystems that occur within the system (up to 3)
C	1990 census district code
D	Lowest altitude of land used for agriculture in the AGSYS (metres)
E	Highest altitude of land used for agriculture in the AGSYS (metres)
F	Slope code
G	Most common fallow type code
H	Significance of short fallows (<12 months) code
I	Length of fallow period code
J	Cropping interval code
K	Most important staple crops code

Source: MASP.

The main fieldwork for the MASP database was conducted 10–15 years ago (1990–1995). Some field checks by other, independent observers have found that, in general, the information remains current. The information that is becoming dated is that on sources of cash income.

Sources

Bellamy, J.A. (1986). *Papua New Guinea Inventory of Natural Resources, Population Distribution and Land Use Handbook.* Natural Resources Series No. 6. Division of Water and Land Resources, Commonwealth Scientific and Industrial Research Organisation, Canberra.

Bellamy, J.A. and McAlpine, J.R. (1995). *Papua New Guinea Inventory of Natural Resources, Population Distribution and Land Use Handbook.* Second edition. PNGRIS Publication No. 6. Australian Agency for International Development, Canberra.

Bourke, R.M., Allen, B.J., Hobsbawn, P. and Conway, J. (1998). *Papua New Guinea: Text Summaries.* Agricultural Systems of Papua New Guinea Working Paper No. 1. Two volumes. Department of Human Geography, Research School of Pacific and Asian Studies, The Australian National University, Canberra.

McAlpine, J.R., Keig, G., Bellamy, J.A., Bleeker, P., Cuddy, S.M., Hackett, C., Hide, R.L., Saunders, J.C. and Freyne, D.P. (1986). A natural resources information system for village agriculture in PNG. *CSIRO Division of Water and Land Resources Research Report 1983–85.* Commonwealth Scientific and Industrial Research Organisation, Canberra. pp. 78–86.

McAlpine, J. and Quigley, J. (c. 1995). *Natural Resources, Land Use and Population Distribution of Papua New Guinea: Summary Statistics from PNGRIS.* PNGRIS Report No. 7. Australian Agency for International Development, Canberra.

Trangmar, B.B., Basher, L.R. and Rijkse, W.C. (1995). *Land Resource Survey, Upper Ramu Catchment Papua New Guinea: Database Attributes.* PNGRIS Report No. 3. Australian International Development Assistance Bureau, Canberra.

Vovola, P. and Allen, B.J. (2001). Computer managed databases relevant to agriculture in PNG. In Bourke, R.M., Allen, M.G. and Salisbury, J.G. (eds). *Food Security for Papua New Guinea. Proceedings of the Papua New Guinea Food and Nutrition 2000 Conference.* ACIAR Proceedings No. 99. Australian Centre for International Agricultural Research, Canberra. pp. 467–475.

PART 2

Food Production, Consumption and Imports

R. Michael Bourke, John Gibson, Alan Quartermain, Kate Barclay, Bryant Allen and Jean Kennedy

2.1 Food in Papua New Guinea: an overview

Food is made up of three major components – proteins, carbohydrates and fats – and each is necessary for growth and healthy living. Although all three provide energy, carbohydrates, which consist of starches and sugars, provide the highest proportion of the *food energy* (or fuel) that human bodies need to function. Protein, used for building and repairing the body, comes from animal products such as meat, fish, and milk, but also from grains and vegetable foods. Small quantities of fats and oils are also important in a balanced diet. They provide more food energy per gram than either carbohydrates or protein.

Staple foods provide most of the carbohydrate (and thus food energy). In PNG the staple foods are starchy root crops, sago and banana. The amount of energy in these foods depends on how much protein, carbohydrate and fat they contain. The main sources of fats and oils in PNG are meat (pigs, fish and other animals), coconut, *marita* pandanus and imported vegetable oil.

This section provides an overview of food sources in PNG and introduces the more detailed sections that follow.

Estimates of energy and protein production and consumption

Most of the food consumed in PNG is produced within the country. In 2006 it was estimated that 83% of food energy and 76% of protein consumed in PNG was produced in PNG. The balance was imported (Figure 2.1.1). In 1996 it was calculated that locally grown food provided 80% of the food energy consumed in PNG; rural villagers obtained 84% of their food energy from locally grown food and urban people obtained 50%.

The 1996 and 2006 estimates are based on different measures, so it is not possible to conclude that the proportion of food energy derived from local sources increased or decreased during that decade. Nevertheless, it is almost certain that people are now obtaining more of their food requirements from food produced within PNG than they did in the mid 1990s. This is because after 1997 the PNG kina lost value against other currencies, the price of imported rice increased and its consumption decreased (see Part 4 and Section 2.7). As a result, consumption of locally grown food increased.

The 1996 Household Survey found significant differences in food consumption between urban and rural households. Urban people consumed more rice, wheat-based foods, soft drink and beer, and less root crops and banana (Table 2.1.1). Consumption estimates from the 1996 Household Survey and estimates of the quantities of the most important foods consumed in 2006 are similar (Table A2.1.1). Most of the differences that exist between the two estimates are probably errors in the data but some reflect real changes, such as the increased consumption of foods prepared from flour.

Plant foods grown in PNG – sweet potato and other root crops, banana, sago, sugar cane, coconut, vegetables, peanuts, fruit and nuts – provide an estimated

76% of food energy and 57% of protein consumed by rural and urban people (Figure 2.1.2). Imported rice, and imported wheat that is manufactured into bread, biscuits and noodles in PNG, provide a further 14% of food energy and 17% of protein. Meat and fish, most of which is produced in PNG, contribute 6% of food energy, but 25% of protein in people's diets. Smaller amounts of food energy and protein come from commercially refined sugar (produced in the Ramu Valley), imported and locally produced vegetable oil, imported animal fat, imported dairy products and other minor food products.

Diet

People's diets vary across the country, particularly between rural and urban areas. The broad pattern is that most food energy in rural areas comes from root crops, banana and sago, with coconut, other nuts and green vegetables making a small but significant contribution to energy and a greater contribution to the intake of other nutrients, particularly protein. Purchased foods based on imported rice, flour and vegetable oil typically provide about a fifth of the food energy in people's diets, considerably more in urban areas and considerably less in remote locations. Diets change with economic development, urbanisation and increasing cash incomes. Imported rice and wheat-based foods become more important and the consumption of cooking oil, tinned meat and tinned fish increases.

Changes in diet have been observed in a number of places in PNG over the past 50 years. For example, the contribution of sweet potato to dietary energy at Yobakogl village in the Sinasina area of Simbu Province fell from 76% in 1956 to 53% in 1981 (Figure 2.1.3). Over this period, the proportion of food energy derived from rice, wheat-based foods and corn increased from 4% to 22%. These changes were associated with increases in cash income from coffee sales. Similarly, on Ontong Java Atoll, the contribution of imported foods increased from 27% to 51% between 1971 and 1986, while the relative importance of locally produced coconut, fish, swamp taro and taro decreased. Changes have also occurred in the consumption of locally grown foods.

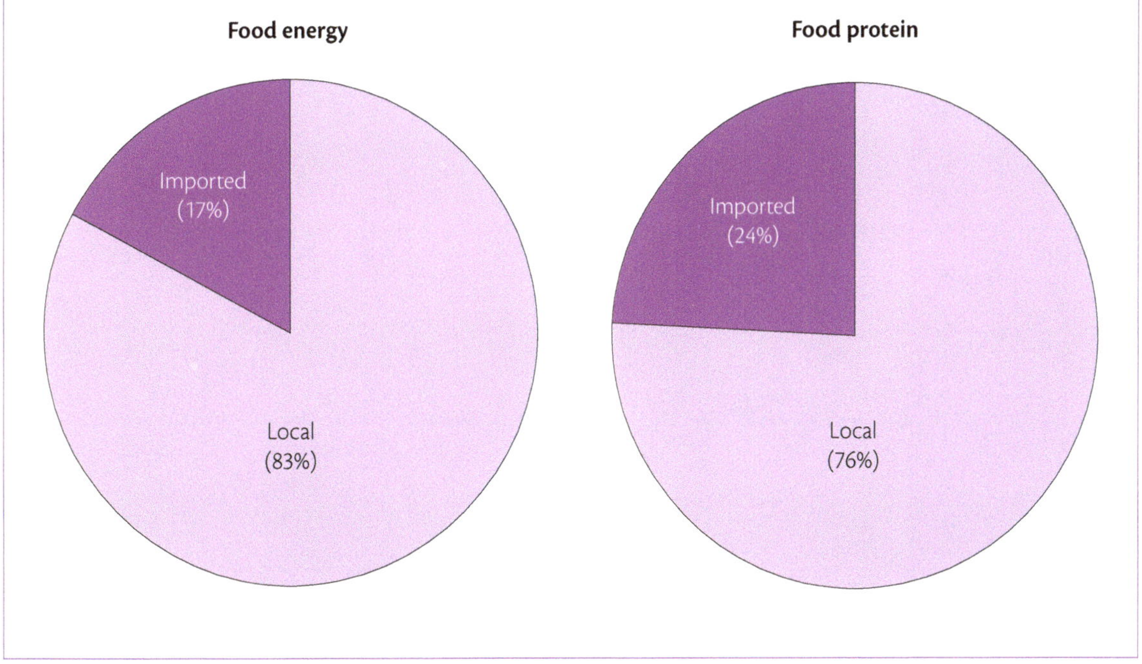

Figure 2.1.1 Proportion of energy and protein derived from locally produced and imported foods, 2006.
Source: Table A2.1.1 and author's calculations.

For example, in the late 1960s taro was the single most important garden food on Karkar Island in Madang Province, but by the early 1990s banana and Chinese taro had also become important foods.

Fresh food markets

Most food grown in PNG is consumed in the producing household. But many people also sell some of the food that they grow at local and regional fresh food markets (see Section 5.3). Fresh food markets are numerous in all urban and many rural locations. They are an important source of food in urban areas and as a source of income in rural areas. The total volume of fresh food sold through all food markets is not known, but the amount of fresh food sold in Port Moresby in 2005 was estimated as 15 000 tonnes. Information from the Mapping Agricultural Systems of PNG Project (MASP; see Section 1.15) illustrates the importance of domestically marketed food. It suggests that the amount of cash earned from selling fresh food is exceeded only by cash earned from coffee sales (Figure 5.1.1). A large number of foods are marketed, the most common being root crops, particularly sweet potato; peanut; banana; coconut; fruit such as mango, pawpaw and pineapple; and many types of green vegetables. Fresh food is not exported from PNG because of uncompetitive pricing, the generally poor presentation of the produce, expensive or unavailable transport, and quarantine barriers.

Table 2.1.1 Proportion of rural, urban and total population consuming different foods during the PNG Household Survey in 1996 (%)

Food	Rural	Urban	Total population
Greens	74.3	78.9	75.0
Sweet potato	65.0	33.6	60.2
Rice (imported)	25.8	87.4	35.1
Banana	33.6	38.7	34.3
Coconut	28.4	34.2	29.2
Biscuit/bread/flour/scone	14.4	74.6	23.5
Taro and Chinese taro	23.9	9.6	21.7
Sago	13.3	18.9	14.2
Tinned meat	5.9	51.7	12.8
Legumes	12.7	7.8	12.0
Tinned fish	9.1	24.5	11.4
Yam	12.5	4.8	11.3
Fresh fish, shellfish	7.1	28.2	10.3
Chicken	4.1	26.5	7.5
Pork, beef, other meat	6.4	9.9	6.9
Cassava	6.9	4.3	6.5
Lamb and mutton	5.0	13.7	6.3
Bush meat	1.8	1.5	1.7

Source: Gibson (2001:47).

Main sources of food

Food in PNG has five main sources:

- Local food plants.
- Imported plant food.
- Local industrial-scale production.
- Local foods of animal origin.
- Imported foods of animal origin.

Local food plants

PNG villagers grow or harvest about 400 plant species for food. Many of these foods are consumed in small quantities or in a limited number of places. Some are grown only for sale while others are grown just to try them out. Most people regularly grow between 30 and 80 species of food crops and many varieties of the most important species (see Section 2.3). Most foods plants are cultivated in gardens or in orchards, but food is also harvested from self-sown plants in food gardens, in fallow regrowth (see Section 3.8), in swamps, in unmanaged forest, or in grasslands. Food plants may be grouped into root crops, sago and banana (the staples) (see Section 3.1), leafy green and other vegetables (Section 3.2), fruit (Section 3.3), nuts (Section 3.4), sugar cane and stimulants (Section 3.5).

Sweet potato is by far the most important staple food in PNG. It provides around two-thirds of the food energy from locally grown food crops and is an important food for 65% of rural villagers. Other important staple foods include banana, sago, cassava, yam, taro and Chinese taro (Figure 2.2.2, Table 2.2.1). Domestic rice production is negligible compared with that of root crops, sago and banana, and is less than 1% of the quantity of imported rice (see Sections 2.5 and 2.7).

Many species of leafy green vegetables are eaten in PNG, and they are consumed daily (Table 2.1.1). The most important greens are pumpkin tips, *aibika*, amaranthus, rungia, *tulip*, oenanthe, cabbage, fern fronds, rorippa, Chinese cabbage, choko tips and taro leaves. Important non-leafy green vegetables include corn, highland *pitpit*, lowland *pitpit*, common bean, cucumber, winged bean, snake bean, pumpkin fruit and spring onion (Table 3.2.1).

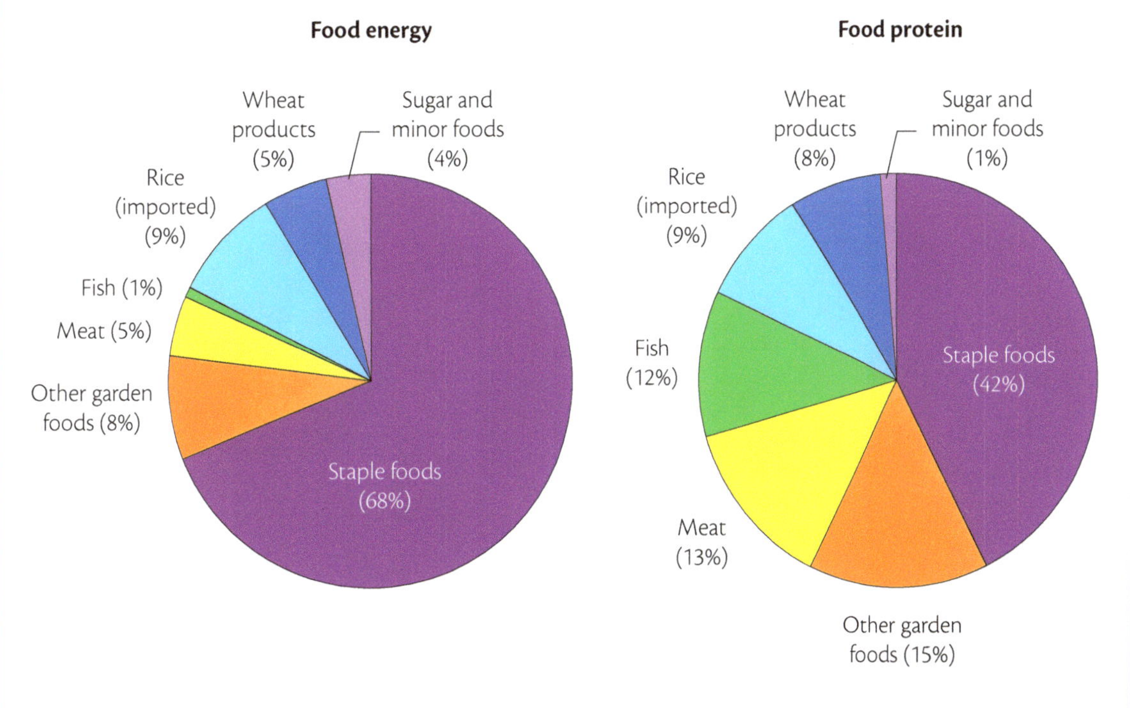

Figure 2.1.2 Source of energy and protein by main food groups, 2006. Source: Table A2.1.1 and author's calculations.

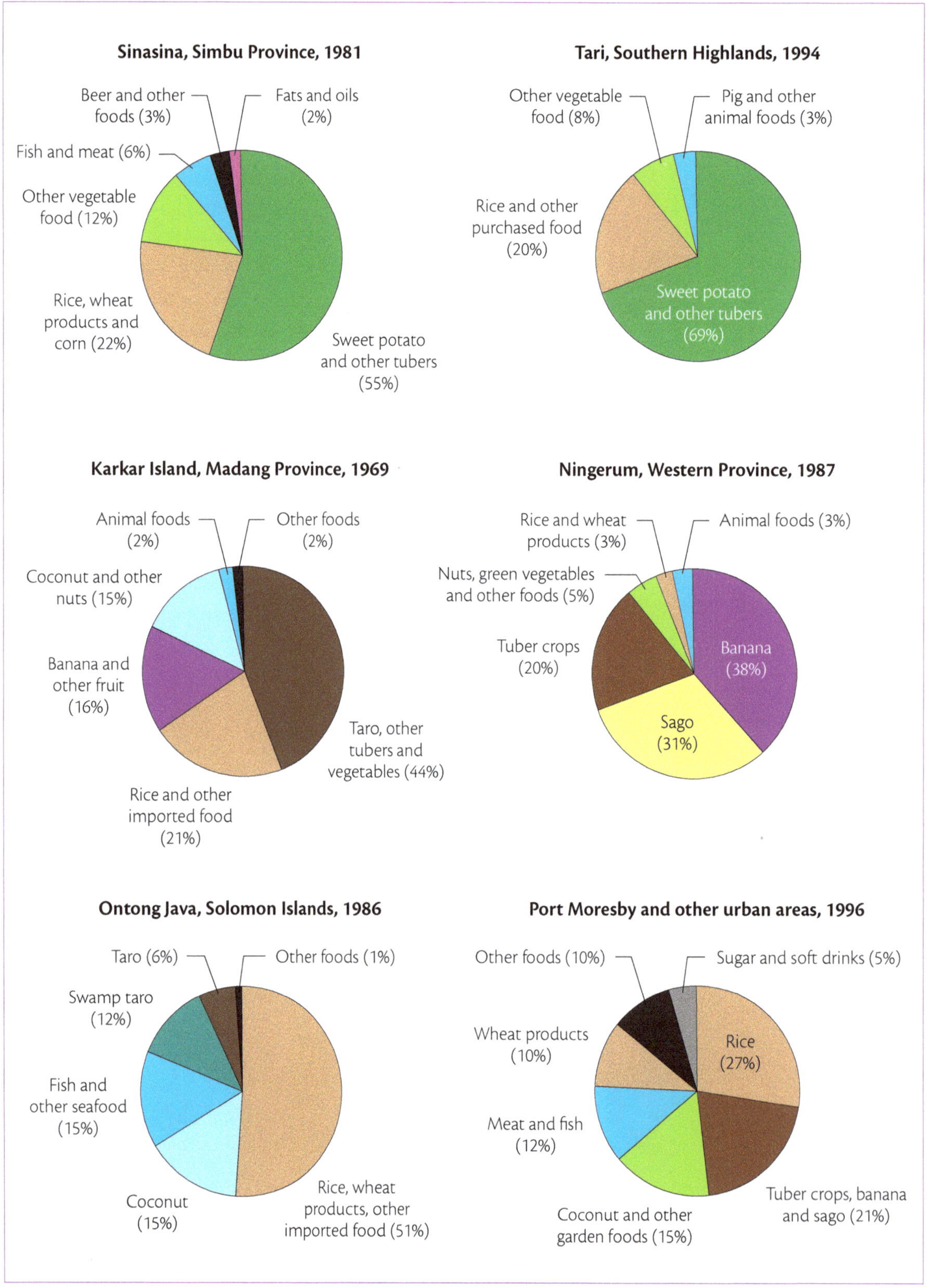

Figure 2.1.3 Proportion of total energy from different foods in selected locations. **Note:** Ontong Java is an atoll group in Solomon Islands near the international border with PNG. This data was used because it is of higher quality than that available for atolls in PNG and is likely to be representative of PNG atolls. Sources: Sinasina: Harvey and Heywood (1983); Tari: Umezaki et al. (1999); Karkar Island: Norgan et al. (1974); Ningerum: Ulijaszek (1992); Ontong Java: Bayliss-Smith (1986); Port Moresby: Gibson (2001:43).

Many fruits are eaten, including pawpaw, *marita* pandanus, pineapple, mango, watermelon, *ton*, Malay apple, guava, orange and passionfruit (Table 3.3.1). The most important nut crops are coconut and peanut.[1] Other nuts, which are commonly eaten seasonally, include breadfruit seed, *karuka* nut pandanus, *galip*, Polynesian chestnut (*aila*), sea almond (*talis*), *pao* and *okari* (Table 3.4.1). Sugar cane is another important food. Consumption of village-grown sugar cane is estimated to be 60 kg of cane per person per year (Table A2.1.1).

A number of stimulants are widely used, particularly betel nut, betel pepper, tobacco and, less commonly, marijuana, but they contribute almost nothing nutritionally (see Section 3.5).

Imported plant food

Rice and wheat are the most important imported foods (see Section 2.7). Rice imports have averaged 152 000 tonnes per year since 1990 and wheat imports 117 000 tonnes per year. It is possible that rice consumption per person in PNG has reached a plateau, although it increased in 2006 (to 184 000 tonnes, or 30 kg/person/year). Wheat is milled in PNG to produce flour that is used to manufacture commercial quantities of bread, biscuits and 'instant' noodles for sale in retail outlets. Individuals also use flour to make 'scones' from unleavened dough deep-fried in vegetable oil. Since 1998, per person wheat imports have been increasing while rice imports have decreased or been static (Figures 2.7.2, 2.7.3).

The 5000–6000 tonnes of vegetable oil imported annually (Table 2.1.2) is supplemented by locally refined coconut and palm oil.

A small quantity of fresh fruit and vegetables is imported from Australia and New Zealand (see Section 2.8). The quantity imported has been declining and is now less than 6000 tonnes per year (about 1 kg/person/year). Onions, potatoes, apples, oranges and peas are the most important commodities imported (Table 2.1.2, Figure 2.8.2). Potato imports declined significantly from around 1980 because of increased production in the highlands, but have increased again following the outbreak of potato late blight disease in the highlands in 2003.

Local industrial-scale production

Six foods are produced on an industrial scale in PNG: coffee, cocoa, coconut (for copra oil), oil palm (for palm oil), sugar and tea. Most of the production is exported (see Part 5), with sugar cane the only industrial crop grown primarily for the domestic market (see Section 5.10). Small quantities of locally grown coffee, tea and refined palm oil are also consumed, but domestic consumption is a very small proportion of total production.

Local foods of animal origin

Fish, shellfish, pigs, chickens and cattle are the most important animal foods consumed in PNG. Much less important are rabbit, duck, sheep, goat and dog. Wild pigs, bandicoots and other small mammals, birds and insects are hunted and collected widely and are important foods in more remote locations.

Fish, shellfish and other marine animals, such as octopus and turtle, are caught and eaten. Fish is a major dietary item for villagers living on the coast and along major rivers. The per person consumption of fish in PNG is much lower than in many other Pacific and Asian countries (see Sections 5.9 and 2.10).[2]

The most important animal food is the domestic pig, with an estimated 1.8 million pigs being raised in PNG villages. Commercial pig production is relatively small. An estimated 1.5 million chickens in PNG villages are used for meat and eggs but, in most villages, chickens scavenge and production is low. The commercial chicken sector is more productive. Village cattle projects that took the form of small-scale

1 In this book coconut is grouped with the staple food crops (Section 3.1) and peanut with the vegetables (Section 3.2).

2 Estimates of fish and other seafood consumption in PNG vary widely. The National Fisheries Authority estimates consumption at 25 000 to 50 000 tonnes per year (5–10 kg/person/year) (see Section 5.9). However, other estimates for fish and seafood caught locally are as high as 120 000 tonnes per year (24 kg/person/year) (see Section 2.10). Figures of 50 000 tonnes (8 kg/person/year) for fresh and smoked fish and 15 000 tonnes (2 kg/person/year) of tinned tuna were used here to calculate the contribution of fish to the national diet (Table A2.1.1).

enclosures planted to pasture grasses collapsed in debt and unmanageable stock in the 1980s. Eighty per cent of the national cattle herd, estimated at around 80 000 head, is today maintained on large holdings, with only 20% owned by villagers (Table 2.6.3).

A significant amount of wildlife is hunted or gathered and consumed each year, particularly in locations where the human population density is low. Somewhere between 0.8 and 1.6 million people derive significant dietary protein from wild animals and consume 4–8 million vertebrate animals a year. The estimated replacement value of this wild or bush meat with imported tinned mackerel or mutton flaps was K75 million in 2005.

Imported foods of animal origin

Animal foods imported into PNG are sheep meat, beef and offal (lungs, liver, kidneys and other internal organs) and mackerel and tuna (see Sections 2.9 and 2.10). Over the past 20 years significant changes have taken place, first in the composition of meat imports, with sheep meat becoming more important than beef (Figure 2.9.2) and, second, in a decline in consumption of tinned meat, fresh meat and tinned mackerel and an increase in consumption of tinned tuna. These changes were driven by the relative costs of meat and fish, with a small tin of tuna retailing for about half the price of a comparable tin of meat.

Table 2.1.2 Volume and value of imports of meat, fish, vegetables, fruit, dairy products, animal fat and vegetable oil, 2002–2004[a]

Imported food	Volume (tonnes)			Value (K'000)		
	2002	2003	2004	2002	2003	2004
Sheep meat	20,528	16,845	22,046	76,501	80,094	83,798
Beef	6,615	3,806	3,028	34,535	26,347	25,679
Offal	1,658	1,706	2,382	6,656	8,493	8,506
Pig meat	85	192	200	481	1,244	2,031
Other meat	77	92	118	302	608	1,037
Fish[b]	7,986	9,324	8,903	26,919	24,673	26,652
Onions	1,263	977	955	2,171	2,355	2,294
Potatoes[c]	161	735	502	471	2,118	2,469
Apples	624	674	428	3,489	3,330	2,806
Citrus	225	306	222	1,235	1,396	1,248
Other fruit and vegetables	784	772	805	5,297	5,250	4,845
Milk and other dairy products	5,920	5,938	4,196	32,273	26,161	31,373
Butter and dairy spreads	678	673	448	6,169	6,644	6,283
Animal fat[d]	–	–	6,182	–	–	12,264
Vegetable oils[d]	–	–	5,477	–	–	23,527

[a] NSO figures for volume and value of food imports are generally lower than those from other sources, including data from the exporting countries, where there is an overlap in coverage (see Tables A2.8.1, A2.9.1, A2.10.1). This suggests that not all import data are recorded in the NSO database.

[b] The figures for fish imports exclude bait fish (average 1150 tonnes/year).

[c] The figures for potato imports exclude seed potato (average 8 tonnes/year).

[d] The source for animal fat and vegetable oils is McGregor (2006: Table 14).

Source: National Statistical Office of PNG.

Around 4000–6000 tonnes of dairy products are imported annually (about 1 kg/person/year), mainly from Australia and New Zealand (Table 2.1.2). UHT (ultra heat treated), fresh and powdered milk account for about 80% of dairy imports. Other imported dairy products are butter, dairy spreads, yoghurt and cheese. About 6000 tonnes of other animal fat, such as lard, is imported each year and is used in baking and other food preparation.

Sources

Allen, B.J., Bourke, R.M. and Hanson, L. (2001). Dimensions of PNG village agriculture. In Bourke, R.M., Allen, M.G. and Salisbury, J.G. (eds). *Food Security for Papua New Guinea. Proceedings of the Papua New Guinea Food and Nutrition 2000 Conference.* ACIAR Proceedings No. 99. Australian Centre for International Agricultural Research, Canberra. pp. 529–553.

Bayliss-Smith, T. (1986). Ontong Java Atoll: population, economy and society, 1970–1986. *South Pacific Smallholder Project Occasional Paper No. 9.* University of New England, Armidale.

Bourke, R.M. and Vlassak, V. (2004). Estimates of food crop production in Papua New Guinea. Land Management Group, Research School of Pacific and Asian Studies, The Australian National University, Canberra.

French, B.R. (1986). *Food Plants of Papua New Guinea: A Compendium.* Sheffield, Tasmania.

Gibson, J. (2001). Food demand in the rural and urban sectors of PNG. In Bourke, R.M., Allen, M.G. and Salisbury, J.G. (eds). *Food Security for Papua New Guinea. Proceedings of the Papua New Guinea Food and Nutrition 2000 Conference.* ACIAR Proceedings No. 99. Australian Centre for International Agricultural Research, Canberra. pp. 45–53.

Gibson, J. (2001). The economic and nutritional importance of household food production in PNG. In Bourke, R.M., Allen, M.G. and Salisbury, J.G. (eds). *Food Security for Papua New Guinea. Proceedings of the Papua New Guinea Food and Nutrition 2000 Conference.* ACIAR Proceedings No. 99. Australian Centre for International Agricultural Research, Canberra. pp. 37–44.

Harvey, P.W.J. and Heywood, P.F. (1983). The research report of the Simbu land use project. Volume IV: nutrition and growth in Simbu. Institute of Applied Social and Economic Research, Boroko, Papua New Guinea.

Mack, A.L. and West, P. (2005). *Ten thousand tonnes of small animals: wildlife consumption in Papua New Guinea, a vital resource in need of management.* Resource Management in Asia-Pacific Working Paper No. 61. Resource Management in Asia-Pacific Program, Research School of Pacific and Asian Studies, The Australian National University, Canberra.

McGregor, A. (2006). Special products and special safeguard mechanism: the Papua New Guinea country study. International Centre for Trade and Sustainable Development, Geneva.

Norgan, N.G., Ferro-Luzzi, A. and Durnin, J.V.G.A. (1974). The energy and nutrient intake and the energy expenditure of 204 New Guinean adults. *Philosophical Transactions of the Royal Society of London. Series B, Biological Sciences* 268(893):309–348.

Quartermain, A.R. (2001). Potential for producing more meat from small-scale livestock production. In Bourke, R.M., Allen, M.G. and Salisbury, J.G. (eds). *Food Security for Papua New Guinea. Proceedings of the Papua New Guinea Food and Nutrition 2000 Conference.* ACIAR Proceedings No. 99. Australian Centre for International Agricultural Research, Canberra. pp. 625–629.

Ulijaszek, S.J. (1992). Dietary and nutrient intakes of 25 Ningerum (New Guinea) adult males at two times of the year. *American Journal of Human Biology* 4(4):469–479.

Umezaki, M., Yamauchi, T. and Ohtsuka, R. (1999). Diet among the Huli in Papua New Guinea highlands when they were influenced by the extended rainy period. *Ecology of Food and Nutrition* 37(5):409–427.

2.2 Staple food crop production

Estimates of staple food crop production

In PNG the staple foods are mainly starchy root crops, sago and banana. Estimates of staple food crop production were made in 1961–1962 as part of the Survey of Indigenous Agriculture (see Section 6.5), and in 1996 from the PNG Household Survey (see Section 2.1). New estimates were made for the year 2000, drawn from data from a nationwide survey of village agriculture conducted in 1990–1995 (MASP; see Section 1.15). The MASP survey identified 342 agricultural systems. To generate estimates of staple food crop production, the proportion of garden area devoted to each food crop in each agricultural system was combined with population data from the 2000 National Census, average garden area planted per person per year, and average crop yield in different environments (Table 2.4.1) (see Box 2.2).

Total staple crop production in 2000 was estimated as 4.5 million tonnes per year, with a food energy content of 18 196 × 10^9 kilojoules (kJ)[1] (Table 2.2.1). This is equivalent to 870 kg/person/year of staple food crops and contains food energy of 9600 kJ/person/day. A significant amount of sweet potato grown in the highlands is fed to domestic pigs, so the estimate of energy produced is not directly comparable to an estimate of the energy consumed by people.

The 2000 estimates were checked by comparing the figures with the daily energy requirements of the population.[2] The check indicated that the new figures are reasonably accurate estimates of production

1 A kilojoule is the internationally recognised unit of measurement for the energy value of food. A common term for unit of energy was the Calorie. One Calorie equals 4.1868 kilojoules.

2 'Daily energy requirement' is the amount of energy that humans need to live and perform active lives. It varies a lot between people and the figure used here (11 160 kJ) is an average one for Papua New Guineans.

Box 2.2

Annual production of each staple crop (tonnes)	=	Proportion of garden land planted with each crop	×	2000 rural population in each agricultural system	×	Average garden area planted per person per year (ha)	×	Average crop yield adjusted for environment (t/ha)

of staple food crops. The figures presented in this section are for *total* production, that is, they include food that is wasted, fed to livestock (especially pigs in the highlands), and kitchen scraps.

Sweet potato dominates food production in PNG. It accounts for almost two-thirds (64%) of production of staple food crops by weight and 63% of food energy production (Table 2.2.1). No other staple food crop contributes more than 10% of the total national production by weight or food energy. The contribution by weight for the more important foods is: banana (9.7%), cassava (6.0%), yam (6.0%), *Colocasia* taro (5.1%), Chinese taro (5.0%), coconut (2.2%) and sago (1.8%). A number of minor crops make a very small contribution to total staple food crop production, with locally grown rice responsible for only about 0.01%. The proportions of food energy are similar to those for production by weight; however, the contribution to food energy is greater for sago (6.8% of energy produced from staples), coconut (3.5%) and rice (0.03%) because these foods have a higher energy value per unit weight than banana or the root crops.

Production of main staple food crops

Sweet potato dominates food production in the highlands, but is also an important food in many lowlands locations, where it has increased in importance over the past 60 years. Sweet potato is grown in almost every agricultural system in PNG, with some

Table 2.2.1 Estimated production of 18 staple food crops in 2000

Crop	Weight (tonnes)	Weight (%)	Energy (kJ x 10^9)	Energy (%)
Sweet potato	2,871,851	63.57	11,422.68	62.77
Banana	436,496	9.66	1,260.98	6.93
Cassava	271,894	6.02	1,115.61	6.13
Colocasia taro	229,088	5.07	748.14	4.11
Chinese taro	226,536	5.01	739.81	4.07
Lesser yam (*Dioscorea esculenta*)	180,370	3.99	656.99	3.61
Coconut	100,929	2.23	633.88	3.48
Greater yam (*D. alata*)	91,358	2.02	294.54	1.62
Sago	82,962	1.84	1,240.00	6.81
Irish potato	18,759	0.42	55.77	0.31
Taro (*Alocasia*)	2,389	0.05	7.79	0.04
Queensland arrowroot	1,431	0.03	4.69	0.03
Taro (*Amorphophallus*)	1,217	0.03	3.98	0.02
Swamp taro	823	0.02	2.68	0.01
Yam (*D. nummularia*)	478	0.01	1.55	0.01
Aerial yam (*D. bulbifera*)	467	0.01	1.51	0.01
Rice	407	0.01	5.82	0.03
Yam (*D. pentaphylla*)	37	0.00	0.13	0.00
Papua New Guinea	4,517,492	100.00	18,196.54	100.00

Source: Bourke and Vlassak (2004).

minor exceptions in parts of East Sepik and Western provinces. It has the widest distribution of any crop in PNG (Table A2.2.1).

Average sweet potato production for all rural villagers is calculated as 685 kg/person/year (Table 2.2.2). Production ranges from 1150 to 1200 kg/person/year in the five highlands provinces. Significant quantities per person are also produced in Bougainville, Morobe, Oro, West New Britain, New Ireland, Central, Madang, East New Britain, Milne Bay and Gulf provinces.

Banana is grown in all parts of PNG. Average banana production for all rural villagers is calculated as 105 kg/person/year. At a provincial level, production is greatest in Morobe, East New Britain, Central and Madang provinces. Production per person is greatest in East New Britain, Central, Morobe, Western and Madang provinces.

Cassava is an important crop in the lowlands. Production for all rural villagers is calculated as 65 kg/person/year. Production is greatest in Milne Bay Province, where it is commonly planted with sweet potato after taro and yam have been harvested, and in West New Britain. The highest production per person is also in West New Britain and Milne Bay provinces.

Taro (*Colocasia esculenta*) (*taro tru*) is grown in most locations in PNG, but often only as a supplementary crop. It was formerly the most important staple food in much of the lowlands, and was the most important food in the highlands before sweet potato was adopted there about 300 years ago (Figure 3.1.2). *Colocasia* taro production is greatest in Madang, East Sepik and Morobe provinces (Table A2.2.2). Production per person is greatest in West New Britain, Madang and Oro provinces. Average *Colocasia* taro production for all rural villagers is calculated as 55 kg/person/year.

Chinese taro total production and per person production is greatest in Morobe, Madang, East New Britain and West New Britain provinces. Average Chinese taro production for all rural villagers is calculated as 55 kg/person/year.

Yam is grown in all provinces and is an important staple food in some locations. Five species are grown, but three of these are unimportant as foods. The lesser yam (*Dioscorea esculenta*) (*mami* or *taitu*) accounts for 66% of yam production, and the greater yam (*D. alata*) (*yam tru*) for most of the rest. The greatest production of yam, especially *D. esculenta*, occurs in the hilly parts of East Sepik Province north of the Sepik River. Significant quantities are also produced in Madang and Milne Bay provinces. Production of yam per person is greatest in East Sepik and Milne Bay provinces. For the five yam species combined, average production for all rural villagers is calculated as 65 kg/person/year.

Coconut is grown in all coastal and many inland locations. The figures generated here are only for nuts consumed by people. Significant quantities of coconut are also fed to pigs or used to produce copra in some coastal locations. Average coconut production for all rural villagers is calculated as 40 kg/person/year.[3] The highest production of coconut for human consumption occurs in East Sepik, Madang, Milne Bay, Bougainville and East New Britain provinces. Production per person is also high in Manus, New Ireland, West New Britain and Sandaun provinces.

Sago is grown and eaten in most provinces. Average sago production for all rural villagers is calculated as 75 kg/person/year. The greatest production occurs in East Sepik, Sandaun, Western and Gulf provinces. Production per person is greatest in Manus, Western, Gulf, Sandaun and East Sepik provinces.

Irish potato is a relatively recent introduction and produced in lesser quantities than the older staples. Some growers produce commercial quantities of Irish potato for sale, but the figures generated here are only for subsistence consumption and produce sold from village gardens. Production is greatest at high-altitude locations in Enga Province. Production dropped greatly following an outbreak of potato late blight disease in 2003, and it is likely that production in 2006 was less than the estimates given here.

---

3 Coconut and sago have a higher food energy content per unit weight (Table 2.2.1), so their average production figures were converted to sweet potato equivalent to be comparable with data on root crops and banana (Table 2.2.2).

Table 2.2.2 Estimated production of staple foods per rural person in 2000, by province (kg/person/year)

Province	1st staple		2nd staple		3rd staple		4th staple		5th staple		6th staple		7th staple		8th staple	
Western	Sago	450	Banana	140	Yam	90	Cassava	85	Sweet potato	65	Coconut	65				
Gulf	Sago	420	Sweet potato	220	Banana	115										
Central	Sweet potato	315	Banana	275	Cassava	125	Yam	115	Coconut	50						
Milne Bay	Cassava	275	Sweet potato	235	Yam	215	Banana	125	*Colocasia* taro	85	Coconut	80				
Oro	Sweet potato	500	*Colocasia* taro	110	Yam	110	Chinese taro	105	Cassava	100	Banana	95	Sago	55		
Southern Highlands	Sweet potato	1175														
Enga	Sweet potato	1200														
Western Highlands	Sweet potato	1150	Banana	80												
Simbu	Sweet potato	1210														
Eastern Highlands	Sweet potato	1195														
Morobe	Sweet potato	550	Banana	180	Chinese taro	165	Yam	75	*Colocasia* taro	65						
Madang	Sweet potato	250	Chinese taro	150	Yam	135	Banana	135	*Colocasia* taro	125	Coconut	75	Sago	65	Cassava	55
East Sepik	Sago	290	Yam	235	Banana	120	Coconut	95	Sweet potato	85	*Colocasia* taro	80				
Sandaun	Sago	375	Sweet potato	150	Banana	110	*Colocasia* taro	95	Coconut	75	Yam	55				
Manus	Sago	490	Cassava	165	Sweet potato	130	Coconut	90								
New Ireland	Sweet potato	375	Cassava	150	Yam	140	*Colocasia* taro	90	Banana	80	Coconut	85	Sago	65		
East New Britain	Banana	320	Sweet potato	245	Cassava	195	Chinese taro	180	Coconut	80	*Colocasia* taro	55				
West New Britain	Sweet potato	415	Cassava	365	Chinese taro	220	*Colocasia* taro	145	Banana	110	Coconut	75				
Bougainville	Sweet potato	555	Banana	105	Coconut	90	Chinese taro	75	Cassava	70						
All rural PNG	Sweet potato	685	Banana	105	Sago	75	Cassava	65	Yam	65	*Colocasia* taro	55	Chinese taro	55	Coconut	40

Source: Figures were generated by dividing the estimated production of each staple food in each province (Tables A2.2.1 and A2.2.2) by the provincial rural village population in 2000 (Table 1.1.1). So that a direct comparison can be made between sago and the other staple crops, production estimates for sago were converted to sweet potato equivalent by multiplying the food energy content per kilogram of sago divided by the food energy content of sweet potato. The same was done for coconut. Data for the five yam species have been combined. Crops where the production is less than 50 kg/person/year at the provincial level have been excluded. Figures have been rounded to the nearest 5 kg.

Rice for local consumption is grown in very small quantities in a number of provinces (see Section 2.5). Only in Bougainville Province did production exceed a few hundred tonnes per year in the mid 1990s, although rice production had virtually ceased there by 2002. Village plantings of rice have increased in a number of provinces since the late 1990s. However, total production in 2006 was probably less than 1000 tonnes per year, despite the increased interest in rice growing by some villagers and significant external support. Total rice production is still unlikely to exceed 0.1% of total staple food production in PNG.

Economic value of staple food production

These estimates of staple food crop production can be allocated a kina value by calculating the value of grain that would have to be imported to feed the population if all production of staple food crops ceased.

The quantity of *additional* rice that would have to be imported into PNG is 1.2 million tonnes. (Existing rice imports have averaged 152 000 tonnes per year since 1990 – see Section 2.7.) The retail value of this additional rice was K2850 million in 2004, indicating the great economic value of food production in PNG. The figure would be even larger if the value of vegetables, fruit, nuts, fish and meat was calculated.

Changes in production patterns

Estimates of production made in 2000 can be compared with estimates made in 1961–62 (Figure 2.2.1). The most striking change over this 40-year period has been the increased significance of sweet potato, cassava, Irish potato and Chinese taro. These are crops that were domesticated in the Americas and introduced into PNG over the past 300 years (in the case of sweet potato) or about 130 years ago (for the other crops).

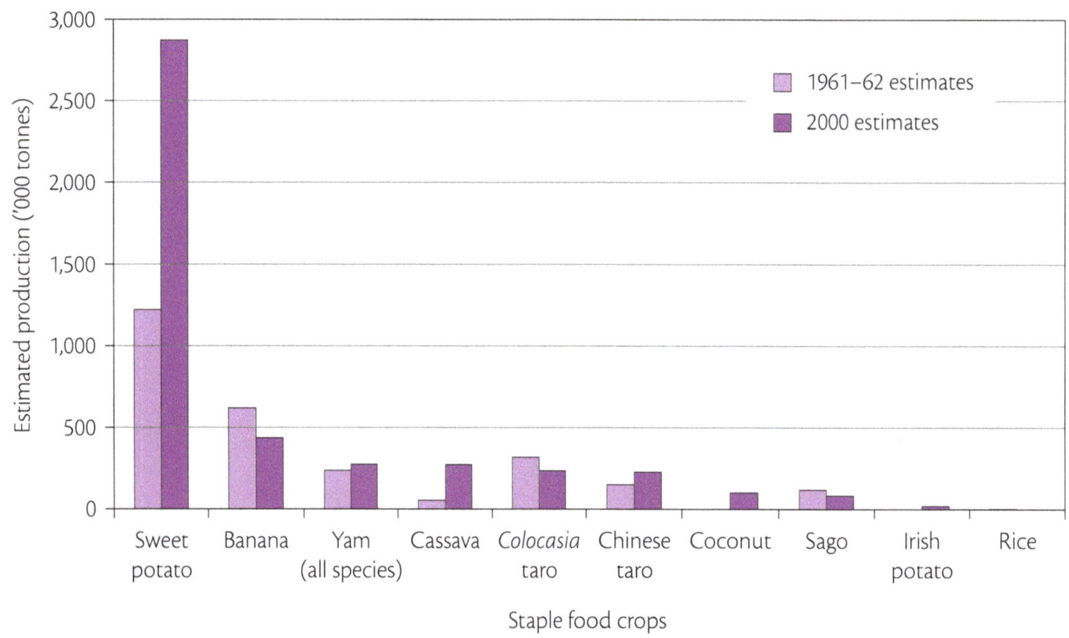

Figure 2.2.1 Comparison of the 1961–1962 and 2000 estimates of production by weight of ten staple food crops.
Note: Coconut production was not estimated in 1961–62. Estimated Irish potato production in 1961–62 was negligible. Rice production was estimated as 3000 tonnes in 1961–62, and 407 tonnes in 2000. The 1961–62 figure for rice is unlikely to be accurate (see Section 2.5). Source: Bourke and Vlassak (2004).

In 1961 sweet potato provided an estimated 45% of the food energy from the staple food crops (excluding coconut). This proportion had grown to 66% by 2000 (Figure 2.2.2). Similarly, the food energy from cassava increased from 2% to 6%. Irish potato production was negligible 40 years ago, but has increased rapidly since then. Production of Chinese taro increased rapidly in the 1960s and 1970s but, from the 1980s onwards, production decreased in many locations because of a disease problem, probably a root rot. Hence the relative contribution of Chinese taro to food energy in 2000 was about the same as it was in 1961–62.

In contrast to crops that originated in the Americas, the proportion of total food energy provided by the food crops that originate in the Asia–Pacific region has declined since 1961. This is the situation for banana, yam, *Colocasia* taro and sago. The estimated total production of these crops in 2000 was similar to the estimates 40 years earlier, but their proportional contribution to total production has dropped relative to that of the crops from the Americas, especially of sweet potato and cassava (Figure 2.2.3).

Sources

Bourke, R.M. and Vlassak, V. (2004). Estimates of food crop production in Papua New Guinea. Land Management Group, Research School of Pacific and Asian Studies, The Australian National University, Canberra.

Gibson, J. (2001). The economic and nutritional importance of household food production in PNG. In Bourke, R.M., Allen, M.G. and Salisbury, J.G. (eds). *Food Security for Papua New Guinea. Proceedings of the Papua New Guinea Food and Nutrition 2000 Conference*. ACIAR Proceedings No. 99. Australian Centre for International Agricultural Research, Canberra. pp. 37–44.

Walters, C.L. (1963). *Survey of Indigenous Agriculture and Ancillary Surveys, 1961–1962*. Territory of Papua and New Guinea Bureau of Statistics, Port Moresby.

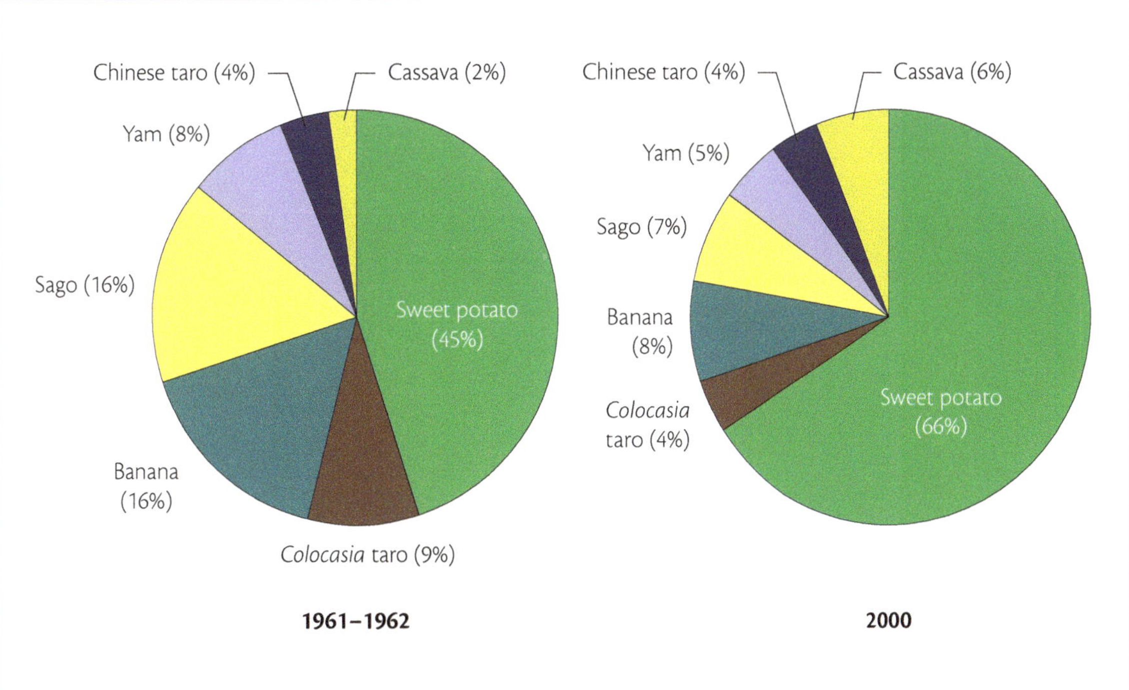

Figure 2.2.2 Estimated production of food energy of staple food crops: Survey of Indigenous Agriculture, 1961–1962 and Mapping Agricultural Systems of PNG Project, 2000. **Note:** Rice was estimated as 0.4% of food energy of the staple food crops in 1961–62, and as 0.03% in 2000. Sources: Walters (1963); Bourke and Vlassak (2004).

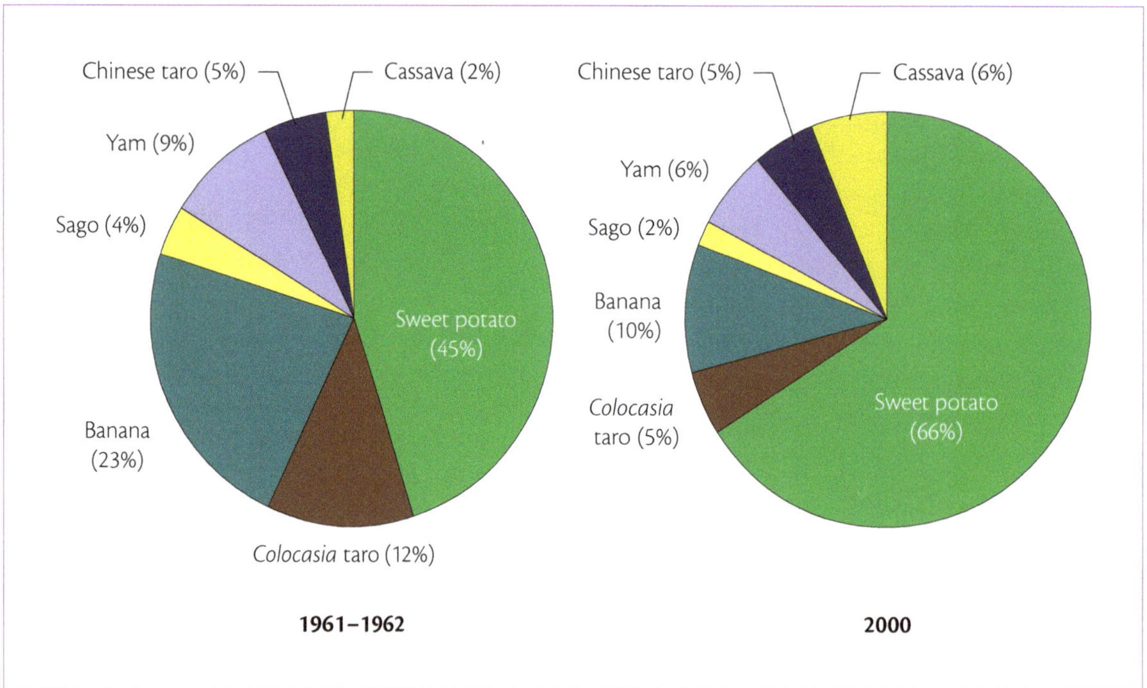

Figure 2.2.3 Estimated production by weight of staple food crops: Survey of Indigenous Agriculture, 1961–1962 and Mapping Agricultural Systems of PNG Project, 2000. **Note:** Rice was estimated as 0.1% of production by weight of staple food crops in 1961–62, and as 0.01% in 2000. Coconut production was not estimated in 1961–62. Coconut contributed an estimated 2.2% of food produced in 2000, but has been excluded from these figures so that the two datasets can be compared. Sources: Walters (1963); Bourke and Vlassak (2004).

2.3 Genetic diversity of food crops

Papua New Guinean villagers pay particular attention to variation among cultivated crops. They try out new variants and retain useful ones which, in suitable conditions, become established as cultivated varieties (cultivars). This is a continuous process and enhances crop genetic diversity. It has been going on in PNG for thousands of years, providing insurance against changing conditions such as drought or new diseases, and adapting plants to new environments. Both indigenous crops such as taro and banana, and introduced ones such as sweet potato, are managed in this way.

Most crops in PNG are propagated vegetatively, that is, by planting suckers, shoots or cuttings. This is known as asexual propagation, and only requires one parent plant. New plants grown by this method are usually identical to the parent plant. Sometimes, however, changes develop in parts of a plant because of abnormal division of somatic (body) cells. This process of somatic mutation is random and often injurious to the new plant. If the plant survives, human judgements about the usefulness or otherwise of the new form will then determine whether the new variant becomes an established variety. If the mutation occurs in buds, suckers or other plant parts that can be propagated, human intervention may result in the establishment of a new cultivar. Variant plants are not necessarily useful, but innovative villagers will try them out. Visible changes such as a new leaf shape or stem colour might be valued for associated desirable fruit or root characteristics, or might be retained just because they look attractive.

Some PNG crops that are planted vegetatively are not completely sterile, and can interbreed with wild forms. For example, some of PNG's many banana cultivars produce occasional seeds and/or pollen. Cross-fertilisation with wild banana plants may produce edible forms with novel characteristics. These may be noticed and tried out by further propagation. This has happened in the past, and has produced hybrid bananas unique to PNG. Cultivars of sweet potato, sago, *aibika* and *marita* pandanus may also have arisen from flowering and seeds, as well as from somatic mutation.

Variation is common in sexually propagated plants. Some PNG crops, such as breadfruit, coconut and other tree crops, grow from seed. Seedlings may or may not have the characteristics of the parent plant, depending on whether the breeding mechanism involves cross-pollination (the transfer of pollen from another plant of the same species) or self-pollination. Breadfruit and coconut are mainly cross-pollinated, and the resulting seedlings are genetically as diverse as the parent population. However, villagers may select and retain or transplant seedlings with desirable characteristics. If a particular trend of selection continues through enough breeding cycles, the characteristics of the population may be progressively altered. A process like this probably produced the large-fruited forms of tree crops like *galip* and *okari*. Breadfruit trees with particularly desirable characteristics are reproduced from rootstocks,[1] to give an identical new tree. Desirable new forms of plants are commonly given to people in other villages and so spread quickly away from their origins.

It is common throughout PNG to find many named cultivars of staples and supplementary fruit, vegetables and nut crops, and villagers are often well aware of prized local cultivars and their origins. Despite this diversity, there have been few studies of food crop variation, and even fewer of how villagers manage it. Conserving the genetic diversity of PNG's crops has so far been largely in the hands of PNG's villagers. In a modernising world, they will need improved institutional support, which cannot be provided without more studies of cultivated varieties and of local management practices.

Sources

Arnaud, E. and Horry, J.P. (eds) (1997). *Musalogue: A Catalogue of Musa Germplasm. Papua New Guinea Collecting Missions, 1988–1989.* International Network for the Improvement of Banana and Plantain and International Plant Genetic Resources Institute, Montpellier and Rome.

Ballard, C., Brown, P., Bourke, R.M. and Harwood, T. (eds) (2005). *The Sweet Potato in Oceania: A Reappraisal.* Ethnology Monographs 19 / Oceania Monograph 56. Oceania Publications, University of Sydney and Ethnology, Department of Anthropology, University of Pittsburgh, Sydney and Pittsburgh.

Brookfield, H.C. (2001). *Exploring Agrodiversity.* Perspectives in Biological Diversity Series. Columbia University Press, New York.

Preston, S.R. (1998). *Aibika/Bele* Abelmoschus manihot *(L.) Medik.* Promoting the conservation and use of underutilized and neglected crops 24. Institute of Plant Genetics and Crop Plant Research and International Plant Genetic Resources Institute, Gatersleben and Rome.

1 Rootstocking involves taking a short piece of root from a tree and growing it into a tree seedling with its own roots and leaves.

2.4 Food crop yields

Yield is an important measure used to compare production between different crops or between different varieties. Yield is also used to compare changes in a crop over time or under different cultivation conditions. Knowing crop yield helps development workers understand why villagers grow certain crops or adopt certain practices, although other factors influence villagers' decisions, including taste and cultural values.

A lot of information exists about food crop yields in PNG, with most of the data from research station experiments. The extensive number of publications consulted (see sources) show that information on crop performance is widely dispersed throughout the literature, some of which is difficult to locate. A summary of these data is presented in tables in this section. However, a complete review of published and unpublished trial data on food crop yields – experimental and village plantings – is beyond the scope of this book.

The four primary measures of yield are:

- **Weight of economic product per unit area.** The most common way to present yield data – the usable portions of tubers, grain or fruit from plants growing within a known area – is weight per unit area. For field crops it is usual to express yields as kilograms per hectare (kg/ha) or tonnes per hectare (t/ha).
- **Weight of economic product per plant.** In the case of tree crops, it is common for yield to be expressed as weight per tree (kg/tree), in which case the weight of fruit from a known number of trees is divided by the number of trees to give an average yield per tree.
- **Weight of economic product per unit area per unit time.** This is a useful measure for some purposes, especially for comparing yields in different environments, where crops may take different times to mature. A hypothetical example of such a comparison is: sweet potato yields are 13 t/ha over 18 weeks at sea level and 25 t/ha over 35 weeks at 1800 m altitude. In this case, the yield per unit time is 0.7 t/ha per week in both environments.
- **Weight of economic product per unit labour input to grow the crop.** Yields per unit of labour estimate how much work is required to produce a crop. This measure of yield is particularly important when looking at human–crop interactions. For example, weeding, pruning and draining will double the yield of coffee per unit area, but labour inputs needed to achieve the higher yield are *more than double* those used by most villagers. The additional labour will significantly reduce the yield per hour worked. This explains why recommended techniques are not adopted by most villagers (see Section 5.20). It is difficult to collect good quality labour data, so this important measure is rarely used.

Yield data

Yield data for PNG comes from two main sources: research stations or experiments by researchers on village land, and village gardens. Experimental data are collected when a researcher grows a crop under controlled conditions, usually on a research station, and then weighs the harvest. Considerable amounts of experimental data are available for rice, sweet potato, sugar cane, corn (maize), peanut, Irish potato, soya bean, cabbage, tomato, winged bean and taro, and for the export cash crops cocoa, coffee, coconut and oil palm. These data come mainly from agricultural research stations (Aiyura, Bubia, Dami, Keravat, Kuk, Laloki and Tambul (see map on page xix)). Little data exists for other food crops such as banana or sago, or for the minor cash crops balsa, cardamom, pyrethrum, rubber, tea and vanilla. Much experimental data is not published and so is not accessible.

Village yield data comes from measuring production from village gardens that were planted using village methods without interference from the researcher. Some village yield data are available for sweet potato and, to a lesser extent, for taro, yam, winged bean and Arabica coffee. For most other food crops, few yield records have been made from village plantings.

Important distinctions must be made between experimental and village yield data and the type of data must always be determined before comparisons are made. It is difficult to measure yield from village-controlled plantings because most crops are grown in mixed plantings. Meaningful yield data per unit area can only be calculated where a reasonably large number of plants are planted together in a plot. Thus one can calculate yield per hectare of sweet potato, taro, yam, cocoa or coffee in PNG, as these crops are often planted as a single species plot. However, meaningful yield per unit area cannot be calculated where plants are grown interplanted with other species. This is how most vegetables and minor staple foods are planted in village gardens.

Most crops in PNG are harvested progressively over some months (yam is an exception) and collecting yield data under these circumstances is difficult. Weighing a single harvest does not properly reflect village production. Village trees are rarely planted together, but are scattered through gardens and fallow land. Thus yield from fruit and nut trees is best expressed as yield per tree per year. Another difficulty is that experimental plots, with regular weeding and maintenance, may produce artificially high yields because the crop management does not accurately reflect village practices.

The best yields of root crops from PNG are as good as, or better than, the best yields reported in other tropical locations, although not as high as those in many subtropical locations where soil fertility is maintained by fertiliser and improved varieties are grown. Yields of subtropical and temperate-climate fruit and vegetables tend to be lower in PNG than in subtropical and temperate-climate locations. Yield data on export cash crops are not reviewed in this section. (See Part 5 for production and export data for cash crops.)

Some sources of yield data are unreliable and care should be taken in their use. The Food and Agriculture Organization of the United Nations (FAO) publishes yield data for food crops that purports to be for PNG, but these data are not based on any recordings in PNG. FAO also presents yield data for export tree crops that appear to be reasonably accurate, but cannot be used to interpret trends over time. Information on the area planted each year is needed to calculate changes in yield over time. The only PNG crops that have reliable long-term data on the area planted each year are oil palm and commercial sugar cane.

Crop yields appear to be changing in some locations. On small islands in particular, crop yields are now very low because population increases have led to intensification of land use, resulting in reduced soil fertility (see Table 1.3.1 and Sections 3.6 and 3.7). In these places, food crops that require high levels of soil fertility, such as taro, yam and corn, cannot be grown, and less demanding crops such as cassava, sweet potato and banana yield poorly. Elsewhere, villagers report that declines in crop yield and the inability to grow certain food crops have taken place over the past 30–50 years.

Staple food crops

A moderate amount of information is available for yields of staple food crops under village conditions, particularly for sweet potato, but much more data is available from experiments. Many of the datasets have very wide ranges, which makes it difficult to generalise. It is sometimes necessary to use a single figure, for example the average yields for the lowlands and highlands (Table 2.4.1). (These are the same figures that were used to generate estimates of staple food production in PNG – see Section 2.2.) These are the best estimates for yields under village conditions but much variation around these averages has been recorded and care must be taken when interpreting results from their use.

Banana

Published data on village banana yields are available from only two studies (Table 2.4.2). The study on banana production in the village north of Nomad in Western Province is one of the most detailed studies of lowland village production for any food crop. Banana yields in the Madang and Nomad areas are similar at 6–8 t/ha (4–5 t/ha edible portion).

Banana yield data comes from four experimental plantings in three lowland provinces. These yields vary between trials and locations, but are generally higher than for village production. Mean fruit bunch weight ranged from 9 kg to 27 kg. Mean bunch yield was in the range 5–15 t/ha/year, although the bearing period for yield recording varied between the different studies. The edible portion of diploid varieties was 60–66% in the two village studies and this was higher than that for triploid Cavendish varieties at Keravat (47%) or mixed triploid varieties in villages near Madang (57%).[1]

[1] Diploid varieties have two sets of chromosomes; triploids have three sets. Triploid banana varieties are usually larger plants and produce larger bunches. Many triploid varieties persist for many years after planting, especially if weeds are removed.

Cassava

No yield data exist for cassava under village conditions. Where soil fertility is high and the growing period is long, for example in the Wahgi Valley, tuber yield is very high and individual tubers have been recorded as heavy as 12 kg. In contrast, where soil fertility has been reduced by long periods of cropping in the lowlands because of population

Table 2.4.1 Average crop yield used for calculating staple food crop production (t/ha)[a]

Crop	Lowlands (0–1200 m)	Highlands (1200–1800 m)
Banana	12	9
Cassava	22	16
Irish potato	–	14
Queensland arrowroot	10	–
Rice	2	–
Sweet potato	13	15
Taro (*Alocasia*)	8	–
Taro (*Amorphophallus*)	6	–
Taro (*Colocasia*)	8	10
Chinese taro	14	11
Swamp taro	6	–
Aerial yam	13	13
Greater yam	13	11
Lesser yam	15	–
Yam (*Dioscorea nummularia*)	13	–
Yam (*Dioscorea pentaphylla*)	9	–

[a] These figures were used with other data to calculate production of staple food crops at the level of agricultural system, and then aggregated to provincial and national level (see Section 2.2). The figures are best estimates only and are more precise for the better-known species, such as sweet potato, *Colocasia* taro, lesser yam and cassava, than for the minor species.

Figures were assembled from published and unpublished records for village production. Where there were few or no village recordings, a figure of 75% of average experimental yield was used. For some of the minor yam and taro species, yields were taken as the same as the better-known species.

Source: Bourke and Vlassak (2004:19).

pressure, for example on some small islands, tubers may weigh only 100–300 g and yield per plant is only 1–2 kg.

A moderate amount of experimental research has been conducted on cassava and most results have been published. Experimental yields are often high and tuber yields up to 45 t/ha are common (Table 2.4.3). In one variety trial conducted at Laloki in 1985–86, six varieties yielded in the range 76–124 t/ha, with a mean tuber yield of 101 t/ha. Crops may be harvested from 10 months after planting. These experimental yields are mainly for crops 12–14 months old.

Irish potato

There are no published data on yield of Irish potato under village subsistence or commercial field conditions. Until the mid 1980s, Irish potato was rarely grown in single species stands and villagers interplanted it with sweet potato, so it was not possible to gather reliable village yield.

Many experiments have been done on Irish potato, particularly in the highlands provinces. There is less information from lowland and intermediate altitudes. Experimental yields have been recorded as low as 1 t/ha and as high as 60 t/ha. Yields in the highlands are typically in the range 10–35 t/ha (Table 2.4.3). As the time to maturity is about 100 days at 1600–1800 m altitude, productivity per unit time is very high for Irish potato, especially when crops are

Table 2.4.2 Village and experimental yield of banana

Location	Bunches (per hectare)	Mean bunch weight (kg)	Bunch yield (t/ha)	Yield edible portion (t/ha)	Source
Village yield					
Amele[a]	1032	5.4	5.6	3.7	King et al. (1989)
Nomad[b]	1313	5.7	7.5	4.5	Dwyer and Minnegal (1993)
Experimental yield					
Lejo[c]	510	9–27	4.6–13.8	–	Heenan (1973)
Keravat[d]	–	–	9.9	–	Gallasch (1980)
Keravat[e]	1240	12	14.9	7.0	Tarepe and Bourke (1982)
Laloki[f]	–	15.3	–	–	Kambuou (2005)

[a] The Amele area is south-west of Madang town. About 90% of bananas in food gardens are diploid. The edible proportion of diploid varieties was 67% and for triploid varieties was 57%. Mean bunch weight and yield of edible portion are based on diploid varieties. Banana is only one component of food gardens, so yield per hectare is less than from plots consisting of mainly banana plants.

[b] In villages north of Nomad, Western Province, diploid bananas are grown in mainly single-species stands. Bunches were available for eating 8–20 months after planting.

[c] Banana yields were recorded over a 16-month period for Giant Cavendish, Dwarf Cavendish and Tui varieties at Lejo Experiment Station in Oro Province. Yield recording commenced two years after planting. Bunch weight increased from 9 kg at two years to about 27 kg by 40 months after planting. The bunch yield given here is derived by multiplying the mean weights by the original planting density (510 plants/ha).

[d] The Yawa variety was planted under coconuts at Keravat, East New Britain Province. The bunch yield of 9.9 t/ha was recorded for the first 32 months of the trial. It takes about a year for this variety of banana to bear, hence the yield is equivalent to 9.9 t/ha over 20 bearing months.

[e] Experimental yields in the early to mid 1970s from five blocks of mature Giant Cavendish banana ranged from 9 to 20 tonnes of bunch per hectare per year, with a mean of 14.9 t/ha/year.

[f] In an irrigated trial at Laloki near Port Moresby, six common varieties and eight varieties derived from breeding lines were evaluated. Mean bunch weight ranged from 1.5 to 37.6 kg/bunch, with a mean of 15.3 kg/bunch.

Table 2.4.3 Village and experimental yield of selected root crops

Crop	Edible part	Village yield range (t/ha)	Typical experimental yield range (t/ha)	Maximum experimental yield (t/ha)	Maximum weight for an individual 'root' (kg)	Source[a]
Cassava	tuber	–	10–45	124.3	11.9	Grant et al. (2004); King (1986b, 1988)
Irish potato	tuber	–	10–35	61.2	1.8	Pitt and Yandanai (1988); Wiles (2001b)
Sweet potato	tuber	2–50	5–30	71.2	9.5	Bourke (1985, 2005)
Taro (*Alocasia*)	corm	–	–	–	40.0	
Taro (*Colocasia*)	corm	3–38	4–12	44.3	10.0	Bayliss-Smith (1982); Gendua et al. (2001); Singh et al. (2006)
Chinese taro	cormel	–	8–25	38.0	3.7	
Swamp taro	corm	8–12	17–48	48.1	15.0	Bourke and Betitis (2003)
Winged bean	tuber	6–12	1–3	14.8	–	
Greater yam	tuber	10–16	5–30	42.2	63.6	King (1986a, 1986b, 1989); Quin (1984); Risimeri (2001)
Lesser yam	tuber	10–21	10–40	95.6	11.5	Allen (1982); Johnston and Onwueme (1999); King (1986a, 1986b, 1989); Quin (1984); Risimeri (2001)
Yam (*Dioscorea rotundata*)	tuber	–	10–45	61.0	–	Risimeri et al. (2001)

[a] This table was adapted from Bourke (1982:56), where the original sources are given. The sources listed here provide supplementary information.

fertilised. Because there are big differences in yield between varieties and Irish potato is responsive to fertiliser, there are often large differences in yield even within a single trial. The only published yield data from the lowlands come from Keravat where different varieties yielded from 0.1 t/ha to 17 t/ha. On the Lelet Plateau on New Ireland Province at 900 m altitude, yields of assorted varieties were in the range 5–37 t/ha.

Sago

Published sago yields are available for 14 locations in PNG (Table 2.4.4). The recorded yield per palm varies widely, even within one location, with a range of 30 kg to more than 500 kg of dry starch per palm and a mean of 180 kg per palm. A feasibility study for production of sago flour in East Sepik Province by Toyo Menka Kaisha Ltd in 1972 recorded an average of 168 kg of dry starch per palm. Planting density is reported as 280–310 palms per hectare. A harvesting rate from planted or wild stands has been reported from five locations, with a range of 7 to 42 logs per year. Combining this harvest density with an average yield of 168 kg dry starch per palm suggests yields of dry starch of 1–7 t/ha/year. This contains the same food energy as between 4 and 27 tonnes of sweet potato. This is the potential yield from current stands of sago where extraction is done manually.

The potential yield from a sago plantation is likely to be much higher, and has been calculated as 25 tonnes of dry starch/ha/year (based on a harvest of 138 logs/ha/year and 185 kg starch/palm). Yields can be increased from existing stands by thinning out palms, removing other tree species and extracting starch mechanically. The long period from planting to maturity (12–15 years) is a consideration in planting sago in a managed plantation, although some varieties mature in less than 10 years.

The quantity of dry starch generated per labour input has been recorded at 10 locations on the PNG mainland at 2.0–3.7 kg/hour. A lower return of 0.5–0.8 kg was reported from south-west Bougainville where a different species of sago (*Metroxylon salomonense*) is used.

Table 2.4.4 Village starch yield per palm and per labour input from sago palms in various locations

Location	Starch yield (dry weight)	
	Per palm (kg/palm)	**Per labour input (kg/hour)**
Fly River, Western Province	79–159	–
Oriomo Plateau, Western Province	29–104	2.0
Ok Tedi area, Western Province	–	1.9
Kikori area, Gulf Province	–	2.6
Purari River delta, Gulf Province	–	3.5
Middle Sepik River, East Sepik Province	–	3.7
East Sepik/Enga provinces border area	–	3.1
Karawari River, East Sepik Province	230	2.9
Ambunti area, East Sepik Province	28–206	2.2
Ambunti area, East Sepik Province	–	2.0
Angoram area, East Sepik Province	150–400	–
Maprik, East Sepik Province	–	2.0
Nuku area, Sandaun Province	70–513	–
South-west Bougainville Island	–	0.5–0.8

Sources: Adapted from Townsend (1982:14–15) and Sowei (2006), who both give original sources.

Sugar cane

No yield data are available for village sugar cane, even though production throughout PNG is substantial (variously estimated as 190 000 to 440 000 tonnes of cane per year). The sugar content of village cane is almost certainly lower than that of commercial production because village sugar cane has been selected for high juice content rather than high sugar content.

Sugar is produced commercially by Ramu Agri-Industries Ltd in the Ramu Valley, Madang Province (see Section 5.10). Cane yields there vary between 50 t/ha and 90 t/ha, and are typically around 55–60 t/ha. The sugar content of the cane is about 9%, so 55–60 tonnes of cane produces 5–6 tonnes of sugar per hectare. The yield achieved in any year is influenced by the incidence of pests and diseases.

Sweet potato

It is difficult to measure sweet potato yields under village conditions because of the progressive harvesting method that villagers use. It is possible to record the weight of all tubers from a plot at one time, but this is not how villagers harvest and figures are likely to underestimate the total yield.

By the mid 1980s, village yields were available for about 30 PNG locations, mostly in the highlands. The range of recorded yields was 2–50 t/ha. However, some of the earlier recordings are unreliable. The range for the more reliable figures in the highlands was 5–31 t/ha. Little information on village sweet potato has been published since then and the more recent figures do not extend the range.

A large number of experimental yields have been recorded for sweet potato, with much of the data unpublished. Experimental yields are typically in the range 5–30 t/ha, but yields of up to 71 t/ha have been recorded (Table 2.4.3). Plantings in larger blocks have also given high yields, for example, the average yield in a fertiliser trial near Goroka was 70 t/ha for a 34-week crop. Yields are often higher in the highlands than in the lowlands, but the ranges cited above have been recorded in both environments. Experimental yields of 20–30 t/ha are common in the highlands and a lower range of 15–20 t/ha is more typical in the lowlands. The growing period in the lowlands is 3–5 months.[2] It is 5–8 months at 1500–2000 m altitude and longer at higher altitudes.

Sweet potato yields can vary greatly, even between different parts of the same garden. Yields often vary considerably from crop to crop, even under the most rigorous experimental conditions. Some of this variation is probably associated with minor variation in experimental technique. For example, it is known that planting different lengths of vine can influence yield. It is often not obvious why sweet potato yield varies so much from crop to crop, but sweet potato is sensitive to soil moisture conditions and nitrogen levels.

Colocasia taro

Colocasia taro yields are low compared with the other root crops. They have declined over the past 50 years probably because of reductions in soil fertility associated with more intensive land use and increased virus, fungus and insect problems. Village yields have been recorded as high as 38 t/ha under irrigated conditions, which are not common in PNG (Table 2.4.3). The highest subsistence yield recorded for non-irrigated conditions is 15 t/ha. The crop matures in 6–9 months in the lowlands but requires more than a year in the highlands. It is relatively easy to estimate taro yields in village plots based on the weight of a sample of corms because the harvest density is typically about 10 000 plants/ha and there is only one main corm per plant. If mean corm weight is 0.8 kg, the mean yield is about 8 t/ha.

Yields under experimental conditions are typically in the range 4–12 t/ha, with the recorded range from complete crop failure (0 t/ha) to 44 t/ha. A valuable review of yield and labour inputs under village and experimental conditions up to 1980 is given by Bayliss-Smith (1982). He recorded yield at Baisu Corrective Institution near Mount Hagen at an average 17 t/ha, which demonstrates that yields of taro on highly fertile soil in the highlands can be higher than in the lowlands, although the growing period is longer.

---

2 It is commonly stated that sweet potato matures in three months in PNG, but three months (90 days) is the minimum period to maturity in the lowlands. Yield commonly increases after 90 days until 150 days after planting in the lowlands.

Chinese taro

No yield data exist for village plantings of Chinese taro. A few experiments have been conducted at Keravat where yields were usually in the range 8–25 t/ha for a 12-month crop. Chinese taro is usually harvested from about 10 months after planting and can continue for some years, although harvesting is reduced after about two years from planting.

Swamp taro

On Mortlock Island east of Bougainville, swamp taro yields were estimated as 8–12 t/ha/year. Mean corm weight was about 8 kg and time to maturity was said by villagers to be 3–6 years, but larger corms are produced if harvesting is delayed.

The only published experimental yield data are from two plantings at Keravat which yielded 48.1 t/ha after 7.5 years and 16.7 t/ha after 2 years. This is equivalent to 6–8 t/ha/year, similar to the lower estimate from village plantings on Mortlock Island. The largest individual corm in the experimental plantings weighed 15 kg.

Taro (*Alocasia* and *Amorphophallus*)

There are no village or experimental yield data for *Alocasia* or *Amorphophallus* taro, except for some recordings on the weight of individual corms of *Alocasia* in village plantings on the Gazelle Peninsula, East New Britain Province. The range of weight recorded was 8.5–40.0 kg per corm. In both species, the edible corm does not mature and can be harvested as needed, sometimes when a plant has been growing for some years.

Yam

Some yam yield data have been recorded from villages in East Sepik and Central provinces. The average of all village recordings for greater yam (*Dioscorea alata*) is about 13 t/ha and for lesser yam (*D. esculenta*) is about 15 t/ha (Table 2.4.3). In one study of 12 villages in the lowlands of Central Province, greater yam tubers were larger on average (365 g) than lesser yam (220 g), but greater yam had fewer tubers per plant (2.9 compared with 9.3), so the yield per area was higher for *D. esculenta* (20.5 t/ha) than for *D. alata* (10.5 t/ha).

There are more experimental data for lesser yam than for greater yam. Like the village data, experimental yields of lesser yam (10–40 t/ha) are usually higher than those for greater yam (5–30 t/ha), with yields as high as 95 t/ha for lesser yam and 42 t/ha for greater yam. Greater yam matures at around 6–7 months and lesser yam at around 6–9 months.

White yam (*D. rotundata*) is an African species introduced to PNG in 1986 and increasing in popularity. Published information on crop performance is limited. In an evaluation at Bubia near Lae, yields of 61 t/ha were recorded.

Grains and grain legumes

Much research has been conducted in PNG on grains and grain legumes, particularly rice. A comprehensive review of published and unpublished literature is not attempted here, but an indication of the yield ranges is given. Almost all yield data are from experiments, although some data are available from commercial maize, commercial sorghum and village rice production in the Markham Valley, Morobe Province.

Corn (maize)

No yield data exist for village plantings. Corn has been grown commercially for more than 30 years in the Markham and Ramu valleys, where the yields were reported as 2.5 t/ha in 1976.

Yields of up to 9.2 t/ha have been reported from small experimental plots of imported varieties but experimental yields are commonly in the range 2–6 t/ha (Table 2.4.5). The best yields of corn in PNG are lower than those obtained in subtropical and temperate climates – in Australia and the United States, for example. Hybrids generally give higher yield than open-pollinated types, but the latter are more suitable for village conditions because the seed can be replanted. Seed from village plantings often gives a low yield when grown under experimental conditions. This is because villagers select seed for replanting from a small number of cobs, which results in 'inbreeding depression', where there is insufficient genetic diversity to obtain good yields. A simple way to increase village corn yield is to select seed for replanting from a larger number of cobs from all over the garden.

Peanut

In the early 1960s average village peanut yields in the lowlands were estimated as 1.2 t/ha of pods (nut-in-shell) (about 0.8 t/ha of kernel). A 2004 National Agricultural Research Institute (NARI) survey reported village pod yields as 0.6 t/ha in Eastern Highlands Province and 1.5 t/ha in the Wahgi Valley of Western Highlands Province. Average village pod yields in the upper Markham Valley in 2005 were reported as 0.9–2.0 t/ha, with a maturity time of 105 days or more.

Experimental peanut yields at Aiyura, Keravat and in the Markham Valley are typically 1–4 t/ha, with pod yields of up to 7.6 t/ha reported (Table 2.4.5). In 2003–2005, a number of varieties introduced from India were evaluated in the upper Markham Valley and at Aiyura. Pod yields of the best varieties in the Markham Valley were 4–5 t/ha and the best of the introduced varieties outyielded local varieties. Short-term varieties matured in up to 90 days and medium-term varieties required more than 120 days. At Aiyura, the best varieties yielded 2–5 t/ha of pods; local varieties yielded 0.5–2.0 t/ha. Short-term varieties took 126–134 days to mature at Aiyura and medium-term ones required 150–164 days.

Table 2.4.5 Experimental yield of six grain and food legume crops (t/ha)

Crop	Location	Experimental yield		Source
		Total range	Typical range	
Corn	Aiyura	1.4–6.0	2–4	Akus (1983)
Corn	Keravat	0.5–7.2	2–5	Bourke (1976b)
Corn	Laloki, Tanubada	0.5–7.6	3–6	King and Bull (1989)
Corn	Markham Valley	0.0–9.2	3–6	King (1987); Vance (1976, 1987a)
Peanut	Aiyura	0.1–7.6	1–4	Rachaputi et al. (2006)
Peanut	Keravat	0.3–3.5	1–3	Bourke (1977); Gallasch (1980)
Peanut[a]	Markham Valley	0.3–4.2	1–2	Vance (1987b)
Peanut	Upper Markham Valley	1.1–5.9	1–4	Rachaputi et al. (2006)
Rice[b]	Five provinces[c]	0.4–8.5	2–5	Wohuinangu and Joo (1982)
Rice	Markham Valley	2.9–10.5	5–8	Lin (1993); Sajjad (1996)
Sorghum	Aiyura	1.6–3.7	2–3	Kimber (1977b)
Sorghum	Keravat	0.9–3.6	1–3	Bourke (1977)
Sorghum	Markham Valley	0.0–11.0	1–5	Vance (1981); Vance and Li (1971)
Soya bean	Aiyura, Goroka, Kuk	–	1–2	Kimber (1977a)
Soya bean	Markham and Ramu valleys	0.2–3.3	1–3	Kambuou (1992)
Wheat	Aiyura	0.0–2.5	0.5–1.5	Kimber (1977b)

[a] Peanut yields reported by Vance are generally expressed as kernel yield, while other authors report pod yield. Kernels constitute about 65% of the weight of pods.

[b] Rice yields are for paddy rice. The recovery of white rice from paddy rice after milling is about 60%, so multiply these figures by 0.6 to convert them to white rice equivalent.

[c] Wohuinangu and Joo (1982) give an overview of a large body of research on rice conducted between 1965 and 1980 in Morobe (Bubia, Markham Valley), Central (Bereina), Milne Bay, East Sepik (Maprik) and Madang provinces.

Rice

Yields of village rice in the Markham Valley are reported as 1.3 t/ha where there is no elaborate land preparation or added fertiliser.[3] Village yields in the Dreikikir area of East Sepik Province were recorded as 1.5 t/ha (range 0.4–2.8 t/ha). Under broadacre farming conditions in the upper Markham and Ramu valleys, commercial yields were reported as about 2 t/ha in the late 1970s. Upland village rice was reported to yield an average of 2–3 t/ha and a maximum of 5 t/ha in the Markham Valley in the 1990s. The highest reported yield from a village planting is 11.7 t/ha. Small plots grown by 50 households in one village in the Markham Valley growing about 10 ha yielded 4 t/ha under high input conditions.

Rice is the most-researched crop in PNG (Table 2.5.1) and a great deal of experimental yield data are available. Experimental yields of more than 10 t/ha of paddy have been reported in the Markham Valley and experimental yields are typically in the range 2–5 t/ha (Table 2.4.5). Yields of irrigated rice are generally higher than those from upland rice. For example, in the 1980s the recommended variety (NG6637) was reported to yield 3.7 t/ha under upland conditions and 5.2 t/ha under irrigated conditions.

Sorghum

Sorghum was grown commercially in the Markham Valley in the 1970s by expatriate farmers where 300–500 ha were sown each year. It is rarely grown in PNG now. Experimental yields are usually in the range 1–5 t/ha, and up to 11 t/ha in the Markham Valley, but yields were lower at Aiyura and Keravat where less experimental work was conducted.

Soya bean

Soya bean is rarely grown in PNG. A reasonably large amount of research has been conducted on soya bean, mainly in the Markham and Ramu valleys and at various highland locations. Average experimental yields in the lower Markham Valley were reported as 2.5 t/ha and in the drier areas of Gusap in the Ramu Valley and the upper Markham Valley as about 1.2 t/ha. Experimental yields of up to 3.3 t/ha have been recorded.

3 All figures are for paddy rice. To convert this to white rice equivalent, multiply by 0.6.

Wheat

Wheat is not grown commercially in PNG. Some experimental work was conducted in the 1970s, when yields of up to 2.5 t/ha were recorded. It was concluded that rainfall in the PNG highlands was too high for successful wheat production. Since about 1999, some experimental work has been conducted by Chinese and later NARI researchers at Kandep in Enga Province, but yield data have not been made available.

Minor pulse crops

A limited number of yield records have been published for seven minor food legumes (Table 2.4.6).

Table 2.4.6 Experimental yield of seven minor food legume crops at various locations (kg/ha)

Crop	Port Moresby[a]	Bubia and Markham Valley[b]	
	Mean	Range	Mean
Bean, adzuki	–	1200–1600	1300
Bean, common	700	–	–
Bean, jack	1600	–	–
Bean, mung	1500	100–1300	600
Bean, rice	900	–	–
Cowpea	900	–	–
Pigeon pea	800	0–6100	1700

[a] The source for the Port Moresby data is Khan et al. (1976). The data are based on a limited number of experiments and observations.

[b] The source for the data from Bubia (near Lae), Gusap Downs and Leron Plains in the Markham Valley is Kambuou (1984). Kambuou (1982) gives average experimental yields for mung bean as 800 kg/ha and for adzuki bean as 1500 kg/ha.

Vegetables

No yield information exists for traditional (pre-1870) or introduced (post-1870) vegetables grown under village conditions.

A limited amount of published data exist from experiments on the traditional vegetables *aibika*, highland *pitpit*, rungia and especially winged bean (Table 2.4.7). Two trials were conducted at Aiyura which evaluated 12 species of traditional and 13 species of introduced vegetables (Table 2.4.8). Some trials have been conducted in villages, but they were managed by researchers.

At least 237 agronomic field trials were conducted on 30 species of introduced vegetables between 1928 and 1978 (Table 2.5.1). Cabbage and tomato have been well studied (over 40 trials on each up to 1978), but many trials were also conducted on cucumber, onion, lettuce, cauliflower, carrot and capsicum.[4] Many more trials on introduced vegetables have been conducted since 1978. While some recommendations have been published, details of most trial outcomes remain unpublished.

4 Cucumber is a traditional vegetable in PNG, but is considered with the introduced species because the cultivars evaluated are all introduced.

Table 2.4.7 Experimental yield of selected traditional vegetables at several highlands and lowlands locations (t/ha)

Crop	Aiyura[a]	Aiyura[b]	Kuk[c]	Bubia[d]	Port Moresby[e]	Various locations[f]
Aibika	2.8	–	–	13.2–63.7	4.7–5.7	3.7–4.7
Ginger	–	–	–	–	–	10–23
Highland *pitpit*	3.1	–	8–18	–	–	3.9–5.0
Rungia	6.4	0–5.0	4	–	–	–
Winged bean						
– pods	3.1	–	–	–	–	–
– seed	–	–	1.6–2.2	–	–	1.1–1.3
– tubers	2.3	0–2.9	–	–	–	1.5–1.9

[a] The data are an average of two trials that evaluated 25 species of traditional and introduced vegetables for their suitability for schools (Akus and Nema 1995).

[b] The data for rungia and winged bean are two time-of-planting trials. Figures given here are yields for the worst and best months to plant (R.M. Bourke unpublished data).

[c] The data are from experiments on three vegetable species at Kuk near Mount Hagen (Powell 1982). Highland *pitpit* yields are total stem yield (edible and inedible portions). About 30% of *pitpit* stem is edible when cooked in an earth oven.

[d] The *aibika* data from Bubia are the range of yields from an evaluation of 41 varieties (Sutherland 1984/85). Mean yield was 36.4 t/ha. The author cautions that plots were not replicated and edge effects may have resulted in a higher apparent yield when plot yields were converted to t/ha.

[e] The *aibika* figure from Port Moresby is the mean yield for a trial that evaluated 12 varieties and for a method of propagation trial (Westwood and Kesavan 1982). The authors also present data on the nutritional value of eight vegetable species.

[f] *Aibika* yields are the averages of the best treatments from a time of weeding trial and a chicken manure fertiliser trial conducted at Laloki near Port Moresby. Each figure is the average for three varieties and two trials (Sowei and Osilis 1995; Sowei et al. 1996). Ginger data are from a coconut interplanting trial at Keravat (Gallasch 1976). The highland *pitpit* data are from a trial with two varieties at Tari in Southern Highlands Province (Rose 1980). Yields are for edible portion. The winged bean seed and tuber yield data are from an evaluation of 10 varieties at three lowlands (Laloki, Lae, Waigani) and three highlands (Aiyura, Kuk, Wapenamanda) sites over two years (Kesavan and Stephenson 1982).

Table 2.4.8 Experimental yield of 12 traditional and 13 introduced vegetable species at Aiyura, Eastern Highlands Province[a]

Crop	Part eaten	Yield over total harvest period (t/ha)	Period to first harvest (weeks)	Period to last harvest (weeks)
Traditional vegetables				
Aibika	Leaves	2.8	24	76
Amaranthus	Leaves	1.5	14	54
Bean, winged	Pods, seed	3.1	20	29
Bean, winged	Tubers	2.3	45	45
Bottle gourd	Young fruit	8.6	19	27
Cyanotis moluccana	Leaves	10.1	14	76
Dicliptera	Leaves	5.0	15	76
Karakap	Leaves	2.8	14	70
Rorippa	Leaves	0.9	12	21
Oenanthe	Leaves	9.7	14	76
Pitpit, highland	Young stem	3.1	17	69
Pitpit, lowland	Inflorescence	0.5	49	82
Rungia	Leaves	6.4	15	76
Introduced vegetables				
Bean, climbing	Pods, seed	3.3	15	25
Bean, dwarf	Pods, seed	2.5	14	22
Bean, lima	Pods, seed	3.1	23	72
Cabbage	Head (leaves)	10.5	20	63
Ceylon spinach	Leaves	6.0	15	68
Choko	Leaves	5.9	15	74
Choko	Fruit	34.5	18	74
Cowpea	Pods, seed	3.5	19	56
Pak choi	Leaves	3.4	20	25
Pea	Pods, seed	0.7	12	14
Pumpkin	Leaves	8.4	16	74
Pumpkin	Fruit	59.8	19	71
Russian comfrey	Leaves	11.6	15	77
Silverbeet	Leaves	16.4	16	63
Spinach	Leaves	0.3	27	34

[a] Figures given here are the average from two trials conducted at Aiyura between 1979 and 1982. The authors also present data for dry weight yield.

Source: Akus and Nema (1995).

Temperate-climate vegetables were produced commercially from 1965 until the late 1990s at Kabiufa High School, west of Goroka. This operation depended on high levels of chemical input and is reported to have closed because of build-up of chemical residues in the soil. In a paper describing the enterprise, average yields for sweet potato and six introduced vegetables were given (Table 2.4.9).

Much of the available information on onion research in PNG is summarised in a paper by Wiles (2001a). The yield range for experiments on onion is crop failure (0 t/ha) to 45 t/ha. The most common range at Laloki was 10–40 t/ha and the most common range at Aiyura and other sites in Eastern Highlands Province was 8–20 t/ha.

An indication of the range of yields for introduced vegetables that have been measured in experiments is presented in Table 2.4.10 for the highlands and Table 2.4.11 for the lowlands.

Fruit

No published yield data are available for fruit trees grown under village conditions in the lowlands or highlands, but some yield data have been recorded at research stations and other locations. Data for 10 species of introduced fruit trees at the Lowlands Agricultural Experiment Station at Keravat are given in Table 2.4.12. The number of trees and duration of recording varies between sources and over time.

A detailed analysis of yield data for durian, mangosteen and rambutan is summarised in Table 2.4.13. Considerable variation in yield occurred between years for the three tree species. Most trees failed to fruit in some years.

Mean fruit weight from 18 avocado trees in one trial at Keravat in the late 1970s was 475 g. Some yield data have been published for pineapple from Keravat and for watermelon in the Port Moresby area (Table 2.4.14).

Yield records from 7 'banana' mango trees at the Pacific Adventist College near Port Moresby in 1994 indicated that trees yielded up to 300 fruit each, or about 75 kg/tree. Planting material of the mango varieties Banana Callo, Kensington Pride and Totapuri ('Rabaul') was made available by NARI at Laloki. NARI reports average fruit yields of 300 fruit per tree for Banana Callo and more than 400 fruit per tree for Kensington Pride and Totapuri. A mean fruit size of 270 g, 470 g and 580 g at Laloki for these varieties and a planting density of 100 trees/ha indicates a potential fruit yield of 8, 19 and 23 t/ha respectively for the three varieties.

Table 2.4.9 Average commercial yield for sweet potato and six introduced vegetable species at Kabiufa High School near Goroka[a]

Crop	Yield range (t/ha)
Sweet potato	20–35
Cabbage	15–25
Carrot	2–5
Cauliflower	2–5
Lettuce	4–13
Radish, red	1–3
Tomato	2–10

[a] Temperate-climate vegetables were produced at Kabiufa High School near Goroka in Eastern Highlands Province from 1965 until the late 1990s. The authors caution that even with a well-managed operation, crop failure can occur and 'yields range from zero upwards'. These data were presented as average yields for their commercial operation.

Source: Dever and Voigt (1976).

Data collected in the late 1970s at Aiyura for experimental yield of pineapple, strawberry, and four types of citrus are presented in Table 2.4.15. Other limited experimental data are available from Aiyura, for example, fruit production of naranjilla was recorded as 7.5 kg/tree (130 fruit/tree; 58 g/fruit) in the first six months of bearing in an observation plot.

Nuts

Yield data from both village gardens and research stations for edible nut species are limited. More village and experimental data for edible nuts have been recorded from Solomon Islands than from PNG.

The most important edible nut in the highlands is pandanus nut (*karuka*). Recordings were made of planted village *karuka* nut in two locations in the Tari basin in Southern Highlands Province from 1976 to 1980. The number of syncarps (fruit that contain the nuts) varied greatly between years, with a range of 0.2–1.5 syncarps per bearing tree. The edible nut was 8% by weight of a pandanus syncarp. Observations of the pattern of *karuka* bearing at six locations from Oksapmin in the west to Kainantu in the east over a 7–10 year period indicated that the size and timing of the harvest varies considerably from year to year and between locations.

Galip (*Canarium* spp.) nut yields under PNG village conditions are not known. However, detailed observations have been made in village and experimental

Table 2.4.10 Experimental yield of 12 introduced vegetable species at various highlands sites (t/ha)

Crop	Three sites in SHP[a]	Various sites, three provinces[b]	Recommended varieties[c]	Average period to the first harvest (days)[d]
Bean, common	7–8	7–24	7–24	95–100
Broccoli	3–9	1–20	9–14	74–84
Brussels sprout	2–14	–	–	–
Cabbage	11	–	–	–
Cabbage, Chinese[e]	11–45	–	39	56–70
Carrot	3–15	0–45	8–40	142–145
Cauliflower	0–15	1–4	–	70–86
Lettuce	–	1–21	13–21	102
Marrow	–	9–25	21	–
Onion	–	20–51	9–16	122–137
Tomato	8–16	9–28	13–15	120–125
Zucchini	–	17–25	19–25	125

[a] Data from Southern Highlands Province were recorded at Piwa (near Tari), Kuma and Wambip (near Mendi). Figures are lowest and highest marketable yield of different varieties in variety evaluation trials, averaged over the three sites (Kanua and D'Souza 1985).

[b] Data were recorded at Yani (near Gumine, Simbu Province), Aiyura (Eastern Highlands Province) and Kuk (near Mount Hagen, Western Highlands Province). Figures are lowest and highest marketable yield of different varieties or fertiliser treatment in variety evaluation and fertiliser trials, averaged over the three sites (Kanua 1990).

[c] Yields for the recommended varieties are based on the trial data from Southern Highlands Province, Yani and Aiyura (previous two columns). Yield ranges are from a more limited number of varieties than the other datasets. Thus they give a clearer picture of the yield range of the best varieties under experimental conditions, as distinct from data on all varieties evaluated (Kanua et al. 1993).

[d] Average period to the first harvest is the range of periods from these six sites.

[e] The term Chinese cabbage here covers a number of types, including those in the Chinese cabbage cultivar group and those in the pak choi cultivar group.

plots in Solomon Islands on three species of *galip* nut. Average tree yields for *C. indicum* were 100 kg of nut-in-shell (NIS)/tree/year (range 50–300 kg) and mean kernel yield was 16 kg kernel/tree/year. The average kernel proportion was 17% of the weight of nut-in-shell. *Galip* is rarely grown as a monoculture but, at a planting density of 100 trees/ha, yields an equivalent of 10 tonnes NIS/ha/year. Tree and kernel yields were lower for *C. harveyi* (50 kg NIS/tree/year and 12 kg kernel/tree/year) and for *C. salomonense* (25 kg NIS/tree/year and 5 kg kernel/tree/year).

Okari (*Terminalia kaernbachii*) yields have been recorded in a village in Western Province, in Oro Province and from experimental plantings at Keravat. They indicate that production of *okari* nut varies a lot from tree to tree and also from year to year. Experimental plantings at Keravat gave a yield of 16–82 kg fruit/tree/year in the late 1970s. Recordings from the same trees over a three-year period in the early 1990s gave a yield of 2.6 tonnes fruit/ha/year. On the Managalas Plateau in Oro Province yields were recorded as 480 nuts per tree,

Table 2.4.11 Experimental yield of 10 introduced vegetable species at various lowlands sites (t/ha)

Crop	Laloki dry season[a]	Mount Diamond dry season[b]	Waigani–Tanubada[c]	Various locations
Bean, climbing	0.4–4.0	–	–	–
Bean, dwarf	0.1–1.0	–	–	–
Cabbage	14–41	32–63	5–10	15–63[d]
Cabbage, Chinese[e]	4–13	28–40	10–22	–
Capsicum	9–19	–	15–27	3–7[f]
Cauliflower	9	–	–	–
Cucumber	4–25	–	0.4–6.0	–
Eggplant	20–75	–	–	–
Lettuce	–	–	3–9	–
Silverbeet	25–33	–	–	–
Tomato	35–45	–	9–18	30–60[g]
Yam bean	–	–	–	13[h]

[a] Figures are lowest and highest yield of different varieties in evaluation trials conducted at Laloki near Port Moresby. These data are for total (not marketable) yields (Blackburn 1976).

[b] The Mount Diamond data are commercial yields from a high school near Port Moresby (Blackburn 1976).

[c] Data from Waigani and Tanubada near Port Moresby are the lowest and highest yield of different varieties in evaluation trials (Kesavan 1977).

[d] Cabbage yield data under 'various locations' are the range of yields for the best varieties from variety and fertiliser trials conducted at seven sites on the Gazelle Peninsula of East New Britain Province (Wiles and Mwayawa 2001).

[e] The term Chinese cabbage here covers a number of types, including those in the Chinese cabbage cultivar group and those in the pak choi cultivar group.

[f] Capsicum yield under 'various locations' is from an experiment conducted near Goroka which evaluated the effect of sheep manure and inorganic fertiliser. Yield figures are the ranges for different treatments (Nukundj et al. 1997).

[g] Tomato yield under 'various locations' is the most common range from many trials evaluating a large number of tomato varieties and breeding lines at Laloki near Port Moresby. The range of yields recorded was 3–135 t/ha (Bull et al. 1985).

[h] The yam bean yield is the mean from an experiment at Keravat, where the best spacing treatment yielded 20 t/ha (Bourke 1982).

with an estimated average yield of 5 kg/kernel/tree. Observations in Solomon Islands indicate that the kernel is 5–10% of the fruit weight.

No yield data are available for sea almond or *pao* nut (*Barringtonia procera*) in PNG. In Solomon Islands, estimated yields of sea almond are 10–50 kg fruit/ha/year, with the kernel 6–12% of the fruit weight. Estimated yields of *B. procera* in Solomon Islands are 10–50 kg fruit/tree/year, with the kernel 9% of the fruit weight, giving an estimated yield of 1–5 kg kernel/tree/year.

Table 2.4.12 Experimental yield of 10 fruit species at Keravat, East New Britain Province

Crop	1970s recordings[a] (kg/tree/year)	1980s recordings[b] (kg/tree/year)	1990–1992 recordings[c] (t/ha/year)
Carambola	–	50	6.1
Durian	50–60	104	5.7
Egg tree	–	–	7.3
Langsat	100–140	39	–
Lime	460–700	–	–
Malay apple, giant	100–130	–	–
Mangosteen	6–9	15	2.5
Pulasan	–	20	1.7
Rambutan	–	29	2.1
Santol	–	15	0.8

[a] The data source is Aburu (1982:117–120), who also gives ranges for the number of fruit per tree per year for 34 fruit and nut species. The duration of data is not stated, but is probably for 1978 and 1979 (and sometimes for 1980). Data for West Indian lime are from a formal trial that ran for about 12 years.

[b] The data source is Woodhouse (1991). Some observations are from the same trees as the 1970s data. He also gives average fruit weight for these seven species. Duration of recording is 10 years for mangosteen, 9 years for durian and one year for the other species. Therefore the figures for durian and mangosteen are more reliable than for the other species.

[c] Source for the 1990–1992 data are recordings by S. Woodhouse (in Bourke et al. 2004:185–186). The data are mostly from the same trees as the 1980s data, and are expressed as t/ha/year. They have been aggregated from monthly records.

Table 2.4.13 Experimental yield and yield potential of three fruit species at Keravat, East New Britain Province[a]

Crop	Sample size and span of data collection	Mean yield per tree (kg/tree/year)	Yield of best trees (kg/tree/year)	Assumed planting density (trees/ha)	Potential yield (t/ha/year)
Durian	14 trees; 1980–1992	67	150	50	7.5
Mangosteen	11 trees; 1982–1992	14	20	280	5.6
Rambutan	24 trees; 1989–1992	12	25	100	2.5

[a] These data are from some of the same trees as those in Table 2.4.12, although the actual trees and period of data collection differs between the datasets. The potential yield is derived by multiplying the yield for the best trees by the assumed planting density and dividing by 1000 to convert from kilograms to tonnes.

Source: Wiles (1997).

Table 2.4.14 Experimental yield of pineapple and watermelon at various lowlands locations

Crop	Location	Yield (t/ha)	Source
Pineapple[a]	Keravat	26.5	Bourke (1976a)
Pineapple[b]	Keravat	16.5	Gallasch (1976, 1980)
Watermelon[c]	Laloki	8.6–21.5	Blackburn (1976)
Watermelon[c]	Mount Diamond	57.4–108.9	Blackburn (1976)
Watermelon[c]	Waigani and Tanubada	0.7–5.8	Kesavan (1977)

[a] These data are from a trial that compared different planting material for the rough leaf type of pineapple. Harvesting commenced 11 months after planting. Total fruit yield was 26.5 t/ha over four bearing years or 6625 kg/ha/year. Mean fruit weight was 1.25 kg.

[b] These data, also with the rough leaf type of pineapple, are from a trial where various food crops were planted with coconuts. The yield of 16.5 t/ha was recorded in the first 28 months after planting.

[c] All watermelon yield data are for the lowest and highest yields recorded in variety evaluation trials.

Table 2.4.15 Experimental yield of six fruit species at Aiyura, Eastern Highlands Province

Crop	Yield (t/ha/year)	Notes
Grapefruit	20.8	Data on citrus species are from a rootstock/scion trial. Heavy bearing for all four citrus species commenced in the third year of fruiting.
Lemon	20.2	
Orange	16.6	
Mandarin	9.8	
Pineapple	22.3	Production commenced 14 months after planting. This figure is for the first 12 months of bearing for the smooth leaf Cayenne type.
Strawberry	2.0	Yield with good management

Source: Tarepe and Bourke (1982).

Sources

Aburu, K. (1982). Notes on some fruit and nut trees in the Keravat area of East New Britain. In Bourke, R.M. and Kesavan, V. (eds). *Proceedings of the Second Papua New Guinea Food Crops Conference: Part One*. Department of Primary Industry, Port Moresby. pp. 101–122.

Akus, W.L. (1983). A new corn variety for the highlands. *Harvest* 9(1):39–40.

Akus, W.L. and Nema, R.K. (1995). Evaluation of twenty five vegetable varieties at Aiyura, Eastern Highlands Province. *Papua New Guinea Journal of Agriculture, Forestry and Fisheries* 38(1):11–16.

Allen, B.J. (1976). Information flow and innovation diffusion in the East Sepik district, Papua New Guinea. PhD thesis. The Australian National University, Canberra.

Allen, B.J. (1982). Yam gardens and fallows in the Torricelli foothills, Dreikikir District, East Sepik. In Bourke, R.M. and Kesavan, V. (eds). *Proceedings of the Second Papua New Guinea Food Crops Conference: Part Two*. Department of Primary Industry, Port Moresby. pp. 236–255.

Anon. (c. 1985). Smallholder Agriculture Development Programme: Barai Okari Estates. Unpublished paper.

Bayliss-Smith, T. (1982). Taro (*Colocasia esculenta*) in the New Guinea Highlands. In Bourke, R.M. and Kesavan, V. (eds). *Proceedings of the Second Papua New Guinea Food Crops Conference: Part One*. Department of Primary Industry, Port Moresby. pp. 134–147.

Blackburn, K.J. (1976). Observations on the selection and management of introduced vegetable varieties under Port Moresby dry season conditions. In Wilson, K. and Bourke, R.M. (eds). *1975 Papua New Guinea Food Crops Conference Proceedings*. Department of Primary Industry, Port Moresby. pp. 163–186.

Bourke, R.M. (1976a). A comparison of five types of planting material for rough leaf pineapples. *Papua New Guinea Agricultural Journal* 27(4):107–113.

Bourke, R.M. (1976b). New corn varieties. *Harvest* 3(1):3–5.

Bourke, R.M. (1977). A long term rotation trial in New Britain, Papua New Guinea. In Leakey, C.L.A. (ed). *Proceedings of the Third Symposium of the International Society for Tropical Root Crops*. International Society for Tropical Root Crops in collaboration with the International Institute of Tropical Agriculture, Ibadan, Nigeria. pp. 382–388.

Bourke, R.M. (1982). Root crops in Papua New Guinea. In Bourke, R.M. and Kesavan, V. (eds). *Proceedings of the Second Papua New Guinea Food Crops Conference: Part One*. Department of Primary Industry, Port Moresby. pp. 51–63.

Bourke, R.M. (1985). Sweet potato (*Ipomoea batatas*) production and research in Papua New Guinea. *Papua New Guinea Journal of Agriculture, Forestry and Fisheries* 33(3–4):89–108.

Bourke, R.M. (2005). Sweet potato in Papua New Guinea: the plant and people. In Ballard, C., Brown, P., Bourke, R.M. and Harwood, T. (eds). *The Sweet Potato in Oceania: A Reappraisal*. Ethnology Monographs 19 / Oceania Monograph 56. Oceania Publications, University of Sydney and Ethnology, Department of Anthropology, University of Pittsburgh, Sydney and Pittsburgh. pp. 15–24.

Bourke, R.M. and Betitis, T. (2003). Sustainability of agriculture in Bougainville Province, Papua New Guinea. Land Management Group, Department of Human Geography, Research School of Pacific and Asian Studies, The Australian National University, Canberra.

Bourke, R.M., Camarotto, C., D'Souza, E.J., Nema, K., Tarepe, T.N. and Woodhouse, S. (2004). *Production Patterns of 180 Economic Crops in Papua New Guinea*. Coombs Academic Publishing, Canberra.

Bourke, R.M. and Vlassak, V. (2004). Estimates of food crop production in Papua New Guinea. Land Management Group, Research School of Pacific and Asian Studies, The Australian National University, Canberra.

Bull, P.B., Moles, D.J., George, J.M. and Blackburn, K.J. (1985). Tomato variety evaluation at Laloki Horticultural Research Station 1974–1983. Technical Report 85/1. Department of Primary Industry, Port Moresby.

Dekuku, R.C. (2001). Pilot phase rice production in Markham Valley of Papua New Guinea shows great promise. *Papua New Guinea Journal of Agriculture, Forestry and Fisheries* 44(1&2):66–75.

Densley, D.R.J. and Cairns, I.J. (c. 1978). Grains and legumes. In Densley, D.R.J. (ed). *Agriculture in the Economy: A Series of Review Papers*. Volume 3. Department of Primary Industry, Konedobu. pp. 1–15.

Dever, K.J. and Voigt, A.L. (1976). Vegetable production methods at the Kabiufa Adventist High School gardens. In Wilson, K. and Bourke, R.M. (eds). *1975 Papua New Guinea Food Crops Conference Proceedings*. Department of Primary Industry, Port Moresby. pp. 205–210.

Dwyer, P.D. and Minnegal, M. (1993). Banana production by Kubo people of the interior lowlands of Papua New Guinea. *Papua New Guinea Journal of Agriculture, Forestry and Fisheries* 36(1):1–21.

Evans, B. (1996). Overview of resource potential for indigenous nut production in the South Pacific. In Stevens, M.L., Bourke, R.M. and Evans, B.R. (eds). *South Pacific Indigenous Nuts*. ACIAR Proceedings No. 69. Australian Centre for International Agricultural Research, Canberra. pp. 10–35.

Food and Agriculture Organization of the United Nations website <http://faostat.fao.org/site/291/default.aspx>.

Gallasch, H. (1976). Integration of cash and food cropping in the lowlands of Papua New Guinea. In Wilson, K. and Bourke, R.M. (eds). *1975 Papua New Guinea Food Crops Conference Proceedings*. Department of Primary Industry, Port Moresby. pp. 101–116.

Gallasch, H. (1980). Growing food crops under coconuts. *Harvest* 6(2):72–75.

Gendua, P.A., Risimeri, J.B. and Maima, J.B. (2001). The effect of planting density on the marketable corm yield of taro (*Colocasia esculenta* (L.) Schott). In Bourke, R.M., Allen, M.G. and Salisbury, J.G. (eds). *Food Security for Papua New Guinea. Proceedings of the Papua New Guinea Food and Nutrition 2000 Conference*. ACIAR Proceedings No. 99. Australian Centre for International Agricultural Research, Canberra. pp. 764–767.

Grant, I.M., Allen, M.G., Wiles, G.C. and Laraki, J. (eds) (2004). *Proceedings of the NARI Cassava Workshop: The Future of Cassava in Papua New Guinea*. NARI Workshop Proceedings. National Agricultural Research Institute, Lae.

Hartemink, A.E. and Kuniata, L.S. (1996). Some factors influencing yield trends of sugarcane in Papua New Guinea. *Outlook on Agriculture* 25(4):227–234.

Heenan, D.P. (1973). Preliminary observations on the growth and production of bananas in the Northern District of Papua New Guinea. *Papua New Guinea Agricultural Journal* 24(4):145–155.

Johnston, M. and Onwueme, I.C. (1999). Productivity of lesser yam (*Dioscorea esculenta*) in Papua New Guinea as influenced by sett weight and staking. *Papua New Guinea Journal of Agriculture, Forestry and Fisheries* 42(1 & 2):27–34.

Kambuou, R.N. (1982). Pulse crop research in the Markham and Ramu valleys. In Bourke, R.M. and Kesavan, V. (eds). *Proceedings of the Second Papua New Guinea Food Crops Conference: Part Three*. Department of Primary Industry, Port Moresby. pp. 568–569.

Kambuou, R. (1984). Pulse crops research at Bubia ARC. In George, J. and Moles, D. (eds). Foodcrops Research Advisory Meeting 2/3rd October 1984. Bubia Agricultural Research Centre, Lae.

Kambuou, R.N. (1992). Soybean work in the Markham and Ramu valleys of Papua New Guinea. In Levett, M.P., Earland, J. and Heywood, P. (eds). *Proceedings of the First Papua New Guinea Food and Nutrition Conference: Changes in Food and Nutrition in the Last Three Decades*. University of Papua New Guinea Press and Department of Agriculture and Livestock, Port Moresby. pp. 90–102.

Kambuou, R.N. (2005). Banana varieties tested for Sigatoka disease resistance under irrigated conditions in Papua New Guinea. *Papua New Guinea Journal of Agriculture, Forestry and Fisheries* 48(1–2):29–34.

Kanua, M.B. (1990). Farming systems research in the highlands of Papua New Guinea: results on introduced vegetable research. Occasional Paper No. 1/90. Department of Agriculture and Livestock, Aiyura.

Kanua, M.B. and D'Souza, E.J. (1985). Evaluation of introduced vegetable varieties on volcanic ash soils in the Southern Highlands of Papua New Guinea. Technical Report 85/13. Department of Primary Industry, Port Moresby.

Kanua, M.B., Freeman, M. and Kamis, P. (1993). Rural Development Series Handbook: No. 22. A guide to successful vegetable production in the highlands of Papua New Guinea. *Harvest* 15(1):15–30.

Kesavan, V. (1977). Potential and problems of vegetable production in lowland Papua New Guinea. In Enyi, B.A.C. and Varghese, T. (eds). *Agriculture in the Tropics*. Papers delivered at the tenth Waigani seminar. University of Papua New Guinea, Port Moresby. pp. 453–462.

Kesavan, V. and Stephenson, R.A. (1982). Genotypic and environmental effects for grain and tuber yield in winged bean. In Bourke, R.M. and Kesavan, V. (eds). *Proceedings of the Second Papua New Guinea Food Crops Conference: Part Two*. Department of Primary Industry, Port Moresby. pp. 359–366.

Khan, T.N., Erskine, W. and Kesavan, V. (1976). Improvement of food legumes at the University of Papua New Guinea. In Wilson, K. and Bourke, R.M. (eds). *1975 Papua New Guinea Food Crops Conference Proceedings*. Department of Primary Industry, Port Moresby. pp. 127–136.

Kimber, A.J. (1977a). Assessment of grain soybean yield in the highlands of Papua New Guinea. In Enyi, B.A.C. and Varghese, T. (eds). *Agriculture in the Tropics*. Papers delivered at the tenth Waigani seminar. University of Papua New Guinea, Port Moresby. pp. 420–432.

Kimber, A.J. (1977b). Cereal grains in the highlands. Technical Bulletin No. 2. Department of Primary Industry, Aiyura.

King, G.A. (1986a). A survey of Central Province yam (*Dioscorea* sp.) gardens. Technical Report 86/7. Department of Primary Industry, Port Moresby.

King, G.A. (1986b). Review of research on bananas, cassava and yam in Papua New Guinea. Technical Report 86/6. Department of Primary Industry, Port Moresby.

King, G.A. (1987). Maize variety recommendation for lowland Papua New Guinea. *Harvest* 12(4):5–7.

King, G.A. (1988). Cassava variety evaluation at Laloki, Papua New Guinea. *Science in New Guinea* 14(3):118–123.

King, G.A. (1989). A systems approach to improving subsistence yam production in Central Province, Papua New Guinea. *Agricultural Systems* 31:157–168.

King, G.A., Banag, J., Kambuou, R.N., Ovia, K., Ovia, A., Heywood, P.F. and Hide, R.L. (1989). *Production of bananas by the Amele in Madang Province and in the Vanapa River and Kabadi areas of Central Province, Papua New Guinea*. Research Bulletin No. 44. Department of Agriculture and Livestock, Port Moresby.

King, G.A. and Bull, P.B. (1989). Maize variety evaluation at Laloki. Technical Report 89/3. Department of Agriculture and Livestock, Port Moresby.

Lin, A.A.C. (1993). Trials on the production potential of Taiwanese rice varieties in Papua New Guinea. *Harvest* 15(1):31–35.

Newcombe, K., Holmes, E.B. and Paivoke, A. (1980). Palm energy: alcohol fuel from the sago and nipa palms of Papua New Guinea. The development plans. Report No. 6/80. Department of Minerals and Energy, Port Moresby.

Nukundj, R., Sivasupiramaniam, S. and Mondurafa, R. (1997). Response of capsicum to application of sheep manure. *Harvest* 19(1–2):39–43.

Pitt, A.J. and Yandanai, H.R. (eds) (1988). Working Papers of the 1987 Potato/Sweet Potato Workshop. Department of Agriculture and Livestock, Port Moresby.

Powell, J.M. (1982). Traditional vegetables in Papua New Guinea: retrospect and prospect. In Bourke, R.M. and Kesavan, V. (eds). *Proceedings of the Second Papua New Guinea Food Crops Conference: Part One*. Department of Primary Industry, Port Moresby. pp. 64–77.

Quin, F.M. (1984). Report on yam research (*Dioscorea* spp.) 1981–84. East Sepik Rural Development Project: agricultural research sub-project. Unpublished paper. Wewak.

Rachaputi, R.C.N., Wright, G., Kuniata, L. and Ramakrishna, A. (eds) (2006). *Improving Yield and Economic Viability of Peanut Production in Papua New Guinea and Australia*. ACIAR Proceedings No. 122. Australian Centre for International Agricultural Research, Canberra.

Risimeri, J.B. (2001). Yams and food security in the lowlands of PNG. In Bourke, R.M., Allen, M.G. and Salisbury, J.G. (eds). *Food Security for Papua New Guinea. Proceedings of the Papua New Guinea Food and Nutrition 2000 Conference*. ACIAR Proceedings No. 99. Australian Centre for International Agricultural Research, Canberra. pp. 768–774.

Risimeri, J.B., Gendua, P.A. and Maima, J.B. (2001). The status of introduced white yam in PNG. In Bourke, R.M., Allen, M.G. and Salisbury, J.G. (eds). *Food Security for Papua New Guinea. Proceedings of the Papua New Guinea Food and Nutrition 2000 Conference*. ACIAR Proceedings No. 99. Australian Centre for International Agricultural Research, Canberra. pp. 783–787.

Rose, C.J. (1980). Optimum replanting stage for two varieties of pit-pit (*Setaria palmifolia*) in the highlands of Papua New Guinea. *Papua New Guinea Agricultural Journal* 31(1–4):23–29.

Rose, C.J. (1982). Preliminary observations on the pandanus nut (*Pandanus jiulianettii* Martelli). In Bourke, R.M. and Kesavan, V. (eds). *Proceedings of the Second Papua New Guinea Food Crops Conference: Part One*. Department of Primary Industry, Port Moresby. pp. 160–167.

Sajjad, M.S. (1996). KK 15-36-C and Ayung, two modern high yielding rice (*Oryza sativa* L.) varieties for irrigated lowland ecosystem of Papua New Guinea. *Harvest* 18(1&2):17–19.

Singh, D., Guaf, J., Okpul, T., Wiles, G. and Hunter, D. (2006). Taro (*Colocasia esculenta*) variety release recommendations for Papua New Guinea based on multi-location trials. *New Zealand Journal of Crop and Horticultural Science* 34:163–171.

Sowei, J. (2006). Upland sago cultivation in the Torricelli foothills: a case study of the process of agriculture change of the Beli and Urat people. Unpublished paper. National Research Institute, Port Moresby.

Sowei, J.W. and Osilis, P. (1995). Horticulture Note No. 31. Delayed weeding reduces aibika yield. *Harvest* 17(1&2):6–8.

Sowei, J.W., Nera, S. and Osilis, P. (1996). Horticulture Note No. 30. Chicken manure increases aibika yield. *Harvest* 18(1&2):14–16.

Sutherland, J. (1984/1985). A short note on yields of forty one varieties of aibika. *Science in New Guinea* 11(1):39–42.

Tarepe, T.N. (1982). Naranjilla: a new fruit for the highlands. *Harvest* 8(4):183–185.

Tarepe, T. and Bourke, R.M. (1982). Fruit crops in the Papua New Guinea highlands. In Bourke, R.M. and Kesavan, V. (eds). *Proceedings of the Second Papua New Guinea Food Crops Conference: Part One.* Department of Primary Industry, Port Moresby. pp. 86–100.

Townsend, P.K. (1982). A review of recent and needed sago research. Sago research in Papua New Guinea: IASER Discussion Paper No. 44. Institute of Applied Social and Economic Research, Port Moresby.

Vance, P.N. (1976). Maize in the Markham Valley. In Wilson, K. and Bourke, R.M. (eds). *1975 Papua New Guinea Food Crops Conference Proceedings.* Department of Primary Industry, Port Moresby. pp. 215–220.

Vance, P.N. (1981). *Agronomic studies on grain sorghum in Papua New Guinea.* Research Bulletin No. 30. Department of Primary Industry, Port Moresby.

Vance, P.N. (1987a). *Agronomic studies on maize in Papua New Guinea.* Research Bulletin No. 39. Department of Agriculture and Livestock, Port Moresby.

Vance, P.N. (1987b). *Agronomic studies on peanuts in Papua New Guinea.* Research Bulletin No. 42. Department of Primary Industry, Port Moresby.

Vance, P.N. and Li, C.S. (1971). Grain sorghum in the Markham Valley. *Harvest* 1(4):119–123.

Westwood, V. and Kesavan, V. (1982). Traditional leafy vegetables of Papua New Guinea: aibika (*Hibiscus manihot* (L)). In Bourke, R.M. and Kesavan, V. (eds). *Proceedings of the Second Papua New Guinea Food Crops Conference: Part Two.* Department of Primary Industry, Port Moresby. pp. 391–395.

Wiles, G.C. (1997). Papua New Guinea: a potential exporter of tropical fruit to Asian markets. In Vijaysegaran, S., Pauziah, M., Mohamed, M.S. and Ahmad Tarmizi, S. (eds). *Proceedings of the International Conference on Tropical Fruits. Volume III Supplementary.* Malaysian Agricultural Research and Development Institute (MARDI), Serdang, Selangor. pp. 47–60.

Wiles, G. (2001a). Bulb onions: the challenge of reducing dependence on imported onions. In Bourke, R.M., Allen, M.G. and Salisbury, J.G. (eds). *Food Security for Papua New Guinea. Proceedings of the Papua New Guinea Food and Nutrition 2000 Conference.* ACIAR Proceedings No. 99. Australian Centre for International Agricultural Research, Canberra. pp. 860–869.

Wiles, G. (2001b). Potato production in PNG: the contribution of research to alleviating constraints to potato production. In Bourke, R.M., Allen, M.G. and Salisbury, J.G. (eds). *Food Security for Papua New Guinea. Proceedings of the Papua New Guinea Food and Nutrition 2000 Conference.* ACIAR Proceedings No. 99. Australian Centre for International Agricultural Research, Canberra. pp. 792–806.

Wiles, G. and Mwayawa, P. (2001). University of Vudal workshop on commercialisation of vegetable production. In Bourke, R.M., Allen, M.G. and Salisbury, J.G. (eds). *Food Security for Papua New Guinea. Proceedings of the Papua New Guinea Food and Nutrition 2000 Conference.* ACIAR Proceedings No. 99. Australian Centre for International Agricultural Research, Canberra. pp. 588–596.

Wohuinangu, J.S. and Joo, M.K. (1982). An overview of rice research in Papua New Guinea. In Bourke, R.M. and Kesavan, V. (eds). *Proceedings of the Second Papua New Guinea Food Crops Conference: Part Two.* Department of Primary Industry, Port Moresby. pp. 396–407.

Woodhouse, S. (1991). Asian fruit in Papua New Guinea. In Bala, A.A. (ed). *Proceedings of the Papua New Guinea First International Fruit Conference.* Department of Agriculture and Livestock and PNG Fruit and Vegetable Advisory Board, Port Moresby. pp. 161–168.

Woodhouse, S. (1992). Horticulture Note No. 16. Mango field management: planting and care of young trees. *Harvest* 14(1–2):31–32.

2.5 Rice production

Rice is the most controversial agricultural crop in PNG. Rice imports have been in the range 120 000–208 000 tonnes per year between 1990 and 2005, mostly from Australia (see Section 2.7). In contrast, domestic rice production has been in the range 60–2200 tonnes over the period 1962 to 2000 (Figure 2.5.1, Table A2.5.1), and has averaged about 400 tonnes per year since 1980. This is around 0.25% of rice consumed per year in PNG in recent years. Claims are made for significant local production from time to time, but these are political statements rather than realistic estimates. Production in 2006 was estimated as 300–800 tonnes.

PNG leaders accuse the former Australian colonial administration of discouraging domestic rice production in PNG in order to protect an important export industry in Australia. Since Independence in 1975, plans for a domestic, import-replacement rice industry have been a feature of every government white paper on agriculture. Yet, since 1977, domestic rice production has never exceeded 1% of the amount of rice imported.

Rice growing in PNG

Rice has been grown in many parts of PNG. Before 1900, rice was grown, mainly by Catholic missionaries, in the Bereina area and on Yule Island in Central Province, at Aitape in Sandaun Province, and probably at other places. In inland Finschhafen in the Sarawaget Mountains of Morobe Province, Lutheran missionaries introduced rice growing in the early 1900s and it is today the only place in PNG where it has become a 'traditional' crop.

After 1918, rice growing in Papua (the Southern Region) was a compulsory village activity under the Native Plantation Ordinance (1918). The Papuan colonial administration sent an officer to India, brought Indian instructors to Papua and established a 'fully equipped rice mill' in an attempt to 'make the Territory self-supporting in rice'. For example, the colonial administration promoted village rice cultivation in the Cape Vogel area of Milne Bay Province in 1923–1926. When this initiative failed, it was concluded that rice growing was too labour intensive and the environmental conditions were unsuitable. Cassava was then promoted and successfully adopted in the Cape Vogel area. In New Guinea, rice growing was promoted at Talasea in West New Britain and in East New Britain. On the Gazelle Peninsula, enough rice was produced for a steam-driven mill to be imported. Rice was grown on Umboi Island in Morobe Province until 1941.

During World War II, Japanese troops grew rice on the Gazelle Peninsula in East New Britain, and on New Ireland, but appear to have concluded that sweet potato was a more productive and reliable crop. For example, in Sandaun and East Sepik provinces (then one province), Japanese troops grew sweet potato and Chinese taro, rather than rice, in an attempt to feed themselves after they were cut off from Japan. In

Papua, the Australian military administration made rice growing compulsory at Bereina and introduced a mechanical harvester.

After the war, in 1947, the New Guinea Nutrition Survey Expedition studied village food production in five locations and concluded that the 'wider cultivation of crops such as peanuts and rice, which can be easily stored and transported, would help eliminate regional and seasonal food shortages'. W. Cottrell-Dormer, who was the agricultural officer on this survey, later became the Director of Agriculture in PNG. He was so convinced that rice could be produced satisfactorily at Bereina that he resigned his post as director to personally supervise the Bereina project. At Bereina, machinery was introduced and tractors were used to cultivate relatively large areas.

In the Sepik provinces, in particular around Maprik and Nuku, villagers began growing rice within the traditional shifting cultivation system in the 1950s as part of an indigenous rural development movement led by Pita Simogun at Dagua. Simogun had visited Australia during the war and observed Australian farmer rice-growing cooperatives in the Riverina.

Similar movements occurred in the Markham Valley and in Oro Province. Some of the villagers involved in these movements brought cargo cult elements into the growing of rice.[1] The colonial agricultural extension service attempted to respond to this movement with the introduction of Rural Progress Societies, hand-powered hullers and subsidised purchases. During this period, village rice production was also promoted by government extension services in Morobe (at Finschhafen), Milne Bay, New Britain, Bougainville, Gulf and Central (at Bereina and Kupiano) provinces.

Rice growing since Independence

Domestic rice production in PNG has fluctuated from year to year but has been less than 1500 tonnes per year since 1975 (Figure 2.5.1). Most production has been unirrigated. Rice has continued to be grown spasmodically at Bereina in Central Province,

1 Cargo cults are movements in which it is believed that economic development and political power can be achieved through supernatural means.

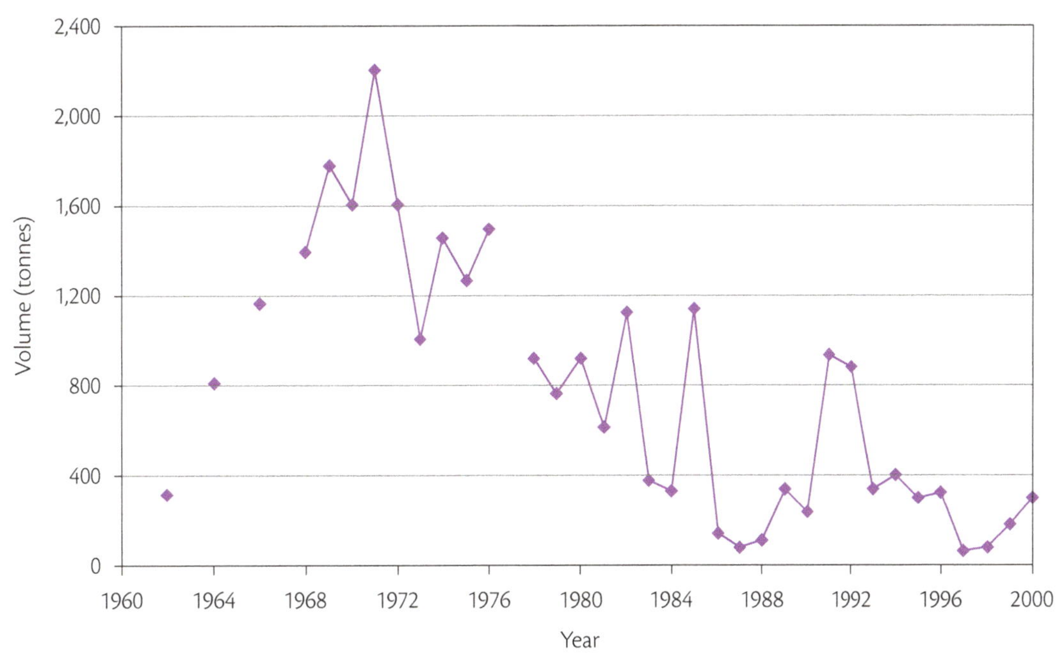

Figure 2.5.1 Estimated rice production, 1962–2000. **Note:** Production figures were not available for 1963, 1965, 1967 and 1977. Sources: 1962–1976: Hale (c. 1978); 1978–1990: DAL (1992:51); 1991–2000: Blakeney and Clough (2001).

as various aid and agricultural investment projects have attempted to make production there sustainable. In the late 1970s the East Sepik Rural Development Project, funded by the Asian Development Bank, made a large commitment to upgrading rice growing and increasing production to 4000 tonnes. However, production in East Sepik Province had almost ceased by 1987.

Small irrigated rice projects have been undertaken near Rabaul using Japanese aid; at Gabmazung near Nadzab in the Markham Valley by the Lutheran Mission; at Bubia with Taiwanese aid; at Cleanwater Creek in the Markham Valley by Trukai Rice; at Erap Research Station, also in the Markham Valley, by DAL; and at Bau near Madang by a Philippines non-government organisation. Rice was grown in Bougainville Province during the civil war (early to mid 1990s), but by 2002 rice growing there had virtually stopped.

From about 2000, production increased in some locations, including in parts of Central, Oro, Morobe, Madang, East Sepik, Eastern Highlands and Simbu provinces. This was in response to the rapid rise in the price of imported rice (Figure 4.3.3). The peak of the recent expansion in rice planting was in about 2001–2003, but production appears to have declined since then. For example, rice production in Madang Province was about 80 tonnes in 2003, 60 tonnes in 2004 and 40 tonnes in 2005. There was little rice being grown in the highlands by 2005. The Trukai Rice depot at Erap in the Markham Valley was able to purchase only 4 tonnes of locally grown rice in 2004 and 7 tonnes in 2005.

Thus locally grown rice remains a minor supplement to the traditional diet in a limited number of locations. At the national level, domestic rice production is still only a small proportion of rice imports and the level of production is a tiny fraction of that of the root crops, sago and banana (see Section 2.2).

Research on rice in PNG

Contrary to assertions that Australia discouraged PNG rice growing, more agronomic field trials have been carried out on rice than on any other crop (Table 2.5.1). Furthermore, many of these trials were done before Independence in 1975. Of the total number of agronomic field trials conducted between 1928 and 1978, 19% were on rice, compared to 11% on sweet potato, the most important food crop for about two-thirds of rural Papua New Guineans. A significant amount of research has also been conducted on rice since 1978.

Table 2.5.1 Agronomic field trials on food crops in PNG, 1928–1978

Crop	Number of trials	Proportion of total trials (%)
Energy and staple crops		
Banana	8	0.6
Cassava	6	0.5
Irish potato	66	5.4
Sugar cane	24	2.0
Sweet potato	136	11.1
Taro	30	2.4
Yam	11	0.9
Other energy crops	5	0.4
Total energy and staple crops	286	23.3
Grain crops		
Buckwheat	1	0.1
Corn (maize)	97	7.9
Rice	234	19.0
Wheat	17	1.4
Total grain crops	349	28.4
Other crops		
Farming systems	30	2.4
Grain legumes	245	20.0
Fruit and nut crops	65	5.3
Vegetables, introduced	237	19.3
Vegetables, traditional	16	1.3
Total other crops	593	48.3
Total trials	**1228**	

Source: Bourke (1982:7–8).

Why rice production has not become sustainable in PNG

A great deal of evidence exists that the colonial administrations of Papua and New Guinea made strenuous efforts to grow rice in PNG in order to offset the costs of imported rice. Given the continued enthusiasm by political leaders and administrators to replace imported rice with domestically produced rice, it is important to understand why rice production has not yet become a sustainable rural industry in PNG.

At least seven detailed investigations on aspects of growing rice in PNG have been undertaken since 1950, a number sponsored by non-Australian-based agencies. A summary of their findings suggests there are three main interrelated reasons why rice has not become a sustainable industry in PNG: these are to do with the environment, with cost efficiencies, and with returns to labour.

Environment

Upland rice has been grown in many parts of PNG, but yields are generally low at around 1000–1500 kg/ha.

Rainfall is too unreliable in some locations in PNG for perennial, unirrigated rice growing. The variation in rainfall from year to year, within the year and in the regularity of the beginning of the wet season is not reliable enough to grow large areas of unirrigated rice (see Section 1.5). This is a major reason why, for example, rice growing has failed to become sustainable at Bereina in Central Province.

Where irrigated rice has been grown, pests, weeds and disease have severely reduced yields. Pests and diseases are not a major problem where fields are shifted every year. Soils in many areas have poor water-holding capacity and are thus unsuitable for irrigated rice.

Economics

The high capital costs of establishing irrigated paddy fields and high production costs per tonne are a severe constraint to the development of a PNG rice industry. The main rice-producing countries of the world have comparatively lower production costs. A number of studies show that the costs of establishing large enough areas of irrigated rice to replace imports would severely distort the PNG economy, would require large subsidies and would result in a substantial increase in the retail cost of rice within PNG. Trukai Industries Limited, the main importer of rice into PNG from Australia, has been growing irrigated rice experimentally in the Markham Valley since 1998, but has been unable to achieve economic yields because of pests, weeds and soil problems.

Until recently, imported rice has been a relatively cheap food. For example, up to 1999 rice gave better value for money than purchased sweet potato, banana or Irish potato in Lae. This position changed with the decline in value of the kina, but taking into account the ease of transporting and cooking rice, it remains a competitive food for urban people in PNG.

Labour

The most important reason that rice cultivation has not become significant in PNG is related to returns on people's labour. Returns to labour are higher in the production of root crops than in rice, both in terms of yield and food energy produced per hour worked (see Section 5.20). Returns from growing coffee or cocoa are also higher than for growing rice in cash income per hour worked. Therefore, after experimenting with growing rice, many villagers decide they are better off growing root crops and export cash crops such as coffee or cocoa. The one place where rice growing has become 'traditional' is in the mountains inland of Finschhafen in Morobe Province, where access is difficult, imported rice is expensive and coffee is costly to market.

Compared with other crops, the cash returns to labour from growing rice for sale are significantly less than for cocoa, oil palm, vanilla, Robusta coffee and sweet potato. Many PNG villagers believe the returns to copra and rubber are too low to make harvesting and selling them worthwhile, so it is not surprising they do not participate in rice growing.

Another reason that rice production has not expanded in PNG is that it does not fit easily with village culture. This is because when a rice crop is ready for harvest, there is a relatively short period when harvest must occur. Unlike the export tree crops or root crops, delays in harvest can result in significant yield loss. Such delays are not uncommon in village communities because of other demands on villagers' time.

Over the last 20 years a number of economists have concluded that PNG is better off to import cheap rice and to export high quality palm oil, coffee and cocoa, than to try to establish a domestic, import-replacement, rice industry. On the basis of these economic analyses, it is unlikely that international aid agencies will provide funds to PNG to establish a rice industry. That does not mean village smallholders should be discouraged from growing rice. But it does mean that import-replacement production levels are unlikely in the foreseeable future.

Sources

Allen, B.J. (1976). Information flow and innovation diffusion in the East Sepik district, Papua New Guinea. PhD thesis. The Australian National University, Canberra.

Anon. (2005). Where is the locally grown rice crop? *Prime* (PNG Resources, Industry, Mining and Energy) (18):36.

Blakeney, M. and Clough, R. (2001). An assessment of grain production and imports in PNG, 1975–2000. In Bourke, R.M., Allen, M.G. and Salisbury, J.G. (eds). *Food Security for Papua New Guinea. Proceedings of the Papua New Guinea Food and Nutrition 2000 Conference.* ACIAR Proceedings No. 99. Australian Centre for International Agricultural Research, Canberra. pp. 23–29.

Bourke, R.M. (1982). Agronomic field trials on food crops in Papua New Guinea 1928 to 1978. Technical Report 82/3. Department of Primary Industry, Port Moresby.

Christian, C.S. (1958). Report on an investigation of rice production in the territories of Papua and New Guinea for the Hon. Minister for Territories, Mr P. Hasluck. Department of Territories, Canberra.

DAL (Department of Agriculture and Livestock) (1992). *Farm Budgeting Manual 1991*. First edition. Policy, Programming and Budgeting Division, Department of Agriculture and Livestock, Port Moresby.

Gibson, J. (1992). Rice self-sufficiency and the terms of trade (why rice is a good thing to import). Policy Working Paper No. 2. Programming, Planning and Budgeting Division, Department of Agriculture and Livestock, Port Moresby.

Gibson, J. (1993). Trends in rice consumption. Policy Working Paper No. 7. Programming, Planning and Budgeting Division, Department of Agriculture and Livestock, Port Moresby.

Hale, P.R. (c. 1978). Rice. In Densley, D.R.J. (ed). *Agriculture in the Economy: A Series of Review Papers*. Volume 3. Department of Primary Industry, Konedobu. pp. 1–53.

Jeffreys, F.J. (1974). The Mekeo rice project. MA thesis. University of Papua New Guinea, Port Moresby.

Leentjes, P. and Emmery, P. (1973). Rice production in Papua New Guinea: an interim report. Department of Agriculture, Stock and Fisheries, Port Moresby.

McKillop, B. (1976). A history of agricultural extension in Papua New Guinea. In *A Brief History of Agricultural Extension and Education in Papua New Guinea*. Extension Bulletin No. 10. Department of Primary Industry, Port Moresby. pp. 5–44.

Mills, F. (ed.) (2002). *The Book of Rice in Papua New Guinea*. Mills Media Limited, Port Moresby.

Mogina, J. (2002). Changing knowledge of plants in transitional societies at Milne Bay, Papua New Guinea. PhD thesis. The Australian National University, Canberra.

Poggendorff, W. (1953). Rice production in Papua New Guinea. Department of Territories, Canberra.

Shand, R.T. and Straatmans, W. (1974). *Transition From Subsistence: Cash Crop Development in Papua New Guinea*. New Guinea Research Bulletin No. 54. New Guinea Research Unit, The Australian National University, Port Moresby and Canberra.

ten Have, H. (1974). Possibilities for increased rice production in Papua New Guinea. Department of Agriculture, Stock and Fisheries, Port Moresby.

Watson, R. (1976). Rice production potential in Papua New Guinea. Asian Development Bank, Port Moresby.

2.6 Animal production

The most important domestic animals in PNG are pigs, chickens, cattle, sheep, goats, ducks and rabbits. Horses and swamp buffalo are used as working animals to a small degree. Pigs and chickens were introduced into PNG about 3500 years ago (see History of agriculture). All other species were introduced from the late 1800s by colonial administrations, settlers and missionaries. According to the 2000 census, some 400 000 households or 47% of total rural households are engaged in some kind of livestock production. About 200 000 households own poultry, but systematic efforts to modernise commercial production of meat and eggs have only been made in the last 50 years and only the last 25 years for broiler chickens.

Pigs

The most notable feature of the PNG pig industry is the 1.8 million village pigs, mainly indigenous breeds, that produce some 27 000 tonnes of pig meat annually (Tables 2.6.1, 2.6.3). These pigs are owned by 60% of all households (1990 census). In the highlands provinces, 77% of the population claimed pig ownership. This figure was more than 80% if Eastern Highlands Province was excluded and was as high as 89% in Enga Province.

The average number of pigs per person in various regions of PNG is summarised in Table 2.6.2. It is clear that there are many more pigs per person in the highlands (above 1200 m altitude) than in the lowlands (sea level to 600 m altitude), with more than one pig per person in the highlands. The ratio of pigs to persons is also higher in the intermediate altitude class (600–1200 m) than in the lowlands. The available data suggest that the ratio is higher in inland lowland locations (0.33:1) than in coastal locations (0.25:1), although the difference is not significant. The data on which Table 2.6.2 is based were recorded between the 1930s and 1990s, but about 75% of the observations were made in the 1960s, 1970s and 1980s. There are indications that the number of pigs per person has declined in recent decades, at least in parts of the highlands. It is possible that there has been an overall decline in the number of pigs per person in PNG and the total number of pigs now is less than these figures suggest.

Commercial pigs are slaughtered in four registered abattoirs. Of the 29 600 pigs slaughtered in 2005, the Lae abattoir accounted for 45% and Abunaka, a private pig farm near Lae, 38%. With an average carcase weight of 48 kg, this gives a commercial annual production of 1420 tonnes. However, this does not allow for the very large number of live sales and there has been a steady but slow increase since the 1000–1200 tonne estimates of the 1990s. There are currently about 32 000 pigs on commercial farms and total production is estimated as 2300 tonnes. Sow numbers on large farms may be declining but village production is steadily increasing.

Table 2.6.1 Pig industry characteristics

Type of holding	Herd size	Number of herds	Estimated number of pigs	Trends	Breeds
Smallholder, traditional	1–20	360,000	1,800,000	Static; may be increasing with human population	Native
Smallholder, penned	1–3	2,000	4,000	Growing rapidly	Native
Smallholder, commercial	10–100	100 (including at prisons and high schools)	6,000	Growing steadily	Modern commercial
Middle-sized commercial	100–500	4 (3 institutional)	2,000	Static	Modern commercial
Large-scale commercial	>500	7	20,000 (2500 sows)	Declining slowly	Modern commercial

Source: Updated from Quartermain and Kohun (2002).

Table 2.6.2 Average number of pigs per person in three altitude classes

Altitude class	Ratio of pigs to people	Number of observations
Lowlands (sea level to 600 m)	0.3:1	37
Intermediate (600–1200 m)	0.5:1	33
Highlands (1200–2800 m)	1.2:1	94

Source: Generated from data presented in Table 5.8 of Hide (2003:39–47). Differences are statistically significant.

Poultry

There are some 1.5 million village or household scavenging chickens owned by 27% of households (1990 census). About 50% of households in Milne Bay, Madang and East New Britain provinces kept chickens, but chicken-keeping was not adopted in the highlands until the 1950s. Commercial poultry production only began in the 1970s, fostered by deliberate government policy and protection from imports. Domestic production has grown from around 4000 tonnes in 1980 to 24 000 tonnes in 2005 (Table 2.6.3). Two large companies, one operating through contracted outgrowers, supply the bulk of the frozen chicken market.

A feature of commercial production is the large number of smallholder farmers, perhaps up to 15 000, who purchase day-old hybrid broiler chicks in lots of 50 or 100 from one of the three hatcheries and sell them when grown, mainly as live birds in local markets. Niugini Tablebirds imports fertile eggs of hybrid broiler grandparent stock from which the final chicks are derived through parental crosses, while Zenag Chicken imports fertile eggs of parent stock. The Christian Leaders Training College (CLTC) at Banz in Western Highlands Province buys parent stock from Niugini Tablebirds and produces 30 000–45 000 day-old chicks per week for highland farmers. Total production of day-old broiler chicks from the three hatcheries is about 400 000 per week.

Commercial egg production, dominated by Zenag Chicken, has also grown from essentially nothing in the 1970s to estimated current production levels of 45 million eggs. About 200–300 day-old layer chicks per week are sold to villagers by the Zenag and CLTC hatcheries.

Muscovy duck ownership is increasing, mainly for household meat and egg production in scavenging systems. There are no estimates of total numbers

Table 2.6.3 Estimated livestock numbers and meat production, 2005

Livestock	Component	Number of animals	Offtake[a] (%)	Dressed carcase weight (kg)	Production
Pigs	Village	1.8 million	50	30	27,000 t
	Commercial	32,000		48	2,300 t
Cattle	Large-scale ranch	63,500	15	200	1,900 t
	Smallholder	16,500	15	200	500 t
Sheep	Smallholder	15,000	30	12	54 t
Goats	Smallholder	25,000	30	12	90 t
Chickens	Commercial broilers				17,000 t (frozen)
	Broilers, live sales[b]			1	7,000 t (carcase)
	Commercial layers	161,000			45 million eggs
	Village[c]	1.5 million		0.8	1,850 t (carcase) 6 million eggs
Rabbits	Village	30,000		1.4	168 t

[a] 'Offtake' refers to the number of animals in a herd that are removed for sale or slaughter in a given time period, usually a year. It assumes the herd is not growing, so the offtake is equal to the potential increase over the period (excluding deaths) if all animals are kept. This is then expressed as a percentage of the base herd, not of the total herd.

[b] The live bird broiler production is based on hatchery sales of 149 000 day-old chicks per week, a 1 kg carcase weight and 12% losses.

[c] For village poultry it is assumed that a hen produces 70 eggs per year, of which 30 are available for consumption. Incubation of the other 40 eggs results in 12 surviving chicks to grow into replacement pullets or be consumed. A standing flock has 66 pullets for every 100 hens. Hence a 1.5 million bird flock has 204 000 hens, 135 000 pullets and 1 156 500 young birds being raised to eat. Actual meat bird output from this flock is 2.3 million birds (1.15 million young birds twice a year).

Sources: Industry sources for cattle, commercial poultry and commercial pigs; author estimates for village production.

but there could be more than 10 000 duck-owning families. The National Agricultural Research Institute has a flock of layer ducks and a few farmers are raising domestic pigeons or Japanese quail.

Despite various attempts by the Australian Administration and by the PNG Government to promote layer ducks, domestic pigeons, Japanese quail, geese, turkeys and guinea fowl, there has been no successful development of commercial production.

Cattle

Beef cattle numbers have been static for the last 20 years, averaging around 80 000 head. During the mid to late 1970s a significant effort was made by government to create a village-based cattle industry. This was largely unsuccessful and most 'cattle projects' had failed by the early 1980s. Cattle numbers declined from a peak of 153 000 in 1976 to the current figure by 1991. However, numbers are now increasing again by about 2000 per year. Current industry estimates for numbers on large-scale ranches are 50 000 in the Markham and Ramu valleys, 4000 in West New Britain Province (Numondo Plantation), 2500 in East New Britain Province (Coconut Products Ltd), 6000 in Central Province and 1000 in New Ireland Province. The remaining 16 500 head are in a large number of small herds containing from one to several hundred animals, mainly in Morobe Province but also scattered throughout Western, East New Britain, East Sepik, Sandaun, Madang and the highlands provinces.

Four registered abattoirs slaughter cattle. Around 9700 animals were slaughtered in 2005; 38% in Lae and 45% by Ramu Agri-Industries Ltd. Possibly another 2000 head were slaughtered for local sales. A total of 12 000 head at an average carcase weight of 200 kg gives an annual production of 2400 tonnes (Table 2.6.3). In addition, there have been eight live cattle export shipments to Asia since 2002 totalling around 8000 head. About 1150 live animals were exported in 2005. There is a surplus of higher-priced beef cuts in PNG because the market demand is for cheaper cuts of meat. An economic solution to this problem is to export the better quality meat as part of live animal exports. It is also convenient to collect a large number of cattle from scattered smallholder herds and hold them at a central location (Trukai Industries Limited in the Markham Valley) until ready for shipment.

Only one small dairy farm in PNG produces fresh milk for sale and, while there have been others in the past, there have never been more than six. While milk consumption appears to be growing, local production is not competitive with production in temperate climates. There is little potential for expansion of milk production from dairy cows in PNG.

Sheep and goats

Sheep and goat numbers are small compared to cattle and pigs but are slowly increasing, especially in highland environments (Table 2.6.3). Only 2% of households claimed to own goats in 1990 and less than 1% owned sheep. The highest numbers of sheep owners were in Enga, followed by Simbu, Eastern Highlands and Morobe provinces. The highest numbers of goat owners were in Eastern Highlands and Simbu, followed by Enga Province. Sheep meat is very popular, driven by the availability of inexpensive imported sheep meat that is affordable by many people (see Section 2.9). Large-scale growth of sheep and goat ownership, especially in the highlands, has only occurred over the past 30 years. This is a consequence of deliberate government policy in the case of sheep, but goat numbers have increased without government encouragement. Goat numbers continue to grow and there is potential for household milk production from goats.

Minor species

Domestic rabbits were only introduced into PNG in 1993. It is estimated there are currently 2000 owners with a total of 30 000 animals. Assuming 6000 breeding does, 20 offspring per doe per year and a carcase weight of 1.4 kg, annual production is around 168 tonnes.

South-East Asian swamp buffalo were originally introduced for draft purposes. However, despite much effort over the years, they have not become popular. There may be 4000 animals in PNG, with 80% estimated to be feral. A few buffalo are used for transport, mainly in East New Britain and Madang provinces, where extension efforts were concentrated. Current efforts at using animals for transport are now focused on cattle.

Horses are used for stock work on cattle ranches and for recreation, but there are no estimates of numbers. Donkeys were once used for transportation, but there are none left today.

Estimating livestock numbers and production

There has never been a complete census of livestock in PNG. The best available estimates of the numbers of animals of the major species and total annual meat production in 2005 are given in Table 2.6.3. The data are industry estimates (cattle, commercial poultry, commercial pigs) or those of the author.

Numbers of village pigs and poultry and smallholder sheep and goats can be estimated using three different sets of information:

- A survey of indigenous agriculture conducted by the Australian Administration in 1961–1962 (see Section 6.5). The numbers of pigs and chickens per 100 persons in the surveyed villages can be extrapolated using rural population data from the 1980, 1990 or 2000 population censuses and assuming 1962 levels of ownership.

- Census questions. During each census, rural householders were asked if they owned pigs, poultry, sheep, goats or cattle. Although there are difficulties in using the census data, for example the problem of multiple ownership of animals, the numbers of animals can be estimated from the numbers of owners, aggregated on a provincial or regional basis, and assumed herd or flock sizes.
- A listing of the ratio of pigs to people observed at various locations in PNG (Table 2.6.2). Again, using census data, it is possible to estimate total pig numbers on a regional basis. Hide (2003) contains a comprehensive summary of all the available observations of pig numbers, distribution and ownership.

While these approaches rely upon population census data, all three methods of calculation for village pig, poultry, sheep and goat populations produce estimates within the same order of magnitude for each livestock species.

Stockfeed

Three Lae-based companies make and sell stockfeed for pigs and poultry. A mill previously serving the Port Moresby market is currently inoperative. The three Lae companies produced 52 200 tonnes of stockfeed in 2005. In addition, one major pig producer (Rumion) in the Markham Valley in Morobe Province makes its own feed from home-grown maize and produced 5800 tonnes in 2005. The Evangelical Brotherhood Church (EBC) produces feed for its own operations and sells a little around Lae.

Apart from Rumion and EBC, production is based on imported grain, mainly sorghum. Sorghum imports averaged 26 000 tonnes per year from 2000 to 2004. Around 10% of feed composition is wheat millrun, a local product derived from imported wheat after milling. Locally produced components of feeds include fish meal from PNG canneries and minor quantities of copra meal and limestone. Imports of fully prepared stockfeed are about 37 000 tonnes or 40% of total usage per year.

Efforts are being made to increase the use of local agricultural and fisheries by-products. Estimated PNG production of potential stockfeed ingredients is 33 000 tonnes of millrun, 21 000 tonnes of copra meal, 31 000 tonnes of oil palm kernel meal and 6000 tonnes of fish meal.

Meat consumption

Imports of meat are dominated by sheep meat and beef from Australia and New Zealand (see Section 2.9). This includes a wide range of products from whole sheep carcases through to cheaper lamb cuts to boned beef for canning. These imports rose from around 25 000 tonnes in 1980 to 60 000 tonnes in 1994, as beef and sheep meat replaced earlier imports of chicken, pork and tinned meat (Figure 2.9.1). Sheep meat imports subsequently declined to around 25 000 tonnes by 2001–2003. Around 90% of imported beef is used by the two commercial canneries to produce corned beef, luncheon meat and meat loaf products.

Total meat production is estimated as 58 000 tonnes. To this can be added 30 000 tonnes in imports, giving total meat consumption in PNG of 88 000 tonnes. Thus average meat consumption is about 15 kg/person/year. However, meat is an extremely variable commodity ranging from whole carcases or bone-in cuts through boned meat of variable fat content, to processed and tinned products. Available statistics are inadequate to enable all this to be expressed on a comparable basis. Regardless of the accuracy and composition of these estimates, consumption is very uneven both geographically and socially, with differences in cash incomes and the importance of meat in feasting and custom. Both rural and urban people will spend income on meat whenever possible. However, most of the traditional or village production of pig, poultry, sheep and goat meat never enters formal markets. For the 2000 census, 175 000 households claimed to sell meat in local markets or on roadsides.

Future prospects

Growth in the production and consumption of commercial pig and poultry products is possible only through increases in cash incomes. Cheaper or more accessible feeding options using a wider range and greater quantities of local feed ingredients will assist this growth. Growth potential is much greater for ruminants (cattle, sheep and goats), with continuing opportunities to increase beef and live cattle exports. The current cattle herd uses around 128 000 ha of grazing land. There is an estimated 445 000 ha of grassland that can be grazed by cattle and capable of supporting 300 000 head. In addition, there are perhaps 100 000 ha of land under tree crops that could be used for sheep and goat production if not by cattle. PNG has an advantage in being free from the major livestock diseases such as foot-and-mouth disease, swine fever and Newcastle disease. However, tuberculosis, brucellosis, fowlpox and anthrax are still present or threats and livestock production systems require improved management of parasitic diseases.

Sources

Askin, D., Grant, I., Kulimbao, D., Taul, P. and Thomas, A. (2001). The productivity of domestic rabbits for subsistence farming families of PNG. In Bourke, R.M., Allen, M.G. and Salisbury, J.G. (eds). *Food Security for Papua New Guinea. Proceedings of the Papua New Guinea Food and Nutrition 2000 Conference*. ACIAR Proceedings No. 99. Australian Centre for International Agricultural Research, Canberra. pp. 652-655.

Hide, R. (2003). *Pig Husbandry in New Guinea: A Literature Review and Bibliography*. ACIAR Monograph No. 108. Australian Centre for International Agricultural Research, Canberra.

Levett, M. and Bala, A. (1994). Agriculture in Papua New Guinea. In Sekhran, N. and Miller, S. (eds). *Papua New Guinea Country Study on Biological Diversity. A Report to the United Nations Environment Programme*. Department of Environment and Conservation and Africa Centre for Resources and Environment, Port Moresby. pp. 125–153.

Quartermain, A.R. (2000). Non-commercial poultry production in Papua New Guinea. *Asian-Australasian Journal of Animal Sciences* 13 (Supplement July 2000 C):304–307.

Quartermain, A.R. (2001). Potential for producing more meat from small-scale livestock production. In Bourke, R.M., Allen, M.G. and Salisbury, J.G. (eds). *Food Security for Papua New Guinea. Proceedings of the Papua New Guinea Food and Nutrition 2000 Conference*. ACIAR Proceedings No. 99. Australian Centre for International Agricultural Research, Canberra. pp. 625–629.

Quartermain, A.R. (2002). Conservation of domestic animal genetic resources in Papua New Guinea. NARI Technical Bulletin No. 4. National Agricultural Research Institute, Lae, Papua New Guinea.

Quartermain, A.R. (2002). Livestock numbers and production in Papua New Guinea. *NARI Nius. Newsletter of the Papua New Guinea National Agricultural Research Institute* 5(1).

Quartermain, A.R. and Kohun, P.J. (2002). Pig production in Papua New Guinea. In Australian Centre for International Agricultural Research (ed). *Priorities for Pig Research in Southeast Asia and the Pacific to 2010*. ACIAR Working Paper No. 53. Australian Centre for International Agricultural Research, Canberra. pp. 40–47.

Vincent, D. and Low, S. (2000). *A Review of Papua New Guinea's Red Meat Industry*. ACIAR Monograph No. 66. Australian Centre for International Agricultural Research, Canberra.

Walters, C.L. (1963). *Survey of Indigenous Agriculture and Ancillary Surveys, 1961–1962*. Territory of Papua and New Guinea Bureau of Statistics, Port Moresby.

2.7 Rice and wheat imports

Total imports

Rice and wheat have the largest share of any foods imported into PNG. These two cereals attract attention because of concerns that the population is becoming dependent on imported food and because of the largely unsuccessful efforts to produce these crops within PNG (see Section 2.5). The time series on import quantities presented here may help to give the debates about these foods a more factual basis.

Figure 2.7.1 and Table A2.7.1 show the annual quantities of rice and wheat imports from 1961 to 2005. Since 1990, rice imports have averaged 152 000 tonnes per year. This figure includes the exceptionally high import of 208 000 tonnes in 1997 when rice replaced food unable to be produced because of the drought and frosts of that year. But total rice imports fell after 1998, and averaged 151 000 tonnes per year between 1999 and 2005.[1]

Wheat imports have averaged 117 000 tonnes per year since 1990. The average quantity of wheat imported has been approximately 80% of the quantity of rice since 1990. However, these ratios change if the comparisons are made in terms of monetary value or food energy supplied.

Rice imports have increased fivefold since the 1960s, when they averaged about 30 000 tonnes per year. The rate of increase in wheat imports has been greater, with a nearly eightfold increase from an average of 15 000 tonnes in the 1960s. Thus, while wheat imports were about half the quantity of rice imports in the 1960s, they were about the same as rice imports for the first time in 2005.[2] This substitution away from rice and towards wheat is a common feature in many countries.

Initially there were no milling facilities in PNG, so bulk flour was imported. Flour imports began to be phased out from 1977 and by the early 1980s the imports were entirely of wheat, which was milled into flour in PNG. An adjustment is made in the figures presented, by using wheat-equivalent quantities in the years when flour was imported. Australia is the source for almost all wheat imported into PNG.

Trukai Industries Limited has historically dominated PNG's rice imports. A number of smaller companies entered the market after 1998 and by 2005 four were importing rice into PNG, but they were responsible

1 Rice imports increased to about 184000 tonnes in 2006, with increased sales in the highlands and in Port Moresby. This presumably reflects higher incomes for coffee producers and an improved national economy.

2 Wheat import figures cannot be compared directly with those for rice imports as virtually all rice imported into PNG is consumed by people. In contrast, wheat is converted into flour and in the process about 30% becomes unavailable for human consumption and is used for livestock feed. As well, some wheat (approximately 5%) is milled for livestock feed rather than to make flour; and some flour-based products, such as biscuits, are exported from PNG to other Pacific island countries.

for only about 3% of imports. Until 1998, the most common brand marketed by Trukai Industries was 'Trukai'. The devaluation of the kina in 1997 (see Part 4) led to a rapid increase in price from early 1998 (Figure 4.3.3) and caused a marked reduction in sales. Trukai Industries countered this by importing cheaper rice, marketed as 'Roots'. The new brand rapidly became popular and accounted for about 90% of sales within a few years of its introduction. 'Trukai' brand declined to less than 5% of sales by 2004, but had recovered to about 15% by 2005.

Until 2002 almost all of the rice imported into PNG came from Australia. A significant drought in southern Australia between 2002 and 2005 greatly reduced rice production there and subsequently less Australian rice was available for export. Over this period PNG imported rice from various countries including Vietnam, the United States, China, Egypt, India, Thailand and Australia. The preferred source for PNG rice imports is Australia because of the high quality product and predictability of shipping arrangements. But this depends on future water supply and rice production levels in Australia.

Per person imports

The most important reason for the increase in rice and wheat imports is the increase in PNG's population. A different understanding of the significance of rice and wheat imports can be gained if the average quantity imported per person per year is examined.[3]

Rice

Changes in average per person rice imports in PNG over the period 1961–2005, along with the trend growth rates for each decade, are shown in Figure 2.7.2. In the 1960s, rice imports averaged about 13 kg/person/year, while by the 1990s they were approximately 2.5 times higher, at 34 kg/person/year. Most of this increase occurred in the 1980s, when the imports jumped from the 1970s figure of 22 kg/person/year to 34 kg/person/year. The rapid increase in rice consumption (and therefore imports) between the 1960s and the 1980s resulted in forecasts that, by 2000, Papua New Guineans would be consuming an average of 50–60 kg of rice per person per year. But since 2000, the actual consumption level has been 27 kg/person/year; about half of what was forecast.

The slowing in the increase in rice imports since 1980 is also apparent in the estimates of trend growth rates. Between 1961 and 1970 per person imports increased by 6.3% per year; from 1971 to 1980 they increased by 5.2% per year; and from 1981 to 1990 by 1.7% per year. Rice consumption per person was static between 1991 and 2000 and fell at 0.5% per year from 2001 to 2005. In recent years average rice imports have fallen behind population growth. The per person level of imports has fallen from its peak of 43 kg in 1997 – when more imports were needed because of subsistence food shortages – to 27 kg per person in 2005. The last time per person imports were lower than they were in 2005 was nearly 30 years earlier, in 1977.

Rice consumption also fluctuates about its trend because of changes in prices and the exchange rate (see Section 4.1) and because of food shortages such as those in 1997 (see Section 1.6). Modelling of average per person rice imports suggests that rice consumption is becoming less sensitive to changes in average income. This change is consistent with patterns in other countries where rice consumption reaches a saturation point and may decline after that point is reached. If a saturation point has been reached in PNG, it is notable that it has occurred at a lower level of consumption than in other countries. The wider availability of other staple foods in PNG could explain this pattern. But the large difference in rice consumption between urban and rural areas (approximately 30 kg/person/year higher in urban areas; Tables 2.1.1, A2.1.1) suggests that the plateau in per person rice consumption also reflects a stagnation in rural incomes. Consistent with this explanation is the fact that when average prices for export tree crops are higher and rural households receive higher incomes, average rice consumption increases. A significant increase in the price of rice since 1997, caused by the fall in the value of the kina, is another explanation of the fall in rice consumption (Figure 4.3.3).

3 Average imports per person per year are calculated by dividing annual rice and wheat imports by the total population for that year. The population for each year is calculated from that in 1980 (3.01 million), that in 2000 (5.19 million) and the growth rate between 1980 and 2000 (2.76% per year).

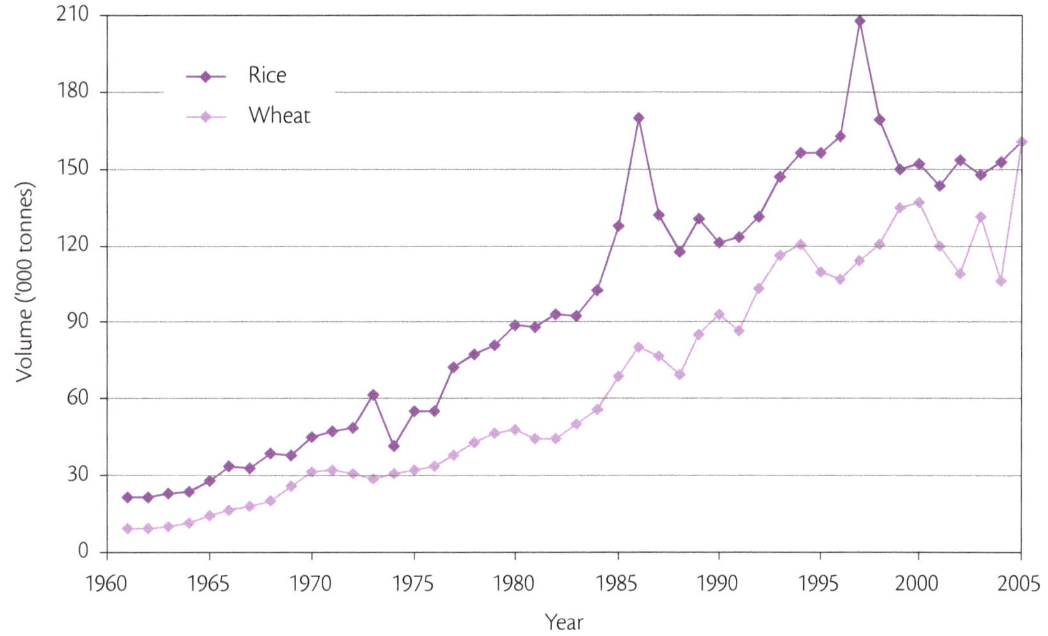

Figure 2.7.1 Volume of rice and wheat imports, 1961–2005. Sources: 1961–1999 Gibson (2001a: Appendix C); 2000–2005 Marketing Department, Trukai Industries Limited, Port Moresby, and Australian Government Wheat Export Authority.

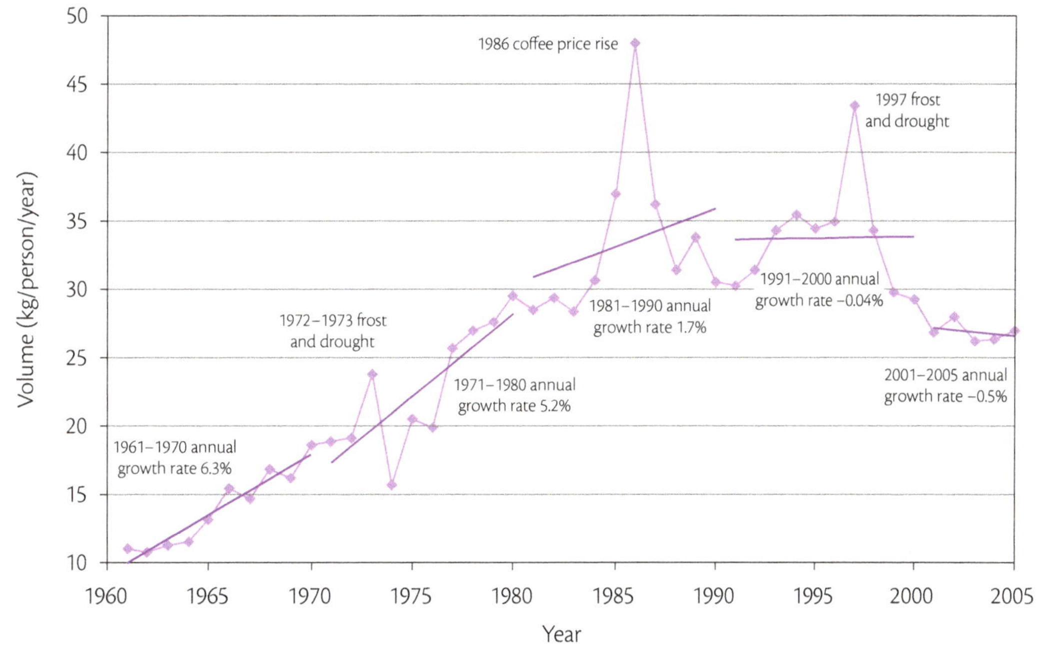

Figure 2.7.2 Average per person rice imports and growth trends by decade, 1961–2005. Sources: 1961–1999 Gibson (2001a: Appendix C); 2000–2005 Marketing Department, Trukai Industries Limited, Port Moresby, and Australian Government Wheat Export Authority.

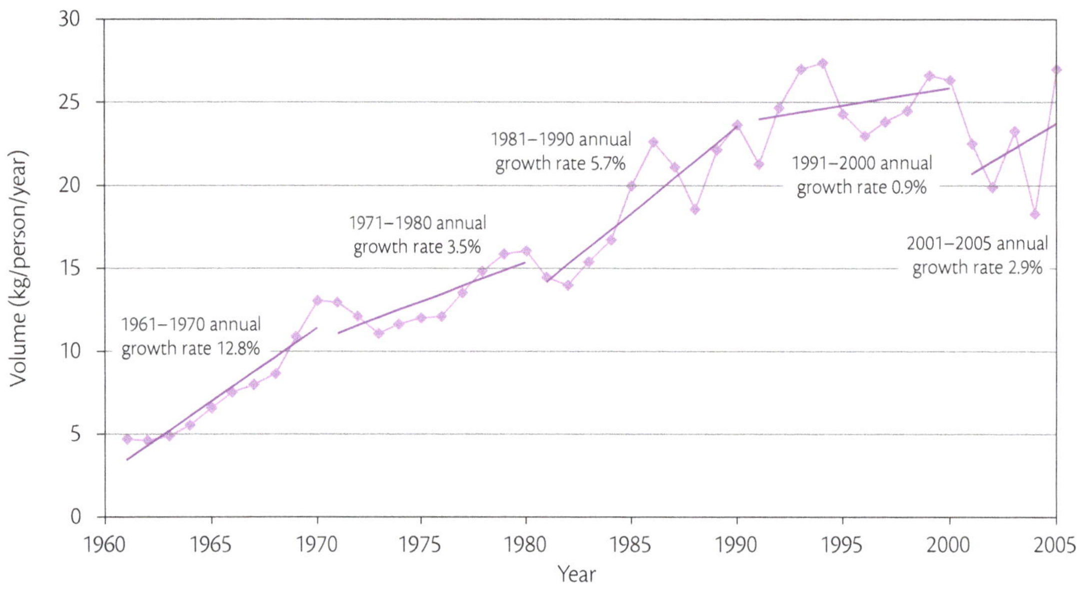

Figure 2.7.3 Average per person wheat imports and growth trends by decade, 1961–2005.
Sources: 1961–1999 Gibson (2001a: Appendix C); 2000–2005 Marketing Department, Trukai Industries Limited, Port Moresby, and Australian Government Wheat Export Authority.

The regional distribution of rice and flour sales also highlights the role of export tree crop prices. In recent years Lae and the Highlands Region have contributed the most to the falling sales of rice and flour, and these are the two sales areas where rural incomes are most affected by the price of coffee. A worldwide decrease in the price of coffee and the increase in the price of imported food in PNG reduced villagers' purchasing power. For example, in mid 1999, the sale of one kilogram of parchment coffee gave a villager sufficient money to buy three kilograms of rice in Goroka. By 2003, a kilogram of coffee could no longer buy a kilogram of rice. Coffee prices recovered to some extent by 2005 but, even so, sale of a kilogram of coffee still only gave enough money to buy about one kilogram of rice. The recovery in coffee prices was associated with an increase in rice sales in the highlands.

Wheat

From 1996 to 2005 wheat imports have been 24 kg/person/year (Figure 2.7.3). Like rice, the per person rate of growth in consumption of wheat products has slowed substantially over the last decade, falling from 5.7% per year in 1981–1990 to 0.9% per year in the period 1991–2000. Since 2001, per person wheat imports have increased at a rate of 2.9% per year. But the instability in previous trend growth rates of wheat consumption, which saw rapid growth in the 1960s and slower growth in the 1970s, means that it is harder to conclude that wheat consumption is also reaching a mature phase, like rice has.

Sources

Australian Government Wheat Export Authority website <http://www.wea.gov.au/>. Accessed July 2006.

Food and Agriculture Organization of the United Nations and PNG Department of Primary Industry (1986). Rice development policy: an analysis of the options for meeting future rice consumption requirements, Papua New Guinea. Mission report. Food and Agriculture Organization of the United Nations, Rome.

Gibson, J. (2001a). *Food security and food policy in Papua New Guinea*. INA Discussion Paper No. 83. Institute of National Affairs, Port Moresby.

Gibson, J. (2001b). Rice demand in Papua New Guinea. *Pacific Economic Bulletin* 16(2):27–35.

2.8 Fruit and vegetable imports

PNG imports fruit and vegetables mainly from Australia and New Zealand. The major items imported are Irish potato, onion, apple and, until recently, citrus. In contrast to cereal and meat imports, fruit and vegetables account for only a small proportion of the total value of imported food and are dwarfed by the scale of domestic fruit and vegetable production (see Sections 2.2, 3.1, 3.2 and 3.3). In recent years, PNG's imports of fruit and vegetables have been valued at around K15 million.

Fruit and vegetable imports from Australia and New Zealand averaged about 8000 tonnes per year in the period 1983–2003 (Figure 2.8.1, Table A2.8.1).[1] However, a distinct change in the trend in imports occurred around 1997. From 1983 until 1997 total imports averaged 8800 tonnes per year and did not keep up with population growth. From 1997 to 2003 total imports declined from 9200 tonnes to less than 6000 tonnes; a decrease in annual per person imports of 2 kg/person to 1 kg/person. This fall was prompted by the 1997 devaluation of the kina (see Part 4), which led to imported food becoming more expensive.

While Irish potato comprises more than one-quarter of the total quantity of fruit and vegetable imports, it is less important in terms of value (Figure 2.8.2). The shares of import expenditure on each of the main fruits and vegetables have been roughly constant over the past decade, with the exception of the decline in the value of citrus imports since the mid 1990s and the decline in the value of 'other' fruit and vegetables since the year 2000.

Sources

Australian Bureau of Statistics website <http://www.abs.gov.au/>.

Statistics New Zealand website <http://www.stats.govt.nz/default.htm>.

1 Statistics from the exporting countries are used (Table A2.8.1) because these data are more timely and reliable than PNG's import statistics (Table 2.1.2). The population numbers used to calculate the average per person imports in Figure 2.8.1 are the same as those used in Section 2.7 for per person rice and wheat imports.

Figure 2.8.1 Volume of fruit and vegetable imports from Australia and New Zealand, 1983–2003.
Sources: Australian Bureau of Statistics; Statistics New Zealand.

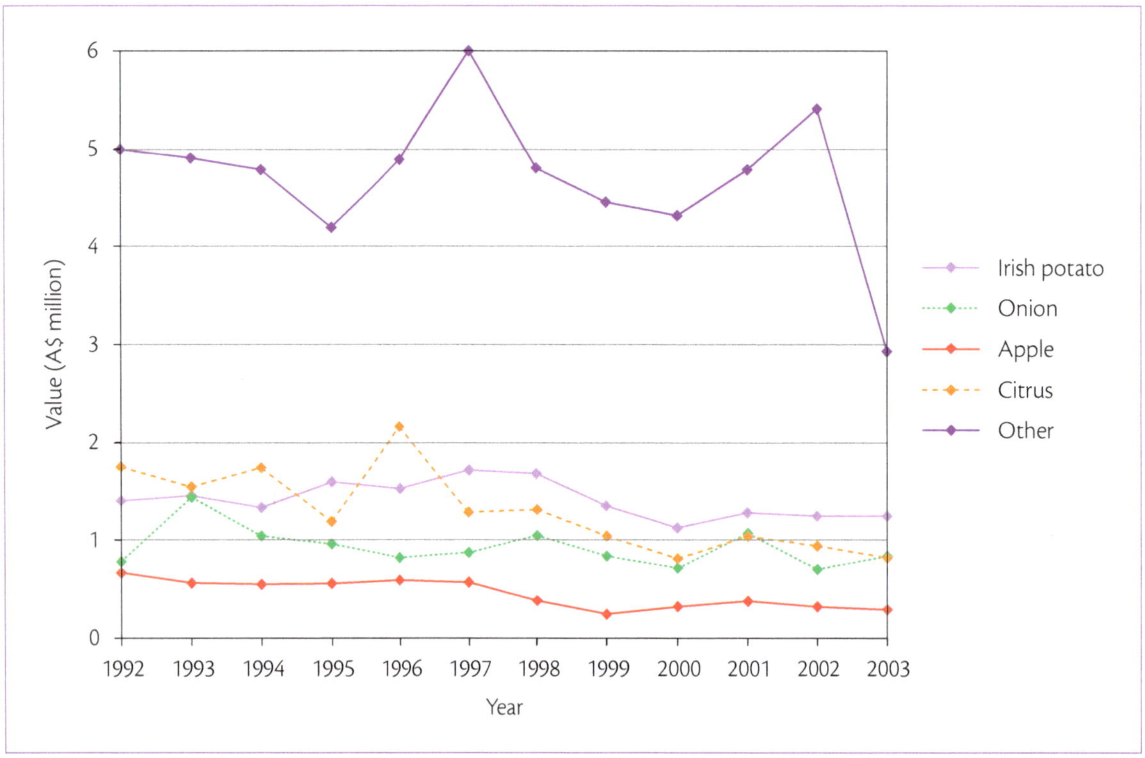

Figure 2.8.2 Composition of fruit and vegetable imports from Australia and New Zealand, 1992–2003.
Sources: Australian Bureau of Statistics; Statistics New Zealand.

2.9 Meat imports

Meat imports into PNG come mainly from Australia and New Zealand, although a limited quantity of tinned pork, chicken, duck and beef is imported from China. The major items imported are sheep meat, beef and offal (lungs, liver, kidneys and other internal organs). Previously, pork, poultry and tinned meat were significant imports but they are now limited by trade barriers (see Section 4.1).

Imported meat together with locally grown meat (see Section 2.6), locally caught fish and imported fish (see Section 2.10) provides scarce protein in people's diets. If the two meat canneries in PNG used only locally grown beef, they would exhaust the entire PNG beef herd in about three months. Meat imports are also important because they are the second most valuable group of food imports, after cereals. In recent years, PNG's meat imports have cost up to K200 million. This has fluctuated with the exchange rate (see Section 4.1), with changing volumes of imports, with changes in the quality of meat imported due to substitution towards cheaper meats, and with consumer substitution of tinned fish for tinned meat. A further reason for interest in meat imports is that claims are sometimes made that meat and meat products (especially lamb flaps) contribute to dietary and health problems.

Meat imports from Australia and New Zealand have averaged about 42 000 tonnes per year since 1983 (Figure 2.9.1).[1] However, from 1983 until 1994 total meat imports rose rapidly, from 25 000 tonnes per year to 60 000 tonnes per year. The amount of meat consumed per person on average almost doubled over this period from 8 kg/person/year in 1983 to 14 kg/person/year in 1994. The annual value of these imports did not rise as fast, however (increasing from about A$40 million in the early 1980s to about A$70 million by the mid 1990s), because over this period the quality of the meat imported was reduced.

The change towards lower quality and cheaper meat is illustrated by sheep meat imports (Figure 2.9.2). In 1983 sheep meat comprised less than one-third of the volume of total meat imports and it was not even the leading individual item imported (tinned meat was). But by 1994 sheep meat had grown to be three-quarters of the total volume of meat imports and it has maintained that share since then.

---

1 Statistics from the exporting countries are used (Table A2.9.1) because these data are more timely and reliable than PNG's import statistics (Table 2.1.2). Because Australia and New Zealand account for almost all meat imported into PNG, the trends shown in Figure 2.9.1 are representative of all meat imports. The population numbers used to calculate the average per person imports are the same as those used in Section 2.7 for per person rice and wheat imports.

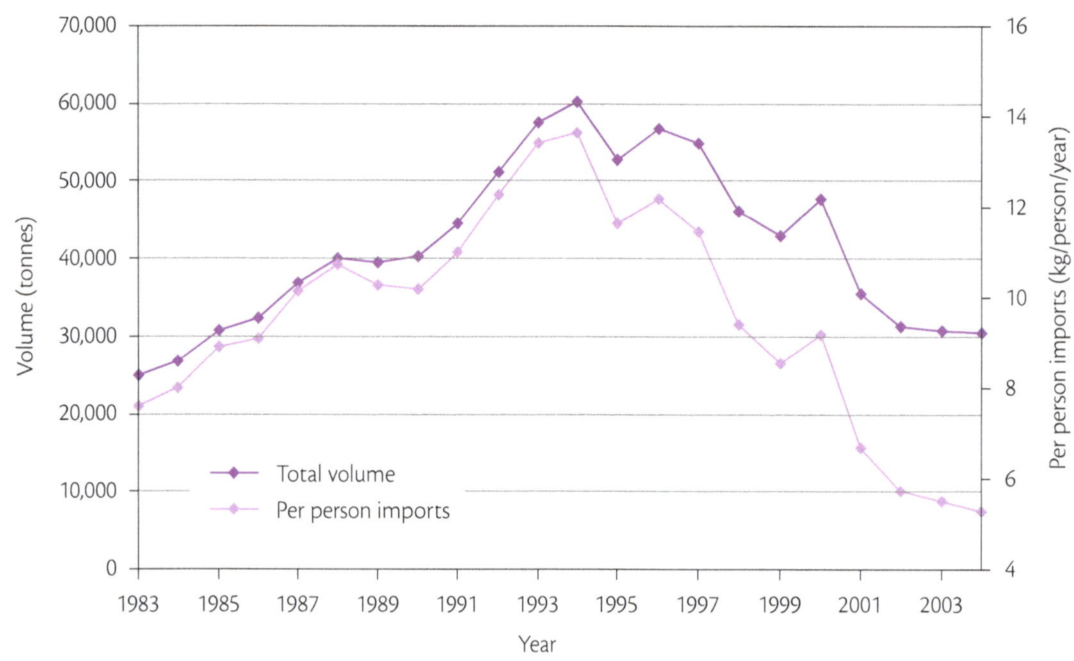

Figure 2.9.1 Volume of meat imports from Australia and New Zealand, 1983–2004.
Sources: Australian Bureau of Statistics; Statistics New Zealand.

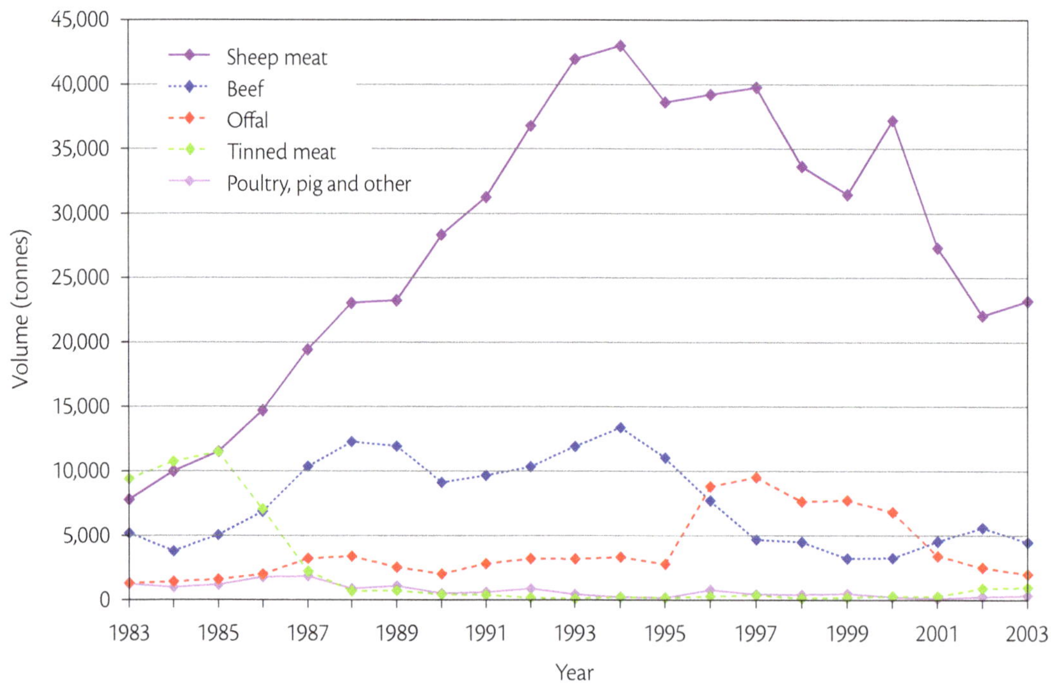

Figure 2.9.2 Composition of meat imports from Australia and New Zealand, 1983–2003.
Sources: Australian Bureau of Statistics; Statistics New Zealand.

Another factor that has caused a reduction in the cost per kilogram of meat imports has been the switch to cheaper cuts of beef and the increased import of offal, which is used as an ingredient in PNG tinned meat. At the same time, imports of the relatively more expensive tinned meat, pork and poultry have been close to zero since the mid 1980s, because of import protection of the domestic livestock industries (Table 4.1.1).

After 1994 the volume of total meat imported fell rapidly (Figure 2.9.1). By 2004, total imports were only 30 000 tonnes, corresponding to less than 6 kg/person/year. This level of meat imports per person is lower than it was 20 years earlier in 1983. The value of these meat imports is also lower, to A$45–A$50 million per year, although the trend in kina terms fluctuates because of the changing value of the kina. The fall in meat import volumes is due to falls in all three of the main imports; sheep meat, beef and offal.

Sources

Australian Bureau of Statistics website <http://www.abs.gov.au/>.

Gibson, J. (1993). *The impact of the exchange rate on meat imports (where's the beef?)*. Policy Working Paper No. 7. Department of Agriculture and Livestock, Konedobu.

Statistics New Zealand website <http://www.stats.govt.nz/default.htm>.

2.10 Fish imports

PNG's annual consumption of fish and seafood is between 12 and 25 kilograms per person. Fish makes up around 1% of PNG's dietary energy supply (see Figure 2.1.2), which is much lower than in most Pacific island countries. The amount of fish and seafood consumed in PNG increased from the 1960s until 1980. By 1998–2000 fish consumption had fallen below 1960s levels of around 15 kg/person/year. Consumption was estimated as 12 kg/person/year in 2006 (Table A2.1.1). Other high quality protein sources such as meat, milk, eggs and nuts make up a low proportion of the overall diet (about 20–30 kg/person/year) compared to starchy roots and fruit and other vegetables (around 900 kg/person/year). The low amount of protein in the diet has been a nutritional concern in PNG, causing problems such as inhibited child growth rates (see Section 6.6). It is nutritionally important to increase access to fish (as well as to high quality cereals, legumes and meat).

According to the United Nations Food and Agriculture Organization, demand for fish products in PNG greatly exceeds supply. It has been estimated that people in Port Moresby alone could consume an additional 2000 tons per year of fresh, smoked and frozen fish. However, in PNG, like other Pacific island countries, lack of cash limits the demand for fish. Buying power, even in urban areas, is limited. In the late 1990s, PNG's estimated total fisheries production was 167 000 tonnes (worth US$161 million), while annual national fisheries consumption is estimated to have been around 16 500 tonnes since the late 1980s. The ratio of exports to domestic consumption supports the argument that lack of buying power is a main cause for low fish consumption.

PNG mostly exports higher-value products such as bêche-de-mer (sea cucumber), trochus shell, prawns and tuna, while it imports low-value fish such as tinned mackerel and the barracouta fillets that are often used in fast food outlets (*kai* bars). Locally produced fish and seafood have long been supplemented by imported fish products. In 1996 fish imports for human consumption were 35 500 tonnes. As with overall fish consumption, PNG's fish and seafood imports peaked in the period 1979–1981 then dropped again and have stayed at low levels. Tinned fish, which can be stored without refrigeration for long periods, has been a particularly important supplement to more perishable local fish products. Tinned fish is vital for urban nutrition in PNG, especially among poorer socioeconomic groups.[1]

In real terms fish imports have shrunk significantly. The United Nations commodity trade statistics database (UNComtrade) shows that the total value of fish imports to PNG has declined in dollar figures

1 People in urban areas tend to consume more fish, rice, biscuits, bread and meat because these products are more accessible (Tables 2.1.1, A2.1.1). The kinds of foods that are more often eaten in rural areas are less accessible in urban areas.

since 1981.[2] In 1981 fish imports totalled US$35 million. By 1998 fish imports were only US$14 million and in 2003 had fallen to US$8 million (Table A2.10.1). This corresponds with the previously noted decline in real per person consumption of imported goods in PNG over the period to the early 1990s. The vast majority of fish imports up to the early 1990s were tinned fish (Figure 2.10.1). Most of this was mackerel from Japan (Figure 2.10.2). The decline in the value of fish imports, however, does not simply mean a decline in consumption of tinned fish, because since the mid 1990s PNG has been canning fish domestically.

In 1995 the International Food Corporation opened a cannery in Lae and started canning mackerel. This is the main reason for the marked shift in the proportions of tinned fish and fresh fish imports between 1990 and 1998 (Figure 2.10.1). By 1998 the value of tinned fish imports had dropped to a fraction of fresh fish imports, with the total value of imports far lower than it had been when imports of tinned mackerel were higher.[3] Because PNG consumers liked the cold water mackerel they had been eating for years, and because it was cheap, the International Food Corporation imported cold water mackerel to use in the Lae factory. Mackerel thus remained by far the largest fish import but, instead of being imported already canned, it was imported frozen (Figure 2.10.3) then canned in Lae. This switch from imports of higher-value tinned mackerel to lower-value frozen mackerel is part of the explanation for the decline in the value of fish imports.

The advent of tuna canning in PNG has contributed to the fall in imports of tinned fish since the late 1990s. From 1995 the PNG Government policy was to tie access to PNG's very large and rich tuna fishing grounds to a commitment to develop onshore processing. In this way several companies that had been fishing for years as foreign fleets in PNG waters were enticed to re-label their fleets as 'locally based' and build canneries and/or loining plants.[4] The first was the Filipino fishing company RD, which established a cannery in Madang in 1997. In 2004 the South Seas Tuna Corporation (SSTC) started a loining plant in Wewak, using fish from a locally based Taiwan-owned fleet. In 2006 another Filipino fishing company, Frabelle, opened a loining plant and cannery in Lae. As well as exporting, RD has been selling tinned tuna domestically. Furthermore, increases in the price of tinned beef have apparently steered consumers towards tinned tuna in recent years (tinned tuna now retails for less than half the price of tinned beef) (see Section 2.9). This has led to the meat canning factory in Port Moresby also canning tuna.

While the trade situation regarding cold water mackerel is straightforward because this fish is only imported, tuna production in PNG is less clear cut in terms of imports versus domestic production. For example, RD and Frabelle are wholly owned PNG subsidiaries of large international companies. While

2 Data for this section are based on datasets available for PNG fishery imports in the commodity trade statistics database compiled by the United Nations (UNComtrade). For the years 1981–1990, the Standard International Trade Classification (SITC) Revision 1, 3-digit codes (which contain only two commodity descriptions – 'fish, fresh & simply preserved' or 'fish, in airtight containers, not elsewhere specified & fish preparations') are the only data available. For figures since 1998, the Harmonized System (HS) 1992 five- and six-digit codes became available (with more detailed commodity descriptions), and for the years 2002–2003 HS 2002 six-digit codes (the most detailed commodity descriptions) are available. There are discrepancies between the sets of data. For example, the general category of 'fish, fresh & simply preserved' (commodity code 031) from a particular country for a particular year should equal the sum of specific categories of chilled or frozen fish from the same country and same year, but occasionally these numbers are not equal across the different datasets. Where there are discrepancies, figures from the set indicating a larger volume of trade have been used.

3 The United Nations trade statistics on value are at odds with the 2002 FAO fisheries country profile of PNG, which uses weight to assert that tinned fish still made up 95% of PNG's fish imports in the early 2000s.

4 Loins are pieces of fish body meat, with the bones, skin and guts removed. Cooked loins are ready for canning. This part of the canning process cannot be mechanised, so is labour intensive. The high labour costs in Europe, the United States and Japan mean these countries are not competitive in fish canning, so they import cooked loins then complete the mechanised stages of the canning process domestically. This enables countries that are no longer competitive in fish canning to keep canneries open for political reasons, even though most of the labour is actually offshore.

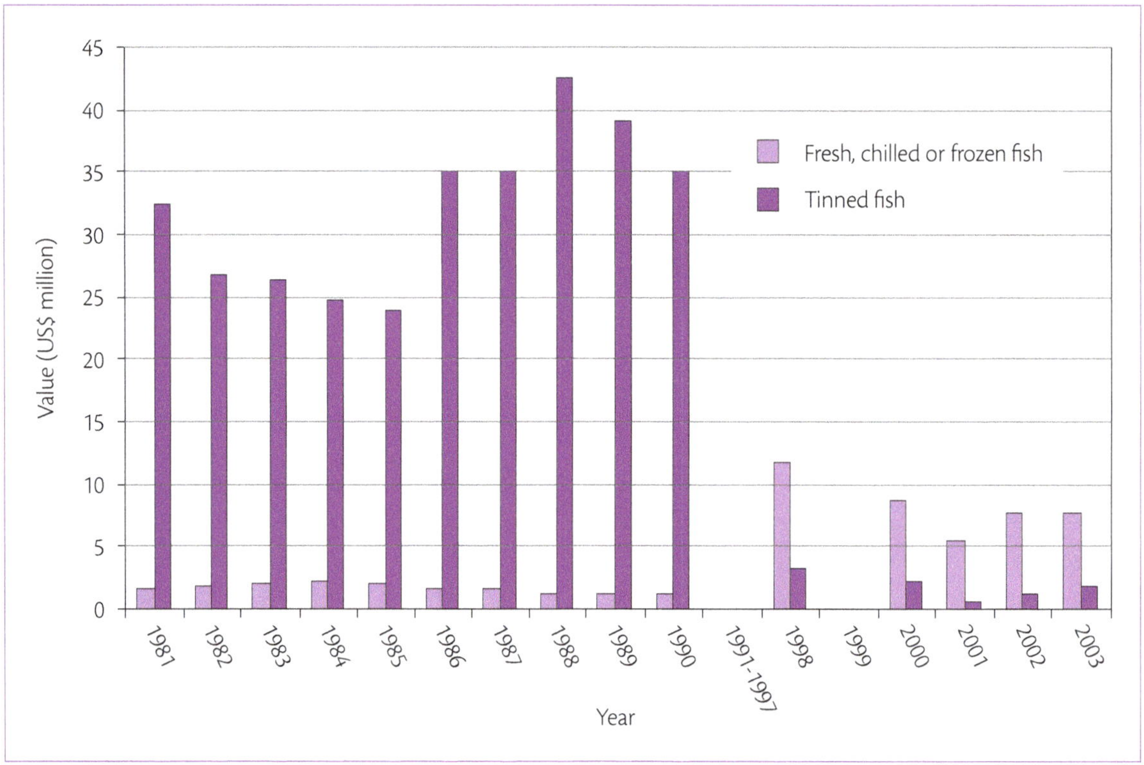

Figure 2.10.1 Value of fresh and tinned fish imports, 1981–2003. **Note:** Data are not available for the years 1991–1997 and 1999. Source: United Nations (2005).

Figure 2.10.2 Value of fishery imports by major source country, 1981–2003. **Note:** A 'major source country' was defined as having import values over US$1 million for one or more years. Data are not available for the years 1991–1997 and 1999. Source: United Nations (2005).

they both have a number of fishing vessels registered as domestic by certain official criteria, these vessels are not really locally owned and operated in a broader sense. RD's vessels are owned by Philippines-based interests who contract them to RD PNG and are crewed mostly by Filipinos recruited through a Philippines-based company. The fleet supplying SSTC is similarly tenuously 'local', as are the purse seine vessels in Frabelle's fleet. (Frabelle is also intending to run a fleet of small 'pump boats', which will have more local input.)[5]

5 A purse seine vessel carries and operates a net called a seine that hangs vertically in the water with lead weights at the bottom and floats at the top. When the vessel finds a school of fish, it encircles the fish with the seine. A wire threaded through the bottom of the net is then drawn closed, trapping the school.

Pump boats are small, wooden, single-hulled vessels with outriggers, used in some parts of the Philippines. They are powered by diesel engines and 5–8 crew can stay out at sea for several days on them, fishing with handlines.

Furthermore, while PNG's fishing grounds are the richest for tropical tuna in the region, tuna is a migratory species, so the fleets tend not to stay in one national area, but roam across the Pacific following the fish. A regional agreement called the Federated States of Micronesia Arrangement gives reciprocal fishing rights to nine countries of the western and central Pacific, prioritising locally based fleets. Thus a substantial proportion of the catch from the PNG-based vessels associated with the onshore processing factories is caught by foreign crew on foreign-owned vessels, and they may have been fishing outside PNG's fishing zone. Notwithstanding these issues that complicate the nationality of PNG's tinned tuna, trade figures show a low rate of imported frozen tuna (Figure 2.10.3), so the tuna used in the RD cannery and SSTC loining plant has been classified as domestically produced.

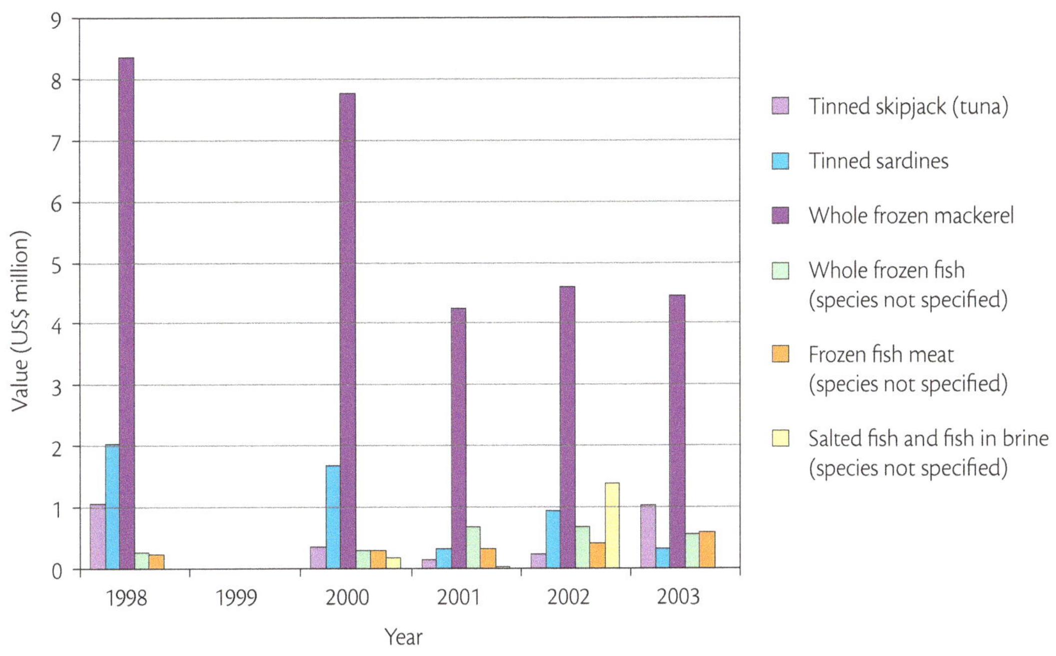

Figure 2.10.3 Value of fishery imports by main commodity, 1998–2003. **Note:** A 'main commodity' was defined as having import values over US$0.5 million for one or more years. The commodity description 'prepared/preserved, not mince' is presumably made up mostly of tinned fish (neither the word 'can' nor 'tin' are used in the commodity descriptions). Data are not available for the year 1999. Source: United Nations (2005).

Sources

Barclay, K. and Cartwright, I. (2006). Capturing wealth from tuna: key issues for Pacific Island countries. Institute for International Studies, University of Technology, Sydney.

Barclay, K. and Cartwright, I. (2006). Capturing wealth from tuna: key issues for Pacific Island countries. Country profiles. Institute for International Studies, University of Technology, Sydney.

Food and Agriculture Organization of the United Nations (2002). Fishery country profile <http://www.fao.org/fi/fcp/en/PNG/profile.htm>. Accessed 22 March 2006.

Food and Agriculture Organization of the United Nations (2003). Nutrition country profiles – Papua New Guinea <ftp://ftp.fao.org/es/esn/nutrition/ncp/png.pdf>. Accessed 22 March 2006.

Gibson, J. (1995). *Food consumption and food policy in Papua New Guinea*. INA Discussion Paper No. 65. Institute of National Affairs, Port Moresby.

Gillett, R. and Lightfoot, C. (2002). *The Contribution of Fisheries to the Economies of Pacific Island Countries*. Pacific Studies Series. Asian Development Bank, Manila, Philippines.

Mueller, I., Vounatsou, P., Allen, B.J. and Smith, T. (2001). Spatial patterns of child growth in Papua New Guinea and their relation to environment, diet, socio-economic status and subsistence activities. *Annals of Human Biology* 28(3):263–280.

United Nations (2005). United Nations commodity trade statistics database (UNComtrade). Standard International Trade Classification (SITC) Revision 1, Harmonized System (HS) 1992, and HS 2002. Accessed 2005.

PART 3

Village Food Production Systems

R. Michael Bourke and Bryant Allen

3.1 Staple food crops

Sweet potato, sago, banana, yam, taro, Chinese taro, cassava, sugar cane, coconut, Irish potato and corn are the main staple foods eaten by rural villagers in PNG. They are grouped into three classes according to their relative importance in any given location – 'most important food', 'an important food', and 'grown for food' (see note, Table 3.1.1). The main vegetables, fruits, nuts and stimulants grown in PNG are described in Sections 3.2 to 3.5. Production per person of the main staple food crops is discussed in Section 2.2.

Sweet potato is grown by almost all rural villagers in PNG (99%), the exceptions being people living in a limited number of locations in East Sepik and Western provinces where land is subject to regular flooding. Sweet potato is the most important food for 66% of the rural population and an important food for a further 15% of the rural population.[1] Thus it is the most important food or an important food for more than 80% of the rural population (Tables 3.1.1, 3.1.2). Sweet potato is especially important for food production at locations above 1500 m altitude. It is also important on Bougainville, New Ireland, New Britain and on coastal and inland areas in Madang, Morobe, Oro, Milne Bay and Central provinces (Figure 3.1.1, Table 2.2.2). Sweet potato is grown in a wide range of environments in PNG: from sea level to 2700 m altitude (Figure 1.13.3); in locations where the mean annual rainfall ranges from 1000 mm to 6500 mm per year; and on most landforms and soils.[2]

Sweet potato was introduced and adopted in the PNG highlands around 1700 AD (see History of agriculture). It displaced taro as the most important food in the highlands so that, by the early colonial period (1880–1940), it provided an estimated 40% of food energy from locally grown staple foods. This proportion grew to 66% by 2000 as production expanded (Figure 3.1.2).

Sago (*Metroxylon sagu*) is grown by a third of the rural population and is the most important food for 11% of people. It is grown in all provinces except East New Britain. Sago grows to an altitude of 1150 m, mostly in places with extended or permanent flooding (Figure 1.11.1), but also in local sites that are poorly drained, such as near creeks and rivers and in depressions in slopes (slump hollows). The total quantities of sago produced did not change greatly between 1960 and 2000 (Figure 2.2.3), but the relative contribution of sago to food energy has

1 The proportion of people for whom a particular crop is the most important or an important food crop can be compared with consumption data from the 1996 PNG Household Survey. The figures from the two sources are reasonably close. For example, sweet potato was consumed by 65% of rural villagers in the Household Survey period (Table 2.1.1), while the MASP data indicates that sweet potato is the most important food for 66% of rural villagers (Table 3.1.1).

2 Altitudinal ranges quoted in Sections 3.1 to 3.5 are mean figures for the usual range in PNG (Bourke 1989). Data were recorded in 1979–1984 before current climate change was significant (see Section 1.8). The ranges are likely to have increased slightly since the early to mid 1980s.

decreased over time. A different species, *M. salomonense*, is grown and eaten on Bougainville, but is only a minor food there.

Banana is grown by most rural people (96%), except by those living at very high altitudes. It is the most important food crop for 9% of the rural population. A further 32% of rural people grow it as an important food. It is eaten cooked and as uncooked fruit, depending on the variety and maturity when harvested. More fruit is eaten cooked than uncooked. Banana is grown from sea level to an altitude of 2150 m. There is a large variation in rainfall between locations where it is the most important food. It is an important food in locations with a marked dry season each year and relatively low annual rainfall, such as in coastal Central Province and the Markham and Ramu valleys in Morobe and Madang provinces (Figures 1.5.1, 1.5.2, 1.5.3). It is also an important food in locations with no rainfall seasonality and very high annual rainfall, such as some places in inland Gulf and Western provinces. Banana is also an important food where the rainfall is neither particularly high

Table 3.1.1 Rural population growing staple food crops

Crop	Most important food		An important food		Grown for food	
	Population	%	Population	%	Population	%
Sweet potato	2,785,005	66	633,791	15	4,142,532	99
Banana	385,748	9	1,341,922	32	4,035,383	96
Colocasia taro	265,094	6	1,026,171	25	3,991,472	95
Greater yam (*Dioscorea alata*)	–	–	167,122	4	2,508,298	60
Cassava	42,847	1	515,140	12	2,318,528	55
Chinese taro	129,061	3	779,783	19	2,244,173	54
Coconut	1,662	<1	1,488,561	36	1,535,066	37
Sago	459,831	11	145,703	4	1,372,004	33
Lesser yam (*D. esculenta*)	271,968	7	237,093	6	1,369,959	33
Irish potato	–	–	120,881	3	668,769	16
Taro (*Alocasia*)	–	–	–	–	315,154	8
Queensland arrowroot	–	–	–	–	184,334	4
Taro (*Amorphophallus*)	–	–	–	–	139,707	3
Swamp taro	680	<1	3,466	<1	31,598	<1
Aerial yam (*D. bulbifera*)	–	–	–	–	21,538	<1
Yam (*D. nummularia*)	–	–	–	–	7,391	<1
Yam (*D. pentaphylla*)	–	–	–	–	3,436	<1

Note: Populations were allocated to the first class ('most important food') where a crop occupies 33% or more of the land devoted to staple food crops in a given agricultural system; the second class ('an important food') where a crop occupies 11–32%; and the third class where a crop occupies 2% or more. The third class ('grown for food') includes the populations in the first two classes. These three classes correspond to 'dominant staple', 'subdominant staple' and 'all staple crops' in the MASP database (see Section 1.15). There can be either one or two crops in the class 'most important food' and up to six crops in the class 'important food'. Sago and coconut are not planted in food gardens, so estimates for the population growing them were based on the estimated food energy derived from these two crops in each agricultural system.

Percentages are the proportion of the total rural population (4 192 561) who grow each crop in each class. Column totals add up to more than 100% because people are counted more than once where they grow more than one crop in that class.

Sources: NSO (2002), MASP.

Table 3.1.2 Rural population growing staple food crops in combinations

Crop	Sweet potato	Sago	Banana	Lesser yam (*D. esculenta*)	Chinese taro	*Colocasia* taro	Cassava	Coconut	Swamp taro	Total population using crop
Sweet potato	2,425,435	19,753	91,047	38,890	33,926	150,615	25,339			2,785,005
Sago	19,753	431,283	8,264			531				459,831
Banana	91,047	8,264	252,203	20,499	11,789		1,946			385,748
Lesser yam (*D. esculenta*)	38,890		20,499	141,473		71,106				271,968
Chinese taro	33,926		11,798		51,474	25,647	6,216			129,061
Colocasia taro	150,615	531		71,106	25,647	17,195				265,094
Cassava	25,339		1,946		6,216		9,346			42,847
Coconut								982	680	1,662
Swamp taro								680	–	680

Note: These figures are for the population growing a staple food crop as their *most important* food crop. There can be either one or two most important (dominant) staple food crops in any agricultural system in the MASP database. The total population in this table is greater than the total rural population as people are counted twice where they grow two most important crops.

Sources: NSO (2002), MASP.

nor low but is moderately seasonal, such as on the Gazelle Peninsula in East New Britain Province. The relative importance of banana has declined over time as has that of the other Pacific staple food crops (taro, sago and yam) because of the adoption of a number of crops from the Americas, particularly sweet potato, cassava, Chinese taro and corn.

Yam species are grown by 60% of the rural population. Greater yam (*Dioscorea alata*) is more widely grown than lesser yam (*D. esculenta*), but greater yam is not the most important food anywhere and is an important food for only 4% of the rural population. Lesser yam is the most important food or an important food for 13% of rural villagers. It is the sole most important food in inland East Sepik Province and the southern part of Western Province (Figure 3.1.1). Greater yam is grown up to 1900 m above sea level and lesser yam up to 1550 m, but most production for both species occurs in the lowlands. Yam is usually more important in seasonally dry climates, such as inland East Sepik, but it is sometimes an important crop in locations where rainfall is well distributed throughout the year, such as the northern islands of Milne Bay Province. Total production seems to have increased between 1960 and 2000 (Figure 2.2.3), but yam's contribution as a proportion of total staple food energy has declined over this period (Figure 3.1.2).

***Colocasia* taro** is grown by most rural villagers (95%) in PNG, but it is now the most important food for only 6% of the rural population, with a further 25% growing it as an important food crop. Despite being grown by many people in most parts of PNG, by 2000 taro was the sole most important food in only a few locations, including the inland Kandrian area of West New Britain Province and most of the Telefomin area of Sandaun Province. Taro is widely grown from sea level to an altitude of 2400 m, although monocultures of taro are not grown above 2200 m. There is a wide range in annual rainfall at these locations (1500 mm to over 7000 mm). In the past, taro was the sole staple food in wetter and less seasonal environments, and where rainfall seasonality was greater it was grown with other staples. (See Figure 1.13.1 for an example of food crops grown in a location where rainfall distribution is moderately seasonal.) Prior to the introduction of sweet potato, *Colocasia* taro provided an estimated half of food energy from the staple foods. Production has declined since 1940, so that by 2000 it provided only 4% of food energy from locally grown staple foods. The use of taro as a food has been affected by taro blight (*Phytophthora colocasiae*), which caused a sudden and severe loss of production in some lowland areas after 1940. Production has also been affected by declining soil fertility associated with more intensive land use, virus infection and taro beetle damage.

Chinese taro is grown by more than half the rural population and is the most important food or an important food for 22% of the rural population. It is the most important food on its own or with one other crop in parts of the Baining Mountains of East New Britain Province, on the north and south of the Huon Peninsula in Morobe Province and in the Adelbert Range and Gogol Valley in Madang Province. Chinese taro grows from sea level to 2000 m altitude. It is a more important crop at intermediate altitudes and lower highland locations (500–1500 m altitude) and in places where rainfall is higher and not seasonal. Chinese taro was introduced into PNG in the late nineteenth century. It increased in importance from 1940 until the 1980s and then declined somewhat as production was affected by a root disease, possibly caused by a fungus (*Pythium* sp.).

Cassava is grown by more than half the rural population, but is the most important food for only 1% of rural people. It is the most important food on its own or with one other crop in some places in West New Britain Province, including the Cape Hoskins and Talasea areas, and in a number of locations in Milne Bay, including the Cape Vogel area and parts of Goodenough Island. In Milne Bay Province, more cassava is produced than any other staple food (Table 2.2.2). Cassava grows from sea level up to 1800 m, but is only an important food in the lowlands. It is an important food in locations with a wide range of rainfall, including weakly seasonal and high annual rainfall, and markedly seasonal and relatively low annual rainfall. Cassava was introduced into PNG in the nineteenth century, but only started to become widely grown after 1950. Production is expanding rapidly, with a threefold increase in its relative importance as a food between 1960 and 2000 (Figure 3.1.2).

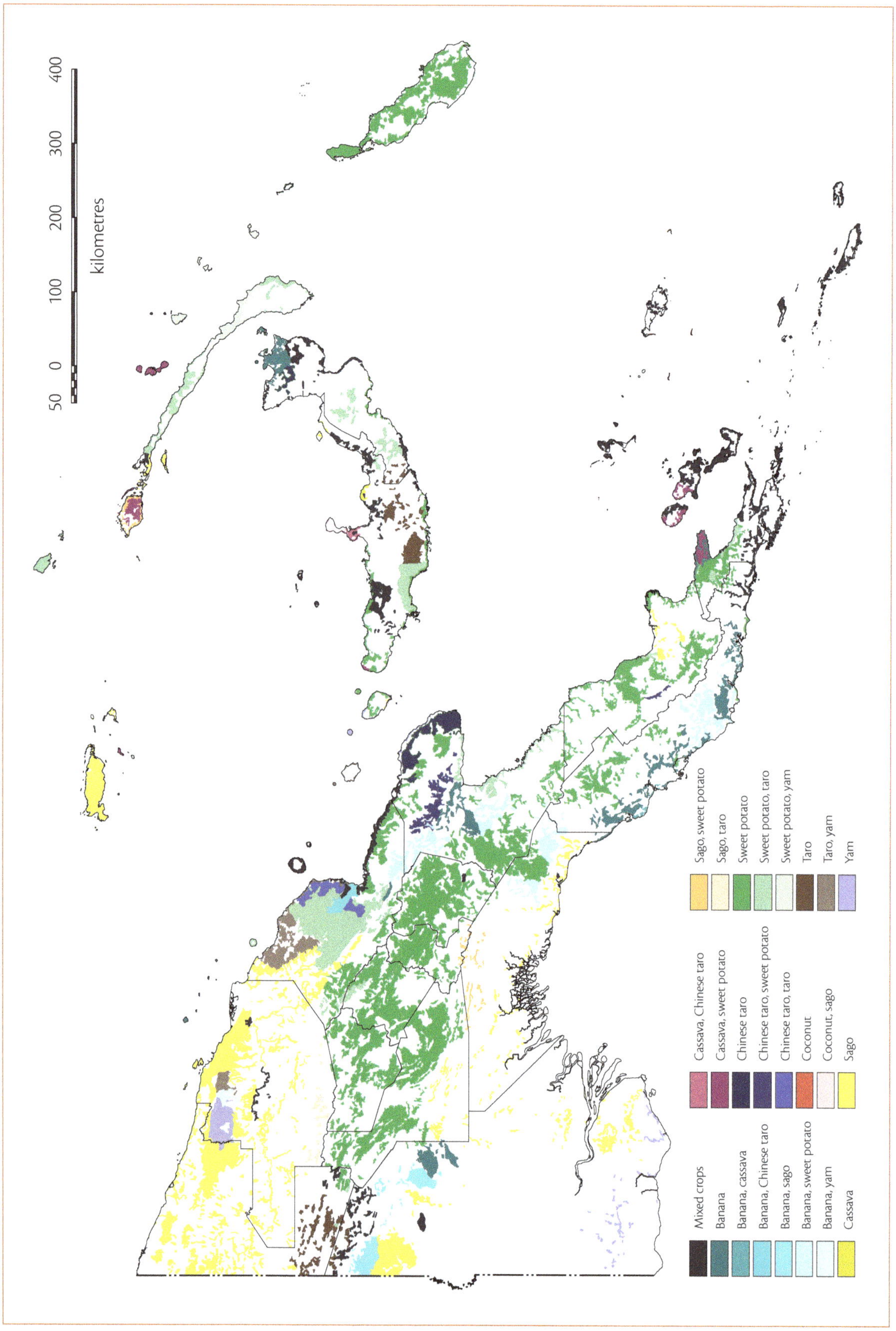

Figure 3.1.1 Distribution of the most important staple food crops. **Note:** There can be either one or two 'most important' (dominant) staple food crops in any agricultural system in the MASP database. Source: MASP.

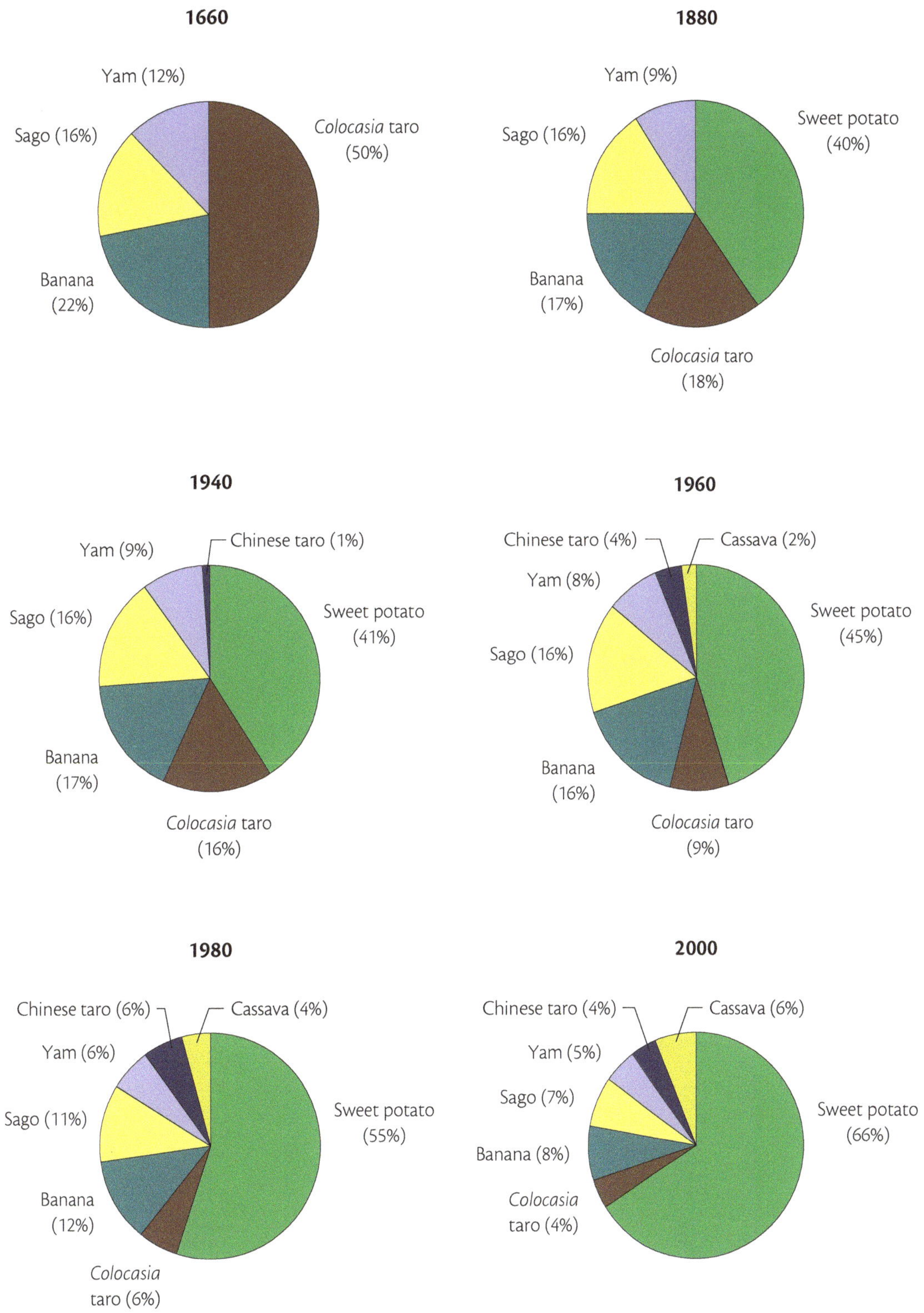

Figure 3.1.2 Estimated contribution to food energy of staple food crops at six different periods, 1660 to 2000 AD. **Note:** The year 1660 was chosen to represent the period prior to the Long Island volcanic eruption and subsequent adoption of sweet potato by about 1700 (see History of agriculture). Sources: 1660, 1880, 1940, 1980: author's reconstruction based on historical information; 1960: Walters (1963); 2000: Bourke and Vlassak (2004).

Sugar cane is one of the most widely grown food crops in PNG, with 99% of the rural population growing it. People suck the juice from the chewed cane. Production estimates are not available that are comparable with estimates for the other staple foods. The 1996 Household Survey estimated production as 190 000 tonnes of cane per year or 35 kg/person/year (Table A2.1.1). Sugar cane is grown almost everywhere in PNG up to its altitudinal limit at 2600 m. There are no reliable data indicating trends in production over time, but this is likely to have increased with population growth.

Coconut is an important food for 36% of the rural population. It is widely grown on islands, coastal locations and some inland locations on the New Guinea mainland up to 950 m above sea level. No reliable data are available for changes over time. However, it is likely that more coconut became available for consumption following widespread planting of palms for copra production from the late 1800s (see Section 5.6).

Irish potato is grown by 16% of the rural population. It produces tubers between 700 m and 2750 m altitude in PNG. It is most commonly grown for subsistence purposes above 2000 m but is also grown for sale in parts of the highlands. Irish potato was introduced into the highlands in the 1930s. Production increased for both subsistence and sale after about 1970, but declined following an outbreak of potato late blight (a fungal disease) in early 2003.

Corn (maize) is not an important food compared with the main root crops, banana and sago.[3] Nevertheless, it is grown by 94% of the rural population (Table 3.2.1). It grows from sea level up to 2450 m. It is often more significant in seasonally dry locations, such as the Henganofi area in Eastern Highlands Province, where it produces food about three months after planting, before other crops mature. Production in the highlands is markedly seasonal (Figure 1.13.2). Corn was introduced into PNG in the nineteenth century. It has increased in importance since then, particularly since the introduction of superior varieties after 1970.

Minor root crops. A number of minor root crops are eaten, including Queensland arrowroot and three species each of yam and taro. The quantities grown and the number of people growing these crops are generally small (Tables 2.2.1, A2.2.2). Swamp taro is the only minor root crop that is a most important food or an important food crop. It is the most important food on three atoll groups in Bougainville Province and important on a few small islands, such as those west of Manus Island.

Other grain crops. Rice is grown by villagers in a number of locations. It has been and continues to be promoted widely, but only a very limited quantity is grown (see Section 2.5). Experimental plantings of a number of other grain crops including wheat, oats and buckwheat have been made, but these crops have not been adopted by villagers.

Sources

Bourke, R.M. (1989). Altitudinal limits of 230 economic crop species in Papua New Guinea. Unpublished paper. Department of Human Geography, Research School of Pacific and Asian Studies, The Australian National University, Canberra.

Bourke, R.M. and Vlassak, V. (2004). Estimates of food crop production in Papua New Guinea. Land Management Group, Research School of Pacific and Asian Studies, The Australian National University, Canberra.

Gibson, J. (2001). The economic and nutritional importance of household food production in PNG. In Bourke, R.M., Allen, M.G. and Salisbury, J.G. (eds). *Food Security for Papua New Guinea. Proceedings of the Papua New Guinea Food and Nutrition 2000 Conference.* ACIAR Proceedings No. 99. Australian Centre for International Agricultural Research, Canberra. pp. 37–44.

NSO (National Statistical Office of Papua New Guinea) (2002). Papua New Guinea 2000 Census: Final Figures. National Statistical Office of Papua New Guinea, Port Moresby.

Walters, C.L. (1963). *Survey of Indigenous Agriculture and Ancillary Surveys, 1961–1962.* Territory of Papua and New Guinea Bureau of Statistics, Port Moresby.

3 Corn was classed as a vegetable in the MASP surveys, so data are not available in the MASP database to generate estimates that are comparable with those for the other staple foods.

3.2 Vegetables

About 150 plant species are grown in PNG that have an edible part which can be classed as a vegetable. These include leafy greens, plants with other edible parts such as beans, stems or flowers, spices and flavourings.[1] Leafy green vegetables are eaten with most main meals in both urban and rural areas. Green vegetables were recorded as being eaten in 75% of households in the PNG Household Survey in 1996 (Table 2.1.1).

Data are presented here on the proportion of rural Papua New Guineans growing vegetables.[2] Thirty species are grown by 1% or more of the rural population (Tables 3.2.1, A3.2.1). Fifteen species that are grown by more than 20% of the rural population are described below.[3]

Pumpkin is grown very widely. More than 70% of the population in the highlands and lowlands eat the cooked tips of the youngest leaves and vines (Table 3.2.1, Figure 3.2.1). Pumpkin fruit is grown by 17% of the rural population and is an important food in some highlands locations, for example on the Nembi Plateau in Southern Highlands Province. Pumpkin grows in a wide range of environments and is tolerant of reduced soil fertility. The most commonly grown species (*Cucurbita moschata*) bears both leaves and fruit from sea level to 2350 m, although other minor species bear higher. Production of tips is non-seasonal. Fruit production in the intermediate and highlands altitudes is seasonal, with the best supply between September and April. Pumpkin was first introduced in 1847 to Woodlark Island in Milne Bay Province and again in 1871 on the Rai Coast of Madang Province (see History of agriculture). It was subsequently introduced to other locations and widely adopted.

Aibika is grown by more than 60% of the rural population. The leaves and stems are eaten cooked. *Aibika* grows from sea level to 1900 m, although it is more common in the lowlands and intermediate

1 See French (1986) for a comprehensive listing of major and minor vegetables in PNG. In this volume Tables 1, 2, 3 and 4 list more than 60 vegetable species according to their likely period of domestication or introduction into PNG.

2 The data are from the Mapping Agricultural Systems of PNG Project (MASP; see Section 1.15). Figures are the number of rural villagers living in agricultural systems where each vegetable was classed as important only (in contrast to the classification of staple crops, for which three categories are used – see Table 3.1.1). Because the number of plants per household is small and the plants are dispersed, relative importance is not easy to assess in the field and the figures are subject to large errors. Nevertheless, the *ranking* of the relative importance of the species is likely to be fairly accurate.

3 Corn was classed as a vegetable in the MASP database. However, it is generally eaten when it is semi-dried and high in food energy in PNG, so it is included with staple foods in this book (Section 3.1).

Table 3.2.1 Rural population growing the most important vegetables

Vegetable	Rural population	%
Corn	3,957,678	94
Pumpkin tips	3,001,792	72
Aibika	2,609,661	62
Amaranthus	2,520,389	60
Highland *pitpit*	2,247,783	54
Lowland *pitpit*	2,197,462	52
Common bean	2,008,173	48
Cucumber	1,832,626	44
Rungia	1,768,336	42
Winged bean	1,654,668	40
Tulip	1,553,334	37
Snake bean	1,537,139	37
Kumu musong leaves	1,393,531	33
Peanut	1,366,074	33
Oenanthe	1,334,096	32
Cabbage	1,043,222	25
Ferns	758,567	18
Rorippa	751,940	18
Pumpkin fruit	691,851	17
Chinese cabbage[a]	671,896	16
Choko leaves	642,636	15
Taro leaves	318,662	8
Karakap	317,321	8
Spring onion	275,192	7
Valangur	229,185	6
Lablab bean	199,520	5
Tomato	101,771	2
Kangkong	92,975	2
Kalava	77,732	2
Lima bean	57,980	1

[a] Chinese cabbage covers all types of oriental cabbage, including pak choi.

Sources: NSO (2002), MASP.

altitude classes than in the highlands. Production is mildly seasonal in the highlands and lowlands, with the best supply between January and March. *Aibika* was probably introduced to PNG thousands of years ago, but may have been domesticated in the New Guinea area. Consumption was estimated as 40 000 tonnes per year in the 1996 PNG Household Survey. No data are available on production trends, but it is likely that production has not expanded as rapidly as population growth has over the past 50 years.

Amaranthus is grown by 60% of the rural population. The leaves are eaten cooked. Three species, all of South-East Asian origin, are common in the lowlands through to the highlands: *Amaranthus tricolor* (0–1950 m), *A. dubius* (0–1800 m) and *A. blitum* (0–2050 m). Two other species, both of which were domesticated in Central or South America, are grown at high altitudes: *A. cruentus* (1350–2300 m) and *A. caudatus* (1600–2400 m). *Amaranthus tricolor* and *A. cruentus* are probably the two most widely grown species in PNG. All species grow best where soil fertility is high and are tolerant of a wide range in rainfall. Production is markedly seasonal in the highlands (Figure 1.13.2). *Amaranthus tricolor* was probably introduced to PNG thousands of years ago. *Amaranthus blitum*, *A. caudatus* and *A. cruentus* are post-1870 introductions. The antiquity of *A. dubius* is not known, but it is probably also a post-1870 introduction. Production of *A. cruentus* has increased rapidly from a small base over the past 60 years, while that of *A. tricolor* has probably not increased as rapidly as population growth.

Highland *pitpit* is grown by over half the rural population. The stem is consumed after cooking. It is important in the highlands, but is also commonly grown in the Momase and Southern regions. Despite its English common name, highland *pitpit* grows from sea level to 2700 m, although it is more prevalent above 500 m. It can be grown in a wide range of environments, including in low fertility soils. Production is usually non-seasonal, but it is somewhat seasonal in some highland locations. Highland *pitpit* was domesticated in the New Guinea area a long time ago.

Lowland *pitpit* is grown by over half the rural population. The edible portion of lowland *pitpit* is the inflorescence (flower), which is eaten cooked. Lowland *pitpit* is widely grown in most environments in lowland and intermediate altitude locations. It is occasionally grown in highland valleys up to 1800 m. Production tends to be seasonal, with crops maturing sometime between November and May, particularly in January to March. It was domesticated in the New Guinea area a long time ago.

Common bean is grown by almost half the rural population, mostly in the highlands. The edible part is the bean, which is commonly harvested and cooked when reasonably mature. It grows from sea level to 2350 m, but it is rarely grown below 400 m altitude and is less common between 400 m and 1200 m. Beans are available throughout the year in most highlands locations, but production is seasonal with the best supply between September and April. Common bean was first introduced into PNG in the 1800s, with the first recorded introduction in 1847. Production has increased in the highlands over the past 50 years.

Cucumber is grown by more than 40% of the rural population in all environments up to 1950 m altitude. The edible portion is the fruit, which is eaten uncooked, often as a snack food. Cucumber is a seasonal crop. Seasonality of production is greater in the highlands than in the lowlands. It is typically planted during the drier months in May–July in the highlands, presumably to avoid fungal damage in the wetter part of the year. The period of best supply varies between locations, but is generally between November and March. Cucumber is a pre-European crop, but because new types have been introduced since 1870 it is sometimes assumed to be a post-European introduction. It was probably introduced from South-East Asia thousands of years ago.

Rungia is grown by more than 40% of the rural population, mainly in the highlands (Figure 3.2.2). The young leaves are edible and are usually cooked, but can be eaten raw. Rungia grows in all highland environments over an altitudinal range of 950–2700 m, although it is more common between about 1400 m and 2300 m. Production is non-seasonal. Rungia is an ancient crop in PNG and was probably domesticated in New Guinea.

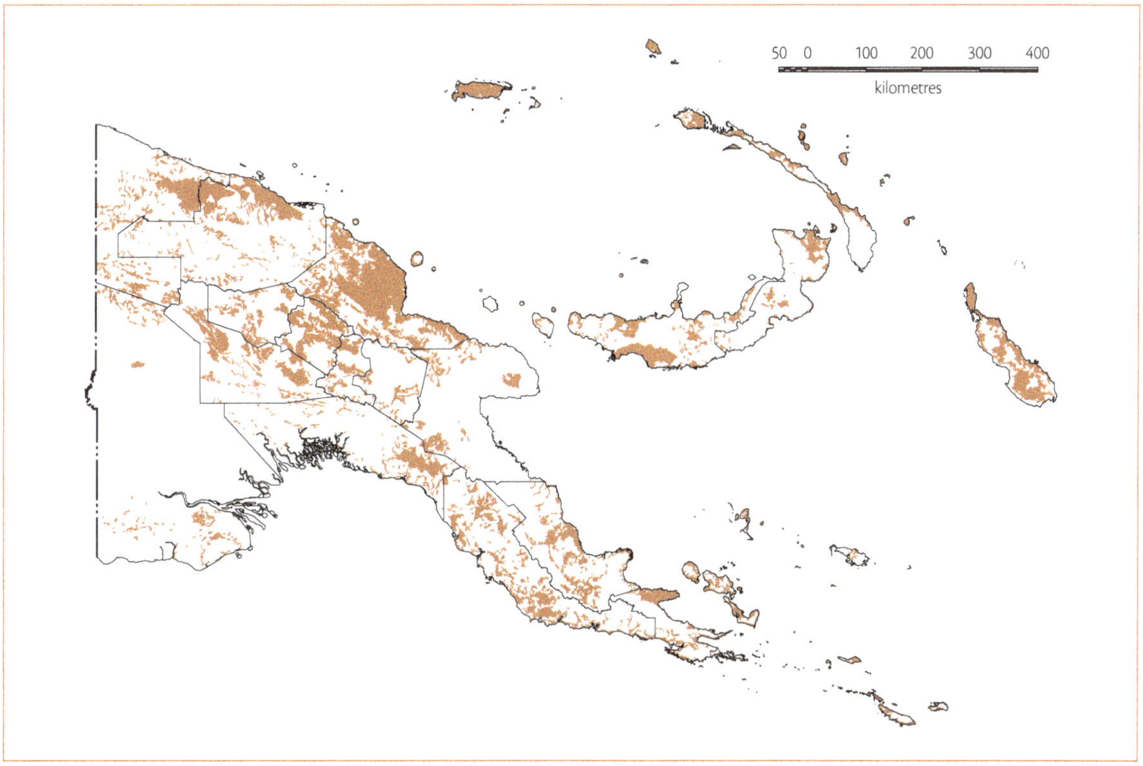

Figure 3.2.1 Distribution of pumpkin tips as an important vegetable. Source: MASP.

Winged bean is grown by 40% of the rural population. The green bean is eaten in the lowlands and the green bean, young leaves and tubers are eaten in the highlands. All edible parts of the plant are eaten cooked. The crop grows from sea level to 1900 m, with tuber production over the range 1200–1900 m (Figure 1.13.3). It is most common in locations that have a seasonally drier climate in Eastern Highlands, Western Highlands, East Sepik and Simbu provinces. Production of tubers, green beans and leaves is seasonal in all environments. In parts of the highlands, tubers are produced from plantings made in the drier months of the year on well-drained sites. The seasonal pattern varies between locations. In the Kainantu area of Eastern Highlands Province, gardens intended to produce tubers are planted in May–August and tubers are harvested in January–March. Plantings intended for bean production (in mixed vegetable gardens) are made in September–December, with young leaves available in January–March and green beans in March–April. Winged bean was introduced from South-East Asia some hundreds to thousands of years ago. Plantings for tuber production in the highlands appear not to be expanding with population growth and production may have decreased in recent decades.

Tulip[4] is a significant vegetable for more than a third of the rural population. The young leaves of this tree are an important vegetable, particularly in sago-growing locations where only limited areas of food gardens are planted. The leaves are eaten cooked; young flowers (cooked) and fruit (raw or cooked) are also eaten. *Tulip* is widely grown in the lowlands (Figure 3.2.3). It bears leaves, flowers and fruit from sea level to 1100 m altitude. The best supply of young leaves occurs about November–December and the best supply of fruit in December–February. *Tulip* was probably introduced from South-East Asia a long time ago.

4 The Tok Pisin name (*tulip*) is derived from 'two leaf' as the leaves are produced in pairs opposite each other. *Tulip* bark is used to make string bags (*bilums*).

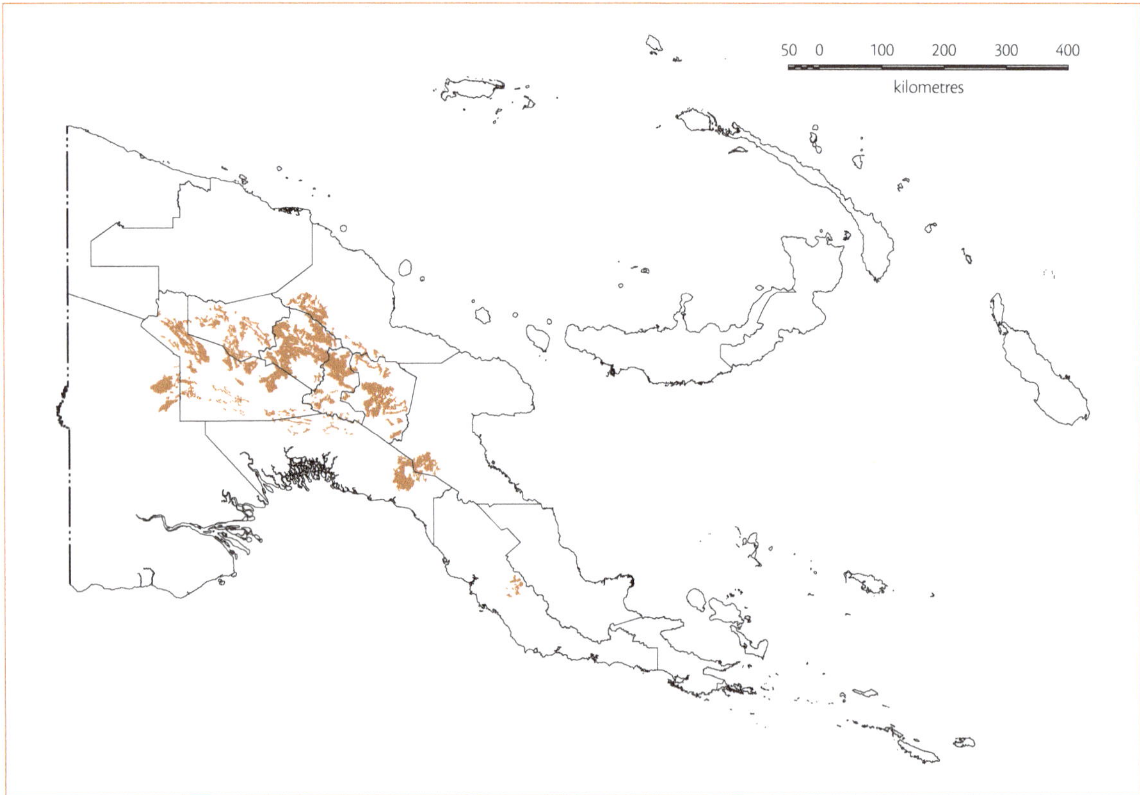

Figure 3.2.2 Distribution of rungia as an important vegetable. Source: MASP.

Snake bean is grown by more than a third of the rural population in a wide range of lowland environments. It is most common in New Ireland, Oro, East New Britain, Milne Bay, Central and East Sepik provinces. It bears from sea level to 1600 m, but is uncommon above about 1000 m. The semi-mature or immature bean is cooked and eaten. The production pattern is not known, but seems to be non-seasonal. Snake bean was introduced to PNG after 1870.

***Kumu musong*[5] leaves** are grown by about a third of the rural population. Young leaves (cooked) and raw fruit of a number of fig-bearing trees are eaten, with *kumu musong* being the most commonly eaten. The cooked leaves are eaten more often than the fruit. The species grows from sea level to 2200 m, but it is less common above about 1200 m altitude. It is found in secondary forest in lowland locations and self-sows in newly cleared gardens, where it is protected from burning. It is most common in Bougainville, West New Britain, East New Britain, Manus, Sandaun, Milne Bay and Oro provinces. In the highlands the young leaves are most abundant in September–November. In Milne Bay Province, leaves are most abundant in January–March and fruit in January–February. This species was probably taken into cultivation in the New Guinea area, although it could have been introduced from Indonesia a long time ago. Consumption of both leaves and fruit has probably declined with the availability of various introduced vegetables and fruit.

Peanut is grown by about a third of the rural population. The edible part is the kernel (nut), which is eaten raw or roasted. Peanut is grown in both the lowlands and highlands, up to the crop's altitudinal limit of 1850 m (Figure 3.2.4). The greatest volume is grown in the Markham and Ramu valleys of Morobe and Madang provinces. Production there is partly mechanised, with tractors used to till large plots. A significant proportion of the crop is sold in local markets. The Markham and Ramu valleys

5 The Tok Pisin name (*kumu musong*) means 'hairy vegetable', named because of raised bristles on the leaves and stems.

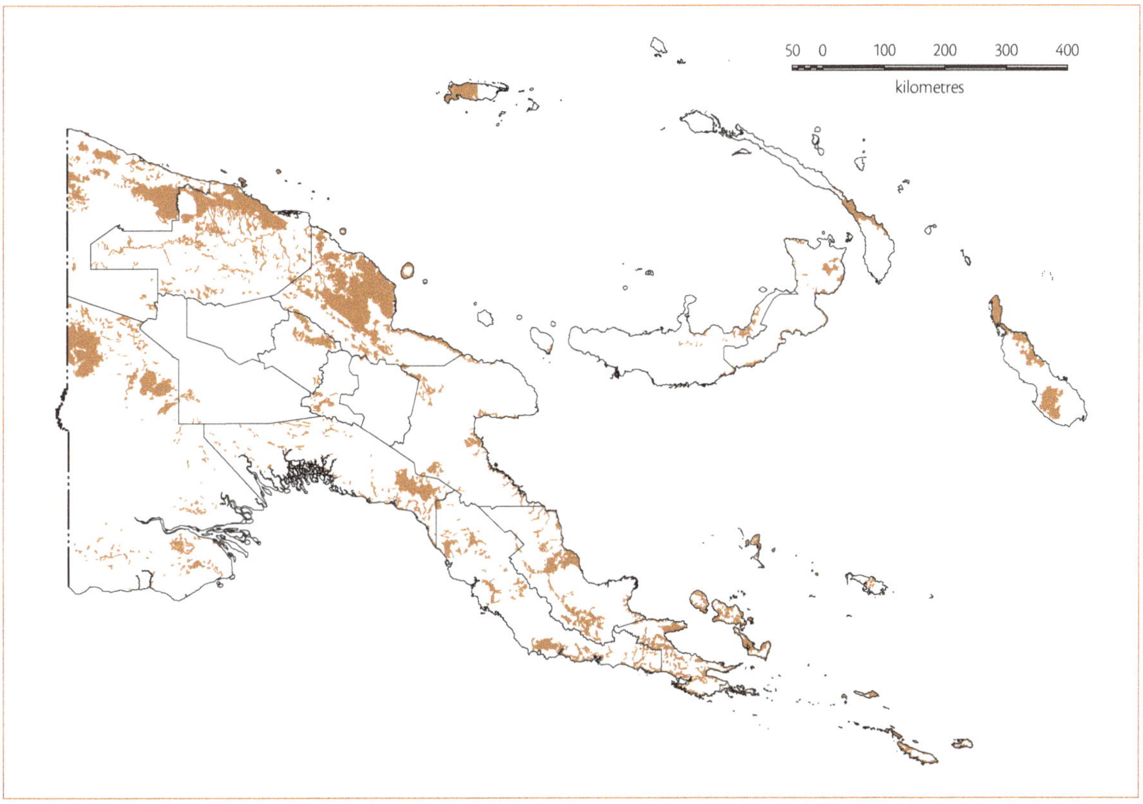

Figure 3.2.3 Distribution of *tulip* as an important vegetable. Source: MASP.

crop is mainly sold to middlemen who retail peanut throughout the highlands. Peanut has replaced winged bean in sweet potato–legume rotations in parts of the highlands where sweet potato is grown for long periods on the same land (see Section 3.10).

The environments where peanut is most common have a seasonally dry climate; mean annual rainfall in the range 1500–2500 mm; flat or gently sloping land; and reasonably fertile and friable soil. Production is mildly seasonal in most lowland locations, with the best supply typically in January–February, but the pattern varies between locations and between years. Production is more seasonal in Eastern Highlands Province, with the best supply in January–March. Peanut was introduced into PNG after 1870. It was grown as an export cash crop in the Markham Valley in the 1950s and 1960s and on a smaller scale in the Goroka area in the 1960s and early 1970s. It was promoted for village production in the Markham Valley as raw material for a peanut butter factory from the early 1970s to the mid 1980s. Peanut was heavily promoted in the highlands in the 1960s to improve the protein intake of villagers. Production was estimated as 21 000 tonnes per year in the 1996 PNG Household Survey. The quantity grown has increased over the past 30 years, particularly in the Markham Valley.

Oenanthe is grown by about a third of the rural population, mainly in the highlands and highlands fringe. The leaves are generally eaten cooked, but occasionally people eat them raw as a snack. It is grown in a wide range of environments, but grows best in moist sites. It was grown over an altitudinal range of 1050–2700 m until about 40 years ago, but highland migrants now grow it in coastal locations. Production is non-seasonal. Oenanthe is an ancient crop that was probably domesticated in New Guinea. There are no data on production trends, but it is likely that production has declined or not kept up with population growth as other vegetables have been adopted in the highlands.

Cabbage is grown by a quarter of the rural population, with most production in the highlands provinces and in the mountains of Morobe Province. The head (leaves) is the edible part and is generally

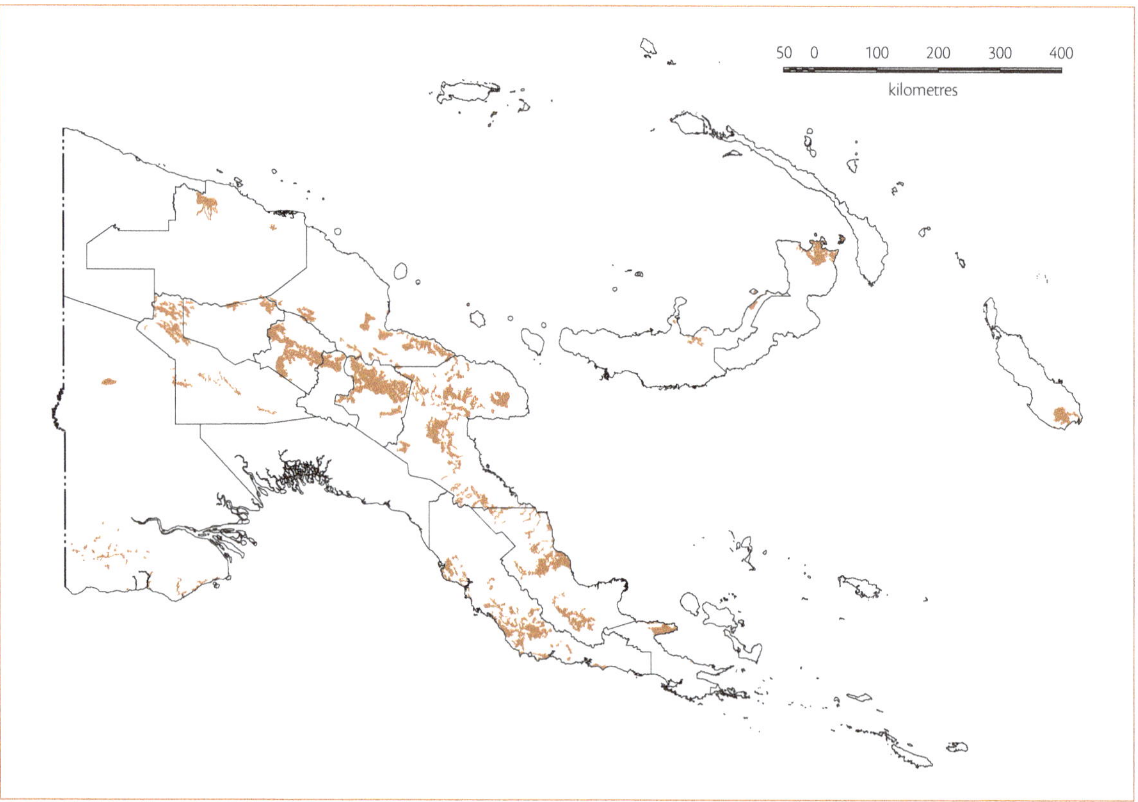

Figure 3.2.4 Distribution of peanut as an important vegetable. Source: MASP.

eaten cooked. In the highlands, cabbage grows over a wide environmental range. It is grown between 700 m and 2700 m altitude, although it is more common above 1700 m. Crops are planted in some lowland locations and sold at nearby urban food markets. These plantings use modern varieties and are grown from seed, in contrast to the highlands where propagation is by the stems of old plants. Production seems to be weakly seasonal in parts of the highlands. For example, in the Kainantu area cabbage is more abundant in September–December. Cabbage was introduced into PNG after 1870. It was promoted as a cash crop for local markets in the highlands from the 1950s onwards. Production has increased rapidly over the past 50 years, particularly at high-altitude locations.

Sources

Akus, W.L. and Nema, R.K. (1995). Evaluation of twenty five vegetable varieties at Aiyura, Eastern Highlands Province. *Papua New Guinea Journal of Agriculture, Forestry and Fisheries* 38(1):11–16.

Bourke, R.M. (1989). Altitudinal limits of 230 economic crop species in Papua New Guinea. Unpublished paper. Department of Human Geography, Research School of Pacific and Asian Studies, The Australian National University, Canberra.

Bourke, R.M., Camarotto, C., D'Souza, E.J., Nema, K., Tarepe, T.N. and Woodhouse, S. (2004). *Production Patterns of 180 Economic Crops in Papua New Guinea*. Coombs Academic Publishing, Canberra.

French, B.R. (1986). *Food Plants of Papua New Guinea: A Compendium*. Sheffield, Tasmania.

Gibson, J. (2001). The economic and nutritional importance of household food production in PNG. In Bourke, R.M., Allen, M.G. and Salisbury, J.G. (eds). *Food Security for Papua New Guinea. Proceedings of the Papua New Guinea Food and Nutrition 2000 Conference*. ACIAR Proceedings No. 99. Australian Centre for International Agricultural Research, Canberra. pp. 37–44.

Kanua, M.B. and D'Souza, E.J. (1985). Evaluation of introduced vegetable varieties on volcanic ash soils in the Southern Highlands of Papua New Guinea. Technical Report 85/13. Department of Primary Industry, Port Moresby.

NSO (National Statistical Office of Papua New Guinea) (2002). Papua New Guinea 2000 Census: Final Figures. National Statistical Office of Papua New Guinea, Port Moresby.

Powell, J.M. (1982). Traditional vegetables in Papua New Guinea: retrospect and prospect. In Bourke, R.M. and Kesavan, V. (eds). *Proceedings of the Second Papua New Guinea Food Crops Conference: Part One*. Department of Primary Industry, Port Moresby. pp. 64–77.

Rose, C.J. (1980). Optimum replanting stage for two varieties of pit-pit (*Setaria palmifolia*) in the highlands of Papua New Guinea. *Papua New Guinea Agricultural Journal* 31(1–4):23–29.

Sowei, J.W. and Osilis, P. (1995). Horticulture Note No. 31. Delayed weeding reduces aibika yield. *Harvest* 17(1&2):6–8.

Westwood, V. and Kesavan, V. (1982). Traditional leafy vegetables of Papua New Guinea: aibika (*Hibiscus manihot* (L)). In Bourke, R.M. and Kesavan, V. (eds). *Proceedings of the Second Papua New Guinea Food Crops Conference: Part Two*. Department of Primary Industry, Port Moresby. pp. 391–395.

3.3 Fruit

About 100 plant species that have an edible fruit are grown in PNG.[1] Many fruits are grown in small quantities in only a limited number of locations. Production of fresh fruit (excluding banana) was estimated as 59 000 tonnes per year by the 1996 PNG Household Survey. More than 20 fruit species are grown in sufficient quantity to be classified as 'important' (Tables 3.3.1, A3.3.1).[2] Half are indigenous species and the others were introduced by foreigners during the nineteenth century. Ten of these 'important' species, which are grown by more than 10% of the rural population, are described below.[3]

Four fruits – mandarin, mango, mangosteen and rambutan – have significant potential for expanded production for the domestic market, particularly for sale in the highlands and in major urban centres.[4] Five well-established species could also be developed and marketed in greater volume. These are avocado, orange, pawpaw, pineapple and *ton*. Another group have some potential for further production and marketing. In the lowlands, these are carambola, custard apple (sweetsop), durian, guava, langsat, longan, pomelo, pulasan, rockmelon (cantaloupe) and watermelon. Highland species in this group are banana passionfruit, cape gooseberry, cherimoya, naranjilla, purple passionfruit, black raspberry, strawberry, *suga prut* and tamarillo (tree tomato).

Pawpaw is an important fruit for about two-thirds of the rural population (Table 3.3.1). The ripe fruit is consumed, often as a snack while people work in food gardens. The immature fruit is occasionally cooked and eaten, particularly as an emergency food. This happened, for example, during the 1997 drought. Pawpaw is less commonly eaten in highlands locations. Many plants grow in newly established food gardens after seed is dispersed by animals, but some trees are deliberately planted. Pawpaw is grown in most lowland environments and bears from sea level to 1700 m altitude (Figure

1 See French (1986) for a comprehensive listing of fruit in PNG. In this volume Tables 2, 3 and 4 list more than 60 fruit species according to their likely period of domestication or introduction into PNG.

2 The data are from the Mapping Agricultural Systems of PNG Project (MASP; see Section 1.15). Figures are the number of rural villagers living in agricultural systems where each fruit species was classed as important only (in contrast to the classification of staple crops, for which three categories are used – see Table 3.1.1). Because the number of plants per household is small and the plants are dispersed, relative importance is not easy to assess in the field and the figures are subject to large errors. Nevertheless, the *ranking* of the relative importance of the species is likely to be fairly accurate.

3 Banana was classed as a staple in the MASP database as banana fruit are more often eaten cooked than uncooked (Section 3.1). Breadfruit is described under nuts (Section 3.4).

4 See papers presented at the Papua New Guinea Fruit and Nut Workshop, University of Vudal in 2005, in particular those on introduced fruit by R.M. Bourke, T. Nevenimo and B.J. Watson. These papers are scheduled to be published by NARI.

1.13.3), although fruit are of poorer quality and less sweet when grown above 1200 m altitude. Production is non-seasonal in the lowlands and intermediate altitude classes. Fruit is available seasonally in Eastern Highlands Province, with fruit ripening in August–October. The first recorded introduction of pawpaw was to the Rai Coast of Madang Province in 1871, where it was rapidly adopted (see History of agriculture). As with other introduced foods, there are likely to have been multiple introductions at different locations. Pawpaw fruit is commonly sold in lowland fresh food markets and sometimes in urban food stores.

***Marita* pandanus** is an important fruit for almost 60% of the rural population. The fruit is cylindrical in shape, up to a metre long, usually red in colour, but sometimes yellow. The pericarp (outer layer) of the fruit is rich in oil. The fruit is cut into pieces then boiled, roasted or cooked in a stone oven. The pulp and seeds are removed from the core, mashed with water and strained to produce a thick, rich red sauce used to flavour other foods such as sweet potato, banana and green vegetables. *Marita* is widely planted on the New Guinea mainland, but is uncommon in the Islands Region (Figure 3.3.1). It is not usually grown near the ocean, but grows from low altitudes in inland locations (10–50 m altitude) up to 1700 m. It is most common over the range 500–1500 m above sea level. It is an important food in intermediate altitude locations where coconut does not bear well and vegetable oil or animal fat in villagers' diets is limited. *Marita* grows best in moist locations, often under shade, and tolerates waterlogged soils. It is frequently grown with other fruit and nut bearing trees in 'orchards' on fallow land in secondary forest.

There is a clear relationship between the length of the *marita* fruiting season and altitude in PNG. Near sea level, production is continuous and non-seasonal. With increasing altitude, the producing period becomes shorter. Near the top of its altitudinal range at 1500–1700 m, fruit ripens over a four-month period, usually January to April (Figure 1.13.4). *Marita* was domesticated in New Guinea a long time ago. It is commonly sold in fresh food markets in the producing areas, particularly in the highlands, although production has probably not kept pace with population growth.

Pineapple is grown by more than half the rural population. It is widely grown in most environments in the lowlands, intermediate altitudes and lower highland valleys. The fruit is eaten raw. Pineapple grows and bears up to 1800 m altitude, but the smooth leaf type is more common in the highlands above 1500 m and the rough leaf type is more common in the lowlands and intermediate altitude classes. The sweetest fruit is grown over the altitudinal range 400–1200 m. Production is seasonal, with the best supply usually between October and March,

Table 3.3.1 Rural population growing the most important fruit

Fruit	Rural population	%
Pawpaw	2,642,804	63
Marita pandanus	2,465,476	59
Pineapple	2,197,921	52
Mango	1,962,731	47
Watermelon	1,090,010	26
Ton	1,046,111	25
Malay apple	985,150	23
Guava	723,650	17
Orange	503,488	12
Passionfruit	327,880	8
Avocado	245,103	6
Bukabuk	231,559	6
Mon	217,778	5
Golden apple	179,985	4
Mandarin	175,972	4
Banana passionfruit	127,472	3
Parartocarpus	102,484	2
Rukam	100,966	2
Tamarillo	74,522	2
Pomelo	61,851	1
Pouteria	56,591	1
Coastal pandanus	55,367	1

Sources: NSO (2002), MASP.

although the period of peak production varies from year to year. Pineapple was introduced after 1870. Fruit is commonly sold in fresh food markets in the lowlands. It has become a popular fruit in urban centres in recent decades and production for sale has subsequently increased. This is partly because pineapple is less easily damaged than softer fruits, the poor handling of which inhibits sales in distant urban markets.

Mango (*Mangifera indica*) is grown by almost half the rural population and is widespread in most lowland locations. The fresh fruit is eaten either partly or fully ripe. Mango bears from sea level to 1600 m altitude, although the quality of fruit is poorer above about 1200 m. Mango bears more, better quality fruit in lowland locations that have a marked dry season each year, such as coastal Central Province, southern Western Province, the Rabaraba–Cape Vogel area of Milne Bay Province, the upper Markham Valley, the Sialum area on the Huon Peninsula, and north-eastern parts of the Gazelle Peninsula of East New Britain Province (Figure 3.3.2). Production is markedly seasonal, with most fruit ripening from October to January.[5] It is commonly sold in fresh food markets in the producing areas. Significant quantities are transported from the Markham and Ramu valleys into the highlands for sale. Mango was introduced to PNG after 1870. Virtually all trees in PNG have been derived from seedlings rather than from selected clones. Thus fruit quality is only moderately high. A number of selected clones are available from research stations, but planting material has not been widely distributed.

An indigenous species, *Mangifera minor*, is also grown and eaten in PNG. It is more common in locations where rainfall is continuous throughout the year. In these places, such as the interior of New Britain, the introduced mango does not bear. Total

5 See Bourke et al. (2004:27–29) for a detailed discussion on variation in the production pattern of *Mangifera indica* between locations in PNG and the relationship with environmental factors, particularly rainfall.

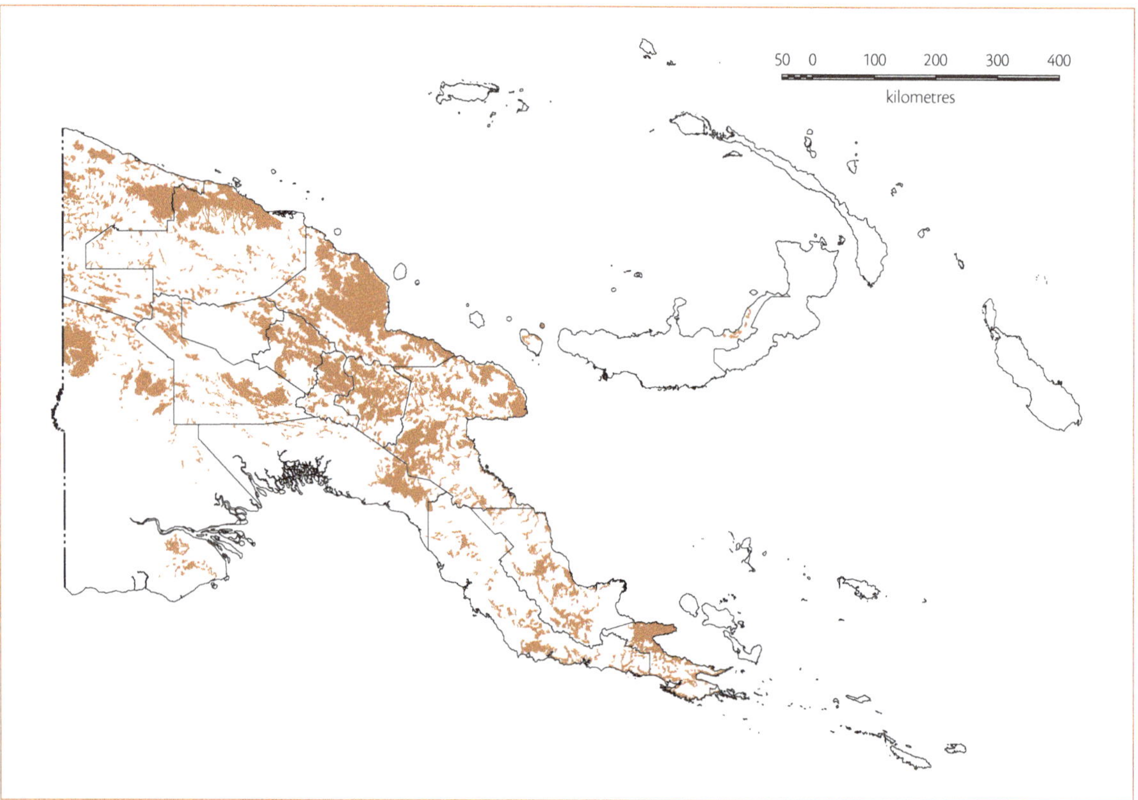

Figure 3.3.1 Distribution of *marita* pandanus as an important fruit. Source: MASP.

production of the indigenous species is much less than that of *M. indica*. The indigenous species has fibrous fruit with a strong turpentine flavour.

Watermelon is grown by about a quarter of the rural population. The flesh is eaten raw. It is most common in lowland locations with a marked dry season each year, such as coastal Central Province, parts of Milne Bay Province, the Gazelle Peninsula in East New Britain Province, and the Markham and Ramu valleys. It is uncommon in locations that are continuously wet throughout the year. Watermelon grows from sea level to 1700 m, but is uncommon above 1200 m altitude. Production is seasonal, with the best supply occurring between November and March. It was first introduced to Woodlark Island in 1847, with the next recorded introduction to the Rai Coast in Madang Province in 1871. It was rapidly adopted in seasonally dry lowland environments. It is commonly sold in lowland markets and significant quantities are transported from the Ramu and Markham valleys to the highlands for sale during the producing season.

Ton[6] is an indigenous fruit from large trees that belong to the same botanical family (Sapindaceae) as the litchi, rambutan and pulasan. *Ton* trees are planted from seed in fallow land, along paths and on the edges of villages. The fruit is eaten raw and has a similar taste and texture to litchi. It is consumed by about a quarter of the rural population. *Ton* is common along the New Guinea north coast and in the Islands Region (Figure 3.3.3). The tree grows from sea level to about 1700 m, but the fruit is eaten only up to 800 m. Above that altitude, villagers say that the tree bears fruit, but the fruit is 'not sweet' and they do not eat it. Fruit is available seasonally for about two or three months sometime between August and April each year, most commonly in the period November–February. *Ton* was domesticated in the New Guinea area a long time ago. A small quantity of fruit is sold in markets in the Momase

6 *Taun* is an alternative spelling for its common name, but *ton* is adopted here as this better reflects the pronunciation in Tok Pisin.

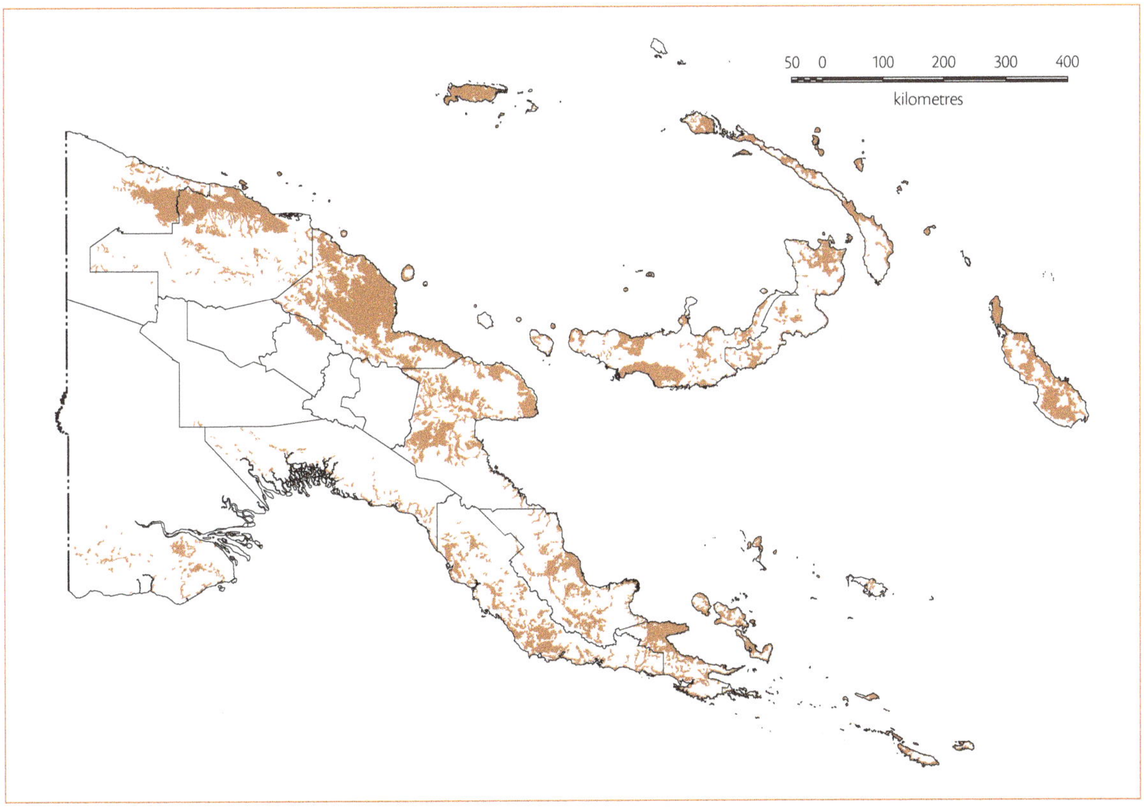

Figure 3.3.2 Distribution of introduced mango as an important fruit. Source: MASP.

and Islands regions. Some observers consider that *ton* has excellent potential for commercialisation for sale within PNG and possibly overseas.

Malay apple is eaten raw and is grown by about a quarter of the rural population. Malay apple grows from sea level to 850 m altitude. It is common in coastal and inland locations in the Islands Region and Milne Bay Province (Figure 3.3.4). A related species, watery rose apple, is grown and eaten occasionally in the lowlands and intermediate altitude classes up to 1600 m, mainly in New Ireland, East New Britain, West New Britain and Milne Bay provinces. Malay apple ripens sometime between September and February, particularly in December–January, but the seasonal production pattern is not well defined. It was probably introduced from South-East Asia some thousands of years ago. In the producing areas it is commonly sold in fresh food markets, where it is a popular fruit.

Guava is grown by about a sixth of the population. Fresh fruit is eaten when partially or fully mature, often by children. Guava grows to 1850 m altitude. It is a minor fruit in many lowland environments and is rarely grown in the highlands. Fruiting is non-seasonal in the lowlands, but seasonal in Eastern Highlands Province, with the best supply in February–May. The first documented introduction of guava to PNG is to the Duke of York Islands in 1875. Minor quantities are marketed in lowland fresh food markets.

Orange and mandarin are grown by about a sixth of the population. Fruit is consumed fresh. Mandarin is a convenient snack food as the fruit is easy to peel and less messy than oranges. Orange is grown in small quantities in the lowlands, intermediate altitudes and lower highland valleys. It is more common in the lowlands south of about 5° latitude. Mandarin is grown in the same broad environment, but is more common at intermediate altitudes. Both species bear from sea level to 1800 m altitude; the best fruit are produced at 800–1400 m.

The main producing season for orange is April to August, but the pattern varies a lot from year to year and fruit is available throughout most of the

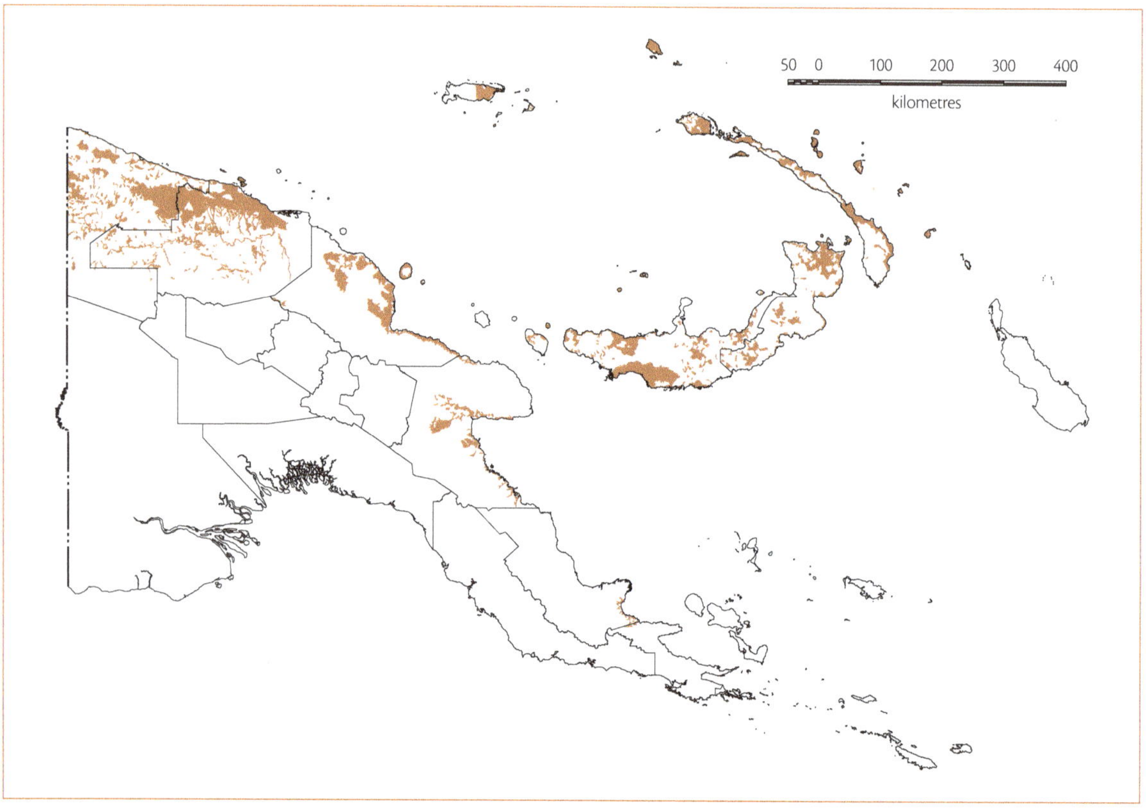

Figure 3.3.3 Distribution of *ton* as an important fruit. Source: MASP.

year. Mandarin has a more well-defined producing season, with most fruit maturing in May–August. Orange, mandarin and other citrus were introduced after 1870, with the first recorded introduction for orange to the Rai Coast in 1873 and to the Duke of York Islands in 1875. Limited quantities of orange are sold in lowland and highland fresh food markets. Greater quantities of mandarin are grown for sale in intermediate altitudes. It is a significant cash crop for villagers in some locations, for example, in the Arona Valley in Eastern Highlands Province, in the Bulolo–Wau area and parts of the Huon Peninsula in Morobe Province, and the Kokoda Trail area of Central Province.

Passionfruit is grown by about a tenth of the rural population, mostly in the highlands. Five types of passionfruit are grown: *suga prut*, purple passionfruit, banana passionfruit, lowland yellow passionfruit and granadilla. The fruit is eaten fresh. All five species were introduced after 1870. *Suga prut* (highland yellow passionfruit) is the most commonly grown. It is grown in the highlands over an altitudinal range of 1350–2350 m. It was a minor fruit until the mid 1970s, but its popularity has increased rapidly since then because the flesh is sweet and sweet fruits are valued by highlanders. Production is non-seasonal. Fruit is commonly sold in highland and some lowland markets.

Purple passionfruit was a village cash crop in the highlands from 1952 to 1974. Around 400–700 tonnes of fruit was purchased each year in Goroka and Mount Hagen and pulp was extracted in a factory in Goroka and exported to Australia. The exports collapsed in 1974 and production declined so that purple passionfruit is now a minor fruit. Purple passionfruit is grown from 800 m to 2300 m altitude. Production is markedly seasonal with fruit available in January–April. Some fruit is sold in highland markets.

Banana passionfruit is a wild species that grows at high altitudes. It fruits from 1850 m to 2800 m altitude. Fruit is gathered for consumption or sale in highland and some lowland markets. Production seems to be non-seasonal.

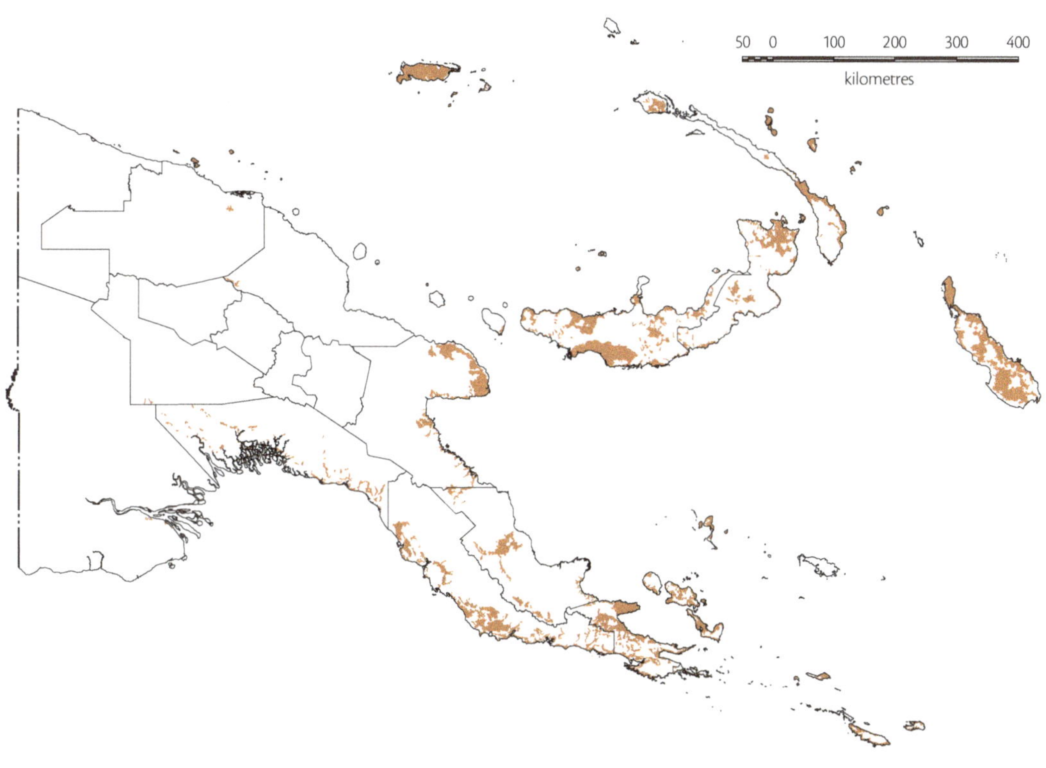

Figure 3.3.4 Distribution of Malay apple as an important fruit. Source: MASP.

Lowland yellow passionfruit is a minor fruit that is grown from sea level to 850 m altitude. Fruit is occasionally sold in lowland fresh food markets. Fruiting is non-seasonal.

Granadilla produces a large fruit, typically 15–25 cm long. It is a minor fruit in the lowlands and grows from sea level to 1000 m. Fruiting is non-seasonal. Small quantities of fruit are sold in some lowland markets.

Sources

Aburu, K. (1982). Notes on some fruit and nut trees in the Keravat area of East New Britain. In Bourke, R.M. and Kesavan, V. (eds). *Proceedings of the Second Papua New Guinea Food Crops Conference: Part One.* Department of Primary Industry, Port Moresby. pp. 101–122.

Bala, A.A. (ed.) (1991). *Proceedings of the Papua New Guinea First International Fruit Conference.* Department of Agriculture and Livestock and PNG Fruit and Vegetable Advisory Board, Port Moresby.

Bourke, R.M. (1989). Altitudinal limits of 230 economic crop species in Papua New Guinea. Unpublished paper. Department of Human Geography, Research School of Pacific and Asian Studies, The Australian National University, Canberra.

Bourke, R.M. (2005). An overview of edible fruit and nuts in Papua New Guinea. Unpublished paper. University of Vudal, East New Britain.

Bourke, R.M. (2005). Potential for further commercial development of introduced fruit. Unpublished paper. University of Vudal, East New Britain.

Bourke, R.M. (2005). Indigenous fruit in Papua New Guinea. Unpublished paper. University of Vudal, East New Britain.

Bourke, R.M., Camarotto, C., D'Souza, E.J., Nema, K., Tarepe, T.N. and Woodhouse, S. (2004). *Production Patterns of 180 Economic Crops in Papua New Guinea.* Coombs Academic Publishing, Canberra.

French, B.R. (1986). *Food Plants of Papua New Guinea: A Compendium.* Sheffield, Tasmania.

Gibson, J. (2001). The economic and nutritional importance of household food production in PNG. In Bourke, R.M., Allen, M.G. and Salisbury, J.G. (eds). *Food Security for Papua New Guinea. Proceedings of the Papua New Guinea Food and Nutrition 2000 Conference.* ACIAR Proceedings No. 99. Australian Centre for International Agricultural Research, Canberra. pp. 37–44.

International Bank for Reconstruction and Development (1965). *The Economic Development of the Territory of Papua and New Guinea.* The Johns Hopkins Press, Baltimore.

Nevenimo, T. (2005). Exotic fruit with potential in the lowlands. Unpublished paper. University of Vudal, East New Britain.

NSO (National Statistical Office of Papua New Guinea) (2002). Papua New Guinea 2000 Census: Final Figures. National Statistical Office of Papua New Guinea, Port Moresby.

Tarepe, T. and Bourke, R.M. (1982). Fruit crops in the Papua New Guinea highlands. In Bourke, R.M. and Kesavan, V. (eds). *Proceedings of the Second Papua New Guinea Food Crops Conference: Part One.* Department of Primary Industry, Port Moresby. pp. 86–100.

Thomas, G.S. (1982). Review of the prospects for food processing in Papua New Guinea. In Bourke, R.M. and Kesavan, V. (eds). *Proceedings of the Second Papua New Guinea Food Crops Conference: Part Two.* Department of Primary Industry, Port Moresby. pp. 408–420.

Walter, A. and Sam, C. (2002). *Fruits of Oceania.* ACIAR Monograph No. 85. Australian Centre for International Agricultural Research, Canberra.

Watson, B.J. (2005). Tropical fruit development in Papua New Guinea – 'the hard yards'. Unpublished paper. University of Vudal, East New Britain.

Woodhouse, S. (1991). Asian fruit in Papua New Guinea. In Bala, A.A. (ed). *Proceedings of the Papua New Guinea First International Fruit Conference.* Department of Agriculture and Livestock and PNG Fruit and Vegetable Advisory Board, Port Moresby. pp. 161–168.

3.4 Nuts

More than 40 plant species that have an edible nut are grown in PNG.[1] Many nuts are grown in small quantities in a limited number of locations. Thirteen species are grown in sufficient quantity to be classified as 'important', all of which are indigenous (Tables 3.4.1, A3.4.1).[2] Ten of these 'important' species, which are grown by more than 8% of the rural population, are described below.[3] A number of nut crops have been introduced since 1870 and three of these – macadamia, cashew and pecan – are grown, but all are very minor foods.

Five of the species discussed here are considered as having significant potential for commercial development for both the domestic and export market. These are *galip*, planted *karuka*, *okari*, *pao* and sea almond. Three of these are sold as processed nuts in modern packaging in Vanuatu. In the medium to long term, the indigenous edible nuts could be worth many hundreds of millions of kina to the PNG economy (see Section 5.3).

Breadfruit is grown by more than half the rural population (Table 3.4.1). The flesh and seed (nut) are eaten after being cooked, commonly in a stone oven, but there are numerous ways of cooking and preparing breadfruit.[4] The tree is grown in all lowland environments in PNG up to 1250 m altitude and it is one of the most common food-bearing trees in the lowlands. Both the flesh and nut are eaten on the smaller islands off the north coast of New Guinea, the Bismarck Archipelago, the Solomon chain, and throughout Milne Bay Province. In contrast, only the nut is eaten on the mainland of New Guinea, with some exceptions, such as the mainland of Milne Bay Province and some coastal locations in Central Province. Types of breadfruit with few or no seeds have been introduced into PNG from Polynesia over the past century, but these remain uncommon.

Breadfruit does not produce fruit in a regular manner in most of PNG and the production period varies from year to year. The exception is in Milne Bay Province at 8–12° south where the producing period commences in October or November.

1 See French (1986) for a comprehensive listing of edible nuts in PNG. In this volume Tables 2, 3 and 4 list 22 nut species according to their likely period of domestication or introduction into PNG.

2 The data are from the Mapping Agricultural Systems of PNG Project (MASP; see Section 1.15). Figures are the number of rural villagers living in agricultural systems where each nut species was classed as important only (in contrast to the classification of staple crops, for which three categories are used – see Table 3.1.1). Because the number of plants per household is small and the plants are dispersed, relative importance is not easy to assess in the field and the figures are subject to large errors. Nevertheless, the *ranking* of the relative importance of the species is likely to be fairly accurate.

3 Coconut is considered under staples (Section 3.1) and peanut is included with vegetables (Section 3.2), although both could be classed as a nut.

4 Other uses from the tree include medicine, timber, fuel wood, canoe construction, clothing, rope, wrapping and adhesive.

Production is likely to be seasonal at locations south of about 8° latitude, but is irregular at most locations in PNG that are nearer the equator. Such a pattern results from the changes in daylength during the year with increasing distance from the equator.

Breadfruit is an ancient crop in PNG and was probably domesticated in New Guinea. Even today, people select the best trees from which to collect seed for replanting. Plants are sometimes propagated from root shoots so as to maintain high-yielding trees. Poor-yielding trees are cut down to prevent them cross-fertilising high-yielding trees. Nuts are commonly sold in markets in Momase Region and fruit in markets on islands. Breadfruit is probably now a less important food than before the wide-spread adoption of sweet potato, cassava and Chinese taro in the lowlands.

Planted *karuka* (*Pandanus julianettii*), and a wild species (*P. brosimos*), are the most important of a number of pandanus species that produce edible nuts in PNG, all of which grow above 1000 m. Planted *karuka* is grown by almost half the rural population. It grows in an altitudinal range of 1800–2600 m (Figure 1.13.3), confined mainly to a band in the central and fringe highlands and the Huon Peninsula (Figure 3.4.1).

Kernels (nuts) of *karuka* pandanus are an important seasonal dietary item for those living at high altitudes in the New Guinea highlands. The kernel is eaten raw or cooked by roasting it in an open fire, baking in hot ashes or steaming in a stone oven. The nuts can be preserved by drying and smoking above a house fire. For longer storage, the kernels are extracted and stored in baskets hung in house rafters. Smoke from the house fires imparts a characteristic flavour to the nuts. When the nuts are in season, entire households and their domestic pigs migrate from villages to high-altitude bush camps for weeks to harvest and eat the nuts. When sweet potato is scarce because of frost damage or other causes, villagers depend upon *karuka* if it is in season. The nuts are highly nutritious and provide both protein and oil, the two components that tend to be deficient in highlanders' diets.

Production is irregular in the western part of the highlands, where rainfall seasonality is slight or absent. In the eastern part of the highlands, where rainfall is seasonally distributed, production is more seasonal, but there is still large year-to-year variation in the harvest size. In any year the producing period also varies between locations. The nuts are most likely to mature during January–March, but nuts may mature during any month of the year. After periods of soil moisture stress or drought, the producing periods coincide at most locations. The biggest harvests tend to follow major droughts.

Planted *karuka* is endemic to New Guinea and was domesticated there, probably from the wild species (*P. brosimos*). Surveys indicate that a significant proportion of village trees are immature, which suggests that people are making new plantings. *Karuka* nuts are commonly sold in highland markets, either cooked or uncooked.

Wild *karuka* (*Pandanus brosimos*) is eaten by about a third of the rural population. The kernel (nut) of wild *karuka* is an important food for villagers living at high altitudes in New Guinea, although it is not as important as planted *karuka*. Kernels are eaten both cooked and raw. This species is widespread in high-altitude locations (2400–3100 m) in the central and fringe highlands and the Huon Peninsula. Although

Table 3.4.1 Rural population growing the most important nuts

Edible nut	Population	%
Breadfruit	2,335,541	56
Karuka, planted	1,998,454	48
Karuka, wild	1,343,205	32
Galip (*Canarium* spp.)	1,314,173	31
Polynesian chestnut (*aila*)	636,749	15
Sea almond (*talis*)	545,993	13
Pao (*Barringtonia procera*)	531,634	13
Okari (*Terminalia kaernbachii*)	501,609	12
Okari (*Terminalia impediens*)	352,130	8
Sis (*solomon*)	323,675	8
Castanopsis	301,214	7
Dausia	90,063	2
Tulip	68,619	2

Sources: NSO (2002), MASP.

its range overlaps with planted *karuka*, wild *karuka* grows above 2800 m, the altitudinal limit of settlement and food gardens (see Section 1.10). When fruit matures, villagers migrate to high altitudes to harvest the fruit and extract the nuts.

Nuts are produced in an unpredictable manner. They are most likely to mature in January–February, but may mature in any month. The producing period may coincide with that of planted *karuka* in nearby locations at lower altitudes, but not always. Wild *karuka* is endemic to New Guinea and is not found elsewhere. Nuts have not been recorded in the main highland markets, but it is possible that they are sold in some high-altitude locations.

Galip (*Canarium* spp.) is grown by almost a third of the rural population. A number of species bear edible nuts in PNG, with *Canarium indicum* being the most important. The kernel of *galip* is generally eaten raw, and occasionally roasted. Trees are grown from protected self-sown seedlings or planted seedlings. It is estimated that there are one million edible *Canarium* trees in PNG, with kernel production of 7200 tonnes per year. *Galip* grows from sea level to 700 m altitude and is widely grown in forested locations below 500 m altitude along the north coast and inland areas of the New Guinea mainland and in the Islands Region (Figure 3.4.2).

Nuts are produced seasonally with the producing period typically about three months long. Latitude has a strong influence on the start of the harvesting season. Nuts usually mature from April–June onwards at 4–6° south, with the harvest commencing progressively later at locations further from the equator.

Canarium indicum is the oldest domesticated nut species in PNG. It has been identified in the archaeological record as early as 14 000 years ago, several thousand years before the start of arable agriculture in New Guinea (see History of agriculture). The distribution of this species in the archaeological record suggests that domesticated varieties were being moved between islands thousands of years ago. It is a direct link to an earlier era of hunting and gathering in New Guinea's past.

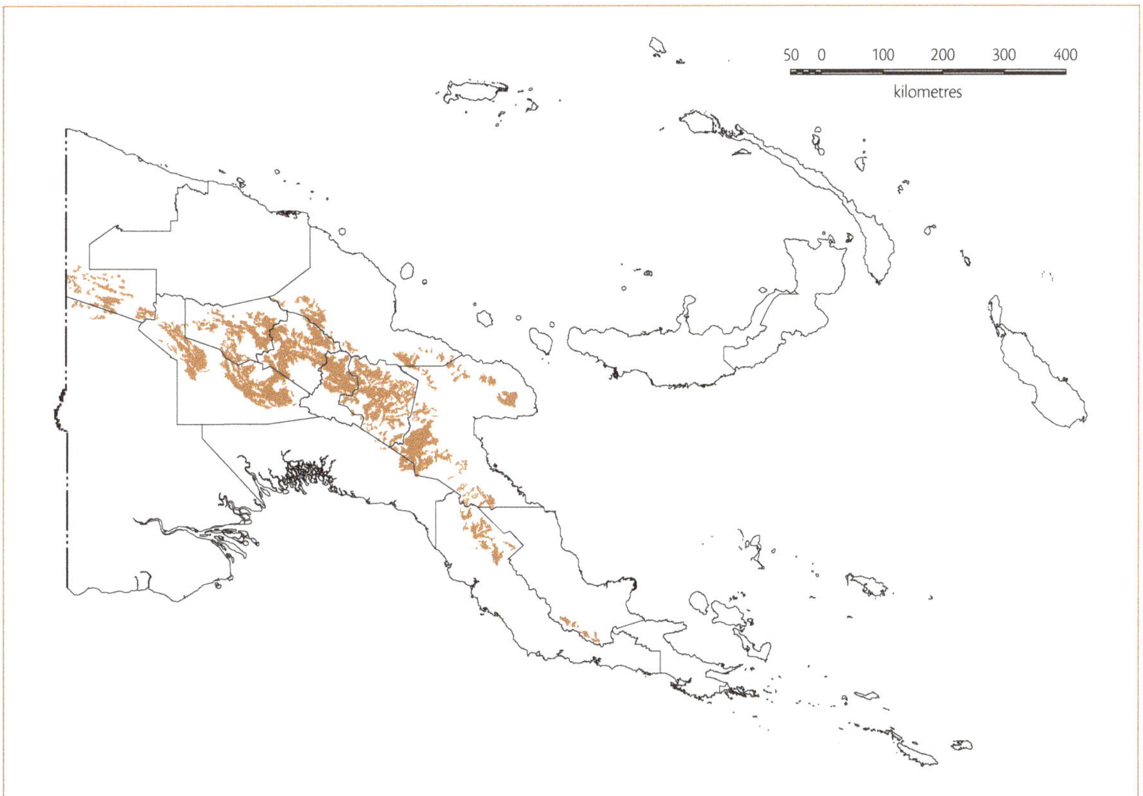

Figure 3.4.1 Distribution of planted *karuka* as an important nut. Source: MASP.

There are indications that consumption is declining. For example, puddings made from *galip* and taro in some locations such as Bougainville Province were once common but are rarely seen now. Nuts are traded in some locations, for example, from Boisa Island to villages near the mouth of the Ramu River in Madang Province (in exchange for sago); on Vokeo Island north-east of Wewak, East Sepik Province; and Siassi Islands in Morobe Province. Again, there are indications that this trade has diminished in recent decades (see Box 5.3).

Potential for commercialisation of *galip* nut is considerable and the potential international market for *galip* nut is large. One unsuccessful attempt to commercialise *galip* was made in the Kandrian area of West New Britain Province. This was done as part of the AusAID-funded Kandrian Gloucester Integrated Development Project in the 1990s and the operation failed as soon as donor support ceased.

Polynesian chestnut (*aila*) is grown by about a sixth of the rural population. It grows from sea level to 400 m altitude on the mainland and island provinces. It is most important in Milne Bay Province, where it is available seasonally when garden food is scarce, and also in the Islands Region. The seed of Polynesian chestnut is cooked prior to consumption by baking the entire fruit or boiling or roasting the nut.

Production in PNG is non-seasonal at locations close to the equator, but fruit is not available in all months. In Milne Bay Province (8–12° south), fruit ripens seasonally over 2–3 months, especially in November–February. Polynesian chestnut has been eaten in PNG for a long time. It seems to have become a less important food since the widespread adoption of sweet potato and cassava over the past 60 years. The decline in importance of the nut as a food is a Pacific-wide phenomenon. Cooked nuts are sold occasionally in markets in the islands.

Sea almond (*talis*) is grown by 13% of the rural population, but this figure overestimates its importance as it is a very minor food in most locations where it grows. The small kernel is eaten raw or roasted, mostly by people who live near

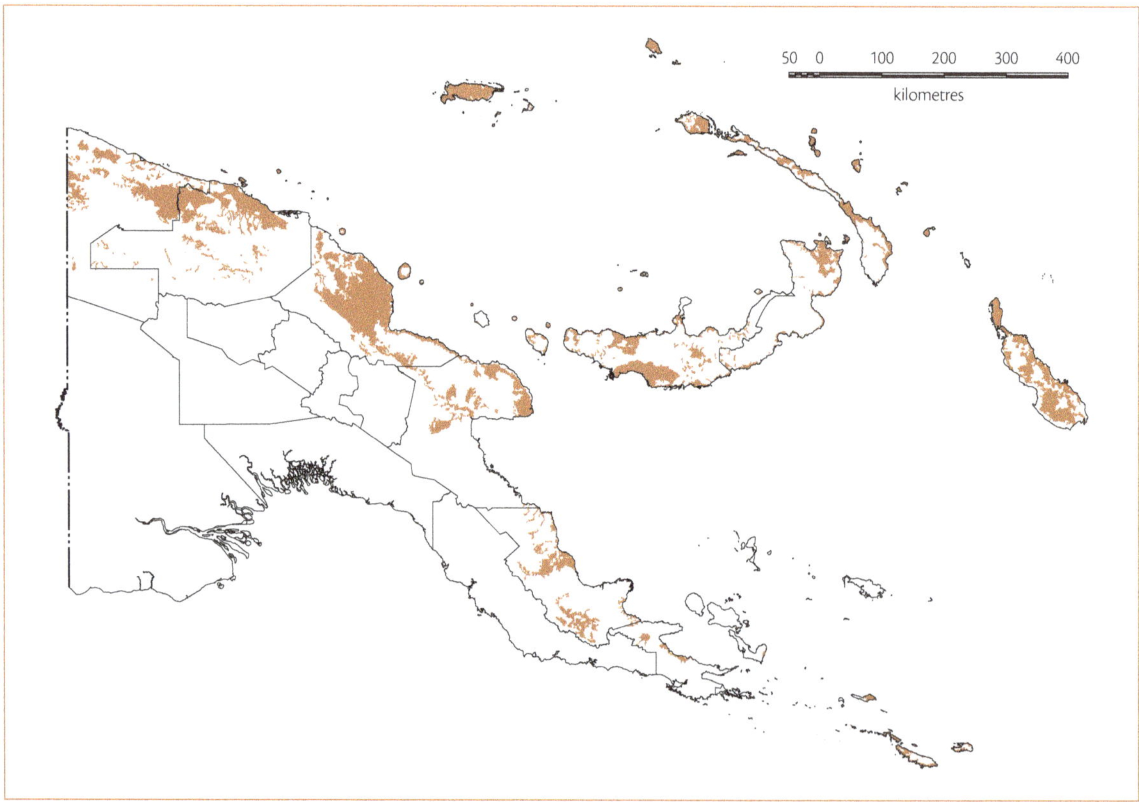

Figure 3.4.2 Distribution of *galip* as an important nut. Source: MASP.

the seashore, particularly children. Sea almond is widely distributed along the coast of all lowland provinces up to 300 m altitude and is most common in Milne Bay Province and the Islands Region. It is sometimes planted in coastal and inland villages and has also been planted as a street tree in a number of PNG towns.

Sea almond fruits sporadically throughout the year near the equator but is seasonal at locations further from the equator, where fruit ripens sometime between November and May. In Milne Bay Province the producing period is reported by villagers as 2–3 months long, with fruiting most commonly around December–February.

Sea almond has a widespread natural distribution in near-coastal areas of the Indian Ocean, throughout coastal tropical Asia and the Pacific Ocean. It has been eaten in PNG for a long time. The fact that superior types exist, including soft-shelled types on Iwa Island in Milne Bay Province and on Mussau Island north of New Ireland, indicate that people have selected superior varieties over a long period. Nuts are not sold in PNG markets. The species has potential for commercialisation for sale within PNG and overseas. It is processed for sale in Vanuatu.

Pao (*Barringtonia procera*) is grown by 13% of the rural population. The kernel is eaten either raw or roasted. *Pao* is planted in coastal and inland locations up to an altitude of 500 m. It has a limited distribution within PNG, being confined mainly to the Islands Region (Figure 3.4.3). *Pao* was probably introduced to PNG from Solomon Islands. It is increasing in importance and since about 1960 has been planted at other locations on New Britain and the New Guinea mainland, including coastal and island locations in Morobe, Madang, East Sepik, Sandaun and Milne Bay provinces.

Pao fruits intermittently in a non-seasonal manner. It is an important edible nut in the Islands Region and is sold in markets there. Edible *Barringtonia* species have been commercialised in Vanuatu, where there is high demand for the nuts. *Barringtonia procera* in particular has considerable potential as a cash crop for both the domestic and export markets, given

Figure 3.4.3 Distribution of *pao* as an important nut. Source: MASP.

the small area required per tree, a relatively short time to maturity and the possibility of growing it as a horticultural crop rather than relying on naturally occurring forest trees. Two related species, *B. novae-hiberniae* and *B. edulis*, bear edible nuts in PNG, but they are very minor foods.

Okari (*Terminalia kaernbachii*) is grown by about 12% of the rural population. The name *okari* is used in Motu (and now English) in the Southern Region for *T. kaernbachii*, but it has been adopted in Tok Pisin as the name for the related *T. impediens*. The former is a more important food and has significant commercial potential, whereas the latter is a minor food with limited commercial potential.

Okari (*T. kaernbachii*) nuts are eaten raw. Trees are preserved in garden land or planted. Villagers either harvest nuts from trees or, more commonly, collect the fallen fruit. *Okari* is mainly distributed in the Southern Region but also occurs in adjacent locations in Southern Highlands, Simbu and Morobe provinces; in Manus Province; and in West New Britain from the Aria River west to Cape Gloucester (Figure 3.4.4).

Okari was domesticated in New Guinea. Over the past 50 years the tree has been introduced to other locations in PNG, including East New Britain and New Ireland. It grows from sea level to 1100 m altitude. The species is uncommon near the ocean in its natural range in southern New Guinea, although it does bear near the ocean at, for example, Keravat on New Britain and Kavieng on New Ireland. It may be that the best production occurs where the diurnal temperature range (the difference in temperature between day and night) is greater.

The producing period is fairly constant from year to year and lasts for 2–4 months, although the size of the annual harvest is variable. There is a clear relationship between latitude and the start of the harvesting period, with the producing period commencing later at locations further south from the equator. This is presumably caused by differences in daylength, although it could also be related to seasonal temperature changes. Nuts mature from about March at 4° south and later at locations further south.

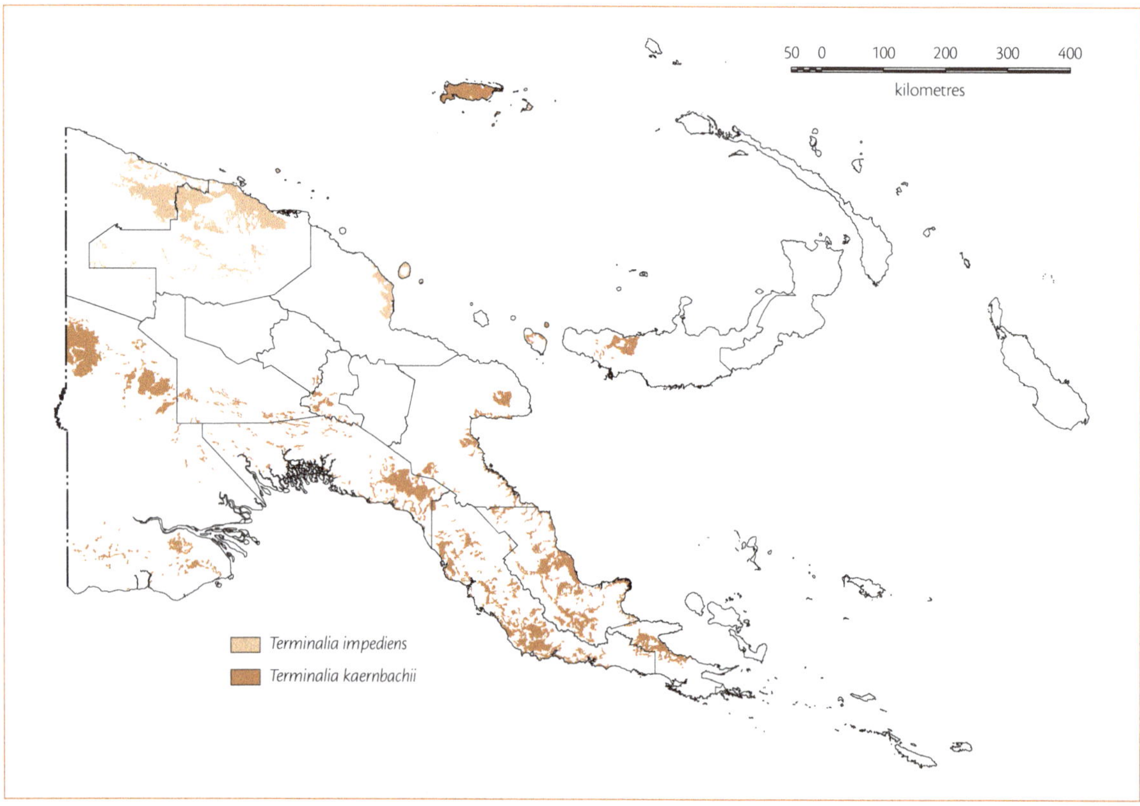

Figure 3.4.4 Distribution of *okari* as an important nut. Source: MASP.

Okari nuts are eaten in the producing region and some are sold in local and regional markets, including in Port Moresby. *Okari* nut has considerable potential for sales within PNG and overseas. A local non-government organisation, Okari Ecoenterprises, attempted to commercialise *okari* nut on the Managalas Plateau in Oro Province in the early to mid 1990s. Nuts were collected from planted and self-sown trees and transported to Port Moresby where they were sold through Associated Distributors. The operation failed after a few years because of an irregular supply of nuts.

Okari (*Terminalia impediens*) is grown by about 8% of rural villagers, but this figure overemphasises its importance as it is a minor food where it is grown. The kernel is eaten raw. Trees are not planted, but are preserved when land is cleared for gardening. This species grows up to 1000 m altitude and is common in East Sepik, Sandaun and Madang provinces (Figure 3.4.4). *Terminalia impediens* was domesticated in New Guinea. Production is not expanding. The species is not sold in fresh food markets and probably has less commercial potential than the closely related *T. kaernbachii* because the nuts are smaller.

Sis (solomon) (*Pangium edule*) is grown by 8% of the rural population and is widely eaten in PNG, despite requiring extensive processing to remove a poison. The seed is eaten only after being washed in water, then roasted and fermented. *Pangium edule* is known by the common name *sis* in some locations in Momase Region and *solomon* in parts of New Britain. It is grown in most provinces, being most commonly eaten in Milne Bay. *Sis* grows from sea level to 1050 m altitude in both coastal and inland lowland locations. Production is seasonal, commonly starting around May or June and lasting 2–4 months. *Sis* is an ancient crop in PNG. It was probably a more important food before the widespread adoption of introduced foods including sweet potato, cassava and Chinese taro. It is occasionally sold in fresh food markets. Because of the extensive processing required to make the nut safe for human consumption, it is unlikely that this species has any commercial potential.

Sources

Aburu, K. (1982). Notes on some fruit and nut trees in the Keravat area of East New Britain. In Bourke, R.M. and Kesavan, V. (eds). *Proceedings of the Second Papua New Guinea Food Crops Conference: Part One.* Department of Primary Industry, Port Moresby. pp. 101–122.

Ase, D. (1996). Production and market development of okari nuts (*Terminalia kaernbachii*) on the Managalas Plateau in Papua New Guinea. In Stevens, M.L., Bourke, R.M. and Evans, B.R. (eds). *South Pacific Indigenous Nuts.* ACIAR Proceedings No. 69. Australian Centre for International Agricultural Research, Canberra. pp. 92–93.

Bourke, R.M. (1989). Altitudinal limits of 230 economic crop species in Papua New Guinea. Unpublished paper. Department of Human Geography, Research School of Pacific and Asian Studies, The Australian National University, Canberra.

Bourke, R.M. (1996). Edible indigenous nuts in Papua New Guinea. In Stevens, M.L., Bourke, R.M. and Evans, B.R. (eds). *South Pacific Indigenous Nuts.* ACIAR Proceedings No. 69. Australian Centre for International Agricultural Research, Canberra. pp. 45–55.

Bourke, R.M. (2005). An overview of edible fruit and nuts in Papua New Guinea. Unpublished paper. University of Vudal, East New Britain.

Bourke, R.M. (2005). Indigenous edible nuts in Papua New Guinea. Unpublished paper. University of Vudal, East New Britain.

Bourke, R.M., Camarotto, C., D'Souza, E.J., Nema, K., Tarepe, T.N. and Woodhouse, S. (2004). *Production Patterns of 180 Economic Crops in Papua New Guinea.* Coombs Academic Publishing, Canberra.

Evans, B. (1996). Overview of resource potential for indigenous nut production in the South Pacific. In Stevens, M.L., Bourke, R.M. and Evans, B.R. (eds). *South Pacific Indigenous Nuts.* ACIAR Proceedings No. 69. Australian Centre for International Agricultural Research, Canberra. pp. 10–35.

Evans, B. (1996). What we don't know about indigenous nuts in Melanesia. In Stevens, M.L., Bourke, R.M. and Evans, B.R. (eds). *South Pacific Indigenous Nuts.* ACIAR Proceedings No. 69. Australian Centre for International Agricultural Research, Canberra. pp. 67–73.

French, B.R. (1986). *Food Plants of Papua New Guinea: A Compendium.* Sheffield, Tasmania.

Henty, E.E. (1982). Some nut-bearing plants in Papua New Guinea. In Bourke, R.M. and Kesavan, V. (eds). *Proceedings of the Second Papua New Guinea Food Crops Conference: Part One.* Department of Primary Industry, Port Moresby. pp. 78–85.

Kennedy, J. and Clarke, W. (2004). *Cultivated landscapes of the southwest Pacific.* Resource Management in Asia-Pacific Working Paper No. 50. Resource Management in Asia-Pacific Program, Research School of Pacific and Asian Studies, The Australian National University, Canberra.

Nevenimo, T., Moxon, J., Wemin, J., Johnston, M., Bunt, C. and Leakey, R.R.B. (2007). Domestication potential and marketing of *Canarium indicum* nuts in the Pacific: 1. A literature review. *Agroforestry Systems* 69(2):117–134.

NSO (National Statistical Office of Papua New Guinea) (2002). Papua New Guinea 2000 Census: Final Figures. National Statistical Office of Papua New Guinea, Port Moresby.

Olsson, M. (1996). Okari Ecoenterprises: a snapshot of participatory rural development. In Stevens, M.L., Bourke, R.M. and Evans, B.R. (eds). *South Pacific Indigenous Nuts.* ACIAR Proceedings No. 69. Australian Centre for International Agricultural Research, Canberra. pp. 94–99.

Pauku, R.L. (2005). *Inocarpus fagifer* (Tahitian chestnut), ver. 1.1. In Elevitch, C.R. (ed). *Species Profiles for Pacific Island Agroforestry.* Permanent Agriculture Resources (PAR), Holualoa, Hawai'i. <http://www.traditionaltree.org>.

Pauku, R.L. (2005). *Barringtonia procera* (cutnut), ver. 1.1. In Elevitch, C.R. (ed). *Species Profiles for Pacific Island Agroforestry.* Permanent Agriculture Resources (PAR), Holualoa, Hawai'i. <http://www.traditionaltree.org>.

Ragone, D. (2005). *Artocarpus altilis* (breadfruit), ver. 1.2. In Elevitch, C.R. (ed). *Species Profiles for Pacific Island Agroforestry.* Permanent Agriculture Resources (PAR), Holualoa, Hawai'i. <http://www.traditionaltree.org>.

Rose, C.J. (1982). Preliminary observations on the pandanus nut (*Pandanus jiulianettii* Martelli). In Bourke, R.M. and Kesavan, V. (eds). *Proceedings of the Second Papua New Guinea Food Crops Conference: Part One.* Department of Primary Industry, Port Moresby. pp. 160–167.

Stone, B.C. (1982). New Guinea Pandanaceae: first approach to ecology and biogeography. In Gressitt, J.L. (ed). *Biogeography and Ecology of New Guinea.* Monographiae Biologicae Volume 42. Dr W. Junk Publishers, The Hague. pp. 401–436.

Thomson, L.A.J. and Evans, B. (2005). *Canarium indicum* var. *indicum* and *C. harveyi* (canarium nut), ver. 1.3. In Elevitch, C.R. (ed). *Species Profiles for Pacific Island Agroforestry.* Permanent Agriculture Resources (PAR), Holualoa, Hawai'i. <http://www.traditionaltree.org>.

Thomson, L.A.J. and Evans, B. (2005). *Terminalia catappa* (tropical almond), ver. 2. In Elevitch, C.R. (ed). *Species Profiles for Pacific Island Agroforestry.* Permanent Agriculture Resources (PAR), Holualoa, Hawaii. <http://www.traditionaltree.org>.

Walter, A. and Sam, C. (2002). *Fruits of Oceania.* ACIAR Monograph No. 85. Australian Centre for International Agricultural Research, Canberra.

Wissink, D. (1996). Galip (*Canarium indicum*) as a cash crop in West New Britain, Papua New Guinea: experiences of the Kandrian Gloucester Integrated Development Project. In Stevens, M.L., Bourke, R.M. and Evans, B.R. (eds). *South Pacific Indigenous Nuts.* ACIAR Proceedings No. 69. Australian Centre for International Agricultural Research, Canberra. pp. 84–91.

Yen, D.E. (1996). Melanesian arboriculture: historical perspectives with emphasis on the genus *Canarium.* In Stevens, M.L., Bourke, R.M. and Evans, B.R. (eds). *South Pacific Indigenous Nuts.* ACIAR Proceedings No. 69. Australian Centre for International Agricultural Research, Canberra. pp. 36–44.

3.5 Stimulants

The term 'stimulant' here means a group of substances that people use to alter their perception, mood, consciousness or behaviour. The substances covered (tobacco, betel nut, marijuana, coffee, tea, locally brewed alcohol, kava and other psychoactive plants) are not all stimulants in a medical sense, but this name is less emotive than terms such as 'narcotic' or 'drug'. The more precise term is 'psychoactive drug'.[1] These include drugs used as anaesthetics, painkillers, psychiatric medication and for recreation. Globally, the most common recreational psychoactive drugs are alcohol (in the form of beer, wine or spirits), nicotine (tobacco), caffeine (coffee, tea and many aerated soft drinks), betel nut and marijuana.

Six plant species are grown in sufficient quantity in PNG to be classified as 'important' stimulants (Tables 3.5.1, A3.5.1).[2] Tobacco and betel nut are the most widely consumed. Because it is illegal to produce or sell marijuana, it was not possible to collect reliable data on its distribution. It is also illegal to sell locally produced alcohol without a licence, but consumption of tobacco, betel nut and kava is not illegal. There is negligible or no use of other recreational drugs such as amphetamines (speed), heroin, cocaine, ecstasy or LSD by Papua New Guineans in PNG.

Tobacco grows in most environments in PNG, from sea level to 2400 m (Figure 3.5.1). Most of the rural population live in locations where tobacco is commonly grown (Table 3.5.1). A large proportion of the adult male population, fewer women and many adolescent children, smoke tobacco. People use home-grown tobacco, and self-rolled and factory-manufactured cigarettes made from imported tobacco. Consumption of home-grown tobacco is probably much greater than that of manufactured cigarettes, but accurate data are not available. Locally grown tobacco is usually smoked as cigarettes rolled in newspaper, or rolled tobacco leaf. It is sometimes smoked in bamboo or wooden pipes and very occasionally eaten (for ritual purposes). It is uncommon on atolls and is not grown where most of the population are members of the Seventh-Day

1 A psychoactive drug is a chemical substance that acts primarily upon the central nervous system, where it alters brain function, resulting in temporary changes in perception, mood, consciousness or behaviour. These drugs may be used recreationally to purposefully alter one's consciousness, for ritual or spiritual purposes, or as medication.

2 The data are from the Mapping Agricultural Systems of PNG Project (MASP; see Section 1.15). Figures are the number of rural villagers living in agricultural systems where each stimulant was classed as important only (in contrast to the classification of staple crops, for which three categories are used – see Table 3.1.1). Because the number of plants per household is small and the plants are dispersed, relative importance is not easy to assess in the field and the figures are subject to large errors. Nevertheless, the *ranking* of the relative importance of the species is likely to be fairly accurate. The term 'narcotic' was used in the MASP database for betel nut, tobacco and kava.

Adventist Church. Towards the top of its altitudinal range it is often grown under the eaves of houses, presumably to protect plants from the cold.

Production is non-seasonal in the highlands. Estimates of production are not available, but it is likely that production has kept pace with population growth in remote locations. It may not have done so where people have greater access to manufactured cigarettes and imported tobacco. Sales in local and distant markets are a significant source of cash income for some people (see Section 5.17).

Betel nut (*Areca catechu*) grows in most lowland environments from sea level to 1100 m (Figures 3.5.2, 1.13.3). It is grown by more than half the rural population. This figure understates the proportion of adults who consume betel nut because significant quantities are traded into the highlands, where it does not grow. Betel nut is usually consumed with the catkins,[3] leaves or stems of the lowland betel pepper plant and less commonly with highland betel pepper (see below), together with slaked lime made from cooked and crushed seashells or coral. People sometimes chew betel nut on its own. It is an everyday substance and is commonly given as a small gift, but is also used in more formal situations, such as ceremonies.

Betel nut was domesticated in South-East Asia and introduced to PNG some thousands of years ago. Production has increased rapidly in recent decades, partly for local consumption, but especially for sale in urban centres and the highlands.

The supply is continuous throughout the year, but varies seasonally. In general the best supply occurs between April and August and the worst supply between September and December. There are some regional differences and the supply is less seasonal in Madang and more seasonal in Lae. The best supply in the Madang area occurs in January–March and this is the period of poorest supply in the Port Moresby area. A disease of unknown cause is destroying betel nut in the Markham Valley. It started to spread in about 2003 and the supply had almost ceased there by mid 2007.

Betel nut is a major cash crop in many lowland locations and significant quantities are transported to Port Moresby, Lae, other urban centres, mine sites and the highlands (Figures 5.1.1, 5.17.2). The 1996 Household Survey estimated production at 49 000 tonnes per year and consumption of betel nut, pepper and lime at 11 kg/person/year, with consumption levels similar in both urban and rural locations (Table A2.1.1).

Lowland betel pepper is grown by more than half the rural population. It is only consumed with betel nut or highland betel nut. The catkin, leaf and vine are used, with the catkin being the part most commonly used in PNG and the leaf and vine only used when the catkin is not available. Lowland betel pepper grows from sea level to 1000 m altitude. Its distribution is very similar to that of betel nut (Figure 3.5.2), although it does not grow to the altitude that betel nut does. Catkins sold in Kainantu market, which come from Karkar Island and the Markham Valley, are available throughout the year, but the supply varies seasonally with the best supply in January–March and the poorest in September–October.

Lowland betel pepper was introduced to PNG from South-East Asia thousands of years ago, almost certainly with betel nut. Production has increased in the lowlands in recent decades and in most locations has probably exceeded population growth as betel nut consumption has increased. In some locations where it has become a significant cash crop there has been a marked increase in production so that catkins can be sold in distant urban or highlands markets.

Table 3.5.1 Rural population growing the most important stimulants

Stimulant	Population	%
Tobacco	4,134,099	99
Betel nut	2,172,390	52
Lowland betel pepper	2,149,154	51
Highland betel pepper	1,272,570	30
Highland betel nut	649,472	15
Kava	33,432	1

Sources: NSO (2002); MASP.

3 A catkin is a spike of unisexual flowers with no petals.

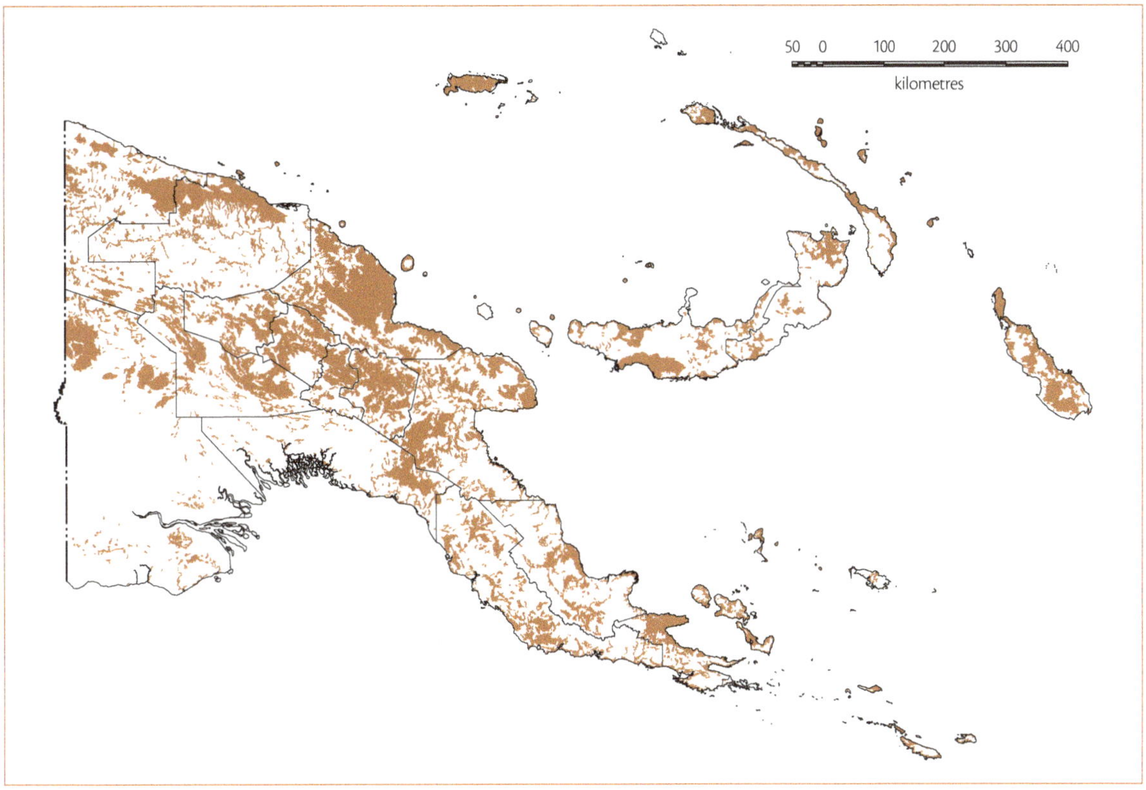

Figure 3.5.1 Distribution of tobacco as an important stimulant. Source: MASP.

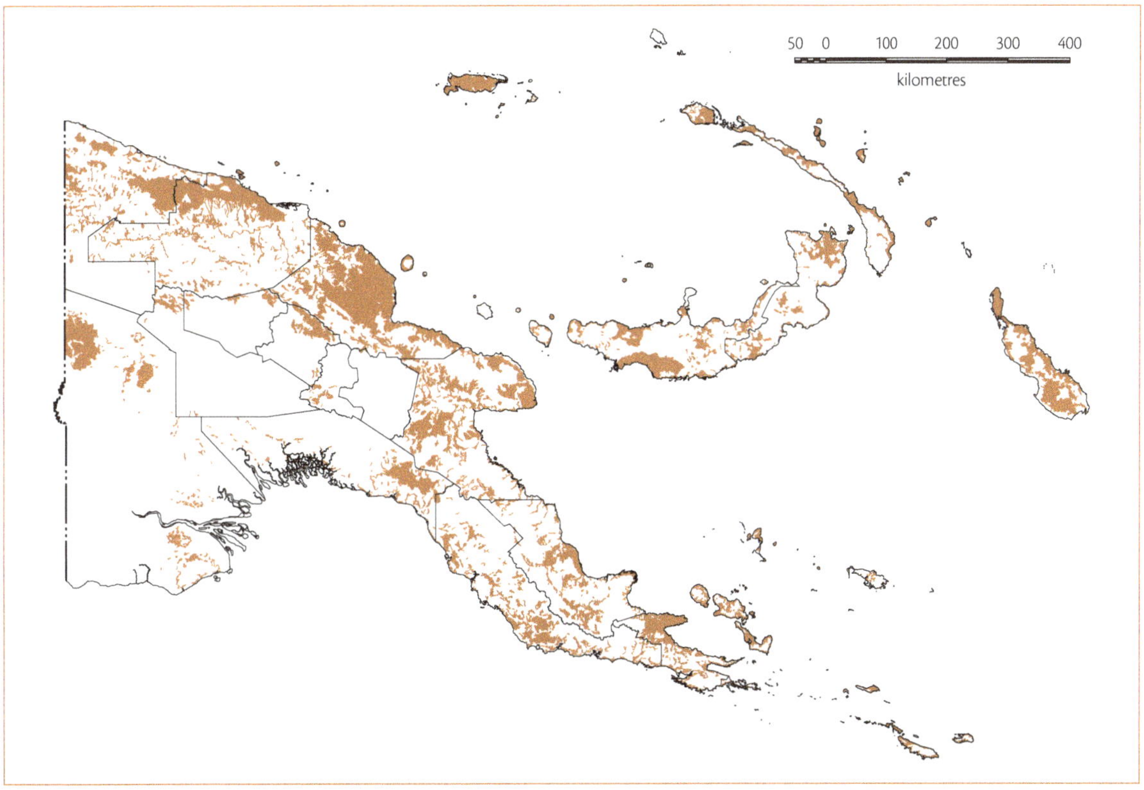

Figure 3.5.2 Distribution of betel nut as an important stimulant. Source: MASP.

Highland betel pepper is grown by about a third of the rural population. The leaves, and occasionally the catkins, are consumed with betel nut or highland betel nut. Highland betel pepper is grown in mountainous areas over an altitudinal range of 1150–2300 m. It is widespread in Eastern Highlands, Simbu, and parts of Southern Highlands provinces, and the Menyamya area and Huon Peninsula of Morobe Province. Its distribution is similar to that of highland betel nut (Figure 3.5.3) and the two items are commonly sold together in highlands betel nut markets. The best supply of highland betel pepper leaves in Kainantu market occurs in September–November. This is the period of poorest supply of lowland betel pepper catkins and it is likely that leaves of the highland species are sold to make up the shortfall of the preferred lowland catkins.

Highland betel nut (*Areca macrocalyx*) is used in a similar way to betel nut, but is considered inferior as the fruit is smaller. About a sixth of the rural population live in locations where it is planted and consumed, two-thirds of whom live in Eastern Highlands and Morobe provinces. It is most commonly planted in Eastern Highlands Province, particularly in the Kainantu area, and in mountainous parts of Morobe Province in the Menyamya area and the Huon Peninsula (Figure 3.5.3). It grows over an altitudinal range of 1100–1950 m. In the Kainantu area nuts are available throughout the year, with the best supply in October–December. This pattern complements that for betel nut. It is likely that more highland betel nut is sold in highlands markets when betel nut from the lowlands is in short supply.

Highland betel nut is an indigenous species and was probably domesticated in New Guinea after the introduction of betel nut from South-East Asia. Consumption of highland betel nut may have declined over the past 20 years as the supply of betel nut from the lowlands has improved. Highland betel nut is commonly sold in Kainantu market and sometimes in other highlands markets. It is occasionally transported to lowlands markets, particularly Lae, when betel nut is scarce.

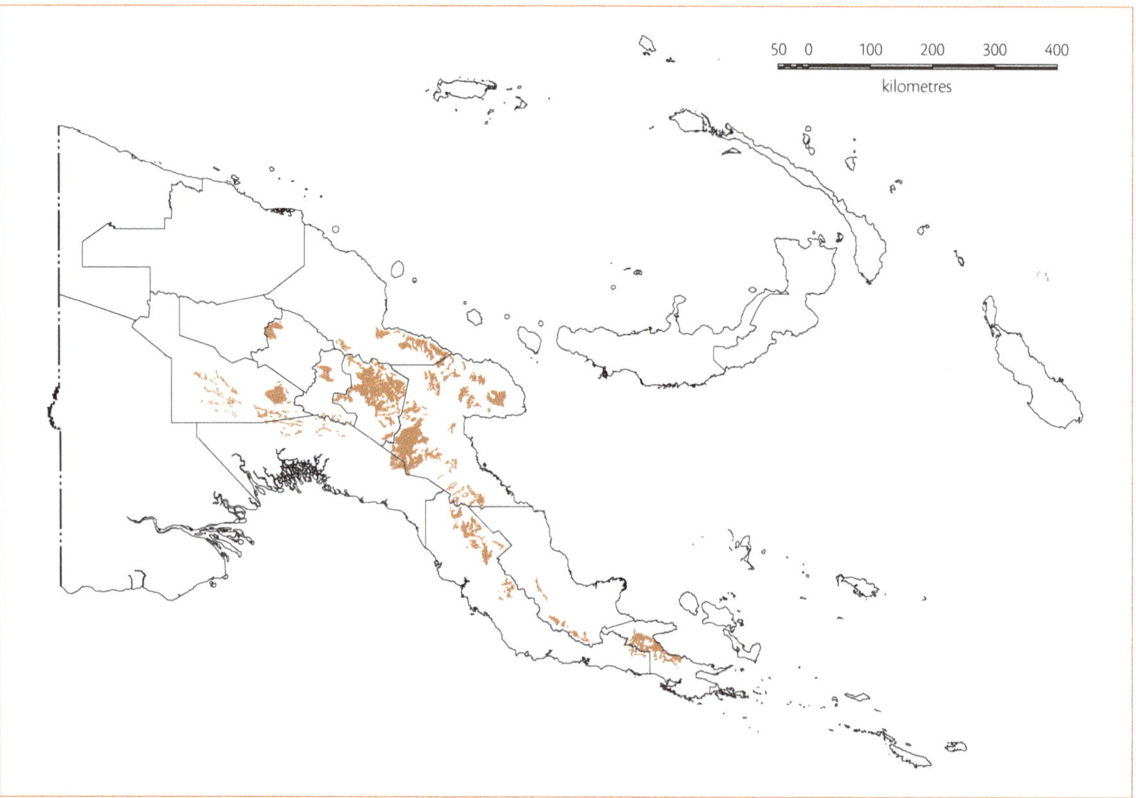

Figure 3.5.3 Distribution of highland betel nut as an important stimulant. Source: MASP.

Marijuana is usually smoked as a cigarette. Both the flowering bud and leaves are used, with the former containing a higher concentration of the active ingredient (THC). Information on production and consumption is limited as production and sale of marijuana is illegal in PNG. Nevertheless, it is likely that a high proportion of young men in the main producing locations grow or consume the plant.[4] Marijuana grows from sea level to about 2600 m in PNG, but is rarely grown in the lowlands. Most production occurs in the highlands from about 1400 m to 2200 m altitude in locations where rainfall varies seasonally. These are parts of Eastern Highlands, Simbu, Western Highlands and Central provinces. Marijuana is particularly common in the driest part of the highlands. Plants grown in seasonally dry highland locations are said to have a higher concentration of the active ingredient than those grown in coastal areas.

Marijuana was probably introduced into PNG in the mid to late 1960s by young expatriates. By the late 1970s it was being used by highlanders.[5] Production and consumption has expanded since then. By the early 1980s it was readily available in Port Moresby and other lowland centres. By the early 1990s expatriate field workers in Eastern Highlands and Simbu provinces were regularly offered marijuana in rural areas; reports of sightings in village plantings became more common; social workers in the highlands commented on the negative impact of heavy use on young men in some highland locations; and some young highland village men spoke openly of how much they used. By 2007, marijuana cigarettes were being offered for sale openly or barely concealed in Goroka, Kundiawa and Mount Hagen markets.

It seems that significant quantities of marijuana are grown in at least four provinces; that many young men use it and some use large enough quantities to have a detrimental effect on their wellbeing; and that it is readily available in urban centres. As a result, marijuana is an important cash crop for some villagers (see Section 5.17). It is unlikely that marijuana is imported into PNG but it is probable that some is exported from PNG to Australia where it is known as 'New Guinea Gold' among some users. Marijuana is implicated in reports of weapons trading from Australia to PNG across Torres Strait.

Coffee, tea and soft drinks. Coffee, tea and many aerated soft drinks contain caffeine, which is a psychoactive substance, although they are so common in most societies worldwide that they are not considered as drugs. Although coffee and tea are produced in PNG for the export market (see Sections 5.4 and 5.12), consumption levels are low. The 1996 PNG Household Survey recorded combined consumption of tea, coffee and Milo (a manufactured chocolate drink) of only 1 kg/person/year in urban areas and a few hundred grams per person per year in rural areas (Table A2.1.1). People consume more soft drink, particularly in urban areas, where consumption in 1996 was 12 kg/person/year and 3 kg/person/year in rural areas. Most coffee consumed is imported 'instant' powder. A very small quantity of locally roasted Arabica coffee is consumed, particularly by expatriates and in up-market restaurants. Some locally produced tea is sold on the domestic market and very small quantities are imported. Aerated soft drinks are manufactured in PNG from locally grown sugar and imported flavours.

Locally brewed alcohol. With the exception of coconut 'toddy' on some atolls, alcohol was unknown in PNG until introduced by Europeans in the late nineteenth century. Consumption by Papua New Guineans was prohibited until the early 1960s. Alcohol is now legally able to be consumed in PNG, with beer being the most common alcoholic drink. Consumption levels are low by global standards, with

4 According to the 2007 World Drug Report, the annual incidence of marijuana abuse in PNG (29.5%) among the population aged 15–64 is the highest of 138 countries surveyed. This can be compared with Australia (13.3%), the Philippines (4.2%), Fiji (0.2%) or Singapore (0.004%). The PNG figure is unlikely to be accurate, but nevertheless indicates a relatively high usage.

5 For example, police prosecuted an expatriate in Arawa on Bougainville Island in early 1970 for possession of marijuana. The prosecution failed because the police could not recognise the product. The author was subsequently requested to grow marijuana in a greenhouse at the Lowlands Agricultural Experiment Station at Keravat from 1970 to 1972 so that PNG police officers could recognise the crop. By the late 1970s, marijuana smoke could be smelt in some public locations in Kainantu that were frequented mainly by highlanders.

beer consumption recorded as 4 kg/person/year in 1996, and use higher in urban than in rural locations. Beer is brewed in Lae and Port Moresby from imported ingredients. Some spirit drinks are made and sold in PNG, based on ethanol (alcohol) manufactured at Ramu Agri-Industries Ltd and imported flavours. Other alcoholic beverages, including wine and spirits, are imported in small quantities.

An alcoholic drink (toddy) is made from the sap collected from the fruit stalk of coconut palms on Mortlock, Tasman and Nuguria islands in Bougainville Province. This appears to be a pre-European custom. The fermented sap produces a refreshing drink with a low alcohol content of about 2–3%. There are reports that coconut toddy was made in the lower Fly River area of Western Province, where it is possible the technique was introduced by Polynesian missionaries in the 1890s.

In a number of locations young men make an alcoholic drink known as 'jungle juice', 'JJ' or 'Yawa', by fermenting fruit, such as banana or pineapple, with yeast and sugar. This custom was possibly first introduced to the Gazelle Peninsula in East New Britain Province by Japanese troops during the 1940s. The Yawa banana variety is commonly used for this purpose because of its high sugar content. Increasingly, the fermented drink is distilled to produce a potent spirit known as *hom bru*. During the civil war on Bougainville in the 1990s the practice was widespread and distilled spirit is now sold at roadside stalls there.[6] It is also sold in some (and perhaps many) betel nut markets in the highlands. If not done correctly the distillation process can result in toxic products and reports of blindness and death among *hom bru* users occur from time to time. The spirit has a high alcohol content and consumption is associated with irrational and sometimes violent behaviour.

Kava is consumed as a beverage made from the chewed, pounded or grated root of a shrub that is related to betel pepper and black pepper. Chewing is the most common method of preparation. Once the kava root has been chewed and mixed with water, it is strained into a bowl or coconut cups and drunk. It is used in a very limited number of locations in PNG and by less than 1% of the rural population.

Kava was domesticated in Vanuatu and introduced from there to some locations in PNG. It was a pre-European crop in the Madang area and possibly in Manus Province. The area of greatest kava consumption in PNG was Manus Province, in particular Lou, Baluan, Pam, Rambutyo and the Fedarb islands. By the 1990s it was used only on Baluan. It was reportedly used in coastal areas north of Madang, the Rai Coast, some inland areas near the Ramu River and on Karkar and Bagabag islands. It is now used only in a limited number of villages in Madang Province. Kava was also consumed in parts of Western Province on the Great Papuan Plateau near Nomad, near Balimo and in the region from Daru to the Papua (west New Guinea) border.

Consumption has declined in recent decades. Some research has been conducted on kava by NARI scientists interested in developing the crop for the export market, but there is no commercial production.

Other psychoactive plants. There are numerous reports of the use of other plants, mushrooms and fungi for their psychoactive effect.[7] Most are indigenous to PNG but some are more recent introductions. They are used during various rituals and to induce visions or dreams. Some have been documented as having hallucinogenic properties. More than 60 species are reported to be used including mushrooms, fungi, and the fruit, leaves, sap, rhizomes, roots, bark and nuts of various plants. Some plants that have edible parts, including ginger, pandanus nut and castanopsis nut, are also reported to be used as psychoactive substances. A *Datura* species that has been used for this purpose has resulted in unintentional poisoning on a number of occasions. There is no reliable information on current usage of other psychoactive plants in PNG. However, it is likely that these plants were used in limited amounts in the past, certainly much less commonly than betel nut and tobacco are now used. It is also likely that their use has declined in recent decades in more accessible parts of PNG.

---

6 The spirit drink is not displayed openly in these stalls, but is signalled by the presence of bottles of Coca Cola, which is consumed with *hom bru*.

7 See Thomas (2000, 2003) for a comprehensive review of the psychoactive flora of PNG.

Sources

Bourke, R.M. (1989). Altitudinal limits of 230 economic crop species in Papua New Guinea. Unpublished paper. Department of Human Geography, Research School of Pacific and Asian Studies, The Australian National University, Canberra.

Bourke, R.M., Camarotto, C., D'Souza, E.J., Nema, K., Tarepe, T.N. and Woodhouse, S. (2004). *Production Patterns of 180 Economic Crops in Papua New Guinea*. Coombs Academic Publishing, Canberra.

Brunton, R. (1989). *The Abandoned Narcotic: Kava and Cultural Instability in Melanesia*. Cambridge Studies in Social Anthropology 69. Cambridge University Press, Cambridge.

Farnworth, E.R. (1976). Betel nut – its composition, chemistry and uses. *Science in New Guinea* 4(2):85–90.

Gibson, J. (2001). The economic and nutritional importance of household food production in PNG. In Bourke, R.M., Allen, M.G. and Salisbury, J.G. (eds). *Food Security for Papua New Guinea. Proceedings of the Papua New Guinea Food and Nutrition 2000 Conference*. ACIAR Proceedings No. 99. Australian Centre for International Agricultural Research, Canberra. pp. 37–44.

Hays, T.E. (1991). 'No Tobacco, No Hallelujah': missions and the early history of tobacco in eastern Papua. *Pacific Studies* 14(4):91–112.

Knauft, B.M. (1987). Managing sex and anger: tobacco and kava use among the Gebusi of Papua New Guinea. In Lindstrom, L. (ed). *Drugs in Western Pacific Societies: Relations of Substance*. ASAO Monograph No. 11. Association for Social Anthropology in Oceania and University Press of America, Lanham, Maryland. pp. 73–98.

Lebot, V., Merlin, M. and Lindstrom, L. (1997). *Kava: The Pacific Elixir. The Definitive Guide to its Ethnobotany, History, and Chemistry*. Healing Arts Press, Rochester, Vermont.

Lepowsky, M. (1982). A comparison of alcohol and betelnut use on Vanatinai (Sudest Island). In Marshall, M. (ed). *Through a Glass Darkly: Beer and Modernization in Papua New Guinea*. IASER Monograph 18. Institute of Applied Social and Economic Research, Boroko, Papua New Guinea. pp. 325–342.

Marshall, M. (1987). An overview of drugs in Oceania. In Lindstrom, L. (ed). *Drugs in Western Pacific Societies: Relations of Substance*. ASAO Monograph No. 11. Association for Social Anthropology in Oceania and University Press of America, Lanham, Maryland. pp. 13–49.

Marshall, M. (1993). A Pacific haze: alcohol and drugs in Oceania. In Lockwood, V.S., Harding, T.G. and Wallace, B.J. (eds). *Contemporary Pacific Societies: Studies in Development and Change*. Prentice Hall, Englewood Cliffs, New Jersey. pp. 260–272.

NSO (National Statistical Office of Papua New Guinea) (2002). Papua New Guinea 2000 Census: Final Figures. National Statistical Office of Papua New Guinea, Port Moresby.

Poole, F.J.P. (1987). Ritual rank, the self, and ancestral power: liturgy and substance in a Papua New Guinea society. In Lindstrom, L. (ed). *Drugs in Western Pacific Societies: Relations of Substance*. ASAO Monograph No. 11. Association for Social Anthropology in Oceania and University Press of America, Lanham, Maryland. pp. 149–196.

Thomas, B. (2000). Psychoactive plant use in Papua New Guinea: a review. *Science in New Guinea* 25(1,2,3):33–59.

Thomas, B. (2003). The psychoactive flora of Papua New Guinea. *Journal of Psychoactive Drugs* 35(2):285–293.

United Nations Office on Drugs and Crime (2007). 2007 World Drug Report. United Nations Office on Drugs and Crime, Vienna.

Wikipedia (2007). Psychoactive drug. Accessed 26 June 2007 at: <http://en.wikipedia.org/wiki/Psychoactive_drug>.

3.6 Intensification of agriculture

Intensification of agriculture occurs when production from an agricultural system is increased, but the area of land in use remains constant. The cause of agricultural intensification is most commonly attributed to the need to produce more food to meet the demands of a growing population, but this is an oversimplification.

The best-known theory of intensification is that of Ester Boserup. In 1965 she argued that the highly influential argument of Thomas Malthus – that population will always outrun food supply, leading to a reduction in food per person, and ultimately to famine and war – was wrong. Rather, said Boserup, as population increases, people will intensify their agriculture by permanently improving land quality (by draining or terracing, for example), changing agricultural techniques, adopting new crops, and working harder, to produce more food.

Boserup's ideas were applied to PNG agricultural systems by Harold Brookfield in the 1970s. Unable to find a good association between population density (the numbers of people per unit area) and the intensity of agricultural systems, Brookfield argued that the production of food in PNG was not solely for consumption by people, but was also used in displays and exchanges, in compensations for death and injury, as sacrifices to ancestral spirits, and in marriages. So food production had a social purpose as well as a nutritional one, which had to be taken into account when agricultural intensity was examined. Large exchanges of pigs and pig meat, some of which was wasted, or the production of yams of extraordinary length for rituals and exchanges, were two examples of Brookfield's 'social production'.

In shifting cultivation systems, land use can be intensified just by using the land more frequently in the absence of any other changes. That is, fallow periods can be reduced and cultivation periods extended (see Section 3.8). However, if fallow periods are reduced and cultivation periods extended to the point at which soil fertility is not restored by the fallow, food production will begin to decline and other symptoms of land degradation will appear. The intellectual and social processes that people go through when they become aware of declining crop yields and environmental degradation are not well understood. But it is assumed that these things trigger a response in the form of changes in the agricultural system that restore crop yields and food production to acceptable levels.

When people adopt new agricultural techniques they are 'innovating'. The sections that follow (Sections 3.7 to 3.11) describe the most important innovations used to intensify agriculture in PNG. They are mostly concerned with the maintenance of soil fertility when long fallow times are reduced or cultivation times are extended. The associations between agricultural intensity and fallowing, planting trees in the fallow, using leguminous crops in rotations with food crops, constructing wooden barriers or terraces, tilling the soil, constructing mounds and beds, and incorporating green manure in the soil (composting) are briefly outlined in Section 3.7. These practices are then examined in detail in Sections 3.8 to 3.11. Lastly, a number of other practices that are not directly associated with agricultural intensification are described in Section 3.12.

Significant changes in social, economic and political organisation, work patterns, land tenure, and gender relations have been observed to occur in association with agricultural intensification. They are not discussed in detail here.

Sources

Allen, B.J. (2001). Boserup and Brookfield and the association between population density and agricultural intensity in Papua New Guinea. *Agricultural Transformation and Intensification. Asia Pacific Viewpoint Special Issue* 42(2/3):237–254.

Allen, B.J., Bourke, R.M. and Hide, R.L. (1995). The sustainability of Papua New Guinea agricultural systems: the conceptual background. *Global Environmental Change* 5(4):297–312.

Allen, B.J., Bourke, R.M. and Hanson, L. (2001). Dimensions of PNG village agriculture. In Bourke, R.M., Allen, M.G. and Salisbury, J.G. (eds). *Food Security for Papua New Guinea. Proceedings of the Papua New Guinea Food and Nutrition 2000 Conference.* ACIAR Proceedings No. 99. Australian Centre for International Agricultural Research, Canberra. pp. 529–553.

Boserup, E. (1965). *The Conditions of Agricultural Growth: The Economics of Agrarian Change Under Population Pressure.* George Allen and Unwin Ltd, London.

Bourke, R.M. (2001). Intensification of agricultural systems in Papua New Guinea. *Agricultural Transformation and Intensification. Asia Pacific Viewpoint Special Issue* 42(2/3):219–235.

Brookfield, H.C. (1972). Intensification and disintensification in Pacific agriculture: a theoretical approach. *Pacific Viewpoint* 13(1):30–48.

Brookfield, H.C. and Hart, D. (1971). *Melanesia: A Geographical Interpretation of an Island World.* Methuen & Co Ltd, London.

McAlpine, J.R. and Freyne, D.F. (2001). Land use change and intensification in Papua New Guinea 1975–1996. *Agricultural Transformation and Intensification. Asia Pacific Viewpoint Special Issue* 42(2/3):209–218.

3.7 Soil fertility maintenance techniques

If crops are planted continuously on the same area, their production declines with every planting, until they produce very little food. This is illustrated in Figure 3.7.1, which shows some results from a long-term trial conducted at Keravat in East New Britain Province. The figure shows yield decline of sweet potato and taro where sweet potato, taro and peanut were grown in a rotation. Three things cause a decline in crop yields:

- Growing crops draw nutrients from the soil (nitrogen, phosphorus and potassium in particular, as well as a number of trace elements) and their harvest removes these nutrients from the soil.
- Soil that is exposed for cultivation loses nutrients when they are leached by rainfall or when soil is physically removed by erosion.
- Pests and diseases that attack crops build up over time to a level where they significantly reduce crop yields.

If agriculture is to be sustainable over the long term, crop production must be maintained at satisfactory levels. In order to achieve this, losses of soil and nutrients must be kept to a minimum. Soil nutrients that are lost during cultivation have to be replaced, physical characteristics of the soil that are altered during cultivation need to be restored or improved, and pests and diseases reduced to acceptable levels.

In PNG a wide variety of techniques are used to slow the loss of soil and nutrients during cultivation and to restore them between times of cultivation. These techniques are:

- Fallowing – long and short fallows.
- Planting trees in the fallow.
- Using leguminous crops in rotations with food crops.
- Constructing wooden barriers or terraces to slow the loss of soil by erosion.
- Tilling the soil and constructing mounds and beds.
- Incorporating green manure in the soil (composting).

The relationship between the use of these techniques and land use intensity, by agricultural system, is shown in Table 3.7.1 (see also Section 1.2). A number of techniques show a clear association with land use intensity; short fallows and composting are found in more than 60% of the highest-intensity systems and are not present in any of the lowest-intensity systems. However, this association is not perfect. It appears that people do not use these techniques, even if they know about them, until crop production falls to an unacceptable level. Many techniques that are used at the highest levels of intensity are also found in systems across the whole range of land use intensities. For example, the use of short fallows, clearly associated with the highest-intensity systems, also occurs in systems with medium and low intensity.

Table 3.7.1 The proportion of systems using selected agricultural techniques, by land use intensity class

Systems in the intensity class (%)	Land use intensity class (R-value)				
	Very low	Low	Medium	High	Very high
81–100				Short fallows	Short fallows
61–100					Composting
41–60				Long beds Small mounds Tillage	
21–40		Small mounds	Short fallows Small mounds Tillage	Composting Legume rotations Medium-sized mounds	Large mounds Medium-sized mounds
6–20	Small mounds Soil retention	Fences Medium-sized mounds Short fallows Soil retention Tillage	Composting Medium-sized mounds Large mounds Legume rotations Square beds	Large mounds	Legume rotations Long beds Small mounds Square beds Tillage
1–5	Long beds Medium-sized mounds Planted tree fallows Tillage	Composting Long beds Planted tree fallows Square beds	Soil retention		
Not present	Composting Large mounds Short fallows	Large mounds Legume rotations	Planted tree fallows	Planted tree fallows Square beds Soil retention	Planted tree fallows Soil retention

Note: Only 'very significant use' is presented in the table. During field surveys to define and describe agricultural systems, a qualitative judgement was made on whether the use of a technique was present and whether its use was 'not significant', 'significant' or 'very significant'.

Source: MASP.

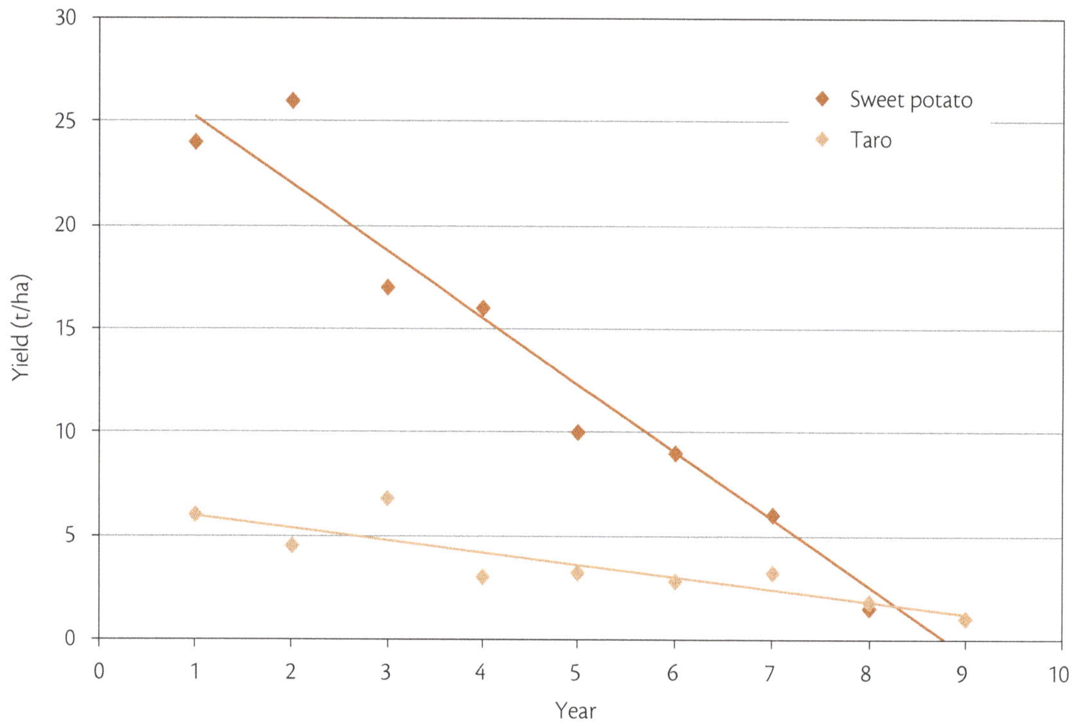

Figure 3.7.1 Effect of ten years of continuous planting on sweet potato and taro yields at Keravat, East New Britain Province, 1954–1964. **Note:** Data are from rotations 4 and 5 of the first ten years of a long-term rotation trial. Treatments were a continuous rotation of sweet potato, taro and peanut. Source: Bourke (1977).

Composting occurs in all land use intensity classes except the very low intensity class. Small mounds are used in some systems at all levels of land use intensity. In contrast, tree planting in fallows occurs only in the low and very low intensity classes.

The sections that follow (Sections 3.8 to 3.11) describe these techniques in more detail.

Sources

Bourke, R.M. (1977). A long term rotation trial in New Britain, Papua New Guinea. In Leakey, C.L.A. (ed). *Proceedings of the Third Symposium of the International Society for Tropical Root Crops.* International Society for Tropical Root Crops in collaboration with the International Institute of Tropical Agriculture, Ibadan, Nigeria. pp. 382–388.

Bourke, R.M. (2001). Intensification of agricultural systems in Papua New Guinea. *Agricultural Transformation and Intensification. Asia Pacific Viewpoint Special Issue* 42(2/3):219–235.

3.8 Fallowing

A very effective and almost universal way of restoring soil fertility in PNG is to use a *fallow*. A fallow is a period when land is not planted with crops, but is left to 'rest', or is planted with a species that assists in restoring soil fertility, such as a leguminous crop (see Section 3.10). In PNG, depending on the need to produce crops, land may be left in fallow from eight weeks to more than 40 years. A distinction can be made between so-called *long fallows*, which last more than twelve months, and *short fallows*, which are less than twelve months and are typically between two and four months.

The importance of short fallows is to restore soil fertility, reduce pests and diseases and, where green manuring (composting) is practised, to provide fallow vegetation that can be incorporated into the soil at the next planting (see Section 3.11). The effect of a 16-month fallow on sweet potato yields in an experiment at Aiyura in Eastern Highlands Province is illustrated in Figure 3.8.1. The solid trend line shows yields falling prior to the fallow and the broken line shows the trend of yields after the fallow. The 16-month fallow restored yields almost to the level they were at eight plantings earlier. Short fallows are commonly used in highlands agricultural systems.

When land is fallowed, naturally occurring plants colonise the previously cultivated area. These plant communities are known as the *fallow vegetation* (see page 237). If the fallow period is long enough and if sources of seed are present, fallow vegetation may progress to a tall tree cover. Tree roots penetrate deep into the subsoil and draw up nutrients to the leaves and branches. Soils become more friable as root systems develop and litter decomposes. The fallow vegetation protects the soil surface from rainfall and sunshine. Pests and diseases that live on particular food crop plants are significantly reduced because their target plants are no longer available to them. When the fallow is cleared most of the vegetation is not removed from the site. Leaf litter and woody material is either burned on site or decomposes, and this provides nutrients to the soil.

Shifting cultivation and fallowing

Shifting cultivation is the basis for most food production in PNG. The exceptions occur where food is derived from tree crops such as sago, coconut or breadfruit and from some sweet potato-based systems in the highlands where land is rarely fallowed. The majority of agricultural systems in the lowlands are shifting cultivation systems that use only long fallows to maintain soil fertility. Other techniques, such as green manuring (composting) and planted tree fallows, are more common in the highlands, but these techniques are still used within shifting cultivation systems. A lot of variation exists but, in general, land under fallow vegetation that is believed to be ready for cultivation again is cleared by slashing, and usually burning, the fallow vegetation and then planting food crops.

Although shifting cultivation is commonly associated with burning, almost 20% of PNG's shifting cultivators do not burn slashed fallow vegetation during the preparation of plots for cultivation. Most of the systems where burning is not important are found in high rainfall areas or where sago is a main food source (Figure 3.8.2). Burning is also unimportant in agricultural systems where the cropping phase is long and soil fertility is maintained by green manuring.

After planting and harvesting crops, the cleared area is returned to fallow. Crops may be planted from one to many times, with 1–3 plantings typical in the lowlands. The cropping phase tends to be 2–5 plantings in the highlands but is even more extended in some highlands locations, particularly where green manuring is practised. The ratio of the time land is cultivated to the time it is left in fallow is one important measure of land use intensity (see Section 1.2).

Just under half of all rural people in PNG use agricultural systems where fallows are between 5 and 15 years long, and 34% use fallows that are longer than 15 years. Of the 11% of people who use fallow periods between 1 and 4 years long, most are located in Southern Highlands and Eastern Highlands provinces (Figures 3.8.3, 3.8.4, Table A3.8.1).

Villagers do not use long fallows in only nine of the 291 agricultural systems identified by the MASP project (see Section 1.15). Agricultural systems where a long fallow is not used have particular characteristics that distinguish them from all other systems: most are highlands systems; they are located between 1600 m and 2800 m in altitude; the lowlands systems all occur on small islands; population densities are high; land use intensity is very high; short fallows are used in all such systems; soils are tilled; in the highlands, mounding and green manuring (composting) are used (see Section 3.11); sweet potato is the most important crop produced; pigs are often tethered in gardens; and burning of fallow vegetation is uncommon. In Enga Province 67% of the rural population do not use long fallows.

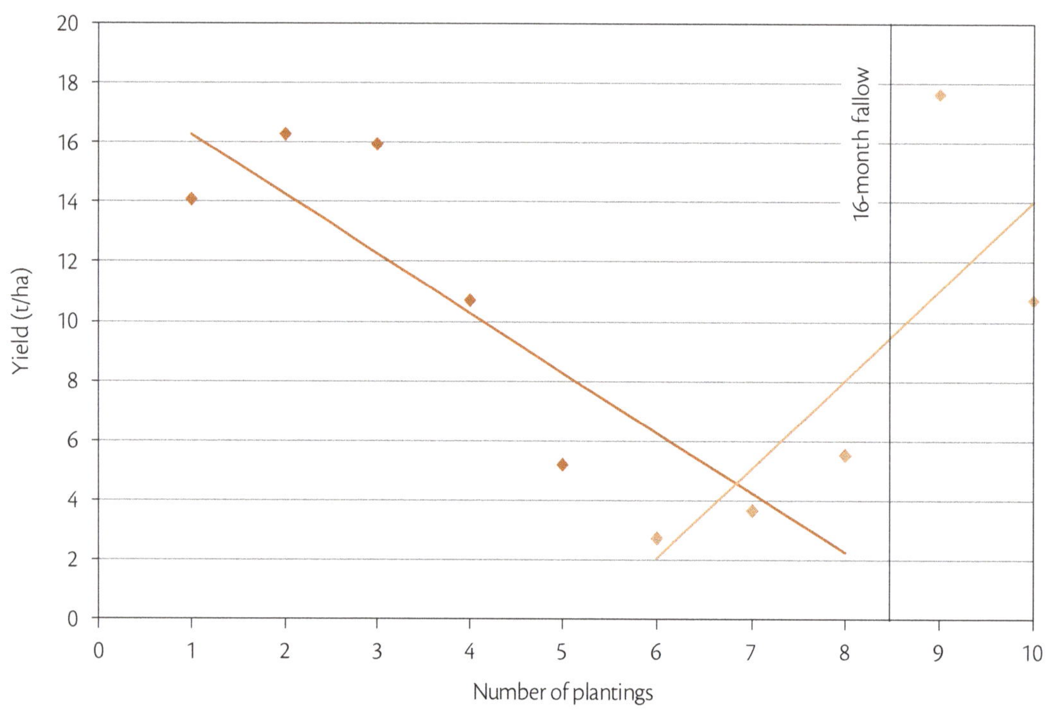

Figure 3.8.1 Effect of continuous planting and a fallow on sweet potato yield at Aiyura, Eastern Highlands Province.
Source: Kimber (1974:78).

Fallow vegetation

When land that has been cultivated is fallowed, a number of vegetation communities follow, or succeed, one another on the site. Hence they are called *fallow successions*. The most common pattern of succession is weeds and grasses first, followed by fast-growing shrubs and bushes, softwood trees, and finally hardwood trees. It may take 30 years to reach a secondary forest succession, with two or three layers of vegetation under a tree cover. Tall secondary forest may appear to be similar to undisturbed forest that has never been cultivated, but the species composition is much less diverse than in undisturbed forest. The rate at which fallow successions proceed depends on the soil fertility of the site, rainfall, sources of seed and whether the succession is disrupted, by a fire for example, or clearing. If fallows are cleared again for cultivation too soon, the fallow successions will not have had time to proceed beyond a certain stage and the replenishment of soil nutrients will not have been completed. Over time, soil fertility will steadily decline, fallow vegetation will not proceed past scrub or tall grass, relatively quick tree successions can disappear, and short grasses or tall grasses can become the dominant fallow vegetation.

Most of the land that is used for agriculture in PNG is cleared from tall woody regrowth[1] that is more than 15 years old. This is particularly marked in Sandaun and Gulf provinces. More than 1.5 million people (37% of the rural population) clear tall woody regrowth fallows for cultivation (Figures 3.8.5, 3.8.6, Table A3.8.2). In the highlands provinces, where fallow periods are often shorter, short and tall grasses and low woody plants dominate fallow vegetation.[2] A further 1.5 million people clear tall grass or grass and woody regrowth fallows.

1 'Tall woody regrowth' is the technical term for tall secondary forest.

2 Common tall grass fallow species are *Saccharum* and *Miscanthus*. Common short grass fallow species include *Imperata*, *Themeda* and *Ischaemum*. Scrub, or low woody regrowth species include *Parasponia*, *Dodonaea*, *Ficus*, *Macaranga*, *Kleinhovia*, *Melanolepis*, *Pipturus* and *Premna*.

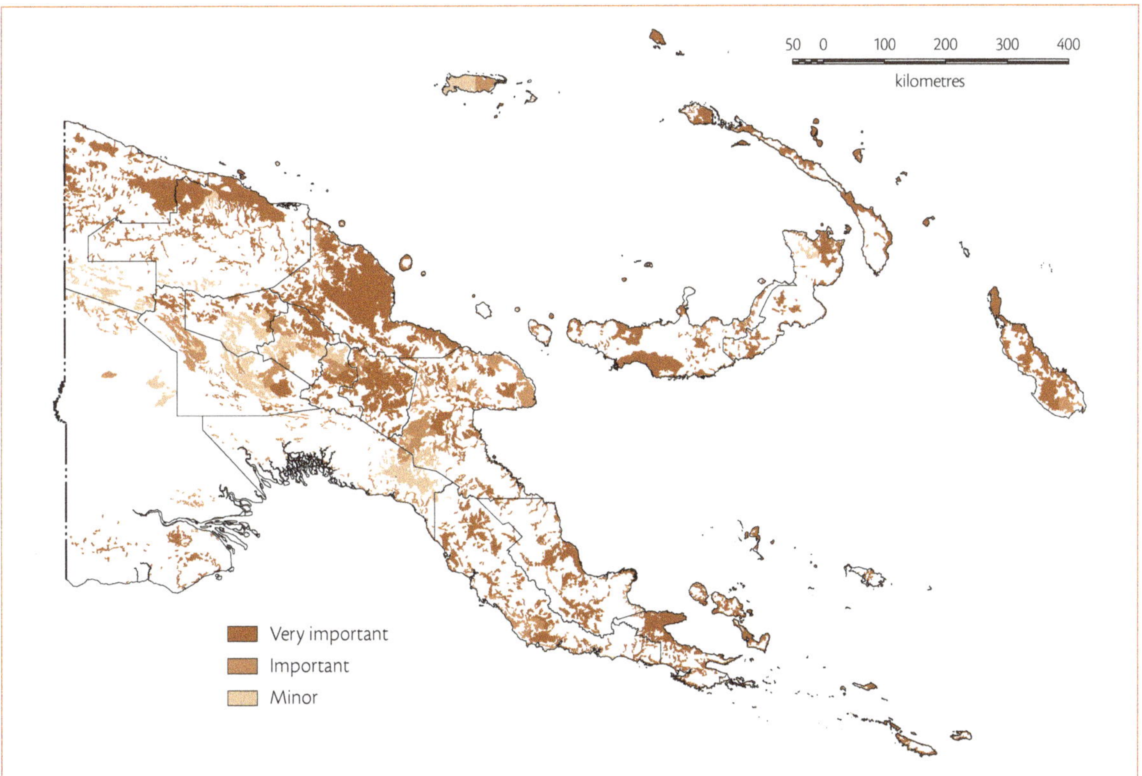

Figure 3.8.2 The use of burning to clear fallow vegetation. Source: MASP.

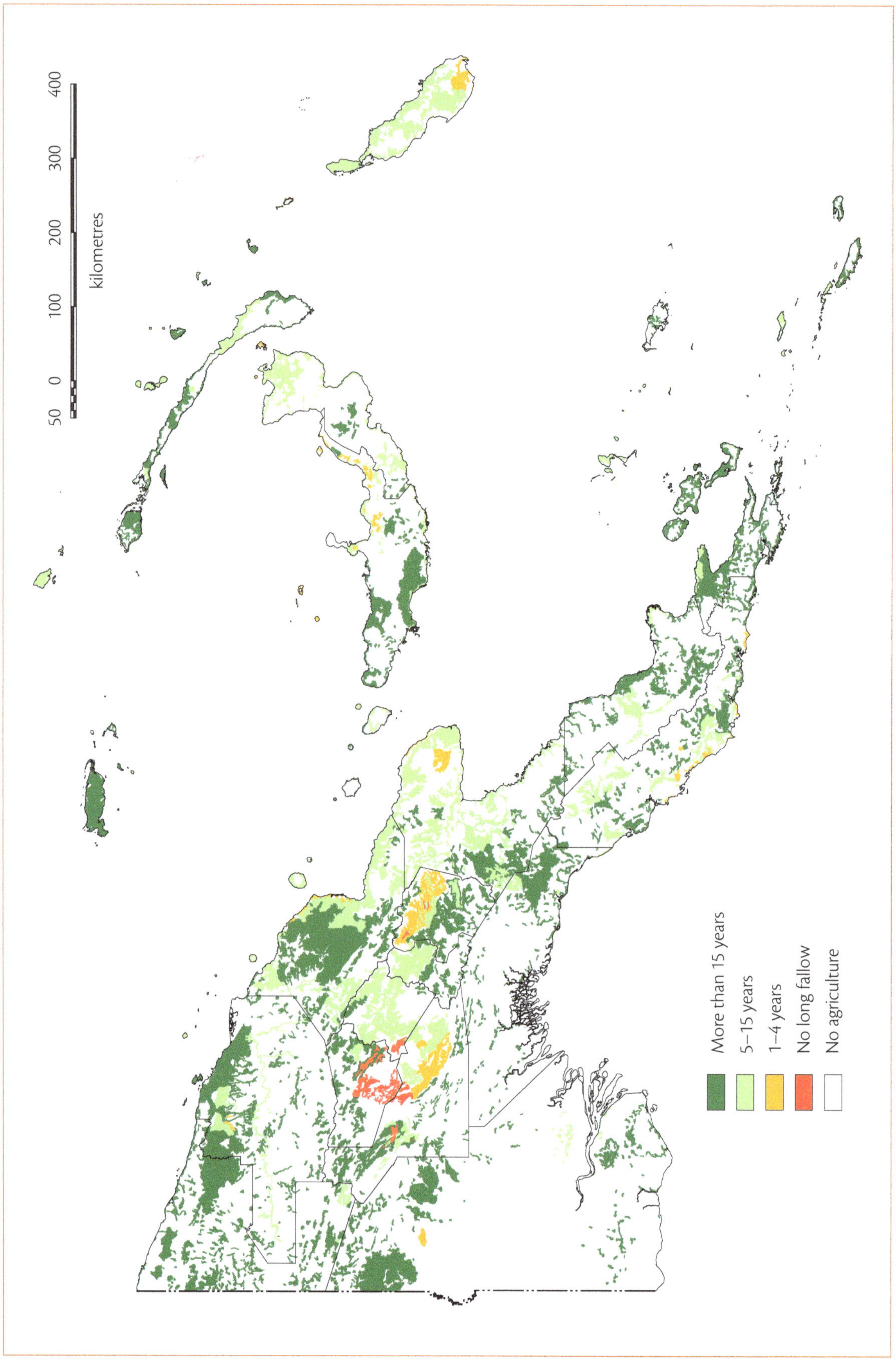

Figure 3.8.3 Length of fallow period. Source: MASP.

In Western Highlands, Simbu, Enga and Southern Highlands provinces, tall grass is the most common vegetation cleared from new garden sites, sometimes with woody regrowth. An important minority of around 10% of the rural population use short grass fallows, particularly in Eastern Highlands Province where 41% of the population use this fallow type.

The influence of rainfall on fallow vegetation is illustrated by the difference between Eastern Highlands, where a distinct dry season is experienced (see Section 1.5) and the four other highlands provinces. In the lowlands provinces, short grasses and savanna are important fallow vegetation only in the coastal areas of Western and Central, where rainfall seasonality is strong.

Although annual rainfall and rainfall seasonality are important influences on fallow vegetation, grass fallow complexes are associated with higher population densities and with fallow periods of less than five years.

Sources

Allen, B.J. (1982). Yam gardens and fallows in the Torricelli foothills, Dreikikir District, East Sepik. In Bourke, R.M. and Kesavan, V. (eds). *Proceedings of the Second Papua New Guinea Food Crops Conference: Part Two*. Department of Primary Industry, Port Moresby. pp. 236–255.

Allen, B.J. (1985). Dynamics of fallow successions and introduction of robusta coffee in shifting cultivation areas in the lowlands of Papua New Guinea. *Agroforestry Systems* 3:227–238.

Hartemink, A.E. (2003). Sweet potato yields and nutrient dynamics after short-term fallows in the humid lowlands of Papua New Guinea. *Netherlands Journal of Agricultural Science* 50(3/4):297–319.

Kimber, A.J. (1974). Crop rotations, legumes and more productive arable farming in the highlands of Papua New Guinea. *Science in New Guinea* 2(1):70–79.

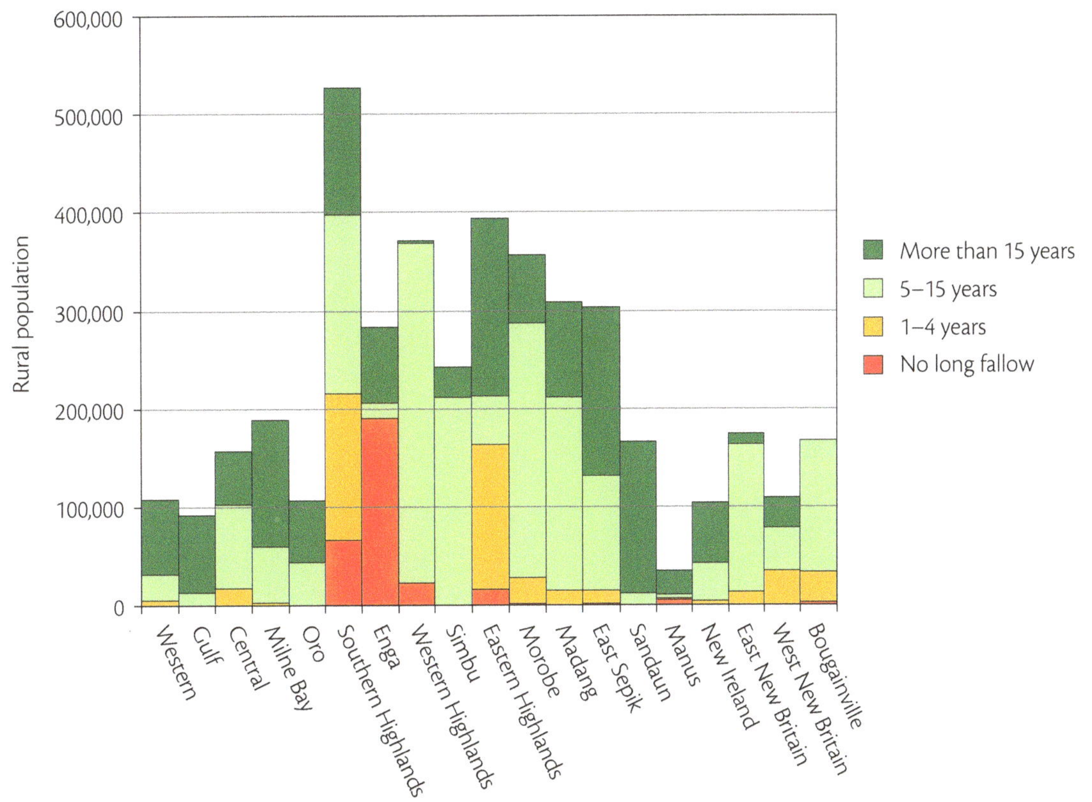

Figure 3.8.4 Rural population using fallows of given periods, by province. Sources: NSO (2002), MASP.

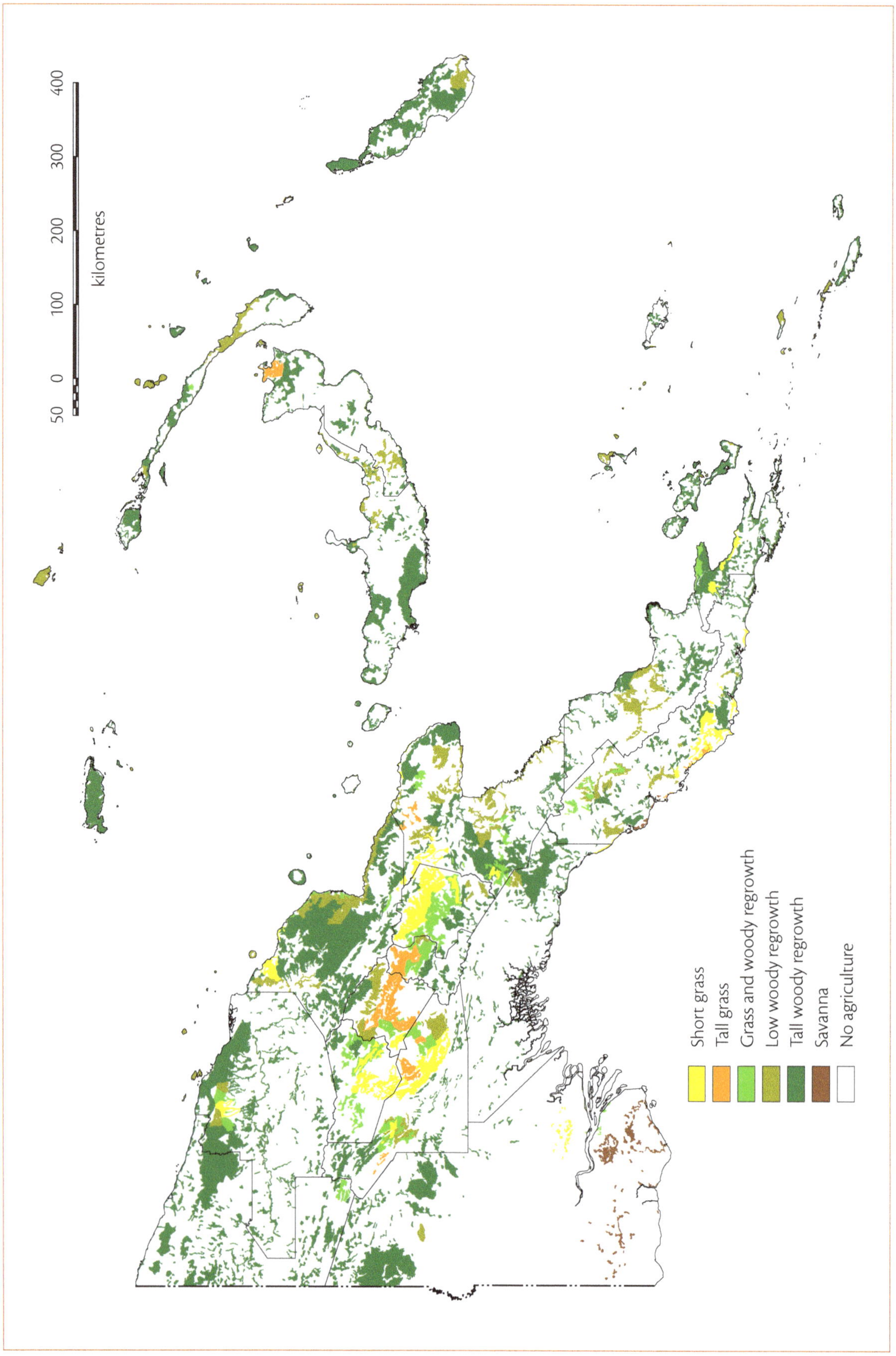

Figure 3.8.5 Fallow vegetation cleared for cultivation. Source: MASP.

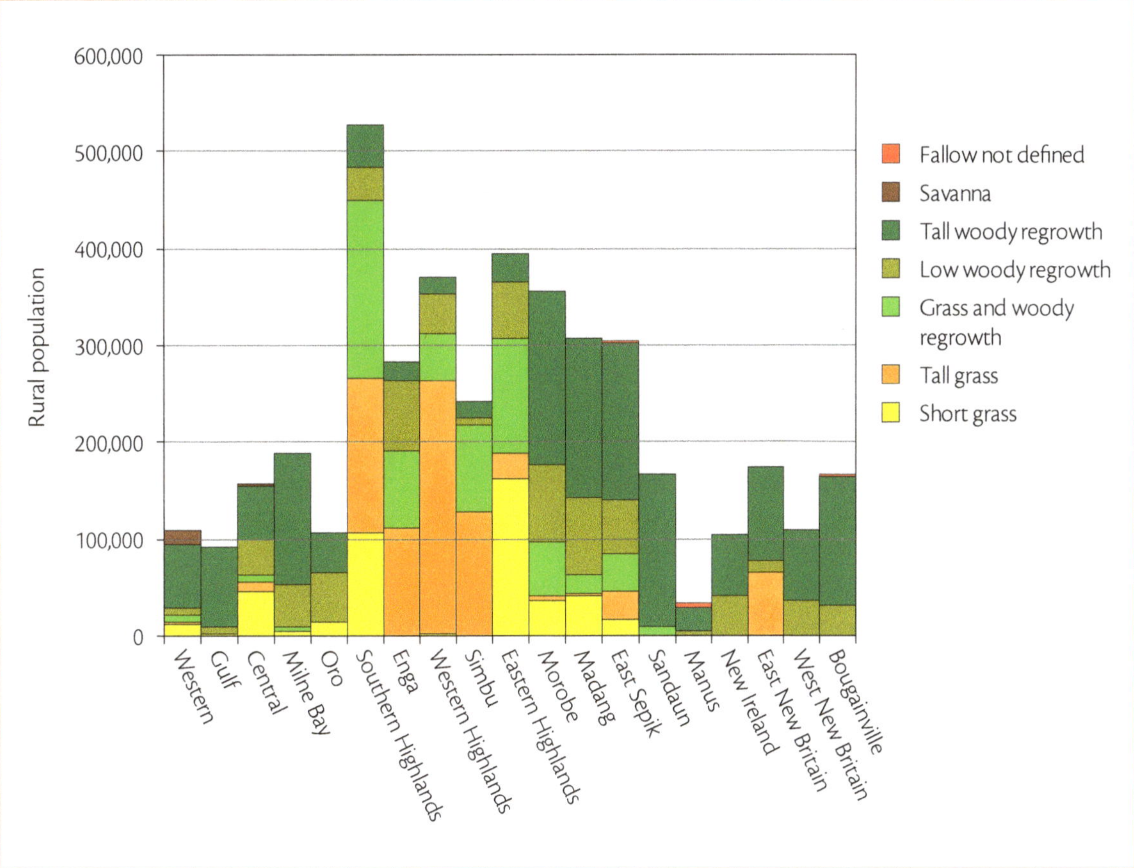

Figure 3.8.6 Rural population clearing different fallow vegetation, by province. **Note:** Agricultural systems with 'fallow not defined' occur in swamps at the mouth of the Sepik River and on some atolls (see Table A3.8.2). Sources: NSO (2002), MASP.

Louman, B.T.M. (1987). Subsistence use of fallow vegetation in the highlands of Papua New Guinea. *Netherlands Journal of Agricultural Science* 35:546–549.

Manner, H.I. (1977). Biomass: its determination and implications in tropical agro-ecosystems: an example from montane New Guinea. In Bayliss-Smith, T.P. and Feachem, R.G. (eds). *Subsistence and Survival: Rural Ecology in the Pacific*. Academic Press, London. pp. 215–242.

NSO (National Statistical Office of Papua New Guinea) (2002). Papua New Guinea 2000 Census: Final Figures. National Statistical Office of Papua New Guinea, Port Moresby.

Paijmans, K. (ed.) (1976). *New Guinea Vegetation*. Australian National University Press, Canberra.

Parfitt, R.L. (1976). Shifting cultivation – how it affects the soil environment. *Harvest* 3(2):63–66.

Sillitoe, P. (1995). Fallow and fertility under subsistence cultivation in the Papua New Guinea highlands: I. Fallow successions. *Singapore Journal of Tropical Geography* 16(1):82–100.

Sillitoe, P. (1995). Fallow and fertility under subsistence cultivation in the Papua New Guinea highlands: II. Soil fertility. *Singapore Journal of Tropical Geography* 16(1):101–115.

3.9 Soil retention and benching

In PNG agricultural systems, two main methods are used to reduce the rate at which soil is lost from a garden site during cultivation. Both techniques have been developed by villagers. Given the widely dispersed use of these practices, it it likely that they were developed independently in a number of locations. The techniques are:

- Construction of physical barriers across the slope to prevent soil loss.
- Creation of benches across the slope to slow run-off and possibly to retain water.

Soil retention

In many parts of PNG, barriers are constructed across the slope to slow or prevent the movement of soil downslope and out of the garden site. The barriers are constructed from wood but, over a number of years, stakes used sometimes sprout and form living, permanent barriers. This is particularly the case where *Cordyline* or *Piper* is used to make the barrier. In the Chimbu Valley for example, carbon dating of soils retained behind a barrier of live *Cordyline* suggest that it was first constructed over 200 years ago.

The size and design of the barriers vary from a simple pole laid along the contour, with stakes hammered into the soil on the downslope side to hold it in place, to a low fence constructed by weaving thin poles in and out of 50 cm tall stakes. In other cases, a 1.5–3.0 m high barrier made from tree branches is held up by substantial props on the downslope side. The distance between the barriers is also highly variable.

Where they are used with shifting cultivation systems, the barriers are often reconstructed in the same places after a long fallow and the soil that builds up behind the barrier becomes the beginning of a terrace. Steep, grass-covered hills south and west of Marawaka in Eastern Highlands Province have numerous ridge lines running along the contour. It is possible these lines are caused by the construction of soil retention barriers during previous cultivations.

Soil retention barriers are most important in agricultural systems in northern Simbu Province, on the Huon Peninsula in Morobe Province, east of Maprik in East Sepik Province, west of the Baiyer Valley in Western Highlands Province, in the headwaters of the Watut, Lakekamu and Waria rivers south of Wau in Morobe Province and in the D'Entrecasteaux Islands and mainland of Milne Bay Province (Figures 3.9.1, 3.9.2, Table A3.9.1).

Vetiver grass has been promoted in Simbu Province as a possible means of soil erosion control. The grass was introduced in the 1980s but was poorly managed and was not widely adopted. In the late 1990s CARE Australia began an extension program using vetiver grass planted behind existing wooden soil retention barriers and also trialled it in association with *Leucaena* trees. It will be some years before the outcomes of this program can be assessed, but there appears to be only limited adoption by villagers about ten years after the program was initiated.

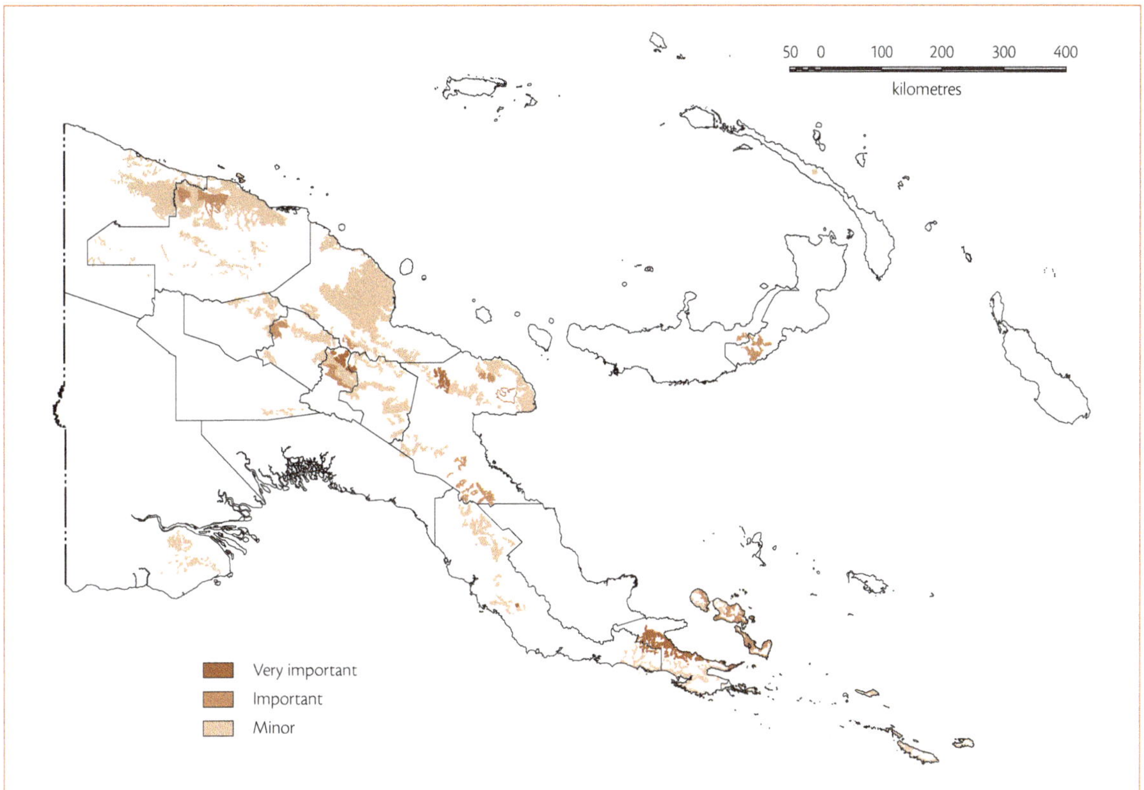

Figure 3.9.1 The use of soil retention barriers. Source: MASP.

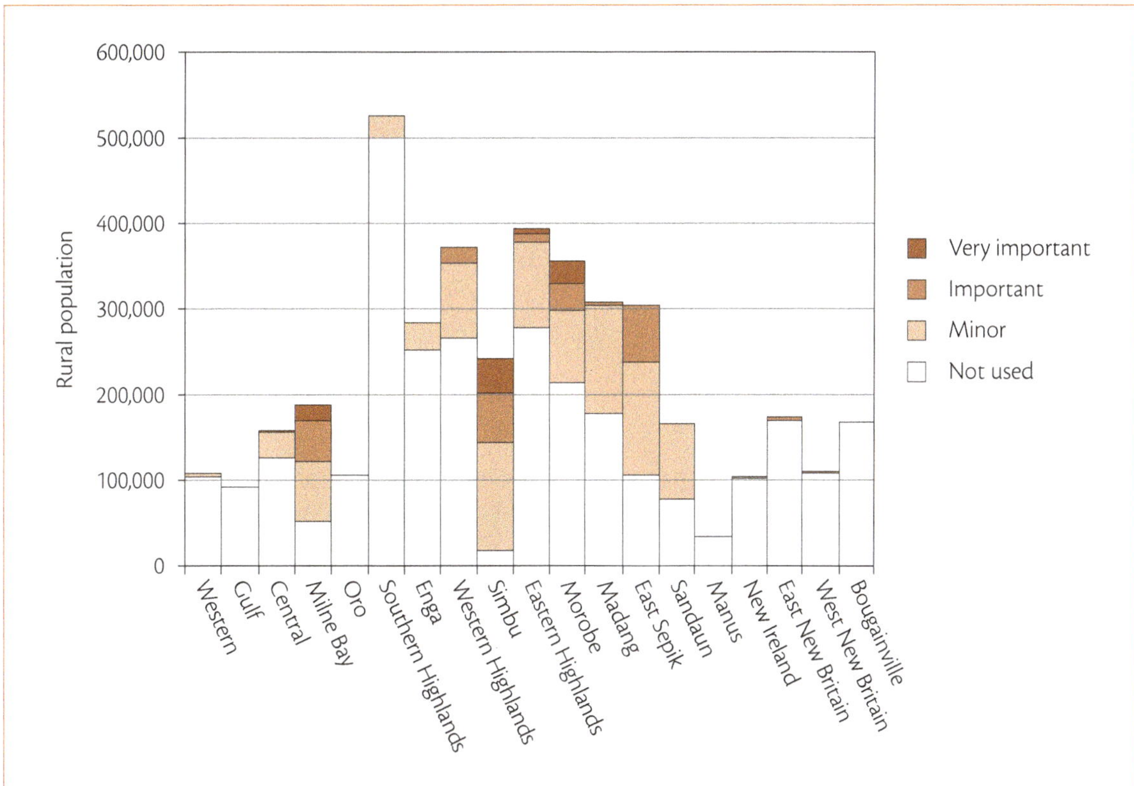

Figure 3.9.2 Rural population using soil retention barriers, by province. Sources: NSO (2002); MASP.

Benching

Another method of retaining soil on steep slopes is benching. Benching involves cutting a narrow bench into a slope along the contour. Benching is not common in PNG and true terracing is not used anywhere. Large areas of hillsides have been benched inland of Rabaraba in mainland Milne Bay Province (Figures 3.9.1, 3.9.2, Table A3.9.1). The benches are about 70 cm wide and 100 cm high. Men construct the benches by working together in a line moving from the bottom of the slope to the top. Wooden digging sticks were previously used but have been replaced by metal spades. The first bench is formed and spoil from the next bench is brought down on top of the first to form a low ridge along the top of the first bench. In this way, the whole slope is benched and ridged. Taro is planted into the ridge of spoil.

Although it seems likely that ground water seeps out of the low wall at the back of the bench, and collects behind the ridges, the practice is described mainly in terms of soil retention. Villagers say that unless this technique is used, severe soil loss will result if substantial rain falls on slopes that have been prepared for taro cultivation. Benches are also used on Goodenough and Wagifa islands in Milne Bay. On the Sogeri Plateau, Central Province, migrants from Kabwum in Morobe Province create small fences along the contour using *Piper aduncum* sticks and move spoil from the slope behind the fences to create a bench and a narrow terrace.

In the Pindiu to Ogeranang area of the Huon Peninsula, large fences 2–3 metres high are constructed from *Piper aduncum* stems. The fences help prevent soil erosion. Soil is moved from higher up the slope down to behind the fence. This creates a small terrace, with the soil deeper just upslope from the fence. Crops that require more fertile sites, such as banana and leafy green vegetables, are planted in the deeper soil with sweet potato planted elsewhere.

Sources

Bourke, R.M. (1991). Indigenous soil conservation practices in Papua New Guinea. *Indigenous Conservation Farming Practices: Report of the Joint ASOCON/Commonwealth Workshop*. Asia Soil Conservation Network Report No. 7. Asia Soil Conservation Network, Jakarta. pp. 67–71.

Carman, K. (1989). Soil loss and runoff from demonstration gardens in Matalau village, East New Britain Province. Technical Report 89/4. Department of Agriculture and Livestock, Port Moresby.

Humphrey, W. (1997). Preventing soil erosion with trash lines and vetiver grass. LAES Information Bulletin No. 82. Department of Agriculture and Livestock, Keravat.

Humphreys, G.S. (1994). Deciphering land use history from hill slopes: an example from New Guinea. *PLEC News and Views* 2:21–25.

Humphreys, G.S. and Wayi, B.M. (1990). Measuring soil erosion on steep lands: the Chimbu experience. In Pushparajah, E. and Elliott, C.R. (eds). *Pacificland Workshop on the Establishment of Soil Management Experiments on Sloping Lands*. IBSRAM Technical Notes No. 4. International Board for Soil Research and Management, Bangkok. pp. 243–269.

Humphreys, G.S. and Brookfield, H. (1991). The use of unstable steeplands in the mountains of Papua New Guinea. *Mountain Research and Development* 11(4):295–318.

Igua, P.B. and Humphrey, W. (1997). Country report on soil erosion research at Keravat, Papua New Guinea. In Dowling, T. and Howlett, D. (eds). *Proceedings of Pacificland Annual Meeting and Workshop on Sustainable Land Management*. Suva, Fiji.

NSO (National Statistical Office of Papua New Guinea) (2002). Papua New Guinea 2000 Census: Final Figures. National Statistical Office of Papua New Guinea, Port Moresby.

Shelton, R.M. and Mondo, C. (2001). The adoption of vetiver grass for soil erosion control in Simbu Province. In Bourke, R.M., Allen, M.G. and Salisbury, J.G. (eds). *Food Security for Papua New Guinea. Proceedings of the Papua New Guinea Food and Nutrition 2000 Conference*. ACIAR Proceedings No. 99. Australian Centre for International Agricultural Research, Canberra. pp. 301–307.

Wayi, B.M. (1981). Soil erosion and conservation research in Simbu Province. *Harvest* 7(1):18–20.

3.10 Tree planting and legume rotations

As intensification of agriculture increases in PNG and cultivation times are extended, methods other than fallowing must be used to maintain soil fertility or to slow the rate at which nutrients are lost from the soil. Two important techniques that villagers use to maintain or improve soil fertility are planting particular trees in fallow vegetation and planting legumes[1] between or prior to planting root crops, particularly sweet potato. These practices provide other benefits, for example, timber for fencing and firewood.

Nitrogen in the form of ammonium is critical to plant and animal growth and life. Although the use of trees and legume rotations in PNG agricultural systems is poorly understood, it is assumed that these plants 'fix' nitrogen, that is, they absorb nitrogen from the atmosphere and convert it to a form that can be used by other plants. It is actually bacteria, which live in small growths or nodules on the roots of leguminous plants, that 'fix' nitrogen and allow the ammonium produced to be absorbed by the plant. Nitrogen is also fixed by certain other plants, including casuarina trees. The nitrogen only becomes available to other plants if the leguminous plant, or part of it, dies and decomposes in the soil.

1 A legume is a plant belonging to the family Leguminosae (peas, beans and pulses). Most legume species contain certain bacteria in their roots which enable them to convert nitrogen from the atmosphere and store it in plant tissue as ammonium, a form of nitrogen that the plant can use.

Tree planting in fallows

During a long fallow, it is usual for plant species to reoccupy the previously cultivated area without human assistance (see Section 3.8). However, in a number of places in PNG, people deliberately cultivate particular trees in fallow gardens in order to speed up the establishment of a tree cover and, frequently, to introduce species into the fallow that will put nutrients back into the soil.

The tree most commonly planted into fallows in PNG is *Casuarina oligodon* (*yar*). Other tree species planted in fallows in PNG are *Albizia* spp., *Parasponia rigida*, *Piper aduncum* and *Schleinitzia novo-guineensis*.

Casuarina oligodon seedlings are planted into fallows in the highlands, mainly between 1400 m and 2100 m in altitude, on sloping land, in hills or mountains, in areas where the annual rainfall is between 2000 mm and 3000 mm and the fallow vegetation is grass. Casuarina occurs naturally along water courses and in disturbed forest areas, such as landslides. Seedlings are transplanted from these sites and from mature stands of trees. Typically men collect seedlings and plant them in sweet potato gardens towards the end of the cropping phase. The seedlings grow during the fallow phase and are mature 8–12 years later.

Planting casuarina trees into fallows is most important in central Simbu Province, in adjacent locations in Eastern Highlands Province, in the Simbai and Kaironk areas in Madang Province, and in the Oksapmin area of Sandaun Province (Figures 3.10.1, 3.10.2, Table A3.10.1). Elsewhere, casuarina is planted around the edges of fields, around houses and in small plantations, but not in fallows.

Villagers say that casuarina trees improve soil conditions. It has not been established scientifically how the tree fixes atmospheric nitrogen, but soil under casuarina trees has been shown to be higher in nitrogen and carbon than the same soils under grasses, food crops or coffee. It has also been shown that nitrogen and carbon levels under casuarina trees increase over time. Soils under casuarina are ready for cultivation again in around 8–12 years. When a fallow that has been planted in casuarina is re-cultivated, the trees are either cut down (the roots are left in the soil to decompose) and used for fencing, ringbarked and eventually used for firewood, or severely pollarded (all but the topmost branches are cut off) to let light into the garden but to keep the tree alive during the cultivation of crops. These casuarinas rapidly form a tree fallow again when cultivation is completed. The pollarded trees grow during the cropping phase and provide seedlings so that the system becomes self-perpetuating.

Casuarina is particularly important in mixed crop gardens where crops are planted that demand higher soil fertility than sweet potato; around villages where the trees provide shade and timber and are also aesthetically important;[2] and as shade in coffee gardens. It is not known for how long people have been planting casuarina trees into gardens to improve soil fertility. It is likely the practice has increased in importance in the last 150 years and that it is still spreading. It is known that casuarina planting has increased in importance in Simbu Province since the 1950s and has been adopted at Oksapmin since about 1920 and at Okapa in Eastern Highlands Province since about 1960.

2 People praise the look of the casuarina tree, its association with settlement and 'home', and the sound that the wind makes blowing through the leaves.

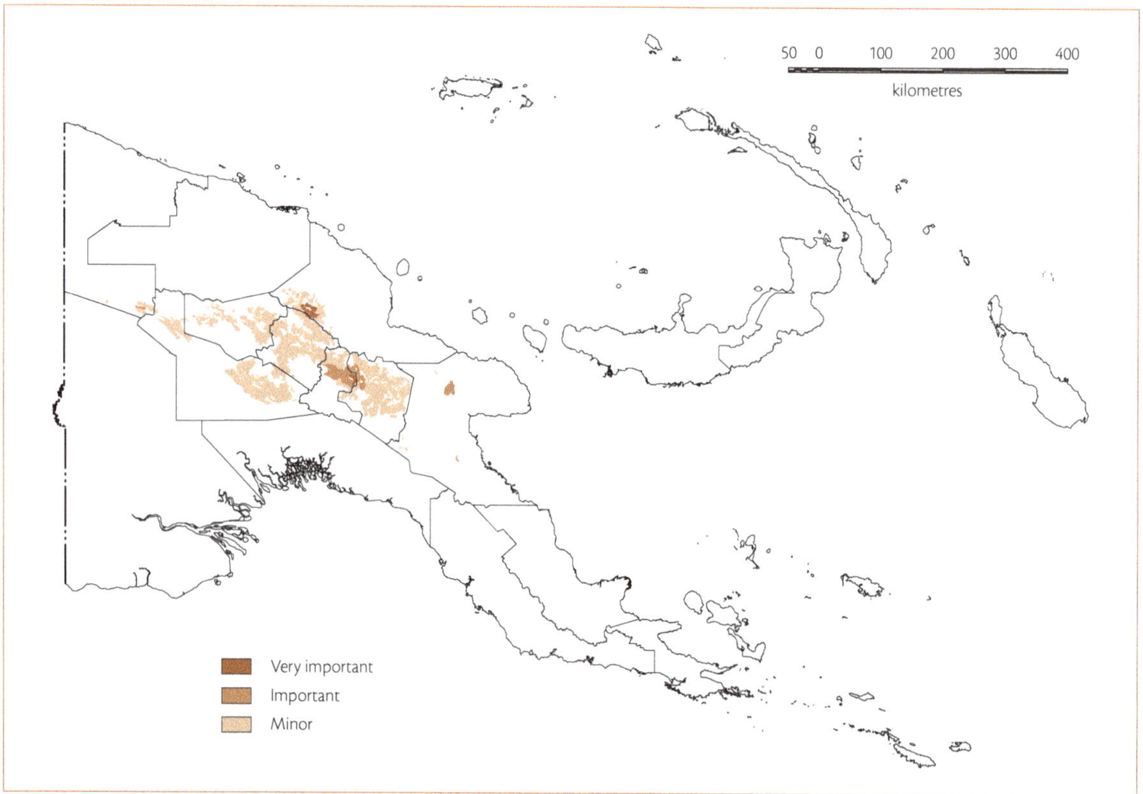

Figure 3.10.1 The planting of trees in fallows. Source: MASP.

Another practice associated with planted casuarina trees is the burning of dried branches on the soil surface and the planting of leafy green vegetables or tobacco plants in the ashes. The fire probably releases nitrogen from the topsoil into the atmosphere. However, fires also kill large numbers of soil micro-fauna that release nitrogen as they decompose, which becomes available to plants.

As well as improving soils, casuarina trees are an important source of wood for firewood, house building and fencing. A 1980 study in Enga Province estimated that 370 000 m^3 of casuarina wood per year was used for firewood (including that used in urban areas), 93 000 m^3 for fences and 26 000 m^3 for house building. This wood was valued at between K2.4 and K3.7 million in 1980 (K15–23 million in 2005 values). One mature casuarina tree was estimated to provide one cubic metre of wood. The study noted that all of this wood was produced by villagers planting trees.

It is reported from the Kaironk and Jimi valleys in Madang Province that, where casuarina plantings do not thrive, *Dodonaea viscosa* is planted in gardens from seed and seedlings.

Albizia spp. is a fast-growing leguminous tree of the subfamily Mimosoideae. It occurs in fallows in a number of locations around PNG, including East Sepik, Morobe, Madang and the highlands provinces. In places it is transplanted into food crop gardens. *Albizia* is used as shade and presumably provides nitrogen in semi-permanent mixed coffee–banana–Chinese taro gardens in the Erap and Wau areas of Morobe Province.

Parasponia rigida is not a leguminous plant, but rhizobia bacteria are found in its root nodules. So although *Parasponia* probably fixes nitrogen, the mechanism of how it is stored in the plant and released to the soil and to other plants is not understood. *Parasponia* is protected from fires and cutting during garden clearing in Enga Province, Southern Highlands Province and in the Kaironk area in Madang Province. It also occurs naturally in lowland fallows across PNG.

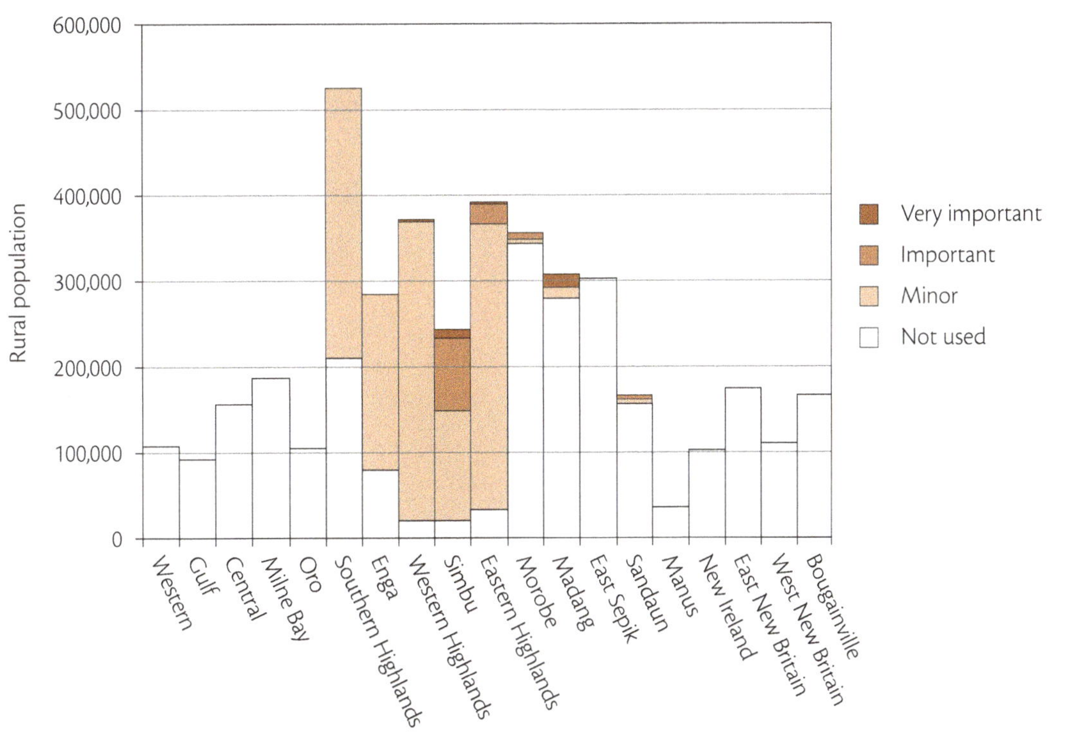

Figure 3.10.2 Rural population planting trees in fallows, by province. Sources: NSO (2002); MASP.

Piper aduncum was introduced to PNG during the 1920s and is spreading in many places, with and without the encouragement of people. On the Sogeri Plateau inland from Port Moresby migrants from the Huon Peninsula in Morobe Province use *P. aduncum* stakes to create soil retention barriers in new sweet potato gardens in grasslands (see Section 3.9). The stakes sprout and after the garden is fallowed *Piper* forms a tree cover that prevents the grass from regrowing.

Schleinitzia novo-guineensis is a leguminous tree, also of the subfamily Mimosoideae, that is found in fallows on a number of small islands in Milne Bay Province. On some islands it is transplanted from natural stands into gardens.

Research is needed to understand why villagers favour these trees and under what conditions the various species perform best. There is much potential for tree planting practices to be spread to other parts of PNG. In some places villagers are experimenting with planting trees. For example, on Pinipel Island in Bougainville Province, some villagers are evaluating the usefulness of *Gliricidia* species to improve sweet potato yields.

Legume rotations

In a number of sweet potato-based agricultural systems, mainly in the highlands, sweet potato plantings are alternated with the planting of leguminous plants. Before the introduction of peanut the most common rotation was sweet potato and winged bean. It is not known how widespread the use of a winged bean rotation was. Following the introduction of peanut between the 1930s and 1950s, a peanut–sweet potato rotation was adopted in parts of the highlands.

Legume rotations are most important in central Eastern Highlands Province and in the Wahgi and Nebilyer valleys in Western Highlands Province.

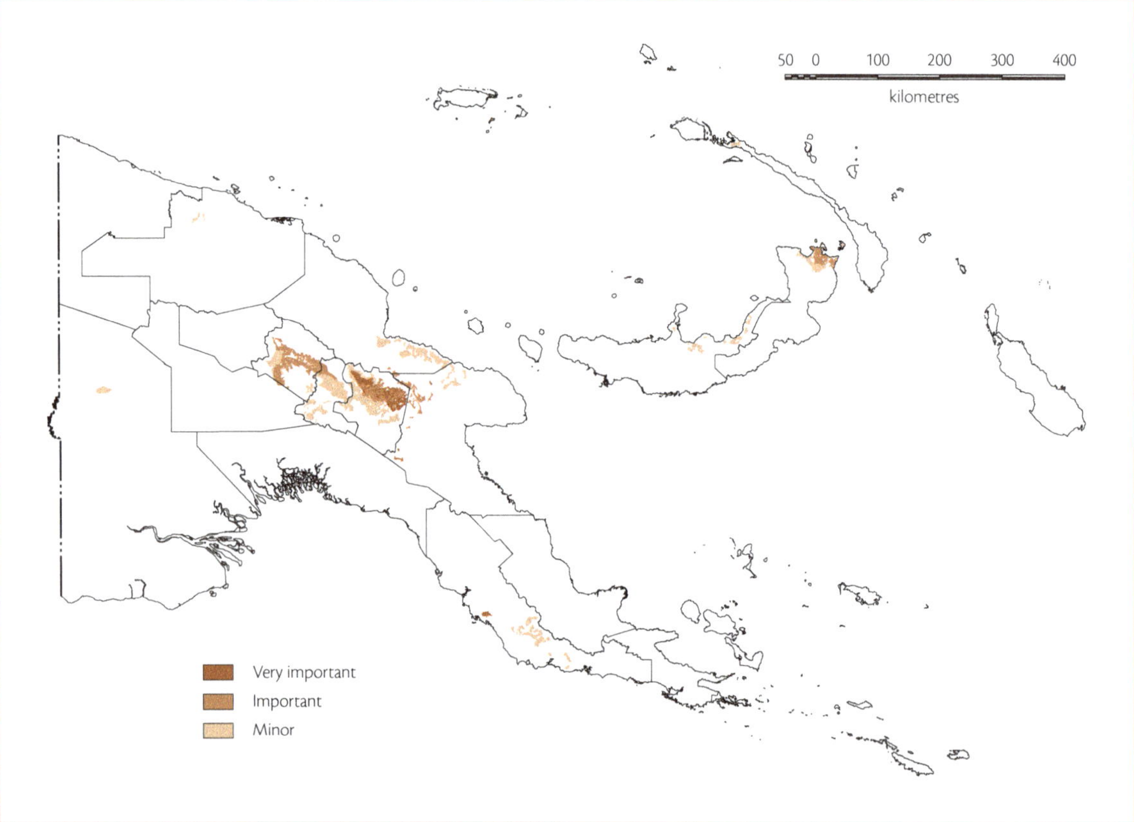

Figure 3.10.3 The planting of legumes (peanut and winged bean) in rotations with root crops. Source: MASP.

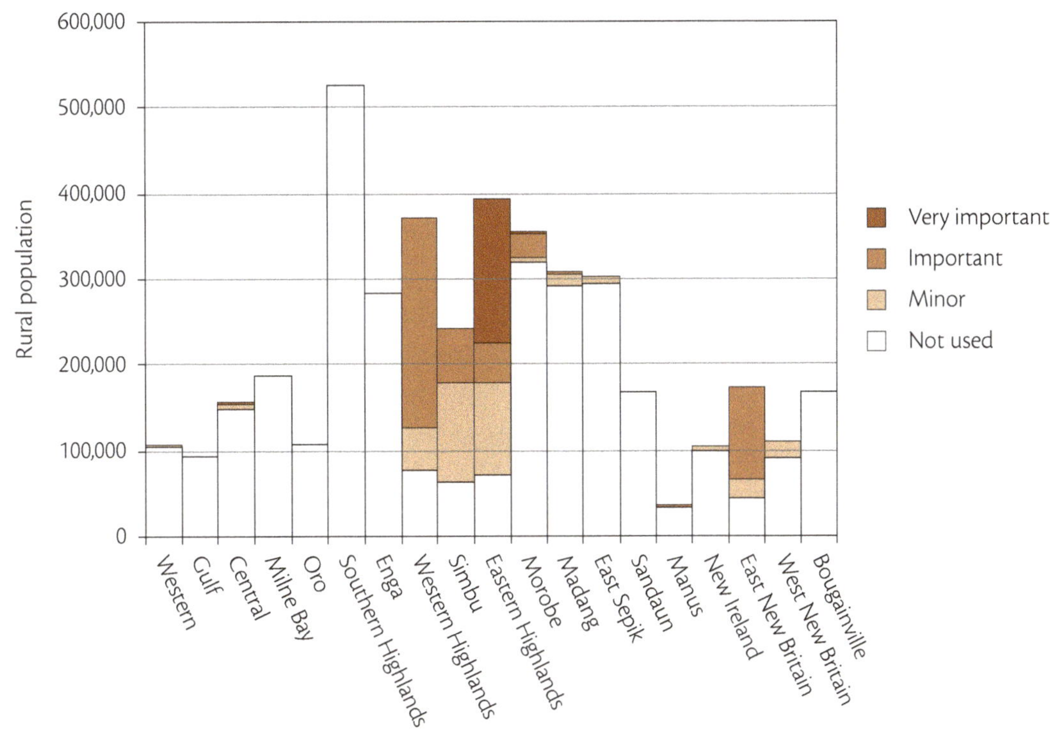

Figure 3.10.4 Rural population planting legumes (peanut and winged bean) in rotations with root crops, by province.
Sources: NSO (2002); MASP.

They have also become important in parts of the Gazelle Peninsula where a sweet potato–peanut rotation is practised, mainly in grassland gardens. Peanut rotations are also used on the Sogeri Plateau in Central Province and in the Ramu and Markham valleys (Figures 3.10.3, 3.10.4, Table A3.10.2).

The rotation of sweet potato with leguminous crops such as peanut or winged bean has two important effects. First, it increases soil nitrogen levels and, second, it reduces the number of sweet potato pests and diseases occurring in the garden soil. Both effects increase yields of sweet potato. Peanut is a widespread supplementary food crop and is also a common locally marketed cash crop. Winged bean is a minor supplementary food in the highlands and is sold in local markets (see Section 3.2).

Sources

Bino, B. and Kanua, M.B. (1996). Growth performance, litter yield and nutrient turnover of *Casuarina oligodon* in the highlands of Papua New Guinea. In Pinyopusarerk, K., Turnbull, J.W. and Midgley, S.J. (eds). *Recent Casuarina Research and Development: Proceedings of the Third International Casuarina Workshop*. Commonwealth Scientific and Industrial Research Organisation, Canberra. pp. 167–170.

Bourke, R.M. (1974). The role of legumes in soil fertility maintenance in the lowlands of Papua New Guinea; a legume evaluation study at Keravat, New Britain; and notes on the use of legumes as food in Papua New Guinea. *Science in New Guinea* 2(1):63–69.

Bourke, R.M. (1985). Food, coffee and casuarina: an agroforestry system from the Papua New Guinea highlands. *Agroforestry Systems* 2:273–279.

Bourke, R.M. (1997). *Management of fallow species composition with tree planting in Papua New Guinea.* Resource Management in Asia-Pacific Working Paper No. 5. Resource Management in Asia-Pacific Program, Research School of Pacific and Asian Studies, The Australian National University, Canberra.

Bourke, R.M. (2001). Intensification of agricultural systems in Papua New Guinea. *Agricultural Transformation and Intensification. Asia Pacific Viewpoint Special Issue* 42(2/3):219–235.

Carrad, B. (1982). The economy. In Carrad, B., Lea, D.A.M. and Talyaga, K.K. (eds). *Enga: Foundations for Development. Volume 3 of Enga Yaaka Lasemana.* Department of Geography, University of New England, Armidale. pp. 146–178.

D'Souza, E., Bourke, R.M. and Akus, W.L. (1986). Intensification of subsistence agriculture on the Nembi Plateau, Papua New Guinea: 3. Sweet potato cultivar trials; crop rotation trials; and crop introductions. *Papua New Guinea Journal of Agriculture, Forestry and Fisheries* 34(1–4):41–48.

Haberle, S.G. and Atkin, G. (2005). Needles in a haystack: searching for sweet potato (*Ipomoea batatas* (L.) Lam.) in the fossil pollen record. In Ballard, C., Brown, P., Bourke, R.M. and Harwood, T. (eds). *The Sweet Potato in Oceania: A Reappraisal.* Oceania Publications, University of Sydney and Ethnology, Department of Anthropology, University of Pittsburgh, Sydney and Pittsburgh. pp. 25–33.

Kanua, M., Brook, R., Maima, R. and Binifa, J. (1989). *Casuarina oligodon* growth and development. In *Agricultural Research Division 1989 Annual Research Report.* Department of Agriculture and Livestock, Port Moresby. pp. 20–22.

Kimber, A.J. (1974). Crop rotations, legumes and more productive arable farming in the highlands of Papua New Guinea. *Science in New Guinea* 2(1):70–79.

NSO (National Statistical Office of Papua New Guinea) (2002). Papua New Guinea 2000 Census: Final Figures. National Statistical Office of Papua New Guinea, Port Moresby.

Parfitt, R.L. (1976). Shifting cultivation – how it affects the soil environment. *Harvest* 3(2):63–66.

Powell, J.M. (1974). Traditional legumes of the New Guinea highlands. *Science in New Guinea* 2(1):48–63.

Talopa, P. (2003). Role of casuarina in the agroforestry systems in Papua New Guinea – our knowledge about *Casuarina oligodon* L. Johnson. *Coffee Industry Newsletter* (2):11–14.

3.11 Tillage, mounds, beds and green manuring

In many PNG agricultural systems, soil is formed into mounds or beds for the cultivation of crops. The main effects of mounding and bedding are to raise crops above the water table, to improve drainage, to increase the depth of topsoil and to thoroughly till the soil. These techniques also allow the application of green manure ('compost') to the soil where this is practised.

Mounds, which are usually roughly circular in shape, may be distinguished on the basis of their dimensions into:

- Small mounds (10–40 cm high and 40–100 cm in diameter).
- Medium-sized mounds (40–70 cm high and 100–250 cm in diameter).
- Large mounds (more than 70 cm high and more than 250 cm in diameter).

Beds can be classified by their shape:

- Square beds.
- Long (or rectangular) beds.

The use of mounds and beds forms reasonably discrete geographical patterns in PNG. The distribution of mounds and beds is influenced by topography, slope, soil type, staple crop, culture and history.

Tillage

Tillage affects the physical, chemical and biological properties of the soil. Tillage influences soil particle size, soil temperature, water infiltration and water retention. Tillage allows air into the soil and so increases the rate of organic matter decomposition, which makes nutrients available to plants. Biological activities of earthworms, insects, bacteria and many other small organisms in the soil are vital to soil productivity.

In PNG, tillage is commonly associated with mounding and bedding. However, there are a number of places in PNG where soil is tilled in the absence of mounds and beds. In these cases, tillage is usually associated with the use of grass fallows (see Section 3.8). For example, grass fallows are tilled in the western Schrader Range along the border between Madang and Western Highlands provinces, and around Wopasali in the south-east of Southern Highlands Province. Hillside gardens are completely dug over to remove cane grass roots in the Tauri River Valley, north and east of Menyamya, Morobe Province. Steep, grass fallow hill slopes are tilled during the building of benches in inland Milne Bay Province from Agaun east towards the Rabaraba coast (see Section 3.9).

Small mounds

Small mounds are used to a varying extent by almost two-thirds of rural people in PNG, mainly to raise crops above a wet soil and to increase the depth of topsoil. With the spread of sweet potato, small mounds have become increasingly widespread because sweet potato yields are reduced by saturated soils.

Small mounds are important or very important in northern Simbu and Eastern Highlands provinces, where they are used by more than 95% of the provincial populations; in Bougainville Province (83%); and in Oro, Central and Milne Bay provinces (more than 75%) (Figures 3.11.1, 3.11.7, Table A3.11.1). Small mounds are also sometimes used to plant crops other than sweet potato, for example, yam (notably in Milne Bay Province), cassava (in Central Province) and green vegetables (in Gulf Province). Small mounds are not used or are a minor practice in Gulf, Southern Highlands, Enga, East Sepik and Sandaun provinces.

Medium-sized mounds

Medium-sized mounds are used in food gardens in the western part of the highlands. They may or may not be constructed over heaps of green manure (composted). Composting is discussed in more detail on page 255. Medium-sized mounds are an important or very important technique for two-thirds of the rural population in Southern Highlands Province and almost half the rural population in Enga Province (Figures 3.11.2, 3.11.7, Table A3.11.2).

Whether or not green manure is incorporated into the mounds, they are constructed in a similar way. Land is cleared from a long fallow of tall grasses and shrubs by removing all vegetation and roots and breaking the soil into a rough tilth. An initial crop of sweet potato is commonly planted on small mounds to further till the soil before medium-sized mounds are created. The garden is then planted and harvested many times, with a short fallow of up to eight months between plantings. The mounds are reconstructed at the beginning of each planting cycle.

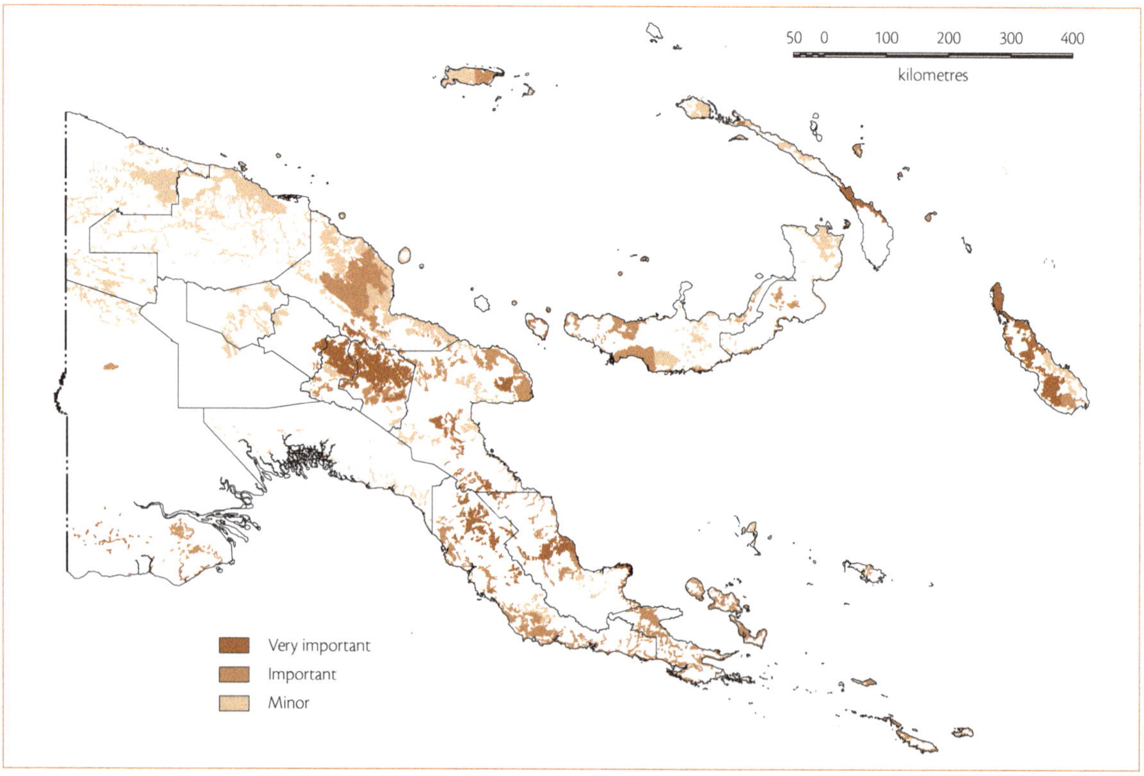

Figure 3.11.1 The use of small mounds in food production systems. Source: MASP.

After a harvest and a short 6–8 week fallow, land is prepared for planting again by breaking down the previous mounds and rebuilding them. Patterns of mound reconstruction include incorporating one-third of the three nearest into a new mound, or using one-quarter of the four nearest mounds to make a new mound. If the first technique is used, the mounds will form a regular triangular pattern across a garden, while the latter technique will create a regular square pattern of mounds. If the mounds are to be composted, the green manure is placed on the soil surface between the old mounds and the new mounds are formed on top of the green material.

Large mounds

Large mounds are used by about 5% of the rural population. They are important in parts of Enga Province at Laiagam, the Lai Valley and Kompiam, and in parts of Western Highlands Province at Tambul and north-west of Mount Hagen town (Figures 3.11.3, 3.11.7, Table A3.11.3). Large mounds are used on soils derived from volcanic ash and are always composted. They are initially constructed in the same way as medium-sized mounds. However, after a harvest the mound is broken open by forming a depression in the centre, so that fallow large mounds look like small bomb craters. During the short fallow, weeds, sweet potato vines and grasses are thrown into the depression where they partially decompose. When the land is prepared for planting, more green material may be added and the mound is reformed by filling in the depression with soil from the outside of the mound. Thus large mounds are a relatively stable feature of the landscape and are not moved during the cultivation cycle, as medium-sized mounds are. Considerably more green manure is placed in large mounds than in medium-sized mounds.

Square beds

Square beds are constructed by digging a grid of shallow drains at regular intervals across the garden site and throwing the spoil from the drains onto the

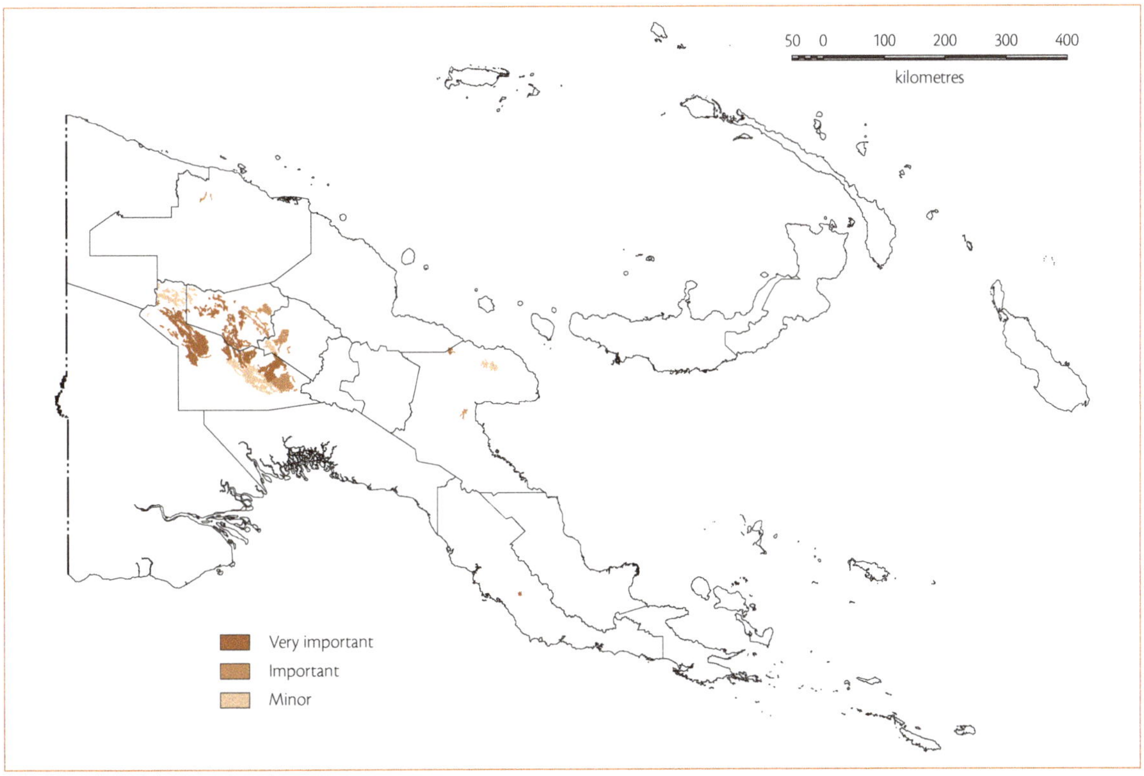

Figure 3.11.2 The use of medium-sized mounds in food production systems. Source: MASP.

surface between the drains. Grasses, weeds and other vegetation are usually left to dry and decompose on the surface and are buried by the spoil from the drains. Additional green material may be added, in which case the beds are composted. After the harvest and a short fallow, the beds are reconstructed by digging a grid of new drains through the middle of the previous beds, filling in the old drains and throwing the spoil onto the surface of what are the new beds. Square beds are usually found on flat or gently sloping land. They form a characteristic 'chequerboard' pattern observable from the air or on air photographs.

Square beds are most important in Western Highlands Province (where they are commonly used by 66% of the rural population) and in adjacent locations in Simbu Province (Figures 3.11.4, 3.11.7, Table A3.11.4). In the western part of Western Highlands Province, square beds give way to large mounds (Figure 3.11.3). Square beds are also important in some locations in Southern Highlands Province. The technique is used in the southern part of Western Province, where large areas of fallow or abandoned beds can be observed beneath secondary forest and savanna. Here beds are sometimes used to remove water from very poorly drained land and to lift crops above the high water table.

Long beds

Long beds, which are found on steeper slopes than square beds, are constructed in a similar fashion to square beds, but with much less regularity. Whereas square beds appear to be primarily a means for lowering the water table, long beds facilitate the passage of surface water across a sloping garden with a minimum loss of topsoil as the water flows between the beds. Both types of beds also result in a deeper topsoil and rooting zone. Long beds in particular are often observed sitting on brown or red clay subsoils.

Long beds are most important in the northern part of Eastern Highlands Province, where they are used by two-thirds of the rural population; in the central part of Southern Highlands Province; and in Western

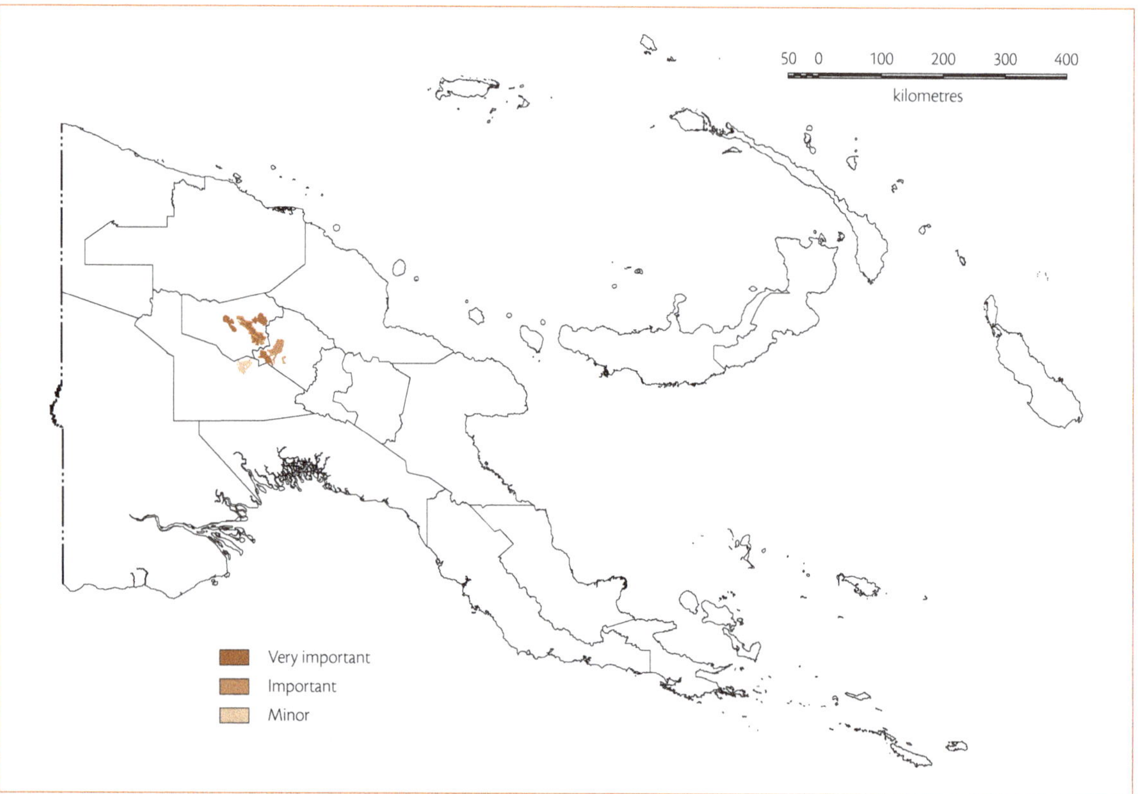

Figure 3.11.3 The use of large mounds in food production systems. Source: MASP.

Province around Balimo. Long beds are also used in a limited number of locations in Central Province and the Menyamya area of Morobe Province (Figures 3.11.5, 3.11.7, Table A3.11.5).

It will be apparent from Figures 3.11.4 and 3.11.5 that square beds and long beds sometimes occur in the same areas. In these locations, square beds give way to long beds as the slope gradient increases.

Green manuring (composting)

In some parts of the highlands, plant material is buried beneath the soil before sweet potato is planted. This is strictly speaking a form of green manuring, but in PNG it has become known as 'composting'. Composting is practised mainly in Southern Highlands Province (using medium-sized mounds and beds); Enga Province (using medium-sized and large mounds); and the western end of Western Highlands Province (using large mounds). The use of composting is notably absent from the eastern part of Western Highlands, Simbu and Eastern Highlands provinces (Figures 3.11.6, 3.11.7, Table A3.11.6). The technique was developed by villagers and not introduced by outsiders. It is possible that composting was adopted in the main valleys of Enga Province about 150–200 years ago and subsequently adopted by other villagers in adjoining locations.

Composting is closely associated with mounding or bedding almost everywhere. It is restricted to areas between 1100 m and 2800 m in altitude; where mean minimum and maximum temperatures are 8–16 °C and 18–27 °C, respectively; and annual rainfall ranges from 2200 mm to 5000 mm. Composting is most common where rainfall is between 2200 mm and 3000 mm. The soils on which composting is most common are derived from volcanic ash. Volcanic ash soils are well structured and friable but have low levels of available phosphate and other essential minerals.

The amount of green plant material placed in mounds varies considerably. Below 1800 m altitude about 5 tonnes/hectare is applied, whereas above 1800 m up to 30 tonnes/hectare can be applied.

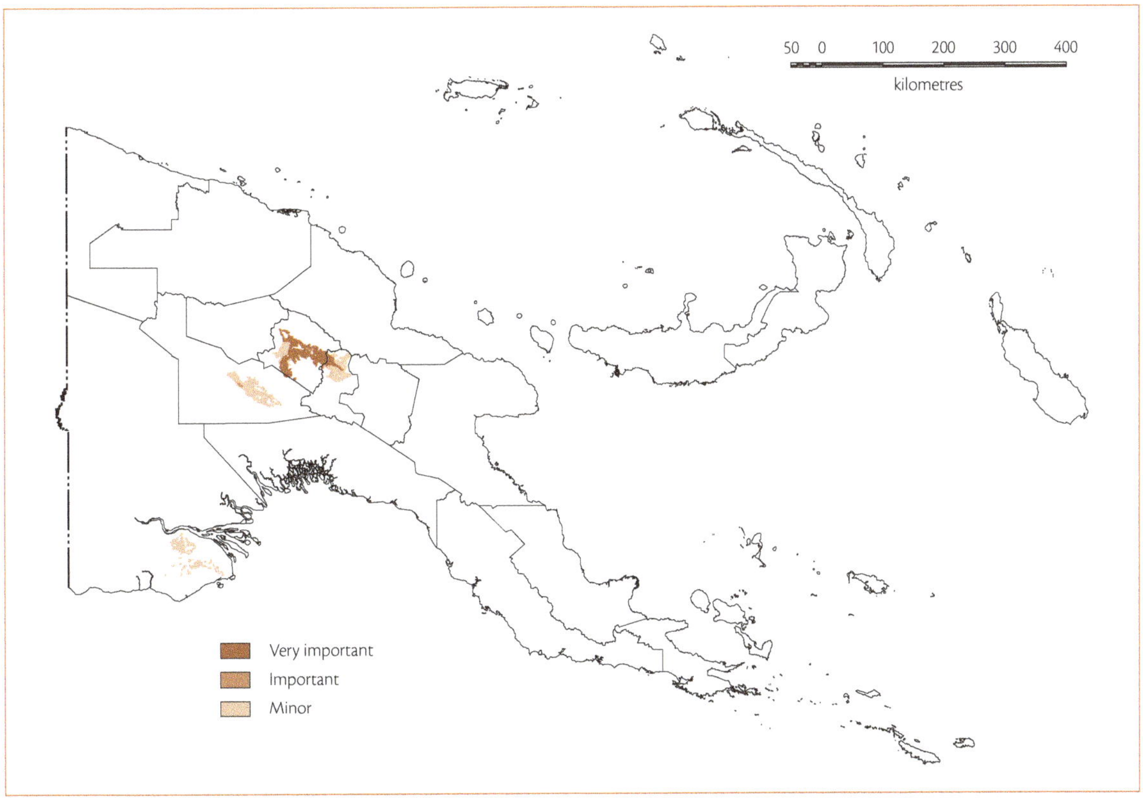

Figure 3.11.4 The use of square beds in food production systems. Source: MASP.

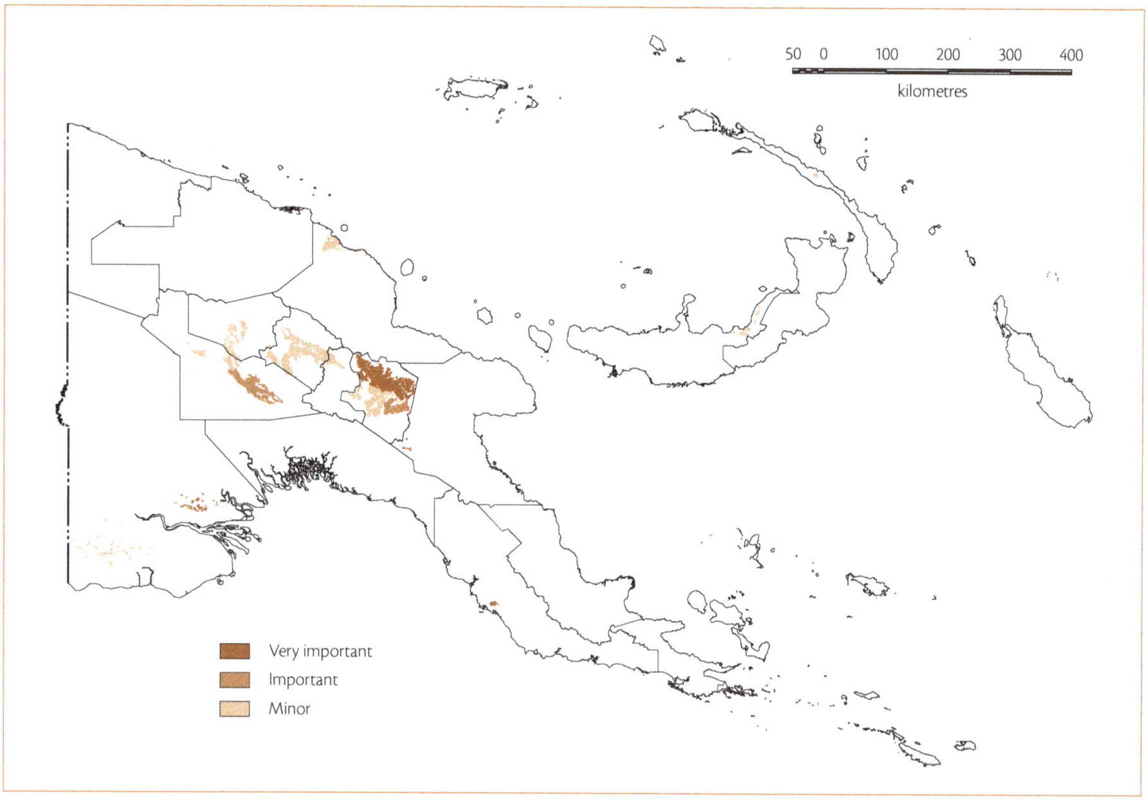

Figure 3.11.5 The use of long beds in food production systems. Source: MASP.

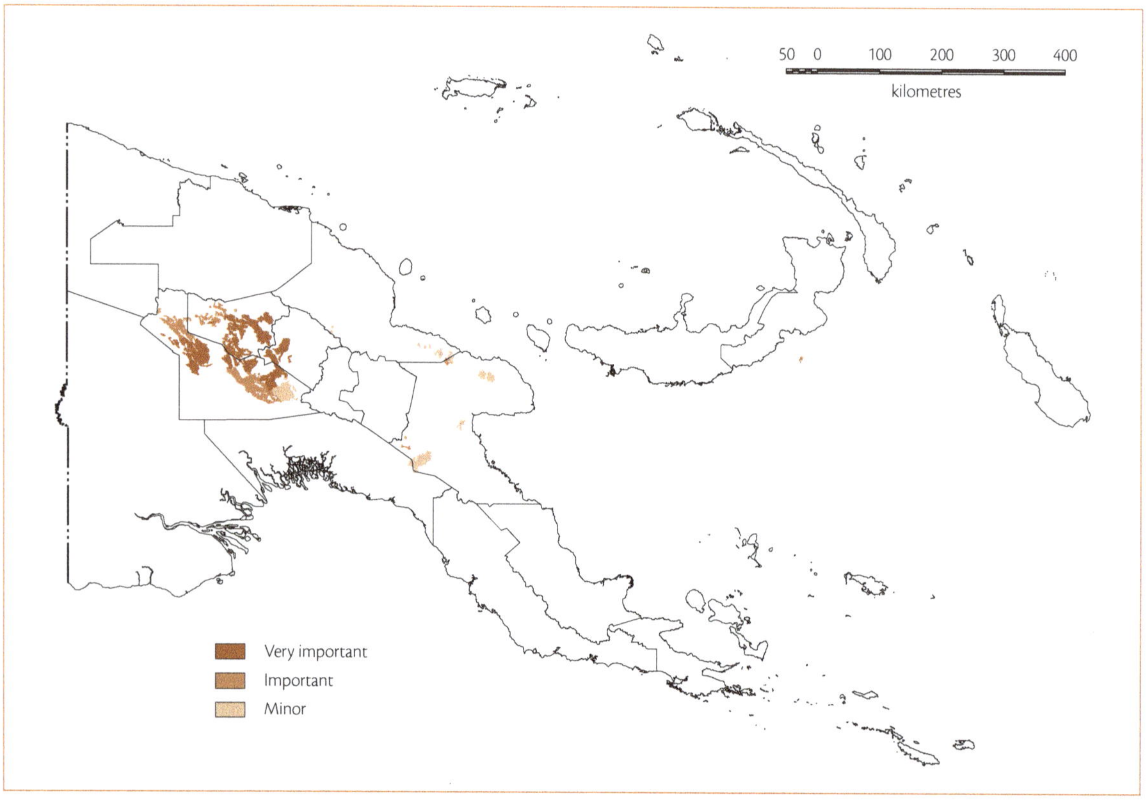

Figure 3.11.6 The use of green manure (compost) in food production systems. Source: MASP.

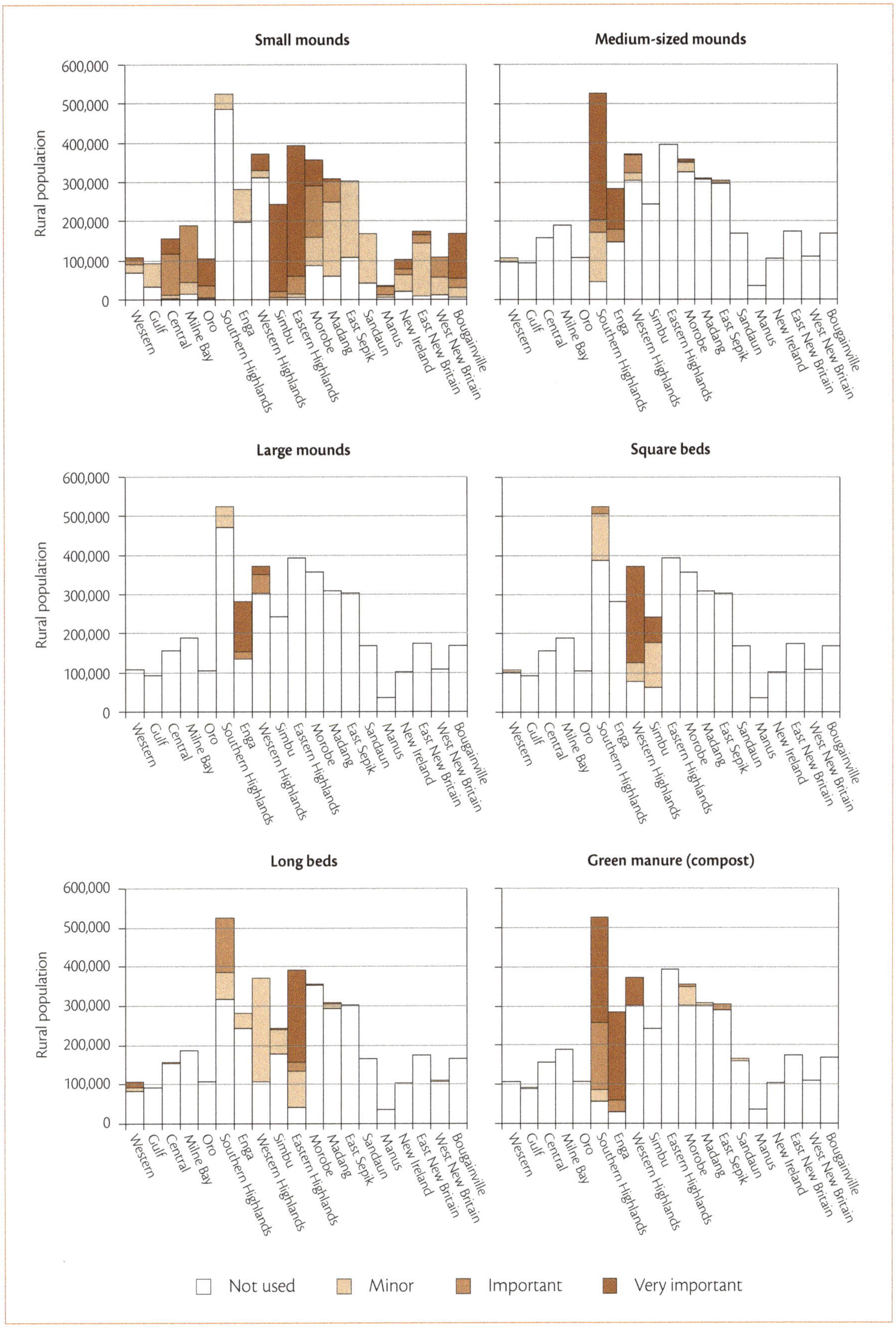

Figure 3.11.7 Rural population using mounding, bedding and green manuring techniques in food production systems, by province. Sources: NSO (2002); MASP.

A study in the Lai Valley in Enga Province found the green manure applied was made up of sweet potato vines (63%), weeds (20%), highland *pitpit* leaves (10%), banana leaves (4%) and sugar cane leaves (3%). In Enga, women sometimes cut grass from nearby fallows and carry it to the garden and place it in mounds.

Composting is almost exclusively associated with sweet potato production. More than 20 agronomic trials have investigated the effect of composting on sweet potato yields. Composting increased sweet potato yields in all the trials but the response varied considerably. Investigations in Southern Highlands Province suggested that compost increased the availability of potassium to the sweet potato plants (other research has indicated that low levels of potassium limit sweet potato yields in the PNG highlands). Compost also increases top growth in sweet potato, increases soil temperature by up to 6 °C (which reduces the time to harvest because time to maturity is temperature-dependent in sweet potato), and reduces the incidence of fungal black rot in tubers. One researcher suggested that composting protects sweet potato from frosts at high altitudes, but this is unlikely.[1] The greatest benefits of composting are probably increased crop yield, reduced time to harvest and reduced incidence of rot.

The advantages of green manuring for the maintenance of soil fertility have been recognised by villagers in a number of locations outside the highlands. At Teptep in the Finisterre Mountains on the border of Morobe and Madang provinces, composted ridges constructed along the contour are used to grow sweet potato on volcanic ash soils at altitudes over 2000 m. The ridges are about 1.5 m wide, 0.6 m high, 20–30 m long and are constructed along the contour about 1.5 m apart. A short fallow follows each harvest. At the end of the fallow, grasses, weeds and sweet potato vines are placed between the ridges and new ridges are formed by moving half of the adjacent ridges into the space between the ridges to cover the green manure.

At Menyamya (Morobe Province, at 1500 m altitude) grasses and weeds are buried in long beds. Around Aseki (Morobe and Gulf provinces, at 1400–1900 m altitude) vegetation slashed during garden clearing is heaped over low wooden hurdles and banana and taro are planted along the edges of the heap. At Wagau in the Snake River Valley (Morobe Province, at 600–1100 m altitude) weeds, grasses and leaf litter are incorporated into medium-sized mounds, which are planted with single yam seed tubers. Near Bogia (Madang Province, 15 m altitude) weeds and grasses are dug into garden beds on drained alluvial flats. Around Dreikikir (East Sepik Province, 300 m altitude) the dry leaves of *Pometia pinnata* trees cut down during the harvesting of *ton* fruit are placed beneath yam seed tubers in holes in the garden. This practice has been adopted at nearby Seim, Sandaun Province.

1 It was argued that planting sweet potato on mounds raised the succulent parts of plants above freezing air flowing over the soil surface. However, observations of frost damage suggest the freezing air is usually deeper than the average height of most mounds. It has been shown that composting raises the soil temperature inside the mound, but this will not prevent succulent vines exposed to freezing air on the surface from being damaged.

Sources

D'Souza, E. and Bourke, R.M. (1986). Intensification of subsistence agriculture on the Nembi Plateau, Papua New Guinea: 2. Organic fertilizer trials. *Papua New Guinea Journal of Agriculture, Forestry and Fisheries* 34(1–4):29–39.

Esekia, I. (1993). Effect of different legume species as green manure on the yield of Chinese cabbage (pak-choi) (revised version). *Harvest* 15(2):1–3.

Floyd, C.N., Lefroy, R.D.B. and D'Souza, E.J. (1988). Soil fertility and sweet potato production on volcanic ash soils in the highlands of Papua New Guinea. *Field Crops Research* 19:1–25.

Kimber, A.J. (1970). Some cultivation techniques affecting yield response in the sweet potato. In Plucknett, D.L. (ed). *Proceedings of the Second International Symposium on Tropical Root and Tuber Crops*. Volume 1. University of Hawaii, Honolulu. pp. 32–36.

NSO (National Statistical Office of Papua New Guinea) (2002). Papua New Guinea 2000 Census: Final Figures. National Statistical Office of Papua New Guinea, Port Moresby.

Powell, J.M., Kulunga, A., Moge, R., Pono, C., Zimike, F. and Golson, J. (1975). *Agricultural traditions of the Mount Hagen area*. Department of Geography Occasional Paper No. 12. Department of Geography, University of Papua New Guinea, Port Moresby.

Preston, S.R. (1990). Investigation of compost × fertilizer interactions in sweet potato grown on volcanic ash soils in the highlands of Papua New Guinea. *Tropical Agriculture (Trinidad)* 67(3):239–242.

Sterly, J. (1977). Research work on traditional plantlore and agriculture in the upper Chimbu region, Papua New Guinea. *Bulletin of the International Committee on Urgent Anthropological and Ethnological Research* (19):95–102.

Swift, J. and Nalu, G. (1981). *Diwai bilong winim gris (Nitrogen-fixing trees)*. Wau Ecology Institute Leaflet No. 3. Wau Ecology Institute, Wau, Papua New Guinea.

Waddell, E. (1972). *The Mound Builders: Agricultural Practices, Environment, and Society in the Central Highlands of New Guinea*. The American Ethnological Society Monograph 53. University of Washington Press, Seattle and London.

Wiessner, P. (2005). Social, symbolic, and ritual roles of the sweet potato in Enga, from its introduction until first contact. In Ballard, C., Brown, P., Bourke, R.M. and Harwood, T. (eds). *The Sweet Potato in Oceania: A Reappraisal*. Ethnology Monographs 19 / Oceania Monograph 56. Oceania Publications, University of Sydney and Ethnology, Department of Anthropology, University of Pittsburgh, Sydney and Pittsburgh. pp. 121–130.

Wiessner, P. and Tumu, A. (1998). *Historical Vines: Enga Networks of Exchange, Ritual, and Warfare in Papua New Guinea*. Crawford House Publishing, Bathurst.

Wood, A.W. (1984). Land for tomorrow: subsistence agriculture, soil fertility and ecosystem stability in the New Guinea highlands (two volumes). PhD thesis. University of Papua New Guinea, Port Moresby.

3.12 Other agricultural techniques

The most important innovations used by villagers in food production systems are described in Sections 3.7 to 3.11. A number of other agricultural techniques used in PNG deserve attention because they may become more important in the future, or because they demonstrate novel solutions to particular agricultural problems. The techniques described in this section are:

- Use of drains.
- Use of mixed crop gardens.
- Use of house gardens.
- The practice of placing pigs in gardens between crop plantings.
- Supporting crops with stakes.
- Cutting down fallow trees onto growing crops.
- Use of fences and other methods of creating barriers around fields or divisions between fields.
- Irrigation.
- Growing yam tubers in deep holes.
- Use of animal manure and inorganic fertiliser.
- Use of silt from floods to increase soil fertility.

Drains

Removing excess water from gardens is an important part of making crop production possible in saturated soils and increasing production in poorly drained soils. Drains have been a feature of PNG agriculture from its beginnings; the archaeological record of what is almost certainly agriculture at Kuk in Western Highlands Province, dated to 9000 years ago, is based on different phases in the draining of Kuk swamp (see History of agriculture). Around 63% of land used for agriculture in PNG is on mountains and hills (see Section 1.10), where steep slopes make drains unnecessary. Drainage has become more important since the introduction and widespread adoption of sweet potato as a staple crop from around 300 years ago.[1]

The use of drains to remove excess water is widespread in highland food production systems and the discussion that follows refers to this practice. Drains are also widely used in coffee, oil palm and cocoa plots. Drains are a very important part of agricultural systems in the northern part of

[1] Archaeological research by Chris Ballard has shown that agriculture began on the swamp edges at Haeapugwa, near Tari in Southern Highlands Province, over 2000 years ago but that the swamp centre was probably not drained until around 300 years ago. Drains were being constructed at Mogoropugua near Koroba before 1670 AD.

Eastern Highlands Province (Figure 3.12.1), where shallow drains are dug directly down the slope, in combination with the construction of long beds (see Section 3.11). In the Wahgi Valley, from western Simbu and through Western Highlands Province, drains are associated with the use of square beds. In the Lai and Nembi valleys in Southern Highlands Province, shallow, downslope drains similar to those used in Eastern Highlands Province are dug in conjunction with long beds. Drained swamps are a feature of agriculture in a number of locations in Southern Highlands and Enga provinces. Drains of impressive dimensions (up to 5 m deep) make it possible to cultivate sweet potato in the Kandep, Koroba, Kopiago, Tari and upper Mendi swamps. Outside the highlands, drains are important in the swampy southern part of Western Province and on a small area of floodplain near Maprik in East Sepik Province, where large ditches help remove floodwaters from gardens.

Mixed crop gardens

In all five highlands provinces, where land use intensity is high (see Section 1.2), smaller gardens of mixed species are planted (Figure 3.12.2). Mixed crop gardens are separated from the larger sweet potato gardens and their management is different.[2] Mixed crop gardens contain up to 15 food crop species and sweet potato is usually excluded; they are smaller in area than the larger sweet potato fields that often enclose them; they are planted for shorter periods of up to twelve months; usually only one planting is made before a long fallow; they may be planted beneath pollarded casuarina trees (see Section 3.10); and the sites chosen are often less well

2 The term 'mixed crop' garden is used here to distinguish this garden type from the main sweet potato gardens in the highlands only.

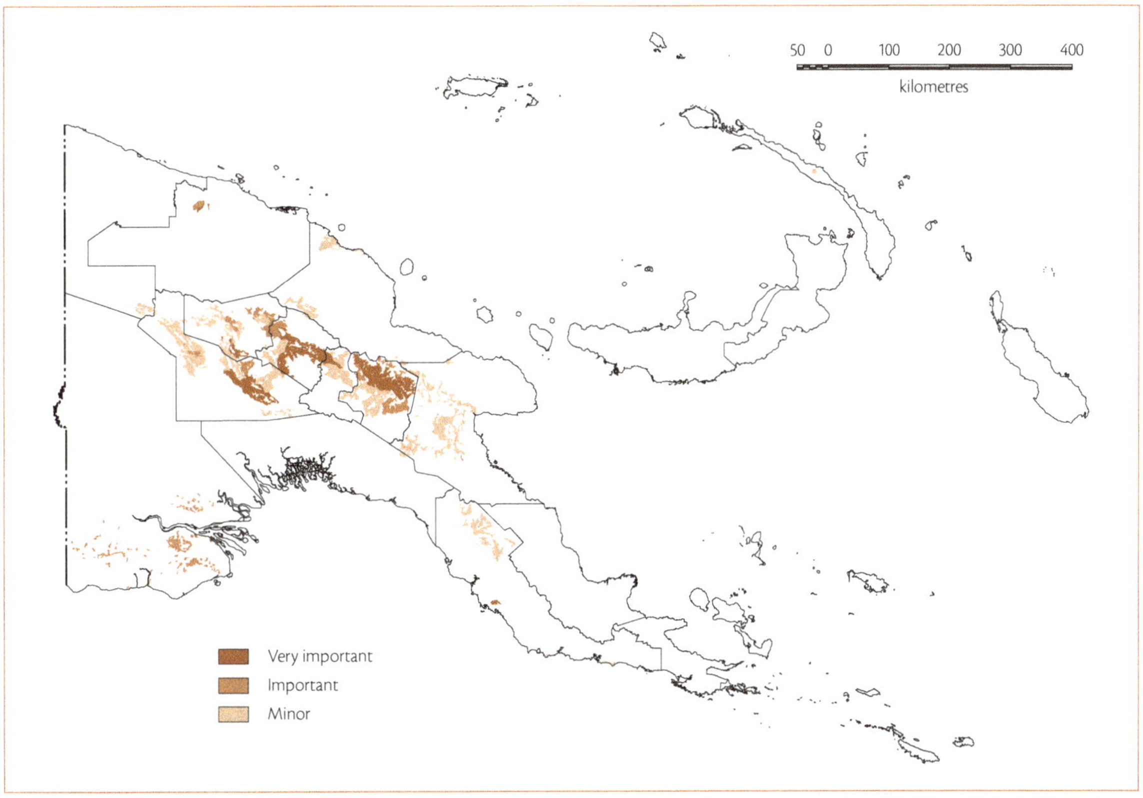

Figure 3.12.1 The use of drains in food production systems. Source: MASP.

drained than those used for sweet potato gardens. Mixed crop gardens are most often observed between 1500 m and 2000 m above sea level.

House gardens

House gardens (also known as 'door yard' or 'kitchen' gardens) are small areas that are cultivated with a number of food species, next to dwelling houses. House gardens are used all over PNG, but are almost universal in the highlands, southern Western Province, the western mountainous areas of Central Province, and parts of Milne Bay and New Ireland provinces (Figure 3.12.3).

House gardens are typically planted with a number of leafy green vegetables, sugar cane and pumpkin (pumpkin leaves are a commonly consumed green vegetable; see Section 3.2). They are maintained for long periods of time by the application of household rubbish: food scraps and peelings, ash from the house fire and sweepings from the house compound.

The placing of pigs in gardens between crop plantings

In restricted parts of PNG, pigs are tethered in gardens after the final harvest and prior to the next planting of sweet potato.[3] This practice represents an intensification of pig management and has a number of benefits. It has been shown at Tari in Southern Highlands Province that pigs tethered in harvested sweet potato gardens gained weight significantly faster than pigs tethered in fallow grassland (probably because of a higher consumption of earthworms than of leftover sweet potato tubers).

[3] Pigs in PNG are tethered by a rope, using a slip knot around the pig's front leg.

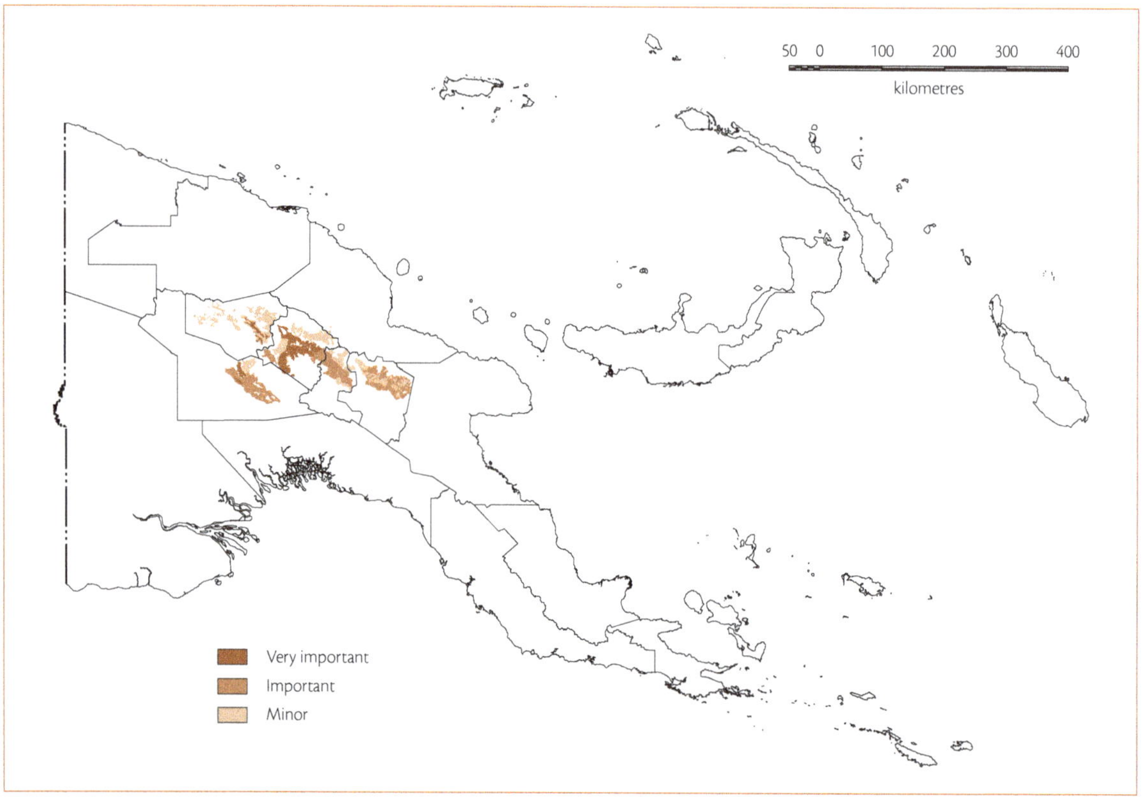

Figure 3.12.2 The use of mixed crop gardens in food production systems. Source: MASP.

In addition, the pigs break up beds and mounds and till the soil. Presumably, their dung and urine also contribute to soil fertility.

The placing of pigs in gardens between crop plantings is almost universal among the Goilala people of Central Province, and is widely observed in the highlands provinces except Eastern Highlands (Figure 3.12.4).

Stakes

The use of stakes to support plants is widespread in PNG agricultural systems, but it becomes significant only where yams are an important crop (Figure 3.12.5). Other crops that are associated with a form of staking are bananas, where large bunches may be propped up or tied to stakes to prevent them breaking the main stem of the banana plant, and sugar cane, which is frequently staked to hold the canes off the ground.

In the Maprik area of East Sepik Province, lesser yam, greater yam and some other less important yam species are staked with stout 2–3 m long sticks. Yams are also trained up vines collected in the forest and tied to dead trees left standing in the garden. In this way, yams may climb up to five metres above the ground. Growers explain how the yam vine grows towards the sun, captures nutrients from the sun, and then, as the vine dries, drains those nutrients into the tubers beneath the ground. Yam staking is also important in southern Western Province, and parts of Central, Milne Bay, Morobe and Madang provinces. However, the staking of yams is not a universal practice. For example, lesser yam grown by Wosera people on the Amogu River floodplains is not staked, but the vines are arranged on the ground to give them maximum exposure to sunlight. Staking has been shown to increase yields of lesser yam.

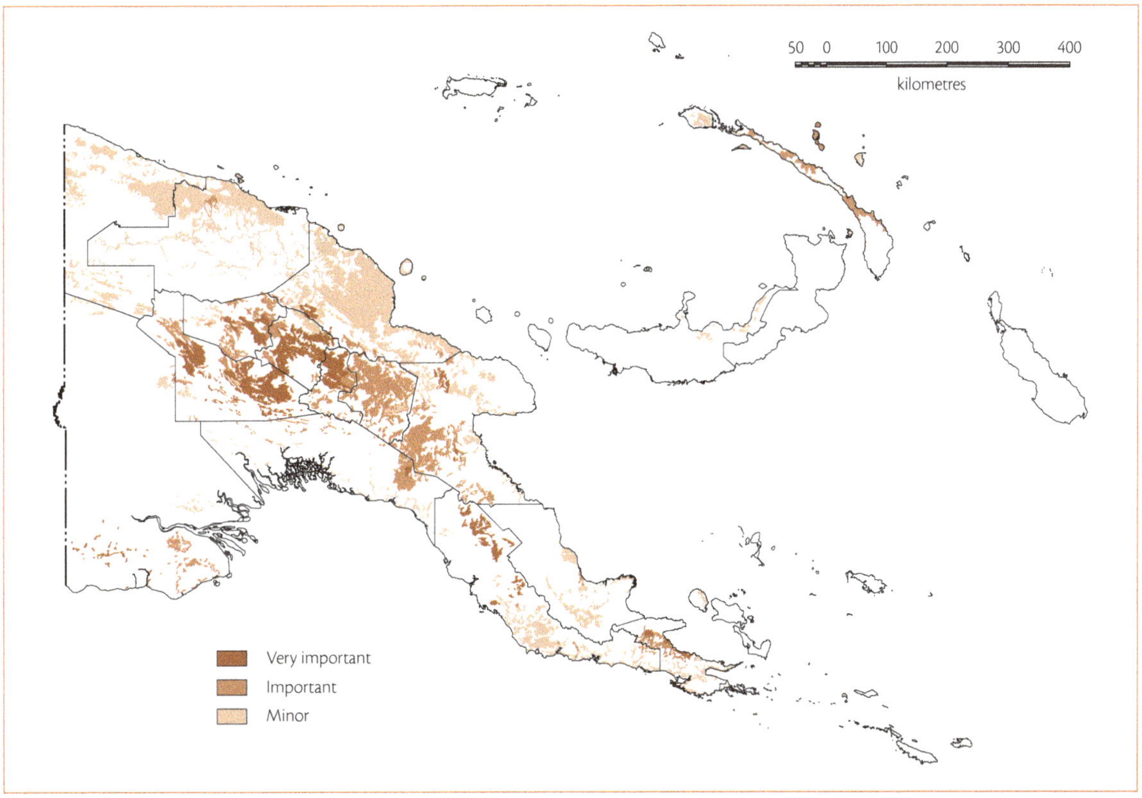

Figure 3.12.3 The use of house gardens in food production systems. Source: MASP.

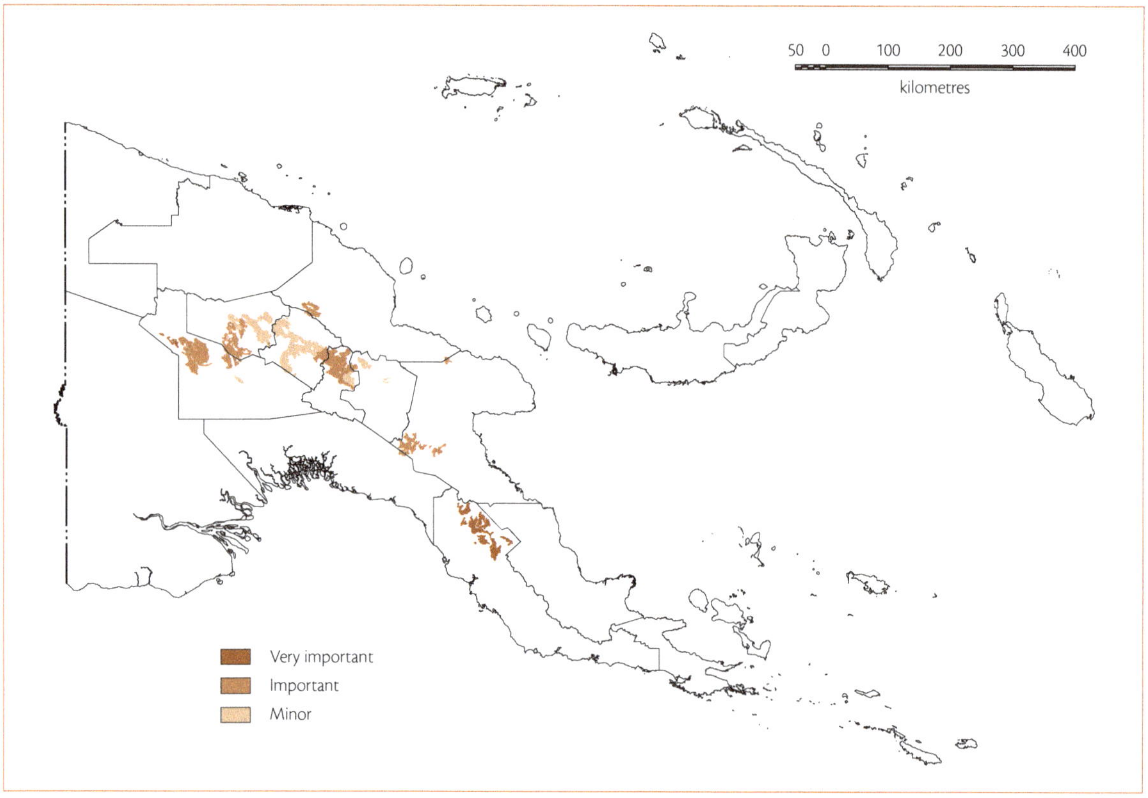

Figure 3.12.4 The practice of tethering pigs in food production systems. Source: MASP.

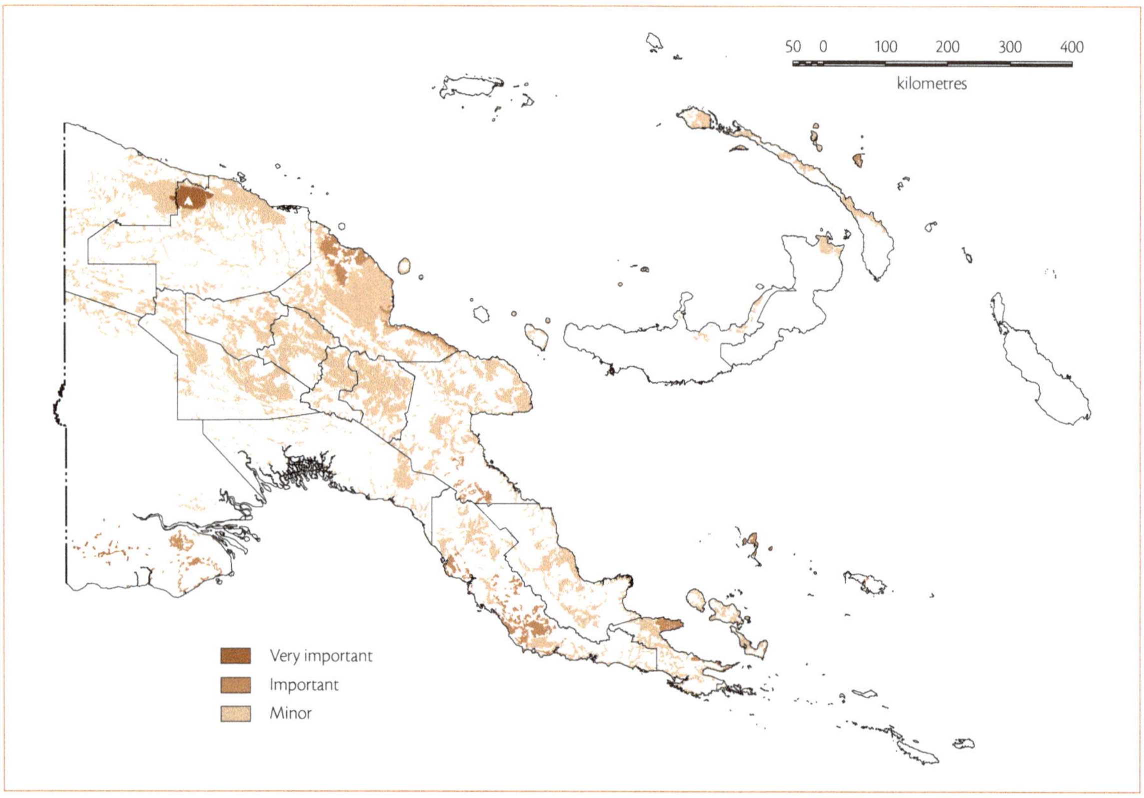

Figure 3.12.5 The use of stakes in food production systems. Source: MASP.

Cutting down trees on top of crops

In a number of areas in PNG, the undergrowth in forest fallows is cleared and food crops are planted beneath standing trees. When the new food plants are established, around 8 to 10 weeks after planting, the trees are cut down on top of them. The areas where this is a typical practice have an average annual rainfall of over 4000 mm; the altitude ranges from near sea level to 1400 m; the soil is relatively infertile; and population density is low. There are two regions where this practice is common. One is on the lower slopes of Mount Bosavi, on the border between Western and Southern Highlands provinces. The other is in the north-eastern end of Gulf Province and nearby locations in Morobe Province (Figure 3.12.6).

Felling large trees onto the developing food plants results in much less damage than might be expected. Immediately after the felling, food plants disappear beneath a mass of twisted branches and tree trunks. However, the high rainfall, humidity and temperatures at this altitude result in the mass of leaves and small branches quickly decomposing and presumably contributing significant amounts of nutrients to the growing food plants. Within a year the trunks of large trees have all but disappeared. This technique reduces exposure of the soil to the very high rainfall and reduces the amount of weeding required. The gardens are planted only once before being fallowed again for up to 20 years.

A similar technique is employed north of Telefomin in Sandaun Province, but most of the trees are not felled. This is another area of very high rainfall (see Figure 1.5.1). The undergrowth beneath tall secondary forest is cleared and planted in taro. As the taro matures, more and more trees are killed by ringbarking or, if they are on the edge of a garden, felled outwards, to allow more light into the garden. This process is extended progressively across a garden site in a strip, exposing the soil to the heavy rainfall for only a short period. An associated technique in this area is the felling of all the trees on a garden site some years before it will be planted

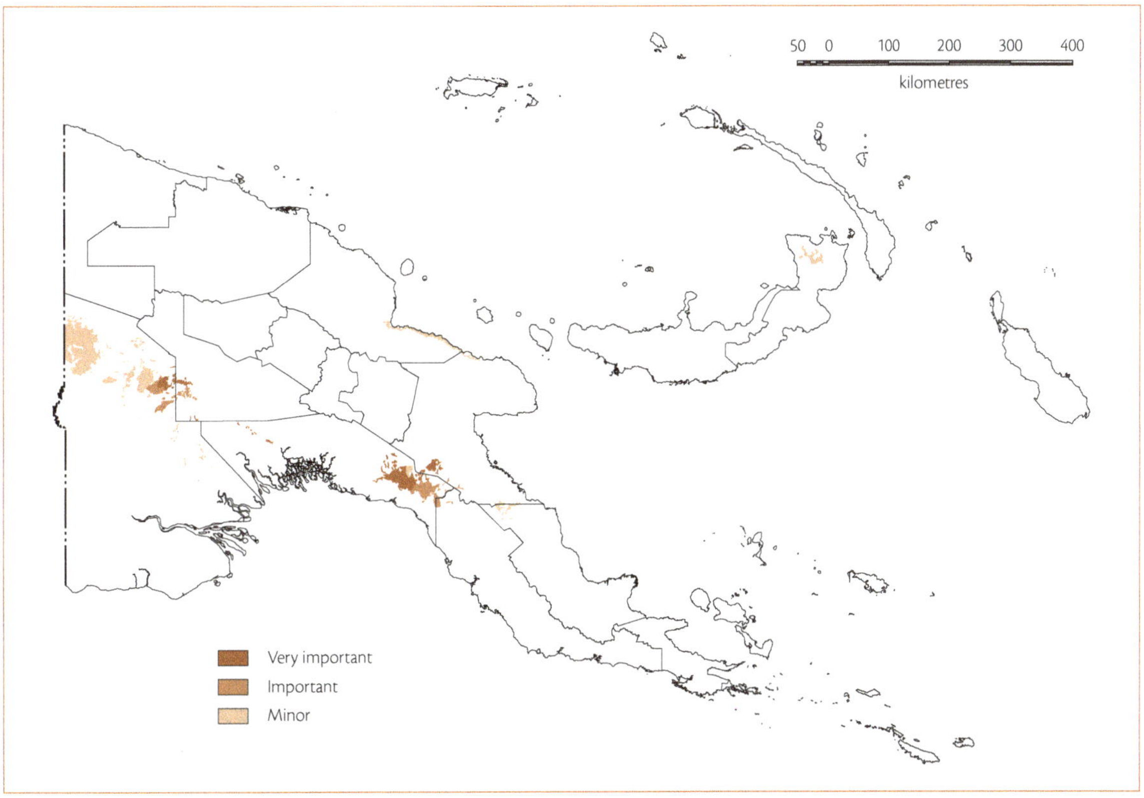

Figure 3.12.6 The practice of cutting down trees on top of growing food crops. Source: MASP.

in taro. The felled forest decomposes rapidly and provides the soil with additional nutrients. A low secondary forest quickly colonises the site again but, when the site is planted in taro a few years later, the size of the trees that have to be thinned is much reduced.

Fencing and other enclosures

Fences and other barriers around garden sites are used throughout PNG (Figure 3.12.7), mainly to keep out domestic and feral pigs and to a lesser extent smaller wild animals such as bandicoots. Fences also create a symbolic barrier that people must cross to enter a garden, which may deter theft of food, for example. In addition to fences, gardens may be surrounded by ditches and earth banks, and stone walls. Fences are commonly constructed from material that becomes available during the clearing of the garden site: cane grass and wood, bound together with split cane.

In the highlands in particular, fences are an integral part of the management of land and domestic pigs. In intensive highlands land use systems, cultivation periods are long, and so fences must last for a long time. They are commonly constructed from split casuarina wood palings, sharpened at each end. The fence is built by pushing one end into the ground and binding the tops together with cane. When the bottom end decays, the palings can be pulled out, reversed and pushed in again. In many parts of the highlands, individual fields are not fenced. Instead, a large area of many gardens, including land in a short fallow, is separated from land in a long fallow by a long perimeter fence. Pigs are allowed to forage in the fallow land and races are constructed to allow them to travel between the fallow land and housing areas without having to enter garden land. A study in Enga in 1980 estimated that 105 000 m^3 of timber per year, mostly casuarina, was used to make fences in the province. Old fences were also an important source of firewood. In the Tari area, deep ditches are constructed around gardens. Pigs and people

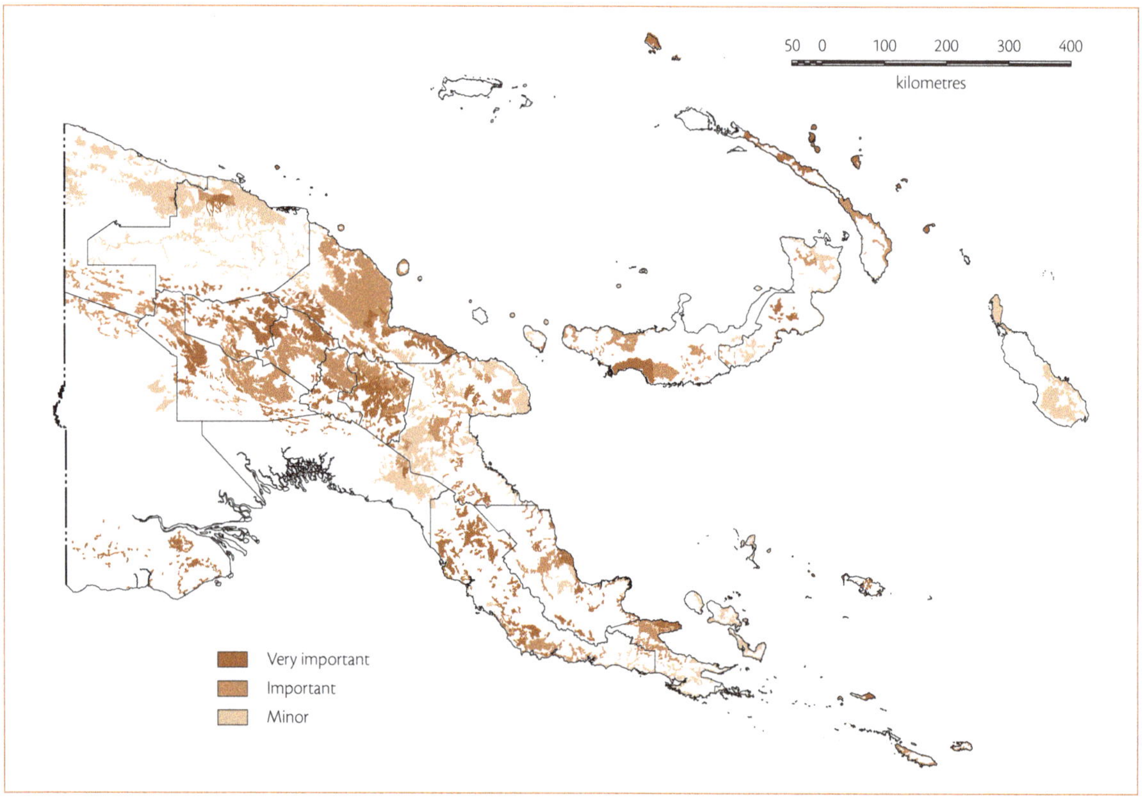

Figure 3.12.7 The use of fences and other barriers in food production systems. Source: MASP.

travel along the ditches between homesteads and to foraging areas, without having to enter garden areas.

In many lowlands locations fences are less important than they were 50 years ago, following the introduction of shotguns and the subsequent reduction in the numbers of feral pigs. In the past, traps for feral pigs were commonly integrated into the garden fence. Fences were constructed every time land was cleared from a long fallow and were maintained only for as long as the land was cultivated, from between one and three years.

Irrigation

Irrigation is not common in PNG, a not surprising situation in view of the very high rainfall received in many parts of the country. Irrigation systems existed where strong seasonal soil water deficits occur (see Section 1.5). But in a number of these places the importance of irrigation systems has decreased following the adoption of drought-tolerant sweet potato and cassava, and a decrease in the importance of taro, which requires more water to produce well.

Irrigation systems existed in the Kaironk Valley in Madang Province, in the Lamari Valley of Eastern Highlands Province, near Kabwum in Morobe Province and on the coast from Cape Vogel to East Cape in Milne Bay Province. At Kabwum irrigated plots were created on narrow terraces beside streams by heaping sods and cane grass roots against wooden barriers. This formed a series of small flooded enclosures in which taro was planted, with corn and other green vegetables. In the Lamari Valley, grassland and forest taro gardens were irrigated with water brought to gardens by bamboo pipes. Pipelines traversed several hundred metres. Use of irrigation had declined by the early 1990s, but in the Lamari Valley during the 1997 drought, water was again led to gardens using bamboo pipes. Two systems of irrigation were described in 1947 in the Wampit Valley, near Mumeng in Morobe Province, both associated with taro production. One involved damming a stream, diverting water into a ditch on the contour and letting water out of the ditch into furrows in a garden below the ditch; the second involved the use of bamboo pipes to bring water into gardens. These systems were not being used by 1992.

The only area where large-scale irrigation systems remain in use in PNG is on the coast near Rabaraba, in Milne Bay Province. Here streams are dammed at the top of large, gently sloping fans and water is led into ditches along the contour, from where it is diverted into gardens below the ditches. A shifting cultivation system producing a mix of staple food crops, including taro, banana, sugar cane, cassava and sweet potato, is maintained in this way. When fallow land is brought into cultivation, water is led through shallow ditches to the new garden. This system is threatened by flash flooding, which destroys the dams and water intakes. The floods are blamed on the clearing of forests in the headwaters of these streams.

Deep holing

In the Maprik area in East Sepik Province and the Wantoat area in Morobe Province, greater yam is planted at the top of deep holes in order to grow tubers to an extraordinary length. In both locations, holes up to 2 m deep and 0.5 m in diameter are dug and refilled with a fine tilth of soil. The seed yam is planted in a small mound on top of the hole and the tuber grows down into the softer soil in the form of a single long tuber. Yams planted without this treatment will produce shorter and possibly forked tubers, depending on the hardness or stoniness of the soil. In both places the yam vines are supported on tall trellises, often made of bamboo.

At Wantoat, the holes are dug around the base of large, circular heaps of stones that have been cleared from the surrounding fields. When a field is cleared after a long fallow of 15–20 years, a search is made for the holes using a long steel rod (previously a long sharpened stick was used). The holes are then re-excavated, refilled with soil and the yams planted. At Maprik, the holes are dug on steep slopes. When the soil is replaced in the hole, a stick is inserted first and then withdrawn when the hole is full, to leave a small empty column down the centre of the hole. The seed yam is planted in a mound, which is supported by a low wooden wall to prevent soil from falling downslope. During the later part of the yam's growth, inspection tunnels are dug horizontally into the side of the hole to monitor the tuber's progress and to assess the potential of the tuber in the competitive displays and exchanges that follow.

Use of fertilisers

The manual addition of animal manure to cultivated areas is unusual in PNG. Occasionally, chicken manure from a chicken-raising project is added to small plots of leafy green vegetables or other food crops. Other materials used occasionally as fertiliser are the skins and pulp from coffee berries left over from processing, ash from wood fires, and pig manure.

The application of inorganic (manufactured) fertiliser on food crops is even less common than the use of organic material. When purchased fertiliser is used on food crops, it is almost always in conjunction with growing vegetables for sale (see Section 5.19).

Use of silt from floods to improve soil fertility

In some low-lying areas, particularly along the Sepik and Ramu rivers, garden land is regularly inundated by floodwaters. These waters deposit a thin layer of fresh silt on the garden surface, which helps improve soil fertility. Another area where this is important is south of Wanigela, Oro Province, on the alluvial flats immediately inland of the coastal mangroves. On the Sepik River, the planting and harvesting of crops, and to some extent the crops used, are governed by the annual rise and fall of the river. For example, greater yam is common in food gardens on the floodplains, whereas lesser yam is rarely planted. This is because lesser yam takes longer to grow and its tubers are not mature before the annual flood covers the garden land.

Sources

Ballard, C. (2001). Wetland drainage and agricultural transformations in the Southern Highlands of Papua New Guinea. *Agricultural Transformation and Intensification. Asia Pacific Viewpoint Special Issue* 42(2/3):287–304.

Bourke, R.M. (2001). Intensification of agricultural systems in Papua New Guinea. *Agricultural Transformation and Intensification. Asia Pacific Viewpoint Special Issue* 42(2/3):219–235.

Boyd, D.J. (2005). Beyond the Ipomoean Revolution: sweet potato on the 'fringe' of the Papua New Guinea highlands. In Ballard, C., Brown, P., Bourke, R.M. and Harwood, T. (eds). *The Sweet Potato in Oceania: A Reappraisal.* Ethnology Monographs 19 / Oceania Monograph 56. Oceania Publications, University of Sydney and Ethnology, Department of Anthropology, University of Pittsburgh, Sydney and Pittsburgh. pp. 137–147.

Carrad, B. (1982). The economy. In Carrad, B., Lea, D.A.M. and Talyaga, K.K. (eds). *Enga: Foundations for Development. Volume 3 of Enga Yaaka Lasemana.* Department of Geography, University of New England, Armidale. pp. 146–178.

Conroy, W.L. and Bridgland, L.A. (1950). Native agriculture in Papua-New Guinea. In Hipsley, E.H. and Clements, F.W. (eds). *Report of the New Guinea Nutrition Survey Expedition 1947.* A.H. Pettifer, Government Printer, Sydney. pp. 73–91.

Coupaye, L. (2004). Growing artefacts, displaying relationships: outlining the technical system of Long Yam cultivation and display among the Abelam of Nyamikum village (East Sepik Province, Papua New Guinea). PhD thesis. University of East Anglia, Norwich.

Gorecki, P.P. (1979). Irrigated taro fields in the highlands. *Oral History (Papua New Guinea)* 7(7):116–120.

Hide, R. (2003). *Pig Husbandry in New Guinea: A Literature Review and Bibliography.* ACIAR Monograph No. 108. Australian Centre for International Agricultural Research, Canberra.

Johnston, M. and Onwueme, I.C. (1999). Productivity of lesser yam (*Dioscorea esculenta*) in Papua New Guinea as influenced by sett weight and staking. *Papua New Guinea Journal of Agriculture, Forestry and Fisheries* 42(1 & 2):27–34.

Kahn, M. (1984). Taro irrigation: a descriptive account from Wamira, Papua New Guinea. *Oceania* 54(3):204–222.

King, G.A. and Risimeri, J.B. (1992). Effects of planting density, height of staking and variety on yield and yield components of the lesser yam (*Dioscorea esculenta*). *Tropical Agriculture (Trinidad)* 69(2):129–132.

Lea, D.A.M. (1966). Yam growing in the Maprik area. *Papua and New Guinea Agricultural Journal* 18(1):5–16.

Quin, F.M. (1984). Report on yam research (*Dioscorea* spp.) 1981–84. East Sepik Rural Development Project: agricultural research sub-project. Unpublished paper. Wewak.

Rose, C.J. (1976). A tethering system of grazing pigs on sweet potatoes in the Tari Valley. In Wilson, K. and Bourke, R.M. (eds). *1975 Papua New Guinea Food Crops Conference Proceedings*. Department of Primary Industry, Port Moresby. pp. 151–158.

Rose, C.J. and Williams, W.T. (1983/84). Ingestion of earthworms, *Pontoscolex corethrurus*, by village pigs, *Sus scrofa papuaensis*, in the highlands of Papua New Guinea. *Applied Animal Ethology* 11:131–139.

Schieffelin, E.L. (1975). Felling the trees on top of the crop: European contact and the subsistence ecology of the Great Papuan Plateau. *Oceania* 46(1):25–39.

PART 4

The Broader Economy

Andrew McGregor and R. Michael Bourke

4.1 The macro-economic environment

PNG has a dual economy. Most people work in the large subsistence sector, producing food for consumption, raising animals and building their own shelter, as well as producing agricultural commodities for sale in domestic and international markets (see Sections 5.2 and 5.3). A smaller number of people work solely in the monetary sector, which is concentrated in urban areas and mining enclaves. The level of national income generated from these activities was, until recently, sufficient to place PNG in the World Bank category of 'Lower Middle Income Countries'. This category includes Thailand, Fiji and Sri Lanka. However, social indicators (for example, literacy rates, life expectancy at birth, infant mortality, school enrolment) in PNG are typical of countries in the 'Lower Income Countries' category. This category includes most of sub-Saharan Africa. In part, this situation can be attributed to the macro-economic environment[1] in which the agricultural sector operates. This environment is determined by a combination of exchange rate, fiscal, monetary and trade policies.

The exchange rate is a key determinant of agricultural competitiveness, both for domestic and export markets. Before 1994, PNG maintained what was known as a 'hard kina' exchange rate policy, in which the exchange rate between the kina and major international currencies[2] was deliberately maintained at a high level, which kept the cost of imports low. The aim of the 'hard kina' policy was to contain wages. This was achieved by improving the purchasing power of urban workers through the low-cost imports. In 1994 the kina was allowed to float and it immediately fell in value (Figure 4.1.1, Table A4.1.1). Following the Asian financial crisis and mismanagement of the economy in 1997, it fell even lower.

The high value of the kina adversely affected the agricultural sector in several ways. People selling locally grown food found it difficult to compete with low-priced imports. For example, in 1993 it cost K0.66 to purchase cabbages worth A$1.00 imported from Australia. But in 2005, cabbages worth A$1.00 cost about K2.40, which encouraged domestic production to such an extent that PNG is now self-sufficient in cabbages.

The hard kina policy also depressed incomes earned by growers of export tree crops. For example, in 1993 the average delivered-in-store (DIS) price of coffee in PNG was K0.57/kg. Ten years later, in 2003, the DIS had increased to K1.23/kg. This occurred despite the average world price of coffee falling from US$1.28/kg to US$0.70/kg during that decade. Although the value of the kina in 2003 was lower than what it

1 The term 'macro-economic' means the economy of PNG as a whole.

2 The exchange rate of the kina is what K1.00 is worth in other international currencies. If the exchange rate of the PNG kina against the United States dollar is 0.35, then one kina can buy US$0.35.

was under the hard kina policy, under international exchange rates, a US dollar bought more kina, which was good for PNG exporters.

The decision to float the kina in 1994 did more to stimulate agriculture than all the direct interventions into the sector by the PNG Government and foreign donors. The decline of the kina swung the terms of trade back in favour of rural areas, where more than 80% of Papua New Guineans live, although it disadvantaged urban dwellers.

The rapid expansion of the mining industry through the late 1970s and 1980s provided a major boost to the PNG economy. However, the income from mining was not managed in a way that encouraged the development of agricultural and other industries. Instead, it was used to create unsustainable government structures that resulted in a large public debt and an over-valued kina. Agricultural commodity earnings fell and the standard of living of growers declined. This would not necessarily have been a problem if other industries had been developed to replace the mines when they eventually close, but replacement industries have not been developed.

The commercial agricultural sector has been hampered by high interest rates and lack of finance. This was caused by the way large budget deficits have been financed. Government expenditure has consistently exceeded income, often by substantial margins. These budget deficits could have been financed by borrowing either internationally or locally. PNG governments chose to borrow locally by selling Treasury bills to the commercial banks.[3] This avoided the build-up of a high level of foreign debt, however, with only a small local market for the sale of Treasury bills, interest rates were driven to exceptionally high levels. In August 1999, the Treasury bill interest rate peaked at 28% per year. This made borrowing for investment in agriculture uneconomical. Since the commercial banks were able to earn very high profits by lending to the government, there was no incentive for them to lend to productive sectors like agriculture. In 2004, for the first time in 13 years, a budget surplus

[3] A Treasury bill is a document that can be bought from a government that promises to pay the buyer the cost of the bill plus an amount of interest, at a given date in the future.

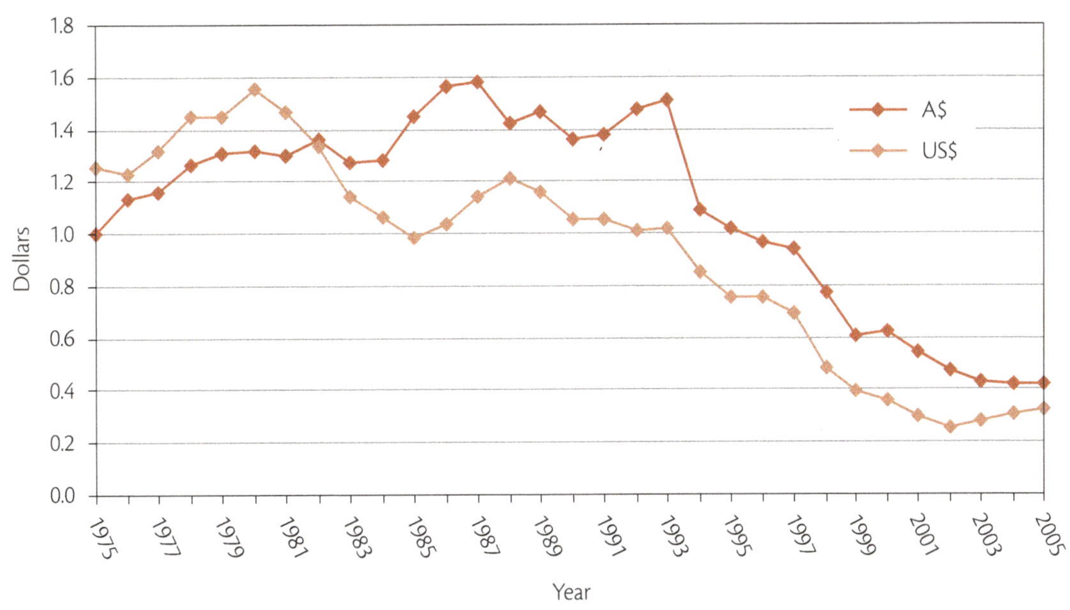

Figure 4.1.1 Exchange rate for one PNG kina against the Australian dollar and United States dollar, 1975–2005.
Source: Bank of Papua New Guinea Quarterly Economic Bulletins.

was achieved. As a consequence, the Treasury bill interest rate fell below 5%, making it worthwhile for banks to consider other lending opportunities. The challenge for PNG fiscal and monetary policy is to balance public debt, to find an appropriate method of financing the debt, and to invest in infrastructure to support productive sectors such as agriculture.

Previous trade policy placed restrictions on the import of food and agricultural products to encourage import replacement. Prior to PNG joining the World Trade Organization in 1995, local fruit and vegetables were protected from imports by a combination of high tariffs, quotas and outright bans.[4] In 1995, these measures were replaced by a 75% tariff on all imports, which has been gradually reduced as a part of the Tariff Reduction Program. The Tariff Reduction Program has had a positive impact on the economy and has reduced the costs of imported food for urban Papua New Guineans.

Despite the Tariff Reduction Program, inconsistencies and inequities still exist in the tariff structure that hinder overall agricultural development. Self-sufficiency in poultry and pig products, which has been achieved in recent years, was the outcome of a high level of protection, with imported poultry products incurring a tariff of 57% (Table 4.1.1). But the additional value to the economy of a local poultry industry is substantially reduced by an almost total reliance on imported feed grains (see Section 2.6). Locally produced maize, copra meal and fish meal must compete with imported feed grains that have no tariff. The pig and beef industries receive a lower rate of protection than poultry (a 15% tariff). The cattle industry, which relies on local pastures and not imported grains, pays high import duties (50%) on essential inputs such as fencing wire. The agricultural sector would be better served if the same low level of protection applied to all industries.

Table 4.1.1 Tariffs applying to selected agricultural imports from January 2006

Product	Proportion of landed value
Potatoes	40%
Vegetables	40%
Fruit	40%
Sugar	70%
Rice	zero
Wheat	zero
Maize	zero
Sheep meat	zero
Beef	15%
Pig meat	15%
Tinned fish	25%
Poultry	K2.20/kg
Eggs	K1/dozen

Source: APEC (2005).

4 A tariff is a tax on imported goods. A tariff may be put in place just to raise money for the government (when the imported good, such as a vehicle, is not produced in the country), or to protect local producers from cheaper imports. A 25% tariff means that an importer must pay the government 25% of the total value of the imported good. The World Trade Organization opposes tariffs because they increase the cost of trading, which in turn reduces trade and output below optimum levels in a country that has no control over the prices of international commodities.

An import quota is a limit on the quantity of a good that can be imported in a particular year. For example, if a government makes a law that only 1000 tonnes of potatoes per year can be imported, that is an import quota on potatoes. A ban on the import of potatoes means no potatoes may be imported at all.

Sources

ADB (Asian Development Bank) (2004). Preparing the agriculture and rural development project, Papua New Guinea: Agricultural markets, marketing and rural enterprise development. ADB TA4055-PNG. Asian Development Bank.

APEC (Asia–Pacific Economic Cooperation) (2005). Tariff Database.

Bank of Papua New Guinea Quarterly Economic Bulletin (various issues).

Corden, W.M. (1984). Booming sector and Dutch disease economics: survey and consolidation. *Oxford Economic Papers* 36(3):359–380.

4.2 Consumer price index

The consumer price index (CPI) is a measure of the cost of living, or inflation. It is calculated by determining the average cost of a standard basket of retail goods and comparing the cost of those goods to the cost of the same goods at a base period. It provides a useful way to compare the relative value of a country's currency from year to year. In PNG, the index is based on the prices of selected goods and services in five towns (Port Moresby, Goroka, Lae, Madang and Rabaul). The CPI in PNG is principally an urban measure: data for the cost of rural living in PNG do not exist. The commodity groups selected are food; clothing; alcohol, tobacco and betel nut; household sundries; and other goods and services, including transport, education, medicine and groceries other than food.

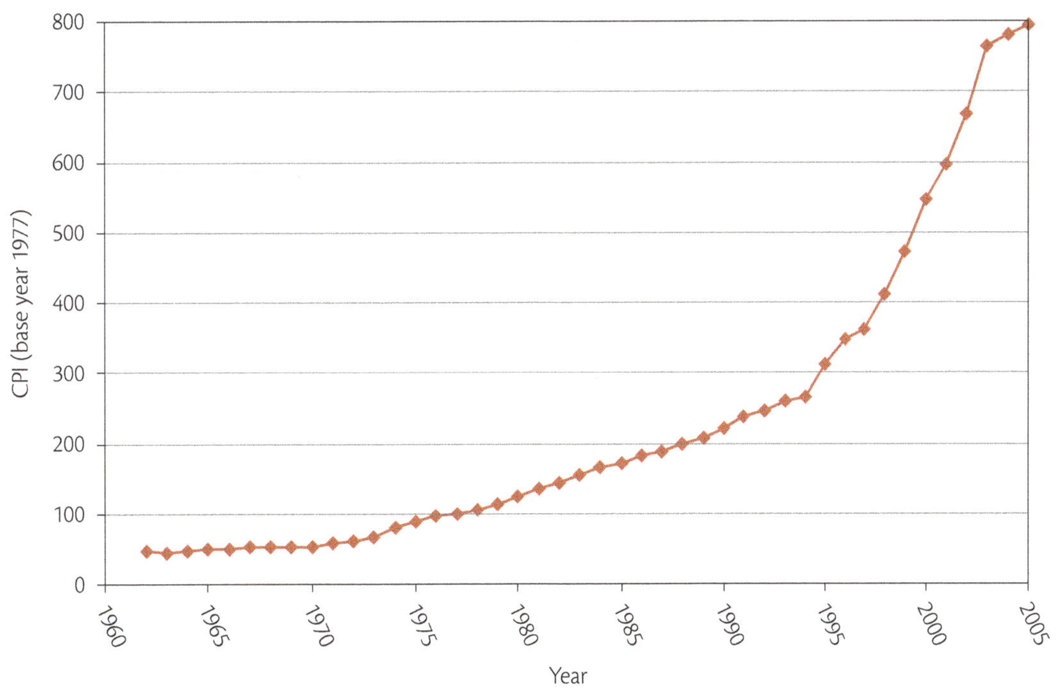

Figure 4.2.1 PNG consumer price index, 1962–2005. Source: National Statistical Office of PNG.

There are three CPI series in PNG, extending from 1962 to the present. The series used now begin in 1977. The values for the three series are listed in Table 4.2.1. The figures have been converted to a common base (100) with 1977 as the base year, and a combined index created.

Prices rose only slightly during the 1960s and early 1970s (Figure 4.2.1). From 1974 until 1994, prices rose at a faster rate. There were large increases in prices each year from 1995 to 2003. This was associated with the devaluation of the PNG kina relative to the United States dollar. The rate of increase slowed in 2004 and 2005. In 1994 the kina was floated, meaning that since then, it has been allowed to move freely against other international currencies. In 2005, K9 had about the same spending power that K1 had in 1975. That is, prices have gone up by a factor of nine over this 30-year period.

The consumer price index can be used to convert prices or income from one year to a value with an equivalent buying power in another year. To convert earlier prices or values to current ones, the following formula is used:

$$\text{Equivalent value in current year} = \frac{\text{Earlier value} \times \text{CPI in current year}}{\text{CPI in earlier year}}$$

Example 1. The value of one kilogram of coffee beans (parchment) for a highland grower in 1980 in 2005 prices.

Equivalent price in 2005

= (Price in 1980 × CPI in 2005) ÷ CPI in 1980

= (2.26 kina/kg × 795.2) ÷ 125.4

= 14.33 kina/kg

Coffee growers would need to have received 14.33 kina per kilogram in 2005 to have the same spending power they had 25 years earlier. (In reality, growers received much less than 14.33 kina in 2005, so they had less spending power in 2005 than in 1980.)

To convert current prices or values to earlier ones, the following formula is used:

$$\text{Equivalent value in earlier year} = \frac{\text{Current value} \times \text{CPI in earlier year}}{\text{CPI in current year}}$$

Example 2. The value of a fortnight's salary in 2005 in 1980 values.

Equivalent salary in 1980

= (Current salary × CPI in 1980) ÷ CPI in 2005

= (300 kina/fortnight × 125.4) ÷ 795.2

= 47 kina/fortnight

Thus a salary of 300 kina per fortnight in 2005 has the same spending power that a salary of 47 kina had in 1980.

Source

Bank of Papua New Guinea (2001). *Money and Banking in Papua New Guinea*. Bank of Papua New Guinea, Port Moresby.

Table 4.2.1 PNG consumer price index, 1962–2007

Year	Retail price index[a]	Consumer price index Base year 1971[b]	Consumer price index Base year 1977[c]	Combined index Base year 1977[d]
1962	100.0[e]			46.1
1963	97.8			45.1
1964	99.5			46.0
1965	105.1			48.5
1966	110.2			50.8
1967	113.0			52.1
1968	114.3			52.7
1969	113.9			52.6
1970	115.7			53.4
1971	123.2	100.0[e]		56.8
1972		106.1		60.3
1973		114.9		65.3
1974		141.6		80.5
1975		156.4		88.9
1976		168.4		95.7
1977		176.0	100.0[e]	100.0[e]
1978			105.8	105.8
1979			111.9	111.9
1980			125.4	125.4
1981			135.5	135.5
1982			143.0	143.0
1983			154.3	154.3
1984			165.8	165.8
1985			171.9	171.9
1986			181.3	181.3
1987			187.3	187.3
1988			197.5	197.5
1989			206.4	206.4
1990			220.7	220.7
1991			236.1	236.1
1992			246.3	246.3
1993			258.5	258.5
1994			265.9	265.9
1995			311.9	311.9
1996			348.1	348.1
1997			361.9	361.9
1998			411.0	411.0
1999			472.4	472.4
2000			546.1	546.1
2001			596.8	596.8
2002			667.2	667.3
2003			765.4	765.4
2004			781.6	781.6
2005			795.2	795.2
2006			814.7	814.7
2007			822.2	822.2

[a] The retail price index is a combined index for Port Moresby, Lae and Rabaul. It is based on prices of food, tobacco and certain household items consumed by expatriates.

[b] The consumer price index base year 1971 is based on prices of a group of goods selected on the basis of an urban expenditure survey of PNG public servant households.

[c] The consumer price index base year 1977 is based on prices of a group of goods and services purchased by urban PNG wage-earning households.

[d] The combined index was generated from the other three series of indices.

[e] The base year for the various indices is shown as 100.

Source: National Statistical Office of PNG.

4.3 Prices of selected items

Local prices of fresh food and store items can demonstrate how international and national economic forces, such as the strength of the currency, have an impact on the daily lives of urban and rural Papua New Guineans.

Four commodities have been chosen to illustrate variation in prices in Port Moresby and Madang over a 35-year period. Sweet potato and *aibika* were selected as representative of fresh foods, while rice and kerosene show price movements for imported items. The pattern for fresh food in Port Moresby is quite different from that in the other main urban areas in PNG. Madang was selected to demonstrate the patterns in the other towns.

Thirty-five years ago, the price of fresh food was somewhat more expensive in Port Moresby than elsewhere, but the difference has increased over time (Figures 4.3.1 and 4.3.2). For example, in 1971 sweet potato and *aibika* were about twice the price in Port Moresby than in Madang, but by 2005 they were around two to three times the price (Table A4.3.1). In contrast, the prices of imported items such as rice and kerosene have not varied much between the main ports during the last 35 years (Figures 4.3.3 and 4.3.4). For many items, prices are higher in remote parts of most provinces and in the highlands because of transport costs and lack of competition. Rice is an exception because Trukai Industries Limited (PNG's primary importer and supplier of rice) maintains the same wholesale price for all main towns.

The prices of goods in PNG towns have risen rapidly since 1997, when the PNG kina fell sharply in value against the United States dollar, the currency in which imports and exports are priced (Figure 4.1.1). This means that it now costs urban and rural consumers much more to buy goods and services in towns than it did prior to 1997. The steep rises in the prices for rice and kerosene from 1997 to 2005 are clear (Figures 4.3.3 and 4.3.4). The prices of fresh foods also increased in Port Moresby and the other urban areas over this period, but not as rapidly as for imported items.

More fresh food has been offered for sale in recent years, probably because the demand has increased as prices of imported foods have risen sharply. This is reflected in the prices of sweet potato and *aibika* in Madang market, which have fallen over the period 2002 to 2005, presumably because of more sellers in the market. Locally grown staple foods are now more competitively priced against imported rice and flour than they were before 1997, for most locations in PNG except Port Moresby. This is good for both producers and consumers in PNG.

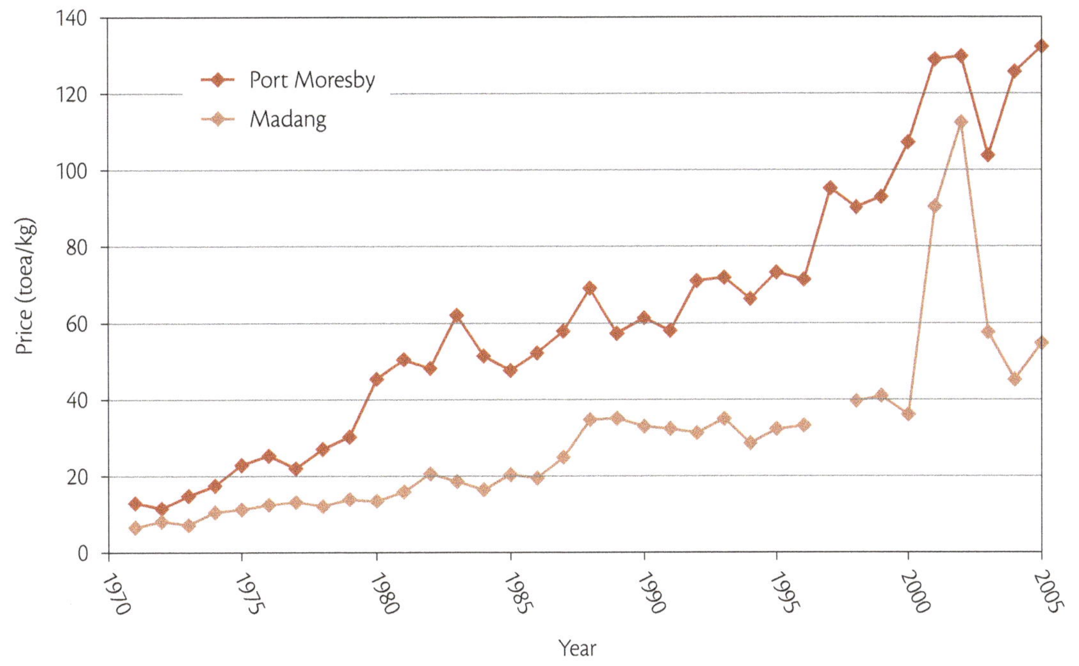

Figure 4.3.1 Average market price of sweet potato in Port Moresby and Madang, 1971–2005. **Note:** The gap in prices of fresh foods in 1997 occurred because there was very little or no fresh produce in many markets during the drought, hence it was not possible to record prices at that time. Source: National Statistical Office of PNG.

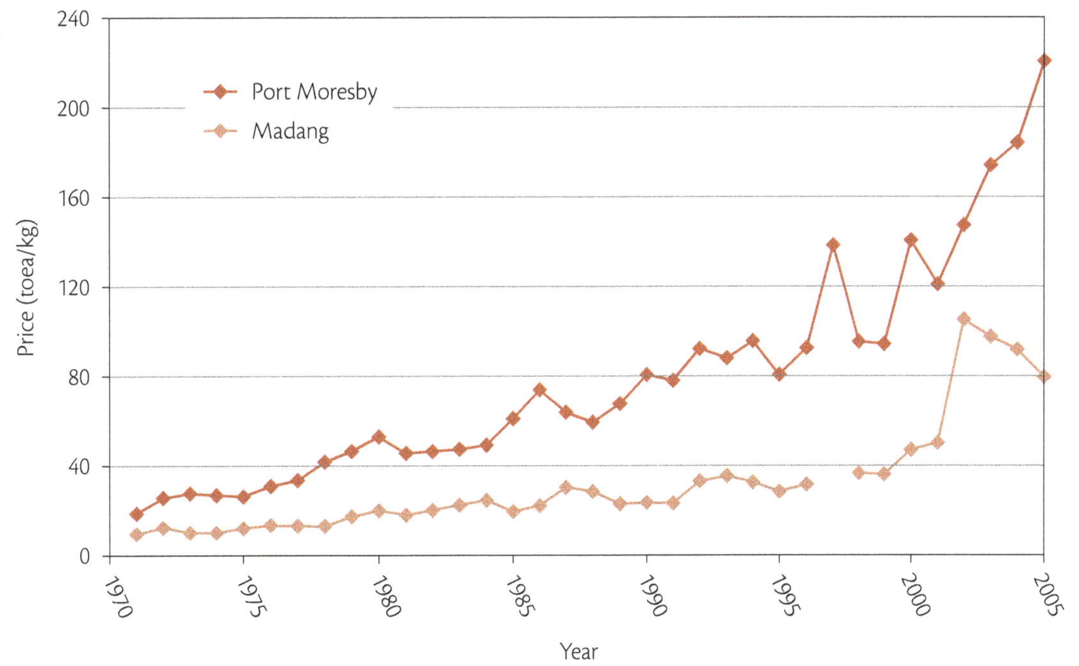

Figure 4.3.2 Average market price of *aibika* in Port Moresby and Madang, 1971–2005. **Note:** The gap in prices of fresh foods in 1997 occurred because there was very little or no fresh produce in many markets during the drought, hence it was not possible to record prices at that time. Source: National Statistical Office of PNG.

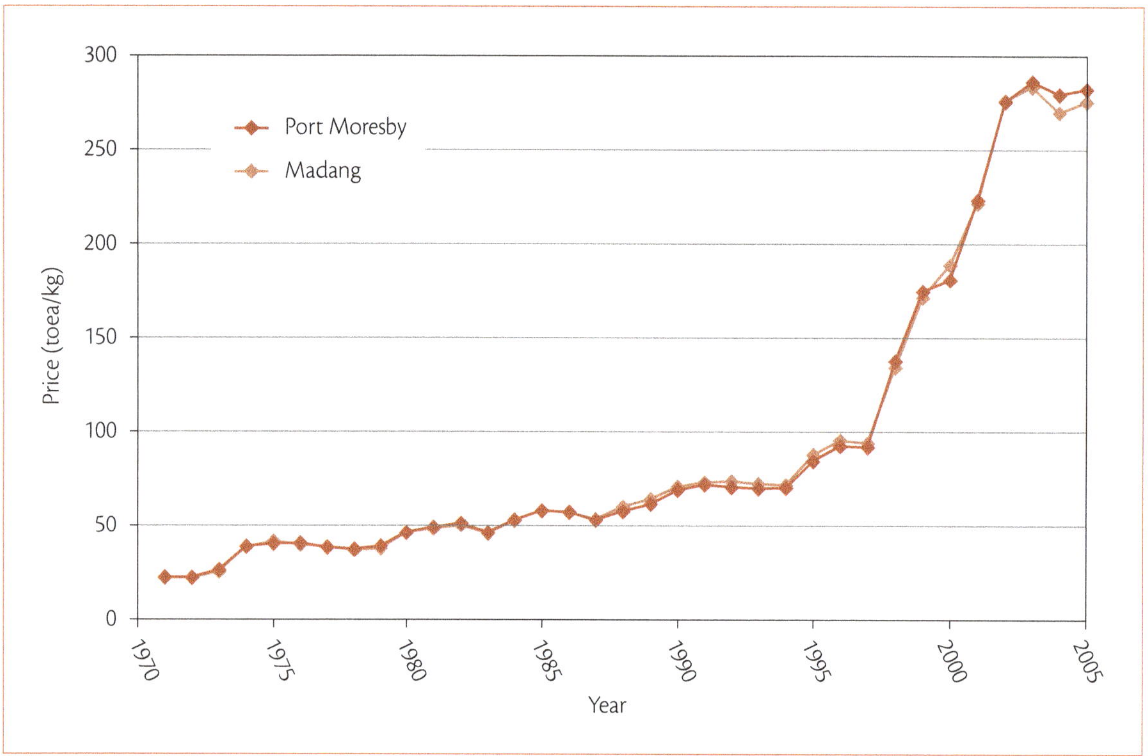

Figure 4.3.3 Average retail price of rice in Port Moresby and Madang, 1971–2005.
Source: National Statistical Office of PNG.

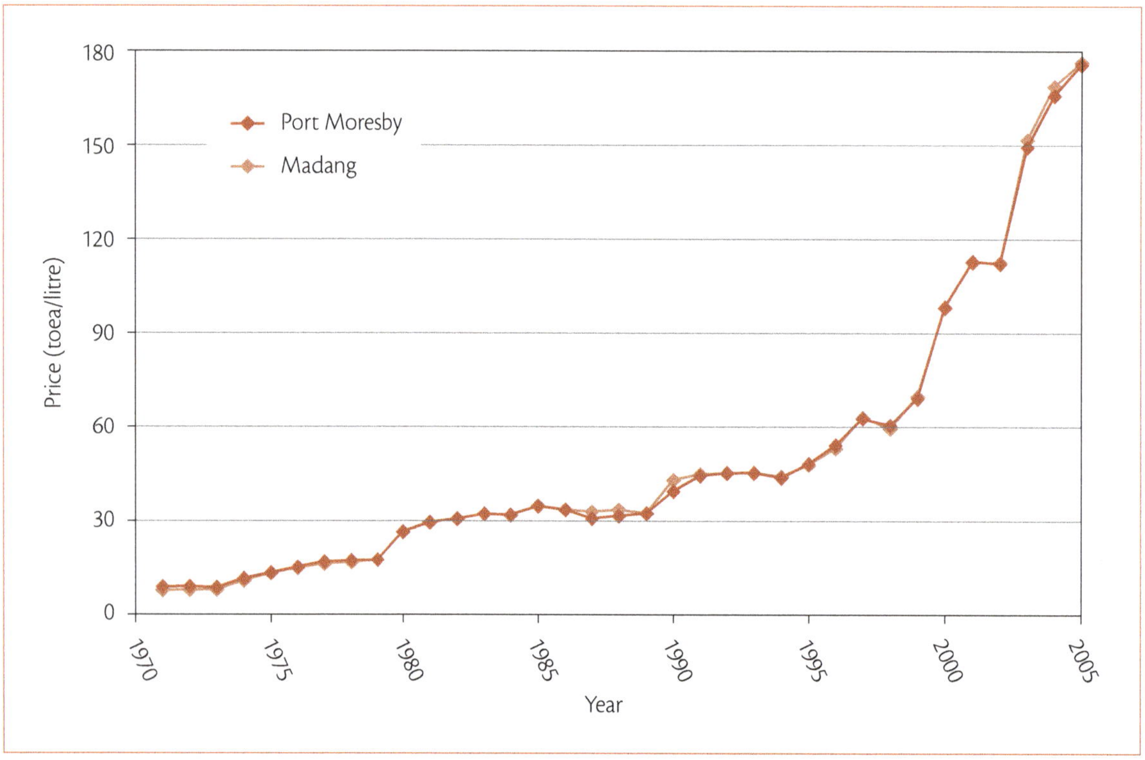

Figure 4.3.4 Average retail price of kerosene in Port Moresby and Madang, 1971–2005.
Source: National Statistical Office of PNG.

4.4 Pay fortnight food market and retail cycles

A regular cycle of retail activity occurs in urban PNG that is closely associated with the fortnightly government pay period.[1] Because of the large number of public servants in urban areas, relatively large amounts of money move into the local economy once a fortnight when public servants are paid. This pattern is less striking where there are more private enterprise employers or other sources of income. For example, in the Kimbe area of West New Britain Province, payment to oil palm growers is a major source of income and the government payday effect is less marked.

Public servants do much of their banking and shopping in local food markets and stores on the Friday and Saturday of the government pay week. Villagers often come to their local town on those days, some seeking handouts from their relatives in the paid work force, others to market fresh food when more money is being spent. Thus there are often more sellers and buyers at local markets on these two days than on any other day in the fortnight.

These patterns are illustrated in Figure 4.4.1 with sales data of fresh food and betel nut in Kainantu market, Eastern Highlands Province, over a two-week period in 1982. The busiest days for the sale of both fresh food and betel nut in Kainantu are the government pay-Friday and the following Saturday. Sales from a retail store in Kainantu over a 10-week period are shown in Figure 4.4.2. At this store, a third of the sales took place on only 2 of the 12 trading days each fortnight. This seems typical of other centres where there are few private enterprise employers.

Private sector payday (known as 'company' payday) falls on the alternate Friday to the government payday, and contributes to a smaller surge in spending on the Friday and Saturday of the non-government pay week (Figure 4.4.2). Although these data were recorded in 1982, the same patterns can still be observed in many urban locations in PNG, including in Port Moresby and the main highlands towns.

An increase in spending prior to the Christmas break in December forms another retail spending cycle in PNG. It probably occurs because teachers and some other public servants spend much of their holiday pay then. This is recorded as an increase in some sales figures in December. Sales of many items are often very low in January and February, presumably because many people have spent their holiday pay. This annual cycle is not as predictable or as large as the two fortnightly cycles described above, and seems not to affect fresh food markets.

1 This section describes retail cycles in urban areas, and involves both urban and rural people. Another retail cycle, associated with coffee harvesting, occurs in the highlands (see Box 5.4).

Source

Bourke, R.M. and Nema, R.K. (1985). *Surveys of three food markets in the Kainantu area, Eastern Highlands*. Technical Report 85/6. Department of Primary Industry, Port Moresby.

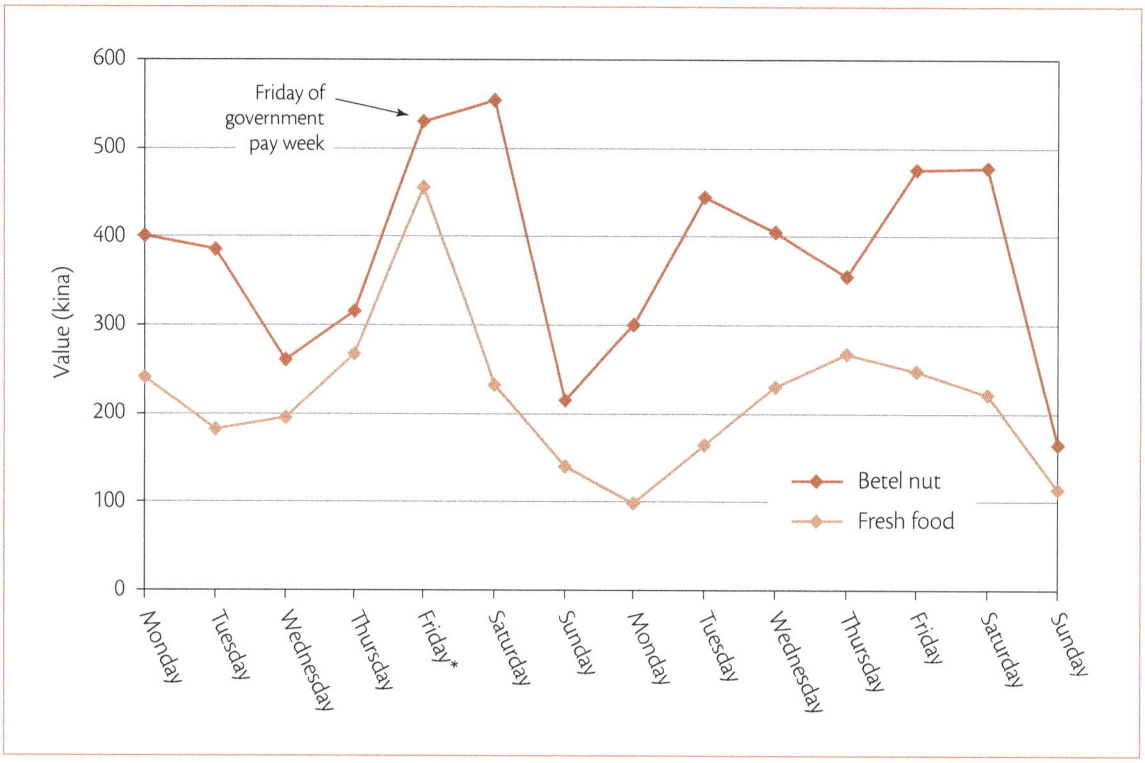

Figure 4.4.1 Value of fresh food and betel nut sold in Kainantu market over a 14-day cycle, July 1982.
Source: Bourke and Nema (1985).

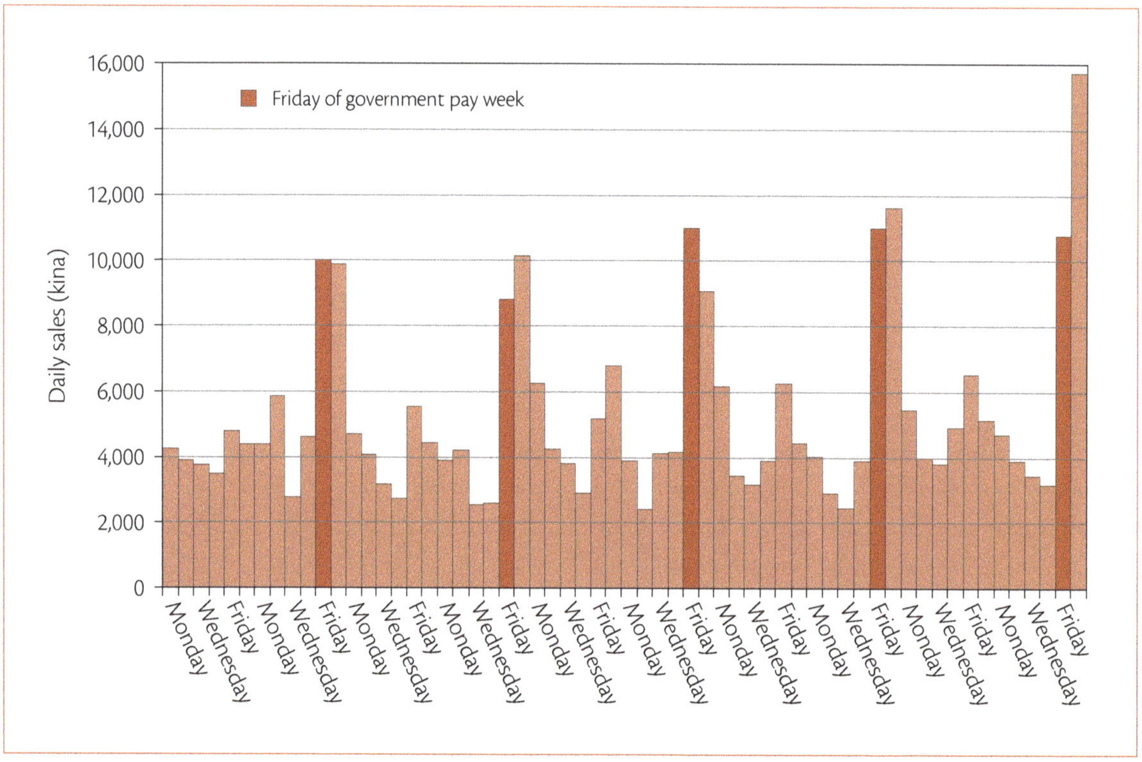

Figure 4.4.2 Daily sales in a retail store in Kainantu over a 10-week period, January–March, 1982. **Note:** Sundays are excluded because the store is closed on Sundays. Source: R.M. Bourke unpublished field notes.

PART 5

Cash Income from Agriculture

Matthew Allen, R. Michael Bourke and Andrew McGregor

5.1 Rural villagers' cash income

Most rural Papua New Guineans earn cash from agriculture and closely associated activities such as selling firewood, fish or animals. This section presents estimates of the amounts of cash earned, how it is earned and the numbers of people who earn it. Rural villagers derive cash income from a number of sources, with sale of agricultural produce the most significant in both total income and the number of people who receive it. The amounts of cash earned are generally low, but there are exceptions. For example, some oil palm producers, people growing betel nut west of Port Moresby, and people in the Markham Valley who sell betel nut and fresh food to highland traders have relatively high incomes.

Agricultural activities

The best available data on cash income for rural villagers comes from surveys conducted as part of the Mapping Agricultural Systems of PNG Project (MASP; see Section 1.15).

During the MASP surveys the importance of various cash-earning agricultural activities was estimated using a four-point scale, from absent to very significant. After completion of fieldwork a cash value was assigned to each rating based on detailed studies of cash income in particular villages. The rating 'absent' was allocated a cash income of K0–10 per household per year, 'minor' was allocated K11–50, 'significant' was allocated K51–250 and 'very significant' more than K250 per household per year. Based on the allocated average cash income for each agricultural activity and the rural village population in each agricultural system, the total cash income for each agricultural activity was calculated at the level of agricultural system, province and nation.

The estimates from the MASP database at provincial and national level were then compared with data from industry sources for coffee and cocoa, two commodities for which reliable export income exists. The MASP estimates of income earned by the rural population from cocoa and coffee were close to industry data and increase the confidence in the accuracy of estimates for commodities sold in the informal sector, such as fresh food, betel nut and firewood.[1]

Since 1990–1995 when the MASP fieldwork was conducted, changes have occurred in the value of the PNG currency relative to the United States dollar, with consequent impacts on prices and income in kina terms (Figures 4.1.1, 4.2.1). World prices for most export commodities have changed, as have some PNG production patterns. Within PNG significant changes have occurred in vanilla and copra production over the past decade, and are continuing in oil palm production. It is probable that a significant increase has occurred in the volume of

1 The MASP estimates for rubber and oil palm were also similar to industry sources, but the MASP data overestimated incomes for copra. See Allen et al. (2001: Figure 5) for a comparison of estimated cash income for coffee and cocoa using MASP, Coffee Industry Corporation and Cocoa Board data.

fresh food sold within PNG since 1995. The MASP database does not include income from sale of sugar by outgrowers to Ramu Agri-Industries Ltd or that from sale of chickens to processing facilities near Lae.

Nevertheless, the broad patterns identified by the MASP surveys remain valid 10–15 years after the data were recorded. More recent information is available for the export cash crops and is presented in the sections that follow. However, the MASP database provides the only reliable estimates of the value of commodities sold in the informal economy.

The estimated cash income to rural villagers from sale of agricultural produce for the period 1990–1995 was about K200 million per year (about the same in US dollars at that time). Arabica coffee generated more income than any other commodity, providing 33% of all income from agricultural activities. This was followed by income from sales of fresh food (22%), cocoa (11%), betel nut and betel pepper (10%), copra (8%), oil palm (3%), firewood (3%) and fresh fish and shellfish (2%) (Figure 5.1.1, Table 5.1.1).

Sales of fresh food provided cash income to more households than any other activity. More than 90% of rural villagers lived in households where income was derived from the sale of fresh food (including Irish potato).[2] This was followed by Arabica and Robusta coffee (53% of rural villagers), betel nut and betel pepper (35%), cocoa (27%), firewood (23%), tobacco (19%), copra (17%), fish and shellfish (13%) and cattle (13%) (Figure 5.1.2, Table 5.1.1).[3]

Sales of oil palm or Arabica coffee gave the greatest returns per person (K47/person/year), followed by copra (K31), cocoa (K26), betel nut and betel pepper (K18) and fresh food (K15). Although kina returns

2 Cash income from sales of Irish potato was recorded separately from other fresh food in the MASP project (see tables), but has been included here with other fresh food.

3 The population data refers to the number of villagers living in MASP agricultural systems (see Section 1.15). So it is accurate to state that X% of rural villagers live in households *in locations* where income is derived from the sale of a certain commodity.

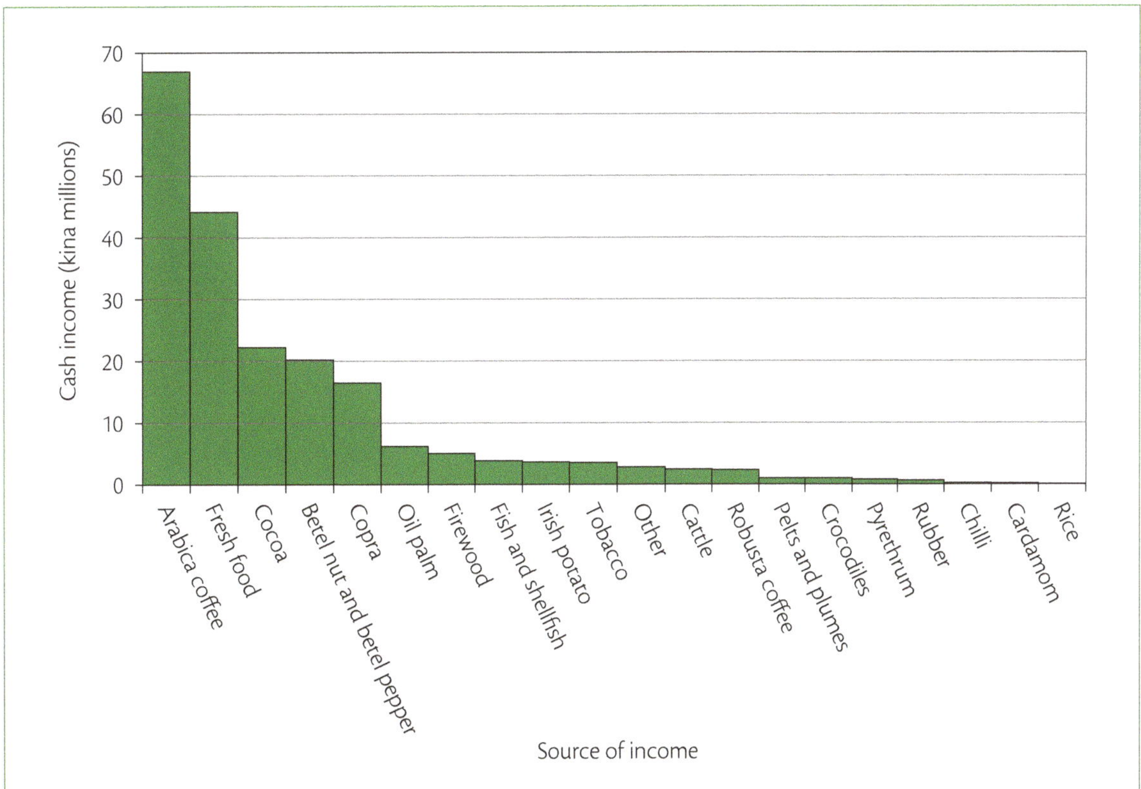

Figure 5.1.1 Estimated annual cash income of the rural population from agricultural activities, 1990–1995. Source: MASP.

Table 5.1.1 Estimated annual cash income of the rural population from agricultural activities, 1990–1995[a]

Product	Income (kina/year)[b]	Proportion of total income (%)	Rural population[c]	Proportion of total rural population (%)	Average cash income (kina/ person/year)
Arabica coffee	66,937,782	32.9	1,416,000	44.5	47
Fresh food	44,136,734	21.7	2,991,000	94.0	15
Cocoa	22,237,432	10.9	850,000	26.7	26
Betel nut and betel pepper	20,216,030	9.9	1,121,000	35.2	18
Copra	16,460,230	8.1	527,000	16.6	31
Oil palm	6,147,153	3.0	130,000	4.1	47
Firewood	5,029,200	2.5	741,000	23.3	7
Fish and shellfish	3,812,507	1.9	412,000	13.0	9
Irish potato	3,593,202	1.8	536,000	16.8	7
Tobacco	3,547,451	1.7	591,000	18.6	6
All other products	2,814,518	1.4	409,000	12.9	7
Cattle	2,417,596	1.2	403,000	12.7	6
Robusta coffee	2,318,227	1.1	270,000	8.5	9
Pelts and plumes	959,931	0.5	160,000	5.0	6
Crocodiles	950,192	0.5	131,000	4.1	7
Pyrethrum	748,667	0.4	125,000	3.9	6
Rubber	619,633	0.3	64,000	2.0	10
Chilli	173,207	0.1	29,000	0.9	6
Cardamom	148,367	0.1	25,000	0.8	6
Rice	39,705	0.0	7,000	0.2	6
Total	203,307,762	100.0			14

[a] This table is derived from data generated by MASP (Bourke et al. 1998; Allen et al. 2001). These estimates differ somewhat from those in Allen et al. (2001: Table 9) due to the use of a different population growth rate to derive the 1990 populations. There are also different assumptions regarding mean income for each class in the MASP database.

[b] These estimates are for income received by villagers. Total revenue from the sale of export crops was greater than this, but some of the total revenue goes to buyers and processors and pays for transport, insurance, levies and other expenses. Hence the income that villagers receive is less than the total export revenue.

[c] This is the population of rural villagers who live in agricultural systems where income is derived from each activity. Villagers may have more than one major source of cash income from agriculture. Hence the population total is greater than the actual rural population. Similarly, the total proportion of the population is greater than 100%.

per household are now generally higher than they were in the mid 1990s, the overall patterns have not altered greatly since then.

The significance of the informal sector is commonly overlooked in economic analyses of the PNG economy. Of the 11 sources listed in Table 5.1.1 that generated more than K1 million per year income in rural areas, six are part of the informal economy. These are fresh food, betel nut, firewood, fish, tobacco and cattle. The other products (Arabica and Robusta coffee, cocoa, copra and oil palm) are exported and appear in official statistics.

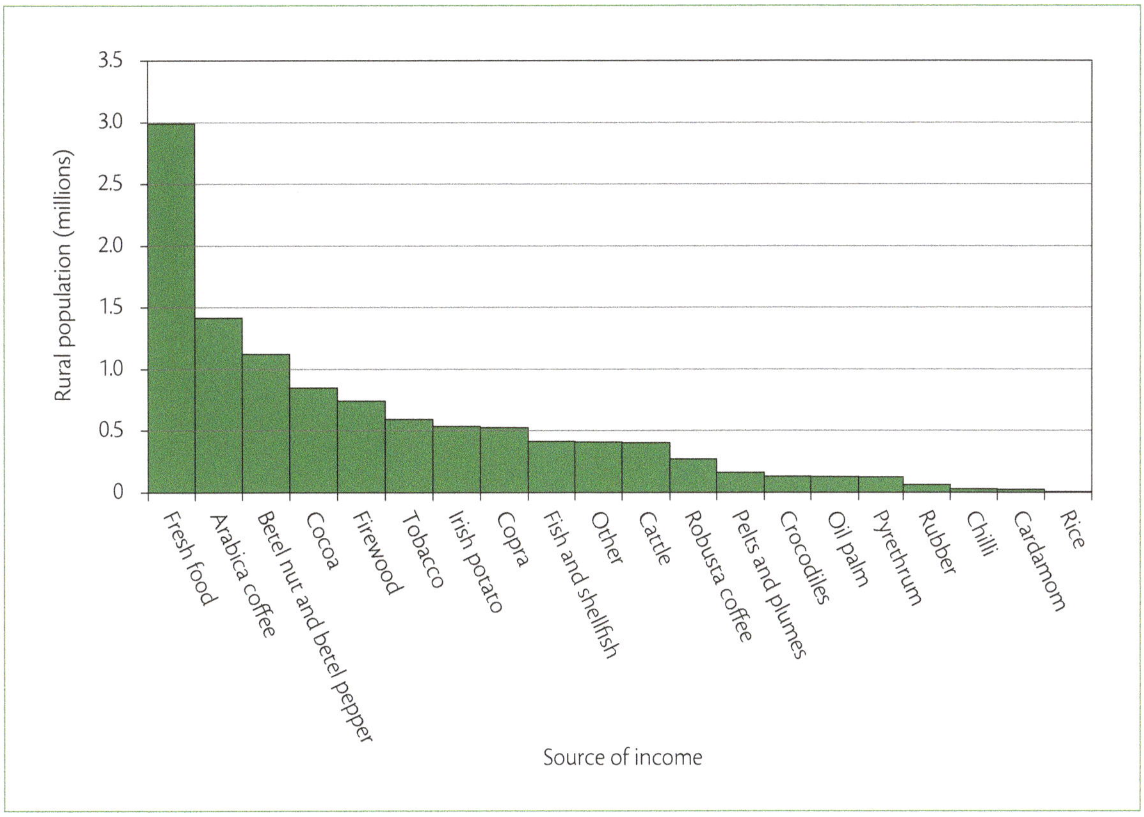

Figure 5.1.2 Estimated rural population who receive cash income from agricultural activities, 1990–1995. Source: MASP.

In reality, most villagers earn cash income from a number of different agricultural activities. People have the highest cash income where they sell a number of commodities into the export and domestic markets, and where the volumes sold are larger. Cash incomes are lowest where people sell only small quantities of one or two commodities.

Distribution of agricultural income

It is not surprising in such a physically diverse nation as PNG where agricultural activities differ widely between locations that the amounts of income generated also vary. The national spatial pattern of income distribution for the most important sources of cash income are given in maps in the sections that follow.[4]

[4] The distribution of cash income from agricultural activities within provinces is presented as maps in Hanson et al. (2001).

Arabica coffee production is confined to the central highlands and some mountainous locations in Morobe and Madang provinces (Figures 5.1.3, 5.4.2, Table A5.1.1). High levels of fresh food sales are more widely distributed, being significant in many locations in the highlands and lowlands (Figure 5.3.1). Cocoa production is concentrated in East New Britain, Bougainville, East Sepik, Madang and New Ireland provinces (Figure 5.5.4).

Two large markets exist for betel nut – the entire highlands region, and urban areas, particularly Port Moresby and Lae. This determines the pattern of income from betel nut, with sales higher in locations in Central and Gulf provinces west of Port Moresby; in Morobe, Madang and East Sepik provinces; and in areas readily accessible to urban centres in the Islands Region (Figure 5.17.1). Copra production is greatest in East New Britain, Madang, New Ireland, Bougainville and West New Britain provinces (Figure 5.6.3). Oil palm is concentrated in West New Britain and Oro provinces. Estates in Milne Bay and New Ireland provinces involve a small number of smallholders (Figure 5.7.1).

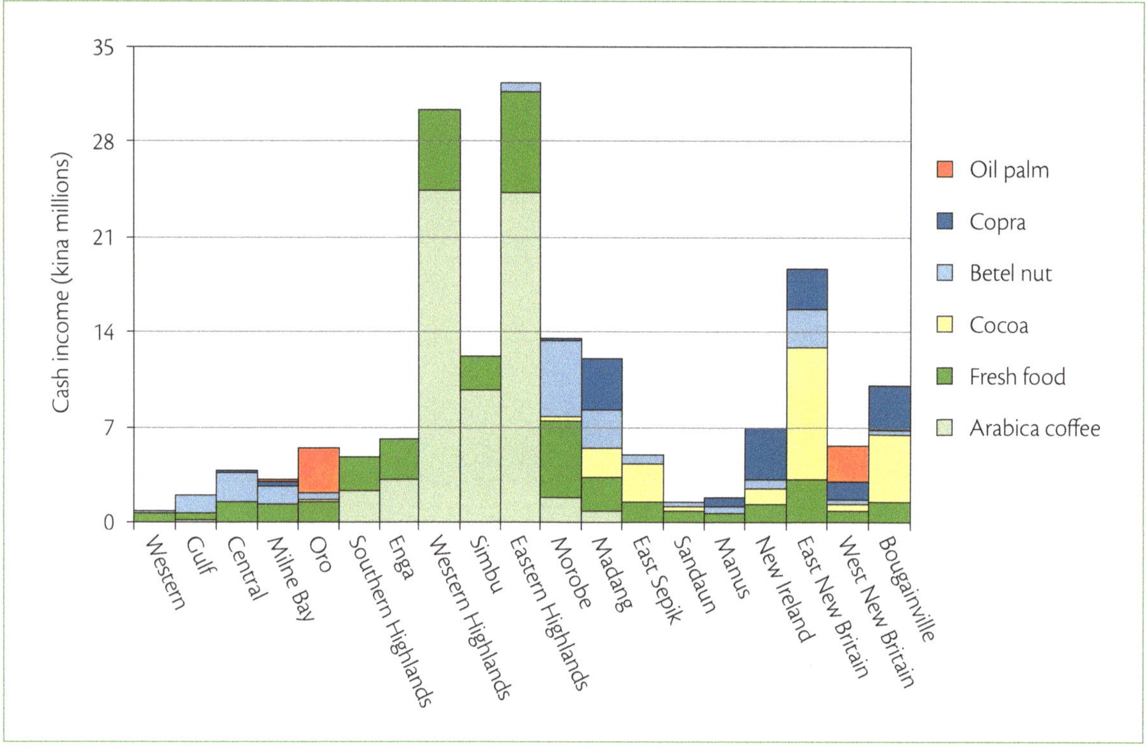

Figure 5.1.3 Estimated annual cash income of the rural population from the six most important agricultural activities, by province, 1990–1995. Source: MASP.

More than a third of the rural population lived in locations where the average cash income from agricultural activities was less than K20/person/year in 1990–1995 (Table 5.1.2). Of these people, 29% live in Southern Highlands Province, 12% in Morobe Province and 10% in Sandaun Province (Figure 5.1.4). People in Sandaun and Southern Highlands provinces had the lowest cash income from agricultural activities, with 92% and 86% respectively of the rural population earning less than K20/person/year. Other provinces with a substantial proportion of the rural population earning low cash incomes were Morobe (53%), Western (48%), West New Britain (45%) and Madang (40%) (Table A5.1.2).

At the other end of the scale, 18% of the rural population lived in locations where average cash income from agricultural activities was greater than K150/person/year in 1990–1995 (Table 5.1.2). Provinces with a higher proportion of the rural population earning relatively large incomes from agriculture were Western Highlands (31% of those earning more than K150 per year), Eastern Highlands (26%) and East New Britain (15%).

Even in a province such as East New Britain, where most rural people earned relatively high incomes from agriculture, a fifth of the population, most of whom lived in Pomio District, earned less than K20/person/year. This illustrates how low-income populations can occur in pockets within a province and that a province is a crude unit with which to examine rural income distribution. The places where rural cash incomes are low are characterised by poor agricultural environments, with excessive rainfall, steep slopes, areas of inundation and soils of low fertility. Road access is also poor or non-existent and access to markets and services is poor. A number of health and education indicators also tend to be greatly below the national average in these locations.[5]

5 These issues are discussed in detail at provincial and district levels by Hanson et al. (2001). See also Section 6.10 in this volume on rural poverty, particularly Figure 6.10.4.

Table 5.1.2 Estimated annual cash income of the rural population from agricultural activities, by income class, 1990–1995

Income class (kina/person/year)	Total rural population[a]	Proportion of total rural population (%)	Total cash earned (kina millions)	Proportion of total cash earned (%)
0–20	1,175,815	37.0	16.3	8.0
21–50	865,030	27.2	30.5	15.0
51–100	300,591	9.4	20.4	10.0
101–150	256,194	8.1	25.3	12.4
151–200	472,550	14.9	71.8	35.3
201–300	111,135	3.5	39.0	19.2
Total	3,181,315	100.0	203.3	100.0

[a] Population figures are for estimated rural population in 1990 extrapolated from the 2000 census data, using the 1980 to 2000 growth rate.

Source: MASP.

Non-agricultural income sources

Sources of rural cash income other than agriculture include:

- Small retail businesses, particularly small retail stores ('trade stores') that sell basic items, especially imported food.
- Transport businesses, including vehicles licensed to carry paying passengers.
- Sales of manufactured foods, including bread and other foods made from flour.
- Trading, as middlemen, of betel nut, betel pepper and lime.
- Paid labour for other villagers, plantations, logging operations and local businesses.
- Remittances of money from relatives working in urban locations or elsewhere in PNG.
- Royalties, compensation and other payments from mines and oil fields.
- Alluvial gold mining.
- Royalties, compensation and other payments from forestry operations.
- Other small businesses, including carpentry, coffin construction and vehicle repairs.

The major mines and oil fields can be sources of large sums of money for nearby villagers. However, payments go to a relatively small number of people. Large mines are located at Ok Tedi in Western Province, at Porgera in Enga Province and on Lihir Island in New Ireland Province. There are smaller gold mines at Tolukuma in Central Province, near Kainantu in Eastern Highlands Province, on Simberi Island in New Ireland Province and at Sinivit in the Baining Mountains of East New Britain Province. A number of new mines are planned, including a large nickel mine in Madang Province. Oil and gas fields are located in the Lake Kutubu area of Southern Highlands Province. Payments are made to villagers who own land in the fields and along a Kutubu–Kikori pipeline. Amounts received vary greatly between groups of villagers, often over relatively short distances. Some individuals, mostly men, receive payments of up to K30 000 per year, but only a few hundred people receive such levels, and most villagers receive much less.

An estimated 60 000 to possibly 100 000 people mine alluvial gold. Alluvial gold mining takes place in all provinces, but the main centres of small-scale mining are the Wau and Bulolo areas of Morobe Province, the Maprik area in East Sepik Province, the

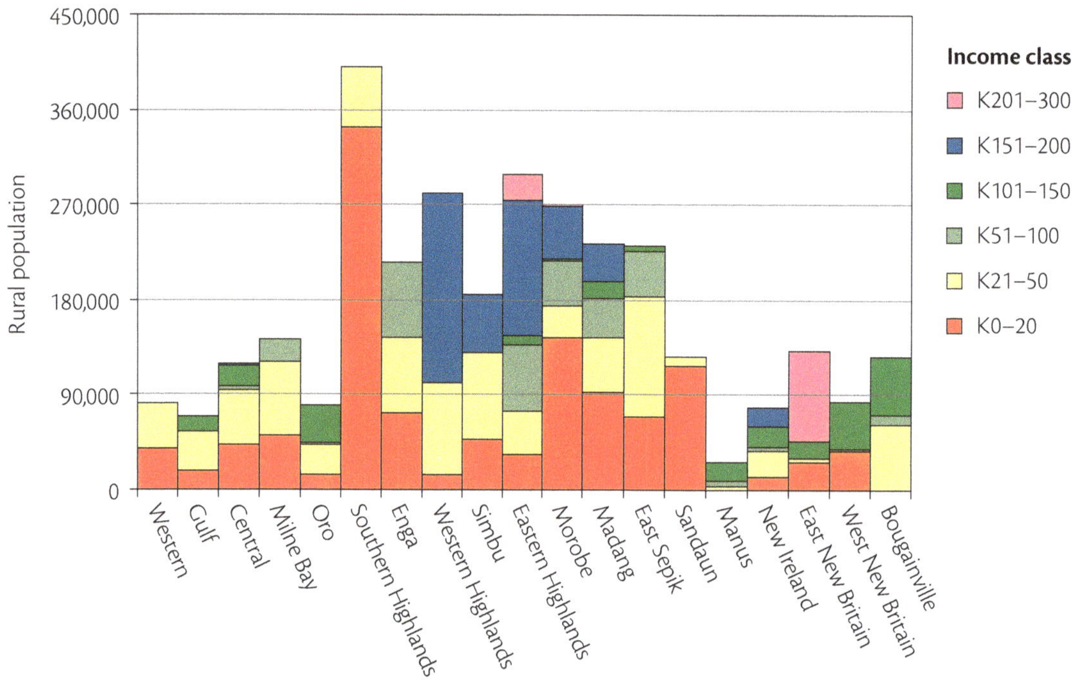

Figure 5.1.4 Estimated annual cash income of the rural population from agricultural activities, by income class per person and province, 1990–1995. Source: MASP.

Mount Kare and Porgera areas in Enga Province and the Kainantu area of Eastern Highlands Province. In 2002 production was four tonnes of gold valued at K100 million.

Royalty payments from logging can be significant. However, discussions with villagers indicate that amounts received are often much less than what is written into logging agreements. Payments to a community tend to be short-term as the logging operations move elsewhere after one or two years. In contrast, payments from mining operations continue over many years and the agreements are better adhered to. Income from casual labour can be moderately high for some rural people in the highlands during the coffee harvest season, particularly when prices are high.

Little numerical data exist on remittances of money from urban areas or enclave developments (mines, oil fields and logging camps). Remittances are particularly large to people in Manus Province, as many well-educated people from that province are in well-paid employment elsewhere in PNG. Remittances are also relatively high to villagers on the Trobriand Islands in Milne Bay Province.[6] The coastal part of Central Province is a third region where remittances from urban-based workers to their rural relatives are relatively high. Transfers commonly occur as gifts when urban people visit their home village for holidays, particularly at Christmas time. They also occur as the payment of school or other education fees for the children of rural-based relatives and purchases of expensive items such as outboard motors for boats.

6 This occurs because of the unique social system in the Trobriand Islands, associated with food exchange systems.

Sources

Allen, B.J., Bourke, R.M. and Hanson, L. (2001). Dimensions of PNG village agriculture. In Bourke, R.M., Allen, M.G. and Salisbury, J.G. (eds). *Food Security for Papua New Guinea. Proceedings of the Papua New Guinea Food and Nutrition 2000 Conference.* ACIAR Proceedings No. 99. Australian Centre for International Agricultural Research, Canberra. pp. 529–553.

Bourke, R.M. (1999). Estimating cash income and access to services for rural Papua New Guinea. Unpublished paper. Department of Human Geography, The Australian National University, Canberra.

Bourke, R.M., Allen, B.J., Hobsbawn, P. and Conway, J. (1998). *Papua New Guinea: Text Summaries.* Agricultural Systems of Papua New Guinea Working Paper No. 1. Two volumes. Department of Human Geography, Research School of Pacific and Asian Studies, The Australian National University, Canberra.

Hanson, L.W., Allen, B.J., Bourke, R.M. and McCarthy, T.J. (2001). *Papua New Guinea Rural Development Handbook.* Land Management Group, Department of Human Geography, Research School of Pacific and Asian Studies, The Australian National University, Canberra.

Susapu, B. and Crispin, G. (2001). Report on small-scale mining in Papua New Guinea. International Institute for Environment and Development, London.

5.2 Cash cropping in Papua New Guinea: an overview

In PNG the production and sale of cash crops is an important source of income at both the national and household level. At the national level, products from the renewable resource sectors of agriculture, forestry and fisheries contribute around a quarter of the total value of exports. However, this figure hides the importance of agricultural products as a source of income for rural villagers, who comprise 81% of PNG's population. The sale of agricultural commodities is the most significant source of income for rural people in PNG, in terms of both total income and the number of people who receive it (see Section 5.1).

This section examines the role of cash crops in the export and domestic sectors of the economy, historical trends in cash crop production levels, and future prospects for cash crop production in PNG.

The export sector

PNG's export economy is presently dominated by minerals (gold and copper) and oil (Figure 5.2.1). These three commodities account for 77% of the total value of exports. However, the contribution of agricultural exports is significant at around 17%, with the remainder from forestry (5%) and marine resources (1%).

In order of importance by value, PNG's agricultural exports are palm oil,[1] coffee, cocoa, copra and copra oil, vanilla, tea and rubber (Figure 5.2.2). These commodities, and other minor agricultural exports, are discussed in the following sections.

The domestic sector

The domestic cash crop economy consists of food (and other commodities such as betel nut) that is sold through PNG's thriving fresh produce markets, sugar produced by Ramu Agri-Industries Ltd (most of which is sold within PNG), and minor quantities of tea and other products. Domestically marketed food was the second most important source of cash for rural villagers after Arabica coffee in the mid 1990s (see Section 5.3). A number of animal products are also sold, including chickens, pigs, fish and cattle (see Section 5.18).

1 Palm oil is the main economic product derived from the fruit of the oil palm. The other products are palm kernel oil, refined palm oil and palm kernel expellent (see Section 5.7).

Trends and changes over time

Significant changes have occurred in the relative importance of different export cash crops since the late 1800s, when the commercial production of copra commenced in the Gazelle Peninsula of East New Britain Province. There was also a decline in the relative contribution of agricultural commodities to the export economy in the 1970s, following the development of a number of large mines. Other important changes have been the decline of the plantation sector and the expansion of the smallholder sector.

In the late 1800s and early 1900s copra was PNG's most important export commodity, accounting for around 90% of all exports in 1921–22. Most copra was produced on plantations that were established throughout the Islands Region in the early 1900s. In 1950 copra still accounted for almost 70% of all exports, but by the mid 1970s its export share had declined to only 5%. This decline was primarily due to development of the coffee, cocoa and mineral industries between 1950 and 1980, and the collapse of the international copra market (Figures 5.2.3, 5.2.4).

A feature that is common to several cash crops is a change in production over time. The relationship between production levels and time tends to produce a characteristic S-shaped curve. The production of coffee in PNG over the past 60 years is a good example of this relationship (Figure 5.2.5). The S-shape of the curve is indicative of a process in which growth initially occurs at a slow rate and then at a rapid rate before slowing down and eventually plateauing. It would appear that each of PNG's main cash crops is at a different stage in this process. Coffee and cocoa are both at an end stage, having experienced periods of rapid expansion (early 1960s to mid 1970s in the case of cocoa and early 1960s to early 1980s for coffee). Oil palm, on the other hand, is still in the rapid growth phase (Figure 5.7.4). It is likely that oil palm production will slow in coming years due to environmental limits on further expansion.

Another characteristic that is common to most export cash crops in PNG is the historical decline of plantation-based production and expansion in smallholder production. The plantation sector has been in decline for all export tree crops over the past 30 years, with the exception of oil palm and tea (Table 5.2.1). On the other hand, smallholder production of oil palm, cocoa, coffee and rubber has been expanding.

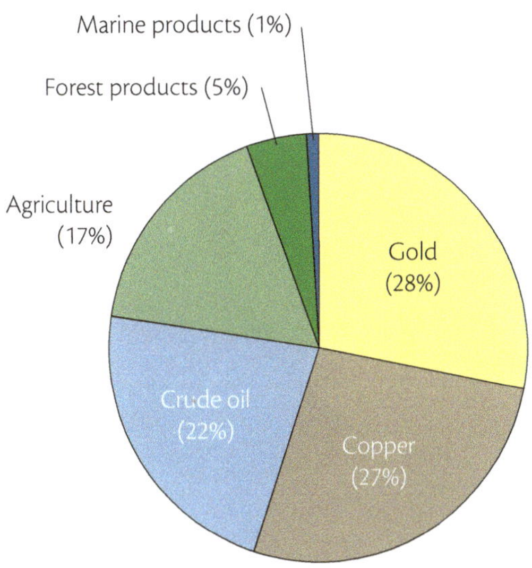

Figure 5.2.1 Value of major exports, 2004–2006 (annual mean). Source: Bank of PNG.

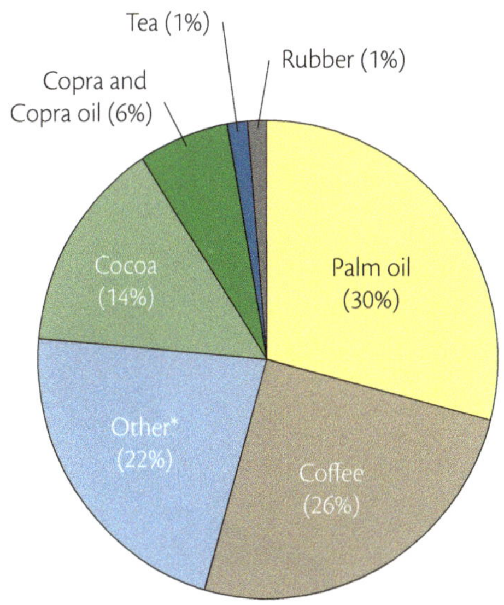

* 'Other' includes tinned tuna, tinned beef, processed tea and coffee, spices, and minor commodities such as artefacts, crocodile skins and butterflies. Vanilla was a major component of the 'other' category in 2004.

Figure 5.2.2 Contribution by value of the main cash crops to agricultural exports, 2004–2006 (annual mean). Source: Bank of PNG.

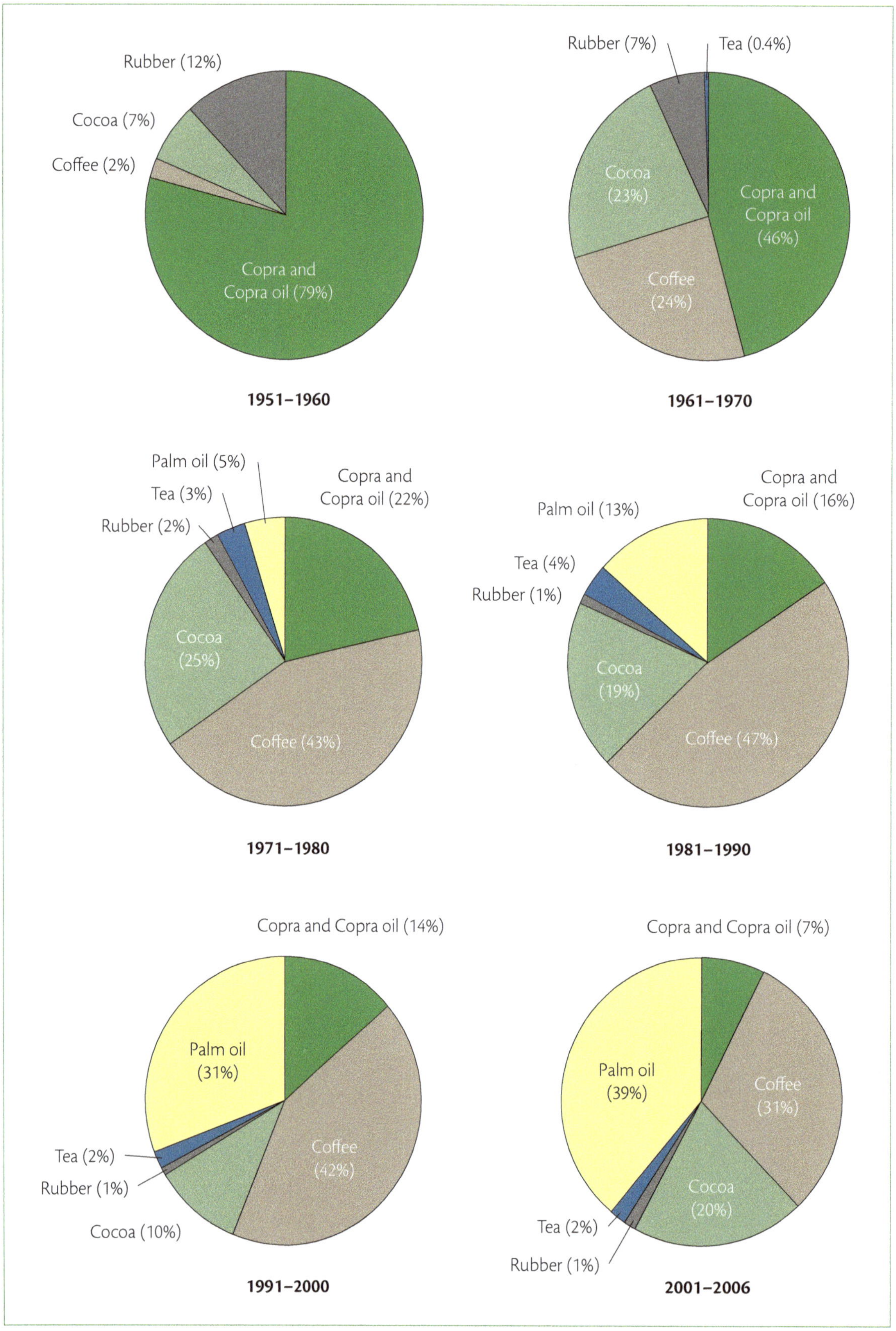

Figure 5.2.3 Contribution by value of the main cash crops to agricultural exports, by decade, 1951–2006.

Sources: Bank of PNG and various industry sources.

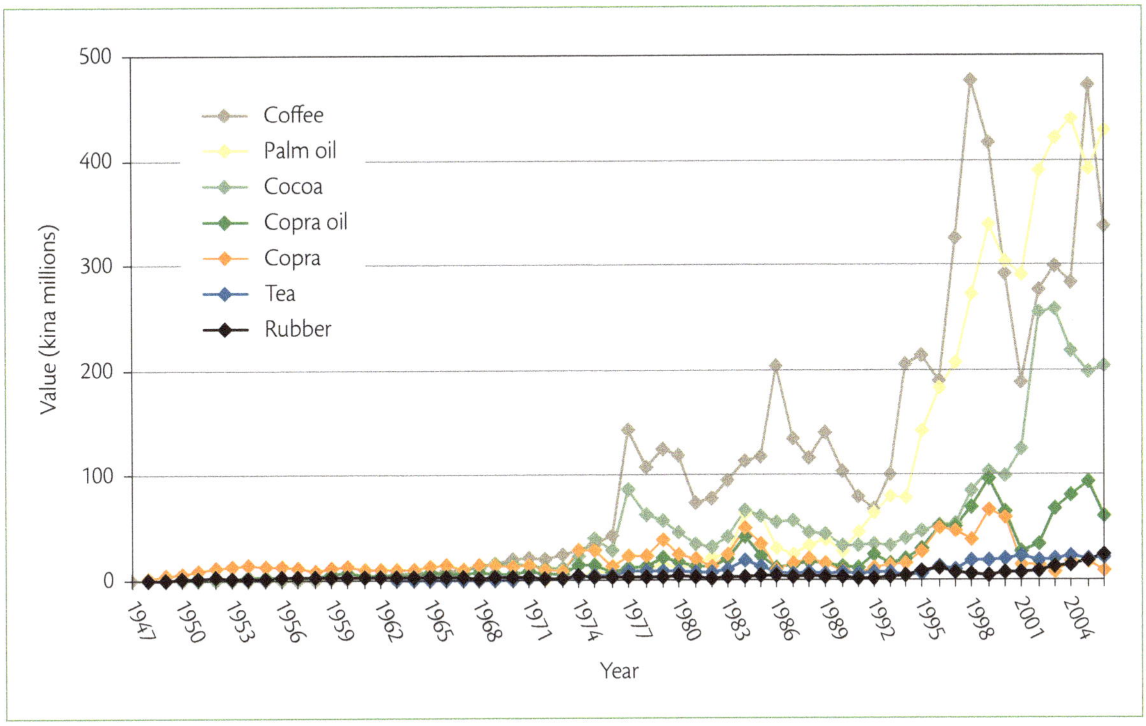

Figure 5.2.4 Export values of main cash crops, 1947–2006. **Note:** The values presented have not been adjusted for changes in the buying power of the kina. This makes the increase in values nearer the present seem greater than they really are. Sources: Bank of PNG and various industry sources.

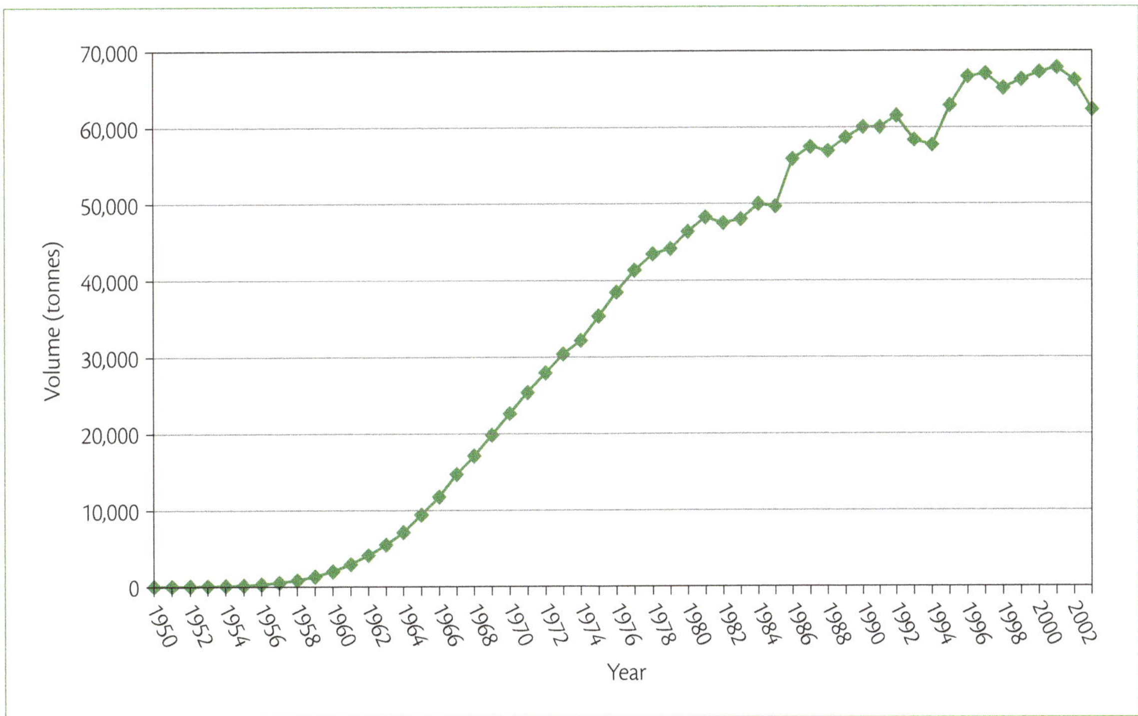

Figure 5.2.5 Volume of coffee exports, 1947–2006. Data presented as a 7-year running mean. Sources: 1947–1976: Munnull and Densley (c.1978); 1977–1991: DAL (1992); 1992–1999: Bank of PNG; 2000–2003: Coffee Industry Corporation Ltd; 2004–2006: Bank of PNG.

The gross annual growth rate data presented in Table 5.2.2 hide this change. For example, in the case of cocoa, overall production grew at 1.3% per year over the period 1986 to 2005, but smallholder production actually increased by 2.7% per year over that period and plantation production declined by 4.1% per year (Table 5.2.1, Figure 5.5.2).

Some of the reasons for the decline in plantation cash crop production are specific to particular crops, such as the age of trees and palms in the cocoa and copra industries. Other factors have affected the plantation sector more broadly. These include the rising costs of inputs, particularly fuel and labour, partly as a consequence of the devaluation of the kina; the declining state of infrastructure, particularly roads and bridges; extreme fluctuations in world market prices; and uncertainties concerning land ownership, particularly around the time of Independence in 1975, when village groups made claims to land that had been previously alienated to the colonial state and was being leased to plantation owners.

The expansion in smallholder production for all of the main cash crops has also occurred for a variety of reasons. In some cases, such as the expansion of smallholder coffee production in the 1960s, government extension and promotion initiatives had a positive effect. However, in many other instances, villagers in PNG have readily adopted cash crops

Table 5.2.1 Annual growth rates of volume of oil palm, cocoa, coffee and rubber from the smallholder and plantation sectors, 1986–2005

Commodity	Annual growth rate (%)[a]		
	Smallholder	Plantation/estate	Both subsectors
Oil palm (fresh fruit bunch)	4.3	8.3	6.6
Cocoa	2.7	−4.1	1.3
Coffee	2.0	−2.9	0.9
Rubber	3.2	−3.3	−0.7

[a] Calculations are based on the growth rate from 1986 to 2005. The 1986 figure is a mean for the three-year period 1985–1987; likewise, the 2005 figure is a mean for the period 2004–2006. For rubber, the period is 1986–2000 (1999–2001).

Sources: Calculated from data in relevant sections in this book.

Table 5.2.2 Production volume and annual growth rates of export tree crop commodities, 1975–2005

Commodity	Production (tonnes)[a]		Annual growth rate (%) 1975–2005
	1975	2005	
Palm oil	18,145	332,167	10.2
Copra oil	26,093	47,000	2.0
Coffee	35,141	61,992	1.9
Tea	4,772	7,181	1.4
Cocoa	31,557	45,857	1.3
Rubber	5,526	4,722	−0.5
Copra	87,320	18,067	−5.1

[a] The 1975 figure is a mean for the three-year period 1974–1976; likewise, the 2005 figure is a mean for the period 2004–2006.

Sources: Calculated from export data in relevant sections in this book.

without any advice or assistance from government. Moreover, contrary to the predictions of some academics and policy makers, the customary land tenure arrangements existing in PNG appear not to have inhibited this expansion.

Future prospects

A number of the long-term trends discussed above can be expected to continue into the foreseeable future. It seems likely that the plantation sector will continue to decline, while the smallholder sector will continue to expand. With the exception of oil palm, smallholders now dominate production of the export tree crops. Prospects for the continued growth of these exports are good, so long as smallholders continue to receive reasonable cash returns (see Section 5.20). However, past experience suggests that a number of factors will influence levels of production in the short term, including international market prices, currency fluctuations and the price of imported inputs, particularly fuel.

Future prospects for cash crops are discussed in the sections on individual crops that follow. Forecasts by the World Bank for future prices for selected commodities are given in Figure 5.2.6, although it is difficult to predict future prices with accuracy.

The domestic cash crop economy – particularly the marketing of fresh produce – has expanded over the past 30 years and is likely to continue to do so. This expansion is being driven in part by the devaluation of the kina, which has seen a significant increase in the price of imported food. Consumers in both urban and rural areas have responded by purchasing less imported food and more local produce.

A common perception of the agricultural economy in PNG is that it has not kept pace with population growth and that export cash crop production has actually been in decline since Independence. The data presented in Tables 5.2.1 and 5.2.2 show that these perspectives are wrong. The official annual population growth rate over the period 1980–2000 was 2.7%. The average annual growth of oil palm production of around 10% has significantly outstripped population growth. While the growth of all other export cash crops has not quite kept pace with population growth, growth rates in the smallholder sector of most cash crops have more or less kept pace with population growth.

Another common perception is that smallholder producers are not responsive to market opportunities and conditions. The rapid expansion of vanilla production since 1998 (see Section 5.14), the steady expansion of fresh food and betel nut production over the past 30 years, and the adoption and expansion of a range of export tree crops over the past 60 years all indicate that smallholders are responsive to market opportunities.

Constraints

New market opportunities are appearing and more will appear in the future, including such things as the sale of organically certified produce. Smallholders will respond to these market opportunities, but certain constraints need to be removed or reduced to allow them to do so. Constraints that inhibit expansion of export agricultural production in PNG include:[2]

- Poorly maintained transport infrastructure, particularly roads and bridges.
- Inadequate access to credit and working capital for middlemen and traders.
- The poor export quality of some commodities, particularly vanilla and coffee.
- Insufficient knowledge by producers of market requirements.
- Inadequate security for traders and producers, who are vulnerable to robbery and assaults.
- Poor dissemination of information and planting material.
- Limited areas of land that do not have major climatic or physical limitations for high productivity, especially for large-scale production of cocoa and coffee.

---

2 A different group of factors constrain further expansion of domestically marketed food (see Section 5.3).

Continuing population growth is placing increasing pressure on land resources. Eventually population pressure and environmental limitations of steep topography, very high rainfall and inundation will constrain the expansion of land under agriculture. Future growth in the value of both food and cash crop production will have to come from greater productivity of land that is already used for cropping and from higher quality produce. These challenges are common to agriculture in most of the world. There are many reasons to expect that PNG producers will meet these challenges as they have met challenges in the past.

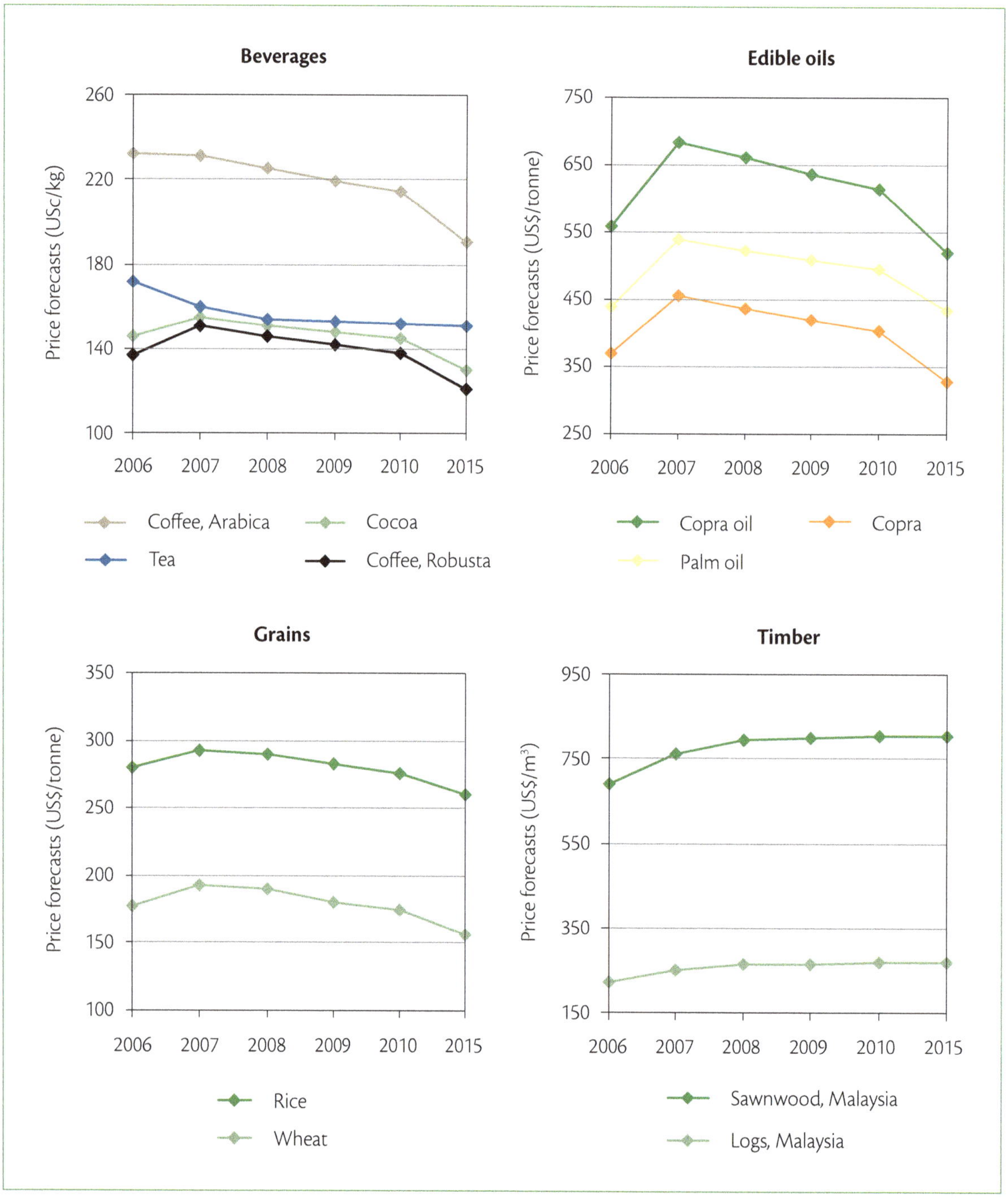

Figure 5.2.6 World Bank real price forecasts for selected commodities. **Note:** Data for 2006 are actual prices and data for 2007–2015 are price forecasts. Real prices are the projected prices in today's spending power, that is, ignoring inflation. Source: World Bank (2007).

Sources

Bai, B. (2003). Performance of major agriculture export commodities and their future prospects. Rural Industries Council, Port Moresby. <http://peb.anu.edu.au/pdf/PNG%20Update%20BAI%202003.pdf>.

Bank of Papua New Guinea Quarterly Economic Bulletin Statistical Tables <http://www.bankpng.gov.pg/> [go to Statistics tab].

Bourke, R.M. (2005). Agriculture in Papua New Guinea: more good news than bad. *Development Bulletin* (67):13–16.

Bourke, R.M. (2005). Agricultural production and customary land in Papua New Guinea. In Fingleton, J. (ed). *Privatising Land in the Pacific: A Defence of Customary Tenures*. Discussion Paper Number 80. The Australia Institute, Canberra. pp. 6–15.

DAL (Department of Agriculture and Livestock) (1992). Exports of important agricultural commodities: quantity, value and average price per unit. Rural Statistics Bulletin 1992/6. Department of Agriculture and Livestock, Konedobu.

Fingleton, J. (ed.) (2005). *Privatising Land in the Pacific: A Defence of Customary Tenures*. Discussion Paper Number 80. The Australia Institute, Canberra.

Munnull, J.P. and Densley, D.R.J. (c. 1978). Coffee. In Densley, D.R.J. (ed). *Agriculture in the Economy: A Series of Review Papers*. Volume 2. Department of Primary Industry, Konedobu. pp. 1–42.

World Bank (2007). Prospects for the Global Economy. Commodity forecasts prepared 21 April 2007 <http://go.worldbank.org/NO5J2C8XJ0>.

5.3 Domestically marketed food

Fresh food grown and sold within PNG is known as 'domestically marketed food'. Sales of fresh food have grown over the past 40 years from a small base, with particularly rapid expansion since 1998 following devaluation of the PNG currency. The term 'fresh food sector' in PNG often refers to introduced temperate-climate vegetables such as Irish potato, broccoli, carrot, tomato and capsicum. However, these vegetables form only a small part of domestically marketed foods that include staple food crops such as sweet potato, banana, sago, taro and yam.

History of marketed fresh food

Prior to colonisation, food was exchanged and bartered in parts of what is now PNG. Bartering was common where significant environmental differences occurred over short distances. For example, people along the Sepik River exchanged fish for sago with those who lived away from the river near sago swamps. This exchange continues on parts of the Sepik River today. Food was also given in competitive exchanges, the most significant being pigs or pig meat in the highlands and yams in parts of the Sepik basin and islands of Milne Bay Province.

Elsewhere, coastal villagers traded coconut, betel nut and marine foods with inland people, who provided taro, yam, banana, green vegetables and other garden produce. This occurred on the Gazelle Peninsula of East New Britain Province; among the Mekeo, Motuan and Koiari people of the Port Moresby area; and on the north coast of the Huon Peninsula, Morobe Province. People on very small islands had limited capacity to grow food because of environmental constraints and pressure on land caused by high population densities. They exchanged marine foods for fresh foods with people on nearby larger islands. A number of pre-colonial trade relationships persist (see Box 5.3), but some have disappeared with the building of roads and long distance travel.

The exchange of fresh food for manufactured goods began as soon as outsiders arrived in PNG. Missions bought food from surrounding areas to feed staff and school children. Marketplaces quickly appeared in new colonial towns. There was a marketplace in Rabaul before 1920, and marketplaces were established in the main highland centres in the 1950s. In the 1970s a number of food marketing schemes were instigated, commencing with the Fresh Food Project in 1973, but most failed after some years.

Intermediate traders or middlemen have been slow to appear in PNG fresh food markets, but numbers are increasing. Middlemen operate at local levels, buying in a main town market and selling within the same town, or moving produce longer distances to mines or to Port Moresby. The growth in the number of middlemen is a sign of a maturing economy and indicates increasing efficiency in the marketing chain.

Box 5.3 Some rural locations where a significant proportion of food comes from trade

- Mid Sepik River and Chambri Lakes, East Sepik Province. Fish is exchanged for sago. About two-thirds to three-quarters of sago consumed is imported from nearby areas.
- Near the mouth of the Sepik River, East Sepik Province. There is no agriculture in this area and sago is imported from surrounding areas in exchange for fish and woven bags.
- Boisa Island, north-west of Manam Island, Madang Province. *Galip* nut, an important seasonal food, is exchanged for sago and yam from the mainland. Sago is the main food during the drier part of the year (May to October) and all sago is obtained from trade.
- Sialum area, north coast of the Huon Peninsula, Morobe Province. Coconut, fish and marine foods are exchanged for banana and root crops from inland locations, particularly during seasonally dry periods.
- Malai and Tuam islands, south of Umboi Island, Morobe Province. In the past, fish, dogs, pigs, wooden bowls and pandanus leaf mats were exchanged for sago, yam, banana and taro from nearby Umboi. This trade continues, but the terms of trade are now less favourable for those on the very small islands and quantities traded have declined.
- Coastal area opposite Daru and Saibai islands, Western Province. Agriculture in this area is limited and dugong, turtle, fish and shellfish are traded or sold to obtain sago. Purchased rice had become a more important source of food energy than traded sago, but the effects of price rises for rice are not known.
- Wari Island, Milne Bay Province. About a third of food needs are obtained from trade with people from Basilaki Island. Clay pots are sold for cash, which is used to purchase food from larger nearby islands.
- Amphlett Group, Milne Bay Province. Clay is obtained from nearby Fergusson Island to make clay pots, which are in turn exchanged with people on Fergusson for about one-third of food needs on the Amphletts.
- Wagifa Island, near Goodenough Island, Milne Bay Province. Wagifa people can only grow a limited amount of fresh food, mostly cassava. They sell fish and other marine foods at a small rural market on Goodenough Island and buy root crops, banana and other garden food. Food is also traded to other small islands in Milne Bay Province including Dobu Island, Brumer Islands, and those in the Engineer Group.
- Numerous islands north and south of Manus Island (Harengan, Arowe, Sori, Andra, Hus, Oneta, Pityilu, Hawei, Ndrilo, Mbuke, Johnston, Big Ndrova, Little Ndrova and Tilianu islands). Fish, crab, lobster, shellfish, octopus, lime, string bags and skirts are exchanged for sago, other vegetable foods, betel nut, tobacco and clay pots.
- Matsungan, Petats, Yame, Pororan and Hetau islands west of Buka in Bougainville Province. Capacity to grow food is limited because of population pressure and reduced soil fertility. Sweet potato, banana and other garden food is obtained from Buka Island in exchange for fish and other marine products. A local saying is: *Sapos yu no gat wanpela kanu bai yu no inap kaikai* – If you do not have a canoe you cannot eat (as you cannot trade for food).

Sources: Various MASP provincial working papers.

The devaluation of the PNG currency in 1994 had a positive effect on marketed food (see Section 4.1). The value of the kina fell from US$1.00 in the early 1990s to about US$0.85 in 1994. The kina further declined after 1997, from about US$0.70 to about US$0.32 by 2005 (Figure 4.1.1). Imported foods, including rice and wheat, became three times more expensive, which resulted in increased demand from both urban and rural dwellers for locally grown food. Rural producers responded by growing more food for sale, particularly in the highlands where low coffee prices had reduced incomes. Both the size and number of food markets in the main towns increased.

Marketing fresh food

Domestically marketed food is sold in the following ways in PNG:

- In small food markets that occur in villages, on roadsides, in small rural centres, at government and church stations, and at logging camps. Larger, permanent markets are present in all cities, towns and smaller urban areas.
- Through supermarkets, restaurants, fast food outlets (*kai* bars), hotels and hostels.
- Directly to institutions and companies that cater for large numbers of people, including to mine and oil field settlements (such as Ok Tedi, Porgera, Lihir and Kutubu), high schools, universities, prisons and defence force establishments.

Most food is sold in open food markets. However, wholesalers also buy fresh food from producers and sell it to retailers or other consumers. Wholesalers include Alele Fresh Farm Produce (Port Moresby, Mount Hagen, Goroka and Lae), Vegmark Fresh Produce (Mount Hagen) and Green Fresh Limited (Port Moresby). Alele and Vegmark also transport fresh food from the highlands to Lae and Port Moresby. Some growers market other people's produce, as well as their own.

Some features of food markets in PNG

Food markets share some common features, despite a wide variety of food and non-food items being sold:

- Items are offered for sale in 'heaps' (or bundles) at a fixed price per heap rather than per unit weight; for example, K1 per heap rather than K1.70 per kilogram.
- The price is varied by adjusting the size of the heap, as well as the price of the heap.
- Considerable variation exists between sellers in the size of heaps on offer and hence the prices of items being sold.
- Premium prices are paid for size for some items, at least in the highlands. For example, the price *per kilogram* of a large cabbage is higher than that for a smaller one.
- There is little obvious competition between vendors. Sellers do not call out or try to attract buyers to their stall.
- Middlemen buy from vendors in one market and resell elsewhere, such as on a street near business offices, for slightly higher prices (smaller heaps).
- Imported goods, uncommon before 1970, are increasingly sold in markets.

Significance of domestically marketed food

People in PNG markets mainly sell fresh foods but, increasingly, cooked foods made from both locally produced and imported ingredients are being marketed. The most commonly marketed fresh foods are sweet potato, other root crops, corn, peanuts, green vegetables such as *aibika*, cabbage, beans and amaranthus, fruit including pawpaw, pineapple, mango and watermelon, and nuts such as *karuka* and *galip*. The most common cooked foods are made from flour.

Statistics on the volume and value of marketed fresh food in PNG are limited but it is clear that this part of the economy is large and is growing. About 83% of the food energy consumed in 2006 was grown within PNG (Figure 2.1.1), while the per person consumption of imported rice has declined since 1998 (Figure 2.7.2). The shortfall of imported food is being met by increased domestic food production and increased sales of fresh food, particularly the staple root crops, banana and sago.

It is not known what proportion of the estimated 4.5 million tonnes of staple foods produced in 2000 (see Section 2.2) was sold. If just 1% of this volume was sold, it would amount to around 45 000 tonnes of food being marketed. However, staple food crops are only a small proportion of all food sold, so the total volume of food moving through PNG markets each year is considerably greater than 45 000 tonnes.

The estimated annual income from food sales in 1990–1995 was K44 million (see Section 5.1). That was second only to income from sales of Arabica coffee (K67 million) (Figure 5.1.1, Table 5.1.1). Prices of fresh foods have risen in local markets, although not as much as the prices of imported foods (see Section 4.3). It is almost certain that income from fresh food sales has increased since 1996, but there is no reliable estimate.

Although higher incomes are earned from coffee sales, more rural people earn an income from selling fresh food than from any other economic activity (Figure 5.1.2, Table 5.1.1). More than 90% of rural villagers live in areas where income is derived from selling fresh food.[1]

The volume and value of fresh food is greatest nearest to the largest towns and where road transport is good. Port Moresby, three times the size of the next largest city, Lae, has an extensive market influence that reaches into the highlands, even though it is not connected by road to the rest of PNG. Fresh food marketing is most important around Port Moresby, including the peri-urban areas and the Sogeri Plateau; in the Markham and Ramu valleys; the northern valleys of Eastern Highlands and Simbu provinces; the Wahgi Valley and other valleys in Western Highlands Province; the north-east lowlands of the Gazelle Peninsula in East New Britain; and near many of the smaller urban areas (Figure 5.3.1).

Marketing systems

The overwhelming characteristic of food marketing in PNG is the high level of individual grower involvement. In the case of the staples, virtually all marketing is undertaken by growers using public motor vehicles for transport. Farmers lack confidence in the marketing chain, have a perception that they can make more money by cutting out middlemen, and use their produce to fund visits to relatives in Port Moresby, Lae and Madang. The high level of farmer involvement in marketing has undesirable consequences. Marketing costs are higher, farmers earn less when all their costs are considered, produce quality is poorer, market development is limited, and consumer prices are increased. As well, production is reduced because producers are travelling and marketing, rather than growing more produce.

The number of food wholesalers is small and they are concentrated in the highlands. The fresh food industry could not have developed to the extent that it has without the tenacity of these few businesses. Wholesale enterprises require considerable support and assistance to offset the substantial constraints under which they operate. A lack of affordable finance is probably the major constraint to fresh produce marketing, both finance for investment in marketing infrastructure and working capital for traders to buy produce from farmers.

1 Similarly, in the 1980 and 2000 censuses, a high proportion of rural people reported that they lived in a household where income was derived from selling fresh food.

Future prospects

Much progress has been made in the marketing of fresh food over the past 40 years. In the past 10 years in particular the sector has expanded with large increases in the volume of food sold, in the range of foods on offer, in the quantity of cooked foods and in the quality of produce. Nevertheless, the considerable difference between rural and urban areas in per person consumption of the various staple foods suggests unmet demand for locally grown food staples in urban areas (Table 2.1.1) and thus potential for expansion in the marketed fresh food sector. The greatest potential is for further sales of the main staple foods, including sweet potato, banana and sago; temperate-climate vegetables from the highlands to lowland urban areas; and sales of sweet fruit, particularly mandarin, mango, rambutan and mangosteen, from coastal or mid-altitude locations to the major urban areas and to the highlands where sweet fruit do not grow (see Section 3.3).

Constraints

A number of major constraints limit expansion of the fresh food sector in PNG. The constraints differ by food. However, the following is a generalised list of problems that need to be addressed:

- Insufficient middlemen and excessive involvement of producers in selling relatively small amounts of food (for example, spending large amounts of time and money travelling), so that efficiencies of scale are not achieved.
- Inadequate market information (for example, on prices, demand and shipping availability).
- Poor communications between producers, middlemen, transporters and retailers.
- Poor marketing skills by producers.
- Lack of consistency in the volume and quality of produce offered for sale.

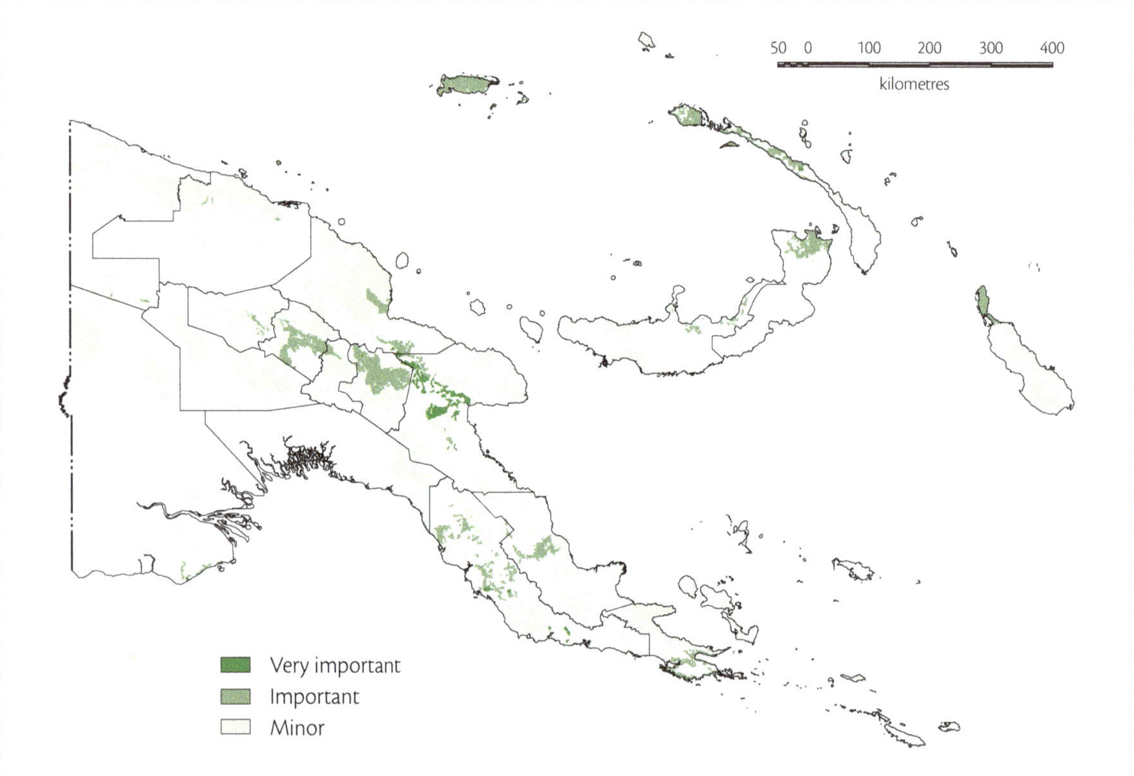

Figure 5.3.1 Locations where sales of fresh food provided income for rural villagers, 1990–1995. Source: MASP.

- Expensive and infrequent sea, river and road transport; a lack of priority by airlines to carry fresh food; and poorly maintained rural roads that cause food to be damaged during transport.
- Inadequate storage facilities at points of transport change; for example, from road to ship.
- Production constraints, including not using the best available varieties; insufficient attention to soil fertility maintenance; production in a less than optimum climatic zone (for example, producing mandarin in the highlands when the best fruit is grown at 800–1400 m altitude); and failure to control major pests and diseases.
- Lack of availability of agricultural inputs, including appropriate planting material, fertilisers and pesticides.

These constraints result in poor quality produce being offered for sale. Produce is handled roughly at every stage of the marketing process from harvesting to retailing; it is inappropriately packaged; is delayed during transport; and is held in unsuitable temperature and humidity conditions.

Indigenous nuts

PNG does not have a competitive advantage in exports of fresh food or flowers (see Section 5.17). A very unfavourable fruit fly situation, poor transport infrastructure and inadequate airfreight links are the main causes. Indigenous nuts are an exception. Nuts with a large export potential include *galip*, *pao* and *okari* (see Section 3.4). These nuts could become for PNG what the Brazil nut is for the Amazon or the macadamia (a nut indigenous to Australia) is for Hawaii. However, despite their inherent qualities, these nuts are largely unknown outside Melanesia. To create a major export industry will require a substantial investment in product and market development. The Hawaiian macadamia nut industry required the injection of substantial equity and risk capital. Such capital has not been forthcoming for new agribusiness ventures in PNG. Until such investment occurs, opportunities exist for small-scale developments supplying local markets, where the nuts are known and are in demand.

Sources

Allen, B.J., Bourke, R.M. and Hanson, L. (2001). Dimensions of PNG village agriculture. In Bourke, R.M., Allen, M.G. and Salisbury, J.G. (eds). *Food Security for Papua New Guinea. Proceedings of the Papua New Guinea Food and Nutrition 2000 Conference*. ACIAR Proceedings No. 99. Australian Centre for International Agricultural Research, Canberra. pp. 529–553.

Benediktsson, K. (2002). *Harvesting Development: The Construction of Fresh Food Markets in Papua New Guinea*. Nordic Institute of Asian Studies, Denmark.

Benjamin, C. (1985). Some food market influences of a large-scale small-holder development in the West New Britain area of Papua New Guinea. *Papua New Guinea Journal of Agriculture, Forestry and Fisheries* 33(3–4):133–141.

Bourke, R.M. (1986). Periodic markets in Papua New Guinea. *Pacific Viewpoint* 27(1):60–76.

Bourke, R.M. (2005). Marketed fresh food: a successful part of the Papua New Guinea economy. *Development Bulletin* (67):22–24.

Epstein, T.S. (1982). *Urban Food Marketing and Third World Rural Development: The Structure of Producer-Seller Markets*. Croom Helm, London & Canberra.

Gibson, J. (2001). The economic and nutritional importance of household food production in PNG. In Bourke, R.M., Allen, M.G. and Salisbury, J.G. (eds). *Food Security for Papua New Guinea. Proceedings of the Papua New Guinea Food and Nutrition 2000 Conference*. ACIAR Proceedings No. 99. Australian Centre for International Agricultural Research, Canberra. pp. 37–44.

McGregor, A., Lutulele, R. and Wapi, B. (2005). Papua New Guinea horticulture industry sector study. Secretariat of the Pacific Community, Suva, Fiji.

Ross, G. (2005). Marketing fresh food in Papua New Guinea: the Alele story. *Development Bulletin* (67):25–26.

5.4 Coffee

Coffee is the most important source of village agricultural income (Table 5.1.1, Figure 5.1.1). It is particularly important for both villagers and urban-based businesses in the highlands where opportunities to earn cash income from other sources are limited. In the early to mid 1990s an estimated 1.7 million people (53% of the rural population) lived in households where cash income was earned from selling Arabica or Robusta coffee (Table 5.1.1, Figure 5.1.2). Coffee generated an average of K364 million per year in export income over the period 2004–2006, which was 26% of the value of agricultural exports in this period (Figure 5.2.2).

Two types of coffee are cultivated in PNG. Arabica coffee, the most important, is usually grown between 700 m and 2050 m altitude, but occasionally as low as 100 m and as high as 2400 m (Figure 1.13.3). Robusta coffee, considerably less important, is usually grown between sea level and 550 m. Both types of coffee are grown in environments where average rainfall is 1700–5000 mm per year.

The production of Arabica coffee is highly seasonal (see Box 5.4). The main season varies a little between years and between locations, but is generally in the period May to September and particularly in June, July and August. A study of Robusta coffee found no definite harvesting season in the Gazelle Peninsula area, East New Britain Province. However, in the Milne Bay area further from the equator, the main harvest occurs between May and August with the peak in June–July.

Adoption and history

Coffee was first introduced to PNG in 1873 (see History of agriculture) and was growing in the Rabaul Botanical Garden by 1890. Between about 1900 and 1940, several plantations were established in Central Province and around Wau in Morobe Province and by villagers in the Sangara area of Oro Province. However, coffee remained an insignificant cash crop until the early 1950s, when commercial production of Arabica coffee commenced on small expatriate-owned plantations and in villages in the central highlands. Before 1960, most of the Arabica coffee exported from PNG was grown in highland regions of Morobe Province, particularly in the Wau, Finschhafen, Kaiapit and Wasu areas, but during the early 1960s a rapid expansion of smallholder coffee production occurred in Western Highlands, Simbu and Eastern Highlands provinces. This expansion occurred as a result of extension activities, the absence of alternative cash-earning opportunities, high export prices for coffee, the construction of the Highlands Highway in the mid 1960s, and the example provided by the plantation developments.[1]

1 In 1954 the Australian Administration took steps to reduce the rate at which land was being alienated for commercial coffee plantation development. By 1961 new plantation development had effectively ceased.

Robusta coffee had been grown successfully in Oro and Milne Bay provinces, and in parts of Central, East New Britain, Madang and East Sepik provinces. However, production grew slowly in comparison to the rapid expansion that occurred in the highlands. Robusta coffee has always attracted significantly lower prices than Arabica.

The Australian Administration was forced to slow the rate of growth of coffee production in PNG to meet quotas imposed by the 1962 International Coffee Agreement. Measures taken included the prohibition of further estate development; a ban on planting coffee on new agricultural leases, including settlement schemes; and reduction of extension and promotional activities in the smallholder sector. Despite these measures, smallholder production continued to increase at an average rate of 28% per year between 1961 and 1968. Most of this expansion occurred in Eastern Highlands, Western Highlands and Simbu provinces, where it is known that many Australian agricultural extension officers chose to quietly ignore the bans on assisting villagers to plant coffee.

Following the abandonment of the quota requirements of the International Coffee Agreement in December 1972, the PNG Government attempted to reinvigorate coffee extension and promotion activities, particularly in the less developed areas of the country, such as Southern Highlands Province. It is possible that the renewed extension and promotion efforts resulted in the significant increase in smallholder production that occurred in the 1970s, although other factors, such as high export prices, were more important.

In the early 1980s the government sponsored the development of smallholder coffee blocks. These were parcels of land removed from customary tenure and owned and operated by families or groups of families, separate from village plantings. The blocks had from 5 ha to 29 ha under coffee and were often managed by a professional organisation.[2] The Coffee Industry Corporation (CIC) (see Section 6.4) estimated that there were 636 coffee blocks in 2007. The current standard of management is highly variable, with many blocks producing poorly.

2 The Coffee Industry Corporation defines a block as a holding with 5–29 ha under coffee and a plantation as having 30 ha or more planted with coffee.

Distribution of production and planting

Western Highlands and Eastern Highlands provinces continue to dominate coffee production in PNG, contributing about 82% of the total quantity of coffee produced in 2006 (Figures 5.4.1, 5.4.2, Table A5.4.1). Other provinces that contributed to coffee production in 2006 were Morobe, Simbu, Enga and Southern Highlands. Robusta production has traditionally been dominated by East Sepik Province, which increased its output steadily during the 1990s, mainly as a result of new plantings in the Wosera and Maprik areas. However, production in East Sepik Province has fallen significantly since 2002 as a result of increased cocoa and vanilla plantings. Historically, Arabica coffee has accounted for about 95% of production, with Robusta the remaining 5%, but Robusta production has declined in recent years to less than 1% of total production.

There is little current information concerning the area of land planted to coffee in PNG. According to a 1999 estimate, approximately 70 000 ha were planted to coffee, of which 57 000 ha were smallholder plantings, with the remainder in the plantation and block sectors. In 2007, according to CIC figures,

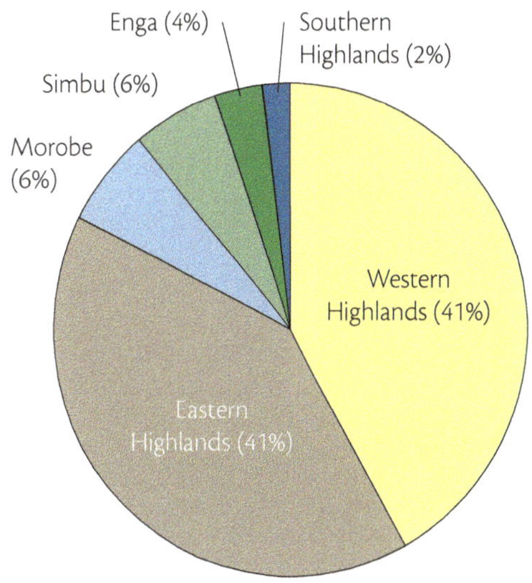

Figure 5.4.1 Coffee production by province, 2006 (by volume). Source: Coffee Industry Corporation Ltd.

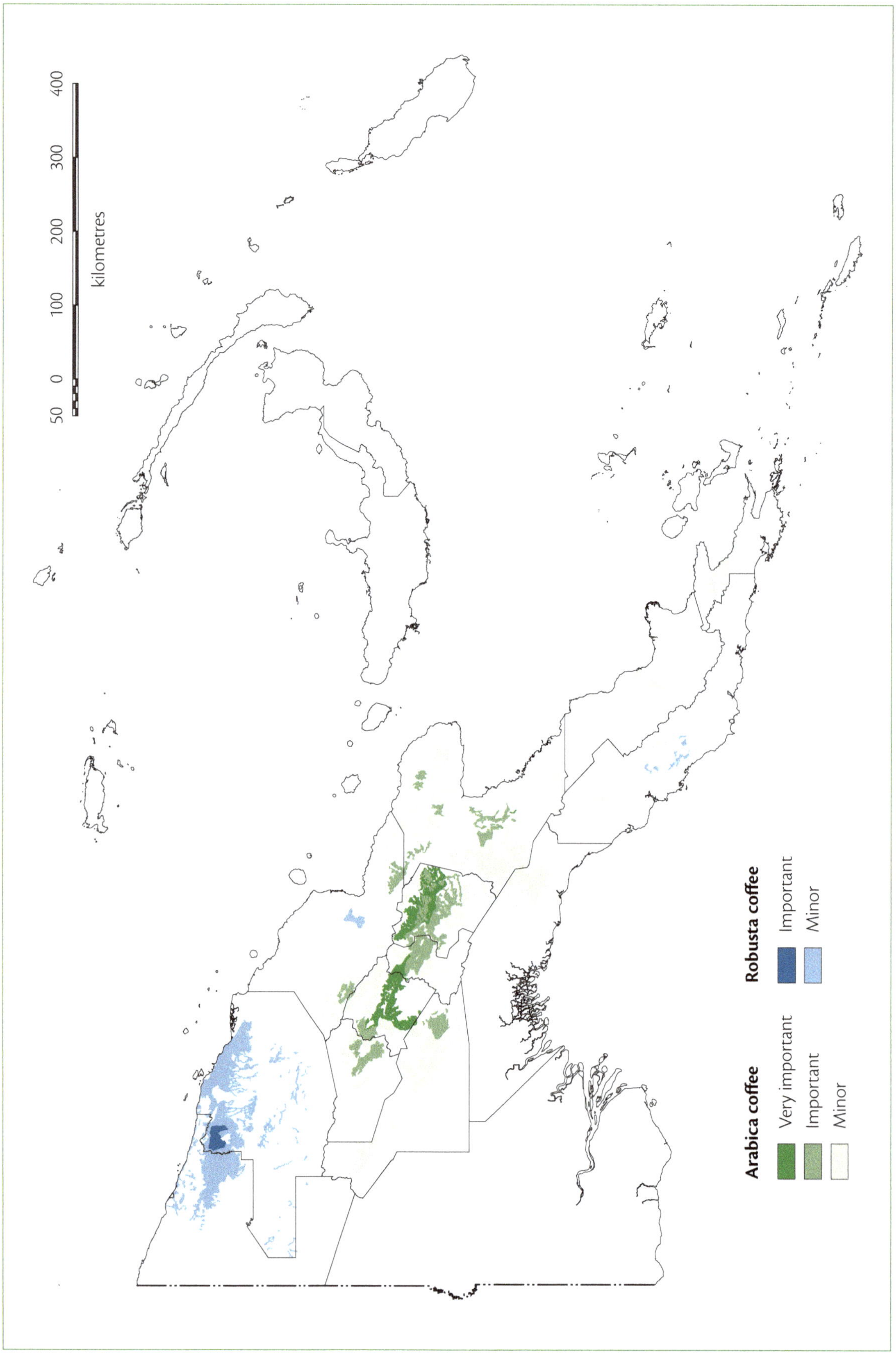

Figure 5.4.2 Locations where coffee sales provided income for rural villagers, 1990–1995. Source: MASP.

plantations had 4400 ha under coffee and the blocks an estimated 12 000 ha. Based on production from 2004 to 2006 and average yields (see below), it is estimated there is 70 000–85 000 ha of smallholder coffee and 12 000–15 000 ha of plantation and block coffee.

Levels of production

Coffee production, measured by the quantity exported, increased rapidly between 1960 and 1980, but the rate of increase has slowed since 1980 (Figures 5.2.5, 5.4.3, Table A5.4.2).[3] There has been virtually no new coffee plantation development since 1961 and the plantation sector has declined since the mid 1980s. There were 107 coffee plantations operating in 1977, but only 33 in 2007. Hence the significant increase in production over the past 40 years has come entirely from the smallholder sector. Between 1985 and 2005 smallholder production doubled and its contribution to overall production increased from 65% to 85% (Figure 5.4.4, Table A5.4.3). Conversely, production from plantations and blocks halved over this period and their combined contribution decreased from 35% to 15%.

Since 1978 the total annual production of coffee has exceeded 40 000 tonnes. Production peaked at 84 000 tonnes in 1989 and again in 1998. The variation in annual production that has occurred since 1980 is mostly explained by fluctuations in the export price and annual rainfall. Smallholder producers are very sensitive to returns on their labour and so are highly responsive to changes in price (see Section 5.20). Growth in production has been limited since 2000 by low prices (until 2004), high rainfall (in 2005 which resulted in a poor harvest in 2006) and deteriorating road access to many producing areas.

3 Roast and ground coffee have been produced commercially in PNG since 1982. Domestic consumption of coffee is less than 100 tonnes per year (about 0.1% of production). Export volumes are therefore a good reflection of total production.

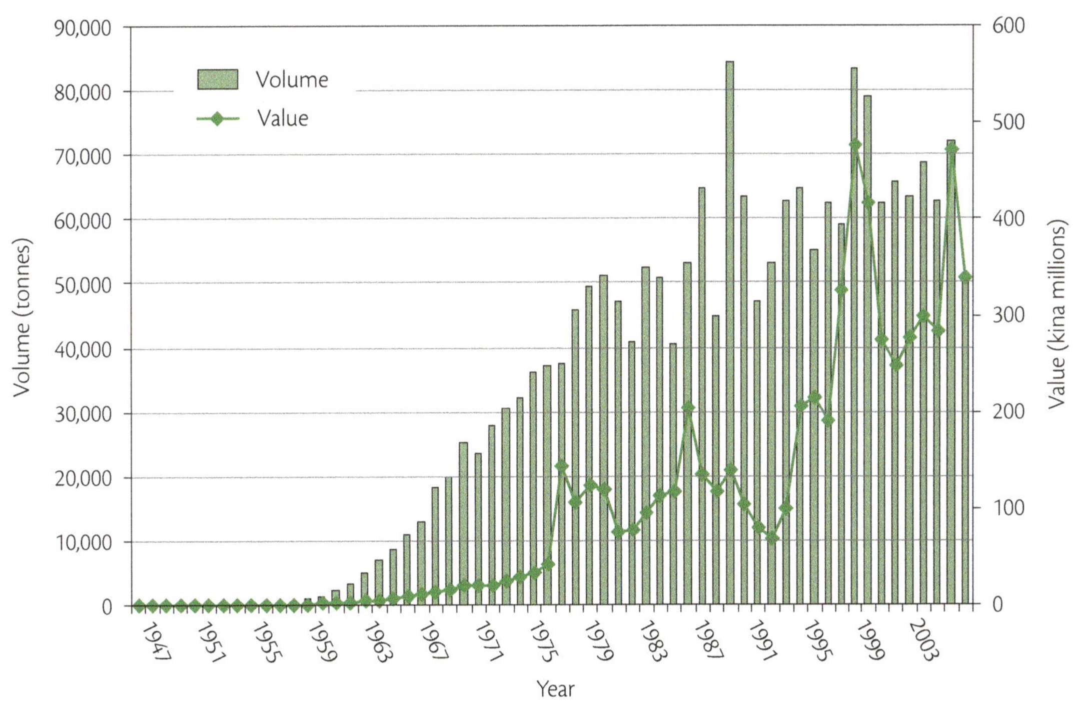

Figure 5.4.3 Volume and value of coffee exports, 1947–2006. Sources: 1947–1976: Munnull and Densley (c. 1978); 1977–1991: DAL (1992); 1992–1999: Bank of PNG; 2000–2006: Coffee Industry Corporation Ltd.

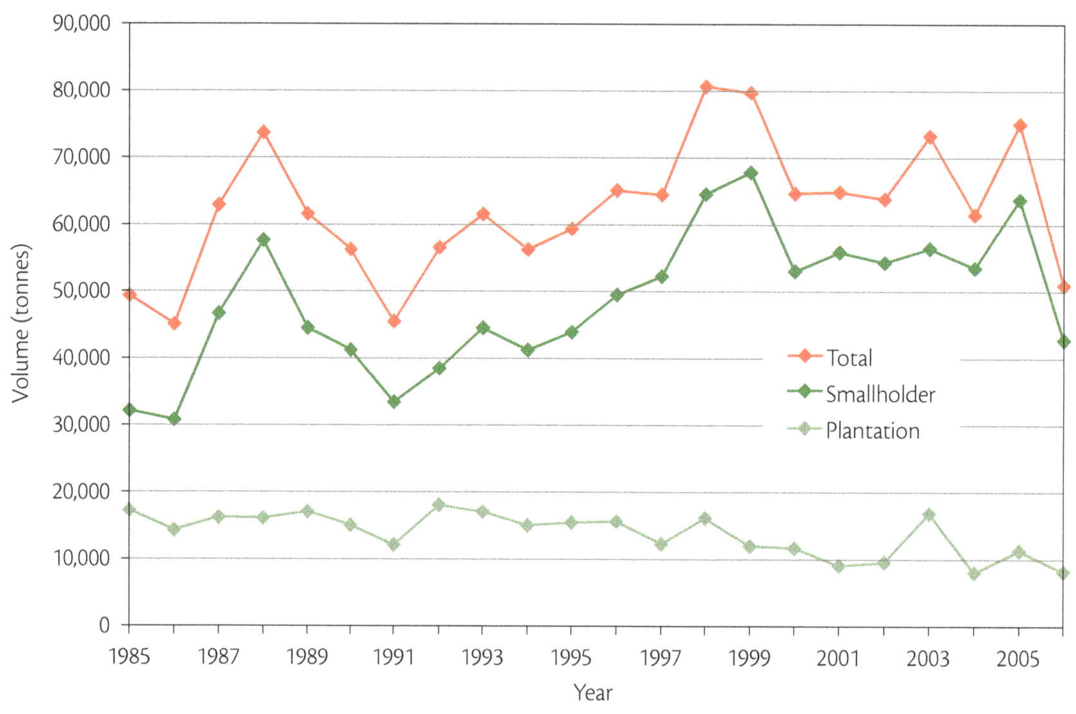

Figure 5.4.4 Coffee production by sector, 1985–2006. Source: Coffee Industry Corporation Ltd.

Survey data from village smallholders and plantations allows yields per hectare to be estimated (Table 5.4.1). For the period 1960 to 1995, the average yield of smallholder Arabica coffee was about 950 kg/ha of green bean, while average plantation yields were almost twice this at 1650 kg/ha. There are significant differences between the highest and lowest reported yields (data not presented in Table 5.4.1), and where the same producers have been surveyed over a number of years, yields may vary considerably between years. On plantations, most of the variation can be attributed to differences in weather conditions between years, rather than greater or lesser attention to harvesting, as is probably the case with smallholders. However, yields on plantations appear to have decreased significantly in recent years; over the period 2000 to 2006, the average yield for plantations and blocks was 600 kg/ha green bean. This is probably more the result of changes in management than of climate.

There is an often-stated belief that coffee has a biennial bearing pattern in PNG (that is, if yields are high in one year, they are lower the following year). However, surveys by the Queensland Department of Primary Industries and by CIC over a number of years do not support this theory (Table 5.4.1). Much published and unpublished data exist on coffee yields under experimental conditions, mostly from Aiyura in Eastern Highlands Province, but these are not reviewed here. Experimental yields range from less than 1 t/ha to more than 4 t/ha of green bean, with means in the range 1.5–2.5 t/ha.

Processing, exporters and markets

The Coffee Industry Corporation registered 17 exporters, 103 processors (58 'dry' factories and 45 'wet' factories) and 5 manufacturers, and estimated there were 5000 itinerant buyers, in PNG in 2007.[4]

4 Raw coffee fruit is called 'cherry'. At wet bean factories the cherry is processed into 'parchment', which is a coffee bean encased in a layer of parchment. At dry bean factories the parchment is removed to reveal the green bean, the main export product.

Table 5.4.1 Average yield of smallholder and plantation Arabica coffee (kg/ha green bean)

Survey year	Average yield (kg/ha)	Location	Source
Smallholder			
1973	750	Six provinces	Munnull and Densley (c. 1978)
1976	910	13 villages, EHP, Simbu, WHP	Anderson (1977)
1982	880	14 provinces	Hassall & Associates (1982)
1982	1170	EHP, Simbu, WHP, Enga	Hassall & Associates (1982)
1987	830	20 village plots, Kainantu area	Harding (1988)
1987–1990	990	14 village plots, over 3 years, Kainantu area	Harding (1991)
1992–93	1090	Benabena District, EHP	Overfield (1994)
Plantation			
1960	1290	30 plantations, EHP, WHP, Wau area	Bureau of Agricultural Economics (1961)
1975	1680	6 provinces	Munnull and Densley (c. 1978)
1980	2220	35 plantations, EHP and WHP	QDPI (1987)
1981	2270	41 plantations, EHP and WHP	QDPI (1987)
1981	2360	10 provinces	Hassall & Associates (1982)
1982	1580	49 plantations, EHP and WHP	QDPI (1987)
1983	2340	49 plantations, EHP and WHP	QDPI (1987)
1984	1610	49 plantations, EHP and WHP	QDPI (1987)
1989	1550	12 plantations, EHP and WHP	Irog (1992)
1990	1100	12 plantations, EHP and WHP	Irog (1992)
1992	1050	20 plantations, EHP, WHP and Morobe	Kufinale (1995)
1993	1400	20 plantations, EHP, WHP and Morobe	Kufinale (1995)
1994	1400	20 plantations, EHP, WHP and Morobe	Kufinale (1995)
1995	1180	16 plantations, EHP and WHP	Stapleton (1997)

Notes

1. Some of the data covering many provinces include lowland areas with Robusta coffee.
2. The assumed factor to convert cherry yield to green bean yield is 6.25:1 (so cherry yield is divided by 6.25 to obtain green bean equivalent).
3. The assumed factor to convert parchment yield to green bean yield is 0.8 (so parchment yield is multiplied by 0.8 to obtain green bean equivalent).
4. Irog (1992) reports mean yields on blocks (5–29 ha) in WHP, EHP and Morobe as 1040 kg/ha in 1989 and 890 kg/ha in 1990.

The majority of exporters and factories are located in Western Highlands and Eastern Highlands provinces. Most smallholder coffee growers process their raw coffee fruit ('cherry') to parchment stage before selling it either directly to factories or, more commonly, to itinerant buyers. A limited amount of coffee is sold to buyers or factories in cherry form. From time to time, attempts are made to ban cherry purchases to prevent theft of cherry from trees.

In 2006/07, six exporters accounted for 85% of the coffee exported from PNG: PNG Coffee Exports (21% of the market), New Guinea Highlands Coffee Exports (19%), Niugini Coffee, Tea and Spice (17%), Monpi Coffee (13%), Kongo (8%) and Pacific Trading (7%). In 2006, PNG exported coffee to 29 countries, with 89% of the exports going to just four: Germany (40%), Australia (20%), the United States (20%) and Japan (9%) (Figure 5.4.5, Table A5.4.4). Minor destinations included Malaysia, New Zealand, South Korea and South Africa.

Future prospects

The PNG coffee industry, despite its problems, has maintained a reasonable degree of international competitiveness. During the early 2000s world coffee prices were at historical lows, but the devaluation of the kina helped maintain kina returns to PNG growers. In 2005, in association with adverse weather conditions in Brazil and an upturn in the normal coffee price cycle, Arabica prices reached their highest level since 1998. Prices have remained firm since. The average export price for PNG coffee in 2006 was 40% higher than in 2004. The World Bank predicts that coffee prices will remain relatively stable until 2010 and will then decline until 2015 (Figure 5.2.6).

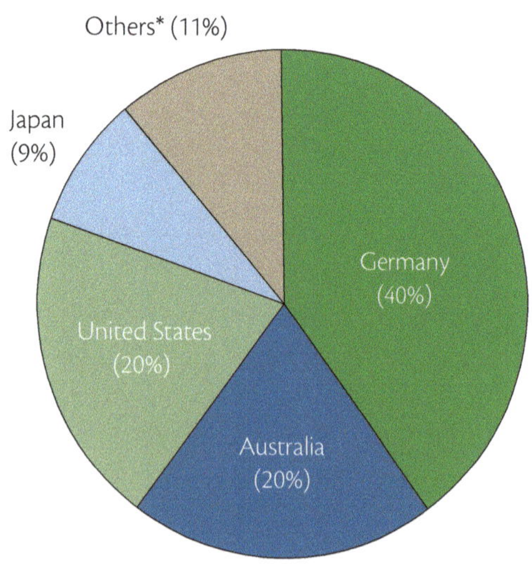

* See Appendix Table A5.4.4 for minor destinations in 2006.

Figure 5.4.5 Destination of coffee exports, 2006 (by volume). Source: Coffee Industry Corporation Ltd.

Fundamental changes in the world coffee market will have major implications for the future of the PNG industry. Over the last decade or so the world coffee industry has boomed at the retail level but remained stagnant at the producer level. In the early 1990s the retail value of the world coffee industry was about US$30 billion; it now exceeds US$70 billion, but the growers' share of profits has fallen from 40% to 10%. The future success of exporting countries, such as PNG, depends on being able to adjust to this new reality.

In addition, there has been a huge increase in the output of low-quality Robusta coffee from Vietnam and medium-quality Arabica coffee from Brazil. The abundant availability of low-priced coffee has brought about a permanent shift in demand. If PNG's coffee industry is to remain viable in the longer term it must produce more, better-quality coffee. This can be done through a number of mechanisms including grower groups where members are committed to producing higher grade coffee and who are rewarded with higher prices for their products.

The PNG Government has set a production target of 90 000 tonnes by 2016. Expansion to these levels is likely to be constrained by limited areas of suitable land in the highlands and ongoing pressure on land; low world prices (Figure 5.2.6); inadequate road access for many producers (see Section 6.9); and possibly by increasing rainfall associated with global climate change (see Section 1.8). A potentially serious insect pest, coffee berry borer, is present in Papua (Indonesian New Guinea) and only urgent quarantine action along the border will prevent it moving into PNG in future years.[5] If this pest becomes established in the PNG highlands it is likely to have a severe effect on coffee production.

5 The clearing of all coffee from a strip of land along the PNG side of the international border may prevent the coffee berry borer from moving into PNG. Most coffee in this area is unproductive, overgrown Robusta and its destruction will have little adverse effects on village incomes.

Box 5.4 Coffee sales and retail spending cycles in the highlands

Coffee is the only major export cash crop in PNG with a strong seasonal production pattern. Arabica coffee is the main source of cash income for most villagers in the highlands.

The seasonal nature of cash income to villagers from coffee sales results in a markedly seasonal pattern of retail sales in the highlands. Sales of alcoholic drinks (mainly beer), rice, other imported food, fuel, soft drinks, clothes and other items are generally greatest during the main coffee harvesting season. Sales are lowest in January, February and March when very little coffee is sold.

This pattern is illustrated with data from the Kainantu area of Eastern Highlands Province. Monthly village coffee production in the Kainantu area, sales of alcoholic drinks from one store in Kainantu and sales of food in another store are shown for a three-year period in Figure 5.4.6 and Table A5.4.5. The close relationship between the quantity of coffee sold by villagers and sales of food and alcoholic drinks is evident. Many other activities in Eastern Highlands, Western Highlands and Simbu provinces show a similar pattern.

It has been claimed that the seasonal nature of coffee harvesting is responsible for food shortages in the highlands. However, a large study of the causes of subsistence food shortages in the highlands showed that there is no relationship between the coffee harvesting period and irregular subsistence food shortages. Rather, these shortages are caused by climatic extremes (frost, excessively high rainfall, drought) and variation in the planting rate of food gardens (see Section 5.20).

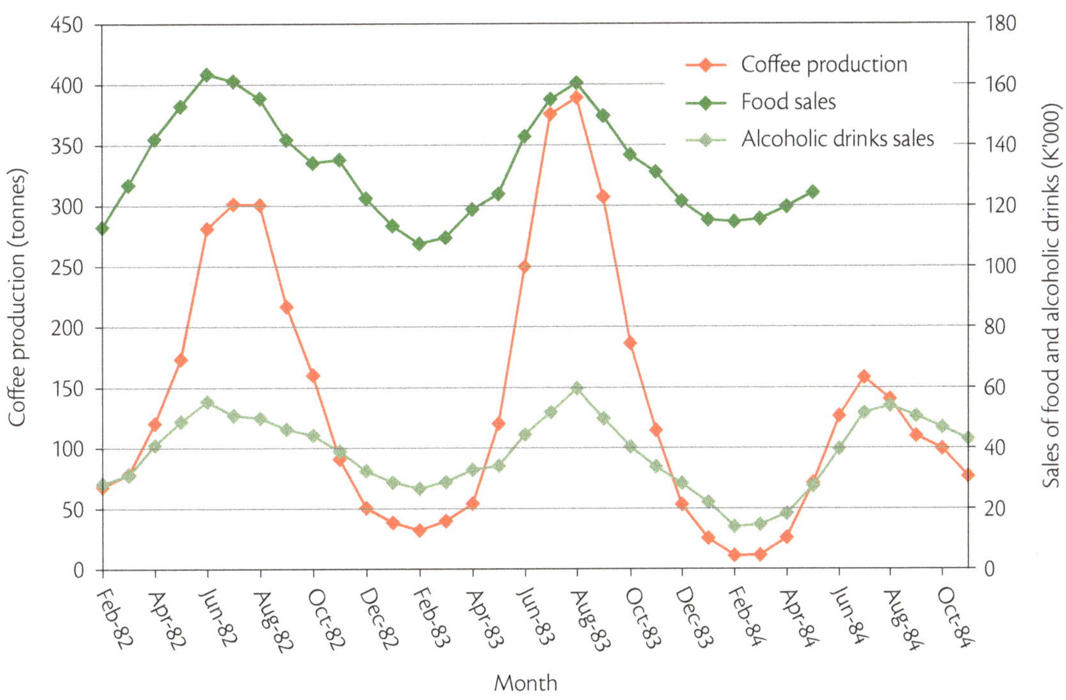

Figure 5.4.6 Smallholder coffee production and sales of food and alcohol in the Kainantu area, Eastern Highlands Province, 1982–1984. Data presented as a 3-month running mean. Source: Bourke (1988:328–9).

The most important issue for the PNG coffee industry is to improve the quality from existing plantings. If it can succeed in doing this, it has a bright future and coffee will continue to make a substantial contribution to highlanders' living standards.

Sources

Anderson, D. (1977). An economic survey of smallholder coffee producers – 1976. Planning Economics Marketing Branch, Department of Primary Industry, Port Moresby.

Bank of Papua New Guinea Quarterly Economic Bulletin Statistical Tables <http://www.bankpng.gov.pg/> [go to Statistics tab].

Bourke, R.M. (1988). *Taim hangre*: variation in subsistence food supply in the Papua New Guinea highlands. PhD thesis. Department of Human Geography, The Australian National University, Canberra.

Bureau of Agricultural Economics (1961). The coffee industry in Papua–New Guinea: an economic survey. Department of Commerce and Agriculture, Commonwealth of Australia, Canberra.

Cartledge, I. (1978). *A History of the Coffee Industry in Papua New Guinea: from inception to the end of 1975*. Papua New Guinea Coffee Industry Board, Goroka.

Coffee Industry Corporation Ltd (ed.) (2005). *Coffee Industry Newsletter*, December 2005.

DAL (Department of Agriculture and Livestock) (1992). Exports of important agricultural commodities: quantity, value and average price per unit. Rural Statistics Bulletin 1992/6. Department of Agriculture and Livestock, Konedobu.

Harding, P.E. (1988). Rehabilitating smallholder coffee gardens in Papua New Guinea: 1. The effects on yields during the first year following rehabilitation. *PNG Coffee* 7(1):79–90.

Harding, P.E. (1991). Foliar nutrient level studies in smallholder coffee gardens in the Kainantu area of Papua New Guinea. Coffee Research Report No. 6. PNG Coffee Research Institute, Kainantu.

Hassall & Associates (1982). Review of the Papua New Guinea coffee marketing system. Final report. Hassall & Associates Agricultural Consultants, Canberra.

Irog, L. (1992). Plantation survey and production costs. Coffee Discussion Paper No. 7. Coffee Industry Corporation Ltd, Goroka.

Kufinale, K. (1995). Plantation cost of production survey. Coffee Discussion Paper No. 18. Coffee Industry Corporation Ltd, Goroka.

Levett, M.P. (2000). Papua New Guinea rural development sector study 2000. Agriculture working paper. Draft final report. Australian Agency for International Development and the World Bank.

Ministry of Agriculture and Livestock (2007). National Agriculture Development Plan 2007–2016. Volume 2: Implementation plan. Ministry of Agriculture and Livestock, Port Moresby.

Munnull, J.P. and Densley, D.R.J. (c. 1978). Coffee. In Densley, D.R.J. (ed). *Agriculture in the Economy: A Series of Review Papers*. Volume 2. Department of Primary Industry, Konedobu. pp. 1–42.

Osorio, N. (2004). Lessons from the world coffee crisis: a serious problem for sustainable development. International Coffee Organization, London. Report to the UNCTAD XI Conference, São Paulo, Brazil, 13–18 June 2004 <http://www.ico.org/documents/pr263.pdf>. Accessed November 2007.

Overfield, D. (1994). The smallholder coffee system: production costs, grower returns, and economic sustainability. Coffee Discussion Paper No. 15. Coffee Industry Corporation Ltd, Goroka.

Queensland Department of Primary Industries, Papua New Guinea Department of Agriculture and Livestock and Papua New Guinea Coffee Industry Board (1987). Papua New Guinea export tree crop study: volume 1 – largeholder coffee. Australian Centre for International Agricultural Research, Port Moresby.

Stapleton, G. (1997). Plantation cost of production survey 1995. Coffee Discussion Paper No. 19. Coffee Industry Corporation Ltd, Goroka.

Wheeler, M. (2002). Papua New Guinea coffee: an integrated market development strategy. Coffee Industry Corporation Ltd, Goroka.

World Bank (2007). Prospects for the Global Economy. Commodity markets briefs – Coffee <http://go.worldbank.org/G37PYVHE00>. Accessed November 2007.

5.5 Cocoa

Cocoa is the third most important source of village agricultural income, after coffee and fresh food (Table 5.1.1, Figure 5.1.1). In the early to mid 1990s, an estimated 850 000 people (27% of the rural population) lived in households where cash income was earned from selling cocoa (Table 5.1.1, Figure 5.1.2). Cocoa generated an average of K218 million per year in export income from 2004 to 2006, which was 14% of the value of agricultural exports in this period (Figure 5.2.2).

In PNG, cocoa is grown up to 800 m altitude (Figure 1.13.3) where annual rainfall ranges from 1800 mm to over 5000 mm. Seasonality data indicate that more cocoa is produced between May and July than in other months, but the pattern varies considerably from place to place.

Adoption and history

Cocoa was introduced to PNG by German settlers around 1900. Commercial cocoa production developed slowly despite a series of subsidies and concessions introduced by the Australian Government during the 1920s. Low copra prices during the Depression also encouraged plantation owners to diversify into cocoa production. The industry was severely affected by World War II, when about two-thirds of PNG's cocoa trees were destroyed.

Following the war, global demand for cocoa increased substantially and the Department of Agriculture, Stock and Fisheries promoted expansion of the industry. Copra plantation owners were encouraged to interplant cocoa with coconut, and soldier settlement cocoa estates (for Australian ex-servicemen) and smallholder blocks were established on the Gazelle Peninsula in East New Britain Province, around Lae in Morobe Province and around Popondetta in Oro Province. These efforts resulted in a major expansion of the industry: the total area planted to cocoa increased from about 3700 ha in 1951/52[1] to around 49 500 ha in 1965/66. Total production increased from 485 tonnes to around 15 500 tonnes over the same period (Figure 5.5.1, Table A5.5.1). The expansion occurred mostly in the plantation sector, which accounted for about 95% of all cocoa produced in the year 1965/66. However, after 1965 smallholder cocoa production increased significantly and by 1979 had surpassed plantation production (Figure 5.5.2, Table A5.5.2). This expansion in smallholder production was concentrated in East New Britain and Bougainville provinces, and these two provinces have continued to dominate the smallholder sector (see below).

---

1 Dates expressed as '1951/52' (for example) represent a cocoa year, which runs from October to September.

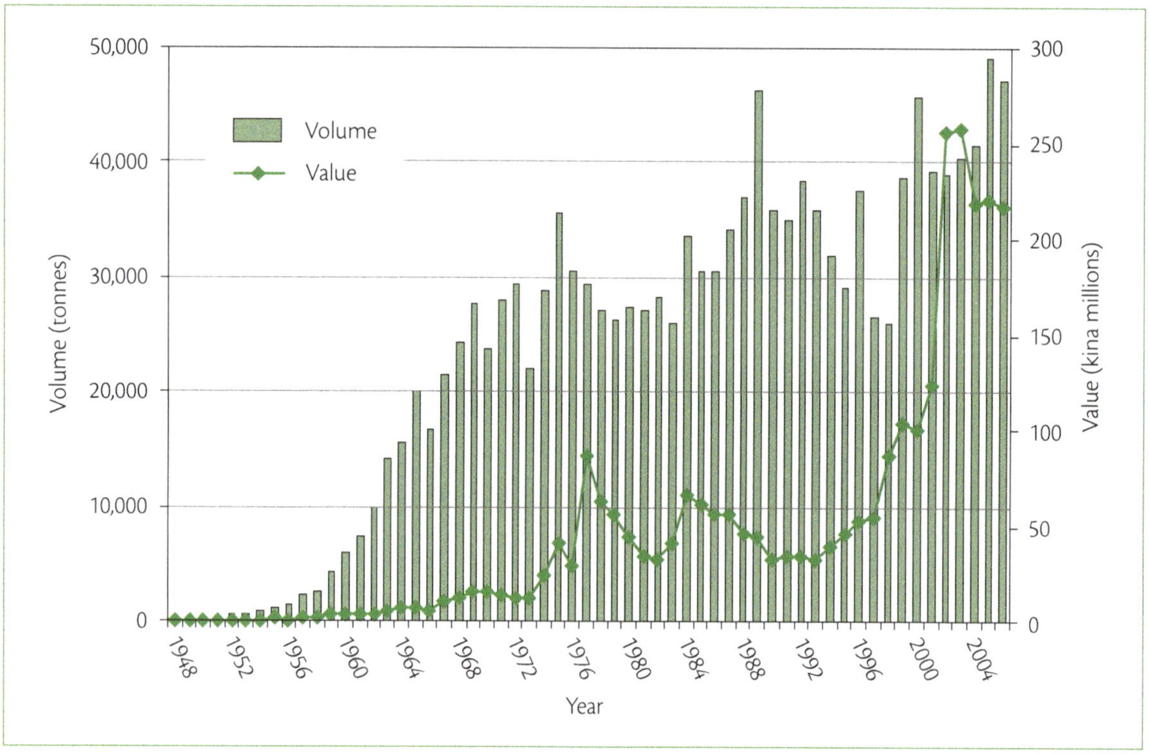

Figure 5.5.1 Volume and value of cocoa exports, 1948–2006. Sources: 1948–1978: Bureau of Statistics, Konedobu; 1979–2006: Cocoa Board of PNG.

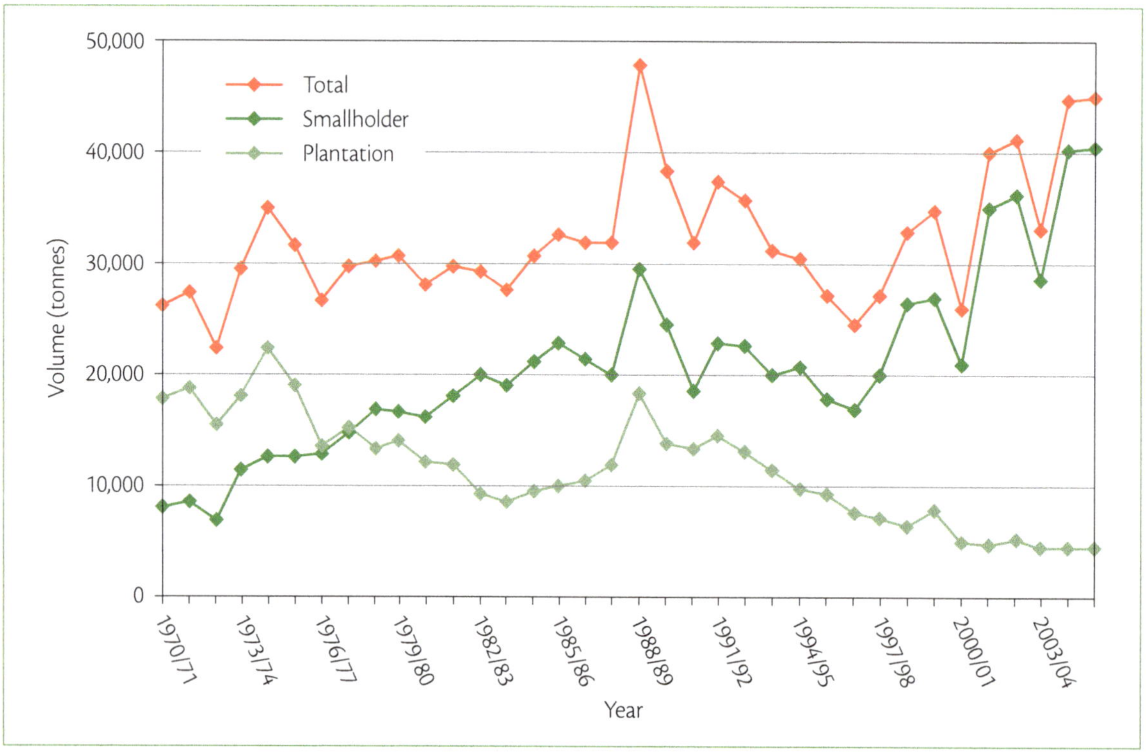

Figure 5.5.2 Cocoa production by sector, 1970/71–2005/06. Sources: 1970/71–1975/76: Department of Primary Industry; 1976/77–2001/02: Cocoa Board of PNG.

Distribution of production and planting

Most PNG cocoa is produced in the north-east lowlands of the Gazelle Peninsula in East New Britain Province and on north-east Bougainville Island (Figures 5.5.3, 5.5.4, Table A5.5.3). Other provinces that contributed to cocoa production in 2006 were East Sepik, Madang, New Ireland, Sandaun, West New Britain and Morobe.

In 2006, smallholders accounted for 90% of national cocoa production, with the plantation sector contributing the remainder. This reflects the continued decline of the plantation sector over the past 30 years (Figure 5.5.2).

There are no accurate estimates of the land area currently planted to cocoa in PNG. Based on production of 40 500 tonnes from smallholders and 4500 tonnes from plantations in 2006, and average yields of 0.25 tonnes/ha and 0.5 tonnes/ha respectively, the area devoted to smallholder cocoa is probably about 160 000 ha and for plantations, 9000 ha. The area of plantation cocoa in 1973 was 55 000 ha.

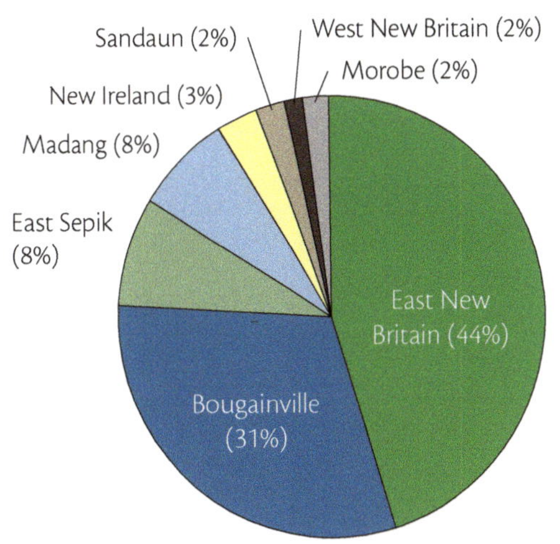

Figure 5.5.3 Cocoa production by province, 2006 (by volume). Source: Cocoa Board of PNG.

Levels of production[2]

The expansion of the plantation sector in the post-war period, coupled with the later increase in smallholder production, saw the overall production of cocoa, measured as total exports, reach 35 500 tonnes in 1975 (Figure 5.5.1). Production then declined in the late 1970s and remained stagnant until the mid 1980s. Smallholder production increased steadily during this period and the decline in production after 1975 occurred in the plantation sector. Factors influencing the decline of the plantation sector included decreases in both yield and production on formerly expatriate-owned plantations that had been returned to local villagers; uncertainties associated with the Land Acquisition Act; and the increasing age of cocoa trees.[3]

The continued growth of the smallholder sector, coupled with a slight recovery in the plantation sector, saw exports reach 46 000 tonnes in 1989 (Figure 5.5.1, Table A5.5.1). During the 1970s and 1980s, Bougainville produced about half of PNG's cocoa. The Bougainville civil war of 1989–1997 caused the closure of plantations and a marked decline in smallholder activity in that province. Production there fell from 18 600 tonnes in 1989, to 7500 tonnes in 1990, and remained in the range 3000–5000 tonnes from 1991 until 2001. National cocoa exports were affected by the decline in Bougainville production, but increased production in East New Britain from 1989 onwards reduced the impact and that province replaced Bougainville as the leading cocoa producer (Figure 5.5.5).

2 There are some large discrepancies between the national cocoa export figures and the national cocoa production figures, with the export figures frequently being higher than those for production. We are unable to account for these differences.

3 The plantation sector experienced some difficulties during the mid 1960s, including the outbreak of vascular streak die-back disease in East New Britain Province, and the devastating effect of weevil borers (*Pantorhytes szentivaiji* and *P. plutus*), which saw the abandonment of large areas of cocoa plantings in the Lae and Popondetta areas. To some extent, the expansion of new plantings in Bougainville and New Ireland provinces compensated for these difficulties.

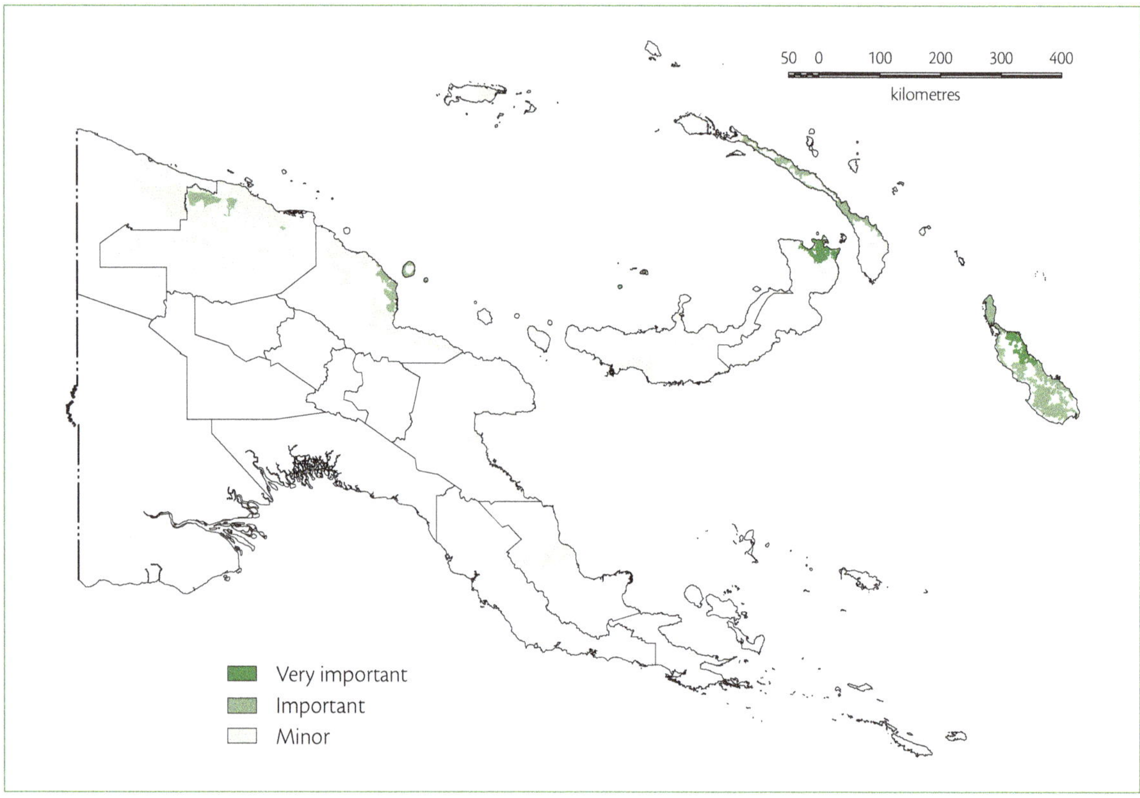

Figure 5.5.4 Locations where cocoa sales provided income for rural villagers, 1990–1995. Source: MASP.

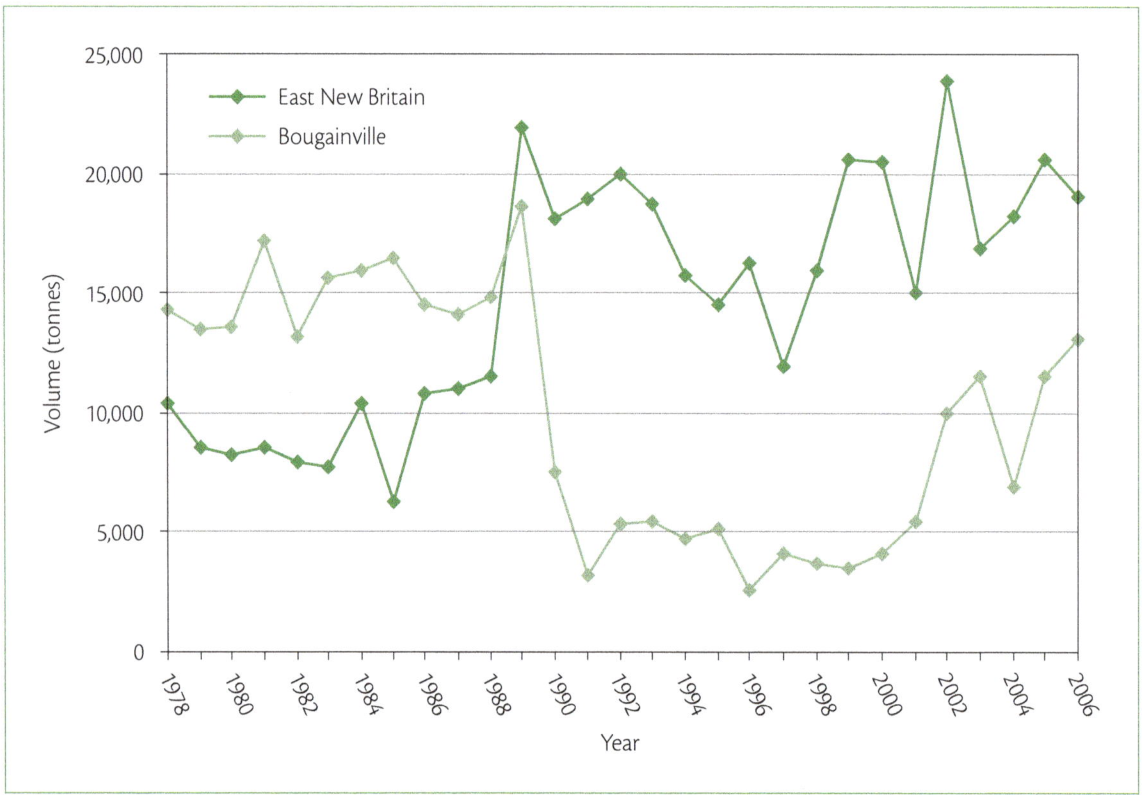

Figure 5.5.5 Cocoa production in East New Britain and Bougainville provinces, 1978–2006. Source: Cocoa Board of PNG.

National production remained relatively low throughout the early to mid 1990s. Three factors contributed to depressed levels of production during this period. First was the continued slump in production in Bougainville Province, particularly in the plantation sector where production had essentially fallen to zero. Second was the decline in the plantation sector in general, due to rising costs of production associated with the devaluation of the kina and increasing wages, and also to land disputes on the Gazelle Peninsula. Third was the volcanic eruptions near Rabaul in 1994, and the drought in 1997, both of which adversely affected production.

Exports increased again from 1999 and reached a peak of 49 000 tonnes in 2005. This recovery was associated with smallholders' response to higher export prices and increased production in Bougainville Province. The recovery in Bougainville was greatly assisted by aid-funded rehabilitation programs and road rehabilitation. Among the smaller cocoa-growing provinces, production has increased in East Sepik, Sandaun and Madang provinces over the past decade. In contrast, it has declined in New Ireland, West New Britain, Manus, Oro, Milne Bay and Central provinces (Table A5.5.3).

Mean smallholder yield of dried cocoa is typically 200–400 kg/ha, and up to 600 kg/ha. The Cocoa and Coconut Institute uses 250 kg/ha as a working average for smallholder yield (Table 5.5.1). Mean yields on plantations are higher at 400–600 kg/ha and are as high as 1500 kg/ha on the best plantations. Much published and unpublished data exist on cocoa yields under experimental conditions, mostly from Keravat and Tavilo in East New Britain Province, but these are not reviewed here.

Most smallholder cocoa trees are more than eight years old and are characterised by a lack of pruning.[4] A lack of maintenance of cocoa leads to a high incidence of pests and diseases and under-harvesting. There is considerable scope for increasing yield per hectare of smallholder cocoa through a number of relatively simple means without expanding plantings. These measures include pruning trees, shade management, and control of the most important pests and diseases.[5]

Processing, exporters and markets

About 2500 cocoa dealers licensed to buy wet bean from producers were registered in PNG in 2003. Dealers process the bean (which involves fermentation and drying) and sell it to exporters. Around 5500 licensed fermentaries were operating throughout the country, with more than half in East New Britain Province and many in villages. More than twenty cocoa exporters were registered but not all were active. In 2005, Agmark Pacific was the major cocoa exporter, accounting for 70% of the market. The next largest exporters were Sepik Coastal Commodities (16%) and Outspan Limited (6%).

In 2006, about 27% of cocoa produced in PNG was exported to the United States, 17% to Belgium, 16% to Malaysia, 16% to Singapore, 11% to Indonesia, and smaller volumes to Thailand and Germany (Figure 5.5.6, Table A5.5.4). The United States and Singapore have been the main export destinations for PNG cocoa since the early 1990s.

Future prospects

The very high prices of the late 1970s and early 1980s resulted in large increases in cocoa plantings worldwide followed by a huge increase in global production over the next decade. The creation of large cocoa stocks dominated the market for more than 10 years, leading to a price slump of unprecedented duration. PNG's smallholder cocoa sector survived the prolonged depressed market essentially intact. A competitive marketing system and the low value of the kina guaranteed a modest regular income for growers that could not be provided by any alternative income source.

4 Plantings in Bougainville Province tend to be younger, with extensive replanting following the civil war.

5 See Curry et al. (2007) for detailed suggestions on how to increase yield per hectare of smallholder cocoa.

Table 5.5.1 Average yield of smallholder and plantation cocoa (kg/ha dry bean)

Survey year	Average yield (kg/ha)	Location	Source
Smallholder			
1961	290	East New Britain	Salisbury (1970:356)
1973	310	East New Britain, Bougainville, Madang	Godyn (1974b:1)
1989	320	East New Britain[a]	Yarbro and Noble (1989:1,15)
1989	170	East Sepik	Yarbro and Noble (1989:1,15)
1989	100	North coast Madang Province	Yarbro and Noble (1989:1,15)
1989	80	Karkar Island, Madang Province[b]	Yarbro and Noble (1989:1,15)
1989	320	Oro[a]	Yarbro and Noble (1989:1,15)
1989	200	Mean all respondents (5 locations)[c]	Yarbro and Noble (1989:1,15)
1994	350	East New Britain	George (1994) cited by Curry et al. (2007)
1998	400	East New Britain	Omuru et al. (2001) cited by Curry et al. (2007)
1999	620	East New Britain	Omuru (2001) cited by Curry et al. (2007)
2000	250	'Average smallholder'	Levett (2000:54)
Plantation			
1971–72	560	East New Britain, Bougainville, Madang	Godyn (1974a:23)
1972–73	440	East New Britain, Bougainville, Madang[d]	Godyn (1974a:23)
1996	590	Various	Omuru (1997) cited by Levett (2000:53)
2000	500–600	'Average plantation'	Levett (2000:54)

[a] Yarbro and Noble (1989) note that yield data for East New Britain and Oro provinces are probably the most reliable.

[b] Low yields on Karkar Island reflect young and newly bearing tree stock at the time of the survey.

[c] Yields (kg) per bearing tree were: 0.9, 0.7, 0.7,1.0 and 1.1 for the five locations surveyed by Yarbro and Noble, with an overall mean of 0.8 kg/tree.

[d] Godyn (1974a:23) notes that yields in 1972–73 were reduced by drought. In 1972–73 the highest production was more than 1500 kg/ha on one plantation, and less than 250 kg/ha on another 12 plantations. Yields per hectare tended to increase with plantation size.

In 2002, when civil unrest began to affect Côte d'Ivoire production (the world's largest cocoa exporter), prices began to move up sharply and have remained at reasonably firm levels since then. Medium-term prices will depend on the political developments in Côte d'Ivoire. In late 2007 the World Bank projected an average price of about US$1.90/kg during 2008 and slightly lower in 2009. Longer term real prices are then projected to decline slightly as supply increases more rapidly than demand (Figure 5.2.6). Even at these prices, PNG smallholders will receive a good return to labour for planting, maintaining and harvesting cocoa (see Section 5.20). However, whether these prices are sufficient to bring about investment in new cocoa plantings remains to be seen.

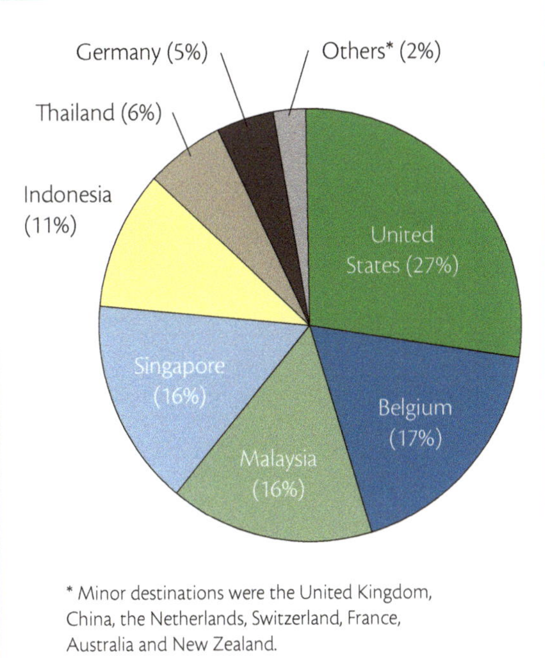

Figure 5.5.6 Destination of cocoa exports, 2006 (by volume). Source: Cocoa Board of PNG.

The PNG Government has set a target of 100 000 tonnes of cocoa exports by 2016. Land in suitable growing environments (fertile soil and rainfall less than 3500 mm/year) is limited. Hence significant expansion will have to come from higher yields per tree rather than an increase in the area devoted to cocoa plantings. Cocoa pod borer, a serious insect pest, has the potential to cause a severe reduction in PNG cocoa production, as it has in Indonesia, Malaysia and the Philippines. It appeared in the Keravat area in East New Britain in early 2006 and has spread to other locations on the Gazelle Peninsula and has been found elsewhere since then.[6]

Further significant increases in export volume will depend on increasing smallholder yield per hectare. The best options for increasing the value of cocoa exports are through improving quality and entering into speciality niche markets.

6 Cocoa pod borer has been detected in the Aitape area of Sandaun Province, from where it could spread to cocoa-growing areas in East Sepik and Madang provinces. It is presumed to have been introduced on logging equipment imported from Malaysia that was not properly cleaned or inspected.

Sources

Bourke, R.M. and Betitis, T. (2003). Sustainability of agriculture in Bougainville Province, Papua New Guinea. Land Management Group, Department of Human Geography, Research School of Pacific and Asian Studies, The Australian National University, Canberra.

Curry, G.N., Koczberski, G., Omuru, E. and Nailina, R.S. (2007). *Farming or Foraging? Household Labour and Livelihood Strategies Amongst Smallholder Cocoa Growers in Papua New Guinea*. Studies in Australia, Asia and the Pacific. Black Swan Press, Perth.

Densley, D.R.J. and Wheeler, M.A. (c. 1978). Cocoa. In Densley, D.R.J. (ed). *Agriculture in the Economy: A Series of Review Papers*. Volume 2. Department of Primary Industry, Konedobu. pp. 1–46.

George, D. (1994). Socio-economic survey of smallholder cocoa growers. Agricultural Research and Extension Project. Asian Development Bank, Manila.

Godyn, D.L. (1974a). An economic survey of cocoa in Papua New Guinea: Part I Introduction and Part II Plantation survey. Department of Agriculture, Stock and Fisheries, Port Moresby.

Godyn, D.L. (1974b). An economic survey of cocoa in Papua New Guinea: Part III Village cocoa. Department of Agriculture, Stock and Fisheries, Port Moresby.

Levett, M.P. (2000). Papua New Guinea rural development sector study 2000. Agriculture working paper. Draft final report. Australian Agency for International Development and the World Bank.

Ministry of Agriculture and Livestock (2007). National Agriculture Development Plan 2007–2016. Volume 2: Implementation plan. Ministry of Agriculture and Livestock, Port Moresby.

Omuru, E. (1997). Cocoa and coconut plantation cost of production survey 1996. Economic Bulletin No. 1. PNG Cocoa and Coconut Research Institute, Rabaul.

Omuru, E. (2001). An estimate of smallholder cocoa and coconut yield profiles and cost of production in Papua New Guinea. Occasional Paper 3: Understanding the smallholder cocoa and coconut sector in Papua New Guinea. PNG Cocoa and Coconut Research Institute and University of New England, Keravat and Armidale.

Omuru, E., Nailina, R. and Fleming, E. (2001). A socioeconomic baseline survey of cocoa and copra smallholders in East New Britain. Occasional Paper 1: Understanding the smallholder cocoa and coconut sector in Papua New Guinea. PNG Cocoa and Coconut Research Institute and University of New England, Keravat and Armidale.

Salisbury, R.F. (1970). *Vunamami: Economic Transformation in a Traditional Society*. Melbourne University Press, Berkeley.

World Bank (2007). Prospects for the Global Economy. Commodity markets briefs – Cocoa <http://go.worldbank.org/G37PYVHE00>. Accessed January 2008.

Yarbro, S. and Noble, S. (1989). Smallholder production, processing and marketing of cocoa and copra in Papua New Guinea: a baseline survey. Designing Monitoring Systems for Smallholder Agriculture in Papua New Guinea, Working Paper No. 10. Department of Human Geography, Research School of Pacific Studies, The Australian National University, Canberra.

5.6 Copra and copra oil

Copra is an important source of village income (Figure 5.1.1). In the early to mid 1990s, an estimated 527 000 people (17% of the rural population) lived in households where cash was earned from selling copra (Table 5.1.1, Figure 5.1.2). From 2004 to 2006 copra and copra oil generated average annual export earnings of K93 million; this was only 6% of the total value of agricultural exports in this period (Figure 5.2.2). Most of this amount (85%) was earned from copra oil exports.

Three economic products are derived from the nut of the coconut palm: copra, copra oil and copra meal. Although coconut will regenerate naturally from seed in coastal locations, almost all coconut palms in PNG have been planted by people. In PNG, coconut is grown in environments where mean annual rainfall ranges from 1000 mm to 6500 mm. It is cultivated from sea level to 1000 m altitude; however, the commercial cultivation of coconut is mostly restricted to coastal locations. Coconut normally bears all year round, but production falls significantly during droughts.

Adoption and history

Although the cultivation and use of coconut in PNG long predates European settlement, the commercial cultivation of coconut palms in PNG commenced in the 1880s, in the Gazelle Peninsula area of East New Britain Province. Coconut meat, dried to copra, was initially in demand for soap manufacture and later for margarine production. Plantation development expanded quickly throughout the New Guinea Islands Region because of high copra prices during World War I. Commercial coconut planting commenced in the coastal areas of Southern Region in 1907, after Australia took over the administration of Papua from the British Colonial Office.

The production and export of copra increased rapidly during the first decades of the twentieth century. Export volumes were recorded as 10 324 tonnes in 1909/10 from a planted area of 16 000 ha, 31 500 tonnes in 1921/22 and 91 500 tonnes in 1936/37 (Figure 5.6.1, Table A5.6.1). Copra was the most important export commodity from PNG during this period. In 1921/22 it contributed 90% of all exports. Copra production and export declined significantly during World War II due to very low prices and the disruption of trade and commerce. Production and exports returned to pre-war levels in the early 1950s when copra exports comprised around 70% of total exports. However, the relative importance of copra and copra oil exports has declined greatly since then (Figure 5.2.3). Until the late 1950s, most copra produced in PNG was grown on plantations. Smallholders produced an estimated 20% of copra in 1954/55.

From the 1950s the Australian Administration adopted policies to develop smallholder and village copra production. Increased extension activities resulted in the establishment of an estimated 75 000 ha of smallholder and village coconut palms

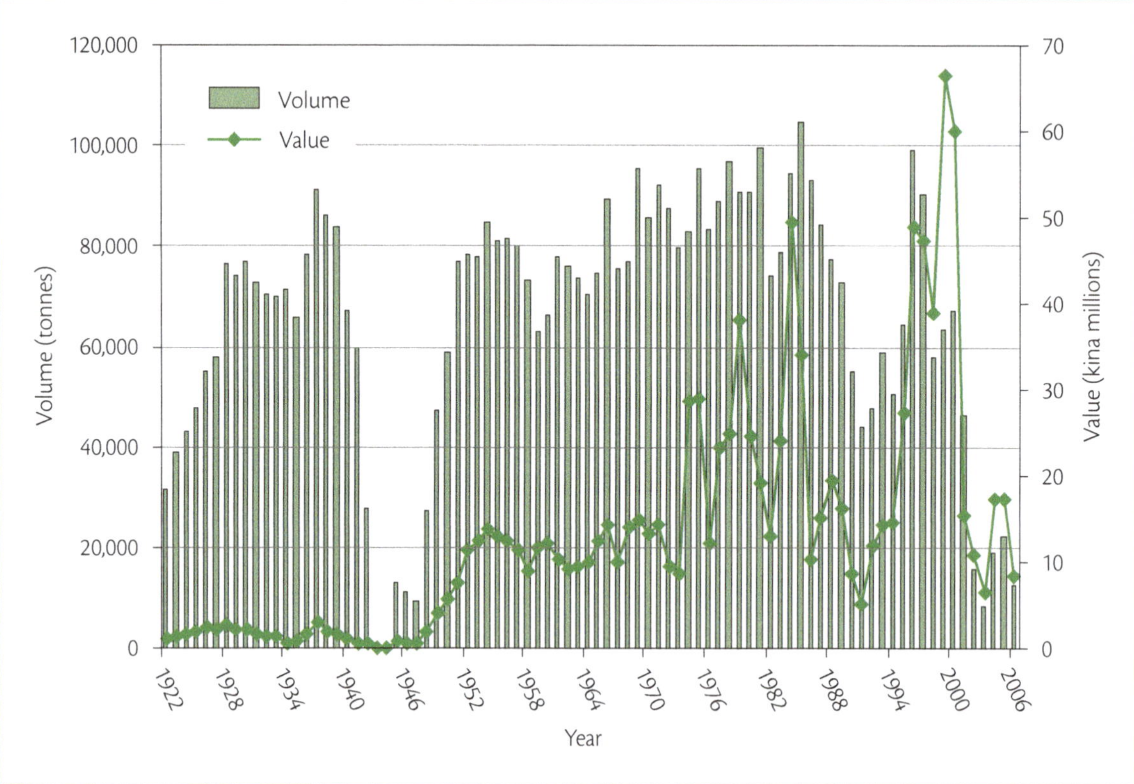

Figure 5.6.1 Volume and value of copra exports, 1922–2006. Sources: 1922–1951: Bureau of Agricultural Economics (1953); 1952–1970: Wheeler et al. (c. 1978); 1971–1978: Bank of PNG; 1979–1989: DAL (1992); 1990–2006: Bank of PNG.

between 1955 and 1965. Most of this development was on village land, not formal settlement schemes. Smallholders produced about 48 500 tonnes of copra in 1971, which was around 35% of national copra production. By 1975/76, smallholders contributed more than 40% of national production, a reflection of both the decline of the plantation sector and the expansion of the smallholder sector. By 1975/76 however, copra and copra oil comprised only 5% of PNG's total exports. This decline in export share was primarily due to the development of the coffee, cocoa and mineral industries during the 1950s, 1960s and 1970s.

Distribution of production and planting

Smallholder copra production is dominated by East New Britain Province, which contributes around one quarter of total production (Figure 5.6.2). Other provinces that make significant contributions to smallholder production are Madang, New Ireland, Bougainville and West New Britain (Figure 5.6.3). Most plantation copra is produced in Madang and East New Britain provinces (particularly in the Gazelle Peninsula area). In 2005, East New Britain Province accounted for 46% of all copra produced and Madang Province 20%. Smaller amounts came from New Ireland, Bougainville and West New Britain provinces (Figure 5.6.4, Table A5.6.2).

It was estimated in 1998 that coconut plantations covered approximately 53 000 ha and smallholder plantings about 128 000 ha. In a large proportion of these areas coconut is interplanted with cocoa, an innovation that was first pioneered in PNG in the 1950s and that increases overall economic productivity of land under coconut for both plantation and smallholder producers. Most coconut palms in PNG are of the 'tall' variety and many are now aged, which reduces productivity. Age is thought to be a greater problem in the plantation sector, where approximately half of coconut palms are 70–80 years old.

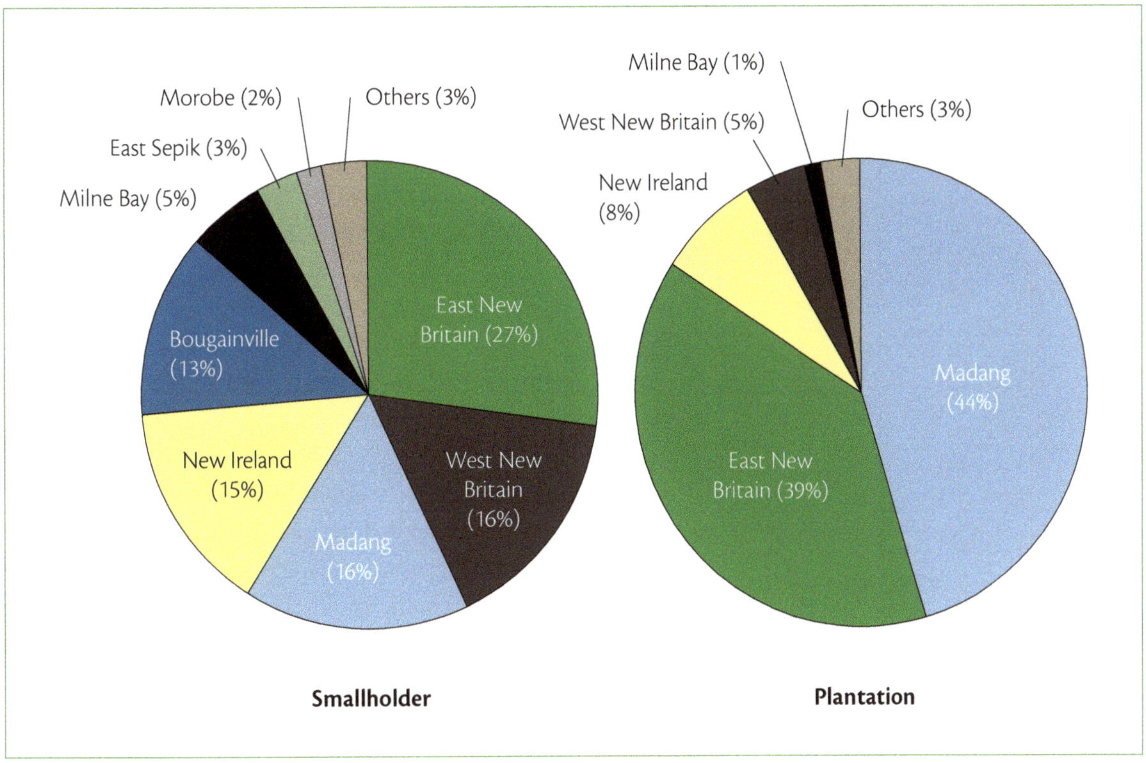

Figure 5.6.2 Copra production by sector and province, 2000 (by volume). Source: Kokonas Indastri Koporesen.

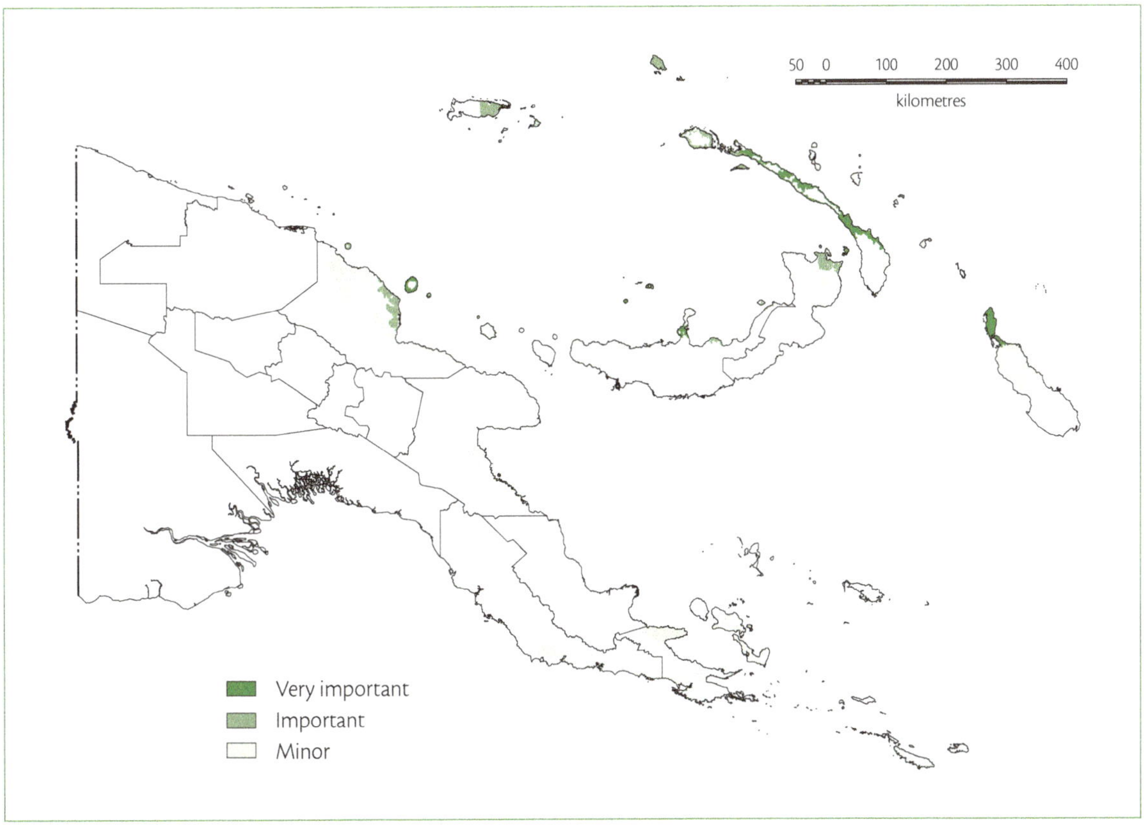

Figure 5.6.3 Locations where copra sales provided income for rural villagers, 1990–1995. Source: MASP.

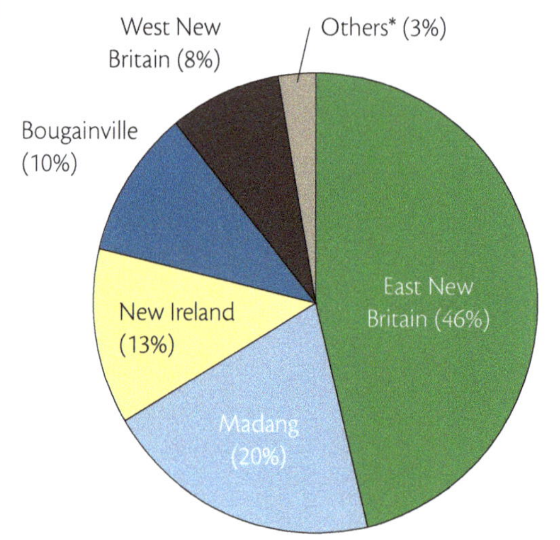

* 'Others' were Milne Bay, Morobe, East Sepik and Manus.

Figure 5.6.4 Copra production by province, 2005 (by volume. Source: Kokonas Indastri Koporesen.

Levels of production

Since the 1970s, the smallholder sector has grown in importance relative to the plantation sector. In 1988, smallholders contributed around 70% of total production, and by 1998 this proportion had increased to 82%. The plantation sector has continued to be adversely affected by extreme fluctuations in world market prices and, more importantly, by the rising costs of inputs, particularly fuel and labour. The sector has also been constrained by investment uncertainty (particularly as a result of the Plantation Redistribution Scheme in the 1970s, which bought back plantation land from owners and returned it to the previous customary owners).

Smallholder producers are sensitive to variations in the export prices paid for copra and copra oil and this largely explains the peaks and troughs that have characterised PNG production levels from the 1970s to the 1990s (Figures 5.6.1, 5.6.5). Very low prices

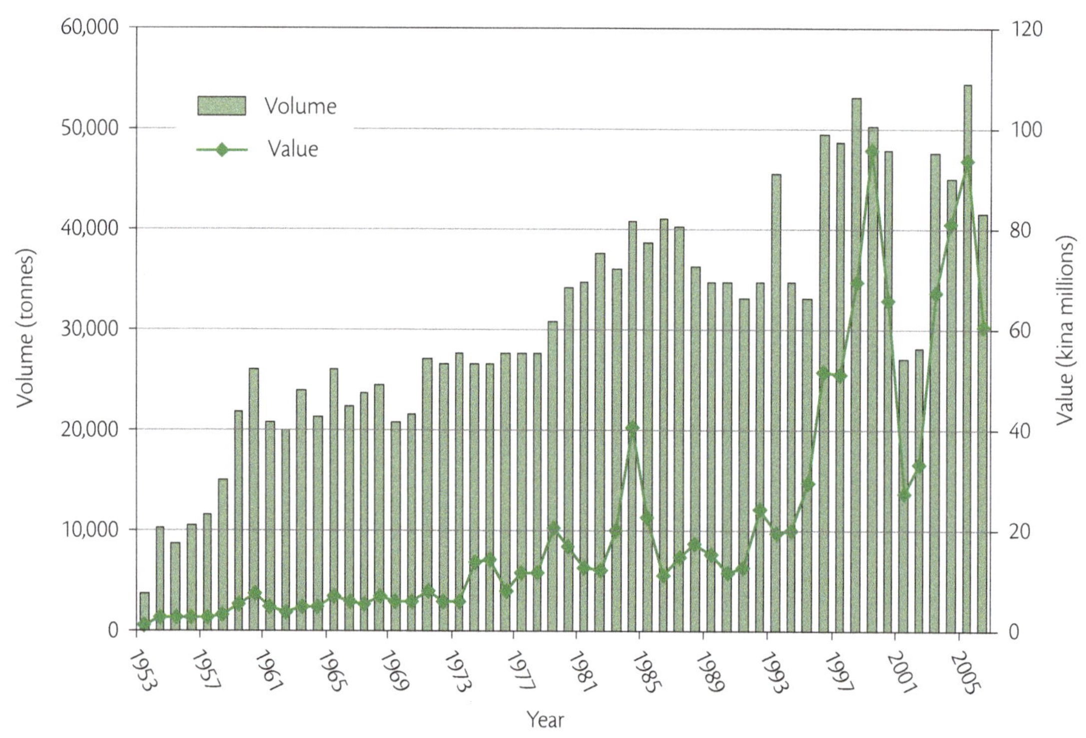

Figure 5.6.5 Volume and value of copra oil exports, 1953–2006. Sources: 1953–1970: Wheeler et al. (c. 1978); 1971–1978: Bank of PNG; 1979–1989: DAL (1992); 1990–2006: Bank of PNG.

between 1985 and 1995 saw copra production reach its lowest levels since the late 1940s (despite K35.4 million of government price assistance and stabilisation funding over the period 1990–95). However, price increases during the 1990s saw production increase again.[1]

The marked decline in copra production in 1998, despite the fact that prices were still increasing, was associated with the opening of a new Copra Marketing Board copra oil mill in Madang. Since the 1960s, the proportion of copra being processed domestically into copra oil has steadily increased. By 1990, copra oil surpassed copra in total export value (Tables A5.6.1, A5.6.3).

Copra production was also adversely affected by significant declines in export prices in 2000 and 2001. In 2002, copra production fell to its lowest level since 1947 and in 2003 copra production reached a historical low of less than 9000 tonnes. Deteriorating infrastructure and increasing transport costs, fewer purchasing depots, and a switch from exporting copra to processing it into oil within PNG have contributed to this decline.

Copra oil production fell in 2001, but recovered in 2003. In 2006, the value of copra oil exports (K60 million) was more than seven times that of copra exports (K8 million).

Copra meal, a low-value by-product of copra oil production, is exported for stockfeed. Copra meal exports averaged 14 000 tonnes/year during the 1990s, with an average value of K1.6 million/year.

Average smallholder yields are typically in the range 400–700 kg of copra/ha, with a mean of about 500 kg/ha (Table 5.6.1). Like other export tree crops, plantation copra yields are higher than those for smallholders, with a range of 700–1000 kg/ha in nationwide surveys and an overall average of about 900 kg/ha.[2]

1 Kina prices of both copra and copra oil trebled between 1992 and 1998 due to a combination of the devaluation of the kina and increasing world market prices.

2 An average full-sized coconut weighs about 1.44 kg. About 6000 full-sized nuts will make one tonne of copra; 1000 full-sized nuts will make 170 kg of copra, which will yield 70 litres of copra oil.

Processing, exporters and markets

Copra production is labour intensive. It involves collecting fallen coconuts, de-husking and splitting them, and removing and drying the coconut meat. Nowhere in the world has coconut harvesting been mechanised. The Copra Marketing Board (CMB), which became the Kokonas Indastri Koporesen (KIK) in 2002, has always regulated the marketing and export of copra in PNG. The CMB enjoyed a monopoly over all aspects of the copra industry. However, in recent years KIK has issued a limited number of private sector export licences, mostly for copra oil.

Until recently, KIK purchased copra from growers at fixed prices at depots and sub-depots.[3] However, between 2001 and 2005 a significant reduction in the number and geographic extent of active purchasing depots occurred, from 22 depots in 11 provinces in 2001, to 15 depots in 10 provinces in 2002, and 10 depots in 9 provinces in 2005 (Table 5.6.2), a reflection of the shift in copra purchasing activity from CMB/KIK to the copra oil mills. Most copra in the Islands Region is now purchased by Coconut Products in Rabaul. This change has disadvantaged smallholder producers who do not have access to the mills and who were previously serviced by CMB/KIK depots, which have now ceased purchasing.

Two large copra oil mills currently operate in PNG: Copra Oil Production Madang Ltd in Madang and Coconut Products at Toboi near Rabaul.[4] The Toboi mill is over 50 years old and has been extensively refurbished. The Madang mill was set up by KIK, but later sold to a private company.[5] In addition to these two mills, a number of very small operations produce copra oil using direct micro-expelling

3 Three different prices were offered depending on quality. Hot Air grade received the best price, Fair Merchantable Standard received the middle price and Smoke grade copra the lowest price.

4 The Toboi mill has a capacity of around 70 000 tonnes of copra oil per year, while the Madang mill has a capacity of about 30 000 tonnes per year.

5 Coconut Oil Production Madang Ltd is currently owned by a New Zealand-based company, Sleepyhead Manufacturing Co Ltd, a mattress manufacturer that has diversified into food production.

Table 5.6.1 Average yield of smallholder and plantation copra (kg/ha)

Survey year	Average yield (kg/ha)	Location	Source
Smallholder			
1961	480	East New Britain Province	Salisbury (1970:356)
1971–74	500	All PNG	Wheeler et al. (c. 1978:12)
1973	400–700	All PNG	Sackett and Williamson (1973) (cited by Wheeler et al. c. 1978:12)
1989	510	East New Britain Province	Yarbro and Noble (1989:22)
1989	570	East Sepik Province	Yarbro and Noble (1989:22)
1989	580	North coast Madang Province	Yarbro and Noble (1989:22)
1989	1210	Karkar Island, Madang Province	Yarbro and Noble (1989:22)
1989	700	Mean all respondents (4 areas)[a]	Yarbro and Noble (1989:22)
Plantation			
1951	710	All PNG	Bureau of Agricultural Economics (1953:38)
1971–72	950	All PNG	Wheeler et al. (c. 1978:12)
1972–73	910	All PNG	Wheeler et al. (c. 1978:12)
1973–74	800	All PNG	Wheeler et al. (c. 1978:12)
1973	1000	All PNG	Sackett and Williamson (1973) (cited by Wheeler et al. c. 1978:12)
1999	800	All PNG	Fleming (1999) cited by Levett (2000:58)

[a] Yield per bearing palm was 7.0, 6.2, 5.0 and 11.6 kg/palm in the four survey areas, with an overall mean of 7.7 kg/palm (Yarbro and Noble 1989:22).

(DME) technology. The Middleton family operation on Karkar Island, Madang Province, produces high-quality copra oil that is used locally to make a number of products including soap, cosmetics and shampoo.

In 2001, 55% of the copra produced in PNG was exported to Europe, particularly to the United Kingdom and the Netherlands; 40% was exported to Japan; and 5% to Singapore (Table A5.6.4). These countries were the major export destinations for PNG copra throughout the 1990s. In 2001 a major shift in the destination of PNG copra took place, with more going to Europe and less to Japan. In 2005, most copra was being exported to Germany (83%) and the remainder to Australia (8%), Singapore (7%), Solomon Islands and India (Figure 5.6.6). This shift in the destination of exports occurred at the same time as the rapid decline in copra export volumes, and fluctuations in the figures need to be viewed in that context. The main export destination for copra oil in 2005 was Australia, with Europe a minor destination (Figure 5.6.6). By 2008, most copra in PNG was being processed into copra oil and exported to Europe.

Future prospects

For many households, copra provides the only source of cash income, but the PNG copra industry is in crisis, with production falling sharply in recent years. Internationally, the PNG copra industry is at best only marginally competitive. Copra oil prices declined for several decades until 2001 when the price for copra increased and reached US$1130 per tonne

Table 5.6.2 Copra purchases by depot, 2005

Depot	Province	Copra purchases (tonnes)	Purchasing agent
Alotau	Milne Bay	1,296	Kumpa Resources Limited & Allen Enterprises
Madang	Madang	20,138	Coconut Oil Production Madang Ltd and Bogia Cooperative
Lae	Morobe	752	Agmark Pacific Ltd
Kavieng	New Ireland	7,327	Coconut Products & NACA
Toboi	East New Britain	46,276	Coconut Products
Wewak	East Sepik	664	Sepik Coffee Agmark JV Limited
Buka	Bougainville	10,161	Coconut Products
Kimbe	West New Britain	7,908	Kimbe Bay Shipping, Island Copra Traders Ltd
Namatanai	New Ireland	5,163	Coconut Products
Manus	Manus	170	Silan Limited
Total purchases		99,855	

Source: Kokonas Indastri Koporesen.

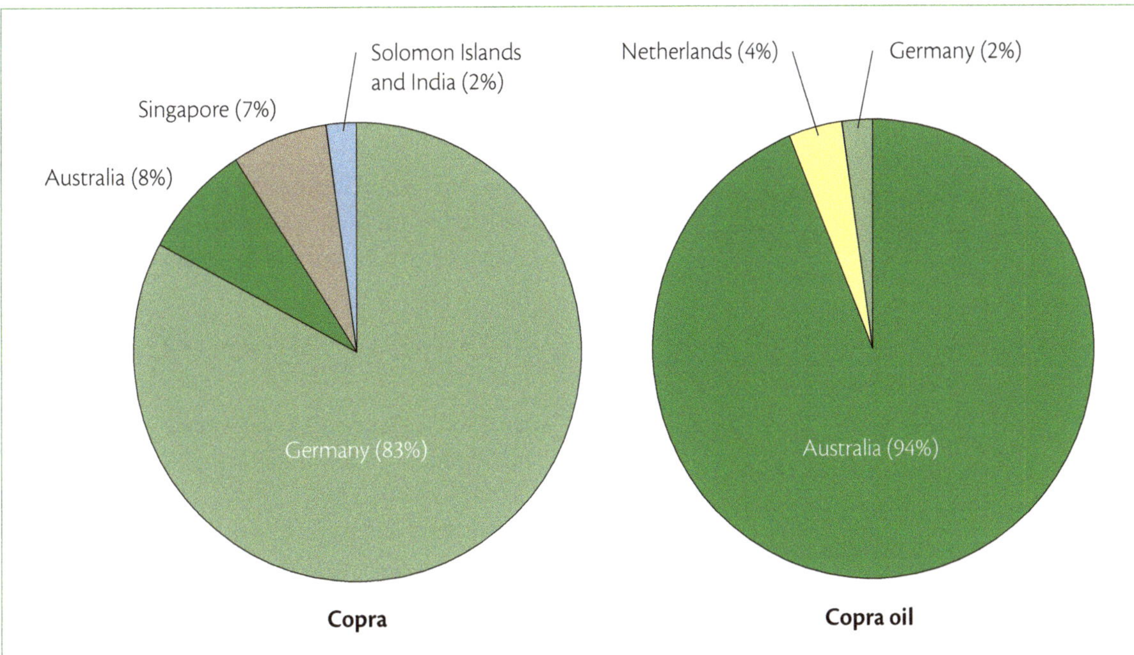

Figure 5.6.6 Destination of copra and copra oil exports, 2005 (by volume). Source: Kokonas Indastri Koporesen.

in late 2007, the highest nominal price since 1984. Price increases have been brought about by structural changes in the world copra oil market. These have been driven by the diversion of other edible oils, particularly palm oil, away from their traditional uses into the rapidly expanding biofuel market. Nevertheless, the World Bank forecasts a significant fall in real copra oil prices through to 2015

(Figure 5.2.6). When prices are poor, the comparatively low yields of PNG copra make it uncompetitive. The once vibrant trade in copra at the village level is now stifled by the closure of buying depots, a deterioration in roads and shipping services, increasing transport costs, and the absence of financial services to support copra purchases by private buyers.

The demand for copra is derived from the demand for copra oil and prices of the two products follow each other closely. Europe was the main buyer for copra. The European Union (EU) applied a zero tariff on oilseeds and meals (including copra), but a much higher tariff on vegetable oils (including copra oil). This resulted in a significant trade distortion that encouraged the importation of copra over copra oil and explains why the EU became the world's largest importer of copra.

The World Trade Organization Uruguay Round led to a reduction of the tariff margin between copra crude oil and refined oil. Since then there has been a steady decrease in Europe's copra crushing capacity and an increase in oil imports. The last Europe-based copra crushing operation (Walter Rau in Germany) closed in 2007 and there is no longer any market for copra in Europe. The copra market is now limited to buyers in Bangladesh, the Philippines and mills in other Pacific island countries (Fiji, Solomon Islands and Vanuatu). The increasing interest in copra from these countries is a reflection of supply problems facing their own industries rather than of growing international demand for copra. These markets, while not sustainable in the long term, provide PNG a few years breathing space to develop a copra industry that is based entirely on oil exports.

Beyond conventional copra and copra oil, some high-value coconut products offer better longer term prospects. These include virgin coconut oil, coconut cream, coconut timber and biofuel.

Sources

Bank of Papua New Guinea Quarterly Economic Bulletin Statistical Tables <http://www.bankpng.gov.pg/> [go to Statistics tab].

Bureau of Agricultural Economics (1953). An economic and cost survey of the copra industry in the Territory of Papua and New Guinea. Bulletin No. 9. Department of Commerce and Agriculture, Commonwealth of Australia, Canberra.

Coconut Quality Improvement Project (1995). Coconut. Manual for Papua New Guinea. PNG Cocoa and Coconut Research Institute, Rabaul.

DAL (Department of Agriculture and Livestock) (1992). Exports of important agricultural commodities: quantity, value and average price per unit. Rural Statistics Bulletin 1992/6. Department of Agriculture and Livestock, Konedobu.

Fleming, E. (1999). Estimates of planted areas and their development costs, yields, export prices and volumes, exchange rates and price support in the tree crop industries in Papua New Guinea. Occasional Paper 12. ACIAR Tree Crops Policy Options Project. University of New England, Armidale.

Levett, M.P. (2000). Papua New Guinea rural development sector study 2000. Agriculture working paper. Draft final report. Australian Agency for International Development and the World Bank.

Omuru, E. (2000). Country paper: thirty years of the coconut industry in Papua New Guinea (1969–1999). PNG Cocoa and Coconut Research Institute, Rabaul.

Sackett, M.A. and Williamson, P.G. (1973). The copra industry in Papua New Guinea. Unpublished paper. Department of Primary Industry, Konedobu.

Salisbury, R.F. (1970). *Vunamami: Economic Transformation in a Traditional Society*. Melbourne University Press, Berkeley.

Wheeler, M.A., Sackett, M.A. and Densley, D.R.J. (c. 1978). Coconuts. In Densley, D.R.J. (ed). *Agriculture in the Economy: A Series of Review Papers*. Volume 2. Department of Primary Industry, Konedobu. pp. 1–44.

Yarbro, S. and Noble, S. (1989). Smallholder production, processing and marketing of cocoa and copra in Papua New Guinea: a baseline survey. Designing Monitoring Systems for Smallholder Agriculture in Papua New Guinea, Working Paper No. 10. Department of Human Geography, Research School of Pacific Studies, The Australian National University, Canberra.

5.7 Oil palm

Palm oil has been PNG's most valuable agricultural export since 2000, when it overtook coffee in this role (Figure 5.2.3). Palm oil exports averaged K420 million per year from 2004 to 2006, which was 30% of the value of agricultural exports for that period (Figure 5.2.2). Oil palm production has expanded at a much greater rate than other export tree crops (Table 5.2.2). However, a smaller proportion of the rural population is engaged in growing oil palm than for the other major export and domestically marketed crops. Approximately 130 000 settlers or villagers derived income from selling oil palm in 1995 (4% of the rural population) (Table 5.1.1, Figure 5.1.2). In 2007, the Oil Palm Research Association estimated that about 166 000 people (3% of the rural population) lived in households that produced oil palm. Many other people derive income directly or indirectly from the PNG oil palm industry, including those working on the nucleus estates.

Four economic products are derived from the fruit of the oil palm: crude palm oil, palm kernel oil, refined palm oil and palm kernel expellent.[1] Of these, crude palm oil is the most significant in terms of export volume and value. Oil palm is grown exclusively in lowland locations, up to a maximum altitude of 200 m. It is cultivated in areas where mean annual rainfall ranges from 2000 mm to 4200 mm. The production of palm fruit is mildly seasonal in West New Britain Province, with about 60% of the crop harvested in January to June each year.

Adoption and history

Although oil palm has been grown in PNG since the 1920s, commercial development did not commence until 1967 with the establishment of a private sector/ government joint venture at Hoskins in West New Britain Province (WNB). This is now the largest oil palm development in PNG. Other large projects are at Bialla (WNB), Popondetta (Oro Province), Gurney and Sagarai (Milne Bay Province), along the coast south-east of Kavieng (New Ireland Province) and in the Ramu and Markham valleys in Madang and Morobe provinces (Figure 5.7.1). All these developments are based on a nucleus estate and smallholder (NES) model, in which a commercially operated estate produces oil palm and also provides a market, processing and technical services for smallholder producers who cultivate oil palm on land adjacent to the nucleus estate.

Initially the smallholder component of the NES model was based on a land settlement scheme (LSS) system (see Section 6.7), which granted settlers 99-year leases over blocks of at least six hectares on land purchased from customary owners. However,

1 Crude palm oil is extracted from the fleshy part of the fruit; palm kernel oil is extracted from the kernel; refined palm oil is processed at Kumbango in West New Britain from crude palm oil; and palm kernel expellent is a residual product exported for stockfeed. Palm kernel is not exported, except for planting material.

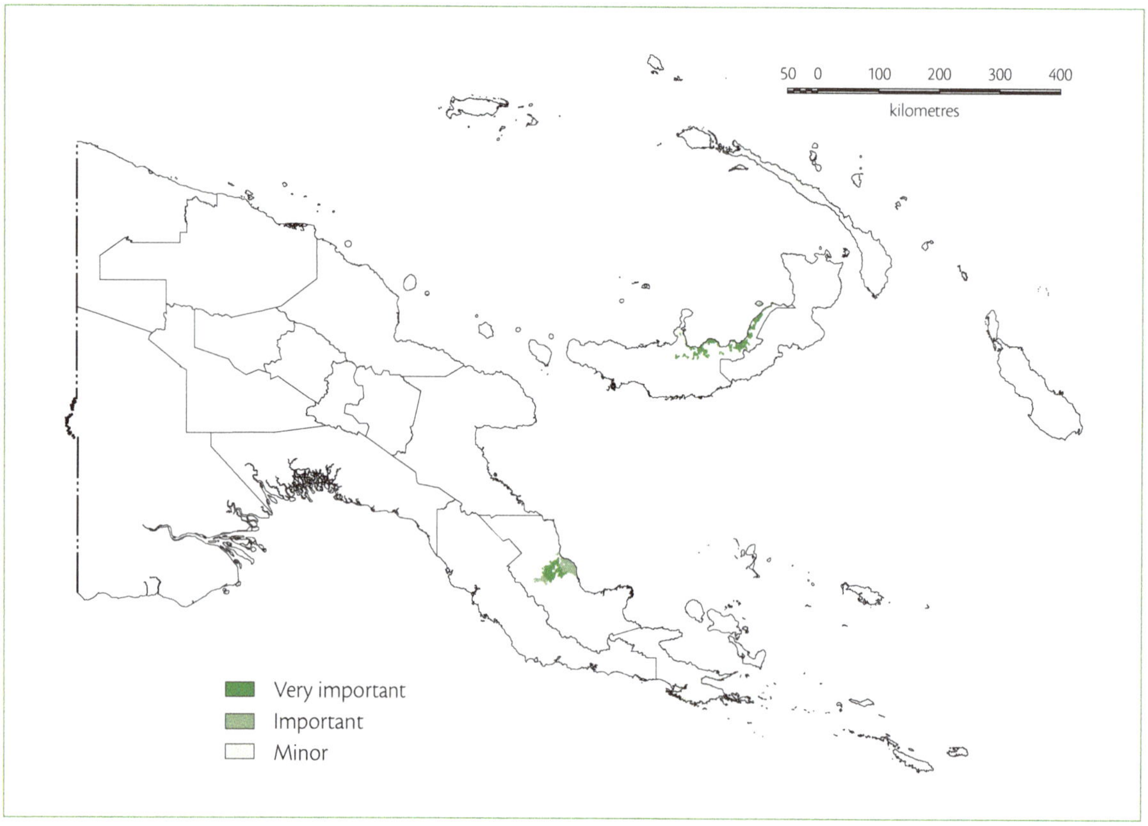

Figure 5.7.1 Locations where oil palm sales provided income for rural villagers, 1990–1995. Source: MASP.

no further land settlement schemes have been undertaken since the mid 1990s due to a shortage of land for further settlement and problems associated with the system. An important issue is the number of people living on blocks of a fixed area. Population density on blocks in the Hoskins LSS has risen from 6 persons/block in the early 1970s to 13 persons/block in 2000, and it is predicted that there will be 20 persons/block by 2011. This rising density is leading to a number of problems, including social instability, conflict over allocation of labour inputs and income, and disputes over inheritance of the blocks.

The existing LSS system has been supplemented by the village oil palm (VOP) system, which provides smallholders with blocks of two or four hectares on customary-owned land, with a Clan Land Usage Agreement giving the blockholder security of tenure and usage rights over the land. The three oldest NES developments, at Hoskins, Bialla and Popondetta, have nucleus estate, LSS and VOP components, while the newer NES developments in New Ireland and Milne Bay provinces have only nucleus estate and VOP components (Table 5.7.1).

With no further land available for land settlement schemes since the mid 1990s, many migrants have entered into informal arrangements to access customary land in the oil palm-growing areas of West New Britain Province. These arrangements are known as customary purchase blocks. Such plantings were providing a growing proportion of fruit from smallholders for the mills in the Hoskins area by 2007.

The latest trend in oil palm development on customary-owned land is for landowning groups to form companies that lease customary land to oil palm plantation companies in exchange for rent and royalties. This system is known as the mini estate system. It is used by New Britain Palm Oil Limited, Higaturu Oil Palms, Milne Bay Estates Limited and Poliamba Limited in West New Britain, Oro, Milne Bay and New Ireland provinces respectively. Most new plantation development since the late 1990s has

Table 5.7.1 Summary statistics for oil palm project areas, 2007

	Hoskins	Bialla	Popondetta	Milne Bay (Alotau)	New Ireland (Lakuramau)	Total
Company	New Britain Palm Oil Ltd	Hargy Oil Palms Ltd	Higaturu Oil Palms	Milne Bay Estates Ltd	Poliamba Ltd	
Area planted (ha)[a]						
Plantation	34,774	8,909	9,009	11,634	5,689	70,015
Smallholder[b]	25,324	14,580	14,285	1,757	2,114	58,060
Total area	60,098	23,489	23,294	13,391	7,803	128,076
Number of blocks[c]						
LSS[d]	2,350	1,851	929	nil	nil	5,130
VOP[d]	4,471	1,593	5,191	536	648	12,439
Production fresh fruit bunch (tonnes)[a]						
Plantation	744,271	180,122	170,206	239,516	117,896	1,452,011
Smallholder	368,729	159,020	162,846	10,376	18,448	719,419
Total production	1,112,999	339,142	333,052	249,892	136,344	2,171,430

[a] Data are from Ian Orrell, Oil Palm Research Association. Figures for plantation areas include mini estates.

[b] Figures for smallholder area do not include oil palm planted on customary purchase blocks, estimated by George Curry and Gina Koczberski to be about 2500–3000 ha in 2007 in the Hoskins area.

[c] Data for Hoskins, Bialla and Popondetta are for 2004/2005 from Curry et al. (2007: Table 1.2). Data for Milne Bay and New Ireland are for 2000 from Koczberski et al. (2001: Table 1.1). The number of VOP blocks in Milne Bay and New Ireland provinces has not significantly increased since 2000.

[d] LSS stands for land settlement scheme; VOP stands for village oil palm.

been on customary land, with extensive plantings in West New Britain Province in particular using this form of land tenure.[2]

Distribution of production and planting

Both smallholder and plantation production are dominated by the Hoskins project area, which, in 2007, contributed more than half of total national production (Figure 5.7.2). Popondetta and Bialla are the next most significant project areas in terms of production, followed by Milne Bay and New Ireland. Although the Hoskins project area accounts for the greatest volume of smallholder production, the Bialla and Popondetta project areas have the largest smallholder components as a proportion of total NES production (Table 5.7.1).

In 2007, 70 000 ha (55% of the total area) was planted to oil palm on plantations (including mini estates) and 58 000 ha planted on smallholdings (distributed between 5100 LSS blocks and 12 400 VOP blocks). In 2007, the Hoskins project area accounted for half of the total area planted to oil palm on plantations (Figure 5.7.3). Hoskins also had the largest area of smallholdings, although Popondetta and Bialla also had significant areas of smallholder oil palm plantings.

2 Customary landowners at Bialla have not entered into mini estate agreements with Hargy Oil Palms Limited. Instead, four Landowner Development Corporations have pooled their resources and are managing their own oil palm plantations with technical support from the company. This arrangement is known as the Community Oil Palm Development (COPD) system.

Levels of production

Exports of crude palm oil have increased exponentially since the early 1970s (Figure 5.7.4, Table A5.7.1) because of an expansion in both smallholder and plantation production (Figure 5.7.5). During the 1980s, smallholder and plantation production of oil palm fruit were approximately equal. However, the establishment of the NES project areas in Milne Bay and New Ireland provinces, which have relatively insignificant smallholder components (4% and 14% of production in 2007 respectively), has caused plantation production to increase at a greater rate than smallholder production since the early 1990s. Plantations continue to dominate production: in 2007 they accounted for two-thirds of total national production.

The volume of fresh fruit harvested has continued to set new records in most years, exceeding two million tonnes for the first time in 2006 (Table A5.7.2). More than 360 000 tonnes of palm oil with an export value of K430 million was extracted from this fruit (Figure 5.7.4, Table A5.7.1). The provisional value in 2007 (K670 million) is much greater, the outcome of particularly high world prices and increased production.

Box 5.7 The Mama Lus Frut Scheme[1]

In 1997, the Oil Palm Industry Corporation, in conjunction with the Hoskins project area and local women's groups, initiated a program for the collection and sale of palm fruit that separates from the bunches during harvesting and transport. Loose fruit is usually very ripe and therefore has considerably higher oil content (about 40%) than the rest of the fresh fruit bunch (about 22%).

Under this scheme, women are issued with their own account by the milling company and the income from the loose fruit that they collect on family blocks is paid directly to their own bank accounts (as distinct from their husband's – or the blockowner's – account).

Between 1997 and 2003, about 4000 women in the Hoskins area, drawn from 88% of all smallholder blocks in that project area, were issued with a harvest card. The loose fruit that drops off the bunch at harvest represents 14% of the total crop, but a greater proportion (about 26%) of smallholder oil palm income is paid directly to women. The additional income consists of loose fruit harvested by the women or by their husbands. In 2003, women collected more than 55 000 tonnes of oil palm and earned an average weekly income of K49. The scheme has increased net household income by a modest 5%. The scheme has also been adopted in the Bialla and Popondetta project areas.

Benefits associated with the scheme include:

- It provides an important source of income for women in an industry that, like other export cash crop industries in PNG, has been traditionally dominated by men.
- The more equitable distribution of income to household members results in fewer conflicts between husbands and wives over money.
- A greater proportion of income from oil palm is spent on family needs.
- There is greater involvement by women in management of the family block.
- There is a more complete harvest resulting in higher overall income for smallholders.
- Provision of a mechanism by which men can contribute to household expense by weighing fresh fruit bunch (which they have harvested) on their wife's card, so that the income goes into the wife's bank account.
- A higher utilisation of fixed milling assets.

1 See Koczberski (2007) for more information on the Mama Lus Frut Scheme.

Yield of fresh fruit bunch per hectare varies between locations and over time. However, accurate comparisons are not possible because data on area planted include both mature and immature palms.[3] Thus yields calculated from the total production and total area planted underestimate yield from mature palms. Nevertheless, available data indicate that smallholder yields are considerably lower than those for plantations (Table 5.7.2). An average yield for mature plantation palms is about 30 t/ha; for settler blockholders it is about 18–20 t/ha; while village oil palm is less again at about 10 t/ha. There are considerable yield data from experimental plantings at Dami in West New Britain Province. Experimental yields are typically 30–32 t/ha and up to 40 t/ha.

Table 5.7.2 Indicative yield of plantation and smallholder oil palm, 2007 (fresh fruit bunch, tonnes/ha)[a]

Project area	Plantation	Smallholder	Total
Hoskins	21.4	14.6	18.5
Bialla	20.2	10.9	14.4
Popondetta	18.9	11.4	14.3
Milne Bay	20.6	5.9	18.7
New Ireland	20.7	8.7	17.5
Total	20.7	12.4	17.0

[a] Yields were derived by dividing production by the area under oil palm in 2007 (Table 5.7.1). The area includes mature and immature palms. The proportion of immature palms is unknown, but it varies between locations and over time. Thus these yield estimates are crude and underestimate yield for mature palms, which is the most reliable statistic to compare yields between sectors or locations and over time.

Source: Ian Orrell, Oil Palm Research Association.

Processing, exporters and markets

Each of the NES project areas has at least one palm oil mill, which process fresh fruit bunch (FFB) produced on both plantations and smallholdings. The FFB is processed to produce crude palm oil and palm kernel. Three of the five project areas also have facilities to further process palm kernel into palm kernel oil and palm kernel expellent.[4]

New Britain Palm Oil Limited (NBPOL), which operates the Hoskins NES, is one of only a handful of oil palm producers in the world that operates commercially viable breeding and germplasm export programs. In 2007, about 4 million seeds were sold by NBPOL. These are supplied to all the plantation operations in PNG and exported to a number of countries, particularly Indonesia, but also to Malaysia, Sri Lanka, Thailand, the Philippines, Honduras, Cameroon, Solomon Islands and Vanuatu. Export of seed generates significant income for the company. Demand from Indonesia is especially large and NBPOL anticipates exporting about 20 million seeds in 2008.

The NES companies maintain a monopoly on purchasing FFB from smallholder producers. Prices are calculated according to a formula devised by the PNG Palm Oil Producers' Association. The formula is based on export prices, and the transport costs and oil palm extraction rates of each NES project area.[5] Prices paid to smallholder producers therefore vary

3 Palms commence bearing within two years of planting and reach maximum yield by about the sixth year. They continue to bear for many decades, but are usually killed by poisoning after 20–25 years when they become too tall to harvest.

4 New Britain Palm Oil Limited opened a palm oil refinery near Kimbe in January 2003. The refinery (which also includes a refraction plant) is able to process 300 tonnes of crude palm oil per day. Crude palm oil is processed into refined palm oil, palm olein, palm stearin and fatty acid distillate. Most of these products are produced for the export market, but small quantities are sold domestically.

5 For example, the price paid to smallholder growers in West New Britain in February 2008 (K362/tonne for FFB) was derived from the following formula: price of crude palm oil delivered to Rotterdam, Holland US$1100/tonne; palm kernel oil delivered to Rotterdam US$1360/tonne; and palm kernel extract delivered to Australia US$57/tonne. Based on average extraction rates from one tonne of FFB of 22.3% for crude palm oil; 2.3% for palm kernel oil; and 3.0% for palm kernel expellent, and after subtracting the cost of transporting the oil from New Britain to Europe, each tonne of FFB was worth US$250/tonne (K704/tonne). Smallholders were paid 57% of the value of one tonne of FFB (K704 × 0.57 = K401). After subtracting the cost of transporting FFB (K34/tonne), and three levies for research and outreach (K5/tonne), the price paid to growers was K362/tonne (K401 – K39 = K362).

among project areas. A large number of smallholder producers have received credit from the Rural Development Bank or from the NES companies, and repayments are deducted from the payments that smallholders receive for fruit.

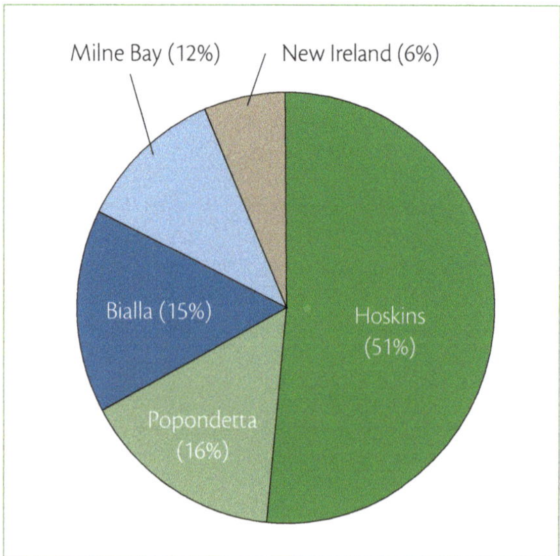

Figure 5.7.2 Oil palm production by project area, 2007 (fresh fruit bunch, tonnes). Source: Ian Orrell, Oil Palm Research Association.

The European Union is the sole export market for PNG's palm oil, with the United Kingdom the biggest buyer.

Future prospects

The area planted to oil palm continues to increase as existing projects expand and new projects commence. New Britain Palm Oil Limited is undergoing an accelerated planting program, with 3000–5000 ha of new plantings planned each year for several years.[6] This company has announced plans to increase its plantings in PNG to about 65 000 ha. The new plantings are on the north coast of West New Britain Province on the Talasea Peninsula and west of there between the Kulu and Via rivers. Land is accessed through agreements with local landowning groups, and palms are planted on both plantations and in village-operated blocks. Hargy Oil Palms Limited is also increasing plantings in the Bialla area of West New Britain. As well as these formal arrangements for new plantations and village oil palm, it is

6 NBPOL had been planting at about 2000 ha/year prior to 2007.

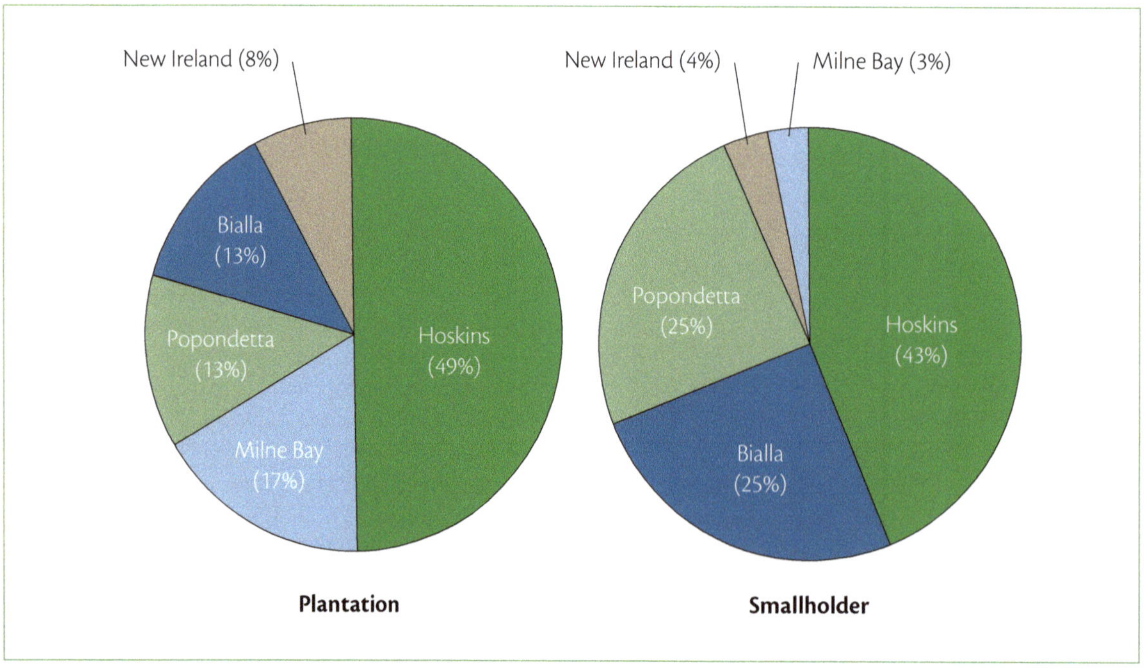

Figure 5.7.3 Oil palm project areas by sector, 2007 (hectares). Source: Ian Orrell, Oil Palm Research Association.

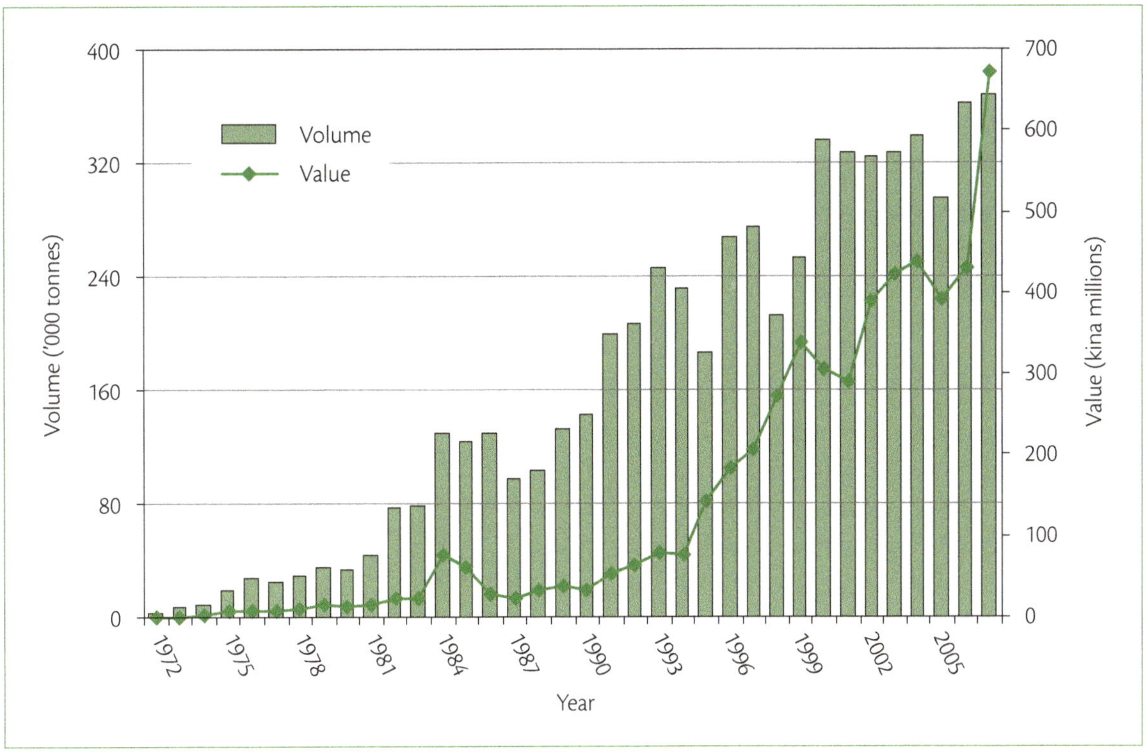

Figure 5.7.4 Volume and value of palm oil exports, 1972–2007. Sources: 1972–1975: Bureau of Statistics, Konedobu; 1976–1979: Bank of PNG; 1980–1983: DAL (1992); 1984–2007: Bank of PNG.

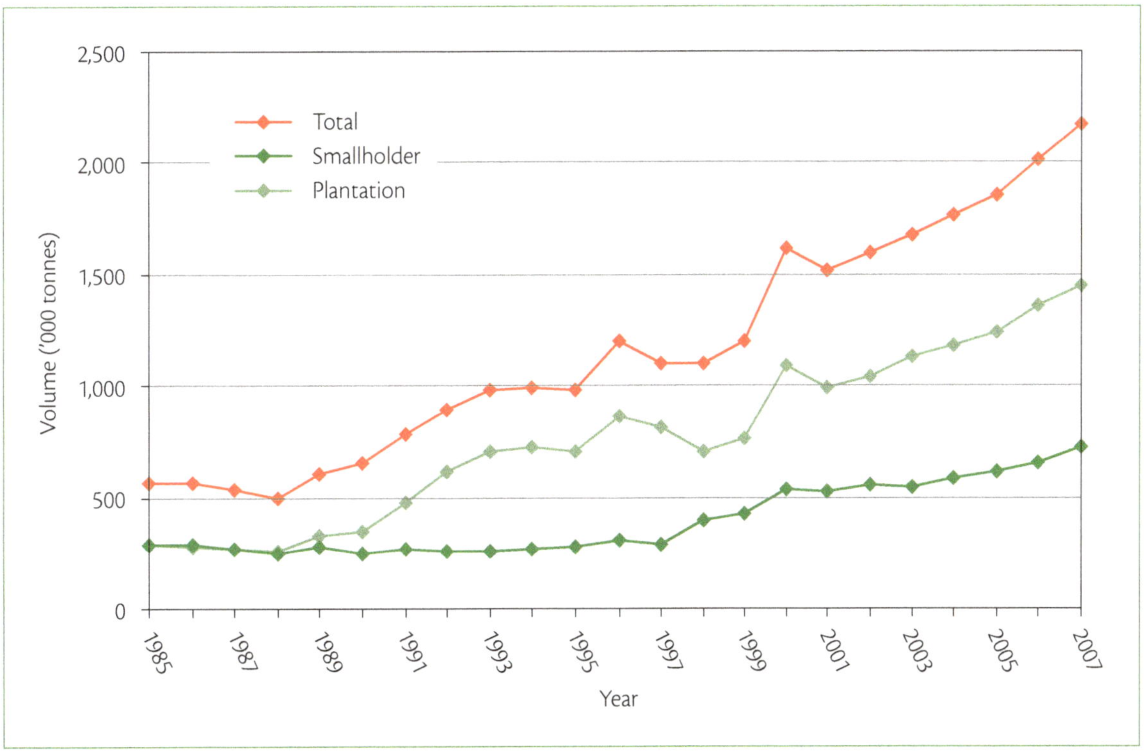

Figure 5.7.5 Oil palm production (fresh fruit bunch) by sector, 1985–2007. Sources: 1985–1995: Oil Palm Industry Corporation; 1996–1999: Palm Oil Producers' Association (compiled by DAL); 2000–2007: Ian Orrell, Oil Palm Research Association.

likely that the area planted under informal tenure arrangements by migrants in West New Britain, and possibly elsewhere, will continue to expand rapidly while the price remains high.

A number of other projects have commenced in PNG in recent years. Ramu Agri-Industries Ltd began planting oil palm in the Ramu Valley in 2003, and had planted 6500 hectares by early 2008. A mill with a capacity of 25 000 tonnes of crude palm oil per year commenced processing fruit in early 2008. An associated village oil palm scheme in the Markham Valley in Morobe Province commenced in 2006, with initial plantings of 100 ha, with 2 ha per household. The village component in Phase 1 of the project consists of 750 ha and can be expanded to 1500 ha.

Plantings commenced by 2007 in several other smaller projects in the Bereina area of Central Province (Mekeo Hinterland Oil Palm) and in the Aitape and Bewani areas of Sandaun Province. Feasibility studies for further projects have been conducted in other locations, including on the Sepik Plains north of the Sepik River in East Sepik Province, near Madang town, and on north-west Bougainville Island.

Oil palm is PNG's most efficient agricultural industry, with yields among the highest in the world. The global price rose rapidly from a low point in 2001, to about US$430/tonne in 2004, and had climbed to US$1100 by early 2008. Recent increases have been in response to strong demand by developing countries, particularly China, India and Pakistan; a slowdown in global production growth in 2007; and increased use of competing vegetable oils for biofuels.[7] The World Bank predicts that the price will peak at about US$900/tonne in 2008 and will then decline until 2015 (Figure 5.2.6). These prices will be more than sufficient to provide good returns to the efficient PNG oil palm industry.

Given current and future plans for expansion, PNG production is likely to continue to increase in the medium term. The high prices being realised in 2007 and 2008, and possibly beyond, mean that palm oil will continue to generate more export income for PNG than any other export crop. However, the limited availability of suitable environments for oil palm is likely to restrict further expansion after around 2030.

7 Biofuels are substitutes for liquid fuels derived from petroleum, and are made from crops such as oil palm, soya bean, canola, coconut, sugar cane or corn (maize).

Sources

Bank of Papua New Guinea Quarterly Economic Bulletin Statistical Tables <http://www.bankpng.gov.pg/> [go to Statistics tab].

Christensen, J. and Densley, D.R.J. (c. 1978). Oil palm. In Densley, D.R.J. (ed). *Agriculture in the Economy: A Series of Review Papers*. Volume 2. Department of Primary Industry, Konedobu. pp. 1–17.

Curry, G.N. and Koczberski, G. (2007). Seeds of discontent. Oil palm and changing production strategies among smallholders in Papua New Guinea. In Connell, J. and Waddell, E. (eds). *Environment, Development and Change in Rural Asia-Pacific: Between Local and the Global*. Routledge, New York. pp. 108–126.

Curry, G.N., Koczberski, G., Omuru, E., Duigu, J., Yala, C. and Imbun, B. (2007). Social assessment report for the Smallholder Agriculture Development Project (SADP) Papua New Guinea. Unpublished report prepared for the World Bank.

DAL (Department of Agriculture and Livestock) (1992). Exports of important agricultural commodities: quantity, value and average price per unit. Rural Statistics Bulletin 1992/6. Department of Agriculture and Livestock, Konedobu.

Hunt, C. (ed.) (2002). *Production, Privatisation and Preservation in Papua New Guinea Forestry*. Instruments for Sustainable Private Sector Forestry Series. International Institute for Environment and Development, London.

Koczberski, G. (2007). Loose fruit mamas: creating incentives for smallholder women in oil palm production in Papua New Guinea. *World Development* 35(7):1172–1185.

Koczberski, G., Curry, G.N. and Gibson, K. (2001). Improving productivity of the smallholder oil palm sector in Papua New Guinea: a socio-economic study of the Hoskins and Popondetta schemes. OPRA Annual Report Supplement. Department of Human Geography, Research School of Pacific and Asian Studies, The Australian National University, Canberra.

Koczberski, G. and Curry, G.N. (2005). Making a living: land pressures and changing livelihood strategies among oil palm settlers in Papua New Guinea. *Agricultural Systems* 85(3):324–339.

Levett, M.P. (2000). Papua New Guinea rural development sector study 2000. Agriculture working paper. Draft final report. Australian Agency for International Development and the World Bank.

New Britain Palm Oil Limited (c. 2003). Kumbango Palm Oil Refinery. New Britain Palm Oil Limited, Kimbe.

New Britain Palm Oil Limited website <http://www.nbpol.com.pg/index.html>.

Oil Palm Industry Corporation (2001). Palm oil industry: a brief background and the smallholder sector. Oil Palm Industry Corporation, Boroko.

World Bank (2001). Study of the smallholder oil palm sector: achievements and potential for future development. Oro Smallholder Oil Palm Development Project. Oil Palm Industry Corporation, Port Moresby.

World Bank (2007). Prospects for the Global Economy. Commodity markets briefs – Palm oil <http://go.worldbank.org/G37PYVHE00>. Accessed February 2008.

5.8 Forest products

Forest products make a significant contribution to export income earned by the renewable resource sector in PNG. They generated an average of K417 million per year in export income between 2004 and 2006, which was 5% of all export income in this period (Figure 5.2.1). The most important forest product exports are logs, which were worth US$187 million (K570 million) in 2007. This is a decline in value of 50% from the logging boom in the mid 1990s when log exports averaged US$385 million between 1993 and 1996. An estimated 10 000 people were directly employed in logging in 2007, a small number compared with the 600 000 smallholders who sell fresh food domestically, for example.

PNG has around 317 500 km^2 of tropical forest covering almost 70% of the national land area. PNG's forests are extremely diverse, ranging from mangroves, swamp forests and eucalypt savanna, to lowland rainforests and montane forests. The most common forest types occur on lowland hills, in lower montane areas and on lowland plains. These three types of forest account for 90% of all forests in PNG and are considered to be the main merchantable forests. Economically valuable trees in PNG number around 400 species, but loggers have tended to focus on only 30–50 species. A characteristic of PNG forests is very high species diversity.

Nearly all forests in PNG are growing on customary-owned land. For this timber to be harvested, the state must first acquire timber rights from the landowners before allocating the rights to logging companies. Prior to 1992, this was done either through the negotiation of a Timber Rights Purchase or a Local Forest Agreement. Since 1992, when a new Forestry Act came into force, state acquisition of timber rights has been through the negotiation of Forest Management Agreements between the PNG Forest Authority and customary owners.

History

The Australian administrations of Papua and New Guinea first attempted to survey the timber potential of parts of Papua in 1908 and parts of New Guinea in 1921. In 1923 and 1924 Charles Lane-Poole completed a survey of Papua and New Guinea and recommended the establishment of a forest service. By the time a forest service had been established in 1938, a number of Europeans had commenced sawmilling operations to service the domestic market, and copra plantation owners in the Islands Region were exporting logs and processed timber to the American veneer market. After World War II a significant expansion of sawmilling operations took place to meet the domestic demand for reconstruction materials. The colonial territories had achieved self-sufficiency in sawn timber production by the time a plywood mill was established at Bulolo in Morobe Province in 1953. The mill resulted in an increase in the export of both processed and unprocessed timber, and timber rapidly became the second most important export commodity after copra. By 1957, forest plantations had been established in the Bulolo Valley and near Port Moresby and Rabaul.[1]

However, the pace of development of the forestry export industry was thought to be unsatisfactory and, encouraged by the World Bank, a shift occurred in policy emphasis towards the clear-felling of natural forest for pulpwood or woodchip production. This strategy culminated in a woodchipping project in Madang conducted by Jant Pty Ltd (New Guinea) and using logs from the nearby Gogol Valley.[2]

After Independence in 1975 the number of logging concessions increased gradually and by 1985 logging operations existed in most lowland provinces. The greatest number of logging concessions were in the Islands Region, particularly Manus, West New Britain and New Ireland provinces. These areas were favoured because they possessed the highest stocking density of commercial timber species and because their coastal locations allowed easy access for shipping raw logs.

In the mid 1990s CSIRO and the PNG Forest Authority created the Forest Inventory Mapping System, a geographical information system containing data about forestry and land use in PNG. The research showed that 35% of the forests of Manus, West New Britain and New Ireland provinces had been logged between 1975 and mid 1996. In comparison, only 10% of the forest area of the mainland lowland provinces was logged over the same period, while in the Highlands Region logging activity was negligible.[3] In the early 1990s, when most concessions in the Islands Region had been allocated, the logging industry shifted its attention to the mainland lowland provinces, particularly Western and Gulf.

1 The Bulolo operation was purchased by PNG Forest Products Pty Ltd in the late 1970s. This company, which is partly owned by the PNG Government, continues to process mostly plantation timber, particularly klinki pine (*Araucaria hunsteinii*) and hoop pine (*A. cunninghamii*), into veneer and then plywood.

2 The Gogol project was originally intended to operate on a cyclical pattern of clearing and replanting. However, about 50% of the landowners refused to allow replanting, which meant that the project had to be expanded beyond its initial boundaries.

3 Logging activity in the Highlands Region focused on sawmilling for the local timber market. Significant plantations were established in the 1960s and 1970s in Eastern Highlands Province and in the Wahgi swamp/Kudjip area of Western Highlands Province.

Forest areas

Information in the Forest Inventory Mapping System (FIMS) suggests the following:

- In 1975, the 'Gross Forest Area' was 330 650 km^2.
- Between 1975 and mid 1996, 19 850 km^2 of forest had been logged and was regenerating.
- Between 1975 and mid 1996, 13 150 km^2 of forest was permanently converted to other land uses. Of this area, 3550 km^2 was logged and then converted; and 9600 km^2 was cleared, but not logged commercially, and subsequently converted.
- Almost all (97%) logging and clearing activities involved the three main types of merchantable forest. These are classified in FIMS as: forest on lowland hills (less than 1000 m altitude), forest on lowland plains (less than 1000 m altitude) and lower montane forest (1000–3000 m altitude). Of the total amount of forest logged or cleared between 1975 and mid 1996, around 70% was forest on lowland hills, 20% was forest on lowland plains and 10% was lower montane forest.
- Over the period 1975 to mid 1996 the 'Gross Forest Area' of PNG was permanently reduced from 330 650 km^2 to 317 500 km^2, or by about 4%, through both logging and land clearance. Of the 317 500 km^2 remaining in mid 1996, 19 850 km^2 comprised forest that had been logged and was regenerating.[4]

4 Not all of the forests included in the Gross Forest Area are suitable for commercial logging activities. Analysis of forest areas, using FIMS, found that of the 1996 Gross Forest Area, 20% was subject to 'extreme physical limitations' and was not suitable for logging by conventional means. A further 40% was subject to 'serious physical limitations' that could constrain commercial logging potential. (An additional category has been considered: 'fragile forests', consisting of forest types and conditions where regeneration is much slower than the normal cycle, or where the forest may not recover at all from commercial forestry.)

Distribution of production

Processed timber and plantations

Logs comprise more than 90% by volume of timber product exports from PNG. However, processed timber has increased in importance over the past decade, rising from 8% of exports (by value) in 2001 to 21% in 2005. Processed timber products include veneer, woodchip, plywood, balsa wood and timber from a number of native or introduced species, with *kwila* the most important of the hardwoods (Table 5.8.1). Veneer is manufactured from forest timber at Panakawa near Balimo in Western Province. Woodchip is processed at a plant in Madang from planted and native forest in the Gogol Valley south-west of Madang, with planted *Acacia mangium* now the main species harvested. Plywood is made in Bulolo from hoop and klinki pine. Balsa wood is processed at four mills from village and plantations plots on the Gazelle Peninsula in East New Britain Province (see Section 5.13).

In 2004, about 52 000 ha of land was under plantation forestry, around 60% of which was managed by the private sector and the rest by the PNG Government. The expected total volume of round wood to be harvested from the current standing trees in the plantations at maturity is 10 million m^3 of hardwood and 4 million m^3 of softwood. As well, there is an estimated one million m^3 of rubber wood (*Hevea brasiliensis*) that could be harvested in the future.

The volume of plantation timber (mostly as round logs) exported in 2004 was 247 214 m^3, worth US$15.5 million (K46 million). The largest plantation areas are at Open Bay in East New Britain Province, Stettin Bay in West New Britain Province, the Wau–Bulolo area in Morobe Province and the Gogol Valley in Madang Province. These four areas account for about 70% of plantation forestry in PNG, with the remaining distributed among 13 locations in 10 provinces.

Logs

The distribution of approved logging concession areas as at mid 1996, with the exception of Bougainville Province, was mapped using FIMS (Figure 5.8.1). The largest concession areas were in Gulf and Western provinces, followed by West New Britain, Sandaun, East Sepik, East New Britain and Central provinces. Gulf Province had the highest proportion of its forest resources in logging concessions (70%). In the Islands Region the proportion was around 30%, except for West New Britain, which had 60% of its remaining forest resources in concession.

Table 5.8.1 Volume and value of processed timber exports, 2004[a]

Product	Volume (m^3)	Value (kina millions)
Veneer	59,249	35.3
Woodchips	51,756	8.0
Mixed sawn	26,586	18.2
Kwila	10,562	13.9
Plywood	3,899	4.5
Balsa wood	3,623	3.0
Teak	1,518	1.9
Other processed products[b]	1,412	0.3
Pencil cedar	1,314	0.5
Ton (taun)	1,218	0.3
Terminalia	1,100	0.2
Malas	1,048	0.2
Red canarium	700	0.1
Rosewood	472	0.7
Pinus	462	0.3
Furniture components (*kwila*)	238	0.3
Mersawa	216	0.1
Plantation	98	0.1
Total processed timber	165,471	87.9

[a] Data are for the first eight months of the year only. The total volume for the full year was 220 000 m^3, valued at K104 million. The volume in 2005 was 250 000 m^3, valued at K106 million, which was 21% of the value of all timber exports in that year.

[b] Other processed timbers are burckella, calophyllum, dillenia, red cedar, red planchonella and white planchonella.

Sources: PNG Forest Industries Association; Bird et al. (2007c: Table 7).

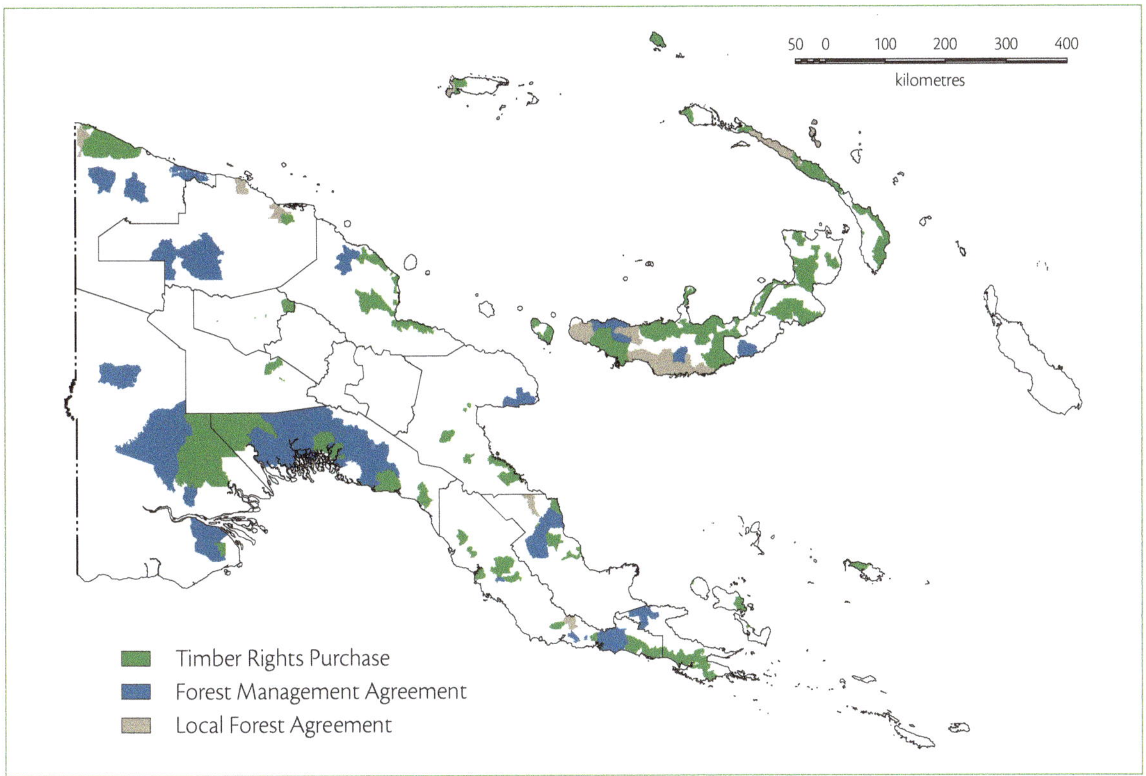

Figure 5.8.1 Distribution of logging concessions, 1996. Source: Forest Inventory Mapping System.

In 2007, West New Britain, Sandaun and Gulf provinces made the largest contributions to national log export volume, accounting for 32%, 15% and 14% of exports respectively (Figure 5.8.2, Tables A5.8.1, A5.8.2). Other provinces that made significant contributions to the total volume of logs exported in 2007 were East New Britain (11%) and Western (8%). Between them, these five provinces accounted for 80% of all log exports in 2007.

There has been a change in the provincial contributions to log exports since the mid 1990s, when the island of New Britain contributed more than half of log exports. Logs from New Britain peaked at 60% of total exports in 1994 (50% WNB, 10% ENB). This declined to about 30% by 2001, but the contribution increased again to 43% by 2007. In 1996, West New Britain Province accounted for 40% of all logs exported from PNG, while East New Britain contributed 16%, Sandaun Province 12%, Western Province 9% and Gulf Province 6% (Figure 5.8.2). The shift in logging activity from New Britain to Western and Gulf provinces follows logically from the distribution of approved logging concessions and from the fact that New Britain was one of the most heavily logged regions until the mid 1990s.

Levels of export

Log exports increased gradually during the 1960s from 136 000 m^3 in 1961 and then remained in the range 360 000–690 000 m^3 a year during the 1970s (Figure 5.8.3, Table A5.8.3). In the early 1980s log export volumes increased significantly, reflecting the expansion in logging concessions that occurred during this period.[5] Between 1979 and

5 There is some concern about the accuracy of log export data during the 1970s and 1980s because of evidence presented by the Barnett Inquiry that illegal and unreported (or misreported) exporting was occurring during this period. The log export data reported since the early 1990s is thought to be more accurate due to improvements in the regulation of the industry and the outsourcing (since 1994) of export monitoring to a private contractor (SGS PNG Ltd).

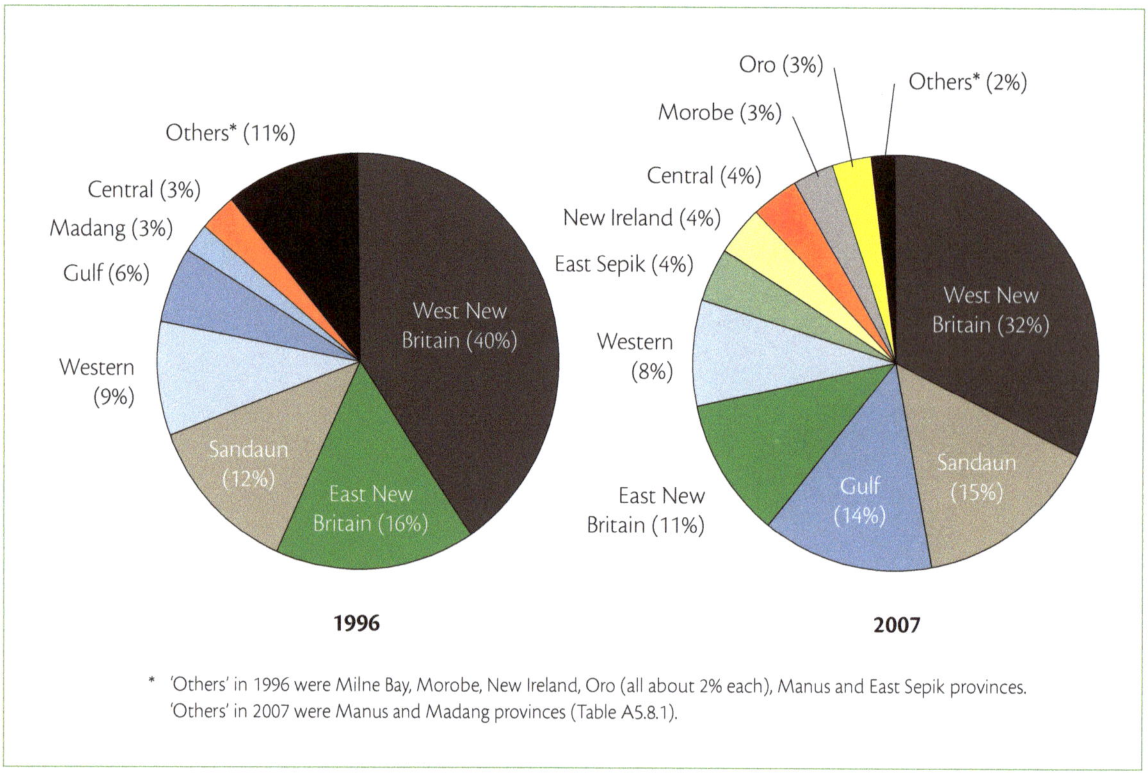

Figure 5.8.2 Log exports by province, 1996 and 2007 (by volume). Source: SGS PNG Ltd.

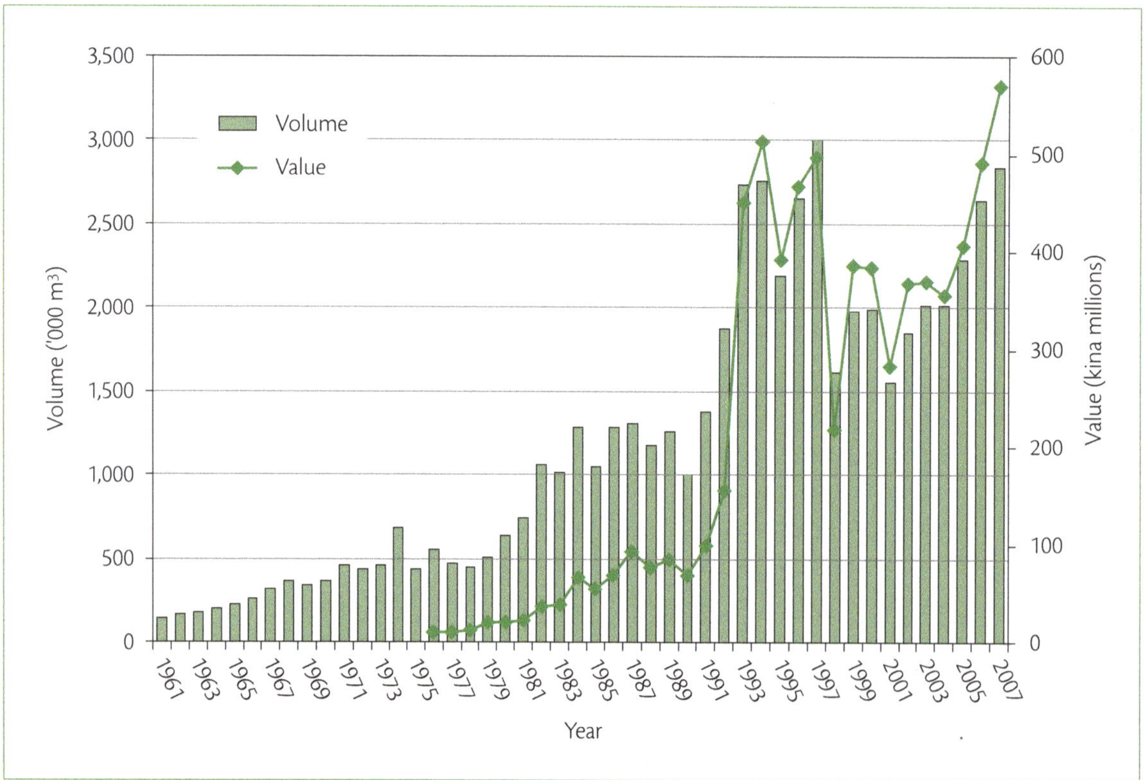

Figure 5.8.3 Volume and value of log exports, 1961–2007. Sources: 1961–1971: PNG Department of Forests; 1972–1979: Bank of PNG Quarterly Economic Bulletins; 1980–1990: PNG Forest Authority; 1991–2007: SGS PNG Ltd.

1982 the annual volume of log exports more than doubled from around 500 000 m^3 to more than one million cubic metres. The volume of log exports increased significantly again in the 1990s, reaching a historical maximum of three million cubic metres in 1997. The highest value of annual log exports was achieved a few years earlier, in 1993, when log exports had a value of US$458 million. At that time, world timber prices were particularly high and export tax rates in PNG were still relatively low. Log exports almost halved between 1997 and 1998, largely as a result of the Asian financial crisis, which caused a decline in demand for timber and lower prices. In 1996, log export taxes were raised substantially and imposed on the basis of value rather than species.

Log export volumes have increased again since 1998, with the volume of 2.8 million m^3 in 2007 returning to the high levels achieved in 1993–1997. The exhaustion of major concessions and the decline in productivity of remaining forest resources contributed to the lower export levels between 1998 and 2002.[6]

Processing, exporters and markets

Around 15% of the log harvest is consumed domestically, mainly for building, with the remainder being exported. Logging companies are responsible for their own marketing and export arrangements, which, since 1994, have been monitored by an independent contractor (the Swiss company, Société Générale de Surveillance, or SGS). As part of its monitoring activities, SGS grades and values logs before they are exported.

There has been much speculation and debate about the extent to which the log export industry in PNG is controlled by a small number of Malaysian companies. In 1998, it was estimated that around 50% of log exports were under the control of a single Malaysian company, Rimbunan Hijau (operating through a large number of subsidiaries), while other Malaysian companies (some with apparent commercial or other links with Rimbunan Hijau) controlled a further 30–35% of the industry.

All of PNG's log exports are to Asian countries. Until the late 1990s, the principal destination for PNG logs was Japan. For example, in 1996 Japan was the destination for 64% of log exports, while South Korea and China made up 18% and 2% respectively. Since 2000, there has been a long-term decline in the Japanese and South Korean markets, and a corresponding expansion into the Chinese market. In 2007, the main destination was China, which was 83% of the export market (Figure 5.8.4, Tables A5.8.4, A5.8.5). Smaller volumes went to Japan (5%), Vietnam (5%), India (3%) and South Korea (2%). A range of factors has caused these changes in export destinations, including the Asian financial crisis in 1997 and material substitutions in the Japanese building industry.

Future prospects

Despite a long and controversial history (see Box 5.8), the PNG forest industry continues to make an important contribution to the national economy. Locally it is a useful source of employment, although nowhere as many rural people earn money from forestry as from agriculture. Although the volume of log exports has returned to levels experienced in the boom period of the mid 1990s, the value is much less because of lower world prices. A number of commentators have expressed concern about the ability of PNG forests to sustain the current rate of logging. Some have suggested that PNG may be following the same path as some other countries in the Asia–Pacific region that have harvested all their easily accessible forest resource. Some of these countries, including Thailand and the Philippines, are now net importers of tropical timber.

6 For example, the timber stock density on the island of New Britain, where much of the forestry activity of the 1980s and 1990s was centred, averages 31.2 m^3 per hectare. In the less fertile Western Province, which in some places is waterlogged and in other places savanna country, and where most of the unlogged forest estate remains, the average timber density is only 24.3 m^3 per hectare.

Box 5.8 A troubled history

From time to time over the past 20 years the logging industry in PNG has been the subject of intense public debate and controversy. In the late 1980s, repeated allegations of impropriety on the part of both private and public sector interests led the then Prime Minister (Paias Wingti) to commission an official inquiry into 'aspects of corruption in the forest industry'. The Commission of Inquiry, which was headed by an Australian member of the PNG judiciary, Justice Tos Barnett, lasted for two years and generated 20 volumes of reports. The reports described a 'forest industry out of control', dominated by foreign investors who had formed 'partnerships' with PNG's political leaders through dubious means, including bribery. The Commission further argued that the failure of the State to adequately and effectively control the forest industry meant that these foreign companies were maximising log exports at the expense of the environment, the sustainability of the forest resources and the domestic processing industry.

Following the Inquiry, a new national forest policy was approved in 1990. This focused on sustainable harvesting and greater decision-making by resource owners. A new Forestry Act was passed in June 1991, but was not gazetted until 12 months later. The legislation established a system of provincial committees and a National Forest Authority under the direction of a board representing a range of stakeholders in the forestry sector, with the aim of reducing the arbitrary power of the Minister for Forests. To support the new policy and legislation, and based partly on the World Bank-led Tropical Forests Action Plan forest sector review, a National Forest and Conservation Action Plan was effected to strengthen capacity in the sector. Extensive concessions were approved during the period leading up to the gazettal of the new Act. Thereafter there was a lull in the issuing of new projects, although a few controversial and major new projects and project extensions were still approved (including Turama Extensions and Vailala Blocks 2 and 3 in Gulf Province, and the Kiunga–Aiambak 'roading' Timber Authority in Western Province).

For a detailed discussion of the Barnett Inquiry, and the history of forest policy in PNG up until the late 1990s, see Filer and Sekhran (1998).

Ongoing concerns remain about the destruction of high-biodiversity forest; loss of income to the PNG Government because of alleged transfer pricing mechanisms by log export companies; and the social disruption caused by payment of relatively large sums of money to small groups of people.

The World Bank forecasts an increase in the real price for logs and sawnwood to the year 2015 (Figure 5.2.6). The forecast for rising prices for logs and sawnwood contrasts with the bank's prediction for falling real prices for tropical tree crops. There is strong market demand for high-value tropical hardwoods and this is likely to continue, particularly as China and India have a large and growing demand for timber. The combination of strong demand and reduced supply, because of the depletion of native forests in Indonesia, Brazil and central Africa, is causing prices to rise. This in turn is causing an increased interest in hardwood plantations.

In the future, the PNG forest industry is likely to rely more on woodlots planted and managed by villagers or plantations rather than the exploitation of native forests. Possible hardwood species include teak, *kwila*, New Guinea walnut (*mon*), *Calophyllum* spp., rosewood, *kamarere* and *ton* (*taun*). As well, there is likely to be larger plantings of fast-growing species such as *Acacia mangium*, *A. crassicarpa* and *Eucalyptus pellita* for chipping, sawn timber or fuelwood. Other multipurpose species with potential to provide timber include rubber, aged coconut palms and *galip* nut.

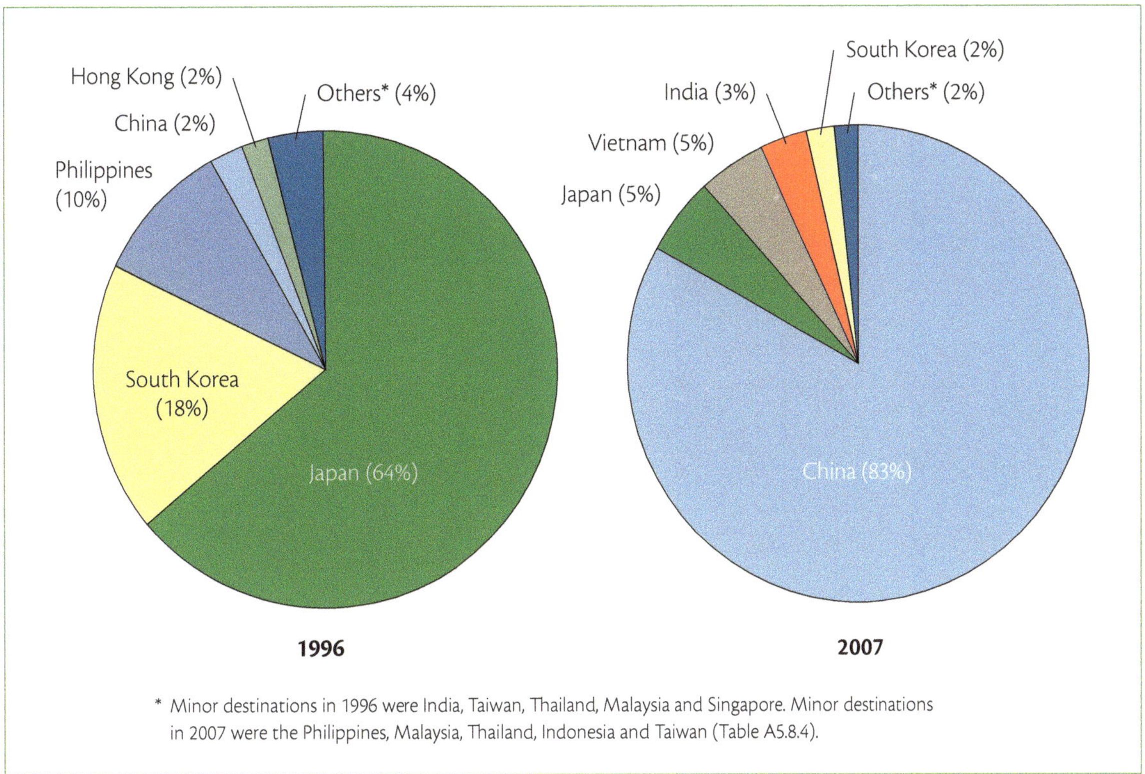

Figure 5.8.4 Destination of log exports, 1996 and 2007 (by volume). Source: SGS PNG Ltd.

A key factor determining which species will be adopted as cash crops by villagers will be whether people are prepared to wait for long periods to harvest and receive income from the sale of timber. Certification schemes guaranteeing that timber comes from a well-managed forest will increase opportunities for access to overseas markets.

Sources

Bird, N., Wells, A., van Helden, F. and Turia, R. (2007a). What can be learnt from the past? A history of the forestry sector in Papua New Guinea. Papua New Guinea Forest Studies 1. Overseas Development Institute, London.

Bird, N., Wells, A., van Helden, F. and Turia, R. (2007b). The current legal and institutional framework of the forest sector in Papua New Guinea. Papua New Guinea Forest Studies 2. Overseas Development Institute, London.

Bird, N., Wells, A., van Helden, F. and Turia, R. (2007c). Issues and opportunities for the forest sector in Papua New Guinea. Papua New Guinea Forest Studies 3. Overseas Development Institute, London.

Bond, A. (2006). Pacific 2020 background paper: forestry. Australian Agency for International Development, Canberra.

Filer, C. and Sekhran, N. (1998). Loggers, Donors and Resource Owners. Policy that Works for Forests and People Series No. 2: Papua New Guinea. Papua New Guinea National Research Institute and International Institute for Environment and Development, Boroko and London.

Hammermaster, E.T. and Saunders, J.C. (1995). *Forest Resources and Vegetation Mapping of Papua New Guinea*. PNGRIS Publication No. 4. Australian Agency for International Development, Canberra.

Hunt, C. (ed.) (2002). Production, Privatisation and Preservation in Papua New Guinea Forestry. Instruments for Sustainable Private Sector Forestry Series. International Institute for Environment and Development, London.

McAlpine, J. and Quigley, J. (1998). Forest Resources of Papua New Guinea: Summary Statistics from the Forest Inventory Mapping (FIM) System. Coffey MPW Pty Ltd for the Australian Agency for International Development and the Papua New Guinea National Forest Service.

Shearman, P. and Cannon, J. (2002). PNG forest resources and the log export industry: a quantitative analysis of forest resources, the impact of commercial logging and the future for the log export industry. The Papua New Guinea Eco-Forestry Forum, Boroko.

5.9 Marine resources

The PNG fisheries zone of 2.4 million square kilometres is the largest in the South Pacific. The coastline and offshore islands of PNG comprise a great diversity of marine environments. The Gulf of Papua contains large delta areas, mud flats and mangrove swamps, while the north coast of the mainland and the coasts of the high islands are characterised by fringing coral reefs and narrow lagoons. Some of the smaller island groups are adjacent to large submerged reef systems or broad shallows. Extensive inland river systems are present in some provinces, such as East Sepik and Western.

PNG's fisheries reflect the diversity of its marine environments. Along the coasts of the mainland and islands, fishing activities include gleaning on reef flats; spear fishing; shallow water hand-lining from dugout canoes, outrigger canoes and outboard-powered fibreglass dinghies; netting; and trapping in the freshwater reaches of the larger rivers. In the swampy coastal areas, fishing activities centre around netting barramundi, catfish and shark, while in the southern part of Western Province a village-based lobster fishery supplies a commercial facility at Daru. Invertebrates are collected throughout the coastal and island areas for commercial purposes, the most important being bêche-de-mer (sea cucumber) and trochus shell. Giant clams are widely harvested for subsistence. Pearl farming is conducted around two islands in Milne Bay Province, with juvenile oysters derived from a hatchery on Samarai Island. Commercial prawn trawling is carried out in the Gulf of Papua.

A domestically based foreign-owned purse seine fishery is rapidly expanding within PNG's Exclusive Economic Zone (see Section 2.10). Aquaculture is practised in many inland locations including parts of the highlands.

Subsistence fisheries

Little reliable information exists on the subsistence fisheries sector, but it is thought to be the most valuable component of PNG's fishing industry in terms of both volume and value. It is estimated that more than 500 000 people participate in both coastal and inland subsistence fisheries, harvesting 25 000–50 000 tonnes of marine produce per year. The best estimate of fish production at the household level comes from the 1996 PNG Household Survey. This survey recorded production of fresh fish, dried fish and shellfish as 50 000 tonnes/year, with an estimated value of K60 million (K142 million in 2007 currency equivalent). Around 60% of the subsistence catch comprises an estimated 20% invertebrates; 30% coastal bay, lagoon and reef fish; and 10% pelagic fish (fish that live in the open sea).

Local area marine food sales

Many people earn some income from selling fresh fish, dried fish and other marine foods including shellfish, lobster, octopus, crab, turtle, prawns and dugong (Figure 5.9.1). Fish is sold in fresh food markets; from informal roadside stalls; and directly to individuals, hotels and other institutions with large boarding populations. Long-distance trade in fresh fish to the larger urban areas is limited, and is constrained by inadequate transport and cold storage facilities, particularly access to ice. Fish and other marine food is sold, and sometimes bartered, in a number of locations, especially from small islands, so that people can buy carbohydrate foods such as sago, sweet potato, imported rice, flour and banana (see Box 5.3).

In the period 1990–1995, sales of fish and shellfish generated an estimated income of K3.8 million for more than 400 000 rural villagers (Table 5.1.1). Over the past decade, the prices of domestically marketed foods have increased (Figure 4.3.1) and the volume of imported fish has decreased (Figure 2.10.1). Hence the value of fish and shellfish sold in 2007 will have increased from the early to mid 1990s, but no recent estimates are available.

Numerous types of marine and some freshwater fish are sold, most of which are native species. Some introduced species are important in a number of locations, particularly tilapia (*Tilapia mossambica* and *T. rendalli*) in the Sepik River area and in parts of Madang and Morobe provinces. Some trout are caught for sale from high-altitude streams (over 2000 m).

Family-based fish farms increased from 5400 in 2001 to an estimated 8000 farms by 2006. As well, an estimated 10 000–15 000 potential fish farmers have constructed earthen ponds, which are not stocked. Most (83%) of the fish farms are in Simbu, Eastern Highlands and Western Highlands provinces, with a further 13% in Morobe, Southern Highlands and Enga provinces. The remainder are scattered in

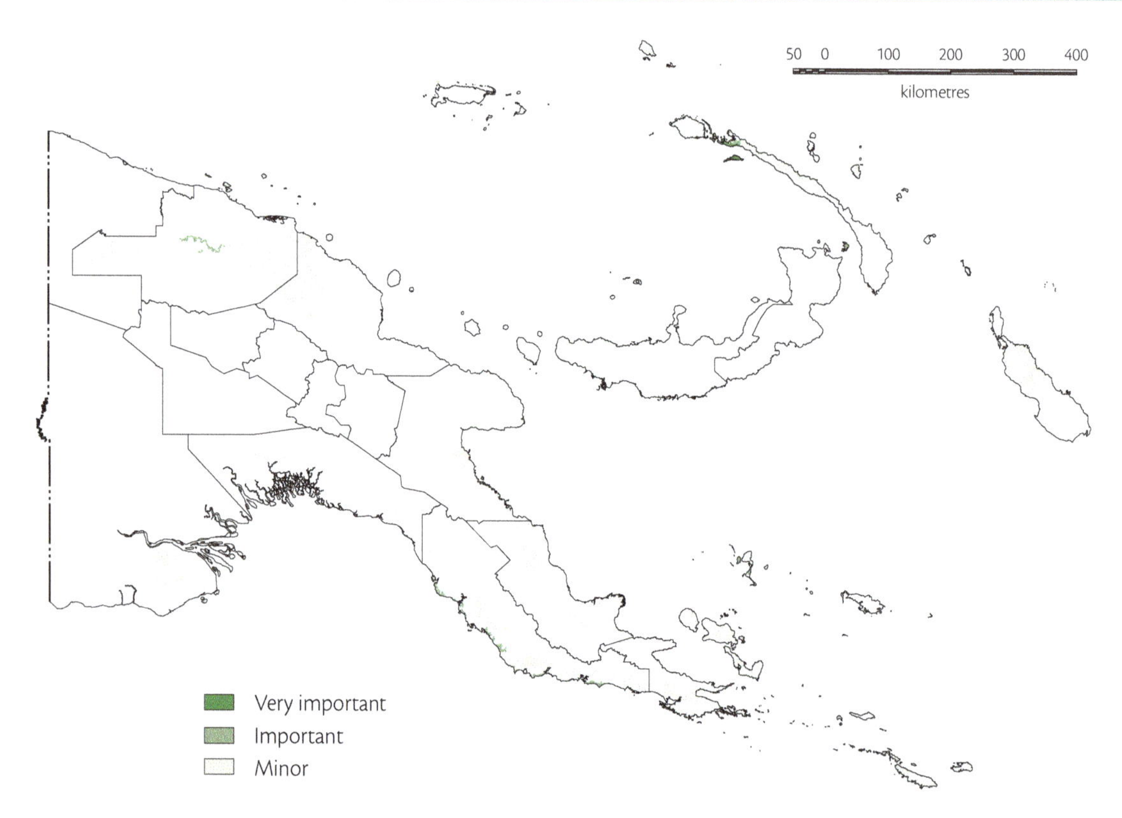

Figure 5.9.1 Locations where sales of fish and shellfish provided income for rural villagers, 1990–1995. Source: MASP.

various lowland locations. There is little information on output and sales from inland fish farms. The three main fish species cultured in inland locations are common carp (*Cyprinus carpio*), rainbow trout (*Oncorhynchus mykiss*) and GIFT tilapia.[1] The Highlands Aquaculture Development Centre at Aiyura in Eastern Highlands Province is the major hatchery in PNG and distributes common carp, GIFT fish and Java carp (*Puntius gonionotus*).

Commercial fisheries

There has been a sharp increase in the overall volume and value of marine exports since the late 1990s (Figure 5.9.2, Table A5.9.1). The major commercial fisheries in PNG are, in order of value, tuna, bêche-de-mer, shell products such as trochus and mother of pearl, shrimp, shark, lobster, and fish and crab (Figure 5.9.3). Prior to the late 1990s, shrimp was the most consistently valuable commercial fishery. However, over the past decade, the tuna fishery has dominated the industry. In 1996 tuna accounted for about 14% of the total value of marine exports but, by 2006, tuna products had increased to around 75% of the total value of marine exports.

Tuna caught in PNG waters accounts for 20–30% of the South Pacific tuna catch and about 10% of the global catch. About 30% of tuna processed in PNG is sold on the domestic market, with local consumption of tinned tuna estimated to be about 15 000 tonnes in 2006 (Table A2.1.1). PNG supplies about 10% of the global market for bêche-de-mer and is the third largest producer.

The increase in the harvesting and export of tuna, and of the value of fisheries exports in general, has occurred as a result of the establishment of the national fisheries policy framework (as articulated

1 GIFT is an abbreviation for Genetically Improved Farmed Tilapia (*Oreochromis niloticus*). This species was introduced into PNG in 1999 and first released to fish farmers in 2002.

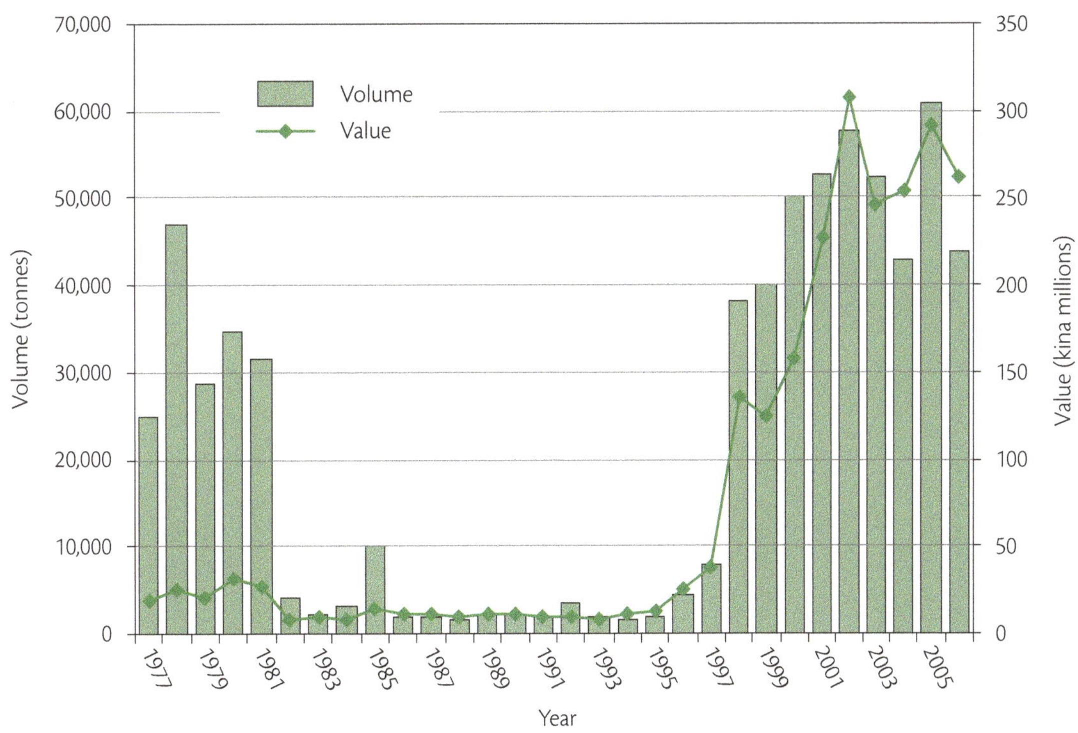

Figure 5.9.2 Volume and value of marine product exports, 1977–2006. Sources: 1977–1991: DAL (1992); 1992–2006: National Fisheries Authority.

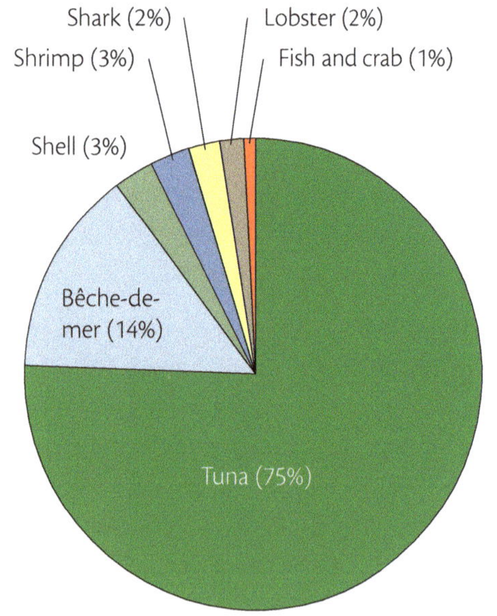

Figure 5.9.3 Value of marine exports by product, 2006. Source: National Fisheries Authority.

in the Fisheries Management Act 1998); the reform of the PNG National Fisheries Authority (NFA);[2] interventions made under successive donor-funded projects; changes to taxation regimes; the development of a domestic tuna fishing industry, including the establishment of canneries in Madang, Lae and Wewak; and the depreciation of the kina. The quality of information collection and monitoring has also improved significantly, but not enough to account for the considerable growth in export volume and value.

It is important to note that the tuna export figures in Tables A5.9.2 and A5.9.3 do *not* include the relatively large amounts of tuna harvested in deep waters by foreign-owned purse seine vessels that pay access and licensing fees to the NFA.[3] Domestically based foreign-owned fleets that operated in PNG during the 1970s left in the early 1980s due to unfavourable duties imposed on tuna exports. Until 1995 distant water fleets paying access fees were the only commercial tuna fisheries in PNG. With policy and governance reform during the late 1990s, the domestic tuna industry expanded rapidly. This includes a domestically based purse seine fleet, which is intended to supply the increasing number of onshore export-oriented processing plants, and a domestic long-line fleet.

Since the domestic long-line fleet is made up of vessels that are smaller, cheaper and of lower technology than the purse seine fleet, this sector was able to incorporate more PNG business people and employees on the vessels. However, with increasing fuel and freight prices, this fishery has proved uneconomic for operators away from the transport hub of Port Moresby, and during 2004–2005 the number of vessels active in the fishery declined by about half. A 'pump boat' handline tuna fishery operated by local fishermen was established near Lae in 2006, with 12 local fishing groups involved. (See footnote 5, Section 2.10 for a definition of pump boat.) There are plans to extend the use of pump boats to New Ireland, Manus and Sandaun provinces.

Processing, exporters and markets

Tuna not only accounts for around 75% of the value of PNG's marine exports, but also represents an important source of revenue in the form of access and licensing fees paid by foreign fishing nations. Most tuna is caught by foreign purse seine vessels. Annual bilateral access agreements are negotiated with a number of countries, particularly Taiwan, Korea, and the Philippines; and a multilateral

[2] The NFA is a non-commercial statutory authority established and operating under the Fisheries Management Act 1998 and related regulations. The NFA is responsible for the management and development of the fisheries sector in PNG. The NFA receives and manages revenue generated from access fees from deepwater fishing nations, licensing fees, donor funding, and penalties from prosecutions under the Fisheries Management Act (see Section 6.4).

[3] Tuna export figures include exports from local processing plants, the domestic long-line catch and the domestic purse seine catch. The export figures do not include catch taken by bilateral access vessels and locally based foreign vessels, despite the fact that some of this catch is transhipped in PNG ports for export. The NFA figures probably underestimate the value of marine exports because of the way the data are compiled.

agreement exists with the United States. Under these agreements, approximately 130 purse seine vessels fish PNG waters each year. Frozen tuna from the foreign fishing vessels is transhipped in the ports of Wewak, Lorengau, Kavieng, Rabaul, Lae and Madang (Figure 6.9.1) for shipment to canneries in Thailand, the Philippines and American Samoa. The foreign fishing vessels catch around 250 000–300 000 tonnes of tuna per year, with an estimated value of US$350–450 million. Access fees paid to the PNG Government by foreign-owned fishing companies increased from US$5.8 million (K15 million) in 1999 to US$13.6 million (K48 million) by 2003.

The domestic long-line industry catches mainly yellowfin tuna and, to a lesser extent, bigeye. Forty licensed locally owned long-line vessels were estimated to be operating in 2003. (According to industry sources, the official figures have consistently underestimated the number of boats because the figures are based on licences and there are chronic problems with the licensing system. There may actually have been between 70 and 80 vessels operating around the peak of the industry in 2003, and about 35 operating in 2005.) Around 1000–2000 tonnes are harvested each year. Most of the fresh catch is exported in chilled form to Japan, while the remainder is exported to Australia. Tinned tuna is exported to the United States and European markets, while fishmeal is mostly exported to Australia and Japan.

The main export destinations for all marine products are Europe and Asia, particularly Germany, the United Kingdom, Hong Kong and Japan (Figure 5.9.4).

Future prospects

Tuna fisheries in particular have the potential to provide substantial long-term economic benefits for PNG. While few Papua New Guineans are employed on, or manage boats in, the purse seine fleets, the domestic fishery generates thousands of jobs in the processing sector onshore and should continue to do so. The domestic long-line fishery has suffered from declining catches as well as freight problems and is unlikely to recover unless these issues are addressed. Small village-based commercial tuna fisheries

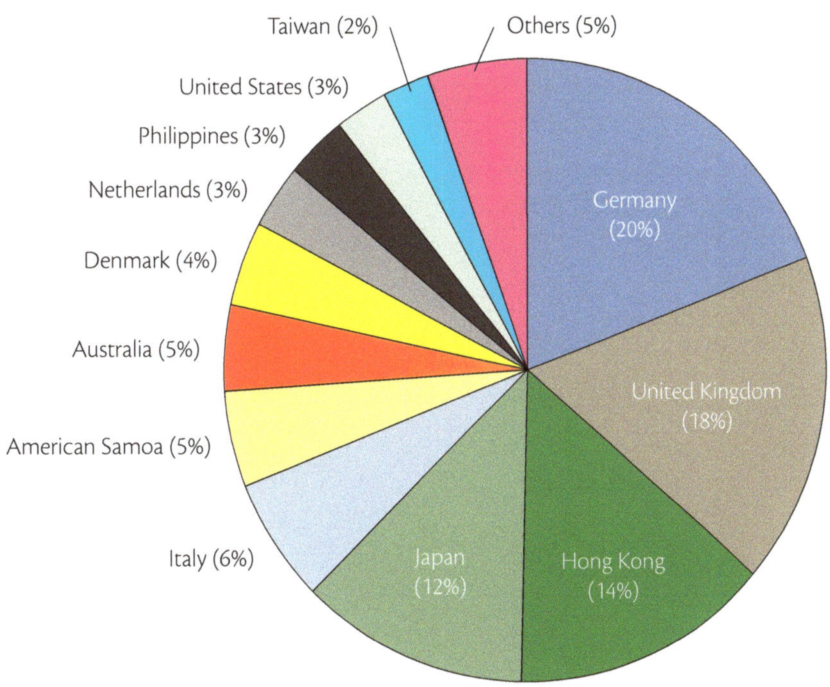

Figure 5.9.4 Destination of marine product exports, 2006 (by value). Source: National Fisheries Authority.

'piggybacking' on large-scale operations in Lae could be an important source of income generation and small business experience, if environmental and economic factors are carefully managed, although similar activities have failed in Madang.

The sea cucumber (bêche-de-mer) fishery can benefit rural people because bêche-de-mer is predominantly export-oriented, although there are concerns over the sustainability of this fishery. The National Fisheries Authority has recently developed a K15 million Fisheries Credit Facility with the National Development Bank to assist rural fisher groups access microfinance.

Future prospects for fisheries in PNG depend upon several factors: the NFA's policy objectives and regulatory ability; greater compliance by all stakeholders; the provision of technical and material support (particularly processing and post-harvest handling); adequate and affordable transport; marketing networks; and a better understanding of the susceptibility of most commercial species to over-harvesting.

Sources

Anon (2005). PNG's robust tuna industry. In Gomez, B. (ed). *Papua New Guinea Yearbook 2005*. Yong Shan Fook, Port Moresby. pp. 122–125.

Barclay, K. and Cartwright, I. (2006). Capturing wealth from tuna: key issues for Pacific Island countries. Institute for International Studies, University of Technology, Sydney.

Barclay, K. and Cartwright, I. (2006). Capturing wealth from tuna: key issues for Pacific Island countries. Country profiles. Institute for International Studies, University of Technology, Sydney.

Kinch, J. (2007). Is management of the beche-de-mer fishery in Papua New Guinea possible? Paper presented at PNG Update, Alotau, April 2007. <http://peb.anu.edu.au/presentations/2007_PNG_Kinch.ppt>.

Kinch, J., Purcell, S., Uthicke, S. and Friedman, K. (2007). Review of the population status, fisheries and trade of commercially important sea cucumbers in Papua New Guinea. Technical Report. Food and Agriculture Organization of the United Nations, Rome.

Lewis, T. (2004). Fisheries – a key economic sector at the crossroads. Unpublished paper.

National Fisheries Authority (c. 2001). National Fisheries Authority Corporate Plan for the three years to 31 December 2004. National Fisheries Authority, Port Moresby.

National Fisheries Authority (c. 2002). PNG country report for workshop on coastal fisheries management, Japan. National Fisheries Authority, Port Moresby.

National Fisheries Authority (2005). National Fisheries Development Plan Framework 2006–2016. National Fisheries Authority, Port Moresby.

National Fisheries Authority website <http://www.fisheries.gov.pg>. Accessed June 2004 and November 2007.

Smith, P.T. (ed.) (2007). *Aquaculture in Papua New Guinea: Status of Freshwater Fish Farming*. ACIAR Monograph No. 125. Australian Centre for International Agricultural Research, Canberra.

5.10 Sugar

Sugar cane is indigenous to PNG. It is widely cultivated in village gardens all over PNG, throughout the year, and is chewed to extract the sweet juice. Modern sugarcane varieties cultivated for refined sugar are complex hybrids (*Saccharum* spp.).[1] Commercial cultivation and production of sugar is confined to the Ramu Agri-Industries Ltd nucleus estate at Gusap–Dumpu in the Ramu Valley of Madang Province.[2] This section focuses on commercial sugar production, with village production for chewing cane covered in Section 3.1. The nucleus estate is located at an altitude of 400 m in an environment where mean annual rainfall is about 2000 mm. The cultivation of sugar at the plantation is seasonal; cane is planted from late February to May to reduce problems associated with insect pests and weeds, and to take advantage of optimum growing conditions.[3] Cane is harvested from April to October using mechanical harvesters.

1 Historically, a number of cane species were used to extract sugar in New Guinea, China and India. The New Guinea species was *Saccharum officinarum*; the Indian species was *S. barberi*; and in China people used *S. sinense*. People in India discovered how to crystallise sugar from cane about 1700 years ago, whereas people in New Guinea and elsewhere in the South Pacific chewed the cane for juice. New Guinea is a centre of diversity for *Saccharum* species, with *S. officinarum*, *S. spontaneum*, *S. edule* and *S. robustum* being endemic. In the 19th century the New Guinea chewing cane (*S. officinarum*) was crossed with a wild cane that is native to South-East Asia and New Guinea (*S. spontaneum*). All modern commercial varieties of sugar cane have been bred from crosses between these two species.

2 The Ramu Agri-Industries Ltd (previously Ramu Sugar Ltd) operation is referred to as a nucleus estate, as opposed to a plantation. Ramu Agri-Industries leases some land from customary landowners, which it uses for sugarcane cultivation. The landowners (known as 'outgrowers') do some manual weed control on the cane, but most operations, including soil tillage, planting, weed control and harvesting, are done by Ramu Agri-Industries. The word 'estate' is used in this section to distinguish the company-operated part of the nucleus estate from the smallholder, or outgrower, component.

Adoption and history

Proposals to establish a sugarcane industry in PNG were first made in the 1930s, with the identification of a possible site for a commercial plantation and processing operation in Oro Province. The establishment of a sugarcane industry was given further consideration in 1951 and again in 1964. The 1964 study concluded that an industry would not be economically viable until domestic sugar consumption reached about 30 000 tonnes which, it was thought at the time, would be achieved by 1975, but was not reached until the early 1980s. An intensive agronomic research program was commenced in 1965, focusing

3 Because sugar cane is endemic to this region, many pests and diseases of the crop exist in PNG and make commercial cultivation more difficult. By comparison, introduced crops, such as coffee, have fewer pest and disease problems.

on the Markham Valley. Other studies followed in the 1970s and a 20 hectare pilot project was initiated by the Department of Agriculture, Stock and Fisheries in the Kemp Welch area of Central Province.

After Independence in 1975, the PNG Government moved ahead with plans to establish a national sugar industry for the purposes of import replacement and export diversification.[4] Several potential sites for a sugarcane plantation were identified, and the Gusap–Dumpu site was eventually selected on the basis that it did not require irrigation or flood protection works. In 1979 a detailed soil survey was undertaken, about 7000 ha of suitable land was identified, and the first sugar cane was planted.

The total area planted to sugar cane on the estate increased rapidly from three hectares in 1979, to 1350 ha in 1982 and 6000 ha in 1989. The area planted to sugar cane on the estate remained at 5000–6000 ha from 1990 to 2006, but expanded in 2007 (Table A5.10.1). The 1990s saw a marked expansion in the area planted to cane by the company on land owned by local villagers ('outgrowers'), from 800 ha in 1995 to 2200 ha in 1999. The area had decreased to 1200 ha by 2007 because of competing claims for payment for use of land by various landowners. The company responded by increasing the area under cane on its estate. In total, about 10 000 ha of company and outgrower land is involved, including that under cane, under short or long fallow, or abandoned.

Levels of production

The aggregate production of sugar from the nucleus estate (including outgrower production) has fluctuated over the past 20 years, particularly between 1982 and 1991, but has nevertheless increased from 11 000 tonnes in 1982 to 49 000 tonnes in 2002 (Figure 5.10.1, Table A5.10.1). The contribution of outgrowers to sugarcane production

4 PNG leaders sometimes accused Australia of reluctance to establish a domestic PNG sugar industry, to protect Australian sugar exports. Given the small proportion of Australian sugar exported to PNG, this is unlikely to have been the case.

Figure 5.10.1 Commercial sugar production, 1982–2007. Source: Ramu Agri-Industries Ltd.

increased during the 1990s. By 2001–2003, about 150 outgrowers provided land which produced around 30% of total sugar cane. However, this proportion had declined to less than 15% by 2007 because of ongoing problems with land ownership.

Because of detailed company records, there are better data on crop yield for commercial sugarcane production than for any other agricultural crop in PNG (Table 5.10.1). Cane yields averaged 58 t/ha between 1982 and 2007, with a range of 28–88 t/ha. This resulted in a mean sugar yield of 5.3 t/ha, with a range of 2.0–8.2 t/ha. The fluctuations in crop yield (and hence in annual production) have been caused by insect pests and diseases and, to a lesser extent, by weeds and climatic variability. For example, the marked decline in production in 1986 was due to the Ramu stunt disease epidemic of 1984–1985, which required that the entire estate be replanted. Other important pests have been moth stem borer, cicadas and white grub.

Efforts to control these problems have been reasonably successful and have resulted in an increase in sugar production since 1992. Crop yields (and total production) were poor in 2006 and 2007 because of an increase in ratoon stunting disease, caused by a bacterial infection. The problem was made worse by inadequate weed control. These problems have been addressed, including heat-treating planting material to kill the bacteria.

Processing, exporters and markets

Most of the sugar produced at the nucleus estate is sold domestically. The domestic sugar industry is protected from imports by a 70% tariff (Table 4.1.1) and Ramu Agri-Industries Ltd enjoys a near monopoly in the PNG market. The import tariff has resulted in high domestic retail prices for sugar in comparison with world prices. The tariff will decrease to 40% in early 2011.

Since 1993, domestic sales of sugar have ranged between 30 000 and 37 000 tonnes per year. Consumption levels of sugar in PNG have declined from about 8 kg/person/year in the early 1980s to less than 6 kg/person/year by 2006 (Figure 5.10.2).

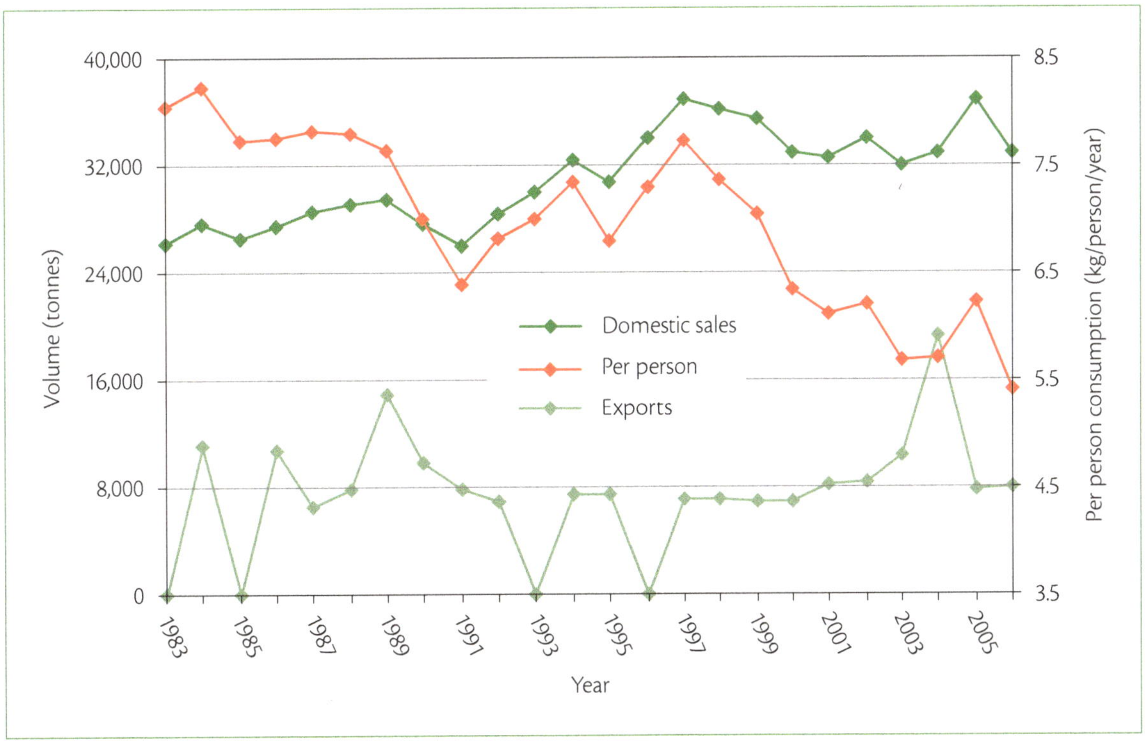

Figure 5.10.2 Volume of domestic sales, per person consumption and sugar exports, 1983–2006.
Source: Ramu Agri-Industries Ltd.

Table 5.10.1 Commercial sugarcane and sugar yields, 1982–2007 (t/ha)

Year	Cane yield			Processed sugar
	Estate	Outgrowers	Total	
1982	88.0	–	88.0	8.2
1983	78.2	79.0	78.3	7.2
1984	64.8	55.9	63.8	6.0
1985	50.5	60.8	52.0	5.3
1986	26.7	31.3	27.6	2.0
1987	64.0	54.6	62.3	4.8
1988	86.0	83.6	85.6	7.6
1989	53.1	51.8	53.0	4.4
1990	48.7	11.4	43.8	4.1
1991	47.7	46.4	47.5	4.2
1992	58.9	63.1	59.4	5.3
1993	50.2	45.4	49.7	4.7
1994	61.1	70.6	62.3	5.9
1995	61.7	69.5	62.6	5.2
1996	57.3	62.5	58.3	4.5
1997	53.1	51.2	52.6	5.2
1998	59.9	62.0	60.4	5.3
1999	63.8	63.6	63.8	5.5
2000	54.7	52.5	54.1	4.8
2001	58.4	57.3	58.0	5.4
2002	70.3	72.9	71.1	6.7
2003	68.9	76.0	70.9	6.6
2004	59.8	56.8	59.1	5.9
2005	58.8	59.1	58.9	5.5
2006	51.2	42.4	49.7	4.4
2007	48.5	40.8	47.3	4.3
Mean	58.5	57.5	58.3	5.3

Source: Ramu Agri-Industries Ltd.

About 7000–10 000 tonnes per year of sugar is exported, but only in years when production is surplus to domestic requirements (Figure 5.10.2). Consequently, no sugar was exported in 1985, 1993 and 1996.[5]

Molasses is a by-product of the sugar refining process and about 15 000 tonnes are produced each year. About 2000–3000 tonnes of molasses is sold for livestock feed. It is also used to produce about two million litres of ethanol (alcohol), most of which is exported and is worth K2–2.5 million per year.

Future prospects

Ramu Agri-Industries anticipates that sugar yields per hectare will increase because of the use of more productive cane varieties that have been selected in the company's breeding program. The company is aiming to achieve yields of 6.5–8.0 t/ha of sugar, compared with 4–7 t/ha in recent years. It plans to increase production to 50 000 tonnes/year by 2010. However, there are potential threats to increased output. Sugarcane smut, a fungal disease that is not in PNG but is present in Indonesia and Australia, is a risk to future production on the estate, as well as to village production of chewing cane.

The company anticipates that sugar consumption in PNG will increase at about 4% per year, which will mean a domestic demand of 45 000 tonnes by 2014. However, given the current rate of population growth, consumption will still be about 6 kg/person/year. Ramu Agri-Industries is seeking to have the maximum allowable quantity exported to the United States increased from 7500 tonnes/year to 11 000 tonnes/year, as the US pays about twice the world price for sugar as an indirect form of aid.

Because import protection on sugar will be reduced and domestic demand is limited, Ramu Agri-Industries has embarked on an ambitious program of diversification, commencing with oil palm and

5 In 2004, exports reached an all-time high of 19 000 tonnes (Table A5.10.1), following several years of record production. In that year, 10 000 tonnes went to Sri Lanka, 7000 tonnes to the United States and the balance to various Pacific islands.

cashew. More than 8000 ha of oil palm will be planted by early 2009, and about 15 000 ha will be planted by 2015. An area of 100 ha will be planted to cashew in 2008, which will be increased to 500 ha by 2010 if this venture is successful. A number of other agricultural products are being examined in detail, including timber for fuelwood and hardwood for export; noni, from which the juice would be extracted for export; edible bamboo for the Asian market; and jatropha oil for use as a biodiesel fuel.

Sources

Asian Development Bank (2003). Preparing the agriculture and rural development project, Papua New Guinea. Inception Report (ADB TA4055-PNG).

Densley, D.R.J. (c. 1978). Sugar. In Densley, D.R.J. (ed). *Agriculture in the Economy: A Series of Review Papers*. Volume 3. Department of Primary Industry, Konedobu. pp. 1–17.

Hartemink, A.E. (2001). Sustainable land management at Ramu Sugar plantation: assessment and requirements. In Bourke, R.M., Allen, M.G. and Salisbury, J.G. (eds). *Food Security for Papua New Guinea. Proceedings of the Papua New Guinea Food and Nutrition 2000 Conference*. ACIAR Proceedings No. 99. Australian Centre for International Agricultural Research, Canberra. pp. 344–364.

Hartemink, A.E. and Kuniata, L.S. (1996). Some factors influencing yield trends of sugarcane in Papua New Guinea. *Outlook on Agriculture* 25(4):227–234.

Kuniata, L.S. (2001). Integrated crop production management of sugarcane at the Ramu Sugar Ltd estate. In Bourke, R.M., Allen, M.G. and Salisbury, J.G. (eds). *Food Security for Papua New Guinea. Proceedings of the Papua New Guinea Food and Nutrition 2000 Conference*. ACIAR Proceedings No. 99. Australian Centre for International Agricultural Research, Canberra. pp. 335–343.

Kuniata, L.S. (2008). Ramu Agri-Industries Ltd: more than just a sugar producer in PNG. Unpublished seminar. Department of Human Geography, Research School of Pacific and Asian Studies, The Australian National University, Canberra.

Levett, M.P. (2000). Papua New Guinea rural development sector study 2000. Agriculture working paper. Draft final report. Australian Agency for International Development and the World Bank.

5.11 Rubber

Rubber is a minor cash crop in a limited number of PNG locations (Table 5.1.1, Figure 5.1.1). From 2004 to 2006 rubber generated average export earnings of K19 million per year, which was 1% of the value of agricultural exports in this period (Figure 5.2.2). In 2007 exports were worth K24 million. The significance of rubber in the economy has declined from the 1950s, when rubber exports made up around 12% of agricultural exports (Figure 5.2.3).

Natural rubber is used for many household and industrial purposes, most importantly the manufacture of motor vehicle tyres and tubes. Other uses include the manufacture of window parts, various items used in engines (belts, hoses, dampeners), gloves, toy balloons, adhesives and rubber bands.

In PNG rubber is grown from near sea level up to about 700 m altitude, in environments where mean annual rainfall ranges from 1500 mm to over 5000 mm. Production is non-seasonal.

Adoption and history

Commercial rubber was first planted in PNG in 1903. Most production was in the Australian Territory of Papua (now the Southern Region of PNG). In German New Guinea, attempts to build a rubber industry were based on assam rubber (*Ficus elastica*), which is inferior to para rubber (*Hevea brasiliensis*). Consequently, very little rubber was produced in German New Guinea.

In Papua, plantations were first developed at Galley Reach, Sogeri and Cape Rodney in Central Province and in the Kerema area of Gulf Province. They subsequently expanded into the Kokoda area of Oro Province and parts of New Ireland Province. The principal exports from Papua in the 1930s and 1940s were copra and rubber, and rubber contributed almost 30% by value in the 1930s.

Prior to World War II smallholder rubber planting was confined to a few villages in Oro Province. The Department of Agriculture, Stock and Fisheries (DASF) promoted the development of village rubber production in the 1960s, particularly in Gulf and Western provinces. Between 1964 and 1970 smallholder rubber planting was also promoted on settlement and resettlement schemes in Central Province; at Gavien near Angoram in East Sepik Province; and at Murua in Gulf Province. As a result, the number of smallholder and village rubber producers increased from less than 2000 in 1970–71 to around 3300 in 1976–77. These schemes were not associated with plantations.

Distribution of production and planting

Rubber has been grown and produced in eight lowland provinces (Table 5.11.1, Figure 5.11.1). In 2005, 4100 ha was planted to rubber on settlement schemes, 4500 ha in villages, 9500 ha on plantations

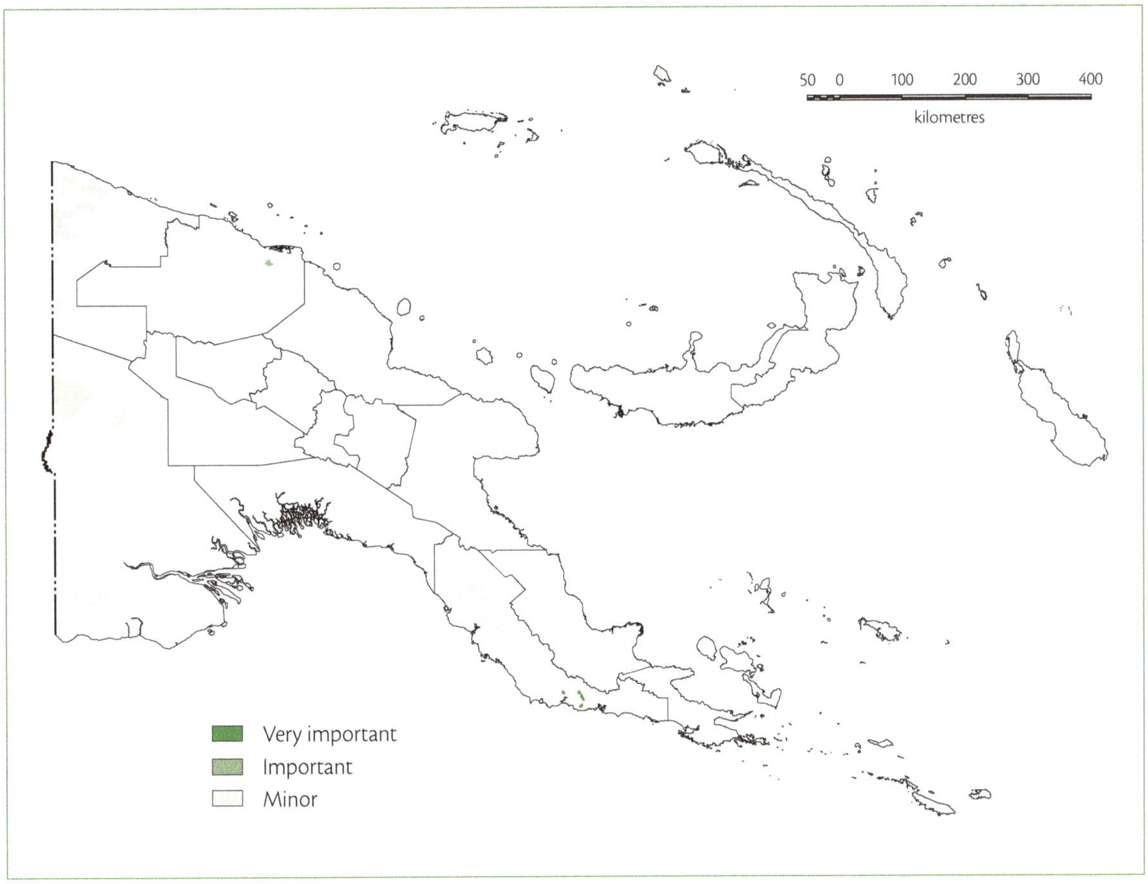

Figure 5.11.1 Locations where rubber sales provided income for rural villagers, 1990–1995. Source: MASP.

and 50 ha at the DASF Bisianumu rubber experiment station.[1] Most production in 2007 was from Central and Western provinces, with smaller amounts coming from East Sepik, New Ireland, Gulf, Oro and Manus provinces. Currently only about one-third of PNG rubber trees are tapped.

Thus, settlement scheme and village-level rubber plantings account for about half the total area of land planted to rubber, with the plantation sector making up the rest. One settlement scheme, Cape Rodney, has 80% of the total area planted to rubber in settlement schemes. Western Province has most village plantings, almost half the total area of village rubber. The Galley Reach Holdings Ltd plantation in Central Province (owned by the Belgian company SIPEF) has 60% of the total area of plantation rubber, and is the only plantation presently producing rubber.

Levels of production[2]

Rubber exports were in the range 1000–2000 tonnes/year between 1936 and 1951, except in 1943 when production was disrupted by World War II. After the war, rubber production continued to be dominated by the plantation sector, with production, measured as exports, rising steadily to over 6000 tonnes in 1970 (Figure 5.11.2, Table A5.11.1).

1 Figures on area planted to rubber are from DAL and the Ministry of Agriculture and Livestock (2007). They underestimate the area planted in Western Province, where about 3800 ha has been planted since 1999 (Table 5.11.2).

2 The production data presented here have been assembled from a number of different sources in order to construct a long-term data run. For some years, there is a certain amount of variation between different data sources. We have presented data that is likely to be the most accurate.

Table 5.11.1 Rubber plantings by area and sector, 2002

Settlement scheme	Number of settlers	Area (ha)
Cape Rodney Agricultural Development Project (Central Province)	2,026	3,400
Bailebo (Central Province)	57	84
Murua (Gulf Province)	36	118
Gavien (East Sepik Province)	154	538
Subtotal	2,273	4,140
Village planting	**Number of growers**	
Western Province[a]	1,999	2,063
Gulf Province	65	100
Central Province	378	359
Oro Province	298	529
East Sepik Province	635	442
Sandaun Province	526	244
Manus Province	128	163
New Ireland Province	333	317
Subtotal	4,362	4,217
Plantation sector	**Number of workers**	
Galley Reach Plantation (Central Province)	600	5,112
Sogeri Rubber Development Corporation (Central Province)	40	1,126
Epo Estate (Gulf Province)	30	942
Other small plantations	30	1,083
Subtotal	700	8,263
DAL Bisianumu rubber experiment station	20	50
Total	7,355	16,670

[a] The figures for Western Province refer to plantings made in the 1960s and 1970s. They do not include 3786 ha planted by 3768 growers between 1999 and 2006 (Table 5.11.2). Most of these figures for other provinces in Table 5.11.1 refer to old plantings, many of which are unproductive and are not being tapped.

Source: DAL.

A significant decline in production followed in the 1970s due to stagnation in the plantation sector and still insignificant smallholder production. The decline of the plantation sector was attributed to the continued use of clonal seedling planting material, rather than higher-yielding bud-grafted material; increased costs of production, particularly labour costs; uncertainty caused by land disputes and the threat of compulsory government land acquisition; and a high turnover of workers and resultant poor tapping standards that reduced the economic life of trees. The area of rubber on plantations declined dramatically after 1970, with several plantations cutting out rubber in favour of other enterprises, such as beef cattle.

Smallholder production increased steadily during the 1980s and by 1990 had overtaken plantation production (Figure 5.11.3, Table A5.11.2). Between

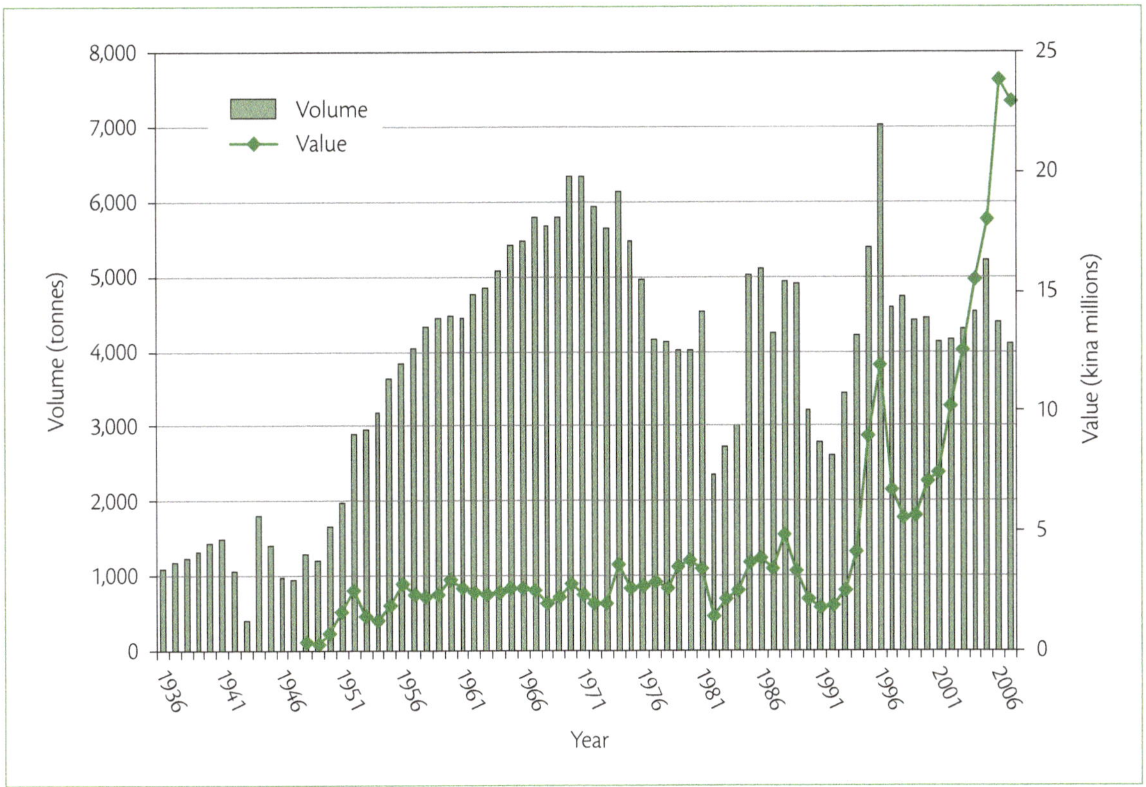

Figure 5.11.2 Volume and value of rubber exports, 1936–2007. Sources: 1936–1947: Bureau of Agricultural Economics (1952:15); 1948–1976: Bureau of Statistics, Konedobu; 1977–1984: DAL (1992); 1985–1999: DAL (2001); 2000–2005: DAL; 2006–2007: Bank of PNG.

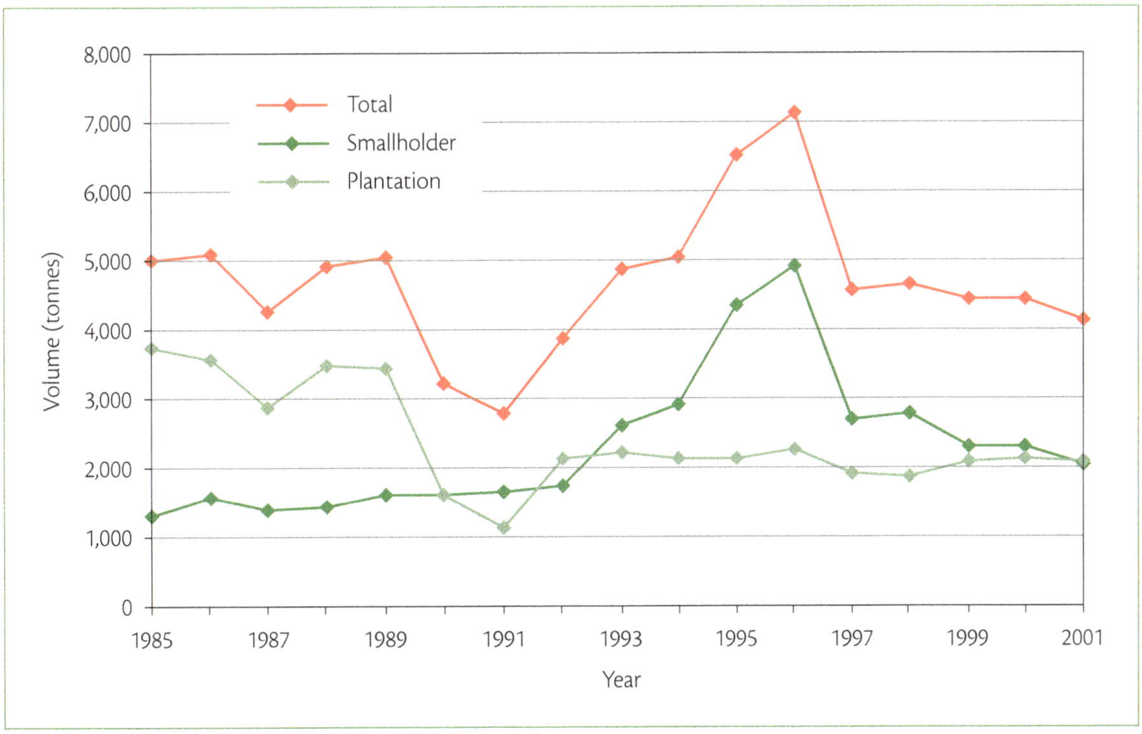

Figure 5.11.3 Rubber production by sector, 1985–2001. Source: DAL.

Table 5.11.2 Rubber plantings in Western Province, 1999–2006

District	Area (ha)	Number of growers
North Fly	2740	2577
Middle Fly	1026	1163
South Fly	20	28
Total	3786	3768

Note: In the 1960s and 1970s, 2063 ha of polyclonal seedlings were planted by 1999 growers and most of this is nearing the end of its economic life.

Production has been in the range 370–1000 tonnes/year of processed rubber between 1994 and 2007. It was 811 tonnes in 2007, the highest figure since 1996.

Source: Chew Boo, North Fly Rubber Limited.

1992 and 1996 a sharp increase occurred in smallholder production that coincided with increasing export prices for rubber. The promotion of rubber in the Kiunga area of Western Province by North Fly Rubber Limited also contributed to the significant rise in smallholder production during this period. Total smallholder and plantation annual production reached an all-time high of 7000 tonnes in 1996.

This was followed by a sharp decline in production in 1997, due to a fall in export prices and other factors which adversely affected smallholder production, including the closure of the processing factory at Gavien in East Sepik Province following damage by fire; low production at Kiunga in Western Province after large compensation payments from Ok Tedi Mining Limited to people living downstream of the mine; and the 1997 drought and very low water levels in the Fly River in Western Province that effectively paralysed the transport of cup lump rubber from Balimo and Lake Murray to the factory at Kiunga, a distance of more than 1300 km. Production remained 4000–5000 tonnes/year from 1997 to 2007, divided evenly between smallholders and plantations.

Although settlers on rubber settlement schemes often produce little or no rubber, they are active agricultural producers and derive significant income from the sale of fresh food and betel nut. Typical are the Murua scheme in Gulf Province, Cape Rodney in Central Province and the Gavien scheme in East Sepik Province, which produce food and betel nut for urban markets.

The average yield of cup lump rubber by village producers in the Kiunga area is about 650 kg/ha (400 kg/ha processed rubber). The best producers obtain up to 1650 kg/ha cup lump (1000 kg/ha processed rubber). Planting material with higher yield potential of over 2000 kg/ha of processed rubber has been distributed by North Fly Rubber Limited in Western Province over the past ten years. A 1949 survey of plantations in Central and Oro provinces recorded average yields of 450 kg/ha of processed rubber, a similar yield to that currently obtained at the Galley Reach Plantation near Port Moresby.[3]

Processing, exporters and markets

Most rubber produced in PNG is initially extracted on-farm as cup lump rubber and is then processed into PNG Certified Rubber 10. Three factories currently purchase and process cup lump rubber; at Doa Plantation (at Galley Reach west of Port Moresby), Moreguina (near Cape Rodney, south-east of Port Moresby) and Kiunga (Western Province). Galley Reach Holdings Ltd buys cup lump rubber from elsewhere in PNG, transports it to Port Moresby and processes it at its Doa Plantation factory. This company produced about 85% of total rubber exported in 2006. Current government policy is that only processed rubber should be exported, but it does not have the power to ban the export of unprocessed cup lump rubber.

Old rubber trees provide high quality timber suitable for making furniture. The timber can be harvested after economic tapping ceases, which is

3 Cup lump rubber contains 40% moisture. To convert cup lump yields to processed rubber, multiply by 0.6. Kiunga smallholder yield data was supplied by Warren Dutton. The 1949 survey of plantations was conducted by the Bureau of Agricultural Economics (1952).

at about 35 years after planting.[4] Sales of old rubber trees for timber increase the economic viability of rubber production.

Most rubber produced in PNG is exported to Europe, particularly to Germany, France, the Netherlands and Belgium (Figure 5.11.4, Table A5.11.3).

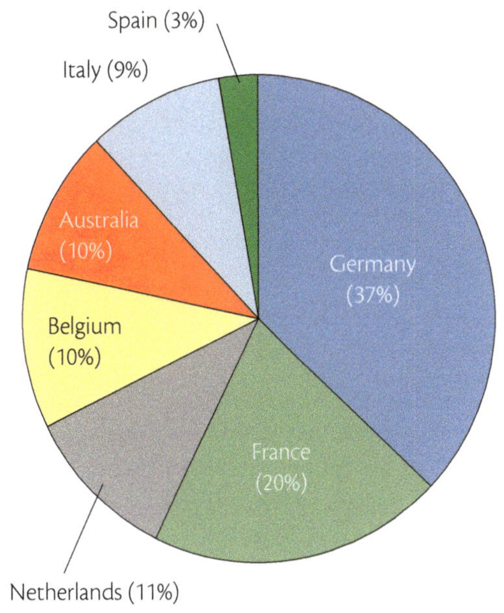

Figure 5.11.4 Destination of rubber exports, 2005 (by volume). Source: DAL.

Future prospects

Following historical lows in world rubber prices in 2001, natural rubber prices staged a remarkable recovery. By 2007, prices were about four times as high as they were in 2001. High oil prices have increased the cost of making synthetic rubber and strong demand for vehicle tyre production has emerged, especially from China. Markets now favour natural rubber and firmer prices can be expected for the next few years. The main future uncertainty in the rubber market is the price of crude oil, which affects the competitiveness of synthetic rubber. Primary constraints on the PNG rubber industry are the poor state of transport infrastructure and the general lack of financing for new planting. Low returns to labour also limit smallholder interest in rubber (Table 5.20.1).

Significant new plantings have been made in recent years only in Western Province. More than 2700 ha of smallholder rubber has been planted in the Kiunga area and over 1000 ha (out of a target of 2200 ha) in the Lake Murray area since 1999 (Table 5.11.2). New plantings are planned in the Bosset, Suki and Balimo areas in the southern part of Western Province. There are plans to increase plantings to a total area of 10 000 ha in the province, with new plantings in the north, middle and south Fly River areas.

The expansion in Western Province is supported by North Fly Rubber Limited, Ok Tedi Mining Limited and the PNG Sustainable Development Program. A total of almost K20 million has been provided, directly and indirectly, from the revenue of the Ok Tedi copper mine in the past decade. The transport costs of moving seedlings from Kiunga to Lake Murray, Suki and Balimo, cup lump rubber from these areas to Kiunga for processing, and processed rubber to a main port for export, is heavily subsidised. Without these subsidies, rubber production in Western Province is unlikely to be economic unless further large areas of high-yielding rubber are planted within the next decade to create economies of scale.

The PNG National Agriculture Development Plan has a goal of rubber production of 29 000 tonnes by 2016, more than seven times production in 2007. If the higher prices of 2007 and 2008 continue, it is likely that rubber production in PNG will increase, but the target of 29 000 tonnes in less than a decade is completely unrealistic. Since 1992, no significant area of smallholder rubber has been planted other than the heavily subsidised plantings in Western Province.

4 Rubber trees are planted at a density of 400 trees/ha. Tapping commences at year 7 and can continue for about 35 years. Tapping can be done every second day. A grower can tap about 500 trees (1.25 ha) each day.

Sources

Bank of Papua New Guinea Quarterly Economic Bulletin Statistical Tables <http://www.bankpng.gov.pg/> [go to Statistics tab].

Bureau of Agricultural Economics (1952). The Papuan rubber industry. Bulletin No. 7. Department of Commerce and Agriculture, Commonwealth of Australia, Canberra.

DAL (Department of Agriculture and Livestock) (1992). Exports of important agricultural commodities: quantity, value and average price per unit. Rural Statistics Bulletin 1992/6. Department of Agriculture and Livestock, Konedobu.

DAL (Department of Agriculture and Livestock) (2001). Handbook on Agricultural Statistics 2001. September 2001 edition. Department of Agriculture and Livestock, Port Moresby.

Densley, D.R.J., Rady, G.L. and Arnold, C.E.W. (c. 1978). Rubber. In Densley, D.R.J. (ed). *Agriculture in the Economy: A Series of Review Papers*. Volume 2. Department of Primary Industry, Konedobu. pp. 1–24.

Department of National Development (1951). *The Resources of the Territory of Papua and New Guinea.* Volume One. Department of National Development, Division of Regional Development, Canberra.

Levett, M.P. (2000). Papua New Guinea rural development sector study 2000. Agriculture working paper. Draft final report. Australian Agency for International Development and the World Bank.

Ministry of Agriculture and Livestock (2007). National Agriculture Development Plan 2007–2016. Volume 1: Policies and Strategies. Ministry of Agriculture and Livestock, Port Moresby.

Ministry of Agriculture and Livestock (2007). National Agriculture Development Plan 2007–2016. Volume 2: Implementation plan. Ministry of Agriculture and Livestock, Port Moresby.

PNG Sustainable Development Program Ltd website <http://www.pngsdp.com/>.

World Bank (2007). Prospects for the Global Economy. Commodity markets briefs – Rubber <http://go.worldbank.org/G37PYVHE00>. Accessed April 2008.

5.12 Tea

In PNG tea is mostly grown in Western Highlands Province between 1200 m and 1950 m altitude in areas where mean annual rainfall is around 2500 mm. A non-producing tea plantation exists at a former government estate at Garaina in Morobe Province at an altitude of around 600 m. Production of tea in PNG is non-seasonal.

Adoption and history

Tea was established as an experimental crop on a government-owned plantation at Garaina in Morobe Province in 1950. A small factory was built at Garaina in 1962 and PNG tea was exported for the first time in 1963. The high yields and good quality that were obtained at Garaina encouraged the Department of Agriculture, Stock and Fisheries to promote the cultivation of tea as a viable cash crop for both plantation and smallholder producers.

The Department's policy was to develop the industry on the nucleus estate model, where a commercially operated estate produces tea and also provides a market, processing and technical services for smallholder producers who cultivate tea on village land adjacent to the nucleus estate. In order to facilitate smallholder development, the government established about 3000 ha of land settlement schemes, mostly in Western Highlands Province. However, tea failed to gain acceptance among smallholders. They preferred other cash crops, especially coffee and fresh vegetables, and disliked the continuous labour inputs required for harvesting ('plucking') tea leaves and the generally high level of skills required to cultivate tea.

Consequently, the tea industry in PNG remained dominated by six foreign-owned estates and factories that were established in Western Highlands Province in the late 1960s and early 1970s. In 1977 these factory estates grew 80% of PNG's tea. The remainder came from five smaller estates (three in Western Highlands Province and two in Simbu Province) and from smallholder producers (mostly in Western Highlands Province). It was estimated that smallholders contributed less than 3% of total tea production in 1975/76.

The Garaina estate in Morobe Province ceased to function as a tea research station in the late 1970s, but people living in the area continued to harvest and sell a limited amount of tea from the site. Smallholder interest in tea cultivation continued to decline to the extent that, by 1992, production of tea from sources other than the factory estates was around 24 tonnes per year, or less than 0.5% of total production. Smallholder tea is no longer produced in PNG.

Distribution of production and planting

In PNG tea is produced and exported only by W.R. Carpenter and Co. The company recently purchased a tea plantation from a second company, and a third company abandoned its operations. This mostly foreign-owned company currently operates five separate estates with a combined area of 2200 ha.

Levels of production

The production of tea, measured in terms of export volumes,[1] increased sharply between 1970 and 1980 and reached almost 8000 tonnes in 1980 (Figure 5.12.1, Table A5.12.1). By the mid 1970s most of the factory estates were fully planted, and production continued to increase as the tea plantings matured. Production levels remained relatively high until 1985, in association with the high world tea prices in the early to mid 1980s. However, prices fell severely after 1985 and in 1991 reached their lowest point (in real terms) since the early 1950s. This saw a corresponding decrease in PNG tea exports, to 4700 tonnes in 1991 and 3400 tonnes in 1994. Exports increased again after the recovery of world market prices during the mid to late 1990s, coupled with the devaluation of the kina, and reached a historical maximum of 9300 tonnes in 1996. Both prices and export volumes have remained relatively stable since then, with volumes averaging more than 7000 tonnes per year.

[1] Because 90% of the tea produced in PNG is exported, exports are a good indicator of total production levels.

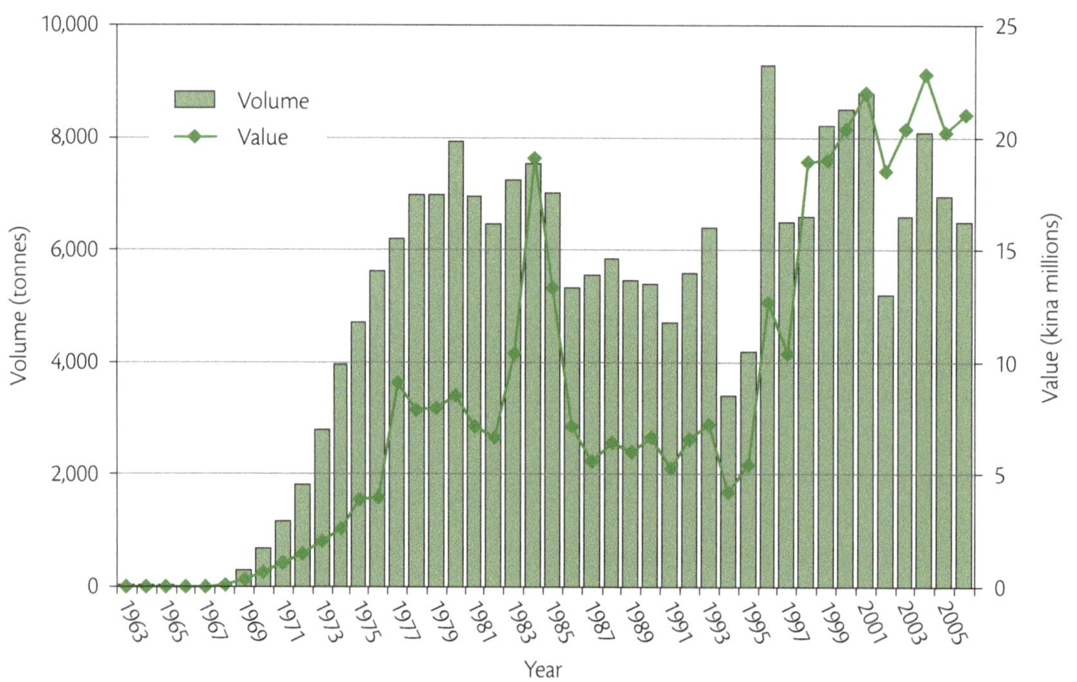

Figure 5.12.1 Volume and value of tea exports, 1963–2006. Sources: 1963–1976: Territory of Papua and New Guinea Bureau of Statistics; 1977–1990: DAL (1992); 1991–2004: Bank of PNG; 2005–2006: Carpenter Estates.

Processing, exporters and markets

About 10% of tea produced in PNG is consumed domestically and the remainder is exported. Domestic sales from the Western Highlands estates are around 700 tonnes per year (equivalent to PNG tea consumption of 0.1 kg/person/year). The export market has changed significantly in recent years. In 2006, 30% of tea exports went to Russia, 15% to Germany, 10% to the United Kingdom, 10% to the United States, 9% to India and 4–6% each to Belgium, Canada, Indonesia and Australia (Figure 5.12.2). Throughout the 1990s, Australia and the United States were the major importers of PNG tea (in most years receiving more than 50% of exports and around 20% of exports, respectively; Table A5.12.2). However, when the two major importers of PNG tea in Australia moved their packing plants to India and Indonesia, PNG exports to Australia collapsed. By 2006, the most important export destination for PNG tea was Russia, a market that only appeared in 2002.

Future prospects

The World Bank predicts that real tea prices will decline by about 12% between 2006 and 2015 (Figure 5.2.6). Global demand for black tea has weakened. This is partly because of a change by consumers from brewed tea to use of tea bags, which require less tea, but also because of a switch by some western consumers from black tea to green tea. Because of the static price of black tea, rising costs of production and the high cost of shipping from PNG compared with that from major overseas ports, PNG's tea producer is struggling to remain profitable. Prospects for expansion of PNG tea production are limited by constant law and order problems, land compensation demands, and the poor outlook for improved global prices.

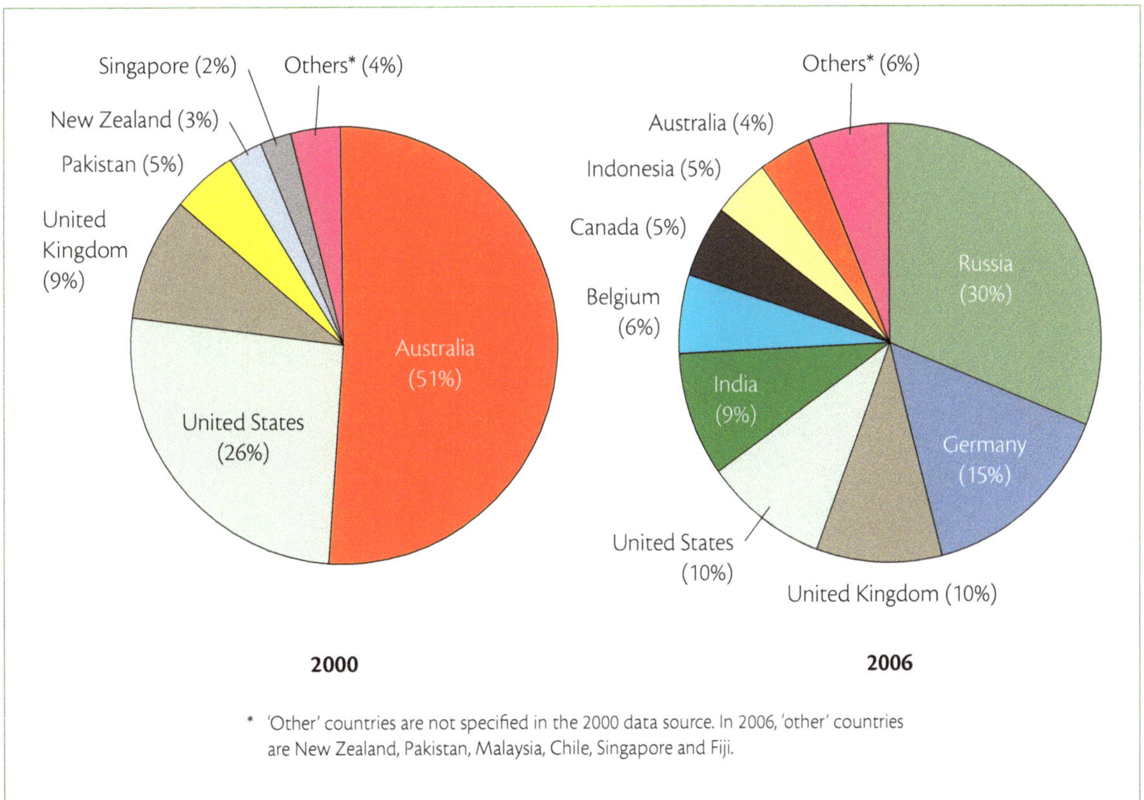

Figure 5.12.2 Destination of tea exports, 2000 and 2006 (by value). Sources: 2000: DAL (2001); 2006: Carpenter Estates.

Sources

Arthur, W.A. (c. 1978). Tea. In Densley, D.R.J. (ed). *Agriculture in the Economy: A Series of Review Papers*. Volume 2. Department of Primary Industry, Konedobu.

Bai, B. (2003). Performance of major agriculture export commodities and their future prospects. Rural Industries Council, Port Moresby. <http://peb.anu.edu.au/pdf/PNG%20Update%20BAI%202003.pdf>.

Bank of Papua New Guinea Quarterly Economic Bulletin Statistical Tables <http://www.bankpng.gov.pg/> [go to Statistics tab].

DAL (Department of Agriculture and Livestock) (1992). Exports of important agricultural commodities: quantity, value and average price per unit. Rural Statistics Bulletin 1992/6. Department of Agriculture and Livestock, Konedobu.

DAL (Department of Agriculture and Livestock) (2001). Handbook on Agricultural Statistics 2001. Department of Agriculture and Livestock, Konedobu.

Levett, M.P. (2000). Papua New Guinea rural development sector study 2000. Agriculture working paper. Draft final report. Australian Agency for International Development and the World Bank.

Simmons, R. and Anoser, K. (1992). The tea industry in Papua New Guinea: problems and policies. Policy Working Paper No. 1. Department of Agriculture and Livestock, Konedobu.

World Bank (2006). Prospects for the Global Economy. Commodity markets briefs – Tea <http://go.worldbank.org/G37PYVHE00>. Accessed November 2007.

5.13 Balsa

Balsa is a fast-growing tree that produces a very light wood used mainly to make model aircraft. The wood is also used in full-scale aircraft, table tennis bats, surfboards, fishing lures and as insulation. Balsa exports from PNG generated an average of K6 million per year from 2004 to 2006, which was about 1% of the value of forest product exports in this period. The value increased to K10 million in 2007. Commercial production of balsa wood in PNG is presently restricted to the north-east lowlands of the Gazelle Peninsula in East New Britain Province. Balsa is currently cultivated in areas below 300 m in altitude and where mean annual rainfall ranges between 2000 mm and 3000 mm, but it may grow well in other environments in PNG.

Adoption and history

Balsa was introduced to PNG in the late 1930s, with further introductions made between 1948 and 1961, mostly to East New Britain Province (ENB). The commercial potential of the crop was investigated in 1952 and 1956 and trial shipments were made to Australia. By the 1960s, a processing mill had been established at Keravat in ENB and a small industry had started to develop on the Gazelle Peninsula. In the 1980s and early 1990s, smallholder participation in the balsa industry on the Gazelle Peninsula was supported by extension activities provided by the National Department of Forestry and the ENB Division of Primary Industry. However, by 1995 extension activities had ceased, which resulted in harvesting rates exceeding replanting rates and a corresponding decline in available balsa. This situation was compounded by the Rabaul volcanic eruption of 1994, which defoliated many trees and caused activity in the industry to cease.

Following the eruption, recognition of the economic potential of balsa, coupled with concerns about the lack of extension services and information about the state of the industry, led provincial and national authorities to request overseas assistance. This resulted in the establishment of the East New Britain Balsa Industry Strengthening Project, funded by the International Tropical Timber Organization (ITTO). The project operated from 1996 to 2003.

Distribution of production and planting

On the Gazelle Peninsula in 2001 there were more than 200 hectares of smallholder balsa plantings and over 80 hectares of private, company-owned balsa plantations. The number of balsa growers in the Gazelle Peninsula area is unknown, but was about 200 in 2001. The area planted, and possibly the number of growers, has increased considerably since then.

Balsa is also grown in other parts of PNG including on New Ireland, Buka, Bougainville and Nissan islands, and in West New Britain and East Sepik provinces. However, much of this balsa cannot be marketed because there are no milling facilities in these areas. Furthermore, balsa matures within four to five years of planting, when it must be harvested or lose its marketability, and many of the trees in these areas are already too old for commercial harvest.

Levels of production

No data is available on balsa production or exports prior to 1996. Annual production, measured in terms of the volume of wood sold to the three balsa mills, averaged 10 000 m³ over the period 1997–2001.[1] Export of milled timber was 2000–4000 m³/year between 1996 and 2004, but increased to 11 000 m³ in 2007 (Figure 5.13.1, Table A5.13.1).

1 Data are from ITTO reports and one of the balsa mills, GS Models at Keravat. The recovery rate of milled timber depends on the product, with thinner sheets having a lower rate. The rate varies between mills in ENB because they produce different products, with an average recovery rate of 29% in 1997–2000, as measured by the volume of exports as a proportion of the volume purchased by the three mills in ENB in that period.

Processing, exporters and markets

Three mills were operating on the Gazelle Peninsula in ENB in 2001 and these processed balsa purchased from both smallholder and commercial plantation producers. By 2007, there were four mills in ENB, and it was anticipated that this would increase to seven mills by late 2008. The East New Britain mills could conceivably cater for balsa produced in other parts of PNG.[2] However, in order to be economic, the balsa would have to be processed into 'flitches' before being shipped to East New Britain for final processing, which means that flitch-processing plants would have to be established in the other balsa-growing areas. Economic potential would also depend upon adequate roads and shipping.

2 The East New Britain mills have purchased small amounts of balsa from West New Britain and New Ireland provinces. However, the balsa industry in these areas is only marginally economic due to the high cost of log haulage.

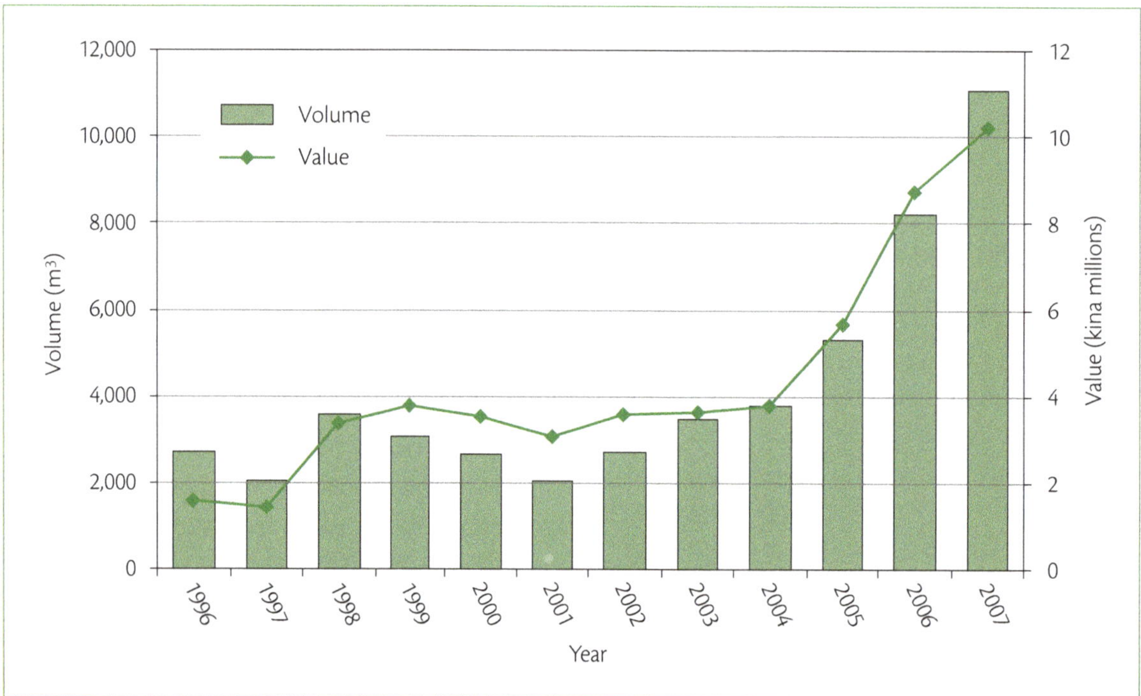

Figure 5.13.1 Volume and value of balsa exports, 1996–2007. Sources: 1996–2002: ITTO ENB Balsa Industry Strengthening Project Phase 2; 2003–2007: PNG Forest Authority.

The balsa mills in East New Britain Province export processed balsa in the form of planed sheets, blocks and strips of various shapes and sizes. Until 2003 balsa was exported to Australia, China, Germany, Italy and the United Kingdom, but since then a greater proportion has gone to China and India. In 2007 the main export destinations were China (43%), India (20%), Australia (14%) and Germany (10%) (Figure 5.13.2, Table A5.13.2).

Future prospects

Because balsa is a fast-growing crop, it can be integrated into village agricultural systems. It has more in common with an agricultural crop than a timber crop. The volume and value of balsa grown in ENB has grown rapidly since 2005. The area planted to balsa has expanded and some plantations have replaced old coconut stands with balsa. While prices remain high, the prospects for further growth are good. There is potential for expansion to other lowland areas, but this would depend on mills being established there.

Sources

Howcroft, N.H.S. (2000). Brief on the ENB balsa industry. ITTO Balsa Industry Strengthening Project Phase 2. International Tropical Timber Organization, Rabaul.

Howcroft, N.H.S. (c. 2001). The balsa industry in Papua New Guinea: its present development and future prospects in the New Guinea islands. Unpublished paper. International Tropical Timber Organization, Rabaul.

Howcroft, N.H.S. (2003). ITTO – East New Britain Balsa Industry Strengthening Project – Phase II. Progress Report 7. ITTO Project PD 7/99 Rev. 2 (F). International Tropical Timber Organization, Rabaul.

Howcroft, N.H.S. and Ohana, J. (2001). The balsa industry: its development and progress in the New Guinea islands and future prospects for further development in Papua New Guinea. Unpublished paper. International Tropical Timber Organization, Rabaul.

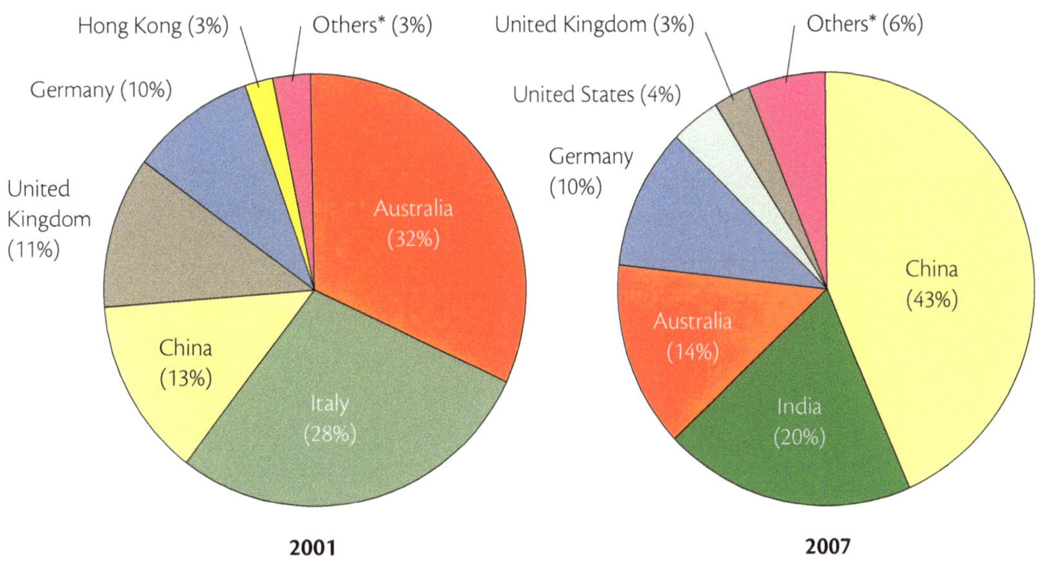

* 'Other' countries in 2001 were Spain, Taiwan, Japan and Malaysia. 'Other' countries in 2007 were Hong Kong, Japan, Malaysia, New Zealand, Taiwan and Vietnam.

Figure 5.13.2 Destination of balsa exports, 2001 and 2007 (by volume). Sources: 2001: ITTO ENB Balsa Industry Strengthening Project Phase 2; 2007: PNG Forest Authority.

5.14 Vanilla

Vanilla is used as a natural food and drink flavouring and as an ingredient in perfume. There are two commercial types of vanilla – Bourbon (*Vanilla planifolia*) and Tahitian (*V. tahitensis*). Both are grown in PNG. Bourbon vanilla is higher yielding, contains more vanillin and has a wider market. Tahitian vanilla needs a shorter period to induce flowering and is thus suited to a wider range of environmental conditions.

Vanilla is successfully grown from sea level to 600 m altitude, although it bears at over 1400 m in PNG. The daily temperature range for its optimal growth is 21–32 °C, with an average around 27 °C. Annual rainfall should be in the range 1700–2500 mm and evenly distributed throughout the year. However, two drier months are required to slow vegetative growth and induce flowering. Areas that do not have this dry season are not suitable for vanilla. The crop requires well-drained soils that are preferably deep and fertile and rich in organic matter. The crop's environmental requirements significantly limit the locations in which vanilla will flower. In PNG, extensive planting has been undertaken in locations where there is no regular dry season. These plantings are not likely to be successful.

Vanilla occupies a very small world niche market. Over the last 20 years, world consumption has varied between 1800 and 3000 tonnes, with production varying between 1200 and 4000 tonnes. This small market is characterised by extreme price fluctuations, made up of high price peaks and prolonged troughs of relatively low prices (Figure 5.14.1). Prices have been particularly sensitive to events in Madagascar, which produces 60–75% of world vanilla. Recent vanilla price fluctuations have been extreme. A severe cyclone that disrupted production in Madagascar in early 2000 triggered a rapid rise in world vanilla prices, which reached more than US$200/kg by the end of 2000. For three years, farmers throughout the vanilla-growing world earned unheard-of returns and responded accordingly. Farmers worldwide began feverishly planting and rehabilitating vanilla. By early 2004, the production from these increased plantings was entering a market that had contracted due to extremely high prices. By July 2004 the inevitable price collapse had begun. Prices have continued to fall since then and were about US$20/kg in mid 2007.

Adoption and history

Vanilla is a very recent cash crop in PNG. The PNG Spice Industry Board believes around 50 000 people were involved in the vanilla industry at the end of 2003. Five years earlier only a few hundred households were growing vanilla. Such meteoric industry growth is unprecedented in PNG agriculture.

The crop was first introduced to PNG in the 1960s at the Lowlands Agricultural Experiment Station at Keravat, East New Britain Province. At about the same time, plantings were made in the Wosera area of East Sepik Province. The original Wosera plantings were mainly of the Tahitian variety and were subsequently abandoned.

The foundations of today's industry were laid in 1993, when Allan Bird (Bangui Bio Products Ltd) planted vanilla on a large block near Maprik in East Sepik Province. He encouraged smallholders around him to also plant. This provided the critical mass upon which a substantial smallholder-based industry could quickly develop once the right price incentives existed.

Nowhere in the world was the response to the huge increase in world vanilla prices as spectacular as in PNG. A combination of factors explains the PNG vanilla phenomenon:

- PNG vanilla farmers had not experienced the previous periods of low prices and expected the high prices to continue in the future.
- The declining value of the kina compared with the US dollar significantly increased the kina price received by PNG growers. The grower price increased 1300% over a two-year period to reach more than K700/kg.
- Environmental conditions in parts of East Sepik Province proved ideal for vanilla production.
- The establishment of the Bangui Bio Products Ltd plantings at Maprik had given surrounding smallholders experience in growing and producing vanilla.
- Vanilla proved very attractive to villagers because it does not demand large areas of land to produce a good income.

Distribution of production and planting

Around 50–75% of PNG vanilla comes from East Sepik Province, with production concentrated in the Maprik, Dreikikir and Wosera areas. Other major producing provinces are Central, Morobe, East New Britain and Madang. Vanilla is now planted extensively in all the lowland and island provinces. While the main production areas of East Sepik Province provide ideal conditions to grow vanilla, the effect of climate was not taken into account in the rush to join the 2000 boom in the development of the industry. Many areas in which vanilla has been planted in PNG are much more humid than the vanilla-producing areas of Madagascar, Indonesia and Tonga. These locations are likely to prove unsuitable, particularly for the Bourbon variety.

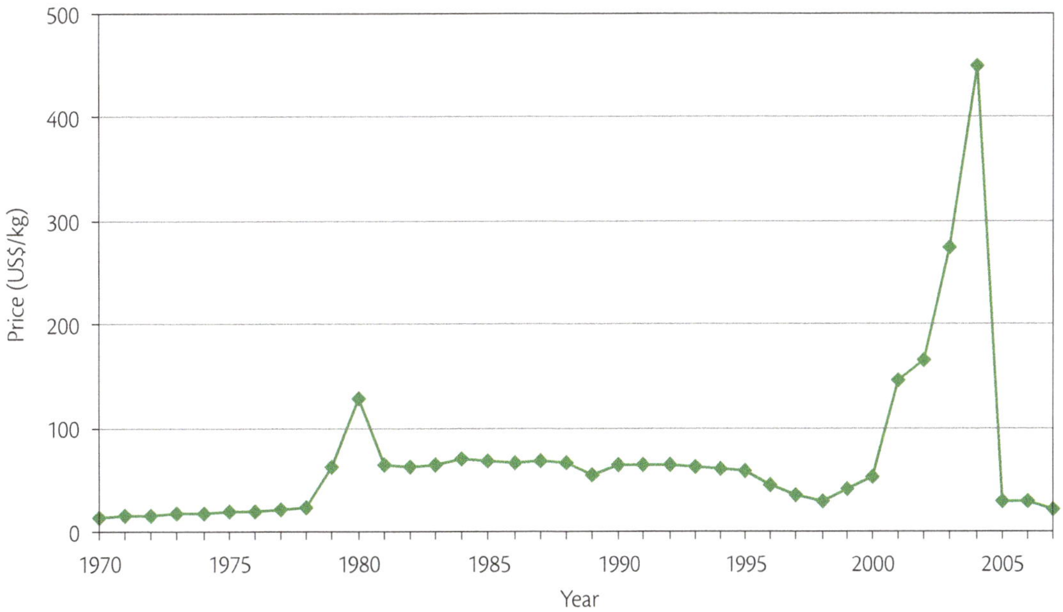

Figure 5.14.1 World vanilla prices, 1970–2007. Source: McGregor (2004) and Public Ledger.

Levels of production

The annual production and export of vanilla in PNG increased rapidly from about one tonne in 1997 to an estimated 200 tonnes by 2003 (Table 5.14.1, Figure 5.14.2). Within six years PNG had become the third largest producer in the world (after Madagascar and Indonesia), contributing about 10% of world production in 2003–2004. PNG production declined rapidly after 2004.

Bank of PNG data indicate that the export value of vanilla was K102 million in 2003 and K50 million in 2004, dropping to K3 million in 2005. However, these figures are probably lower than actual exports because of vanilla smuggled across the border to Indonesian Papua.[1]

[1] In 2004 Indonesia exported around 350 tonnes of vanilla, of which about 50 tonnes was likely to be re-exports of vanilla originally imported from PNG. Estimates of exports from PNG are not precise.

Processing, exporters and markets

The distinctive flavour and fragrance of good quality vanilla is developed by a slow curing process that is labour intensive and takes three to six months to complete. In PNG, villagers do their own curing. In other major vanilla-producing countries, curing is undertaken by specialist businesses and not by farmers. In PNG, grower curing has helped spread the benefits of the industry widely and has allowed for the participation of people in the most isolated of locations, but it has seriously lowered the quality of PNG vanilla. Many growers have little understanding of the slow and demanding requirements for successful curing. They tend to confuse the complex fermentation of vanilla curing with a drying process that they are familiar with for cocoa and copra. A necessary condition for producing quality vanilla is that the beans be harvested fully ripe. Beans that are immature when harvested will have low vanillin content and will quickly go mouldy regardless of how well they are cured.

The lack of knowledge of processing requirements is a consequence of the rapid expansion of the industry – and the lack of extension support provided. During

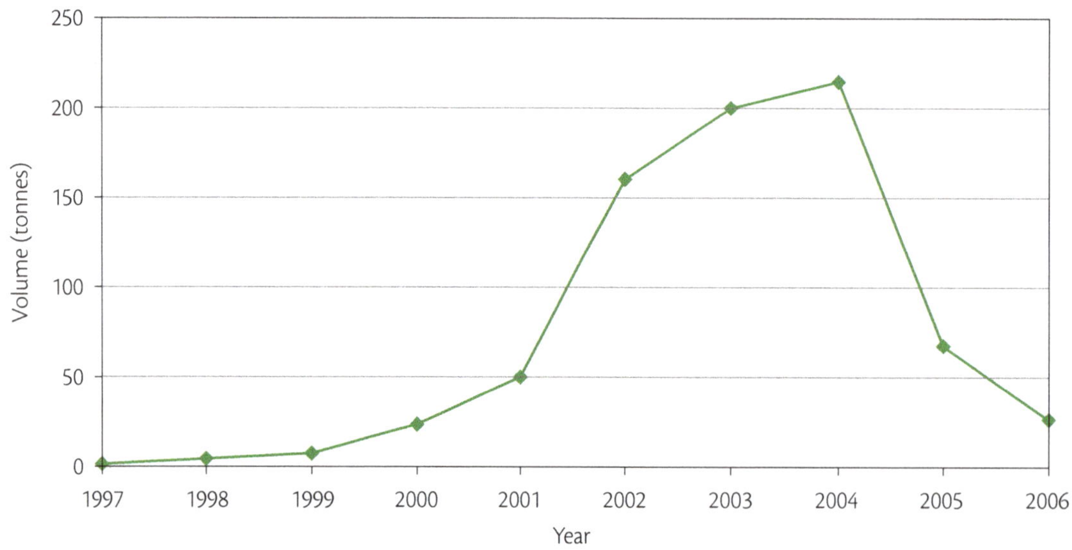

Figure 5.14.2 Estimated volume of cured vanilla bean exports, 1997–2006. Sources: Based on data from Agmark Pacific Ltd, PNG Spice Industry Board and Bank of PNG. These estimates are crude.

Table 5.14.1 Estimated volume of cured vanilla bean exports, 1997–2006

Year	Estimated exports (tonnes)
1997	1
1998	4
1999	8
2000	24
2001	50
2002	160
2003	200
2004	215
2005	68
2006	26

Sources: Based on data from Agmark Pacific Ltd, PNG Spice Industry Board and Bank of PNG. These estimates are crude.

the price boom, the practice of some traders and exporters of paying the same price regardless of quality provided little incentive to growers to learn and adopt correct curing techniques.

Vanilla marketing in PNG is disorderly and largely unregulated. At the end of 2004, 70 vanilla exporters were licensed with the Spice Industry Board. However, only 45 of these actually recorded exports and the ten largest exporters accounted for more than 90% of shipments. Several of these larger companies are representatives of overseas vanilla and spice companies, although most exporters are PNG-based companies. Some exporters apply strict standards and pay significant price premiums for quality. These exporters conduct their own farmer training programs. The results of these efforts in terms of quality have been outstanding and provide an example of how the industry can progress.

Vanilla buying is conducted over a three-month period, starting in February. All vanilla purchases are on a cash-on-delivery basis. Most East Sepik growers bring their cured bean to Wewak for sale. Buying is also conducted at Maprik, either by agents, middlemen or by exporters themselves. For security reasons large volumes are often purchased on-farm. Larger companies fund purchases from their own resources. Those linked to overseas spice companies have been financed by their overseas parent companies. Small companies found it difficult to generate a timely cash flow and many are no longer operating.

In the short term, villagers benefited from the competition created by the large number of exporters. The intense competition helped increase buying prices. However, many traders had little understanding of the product and were willing to purchase inferior quality vanilla at inflated prices. This has had a harmful effect on the quality of PNG vanilla and its overall reputation in the world market. The United States, France, Germany and Indonesia have been the main markets for PNG vanilla.

Future prospects

The vanilla industry that existed in PNG at the beginning of 2004 was based on unrealistic expectations and was not sustainable. By July 2004 prices had fallen to K125–140/kg for Grade 1 vanilla. Growers were starting to realise that poor quality (low vanillin content, over-dried, off-flavour and mouldy) product is unmarketable.

K70–80/kg is a reasonable price to plan for over the next few years. This price is unacceptably low for many present growers, particularly in light of the unrealistic expectations that were created by the boom. These growers will choose to stop producing. Some growers who have already invested in a vanilla plantation will continue to produce.

Vanilla's high unit value and non-perishability (when cured) make it particularly attractive to remote locations with poor or non-existent road access. Vanilla fits well into PNG agricultural systems and is particularly compatible with cocoa in East Sepik Province. Cocoa provides a regular low cash return throughout much of the year and vanilla gives a significant return once a year. Cocoa, with appropriate pruning, can even be used as a support tree for vanilla.

A sustainable PNG vanilla industry could have an initial annual production base of 40–50 tonnes. This would constitute a minor but significant export industry, equivalent to rubber. Like rubber, the vanilla industry is well suited to isolated, poor, lowland areas. The industry could earn high income windfalls similar to the 2000–2004 boom, but these will be infrequent exceptions rather than the norm. The PNG vanilla industry must also make the transition to centralised curing if it is to have a sustainable future.

Even at a price of K70/kg, vanilla provides a good return to labour and land, in suitable growing areas. Farm management models indicate that 0.5 ha of vanilla earns an average of nearly K5000 per year, with returns to labour of over K50 per person-day, provided that growers produce quality bean. This is considerably higher than for most other cash crops in PNG (Table 5.20.1).

Sources

Bianchessi, P. (2004). *Vanilla: Agriculture and Curing Techniques: a photographic handbook for vanilla farmers*. Venui Vanilla Co., Santo, Vanuatu.

McGregor, A. (2004). Diversification into high-value export products: case study of the Papua New Guinea vanilla industry. Food and Agriculture Organization of the United Nations, Rome.

Ngere, O. (2001). Vanilla. NARI Information Bulletin No. 1. National Agricultural Research Institute, Lae.

5.15 Spices and flavourings

A number of crops that yield spices or flavourings are grown in PNG and have potential as export cash crops. Vanilla has been the most significant in recent years (see Section 5.14). Chilli and cardamom were reasonably significant export crops in the 1970s and 1980s, but production of both has declined to low levels. Minor or potential export spice crops include annatto (bixa), black pepper, cinnamon, citronella grass, ginger, Japanese mint, lemon grass, nutmeg, patchouli and turmeric. The export of organically certified essential oils from cardamom, cinnamon, citronella grass, lemon grass, nutmeg and patchouli began in East New Britain Province in 2006.

The value to the national economy of these spices and flavourings is small. However, they provide useful income to villagers, particularly in more remote locations where production and marketing of coffee, cocoa, oil palm or fresh food is difficult. The potential for expansion, or even a return to past production levels, is limited, mainly by poor world prices and consequent low returns to labour inputs. Other factors that limit potential include lack of technical information for growers, poor marketing and inadequate transport.

The most important spice export crops

Chilli

Chilli is a type of capsicum used to flavour food in many parts of the world. The most common variety grown in PNG is the particularly pungent 'birds eye'. It is rarely used in cooking in PNG but is an export cash crop. Birds eye chilli grows between sea level and 1800 m and occasionally as high as 2400 m. Production is non-seasonal. Chilli was introduced in the early colonial period. It was first grown commercially on Aropa Plantation in Bougainville Province around 1955. Village plantings were made in the Dogura area of Milne Bay Province from 1959 and production reached 2 tonnes/year between 1959 and 1964. In the 1960s birds eye chilli was promoted as a cash crop in the Popondetta and Tufi areas of Oro Province and around Erave in Southern Highlands Province.

By the early 1970s the main chilli-producing provinces were Oro, Southern Highlands and Milne Bay. The volume exported increased during the 1970s, peaking in 1978–1981 at 190–265 tonnes/year (Figure 5.15.1, Table A5.15.1). Production dropped significantly from 1982. Export prices doubled between 1981 and 1982 and peaked at K6700/tonne in 1986 (equivalent to K29 400/tonne in 2005 buying power), but production fell because of problems with provincial governments' buying systems. A second factor was the poor quality of the PNG product and consequent loss of reputation of PNG chilli on the world market.

Production continues in a number of lowland and highland provinces, with several companies exporting. Production levels slowly declined from 1990, to less than 10 tonnes/year. Production has increased somewhat in recent years, with purchases from villagers in East New Britain Province of 19 tonnes in 2001 and 15 tonnes in 2002. Most chilli is grown in village plantings.

In the past, chilli has been grown by smallholders, either before export tree crops commenced bearing or in more remote locations where other cash crops cannot be grown or marketed. Smallholders were prepared to grow chilli when the marketing infrastructure was working, but it was never a popular crop. It provides low returns to labour and irritates the skin and eyes when the fruit is harvested and handled. However, demand exists and exporters struggle to meet it. Chilli production is likely to continue at low levels provided that marketing arrangements continue, but it is unlikely that production will return to that experienced in the late 1970s and early 1980s unless prices, and hence returns to labour, increase greatly.

Cardamom

Cardamom is a spice that is a common ingredient in cooking, particularly in South Asia and the Middle East. It is used to flavour tea and coffee in the Middle East and is also used as a medicine. In PNG cardamom grows from sea level to 1900 m but its usual altitude range is 550–1700 m. It was introduced into PNG in the mid 1960s. It is an ideal cash crop for intermediate altitude areas where road access is poor or non-existent, provided that prices are high.

Commercial production commenced in 1973, with early plantings in the Afore area of Oro Province and the Karimui Plateau in Simbu Province. Other plantings were on the Huon Peninsula of Morobe Province, particularly in the Pindiu area; the Baining Mountains and inland Pomio areas in

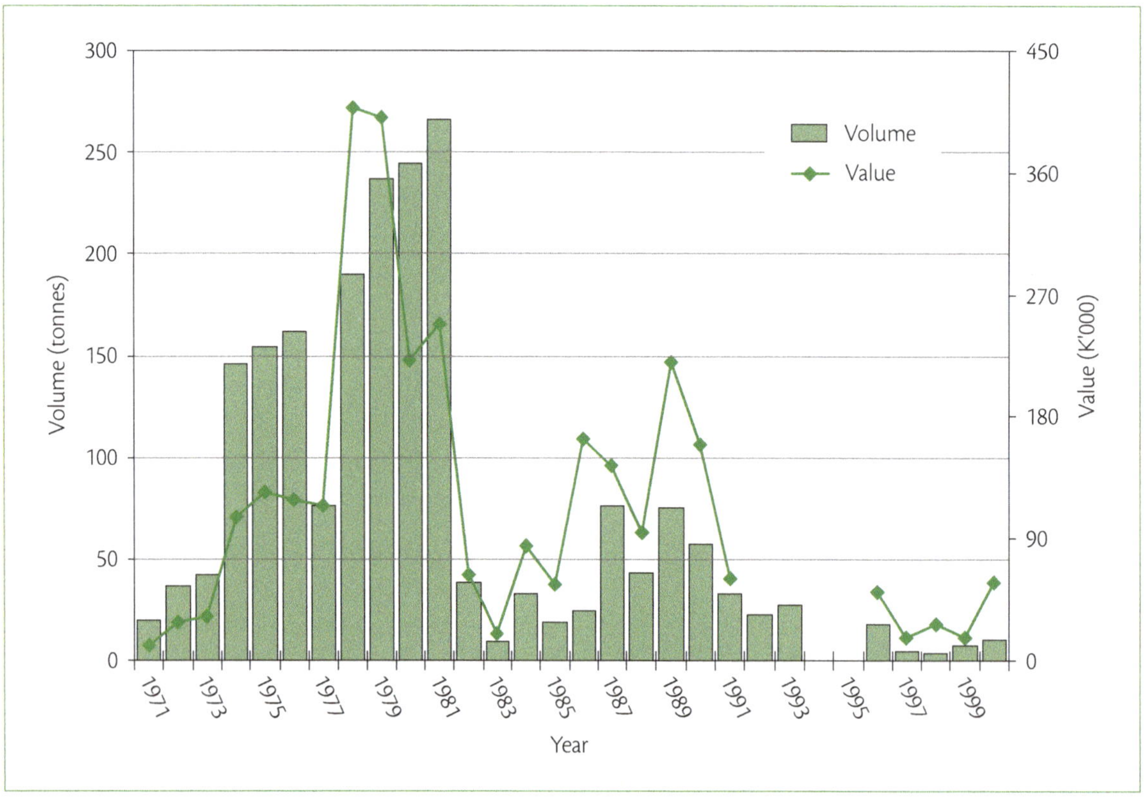

Figure 5.15.1 Volume and value of chilli exports, 1971–2000. Sources: 1971–1976: Wyatt (c. 1978); 1977–1991: DAL (1992); 1992–1993: Waisime (2000); 1996–2000: National Statistical Office of PNG.

East New Britain; and the inland Wakunai area of Bougainville Province. Plantations were established by international companies near Karimui (300 ha) in the late 1970s and near Bundi in Madang Province (160 ha) in the early 1980s. Exports commenced around 1974 and peaked at 320–390 tonnes/year from 1985 to 1987 (Figure 5.15.2, Table A5.15.1). The maximum production followed peaks in the world price from 1983 to 1985. Prices dropped in the late 1980s and PNG exports followed the price down. Both plantations ceased operating at about this time and all production since then has come from village plantings.

Renewed interest in cardamom occurred from 1998, with most production coming from the Baining Mountains, and lesser amounts from the Pomio and Wakunai areas, Karimui Plateau, the Huon Peninsula, and Jimi Valley in Western Highlands Province. Between 1998 and 2003, purchases from villagers in East New Britain Province were in the range 32–60 tonnes/year, averaging 48 tonnes/year. But world prices declined steeply in 2003, resulting in low returns to village growers. In 2004 the export price was less than K2/kg, compared with the peak price of more than K10/kg in 1984 (equivalent to K49/kg in 2004 kina value). Total exports were about 30 tonnes in 2003 and 20 tonnes in 2004, with 21 registered exporters.

Pacific Spices, a company based near Rabaul in East New Britain, commenced buying cardamom again in 2006 and has since exported 80 kg of cardamom oil to Japan, the United Kingdom and Australia. A Port Moresby-based company, Paradise Spices, has been buying cardamom from Simbu and Western Highlands provinces. Purchases from village growers in 2006 and 2007 were less than the period 1998–2003, which in turn was much less than in the mid to late 1980s.

Prospects for further expansion or even maintenance of cardamom exports are constrained by two main factors. The first is low world prices, and hence low prices paid to growers and poor returns to their labour. The second is the poor state of roads or lack

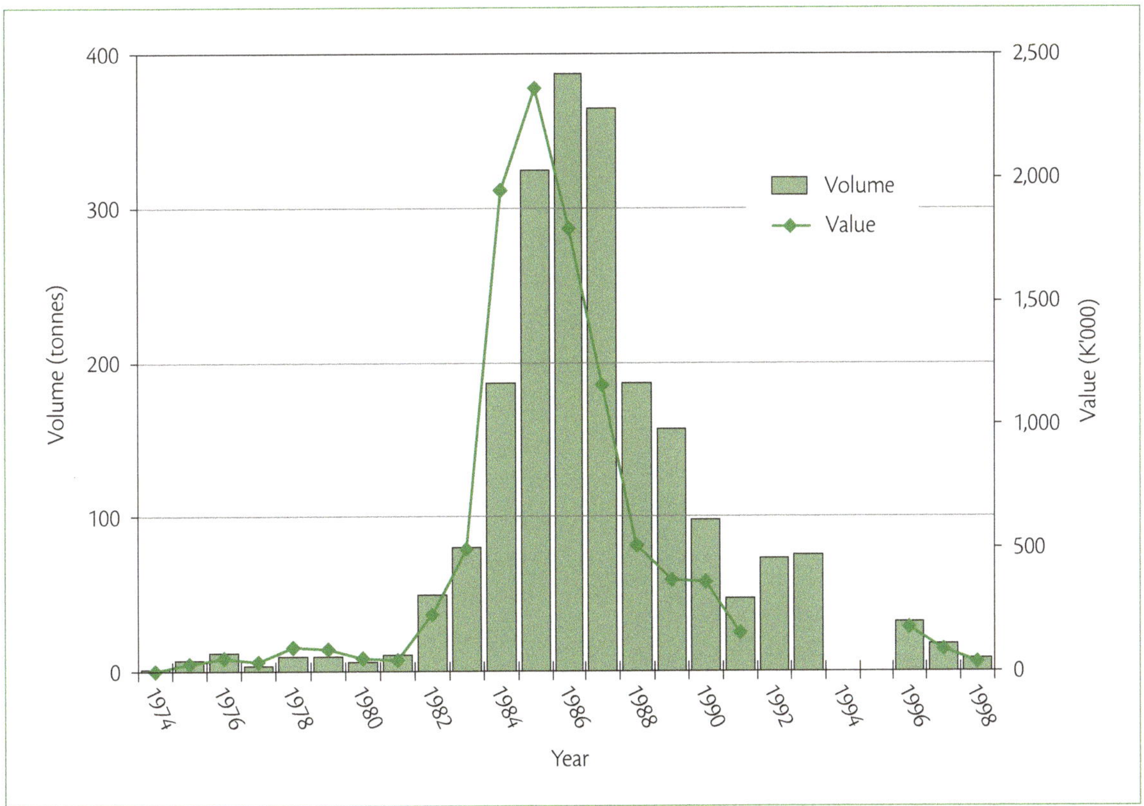

Figure 5.15.2 Volume and value of cardamom exports, 1974–1998. Sources: 1974–1976: Wyatt (c. 1978); 1977–1991: DAL (1992); 1992–1993: Waisime (2000); 1996–1998: National Statistical Office of PNG.

of roads in the intermediate altitude locations where cardamom grows. It is difficult to see a significant increase in production while prices remain depressed and road access is difficult to most producing areas.

Minor and potential spice export crops

Annatto (bixa)

The red dye annatto is extracted from the pulp surrounding the seeds of bixa and yields a red food colouring and a food flavouring. In PNG bixa usually grows between sea level and 1650 m and occasionally as high as 1900 m. The plant is a native of tropical America and was probably introduced into PNG from Indonesia between 1600 and 1870 (Table 1). A few plants are occasionally grown in lowland and intermediate altitudes and the dye is sometimes used as body paint. Bixa has potential as a cash crop, but has not been grown commercially in PNG.

Black pepper

Black pepper is one of the most commonly used spices worldwide. It was introduced into PNG by German settlers in the early colonial period. It grows and bears from sea level to about 700 m and occasionally as high as 1100 m. Pepper was grown experimentally at the Lowlands Agricultural Experiment Station (LAES), Keravat, from 1932, but was not promoted as a cash crop on the Gazelle Peninsula of East New Britain Province until the early 1970s. It was not widely adopted by villagers because labour requirements are high and returns to labour are very much less than for cocoa, betel nut and fresh food. A significant export market did not develop and exports averaged only 1.2 tonnes of peppercorn per year from 1971 to 1976. A very small amount of pepper has been grown for local sale or export since then; for example, 900 kg was sold by two growers in 1991.

Interest in pepper production has been renewed recently. Paradise Spices in Port Moresby is buying pepper for export from growers on the Gazelle Peninsula in East New Britain Province and the Bereina area of Central Province. Pacific Spices has planted about 5 ha and has developed an export market for organically grown pepper. Exports have grown from about one tonne in 2001 to about four tonnes in 2007.

Cinnamon

Dried leaves and dried inner bark of cinnamon are used in many parts of the world as a food flavour. Bark from native cinnamon trees is sometimes used in PNG as a food flavouring, a traditional medicine and for ritual purposes. The species that is commonly grown in the tropics (*Cinnamomum verum*) was introduced to PNG early in the colonial period and has been grown at LAES for many decades. Pacific Spices has purchased several tonnes of bark from the Pomio area, presumably harvested from a native species. About 500 litres of oil were extracted from the bark and exported. Small quantities of dried bark have also been exported. Paradise Spices has purchased cinnamon in Milne Bay Province and exported this as ground cinnamon bark.

Citronella grass

Citronella oil is distilled from citronella grass and is used as an insect repellent, as a perfume in soaps and cosmetics, and as a flavouring. Citronella was introduced to PNG in the late 1960s and some small experimental plots were planted. It is being grown commercially on a small scale on a number of plantations and in village plots on the Gazelle Peninsula. Pacific Spices commenced exporting oil in 2007.

Ginger

The underground stem (rhizome) of ginger is used as a spice throughout the world. It is also used for medicinal purposes. Ginger is an ancient introduction to PNG and grows between sea level and 1950 m and occasionally as high as 2200 m. It is a widely grown minor garden crop in PNG and used in cooking, as a medicine and in magic. Small quantities are sold in local food markets. Ginger yields well, even under coconut shade (Table 2.4.7). Paradise Spices is buying ginger in a number of lowland and highland provinces and exporting small quantities of ginger root flakes. Pacific Spices is growing ginger on a plantation with a view to extracting the essential oil for export.

Japanese mint

An essential oil is extracted from Japanese mint and used to flavour sweets and beverages, including herbal tea. The mint was introduced from Japan in 1968 when there was interest in establishing an industry in PNG. Agronomic trials were conducted at LAES in the early to mid 1970s. No commercial production has followed the experimental work.

Lemon grass

Lemon grass is grown to produce an essential oil that is used as a perfume. Lemon grass is also widely used in cooking to flavour food and drinks. In PNG lemon grass grows from sea level up to about 2000 m and occasionally up to 2100 m. It is an ancient introduction and is sometimes grown in villages. It is used to make a herbal drink (lemon grass tea) and is said to have medicinal and magical properties. It is grown commercially in a number of plantations and in village plots on a small scale on the Gazelle Peninsula. Pacific Spices has exported small quantities of lemon grass oil.

Nutmeg

Two spices are produced from the fruit of nutmeg trees: nutmeg and mace. The former is the tree seed and the latter is the dried reddish seed covering. They are used throughout the world to flavour foods and an essential oil is extracted for the perfume and pharmaceutical industries. *Myristica fragrans* is native to east Indonesia and is the nutmeg of commerce. It was introduced into PNG in the early colonial period and has been grown experimentally at LAES for many decades. A naturally occurring species of nutmeg in the island of New Guinea, *M. argentea*, has been exported in the past, at least from west New Guinea; for example, in 1894, 77 tonnes of nutmeg recorded as originating in 'New Guinea' was sold in Holland.

In 2007 there was about 15 ha of nutmeg in village plantings and 10 ha on two plantations on the Gazelle Peninsula. Pacific Spices processed and exported 1–2 tonnes of nutmeg per year between 2001 and 2007. The company has also extracted and exported some nutmeg oil. Paradise Spices has purchased some nutmeg in Milne Bay Province and exported this as whole or ground nutmeg seed.

Patchouli

An essential oil is extracted from the dried leaves of patchouli and used in perfumes, incense and as a scent in household products. Patchouli was introduced to PNG for experimental purposes in the late 1960s. It is grown commercially on a small scale on the Gazelle Peninsula and Pacific Spices has exported small quantities of oil.

Turmeric

The underground stem (rhizome) of turmeric is boiled, dried and ground into a deep orange-yellow powder. It is commonly used as a spice in curries and other South Asian cooking, to colour food, and as a dye. It is also used in traditional medicine in Asia. Turmeric is an ancient introduction to PNG. It is widely grown up to about 1000 m altitude and is mainly used as a dye rather than as a food flavouring. Turmeric has been grown for export in Madang, East Sepik and East New Britain provinces. Between 2001 and 2007, Pacific Spices purchased and exported about 4 tonnes of turmeric per year. Paradise Spices recently purchased turmeric for export from Central, East New Britain and Milne Bay provinces.

Sources

Arek, T. (2005). Development of Pacific Spices: a small producer, buyer and exporter spice venture. *Development Bulletin* (67):19–21.

Breay, G.S.J. (1971). Cardamom. *Harvest* 1(1):15–19.

DAL (Department of Agriculture and Livestock) (1992). Exports of important agricultural commodities: quantity, value and average price per unit. Rural Statistics Bulletin 1992/6. Department of Agriculture and Livestock, Konedobu.

DAL (Department of Agriculture and Livestock) (1994). Annual review of: export crops, programmes and projects. Export Crops Branch, Department of Agriculture and Livestock, Konedobu.

Hollingsworth, R. and Woodhouse, S. (2002). Potential for a pepper industry in Papua New Guinea. Technical Bulletin No. 5. National Agricultural Research Institute, Lae, Papua New Guinea.

Levett, M.P. (2000). Papua New Guinea rural development sector study 2000. Agriculture working paper. Draft final report. Australian Agency for International Development and the World Bank.

Lowlands Agricultural Experiment Station (1978). *Lowlands Agricultural Experiment Station 1928–1978*. Lowlands Agricultural Experimental Station, Keravat.

Magei, T. (1975). Chillies. *Rural Development Series Handbook No. 6*. Department of Agriculture, Stock and Fisheries, Konedobu.

Powell, J.M. (1976). Ethnobotany. In Paijmans, K. (ed). *New Guinea Vegetation*. Australian National University Press, Canberra. pp. 106–183.

Waisime, M. (2000). Prospects for spice industry in Papua New Guinea. *Papua New Guinea Journal of Agriculture, Forestry and Fisheries* 43(1):108–119.

Wikipedia Encyclopedia (2007). Culinary herbs and spices <http://en.wikipedia.org/wiki/List_of_herbs>. Accessed October 2007.

Wyatt, S.L. (c. 1978). Spices. In Densley, D.R.J. (ed). *Agriculture in the Economy: A Series of Review Papers*. Volume 2. Department of Primary Industry, Konedobu. pp. 1–14.

5.16 Pyrethrum

Pyrethrum produces a daisy-like flower that is used to make a natural insecticide. It grows best at high-altitude locations in the equatorial tropics, including in PNG and East Africa, and in some temperate-climate locations, such as Tasmania, Australia.

The active ingredient, pyrethrin, is extracted from dried pyrethrum flowers and used to make insecticides for household, agriculture, public health and food industry uses. These include aerosols, sprays, pet shampoo and mosquito coils. Pyrethrum is valued because it is highly effective at repelling or killing a broad range of insects, but is not toxic to mammals, including humans, and breaks down quickly in sunlight, leaving no residues.

Pyrethrum thrives at very high altitudes in the PNG highlands and flower production rises steeply with increasing altitude (Figure 5.16.1). Production of pyrethrum is weakly seasonal, being slightly higher in September–March and lower in April–August.

Adoption and history

Pyrethrum plants were first introduced into PNG in 1938. A number of other introductions were made in the 1950s, with plants from Kenya in 1957 forming the basis for selection of planting material in PNG. Agronomic research commenced at Aiyura in Eastern Highlands Province in 1961, but it was quickly found that the station (1600 m altitude) was too low for pyrethrum production. A new research station was established in 1966 at Tambul in Western Highlands Province (2300 m) and the pyrethrum selection program was moved there. The current germplasm collection is maintained at the Taluma Research Station on the Sirunki Plateau in Enga Province.

A processing facility was established in 1964 by Stafford Allen Ltd (PNG) at Kagamuga near Mount Hagen in Western Highlands Province. The enterprise was not profitable and the Australian Administration purchased the facility in 1973 through the company Kagamuga Natural Products Pty Ltd. The factory operated for another 20 years, but closed in 1994. The plant has a capacity to process 420 tonnes of dried flower per year.

Distribution of production and planting

All crop production was done by villagers and pyrethrum was usually interplanted among sweet potato and other crops. The Department of Agriculture, Stock and Fisheries purchased dried flowers from growers. In the late 1960s production was encouraged in many places throughout the highlands, including the Lagaip, Kandep, Wabag and Wapenamanda areas of Enga Province; the Tambul area of Western Highlands Province; the Nipa, Margarima, Ialibu and upper Mendi areas of Southern Highlands Province; the Kerowagi,

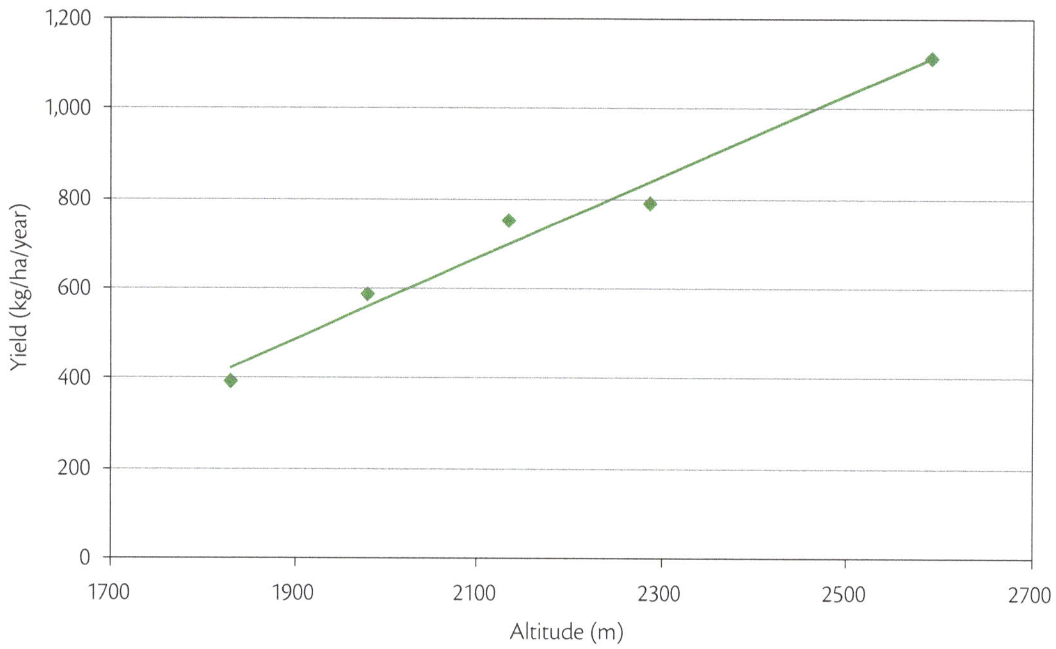

Figure 5.16.1 Relationship between dried pyrethrum flower yield and altitude in village plantings, Enga Province.
Source: Quinlan (1968).

Gumine and Gembogl areas of Simbu Province; and the Okapa, Henganofi and Lufa areas of Eastern Highlands Province.

Pyrethrum was initially grown by villagers at altitudes as low as 1800 m, but within a few years the producing areas shrank to a limited number of very high altitude locations where the crop was most productive. By the early 1970s much of the production was concentrated in the Laiagam area (Enga Province), with significant amounts also grown in the Tambul and Gembogl areas. In 1974 an estimated 22 300 villagers grew pyrethrum. By the late 1970s production was confined to a narrow very high altitude band, mostly in Enga Province (96–99%), and the high-altitude upper Nebilyer Valley in Western Highlands Province.

Levels of production

During the period 1965 to 1993, between 200 tonnes and 400 tonnes of dried flowers were purchased per year, with a peak of just under 600 tonnes in 1967 (Figure 5.16.2, Table A5.16.1).[1] The volume of pyrethrum extract exported followed the pattern of flower purchases (Figure 5.16.3, Table A5.16.2). The pyrethrum processing factory closed in 1994 and no flowers were purchased between 1995 and 1999.

In 1995, the Enga provincial government, concerned about the lack of cash-earning opportunities for villagers in high-altitude parts of the province, formed the Enga Pyrethrum Company to revive the industry. The company initially operated erratically but was revived in 1999 when the company took possession of the Kagamuga processing plant near Mount Hagen. Because pyrethrum was no longer being grown by villagers, it was necessary to multiply and distribute planting material again. This commenced in 1999 and dried flowers were

1 Under optimum conditions, pyrethrum yields about 1200 kg dried flower/ha/year. This contains about 22 kg of the active ingredient (pyrethrin) at 1.8% pyrethrin content. In practice, average grower yields were about 650 kg dried flower/ha/year and PNG pyrethrum has averaged 1.4–1.5% pyrethrin content. A yield of 650 kg dried flower/ha/year containing 1.4% pyrethrin would produce 9 kg of pyrethrin/ha/year.

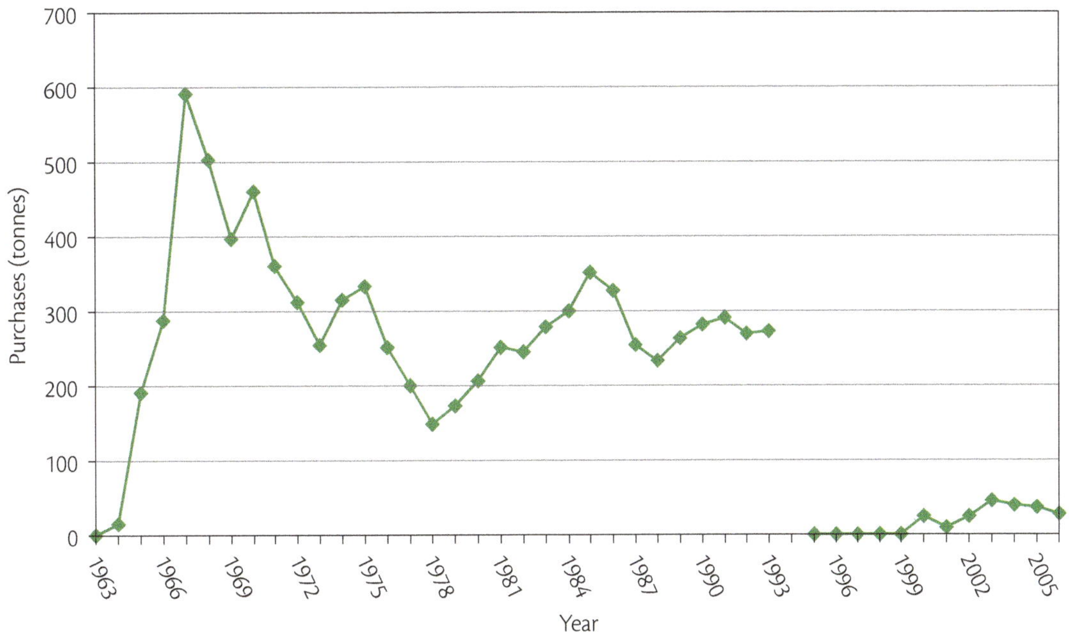

Figure 5.16.2 Purchases of dried pyrethrum flowers, 1963–2006. Sources: 1963–1977: Anderson (c. 1978: Table 1) (data are for Australian financial years, that is, 1962–1963 to 1976–1977); 1978–1979: Carrad (1982:156); 1980–1993: DAL (1994); 2000–2006: Wakasa Mecksaene, Wabag (pers. comm.)

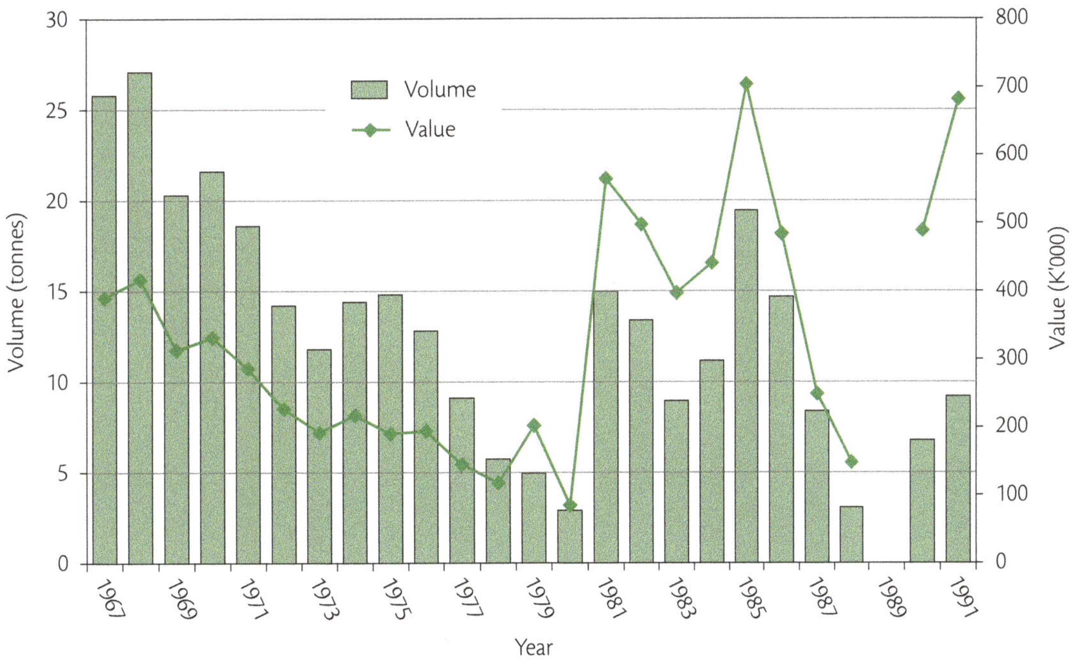

Figure 5.16.3 Volume and value of pyrethrum extract exports, 1967–1991. **Note:** Pyrethrum extract was not exported in 1989. Sources: 1967–1976: Anderson (c. 1978: Table 2) (data are for Australian financial years, that is, 1966–67 to 1975–76); 1977–1991: DAL (1992).

purchased from 2000 onwards. In 2003–2006 the Enga Pyrethrum Company produced pyrethrum extract which it exported to the United States.[2] The volume of exports has been small compared with volumes in the mid 1960s to early 1990s.

Processing, exporters and markets

One of the most important issues for regeneration of the PNG pyrethrum industry is to ensure that production takes place only at high altitudes (2400–2800 m) so that growers achieve the highest possible yields and the best returns on their labour.

Botanical Resources Australia Pty Ltd, a company based in Hobart, Australia, has signed an agreement with the Enga Government to import PNG's pyrethrum extract from 2006 to 2008 with an option to extend this arrangement beyond 2008. This company is one of the largest pyrethrum producers in the world and has supplied about 40% of global natural pyrethrum products in recent years.

Enga Pyrethrum Company paid K1.50/kg for dried flowers in 2006 (K2/kg delivered to the Kagamuga factory). This was a much lower price in real terms than was paid in the past. For example, in 1989–1993, the price received by growers was K1.50/kg. However, the kina now has much less purchasing power than in the 1980s and early 1990s because of depreciation of the currency and inflation (see Section 4.2). At current prices, growers are receiving a gross payment of about K2 per day's labour input. With such a low return to labour, production may not return to the levels experienced from the mid 1960s to the early 1990s (see Section 5.20). The PNG pyrethrum industry may become viable again but this will depend on improved productivity by the growers and the factory producing a reliable supply of good quality product for the export market.

2 The Kagamuga factory processes dried flowers into crude pyrethrum extract, which contains about 26% pyrethrin.

Sources

Anderson, D. (c. 1978). Pyrethrum. In Densley, D.R.J. (ed). *Agriculture in the Economy: A Series of Review Papers*. Volume 2. Department of Primary Industry, Konedobu. pp. 1–23.

Carrad, B. (1982). The economy. In Carrad, B., Lea, D.A.M. and Talyaga, K.K. (eds). *Enga: Foundations for Development. Volume 3 of Enga Yaaka Lasemana*. Department of Geography, University of New England, Armidale. pp. 146–178.

DAL (Department of Agriculture and Livestock) (1992). Exports of important agricultural commodities: quantity, value and average price per unit. Rural Statistics Bulletin 1992/6. Department of Agriculture and Livestock, Konedobu.

DAL (Department of Agriculture and Livestock) (1994). Annual review of: export crops, programmes and projects. Export Crops Branch, Department of Agriculture and Livestock, Konedobu.

International Bank for Reconstruction and Development (1965). *The Economic Development of the Territory of Papua and New Guinea*. The Johns Hopkins Press, Baltimore. pp. 118–120.

Mengge-Nang, A. (1980). Growing pyrethrum in the highlands. *Harvest* 6(2):47–52.

Nagle, W.G. (1981). Notes on pyrethrum. *Harvest* 7(1):43–46.

Quinlan, T.J. (1968). Cultivation of pyrethrum in Papua and New Guinea. *Rural Digest* 9(6):7–13.

Scoullar, B. (1973). Pyrethrum and the highlander. Extension Bulletin No. 3. Department of Agriculture, Stock and Fisheries, Port Moresby.

5.17 Other income from plants

Sales of fresh food and export cash crops are the major income sources for most rural Papua New Guineans. In addition to the major sources (described in earlier sections of Part 5) many other, non-food, products derived from plants are sold within PNG and sometimes exported.[1] The most important are betel nut, firewood and tobacco. These and some other, mostly small volume, plant-derived income sources are described in this section.[2]

Betel nut and betel pepper

Production, trading, transport and retailing of the stimulants betel nut and betel pepper is a significant economic activity. In the period 1990–1995, sales of betel nut and betel pepper generated an estimated K20 million income for rural villagers, which was 10% of the total income earned by villagers in PNG from agricultural production. It was the fourth most important source of income from agriculture for rural villagers after Arabica coffee, fresh food and cocoa (Figure 5.1.1, Table 5.1.1). About a third of the rural population lived in locations where betel nut was sold, a proportion exceeded only by those involved in selling fresh food or coffee (Figure 5.1.2). The amount earned from betel nut sales has increased since 1995 because of price rises over the past 10–15 years and increased sales in the highlands. The amount earned by retailers of betel nut is much higher than that received by producers because large profit margins occur in betel nut trading.

Betel nut grows in the lowlands and at intermediate altitudes (see Section 3.5). It is generally consumed with the catkins, leaves or stems of the lowland betel pepper plant and with slaked lime. Highlanders began consuming betel nut after the Highlands Highway was upgraded to an all-weather road in the mid 1960s and highland men began working as labourers on lowland plantations. Consumption has increased over the past 40 years and significant quantities are now transported into the highlands to meet the demand. Growth of urban populations has also created a large demand for betel nut. Thus there is a demand for betel nut, pepper and lime from highland rural villagers, who cannot grow betel nut (35% of the national population), as well as from the 19% of the national population who are not rural villagers.

Lowland villagers have responded to the demand for betel nut, betel pepper and lime by expanding plantings. Large numbers of people are involved either in growing, trading and transporting betel nut or retailing it (Figure 5.17.1). In the major urban centres betel nut is sold in markets separate from fresh food markets. Betel nut is moved long distances to markets in the highlands, to Port Moresby, and

1 Two food crops were previously grown as cash crops and exported from PNG. These were peanut and purple passionfruit (see Sections 3.2 and 3.3).

2 Clay pots and salt, although not made from plants, are included in this section for convenience.

to mining settlements such as Lihir, Ok Tedi and Porgera. It is also moved shorter distances and sold in all urban centres and numerous other locations (Figure 5.17.2). Most betel nut is moved by road, river or sea transport, but some is flown within PNG, as well as to Australia.

Up to the mid 1970s the demand for betel nut and betel pepper in the highlands was met from production in the Markham Valley and other locations in Morobe Province with road access to the Highlands Highway. With the growth in consumption in the highlands, the area from which betel nut is drawn has expanded greatly to include coastal locations in Sandaun, East Sepik, Madang, Morobe, Oro and Milne Bay provinces; along the Sepik, Keram and Yuat rivers in East Sepik Province; in the lower Ramu River area and Karkar Island in Madang Province; and the west coast of New Britain.

Highlanders travel to lowland locations to buy betel nut from individual producers or from markets. They on-sell to intermediate traders who in turn retail it in highland towns and villages. There are wholesale betel nut markets in Popondetta, Wewak and near Madang town. Betel pepper is sold by Karkar Island traders to highland traders at Kabugum village on the mainland. Lime, manufactured from shells in coastal villages near Lae, is purchased by highland traders in Lae market. 40-Mile market in the Markham Valley was a significant trading point for betel nut producers from the valley and nearby areas. Producers sold betel nut to highlanders who moved it to their home areas, as well as to other traders who sold it in Lae city. An unknown disease has been killing betel nut palms in the Markham Valley since 2003 and the 40-Mile market closed in mid 2007 when supply in this area ceased.

Betel nut and betel pepper sold in the large markets in Port Moresby come from the Bereina area in Central Province and coastal and riverine parts of Gulf Province as far west as Kerema. It is also transported from coastal locations south-east of Port Moresby in Central and Milne Bay provinces.

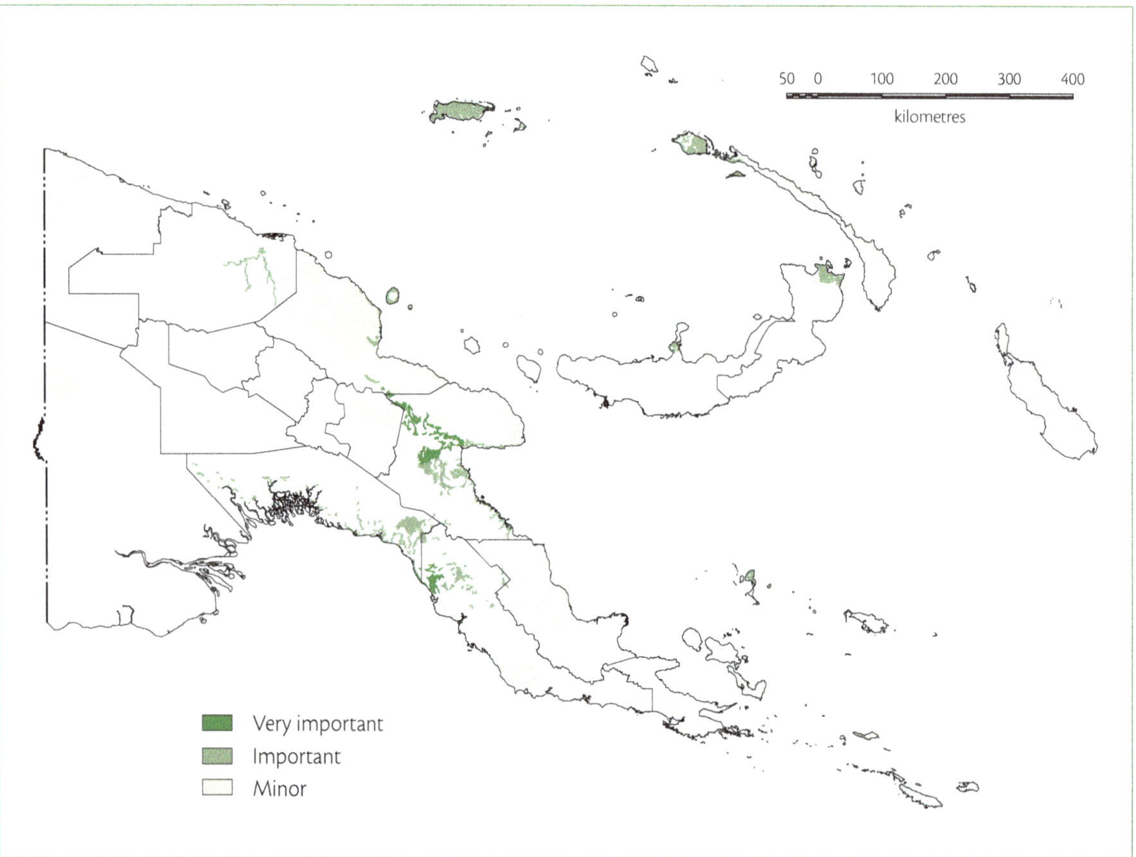

Figure 5.17.1 Locations where sales of betel nut provided income for rural villagers, 1990–1995. Source: MASP.

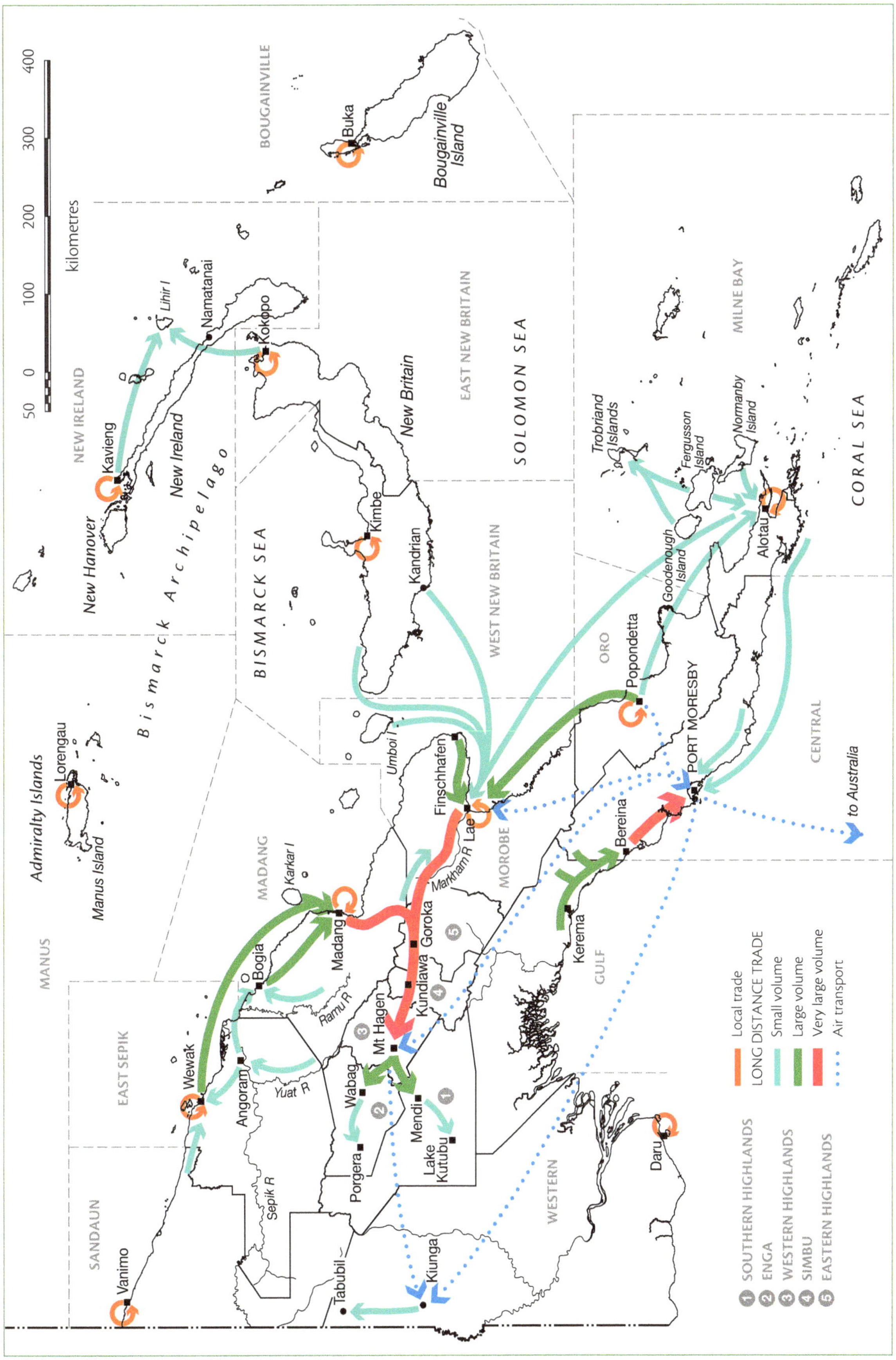

Figure 5.17.2 Betel nut trade routes, 2007. Source: Author's observations and Tim Sharp (pers. comm.).

Producers or traders sell betel nut and betel pepper to highland traders at a number of locations in Port Moresby and several sites along the Hiritano Highway. These traders sell to other traders or directly to consumers in the city.

Firewood

Firewood is sold in many urban markets and some rural markets in PNG, particularly in the highlands, but also near major lowland urban centres (Figure 5.17.3). In the period 1990–1995, sales of firewood generated an estimated K5 million income for rural villagers (Figure 5.1.1). Almost a quarter of the rural population (23%) lived in locations where firewood was sold (Figure 5.1.2, Table 5.1.1).

In the highlands, cool night-time temperatures necessitate the use of open fires for heating, as well as cooking. Timber is the most common source of energy used by villagers for cooking. It is also used in urban areas, with much of the firewood purchased in local markets, although kerosene stoves are commonly used in towns. *Casuarina oligodon* is the most common tree species used for firewood in the highlands. A 1978 study in a Simbu Province village recorded average firewood consumption of 1.3 kg/person/day. A 1980 study in Enga Province estimated that average firewood consumption was 2.25 m^3/person/year for rural people and 1.9 m^3/person/year for urban people. In rural lowlands villages, most firewood comes from trees cleared during shifting cultivation. Fuelwood plantings to supply PNG's biggest cities have been promoted sporadically in the Atzera Hills near Lae and in the Port Moresby area.

Firewood is also used for processing tea, sugar cane, copra and cocoa; for small-scale bread baking; and for making slaked lime for use with betel nut. Some of this timber is purchased from villagers; for example, Ramu Agri-Industries Ltd buys firewood locally. Firewood is scarce in some lowland rural

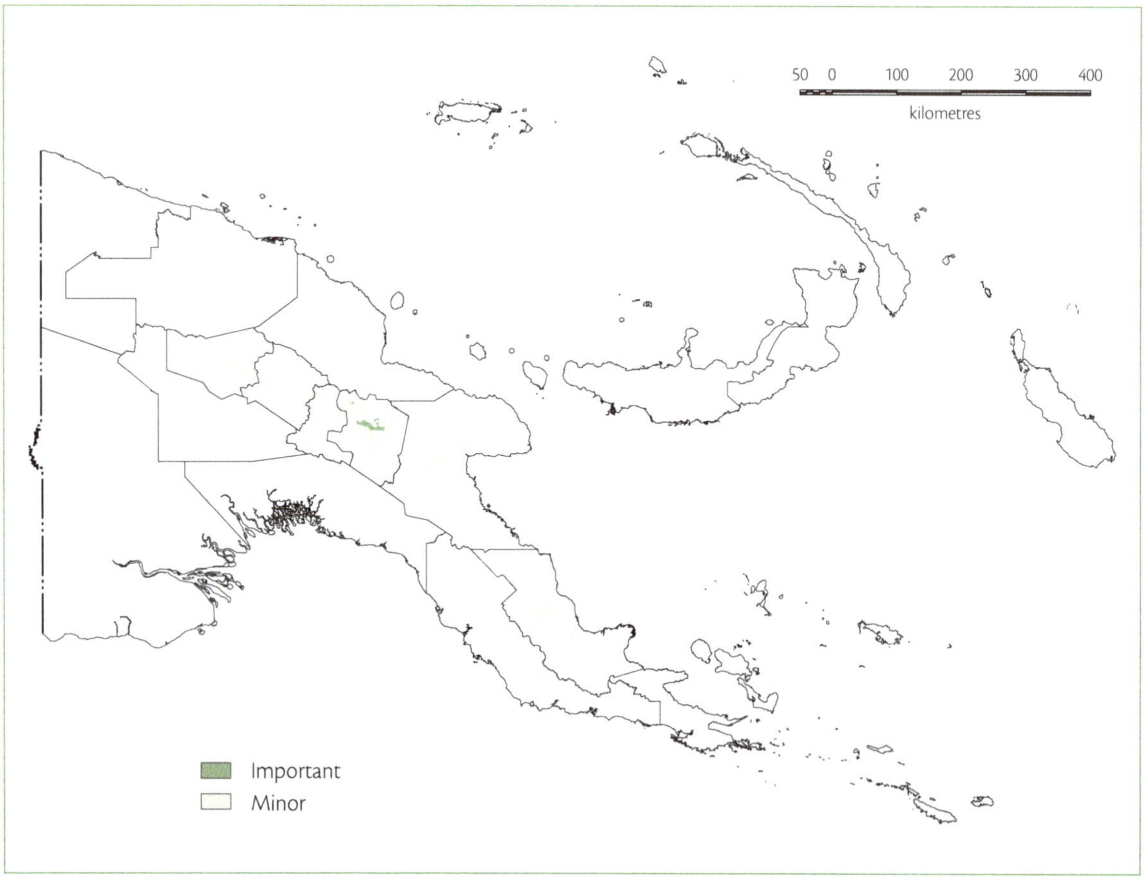

Figure 5.17.3 Locations where sales of firewood provided income for rural villagers, 1990–1995. Source: MASP.

locations, including in some villages on the Gazelle Peninsula, but it is not usually traded for cash. Given the increasing population pressure in some rural locations, ongoing urbanisation and the high cost of imported kerosene (Figure 4.3.4), demand for firewood is likely to continue to increase. Expanded production of timber for firewood, construction and industrial use has the potential to generate good income for rural villagers.

Tobacco

Tobacco, grown for own use or sale, is one of the most widely grown plants in PNG villages (Figure 3.5.1). Leaves are fire-cured and rolled in newspaper to form a cigarette or sometimes smoked in a pipe. Cured tobacco leaves are sold in many markets or directly to consumers (Figure 5.17.4) and are sometimes sold as cut leaf or as cigarettes rolled in newspaper in some highland markets. In the period 1990–1995, sales of tobacco generated an estimated K3 million income for rural villagers (Figure 5.1.1). About a fifth of the rural population lived in locations where tobacco was sold (Figure 5.1.2, Table 5.1.1). The quantity sold per person was generally low, as was income per person.

Commercial tobacco production has a long and not particularly successful history in PNG. The German-owned Neuguinea Kompagnie grew up to 240 ha of tobacco for export at several sites near modern Madang between 1888 and 1902, but rapid losses of soil fertility and pests and diseases caused the crop to be abandoned. Tobacco was also grown as a minor export crop on a number of plantations in East New Britain and Bougainville in the 1930s; for example, 10 ha of plantation tobacco was grown in the Territory of New Guinea in 1933.

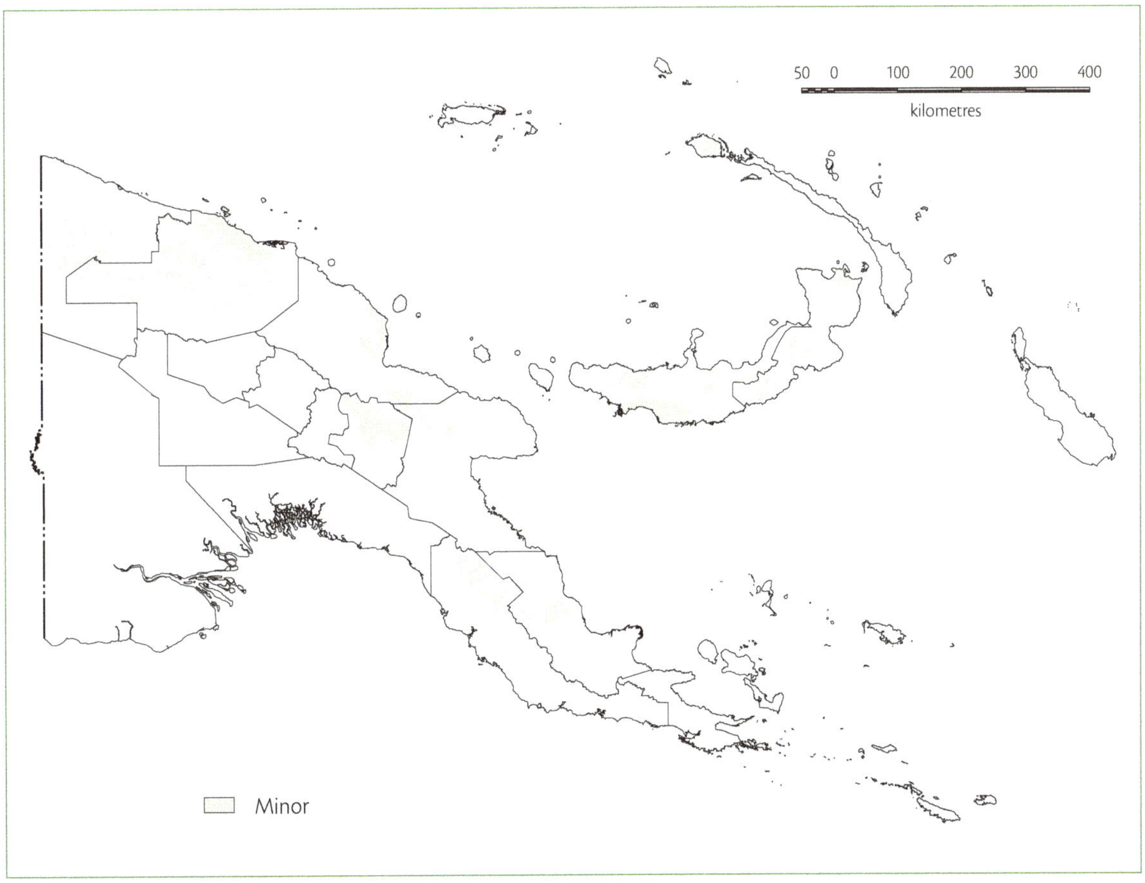

Figure 5.17.4 Locations where sales of tobacco provided income for rural villagers, 1990–1995. Source: MASP.

Tobacco was grown by two companies in the Madang and Port Moresby areas from 1958 and in the Goroka area from 1960. In 1977 a total of 90 ha of estate tobacco was grown in the Goroka area, the Markham Valley, and on a small estate in the Saidor area of Madang Province to supply a factory in Goroka manufacturing cigarettes, twist tobacco and coarse-cut tobacco products for the domestic market. Estate production in the Goroka area ceased in the late 1970s, and large-scale commercial tobacco production in the Markham Valley ceased in the 1980s.

Tobacco was promoted as a cash crop for villagers in Eastern Highlands Province between 1970 and about 1978. By 1971 there were about 150 growers in the Bena Bena and Asaro valleys. In 1975 an area of 97 ha was planted to produce flue-cured tobacco (used in cigarette manufacture) by 500 village growers in the Goroka and Asaro areas. Village growers, dissatisfied with prices paid by the buying companies, ceased production for cigarette manufacture in the late 1970s. The proportion of locally grown tobacco used in manufacture was small, even in the mid 1970s. In 1977 only 3% of PNG requirements were grown locally, with the rest imported. Cigarettes for domestic sale are currently manufactured by British American Tobacco (PNG) Limited at a plant in Madang using imported leaf.

Marijuana

Marijuana is an illegal cash crop that is grown in at least five provinces (see Section 3.5). The quantity grown and value are not known, but anecdotal information and a limited number of published papers indicate that significant quantities are sold, particularly in the larger urban centres; that marijuana cultivation is an income source for some people in rural locations; that its sale is used to fund purchases of guns and ammunition; and that the returns for some middlemen are large. Marijuana is reportedly exported via Torres Strait or other routes to Australia, where it has a reputation for a high content of the active ingredient (THC). However, the extent of international marijuana trade, including the guns-for-drugs trade to Australia, is probably overstated. Prices in the producing areas are much less than in Australia.[3] By 2007 marijuana was being sold more or less openly in markets in at least three highlands provincial capitals and by many street sellers in Port Moresby. Arrests for growing are sporadic.

Kava

Kava was grown and used in a limited number of places in PNG, including locations in Western Province, some small islands in Manus Province and near Madang town. It is still used occasionally in some of these locations, but its use and consumption remain limited. An export market for kava developed in the Pacific in the late 1990s, with demand in Europe and the United States, but kava was exported from PNG only on a very small scale. The PNG Spice Industry Board reported in 2001 that small quantities were grown and processed in several locations in Madang and East Sepik provinces. The dried kava was exported to Japan and Australia. Reports of liver damage from overconsumption of dried kava have caused European and American markets to be restricted.

Building material, canoes, carvings, clothing, rope, string bags, tools and weapons

Building material. Most houses in rural PNG are made from locally grown plant material, including timber, woven cane (used for walls and partitions) and thatch for roofing. People normally gather the material for their houses themselves, but some building material is sold at roadside markets or within villages, particularly sheets of woven cane.

3 In the early 1990s marijuana was regularly offered in rural parts of the highlands at K5–7 for a cigarette pouch, containing 30–50 grams. The same quantity was worth several hundred dollars on the street in Australia at the time. The price in highland markets in 2007 was 50 toea per cigarette, or K15 for a one kilogram rice bag, containing perhaps 300–500 grams. These weights are estimates.

Canoes are widely used to transport people and goods in rivers and the ocean. Timber with particular characteristics is needed to construct the various parts of canoes, including hulls, outriggers and masts. Where this is not available locally, it is sometimes purchased from villagers in other locations. Making and selling canoes is a specialised activity on, for example, Panaeati and Gawa islands in Milne Bay Province. Large ocean-going canoes from these two islands are exported to nearby islands in south-east and north-east Milne Bay Province respectively.

Carvings, bowls and other artefacts are commonly made from wood in PNG. They are sold for local use, in urban markets, to tourists, and some are exported. Occasionally people sell timber to others for carving; for example, on Kiriwina Island in Milne Bay Province, timber for carving is sold by people on the south of the island to people in the north. People from Ialibu in Southern Highlands Province and south-west Bougainville Island are renowned for the high-quality items they make from woven cane, such as trays, table mats, baskets and handbags. People from other highlands locations have copied these products and are also weaving items for sale. The volume and value of trade in carvings, woven items and other wooden artefacts is unknown, but is significant. Such items are sold in all major and many smaller urban centres.

Clay pots are made in a number of locations, particularly along the New Guinea north coast, parts of Milne Bay Province such as the Amphlett Islands and in parts of East Sepik Province. They are traded for food in a number of locations in Milne Bay and Manus provinces (see Box 5.3) and are sometimes sold for cash.

Clothing in the pre-European period was made from plant material, including reeds (for skirts), bark (for belts), gourds and shrubs. In parts of the highlands, men covered their buttocks with *Cordyline fruticosa* leaves. In the western part of the highlands, women's skirts were made from the *Eleocharis dulcis* reed. Traditional clothing has become uncommon, however, and has been largely replaced by second-hand clothing imported from Australia. Traditional clothing is now worn mostly for ceremonial occasions or in some remote locations by older people. Some clothing made from local plant material is sold occasionally.

Furniture. High-quality furniture is made from PNG hardwood timber and cane, but domestic demand is small and little furniture is exported. Most hardwood timber is exported as round logs to China, Japan and Korea (see Section 5.8).

Kapok was introduced to PNG in the 1880s as a potential export crop. It grows from sea level to 1250 m altitude. Kapok was exported from about 1900 until about 1940, with most production from plantations in East New Britain and Madang provinces and small amounts from most other lowland provinces.

Mats and brooms. Floor and sleeping mats are made from pandanus leaves, particularly on some small islands. These have been largely replaced by imported manufactured items, but continue to be made in some locations and may be occasionally sold. Brush brooms are made from locally grown plants and are commonly sold in markets.

Ropes were made in PNG from fibres from a number of plants. Imported rope has largely replaced locally produced rope. Ropes for tethering pigs are sold in highlands markets, but even they are mostly made from imported fibre except in remote locations.

Sisal was introduced into PNG as a potential export crop in the early 1900s. By 1917, 2400 ha was being cultivated in Central Province and processing equipment had been set up in four locations. Some sisal fibre was exported until the mid 1920s, when production ceased.

String bags (*bilums*) were made from fibre from the bark of a number of plants, including the *tulip* tree, a species of *Ficus*, a *Syzygium* species and *Hibiscus tiliaceus*. Nowadays, *bilums* are mostly made from imported wool or string because these fibres last longer. Nevertheless, *bilums* made from locally grown plant fibre are still widely used in some parts of PNG, including in the Telefomin and Oksapmin areas of Sandaun Province. A large number of *bilums* are sold in urban markets each year. They are widely used within PNG by women, men and children, are increasingly used in Solomon Islands, and a small number are exported to Australia.

Tools are made from a range of material, including timber, bamboo, fibre and bone. Metal axe heads and spade blades have almost completely replaced those made from stone and timber. Wooden digging sticks are still widely used in the highlands. Axe handles are homemade and are sometimes sold in highlands markets.

Weapons, including bows, arrows, spears, shields and axes, were very widely used in PNG for hunting birds and game animals and for fighting. These have been partially replaced by firearms, including shotguns and high-powered rifles. However, traditional weapons are still made and used, and a small proportion are sold. Stocks for homemade shotguns are commonly made from wood in the highlands, and presumably some are occasionally sold.

Other items made from plants include musical instruments, lime containers (for use with betel nut), bamboo smoking pipes and household vessels. These are not seen in markets, but some are probably sold occasionally.

Salt

The production and trading of salt has been reported from many parts of PNG. Salt was extracted from sea water, springs and plant ash. To extract salt from springs, grass, leaves or sticks were placed in the spring for a period, then dried and burnt and the ash collected. Where saline springs were absent, people made salt from a number of plants, for example, from the perennial cane grass, *Coix gigantea*, in the Wonenara area of Eastern Highlands Province. Bars of salt were traded long distances in some inland locations. It is not known whether these practices still persist as imported salt is widely available, but it is possible that locally manufactured salt is still traded in some places.

Minor forest products

Cajuput oil is produced from distilling the leaf of *Asteromyrtus symphyocarpa*, a tree that occurs naturally in seasonally inundated areas of Western Province south of the Fly River. Stills are operated by village communities. The oil produced is sold in bulk to local buyers including trade stores in Daru. Trade store owners ship the oil to Port Moresby where it is bottled as Waria Waria oil and retailed through pharmacies and supermarkets or through direct orders. The oil is used externally for the relief of coughs and colds, aches and pains and as a mild antiseptic for wounds. The village industry started in 1996 with the first commissioned still. Current production is around 200 litres per year and is worth about US$8000.

Copal gum is a resin extracted from the bark of kauri trees (*Agathis* spp.), used chiefly for making varnishes. Natural stands occur in Sandaun and East Sepik provinces with some smaller stands elsewhere in the lowlands. Copal was exported from Sandaun Province, at least until the late 1970s.

Dammar resin is extracted from the sapwood of a tree species (*Vatica papuana*) that is present in southern Western Province and some islands in Milne Bay Province. Exports from Tagula (Sudest) Island in Milne Bay Province commenced in 1887 and continued intermittently until the 1970s.

Eaglewood trees grow in New Guinea rainforest below 600 m altitude. The tree produces a fragrant resin that has been exported from PNG since 1998 (see Box 5.17). Over the period 1999–2007 official exports averaged 5.5 tonnes/year, with a mean value of US$438 000/year (Table 5.17.1). The volume peaked in 2001–2003.

Ivory nut is the seed of a plant related to sago and was once used to make buttons. It was exported from PNG between about 1900 and the 1930s.

Massoi bark (*Cryptocarya massoy*) is an aromatic bark used as a medicine in South-East Asia. The tree is widely distributed in the lowlands of mainland PNG provinces. Some massoi bark was exported from Sandaun Province from near the Indonesian border, at least until 1990.

Box 5.17 Eaglewood: an emerging industry

The eaglewood tree produces a fragrant resinous wood that is highly sought after in parts of Asia and the Middle East where it is used for religious and medicinal purposes. High-quality resin can attract very high prices per kilogram. In PNG, a species of eaglewood, *Gyrinops ledermannii*, was first harvested for resin production in about 1998 in the Yapsiei, May River and Ama areas of Sandaun Province. Since then, there has been a significant expansion in the rates of harvesting and export of eaglewood resin (which is often referred to in PNG by its Indonesian name 'gaharu'). Gaharu is primarily harvested in parts of Sandaun, East Sepik and Madang provinces, although a limited amount of harvesting occurs in other parts of the country. The presence of eaglewood has also been confirmed in Gulf, Central, Southern Highlands and Western provinces and there are anecdotal reports of its presence in Enga, Milne Bay and Oro. In PNG, eaglewood grows in rainforest between 70 m and 600 m altitude (but has been recorded at over 1000 m), and where mean annual rainfall ranges between 1700 mm and 5200 mm and is not strongly seasonally distributed. It is important to note that only a small percentage of eaglewood trees contain gaharu. The resin is produced as a chemical defence following injury to the tree. The ability to recognise which individual trees contain gaharu is important for the sustainable management of the resource.

The natural distribution of eaglewood is associated with some of the poorest and remotest rural communities in PNG. This presents both the opportunity to increase income levels in these communities and the challenge of overcoming problems associated with poor access to markets and information. Access to information is particularly important because destructive and inappropriate harvesting and management practices have already led to the severe depletion of trees in some areas. WWF (formerly The World Wide Fund for Nature) has developed a plan, already implemented in some communities, that assists villagers to manage their eaglewood resources through '14 steps for sustainable gaharu harvest'.

According to official records, there are two main gaharu exporting companies operating in PNG, which between them employ about 24 agents who purchase gaharu from several hundred rural households. Prices are determined by the grade of gaharu. Currently there are five grades designated according to the colour, shape and density of the wood. The PNG Forest Authority introduced pricing guidelines in early 2001 to prevent unscrupulous buyers from exploiting villagers who may not be aware of the true value of gaharu. The top grade of gaharu currently attracts a premium price of US$560/kg.

The unreported and unlicensed trade in gaharu is thought to be much larger than the legal trade and is a threat to the sustainable management of eaglewood. The main destination for gaharu from PNG is Singapore, with smaller amounts exported to Malaysia and Indonesia.

For more information on eaglewood in PNG see Gunn et al. (2004).

Table 5.17.1 Volume and value of eaglewood and sandalwood exports, 1997–2007

Year	Eaglewood		Sandalwood	
	Volume (kg)	Value (US$)	Volume (kg)	Value (US$)
1997	–	–	22,000	46,956
1998	–	–	20,000	47,000
1999	1,011	136,888	19,000	47,930
2000	2,670	541,442	42,000	36,269
2001	10,529	1,644,687	53,259	47,554
2002	9,380	521,838	126,092	153,387
2003	12,127	502,982	76,760	117,718
2004	4,560	205,462	52,160	67,560
2005	5,046	208,452	50,870	40,346
2006	3,556	142,597	2,000	2,000
2007	928	33,690	15,354	22,419

Source: Anna Martin, PNG Forest Authority.

Noni (*Morinda citrifolia*) is common in many coastal locations and interest in its commercial production is increasing. A number of products made from noni, particularly noni juice, have become popular as herbal medicines in Western countries and the Pacific since the 1990s. The juice and other products are exported from a number of Pacific island countries. One business has been buying noni fruit in Lae in recent years.

Rattan. There are approximately 60 species of rattan cane in PNG. Rattan cane was exported from PNG in the 1980s and 1990s and has been used for local furniture manufacture. The current status of exports is not known.

Sandalwood (*Santalum macgregorii*) is a tree endemic to PNG that is found in the seasonally dry lowlands of Central and Gulf provinces together with a separate population in Western Province. The natural distribution of the species has been dramatically reduced by fire and harvesting to the point where it is under threat of extinction in the wild. Sandalwood is highly sought after, particularly in Asia and the Middle East, for carving, incense, perfume and medical preparations derived from the oil. Exports of PNG sandalwood are recorded from the 1890s onwards. Over the period 1997–2007 exports averaged 43.6 tonnes/year, with a mean value of US$57 000/year (Table 5.17.1).

Seeds of a number of native plants, orchids, trees and palms have been exported from PNG for use in horticulture and forestry. Several tree species including *Acacia mangium*, *A. crassicarpa*, *Eucalyptus pellita* and *E. deglupta* (*kamarere*) have formed the basis of major plantation developments in South-East Asia since their introduction to commercial forestry there in the mid 1980s.

Tannins are extracted from the bark, roots or leaves of plants. A factory to extract tannin from mangroves was established at Aird Hills in Gulf Province in 1954, but closed after operating for three years.

Tigaso oil (sometimes spelt digaso oil) is the dark brown oily sap of a tree (*Campnosperma brevipetiolata*). People in the Lake Kutubu to Mount Bosavi area of Southern Highlands Province extract the oil by hollowing out one side of the trunk of a mature tree and allowing the sap to accumulate in the hole. They then trade the oil long distances to higher altitude locations in Southern Highlands and Western Highlands provinces where these trees do not grow. People mix the oil with sweet-smelling leaves and rub the mixture on their bodies and hair as a cosmetic and a perfume before ceremonies. A related species (*Buchanania* sp.) is used for the same purpose. It is reported that petroleum-based oils, such as engine oil, and vegetable cooking oil are now being used as a substitute, but the trade in tigaso oil continues. The oil was carried in bamboo tubes until the late 1980s, but by the mid 1990s it was being stored in plastic containers and was flown out from Pimaga airstrip for sale in the highlands.

Ornamental horticulture

Cut flowers are rarely sold in fresh food markets in PNG, unlike some other Pacific island centres, such as Honiara. Some flowers are sold at a craft market in Port Moresby, where they sell readily. For many years villagers have sold rings of alpine flowers to motorists near Daulo Pass in Eastern Highlands Province. These are worn or placed on vehicles to indicate that someone has been on a journey via Daulo Pass. Everlasting daisies are used as hair decoration in parts of Southern Highlands Province and may be sold occasionally.

There is potential for significant expansion of ornamental horticulture in PNG, with possible markets being:

- Domestic markets, particularly in the larger urban areas.
- Export of domesticated indigenous orchids and palms.
- Export of temperate-climate flowers, such as roses, from the highlands.

Domestic demand is unsatisfied and production could be expanded. The market includes restaurants, hotels, business houses and funeral-related sales. Currently plastic flowers are used in many PNG funerals, but this could change if a good supply of fresh flowers was available. Cut flowers for hotels and restaurants are sometimes supplied from within PNG, but imported flowers are mostly used, including orchids from Malaysia and Singapore. The value of flowers imported in 2004 was K139 000 (Internal Revenue data).

The island of New Guinea has the greatest concentration of orchid species in the world, with more than 10% of an estimated 30 000 world orchid species. This presents an unrealised potential for development of a horticulture industry to export orchids. Self-sown orchids are said to be exported illegally. Temperate-climate flowers, like roses, grow well in the highlands. Export potential, for example to Japan, is severely constrained by a lack of air cargo capacity and reliability.

Sources

Allen, B.J. (1981). The north coast region. In Denoon, D. and Snowden, C. (eds). *A History of Agriculture in Papua New Guinea: A Time to Plant and a Time to Uproot*. Institute of Papua New Guinea Studies, Port Moresby. pp. 105–127.

Carrad, B. (1982). The economy. In Carrad, B., Lea, D.A.M. and Talyaga, K.K. (eds). *Enga: Foundations for Development. Volume 3 of Enga Yaaka Lasemana*. Department of Geography, University of New England, Armidale. pp. 146–178.

Dennis, M. (1981). Plantations. In Denoon, D. and Snowden, C. (eds). *A History of Agriculture in Papua New Guinea: A Time to Plant and a Time to Uproot*. Institute of Papua New Guinea Studies, Port Moresby. pp. 219–248.

Doran, J.C., Brophy, J.J. and Niangu, M. (2007). Chemical variation in the oils of *Santalum macgregorii* (PNG sandalwood). In Thomson, L., Bulai, S. and Wilikibau, B. (eds). *Proceedings of the Regional Workshop on Sandalwood Research, Development and Extension in the Pacific Islands and Asia*. Secretariat of the Pacific Community, Nadi, Fiji. pp. 101–107.

Evans, B. (1998). Non-timber forest products with commercial potential in the Kikori River watershed of PNG: Mt Bosavi and upper Kikori. WWF-US Kikori Integrated Conservation and Development Project. WWF, Moro, Papua New Guinea.

Gunn, B.V., Stevens, P., Singadan, M., Sunari, L. and Chatterton, P. (2004). Eaglewood in Papua New Guinea. Resource Management in Asia-Pacific Working Paper No. 51. Resource Management in Asia-Pacific Program, Research School of Pacific and Asian Studies, The Australian National University, Canberra.

Haley, C.G. (1976). Eastern Highlands Province. *Harvest* 3(1):22–29.

Halvaksz, J. (2006). Marijuana in Papua New Guinea: understanding rural drug consumption and production. *Development Bulletin* (69):56–58.

Hughes, I. (1977). *New Guinea Stone Age Trade: The Geography and Ecology of Traffic in the Interior*. Terra Australis 3. Department of Prehistory, Research School of Pacific Studies, The Australian National University, Canberra.

King, B. (1987). Fuelwood planting in urban Papua New Guinea. *PLES: An Environmental Education Journal for the South Pacific Region* (3):1–9.

Langlas, C.M. and Weiner, J.F. (1988). Big-men, population growth, and longhouse fission among the Foi, 1965–79. In Weiner, J.F. (ed). *Mountain Papuans: Historical and Comparative Perspectives from New Guinea Fringe Highlands Societies*. The University of Michigan Press, Ann Arbor. pp. 73–110.

Lemmens, R.H.M.J., Soerianegara, I. and Wong, W.C. (eds) (1995). *Plant Resources of South-East Asia No 5(2). Timber Trees: Minor Commercial Timbers*. Backhuys Publishers, Leiden.

Levett, M.P. (ed.) (1992). *A survey of the north-east lowlands of the Gazelle Peninsula of East New Britain Province with particular reference to the potential and acceptability of new agroforestry projects*. Research Bulletin No. 53. Department of Agriculture and Livestock, Port Moresby.

MacKenzie, M.A. (1991). *Androgynous Objects: String Bags and Gender in Central New Guinea*. Volume 2. Studies in Anthropology and History. Harwood Academic Publishers, Chur, Switzerland.

McDonald, D. (2005). A rapid situation assessment of drug use in Papua New Guinea. *Drug and Alcohol Review* 24(1):79–82.

McGregor, A. (2006). Solomon Islands Smallholder Agriculture Study. Volume 3: Markets and Marketing Issues. AusAID, Canberra.

McGregor, A. (in press). Report on potential for ornamental horticulture in the Pacific.

McKillop, R. and Firth, S.G. (1981). Foreign intrusion: the first fifty years. In Denoon, D. and Snowden, C. (eds). *A History of Agriculture in Papua New Guinea: A Time to Plant and a Time to Uproot*. Institute of Papua New Guinea Studies, Port Moresby. pp. 85–103.

Moulik, T.K. (1973). *Money, Motivation and Cash Cropping*. New Guinea Research Bulletin No. 53. New Guinea Research Unit, The Australian National University, Port Moresby and Canberra.

Newcombe, K. and Bayes, T. (n.d.). Energy in a Simbu village. Report No. 5: Papua New Guinea Human Ecology Programme. UNESCO/UNEP, MAB 11 'Integrated Ecological Studies of Human Settlements'. Centre for Resource and Environmental Studies, The Australian National University, Canberra.

Papua New Guinea Orchid News website <http://www.orchidspng.com/>. Accessed July 2007.

Parsons, N.J. (c. 1978). Tobacco. In Densley, D.R.J. (ed). *Agriculture in the Economy: A Series of Review Papers*. Volume 3. Department of Primary Industry, Konedobu. pp. 1–15.

Ryan, P. (ed.) (1972). *Encyclopaedia of Papua and New Guinea*. 3 volumes. Melbourne University Press in association with the University of Papua and New Guinea, Melbourne.

Saulei, S.M. and Aruga, J.A. (1994). The status and prospects of non-timber forest products development in Papua New Guinea. *Commonwealth Forestry Review* 73(2):97–105.

von Fleckenstein, F.W. (1974). An analysis of the tobacco industry in Papua New Guinea. Unpublished paper. New Guinea Research Unit, The Australian National University, Port Moresby.

5.18 Income from animals

Small amounts of income are derived from the sale of a wide range of animal products in PNG. This is illustrated by the variety of animal products sold in one small highlands urban market (Table 5.18.1). The income received is usually small and irregular, but larger amounts can be earned, particularly from the sale of chickens. The quantities sold and income earned from animal products are not well known. Changes have occurred in recent decades in the type of product sold; for example, there has been an increase in the total number of live chickens sold and a probable decrease in other items, such as meat from hunted wild animals. Income from fish and other marine and freshwater animals, other than crocodiles, is covered in Section 5.9.

Domestic animals

The most important domestic animals marketed live and as meat are pigs, chickens and cattle (Table 2.6.3). Minor domestic animals sold for meat include goats, sheep, ducks and rabbits. Dogs, cats, cassowaries and other captured birds and mammals are occasionally sold as pets. Future prospects for expanded production of domestic animals are discussed in Section 2.6.

Pigs are the most important domestic animal in PNG, but the sale of pig meat or live animals is only a minor source of income in most locations, even in the highlands where pigs are most numerous (Table 2.6.2).[1] In some places where other sources of income are unavailable, for example in the Tari basin in Southern Highlands Province, sale of pork or live pigs is the single most important income source. The price of large animals is often very high, particularly in the highlands, where a large pig may fetch more than K1000. Sale of pig meat is a common response to the need for cash. For example, during the 1997 drought and associated food shortages villagers in parts of the highlands killed pigs, which were becoming difficult to feed, and sold the meat to raise cash to buy imported rice and flour. Pigs and pig meat are significant in customary exchanges in parts of the highlands.

As well as significant village production, there is a small commercial pig sector in which animals are raised to be sold to abattoirs or directly to individual buyers for local slaughter.

Chickens are sold as live animals and occasionally as meat. The chickens are either from free-range village production or from peri-urban and village small-scale 'chicken projects'. Large numbers of broilers are grown by smallholders and sold to one large firm near Lae. Chicken eggs are a minor source of income. Peri-urban chicken production has grown rapidly over the past 20 years and is a profitable enterprise for people who live near an urban area or other market. Prices of live chickens in local markets

1 See Hide (2003:102–110) for a review of the scattered information on trade, exchange and sale of pigs and pork in PNG.

Table 5.18.1 Animal products sold in Kainantu market, 1979–1982

Live animals	chicken, pig, duck, cuscus, dog, cat, bandicoot, wild birds
Meat	pig, chicken, cuscus, beef, carp fish, bat
Eggs	chicken, duck, wildfowl
Other	cuscus fur, insect larvae

Note: The most commonly sold products were live chickens, pig meat and chicken eggs. Other products were sold only occasionally over the survey period. The volume of live chickens has increased greatly since these surveys were conducted and some of these items may no longer be sold in Kainantu market.

Source: Unpublished markets surveys, R.M. Bourke and K. Nema. Surveys conducted every two weeks over a 36-month period (June 1979 to May 1982).

are generally higher than for equivalent-sized frozen birds produced by two large companies and sold in retail stores and supermarkets.

Cattle are sold as live animals for slaughter through abattoirs, directly to consumers or as meat in food markets. In the period 1990–1995, sales of cattle and beef meat generated an estimated income of K2.4 million per year for rural villagers (Figure 5.1.1, Table 5.1.1). It was estimated that 13% of the rural population lived in locations where cattle were sold (Figure 5.1.2). There are small village herds in most lowland and highland provinces, with the greatest concentration in Morobe Province.

Crocodiles[2]

There are two species of crocodile in PNG. Saltwater crocodiles (*Crocodylus porosus*) inhabit both fresh and salt water and are endemic to the western Pacific, South-East Asia and Australia. Their skins are very highly valued for making fashion accessories such as handbags, shoes, purses, wallets and belts. The second species, the New Guinea freshwater crocodile (*C. novaeguineae*), is endemic to the island of New Guinea only and lives in fresh water. Freshwater crocodile skins are not as highly valued as the saltwater species.

Crocodile skins have been exported since the 1950s (Table A5.18.1). In the mid 1960s large animals became over-harvested and the stock of wild animals rapidly declined. In 1966 it became illegal to sell crocodile skins with a belly width greater than 51 cm. This law reversed the decline of large wild animals, but resulted in greater harvesting of smaller animals.

Crocodiles in PNG are hunted in the wild and are farmed. In places with extensive wetlands, villagers hunt wild crocodiles and sell the skins to traders. The skins are sold to export companies and sent to tanneries overseas. Young animals are sold to large commercial crocodile farms or raised in small village enclosures.

In the period 1990–1995, sales of crocodiles or crocodile skins generated income of almost one million kina per year for rural villagers (Figure 5.1.1, Table 5.1.1). The value to villagers is likely to have increased considerably since then. It was estimated that 4% of the rural population lived in locations where crocodiles or skins were sold (Figure 5.1.2). Most young crocodiles sold to large farms come from Ambunti District in East Sepik Province. The bulk of skins from wild animals come from Gulf and Western provinces, but crocodile skins are produced in most lowland provinces (Figure 5.18.1).

The value of crocodile skin and crocodile meat exports was estimated in 2005 as K35 million. In 2006 the Department of Environment and Conservation (DEC) issued 68 licences for crocodile traders, 24 for buyers, 18 for exporters and 3 for crocodile farming.

Most skins exported from PNG are from freshwater crocodiles (60–70% since 1998) (Figure 5.18.2). The number of skins exported from freshwater and saltwater wild animals has been 20 000–25 000 skins per year since 2001. Specialists in DEC consider that this level of export does not present a threat to the number of wild animals. As well, 5000–8000 skins from farmed saltwater crocodiles are exported each year. The total number of skins exported has been 25 000–33 000 per year (average 29 000) since 1998 (Table A5.18.2). There was a surge in the number

[2] David Wilken, Manager, Mainland Holdings Crocodile Farm, Lae, provided much information for the section on crocodiles. Other sources were a presentation by the Department of Environment and Conservation and published papers.

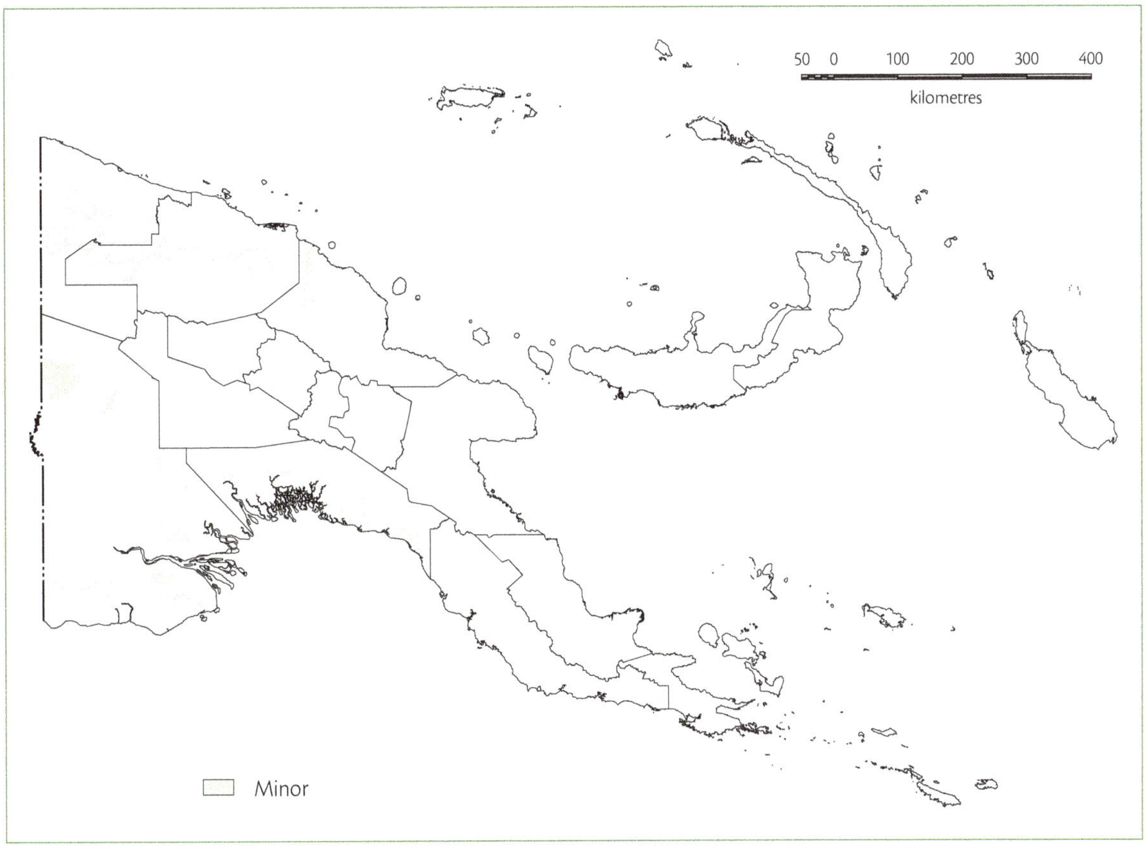

Figure 5.18.1 Locations where sales of crocodiles provided income for rural villagers, 1990–1995. Source: MASP.

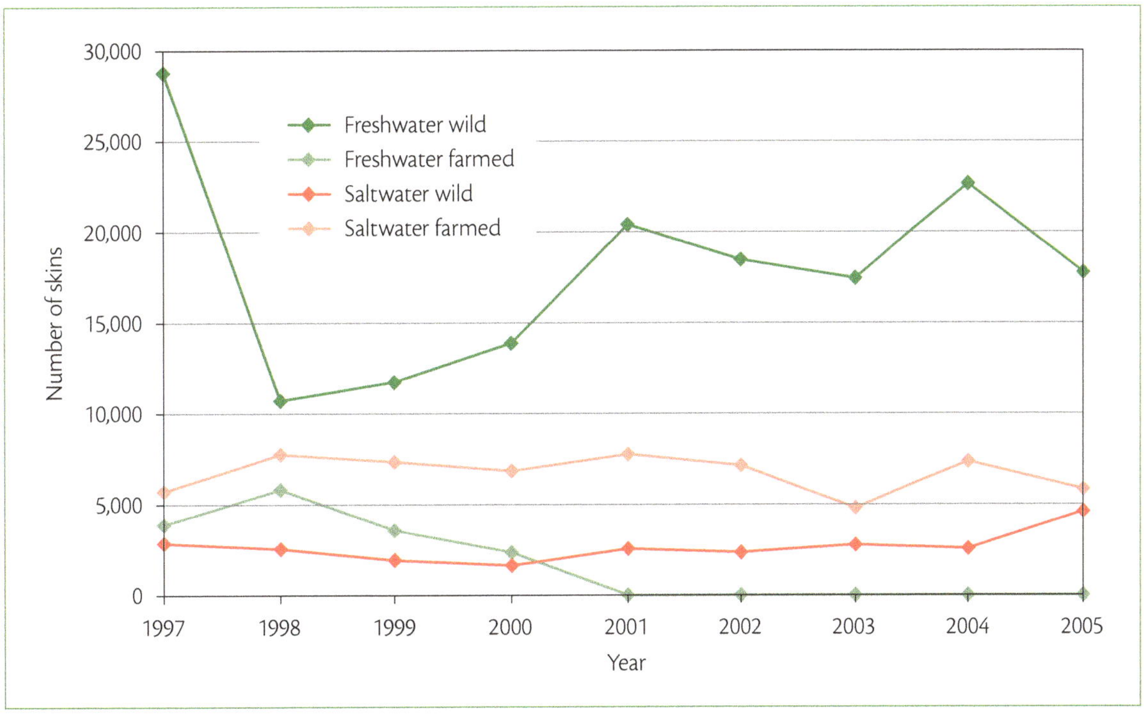

Figure 5.18.2 Number of freshwater and saltwater crocodile skins exported, 1997–2005. Source: Department of Environment and Conservation (2007).

of wild freshwater crocodiles caught and exported in 1997, presumably because they were easy to catch when stream levels were low during the major drought in that year, and people needed cash to buy food. Most exports go to Japan (about 70%), Singapore and France.

Wild crocodile numbers are closely monitored by staff from DEC. If this monitoring is not conducted, exports of crocodile skin from PNG will be prohibited under the Convention on International Trade in Endangered Species of Wild Fauna and Flora (CITES). Monitoring by DEC staff over the period 1982–2006 indicates that the number of saltwater crocodiles is increasing and numbers of the freshwater species are stable in the middle and upper Sepik River region.

The largest crocodile farm in PNG is owned by Mainland Holdings near Lae. It had 51 000 saltwater crocodiles in 2008, which will generate about 10 000 skins per year for export. The crocodiles are fed in part on chicken carcasses from a poultry farm. For the past 20 years this company has sourced about half of its crocodiles from wild juveniles, with the rest from its own eggs and a limited number of eggs from wild animals. Since 2004 however, its stock has come from eggs taken from wild crocodile nests (about half), from farmed eggs (a third) and from wild juveniles (about one sixth). The wild eggs are taken from nests in the upper Sepik River in Ambunti District of East Sepik Province by Mainland Holdings and DEC staff. Villagers are paid K10 per viable egg. The wild juveniles come from a number of lowland provinces, including Oro (Musa Valley), West New Britain, Morobe (Umboi Island, Morobe coast, Watut River), Madang and Bougainville.

Small amounts of crocodile meat are sold to up-market restaurants in PNG and about 40 tonnes/year is exported by Mainland Holdings to Australia.

Honey

The European honey bee (*Apis mellifera*) was first introduced to PNG from Australia in the late 1930s by expatriates, for their own use. Swarms from existing hives and further introductions in the 1960s and 1970s has led to this species becoming widespread in PNG. These bees are most productive at 1500–2000 m altitude in PNG and a small honey industry was established in Eastern Highlands Province (EHP) and other highlands provinces in the early 1970s. Honey production in 1975 was about 25 tonnes. Growth continued until 1986 when the number of hives peaked at about 4000, with 500 producers harvesting 120 tonnes of honey in that year. The honey was processed and packed at a plant near Goroka by the Highlands Honey Producers Company. Some was sold on the domestic market and some was exported to Europe where it was marketed as 'organic', but exports ceased in the early 1990s. Production per colony declined from 30 kg to 20 kg in the 1990s due to poor management, low prices, the closure of the honey producers cooperative, competition from the invading Asian bee (*A. cerana*) and changes in nectar sources following the 1997 drought (Figure 5.18.3, Table A5.18.3).

Recently the industry has made a slow recovery. Prices have improved and the Eastern Highlands Provincial Government has supported honey producers with training. Most honey production is concentrated in EHP, with about 370 producers who typically have 5–10 hives each. The New Guinea Fruit Company is the principal buyer, processor and marketer of honey in EHP. There are a limited number of hives in the other four highlands provinces; near Oksapmin in Sandaun Province; and in Bougainville Province. The number of producers is expanding slowly and interest is increasing from potential producers elsewhere in the highlands.

Honey is sold on the domestic market, with about half through informal arrangements and the remainder through New Guinea Fruit Company. Production in 2007 of about 40 tonnes was worth an estimated K500 000 to growers. Domestic consumption is about 200 tonnes/year, so about 160 tonnes of honey is imported each year.

Both the domestic and export market offer potential for expanded production. Eastern Highlands Province has capacity for an estimated 20 000 colonies (from about 3000 currently). There is demand for 500–600 tonnes/year of PNG organically certified honey from a number of overseas buyers. PNG honey is able to be marketed as 'organic' because antibiotics, miticides, pesticides

Figure 5.18.3 Estimated honey production in Eastern Highlands Province, 1975–2007. Sources: Hardie et al. (2005) and Tella Loie, Department of Primary Industry, Goroka.

or fumigants are not used. A fledgling industry in the Oksapmin area was destroyed by the varroa mite in the 1980s. Fortunately this serious pest has not reached the Highlands Region, but it remains a potential threat to honey production there.

Insect farming

An international market exists for dried insect specimens. 'Farming' insects provides income to small numbers of people in PNG. Butterflies are the main insect exported, with lesser numbers of beetles, stick insects and other insects. Specimens must be undamaged, so they are produced by 'farming' rather than collecting adults from the wild. 'Butterfly farming' involves the farmer planting foods that butterflies eat. Butterflies lay eggs on the food plants, which support caterpillars until they form pupae. The farmer collects the pupae and either sells it to traders or hatches the pupae, kills the newly hatched butterfly and sells the butterfly.

Interest in buying PNG butterflies began in the early 1970s, with trading dominated by a few foreigners. In 1974 the Australian Administration started an Insect Farming and Conservation Project, which became the Insect Farming and Trading Agency (IFTA) in 1978, based in Bulolo. It is currently part of the University of Technology's commercial arm, University Development Consultancy.

Three other organisations are involved in marketing insects from PNG, but the two organisations that dominate this market (IFTA and Wau Ecology Institute) are both based in Morobe Province. A study over an eight-year period (1995–2002) found that insects were farmed in all provinces except Enga in 1995 but, by 2002, farming was reduced to eight provinces. The number of insect farmers was stable at 120–130, but there was a greater concentration in the Wau and Bulolo areas of Morobe Province. The total amount earned per year from insect farming and collecting was in the range K60 000–K120 000 for the period 1995–2002. Average income per farmer or collector over this period was K350/year. However, income has declined over the period 2002–2008 because of shrinking international demand, reduced profitability for traders and problems with obtaining PNG export permits.

Silk

A small village-level silk industry existed in Southern Highlands and Western Highlands provinces between 1979 and 1983. Cocoons of silk were produced by silkworms fed on mulberry leaves and the cocoons were exported. The fledgling industry collapsed when government funding was withdrawn. Production was greatest in the Lake Kutubu area of Southern Highlands Province.

Other animal products

Plumes and pelts. The plumes of many birds, especially birds of paradise, parrots, hornbills, hawks and cassowaries, and the pelts of a number of animals, including tree kangaroo and cuscus, are used as ornaments and decoration in customary dress. Such items were important in local trade and gift exchanges in the recent past. They are sometimes sold for cash. The volume sold or traded is probably declining. In the period 1990–1995, sales of plumes and animal skins generated income of almost one million kina per year for rural villagers (Figure 5.1.1, Table 5.1.1). It was estimated that 5% of the rural population lived in locations where these were sold (Figure 5.1.2).

Sales of plumes and pelts are greatest in locations with low population densities, especially on the northern and southern fringes of the central highlands (Figure 5.18.4). The value of sales is very much less than that for fresh food or the main export cash crops, which tend to be grown in areas with better market access.

Live wild animals. Some wild animals are occasionally sold live. These include parrots, hornbills, other birds, cuscus and bandicoot. Young cassowary birds raised from wild chicks are sold to highlanders by people who live on the fringe of the highlands where the human population density is low and wild birds can be found. Adult birds are highly priced.

Game meat. Small amounts of wildlife meat or animals intended to be killed for meat are sold in markets or directly to consumers. Animals sold include cuscus, wild pig, tree kangaroo, wallaby, bandicoot and birds. Larvae of certain beetles and moths are also sold for food in some places.

Wildfowl eggs. Wildfowl eggs are sometimes sold for food, particularly on the Gazelle Peninsula in East New Britain Province and in a number of locations between Talasea and Bialla in West New Britain Province. Eggs were being sold in West New Britain Province for K4 each in 2007.

Sources

Aisi, C., Hudson, M. and Small, R. (2007). How to ranch and collect insects in Papua New Guinea. Sustainable Insect Collecting and Farming in Papua New Guinea Project Training Manual. Port Moresby.

Bulmer, R.N.H. (1972). Hunting. In Ryan, P. (ed). *Encyclopaedia of Papua and New Guinea*. Melbourne University Press in association with the University of Papua and New Guinea, Melbourne. pp. 543–546.

Cox, J.H., Gowep, B., Mava, A., Wana, J., Genolagani, J.-M., Kula, V., Solmu, G., Sine, R., Wilken, D. and Langelet, E. (2006). The saltwater crocodile *Crocodylus porosus* egg harvest program in Papua New Guinea: linking conservation, commerce and community development. Paper presented at the 18th meeting of the IUCN Crocodile Specialist Group.

DAL (Department of Agriculture and Livestock) (1992). Exports of important agricultural commodities: quantity, value and average price per unit. Rural Statistics Bulletin 1992/6. Department of Agriculture and Livestock, Konedobu.

Department of Environment and Conservation (2007). Role of DEC in wildlife trade – crocodile perspective. Presentation given at Sustainable Use of PNG's Wildlife Workshop, Madang, 20–21 February 2007.

Downes, M.C. (1971). Crocodiles. *Harvest* 1(1):6–9.

Hardie, P.A., et al. (2005). Sustainable plan for the honeybee industry of Eastern Highlands Province, 2005–2015. Unpublished paper. Division of Primary Industry, Goroka.

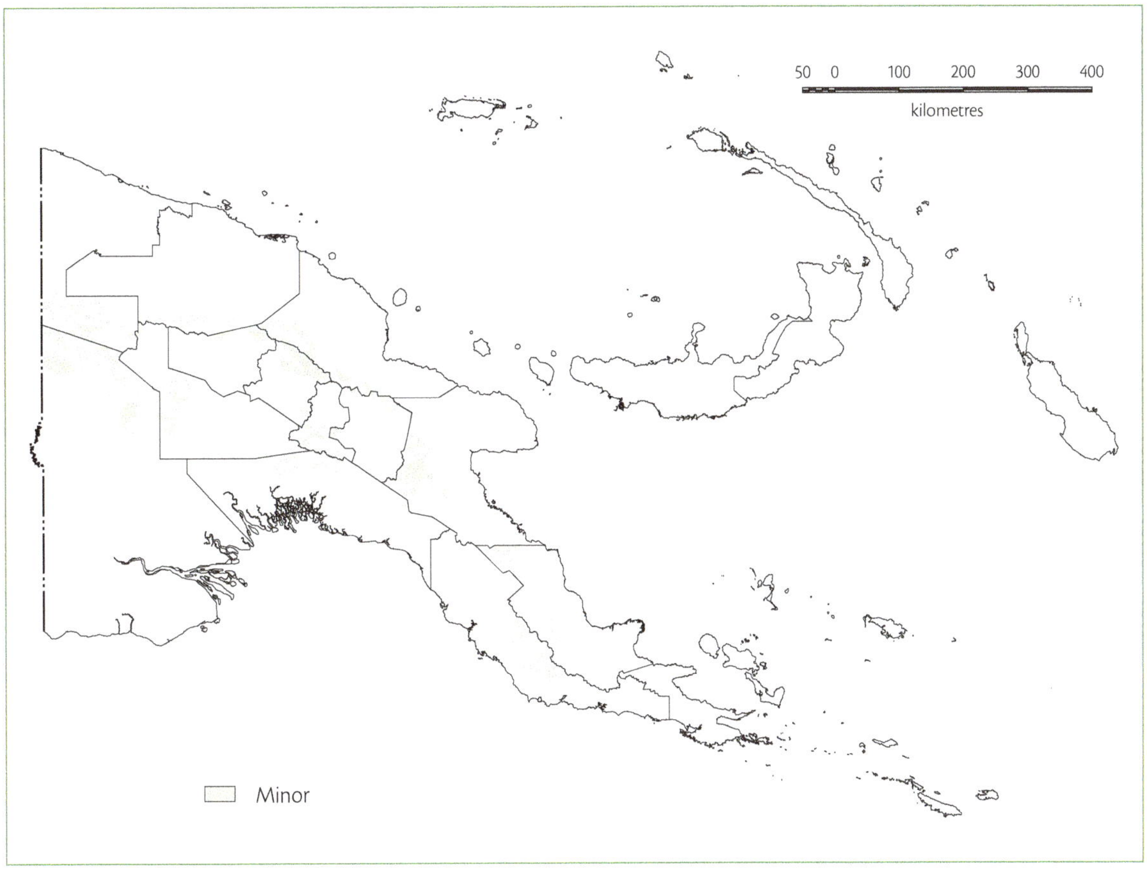

Figure 5.18.4 Locations where sales of plumes and animal skins provided income for rural villagers, 1990–1995.
Source: MASP.

Herington, J. (c. 1978). Wildlife use and conservation. In Densley, D.R.J. (ed). *Agriculture in the Economy: A Series of Review Papers*. Volume 1. Department of Primary Industry, Konedobu. pp. 1–25.

Hide, R. (2003). *Pig Husbandry in New Guinea: A Literature Review and Bibliography*. ACIAR Monograph No. 108. Australian Centre for International Agricultural Research, Canberra.

Small, R. (2004). Uptake and the success of insect farming projects in Papua New Guinea: implications for biodiversity conservation. Master of Philosophy thesis. University of Cambridge, Cambridge.

Small, R.D.S. (2007). Becoming unsustainable? Recent trends in the formal sector of insect trading in Papua New Guinea. *Oryx* 41(3):386–389.

Weiner, J.F. (c. 1985). The social organization of Foi silk production: the anthropology of marginal development. Unpublished paper. Department of Anthropology, Research School of Pacific Studies, The Australian National University, Canberra.

5.19 Purchased inputs for agricultural production

Land and human labour are the main inputs for most agricultural production in PNG. A minimum use is made of purchased inputs, such as fertilisers and pesticides, except in the plantation sector. Even there, the volumes used are low by world standards. The most common purchased inputs in PNG agricultural production are basic tools, especially spades, machetes ('bush knives') and axes. Coastal- and river-dwelling villagers engaged in subsistence and commercial fishing also buy nets, lines and other fishing equipment.

The use of mechanical tillage for village agriculture is limited. The soil is tilled using tractor-drawn ploughs in the upper Markham and Ramu valleys for peanut crops. Occasionally villagers use tractors or small machines to till the soil for vegetable production in the highlands, but this is uncommon. Boats powered by outboard motors are used for fishing in some locations, but non-motorised canoes are more common. Vehicles and boats are used to transport food and cash crops in many locations, and buffalo-drawn carts in a few places. However, almost all agricultural produce is carried by people or canoe 'on the farm'.

Planting material

Little seed or other planting materials are purchased for PNG agriculture. The most significant is oil palm seed, which is produced by New Britain Palm Oil Limited (NBPOL) at Dami Oil Palm Research Station, West New Britain Province. Seed from Dami is used throughout PNG oil palm growing areas and is also exported (see Section 5.7).

Coffee seed and seedlings are sold by the Coffee Industry Corporation Coffee Research Station at Aiyura in Eastern Highlands Province. Some coffee growers buy this seed to produce seedlings which they sell to other coffee growers. Improved cocoa seed and seedlings are sold by the Cocoa Coconut Institute (CCI) Research Station at Tavilo in East New Britain Province. Coconut seedlings are sold by the CCI Stewart Research Station in Madang Province. Small quantities of temperate vegetable seed are imported and sold through retail outlets. Seed potato is imported from Victoria in Australia and is multiplied by the Fresh Produce Development Agency and National Agricultural Research Institute before it is sold to commercial potato growers.

Livestock inputs

Three Lae-based companies produce stockfeed for poultry and pig production from imported grain, particularly sorghum from Australia (see Section 2.6). Other stockfeed ingredients include residues from wheat milling, fish meal and copra meal. One major pig producer in the Markham Valley makes feed from home-grown maize. Locally produced stockfeed totalled 58 000 tonnes in 2005. An additional 37 000 tonnes of fully prepared stockfeed was imported in

2005. Feed grain imports were valued at K18 million in 2004. About 80% of stockfeed is used for poultry production with the rest going to pig production.

The most important purchases of young animals are day-old broiler chickens, with three hatcheries providing more than 400 000 per week. Purchases of other young livestock, including pigs and cattle, are minimal. Fencing materials are an important input for cattle production but attract a high import tariff (see Section 4.1). Very small quantities of chemicals are used in the livestock industry to control parasites.

Fertilisers[1]

Most rural villagers engaged in agriculture in PNG do not use inorganic (manufactured) fertiliser. Small quantities are used by some people growing Irish potato and introduced temperate-climate vegetables, but the total quantity used is negligible. Some villagers use organic materials as green manure fertiliser, particularly in the 'composting zone' in the western part of the central highlands (Figure 3.11.6). Minor amounts of chicken manure and other organic fertilisers, particularly household waste, are used, especially in the highlands (see Section 3.12). The major oil palm estates compost fruit bunch waste to produce significant quantities of mulch and organic fertiliser for their plantations. For example, Higaturu Oil Palms in Oro Province applied 45 000 tonnes of effluent fruit bunch at a rate of 17.6 t/ha in 2004 while NBPOL applied about 36 000 tonnes.

The oil palm industry uses 83% of the inorganic fertiliser imported into PNG (Table 5.19.1, Figure 5.19.1). The coffee, Irish potato, sugar cane and cocoa industries take most of the rest (13% of the total). The balance is retail sales (4%) to villagers, schools, missions, research institutions and other bodies, most of which is used on introduced vegetables. About 70% of retail sales of inorganic fertiliser were made in Mount Hagen and are used for temperate vegetable and Irish potato production in Western Highlands and Enga provinces.

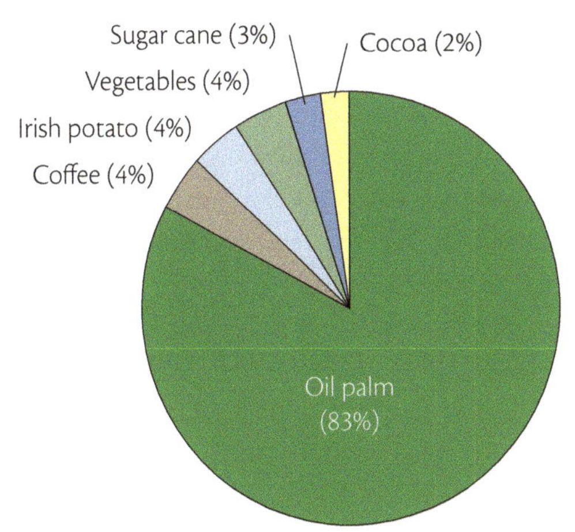

Figure 5.19.1 Inorganic fertiliser use by crop, 2002.
Source: Steve Woodhouse, Farmset Limited, Rabaul.

1 Information on fertilisers and pesticides is largely based on that provided by Steve Woodhouse, Farmset Limited, Rabaul.

Table 5.19.1 Estimated chemical use by crop, 2002

Crop	Fertiliser (tonnes)	Herbicide (litres)	Insecticide (litres)	Fungicide (litres)
Oil palm	14,500	396,000	5,000	0
Sugar cane	500	80,000	1,000	0
Coffee	800	70,000	0	2,000
Vegetables	700	0	4,000	0
Cocoa	300	0	1,000	0
Irish potato	700	0	0	0
Total	17,500	546,000	11,000	2,000

Source: Steve Woodhouse, Farmset Limited, Rabaul.

The most important inorganic fertilisers used in PNG are ammonium chloride and urea. Other fertilisers include potassium chloride, ammonium nitrate, mono ammonium phosphate, mixed fertiliser (the most common being nitrogen–phosphate–potassium (NPK) mixes) and other types, including micronutrients, dolomite, kieserite and sulfur. Many fertiliser mixes now contain 0.3% boron. This follows research in the highlands in the late 1970s and 1980s which showed that application of very small amounts of boron fertiliser increased the yield of many crops in the highlands, including introduced vegetables.

Pesticides

Village production of food and export crops uses negligible amounts of herbicides, insecticides or fungicides. Small amounts of insecticide are applied by a few people to temperate-climate vegetables, particularly brassicas, and very small quantities of herbicide are used by some villagers. Most of the agricultural chemicals imported are applied to oil palm (72%), with most of the balance used on large-scale sugar cane (14%) and coffee (13%). Small quantities are used in vegetable growing (1%) and cocoa production (Figure 5.19.2).

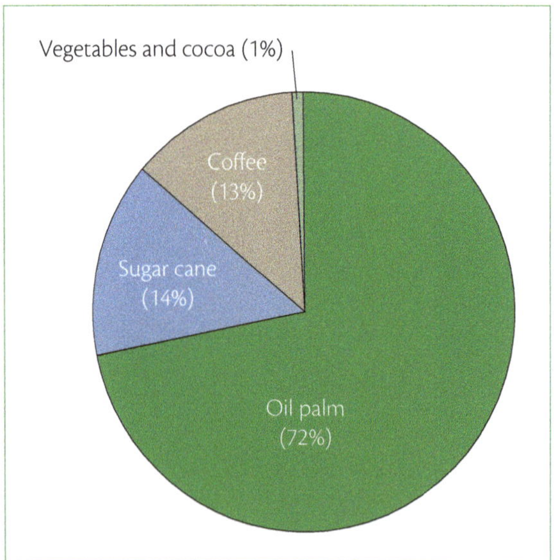

Figure 5.19.2 Herbicide, insecticide and fungicide use by crop, 2002. Source: Steve Woodhouse, Farmset Limited, Rabaul.

Herbicides make up 98% of non-fertiliser chemicals used in PNG agriculture. Only very small quantities of insecticide and fungicide are used (Table 5.19.1). The oil palm industry is the largest user of herbicides (about 400 000 litres/year or 73% of total herbicide used). New Britain Palm Oil Limited is the biggest user of herbicide in PNG. However, consumption by oil palm estates has dropped significantly in recent years because bunch waste is applied on oil palm plantations as compost, reducing weed growth, and herbicide application methods have improved. Some herbicide is also used against weeds for sugar cane and coffee production. The cocoa industry uses almost no chemicals. The most common herbicide used in PNG is glyphosate (marketed as Roundup® and other trade names) and is about 55% by volume of all herbicide use. Gramoxone® is also commonly used (about 20% by volume). Other herbicides and surfactants[2] make up the remainder.

Some insecticide is used in the production of oil palm, vegetables, cocoa and sugar cane. Small quantities of insecticide are used to control green scale and coffee borer in coffee. Some of the larger agricultural companies import insecticide to control malaria-carrying mosquitoes. The total quantity of insecticide used in PNG agriculture is about 11 000 litres per year. The only significant use of fungicide is for the control of coffee rust in the highlands, but the quantity used is small. About 16 tonnes of copper-based chemicals are used per year to control potato late blight disease in commercial Irish potato production in high-altitude areas, especially in Western Highlands Province.

Source

Anderson, T. (2006). Oil palm and small farmers in Papua New Guinea. A report for the Centre for Environmental Law and Community Rights on the economic prospects for small farmers in PNG's oil palm industry. Unpublished report. Australian Centre for Environmental Law, The University of Sydney, Sydney.

2 A surfactant is a substance that when added to a liquid increases the liquid's spreading or wetting properties. It is commonly used with herbicides to increase their coverage and therefore effectiveness.

5.20 Fluctuations in village cash crop production

Village production of crops for sale in domestic and export markets varies from year to year, and often the reasons why are not immediately clear. These fluctuations can be observed in the annual exports of coffee (Figure 5.4.3), cocoa (Figure 5.5.1), copra (Figure 5.6.1), copra oil (Figure 5.6.5), oil palm (Figure 5.7.4), rubber (Figure 5.11.2) and vanilla (Figure 5.14.2). Fluctuations in village cash crop production are influenced by a number of factors, which can act alone or in combination, and which may not occur beyond a local area. It is therefore sometimes difficult to understand or predict what production outcomes will be at any particular time. The most important influences on village cash crop production are:

- Returns to labour.
- Changes in access to markets.
- Marketing arrangements.
- Weather conditions.

Returns to labour

'Returns to labour' refers to the amount of money growers receive in return for the work they do to produce a commodity for sale. If they believe that their returns to labour are not adequate, they will stop producing a commodity or will switch production to another commodity that they think will provide a higher return. Although this phenomenon is difficult to study among PNG villagers, it is clear that they are very sensitive to returns to labour. Their ability to switch in or out of cash cropping is cushioned by subsistence food crop production. This sustains them and their families even if they do not have a cash income for some time.

Many PNG villagers have an amount of money that they believe they should receive when they sell a cash crop. They will stop production if they cannot obtain an income above that amount. This minimum acceptable return is different from place to place and even from person to person. It depends on what crops can be grown in the local environment, experiences of returns that have been received in the past, the existence of alternative money-earning activities (including off-farm employment)[1] and the need to earn cash, for example, for school fees or medical treatment. A rough guide to the minimum return to labour that many people find acceptable is the

1 Most off-farm wage-earning opportunities are local and limited in time, such as road construction work. Local off-farm employment can affect cash crop production in an area because people may choose wage labour over cash crop production. However, local off-farm employment opportunities are not common. Migration is greatest from areas where cash cropping opportunities are poorest, and only a small proportion of the total population is employed in mines or elsewhere in the formal economy. So the number of people employed off-farm does not have a major influence on year-to-year fluctuations in village cash crop production.

minimum rural wage, that is, the amount of money they could earn if they had wage-earning employment. This was about K8 per day in 2008.

The main cash crops produced in PNG provide returns to labour that range from K100/person/day to less than K5/person/day (Table 5.20.1). The reasons that the crops that provide the highest returns to labour (Irish potato and oil palm) are not grown by everyone in PNG is because local environments do not support their production, there is not a local market for them, or other inputs needed are not available. Successful Irish potato production, for example, depends on access to disease-free planting material, fertiliser and fungicide, and a high-altitude location. Oil palm fruit can only be processed in large mills and must be transported to the mill soon after harvest, so production is only by registered growers who live close to a mill. On the other hand, betel nut, which can be grown in most places up to 1000 m above sea level and gives a high return to labour, is produced almost everywhere where transport is available to carry the nuts to a market (see Section 5.17).

Arabica coffee, cocoa and most fresh foods, including sweet potato, provide returns to labour in the acceptable range of K20–K30/person/day. These commodities are produced by most families where environmental conditions and access to a market allow. Most sweet potato marketed in PNG is produced in the highlands valleys wherever there is good access to

Table 5.20.1 Gross returns on labour inputs for selected cash crops

Crop	Mean yield[a] (kg/ha)	Price[b] (kina/kg)	Gross return[c] (kina/ha)	Labour inputs[a, d] (days/ha)	Return[e] (kina/person/day)
Irish potato (tubers)	20,000	2.00	40,000	400	100[f]
Oil palm (fresh fruit bunch)	15,000	0.25	3,750	70	54
Sweet potato (tubers)	14,000	0.80	11,200	350	32
Cocoa (wet bean)[g]	800	1.00	800	40	20
Arabica coffee (parchment)	900	4.50	4,050	275	18
Vanilla (cured bean)	240	15.00	3,600	290	12
Rubber (cup lump)	650	1.60	1,040	100	10
Coconut (copra)	500	1.30	650	65	10
Robusta coffee (parchment)	900	1.60	1,440	275	5
Rice (paddy)	1,300	0.80	1,040	215	5
Pyrethrum (dried flowers)[h]	650	1.50	975	430	2

[a] Yield and labour input data are from different sections of this book, with some data adapted from Hale (c. 1978).

[b] Price data are farmgate prices for 2007 or an average of 2005–2007.

[c] Mean yields are not precise and prices vary greatly over time. Hence gross returns may differ significantly from these figures. Nevertheless, the broad pattern illustrated here will hold under different yields and prices.

[d] Labour inputs to establish tree crops are not included, but are not large when averaged over the life of the crop.

[e] Land and labour are the main inputs for most smallholder production in PNG, but there are some other inputs that require cash, including tools and sometimes planting material. These have been ignored for this comparison.

[f] Cash outlays for disease-free planting material, fertiliser and fungicide are required to achieve high Irish potato yields, so net returns (after cash expenses) to labour inputs are about K60–K70/person/day.

[g] A cocoa wet bean yield of 800 kg/ha is equivalent to 300 kg/ha of dry bean at a conversion rate of 37.5%.

[h] Yield and labour input data for pyrethrum follow Anderson (c. 1978).

the Highlands Highway. Similarly, Arabica coffee is produced by many highlands families (Figure 5.4.2). Cocoa is produced by many households on the Gazelle Peninsula in East New Britain Province, in Bougainville Province, and by some families in other lowland locations (Figure 5.5.3).

Returns to labour for rubber and copra are only about K10/person/day. These commodities are produced where no alternative sources of cash income are available. Their production is limited because producers harvest and process the crop only until they have earned enough money to satisfy their immediate needs.

Robusta coffee, rice and pyrethrum provide returns to labour around K5 or less per day (Table 5.20.1). Only small quantities of Robusta coffee are harvested and sold (Section 5.4) and production is almost solely to meet immediate cash needs. Rice production is also very low. It fluctuates over time because of bursts of extension activity and subsidies by donors, NGOs and some provincial governments who think it is important that PNG produces its own rice to replace imports (Section 2.5). When growers realise the low levels of returns that rice provides, they cease growing it. Most previous studies of rice growing in PNG have concluded that returns to labour are unacceptably low, even compared to growing root crops for subsistence. These studies concluded that it is more logical for PNG to import rice and to export commodities that provide a higher return to labour, such as coffee or cocoa.[2] Other grain crops also give low returns to labour, including corn (maize) and sorghum sold for stockfeed or industrial use.[3] Likewise pulse crops, such as soya bean, also give poor returns to labour when they are sold for processing.

The sensitivity of people to returns to labour results in two broad patterns of village commodity production. The first pattern is found where a commodity provides low returns to labour and there is no alternative crop. Here villagers work for a 'target income', an amount of money that is needed for a particular reason. For example, they might produce enough copra, rubber or bêche-de-mer to earn money to pay their children's school fees. Once they have this amount of money, they stop production.

The second pattern occurs where returns to labour are high. Here villagers are very responsive to price, and will increase production when prices are higher and reduce production when prices are lower. This behaviour is seen in the production of betel nut, fresh food and oil palm. Consequently, production of these crops fluctuates with price. National export figures do not always reflect a production response to price fluctuations because environmental variation, such as rainfall, can result in an overall decrease or increase in production that hides the returns to labour response. However, the response can be readily observed at the local level.

Returns to labour do not always prevent people from adopting a cash crop, largely because the amount of labour involved to produce a crop is not known during the adoption phase. So people frequently enthusiastically adopt a new crop that has been introduced by an extension program, or which is reported to earn high prices. However, it is not until they have actually experienced how much work is required to produce the crop that they decide whether to continue production. It is very common to hear people say, in answer to the question 'Why are you growing this crop?': '*Mipela traim tasol*'

2 Hale (c. 1978: Table 10) compared returns to labour inputs for six lowland cash crops. He showed that returns on rice in 1977 were K0.53/person/day, much less than alternative cash crops, including sweet potato (K7.93), Robusta coffee (K6.24), cocoa (K5.27), copra (K4.82) and rubber (K3.78). He concluded that a major constraint to expanded rice production in PNG was low production at village level and low returns to labour. At the time of writing that paper, Peter Hale was responsible for promoting rice production in PNG. More recently, Mills (2002:69) compared cash returns on growing rice in Japan, Taiwan, Java (Indonesia) and PNG and found that returns in PNG of K2.40/person/day were very low compared with these other countries. Gibson (1994) examined labour inputs for rice and export tree crops. He concluded that increased rice production would result in reduced production of tree crops and would reduce employment rather than increasing it.

3 Corn sold as a fresh food has a much higher price and gives a high return to labour. The same applies to coconut. When sold in food markets to be used for cooking, coconut commands a high price and hence a good return to labour, but when sold as copra for industrial uses, returns to labour are low.

(we are just trying it). If they find that the labour required to produce the crop is high relative to the income received, they are likely to lose interest and either cease production or only produce the commodity when they need cash for a particular reason. Plants of former cash crops that are no longer harvested and sold can be seen in many villages. These include cardamom, chilli, kapok, purple passionfruit, Robusta coffee, rubber and sisal.

A moderate amount of information exists on villagers' labour inputs for a number of important export and food crops (Table 5.20.2).[4] However, much more information is needed for some crops so that informed decisions can be made about what crops are likely to be adopted by villagers.

It has recently been recognised that the amount of labour required to produce a crop can alter over time. This can change the returns to labour and result in lower production or even cessation of production. Crop yields can decline significantly over time because of disease, the ageing of trees or lack of crop maintenance. Furthermore, if regrowth is not kept clear, harvesting becomes heavier work. As the amount of labour required to harvest the crop increases, people will harvest less or will stop harvesting altogether. This has been observed for a number of cash crops, including oil palm, cocoa, coffee and coconut.[5] A similar pattern occurs in food gardens, with less and less food being harvested from old gardens as weeds and regrowth make harvesting more difficult and labour-demanding.

Changes in access to markets

Access to markets is an important influence on cash crop production (see Section 1.14). When road access is good, it is less work and cheaper for producers to market their goods. An example of how production improves with better road access is provided by the road repair program that followed the civil war in Bougainville Province. Cocoa production increased steeply in 2002 following 11 years of much lower production caused by the fighting (Figure 5.5.5). During the war, road maintenance stopped and the ring road around Bougainville Island became almost impassable. However, after the road had been repaired (with AusAID funding), cocoa growers responded with a big increase in production, especially in the main cocoa-growing areas of Wakunai and Tinputz.

The opposite is also true, that is, poor roads reduce growers' access to markets. Unfortunately, there are many examples of this over the past 15 years, especially in the highlands where lack of road maintenance means that places once easily accessible from the Highlands Highway are now isolated.

Marketing arrangements

Changes in the arrangements for marketing produce can have a positive or negative influence on the amount of produce sold. For example, prior to 2002, organisational and financial problems within the Copra Marketing Board, the only buying agency in Bougainville, resulted in low copra production there. In late 2002, Coconut Products began buying copra in Buka. This company processes copra into copra oil at its factory near Rabaul, East New Britain Province (see Section 5.6). Coconut growers on Buka Island and the northern tip of Bougainville Island, the main producing areas, immediately responded. Villagers started making copra again because they could sell it to agents of Coconut Products.

The closure of a number of copra-purchasing depots in recent years has led to a loss of market access for copra producers in some more remote locations in PNG. For example, 22 depots existed in 11 provinces in 2001 but this had become 10 depots in 9 provinces by 2005 (Table 5.6.2). Recently, the closure of depots has been offset to some extent by copra purchases by the ship MV *Coconut Trade* operated by Coconut Oil Production Madang Limited.

---

4 There is a reasonable amount of information on labour inputs for sago starch extraction, which is in the range 2–4 kg dry starch per hour for New Guinea mainland locations (Table 2.4.4).

5 See Curry et al. (2007:68, 126) for discussion on the influence of the quantity of accessible crop on cocoa harvesting levels.

Table 5.20.2 Labour inputs for smallholders for selected export and food crops

Crop	Location	Labour input (days/ha)	Source	Notes
Cocoa	Gazelle Peninsula Bougainville Karkar Island Madang Province	68	Godyn (1974: Part 3, Table 8)	Inputs were for maintenance (32 days/ha), harvesting (31), and bagging and carrying wet bean (5). Time for waiting (6 days) and walking (1) are not included in the total.
Cocoa	Gazelle Peninsula East Sepik North coast road, Madang Karkar Island Popondetta area	39	Yarbro and Noble (1989)	Labour inputs were for clearing land (32 days/ha), planting shade (17), planting cocoa (31), planting coconut (3), pruning cocoa and shade (5), weeding (20), applying fertiliser and chemicals (2), and harvest (12). Comparable data not given for processing. The total of 39 days is to produce wet bean from established cocoa.
Coffee	Eastern Highlands Simbu Western Highlands	274	Anderson (1977: Table 24)	Anderson surveyed coffee producers in 13 villages in 3 provinces in 1976. Inputs were harvesting (174 days/ha), weeding (21), pulping (37), washing/drying (26) and marketing (16). He presented data for each activity for adult men, adult women and children. In total, 45% of labour inputs were by men, 42% by women and 13% by children (Table 25).
Coffee	Benabena area, Eastern Highlands	324	Overfield (1994: Table 3)	Overfield surveyed 18 households over a two-year period. Data were originally presented as days/tonne. Inputs were for harvesting (196 days/ha), weeding/spraying (26), pruning/maintenance (62), processing (33) and marketing (7).
Mixed root crops	Simbai Valley, Madang Province	363	Clarke (1971:173)	
Peanuts	Keravat Research Station	175	Bourke (1977: Table 2)	Soil was tilled mechanically; all other operations were by hand.
Rice	Lowlands	214	Mills (2002:69)	Inputs are for land preparation (78 days), transplanting (10), weeding (40), harvesting and thrashing (80), and other (6).
Sweet potato	Aiyura Research Station	244	Kimber (1976: Table 1)	Labour inputs are for mounding (80), planting (30), maintenance (46) and harvesting (88). Also 12.7 tractor hours/ha for soil tillage.
Sweet potato	Goroka area	240	Anderson (1976)	Inputs are for commercial vegetable growers. This figure does not include initial opening of land after fallow.
Sweet potato	Keravat Research Station	280	Bourke (1977: Table 2)	Soil was tilled mechanically; all other operations were by hand.
Taro	Popondetta area	300	Waddell and Krinks (1968:42, 57, 85)	Labour input figure generated from data from two years' survey (1962–1964) in two villages. Data converted from hours to days by dividing by 8.
Taro	Paniai Lakes, Papua (Indonesia)	338	Bayliss-Smith (1982: Appendix 2)	Data extracted by Bayliss-Smith who summarises labour input data for taro production from 12 Pacific Island studies. Data converted from hours to days by dividing by 8.
Vegetables	Goroka area	233	Anderson (1976)	Labour inputs for commercial vegetable growers are given for pumpkin, cabbage, radish, marrow, tomato, Irish potato, beans, carrots, lettuce, pak choi, silverbeet and sweet potato. The figure here is for beans and carrots, for plantings that follow previous crops ('medium re-establishment'). Inputs for other crops are similar.

The involvement of provincial governments or their business arms in marketing agricultural produce has generally had poor outcomes in PNG. One example is the involvement of the Simbu Provincial Government in buying cardamom from the Karimui area. After the collapse of the producing company there, the provincial government took over marketing. However, cardamom production at Karimui ceased partly because the government financial system did not allow cash advances to be made to public servants to buy cardamom. Another example is the collapse of chilli growing and marketing when a number of provincial governments became involved in marketing in the early 1980s (see Section 5.15). There is a long history of government-run fresh food marketing operations, going back to the mid 1970s (see Section 6.3). Almost without exception, these operations have lost money and have closed down after a few years, leaving producers with unsold produce.

Villagers expect a cash payment for their produce – receiving cash in the hand can be more important to them than the price they receive. Small, wholesale businesses and government agencies in PNG often face cash-flow problems when purchasing produce from villagers. Small marketing organisations do not have access to overdraft facilities from banks. Any delay in payments reduces the confidence of villagers, who may choose to supply to somebody else or not to supply at all. If the villagers do not sell, buyers cannot sell to the wholesalers, and so do not have cash to pay villagers for the produce, a vicious cycle that undermines development of fresh produce marketing in many places.

Weather conditions

Changes in weather conditions, especially rainfall, influence crop yields (see Sections 1.6, 1.13). Because the influence of weather is felt over a wide area, adverse or favourable conditions can influence regional or national production. However, knowledge of how rainfall influences crop production is poor. For example, it was widely assumed that the severe drought in 1997 would adversely affect coffee production and exports. But the opposite happened and the drought conditions in 1997 actually resulted in increased production in the following year (Figure 5.4.3).[6] In contrast, it appears that the 1997 drought reduced cocoa production, which was low in 1997 and 1998, but increased by about 50% in 1999 and subsequent years, even though prices did not vary greatly between 1997 and 2001 (Figure 5.5.1).

For some crops there is better knowledge about their response to water stresses. A minor or moderate drought results in increased production of mango and breadfruit, but reduced production of other fruit, such as pawpaw. Similarly, sweet potato and cassava are tolerant of minor drought, whereas the same drought can negatively affect taro production. Conversely, sweet potato production is often badly affected by very wet periods, while taro is not.

Other environmental conditions can interact with rainfall to affect crop yield. For example, the 1997 drought had a more severe impact on oil palm production on the well-drained limestone soils on New Ireland than it did on the volcanic ash soils of West New Britain Province.

Other local factors

Factors that operate at a local level commonly influence production from a particular area, but generally do not influence national-level output. Localised unrest and fighting in the highlands can cause people to take refuge some distance from their villages, or result in road blocks. For example, in 2005 fresh food producers near Porgera mine in Enga Province were prevented from travelling to the mine site to sell food because of a dispute with local groups. However, people from Mount Hagen in Western Highlands Province were allowed to use the road and were able to sell their produce at the goldmine.

Other things that operate at a very local level and influence production and sale of cash crops include feasts, funerals and other rituals. The need for

6 See Hombunaka and von Enden (2001) for a detailed explanation as to why this happened.

cash to pay school fees has a significant influence on villagers' behaviour, with many people selling produce for a short period so they can pay school fees at the start of each year.

Fluctuations in sweet potato supply in the highlands

The supply of sweet potato from the highlands to local urban markets (and to Lae, Madang and Port Moresby) can fluctuate significantly from time to time. The cause is a complicated relationship between three factors: (1) the supply of sweet potato influenced by rainfall; (2) the area planted; and (3) the way in which families assess how much sweet potato to plant on the basis of present sweet potato supply.

Total sweet potato supply is determined by the area planted and the yield. The yield is in turn influenced by rainfall. The area planted depends on the rate at which sweet potato is planted (that is, the average area planted per day) and this is inversely associated with the existing supply of sweet potato. If sweet potato is abundant *now* the average area planted per day per family *now* will fall below the average long-term planting rate. Conversely, if sweet potato supply *now* is below that which is considered satisfactory, the rate of planting *now* will increase above the average rate (Figure 5.20.1). This pattern has been found to hold true in a number of areas in Eastern Highlands and Southern Highlands provinces.

Sweet potato yield is influenced by rainfall in the period after planting and in the period before harvesting. Tuber formation is reduced if rainfall is particularly heavy during the first 10 weeks after planting. Periods of low rainfall during the tuber bulking phase (when tubers grow large), some five to six months after planting, can reduce tuber size. This combination of particularly high rainfall early in a year followed by a period of low rainfall later in the year is common during El Niño climatic events (see Section 1.6).

The combination of these two rainfall extremes severely reduces sweet potato supply and can result in food shortages. This is reflected by both a

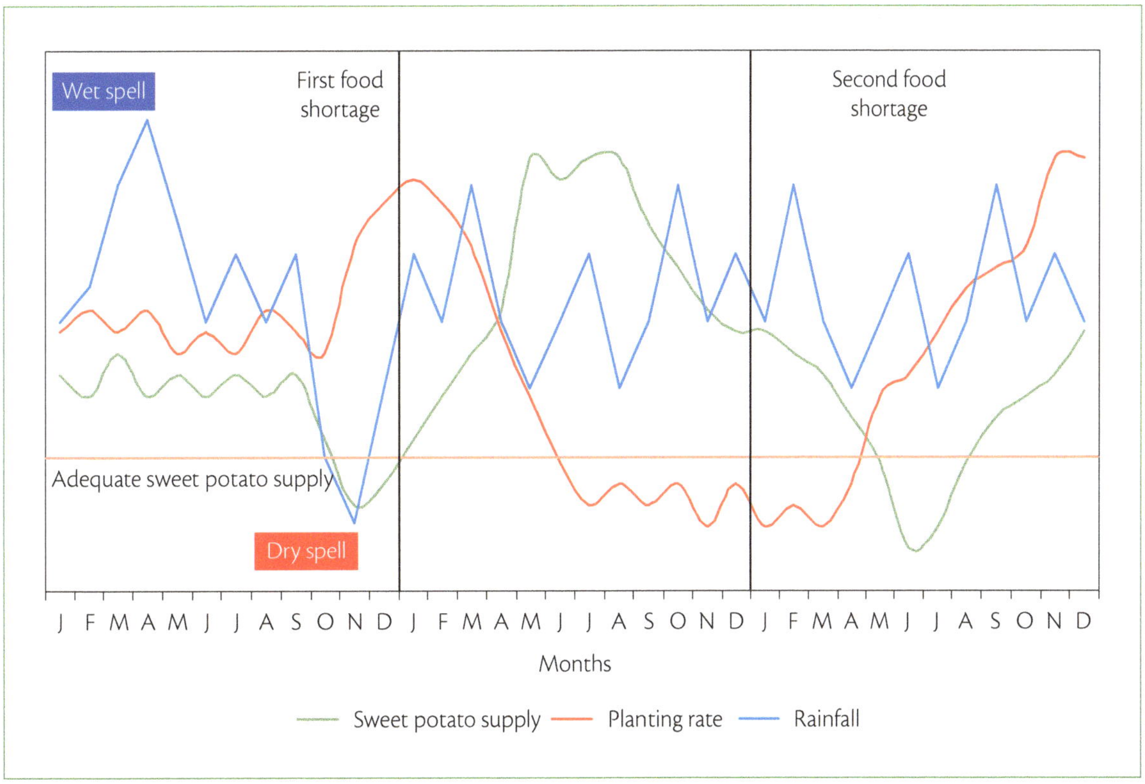

Figure 5.20.1 Model of paired food shortages in highland sweet potato agricultural systems. Source: Bourke (1988).

reduction in sweet potato marketed and an increased demand for sweet potato in markets by rural people. This combination of reduced supply plus increased demand results in a marked increase in the price of sweet potato in food markets.

However, this is not the end of the fluctuations in supply. As a result of a reduction in supply, families immediately increase the area planted per day. Planting rates continue to be higher than normal for up to six months and fallow land may be brought into cultivation. After six months sweet potato supply is high again. In response to this oversupply of sweet potato, families plant less and the planting rate falls sharply (and in local markets the price also falls sharply). Over the next 10 months or so, sweet potato supply remains more than adequate and so the planting rate remains well below average. This results in a second, more severe, food shortage. This event is about two-and-a-half years removed from the initial high rainfall episode and two years after the first food shortage. It is usually blamed on a number of more immediate possible causes, all of which are merely coincidental.[7] This second food shortage causes sweet potato prices to increase once again in response to a significantly reduced supply.

Summary

Many things influence the production of commodities for sale in PNG. The most important influence is the returns to labour. A clear association exists between high returns to labour and high levels of participation in production of a commodity for marketing. Where returns to labour for a commodity are below about K10 per person per day, participation in production of that commodity for marketing takes the form of 'target' earning, that is, when a certain amount of cash has been earned, production of the commodity stops. In order to participate in commodity production, access to markets is necessary and changes in access, due to road conditions or the closure of buying points, will influence production. The direct influence of weather on the crop, particularly excessive or inadequate rainfall, also causes fluctuations in production. The most interesting cause of fluctuations in a commodity is the way in which rainfall and villagers' planting choices combine to cause significant fluctuations in sweet potato supply in the highlands.

7 These have included the demands of the annual coffee harvest, temporary absences from gardening to harvest *karuka* pandanus nuts, temporary migration to towns, labour migration, cattle projects, large ceremonies, and too much card playing or 'laziness' on the part of villagers, none of which are the actual cause (Bourke 1988). In a number of places in the highlands, these food shortages have been associated with significant falls in birthweight and weight loss in children.

Sources

Anderson, D. (1976). Costs and returns of smallholder vegetable growing in the highlands. In Wilson, K. and Bourke, R.M. (eds). *1975 Papua New Guinea Food Crops Conference Proceedings*. Department of Primary Industry, Port Moresby. pp. 299–311.

Anderson, D. (1977). An economic survey of smallholder coffee producers – 1976. Planning Economics Marketing Branch, Department of Primary Industry, Port Moresby.

Anderson, D. (c. 1978). Pyrethrum. In Densley, D.R.J. (ed). *Agriculture in the Economy: A Series of Review Papers*. Volume 2. Department of Primary Industry, Konedobu. pp. 1–23.

Bayliss-Smith, T. (1982). Taro (*Colocasia esculenta*) in the New Guinea Highlands. In Bourke, R.M. and Kesavan, V. (eds). *Proceedings of the Second Papua New Guinea Food Crops Conference: Part One*. Department of Primary Industry, Port Moresby. pp. 134–147.

Bourke, R.M. (1977). Food crop farming systems for institutions in the lowlands. In Enyi, B.A.C. and Varghese, T. (eds). *Agriculture in the Tropics*. Papers delivered at the tenth Waigani seminar. University of Papua New Guinea, Port Moresby. pp. 88–103.

Bourke, R.M. (1988). Taim hangre: variation in subsistence food supply in the Papua New Guinea highlands. PhD thesis. Department of Human Geography, The Australian National University, Canberra.

Clarke, W.C. (1971). *Place and People: An Ecology of a New Guinean Community*. Australian National University Press, Canberra.

Curry, G.N., Koczberski, G., Omuru, E. and Nailina, R.S. (2007). *Farming or Foraging? Household Labour and Livelihood Strategies Amongst Smallholder Cocoa Growers in Papua New Guinea*. Studies in Australia, Asia and the Pacific. Black Swan Press, Perth.

Gibson, J. (1994). Rice import substitution and employment in Papua New Guinea. *Pacific Economic Bulletin* 9(1):46–52.

Godyn, D.L. (1974). *An economic survey of cocoa in Papua New Guinea: Part III Village cocoa*. Department of Agriculture, Stock and Fisheries, Port Moresby.

Hale, P.R. (c. 1978). Rice. In Densley, D.R.J. (ed). *Agriculture in the Economy: A Series of Review Papers*. Volume 3. Department of Primary Industry, Konedobu. pp. 1–53.

Hombunaka, P.H. and von Enden, J. (2001). The influence of available water in 1997 on yield of Arabica coffee in 1998 at Aiyura, Eastern Highlands Province. In Bourke, R.M., Allen, M.G. and Salisbury, J.G. (eds). *Food Security for Papua New Guinea. Proceedings of the Papua New Guinea Food and Nutrition 2000 Conference*. ACIAR Proceedings No. 99. Australian Centre for International Agricultural Research, Canberra. pp. 242–247.

Kimber, A.J. (1976). Costs and returns from growing sweet potato. In Wilson, K. and Bourke, R.M. (eds). *1975 Papua New Guinea Food Crops Conference Proceedings*. Department of Primary Industry, Port Moresby. pp. 313–316.

McGregor, A. (2004). Diversification into high-value export products: case study of the Papua New Guinea vanilla industry. Food and Agriculture Organization of the United Nations, Rome.

Mills, F. (ed.) (2002). *The Book of Rice in Papua New Guinea*. Mills Media Limited, Port Moresby.

Overfield, D. (1994). The smallholder coffee system: production costs, grower returns, and economic sustainability. Coffee Discussion Paper No. 15. Coffee Industry Corporation Ltd, Goroka.

Waddell, E.W. and Krinks, P.A. (1968). The organisation of production and distribution among the Orokaiva. New Guinea Research Bulletin No. 24. The Australian National University, Canberra.

Yarbro, S. and Noble, S. (1989). Smallholder production, processing and marketing of cocoa and copra in Papua New Guinea: a baseline survey. Designing Monitoring Systems for Smallholder Agriculture in Papua New Guinea, Working Paper No. 10. Department of Human Geography, Research School of Pacific Studies, The Australian National University, Canberra.

5.21 Marketing agricultural exports

Agricultural marketing in PNG ranges from the well-organised structures for marketing oil palm, tea and rubber to the informal, dispersed and disorderly marketing of fresh food, betel nut and vanilla. Coffee and cocoa marketing are very competitive with a high degree of self-regulation. A common feature of all these industries is that marketing functions are carried out entirely by the private sector. This contrasts markedly with the situation in a number of other Pacific island countries where government agencies have been actively involved in agricultural marketing, usually to the detriment of industry development.

This section reviews PNG's agricultural marketing of the major tree crop commodities. Marketing of domestic fresh foods is discussed in Section 5.3. Marketing for minor export crops is discussed in the relevant sections.

Tree crop commodity marketing

Markets

PNG's tree crop industries are entirely export-oriented. The small domestic PNG market restricts the opportunities to add value by processing a primary product into a marketable product such as, for example, making chocolate for local sales from cocoa. Despite the importance of the tree crop industries to the national economy, PNG has a very small share of world markets and so has no influence over price. PNG produces around 3% of the world's palm oil, less than 2% of coffee and cocoa, 0.1% of rubber and 2–5% of copra oil.

Marketing systems

In the mid 1960s the World Bank advised the Australian administration on the importance of competitive marketing for coffee and cocoa. This advice was heeded, and the newly formed Coffee Marketing Board and Cocoa Marketing Board did not become directly involved in marketing. Rather, they focused on improving quality standards, price stabilisation, funding, and directing research and development. Competitive and efficient marketing systems developed, to the benefit of growers. However, more centralised and less competitive marketing systems evolved for other commodities.

Coffee

Coffee is the second largest agricultural export by value in PNG (see Section 5.4). Although quality and productivity have declined, the coffee industry has maintained a reasonable degree of international competitiveness. This is because of the low level of purchased inputs used (see Section 5.19) and a fairly efficient marketing system.

The coffee industry's marketing structure has remained virtually unchanged since the Coffee Marketing Board, now replaced by the Coffee Industry Corporation (CIC), was established in 1964 (see Section 6.4). Coffee production has been dominated by the smallholder sector since the early

1960s. Smallholder coffee beans move through a marketing chain of itinerant buyers, dried bean factories and exporters. Virtually all of PNG's coffee is exported as dried green bean. All parts of the marketing chain are competitive to some degree. However, growers have not been offered financial rewards for producing better-quality product, and the result is generally poor quality of smallholder coffee. If the industry is to remain viable in the longer term, the quality of the coffee produced must be improved.

Buying is highly competitive, with several thousand itinerant buyers competing to purchase parchment coffee from village producers at the 'farm gate'. Marketing at this level is unregulated and sometimes chaotic. Buyers provide an important marketing service, particularly for growers who do not have ready access to a processing factory. But the price paid is usually significantly lower than what growers would receive at the 'factory door'. The more remote the grower is from the factory and the poorer the access, the greater the price differential between the 'farm gate' price and the 'factory door' price. Most roadside buyers pay the same price per kilogram regardless of bean quality and keep very poor or no records of their purchases.

Because financial viability depends on the volume of throughput, many of the wet bean factories accept whatever coffee is offered to them, and also pay no price incentives for quality. Under CIC rules, all processors buying coffee at the factory door are required to display prices for different grades. In some locations it has become common practice to offer just one price. Grades are ignored to increase throughput to achieve short-term viability, despite the dry bean processors being registered and more closely regulated. Some processing factories are owned wholly or partly by exporters, although most are owned by small local companies and individuals.

Coffee exporting from PNG is dominated by a small number of large companies. Exporter competitiveness is restricted by the way the industry is financed, with the larger exporters being owned partly or wholly by the larger international coffee trade houses, or tied to them through financing arrangements.

The share of export earnings that goes to PNG coffee growers is comparable to what growers in other countries producing a similar type of coffee receive. However, PNG growers receive considerably lower prices (Figure 5.21.1). This occurs because PNG exporters receive lower returns as PNG coffee exports are dominated by Y-grade coffee (57% of exports in 2006). The average price received in 2006 by PNG exporters for Y-grade coffee was K5.90/kg, compared with K7.10/kg received for premium smallholder grade.

The failure to offer price incentives for quality coffee is to some extent the outcome of the way coffee buying is financed. Commercial banks will not lend working capital to buy coffee without secure guarantees or collateral, so many processors are forced to seek advances from exporters. In turn, exporters have difficulties raising funds locally and so look to their overseas buyers or parent companies to provide bank guarantees or offshore lines of credit. This lack of easily accessible credit creates a substantial barrier to entry into coffee exporting and, as a result, the sector is dominated by a handful of exporters. Exporters must sell to the international buyer that provides their finance and are not in a position to sell to the highest bidder.

Cocoa

Cocoa is the third largest agricultural export earner in PNG (see Section 5.5). The cocoa marketing system has competitive pricing at each stage of the marketing chain. Growers sell wet bean to dealers or directly to fermentaries. Cocoa Board regulations make it mandatory for fermentaries and exporters to display wet and dry bean prices respectively. The PNG competitive marketing system has resulted in low marketing margins and relatively high returns to growers.[1] This contrasts with the situation in other Pacific island countries, where state-owned or controlled monopolies have dominated cocoa marketing and growers receive a lower proportion of the export price. The impact of competition on marketing margins is seen in the differences between Fiji and PNG grower prices during the 1980s cocoa price boom (Table 5.21.1).

1 The marketing margin is the share of the export price that goes to those involved in marketing and production.

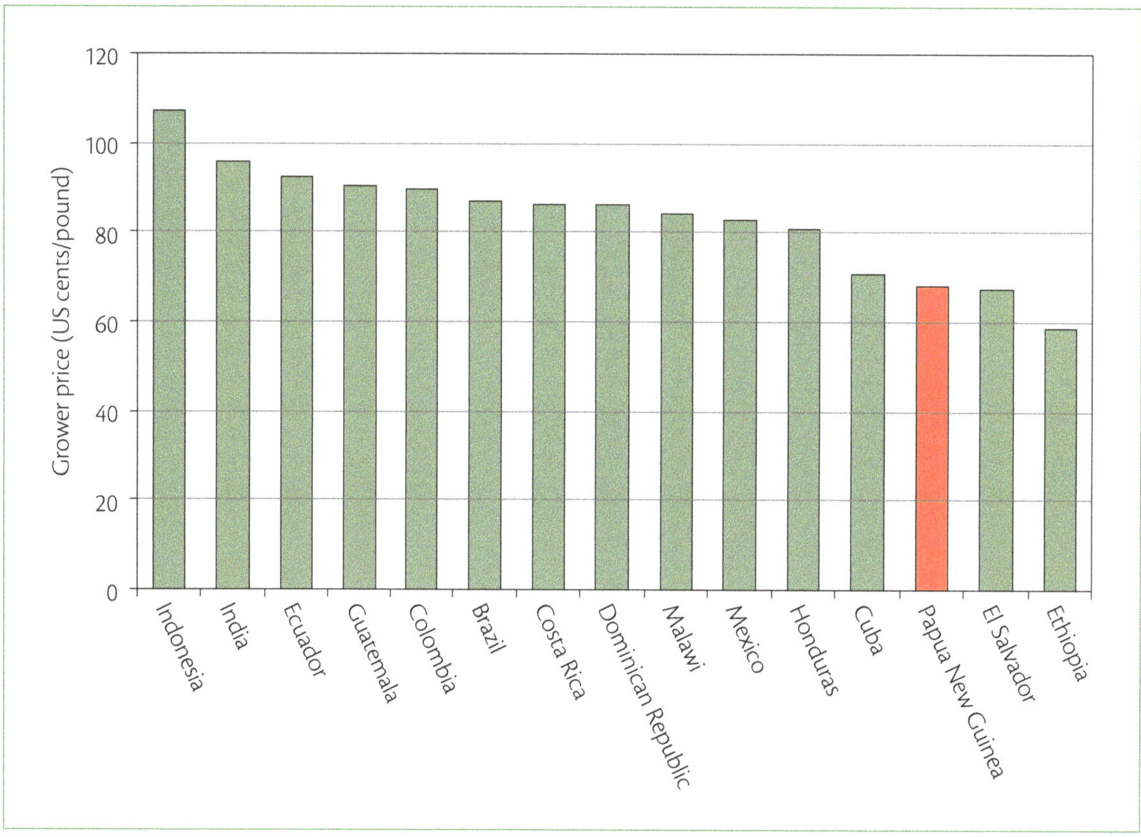

Figure 5.21.1 Average grower prices for Arabica coffee by country, 2006.
Source: International Coffee Organization (2006).

In the past, PNG cocoa had a quality and fine flavour reputation in international markets that earned significant market premiums. This quality advantage has been eroded with the decline of the plantation sector and increasing problems of smoke contamination in drying and under-fermentation in smallholder cocoa. However, some buyers still seek out better-quality PNG cocoa as part of their blends and are prepared to pay higher prices for it.

The Cocoa Board's efficiency and performance as a regulator has declined (see Section 6.4), but the competitive marketing structure that has served cocoa growers well for more than three decades remains largely intact. Despite relatively low yields, the PNG industry has remained internationally competitive due to a low level of purchased inputs and a competitive marketing system. There is scope to increase yields and to raise farm gate prices by improving quality.

Copra

Copra provides a low income for many people living in coastal mainland and island areas (see Section 5.6). Low yields mean that the PNG copra industry is barely internationally competitive.

The Copra Marketing Board (CMB) was established in the 1950s (see Section 6.4). There were good economic and social reasons for creating a monopolistic marketing body for copra. Unlike coffee and cocoa, copra is consistent in quality and sold to few buyers worldwide and substantial economies of scale can accrue in marketing. In the 1950s many copra growers were located in isolated places and transport costs were cross-subsidised from the earnings of better-located buyers. Most copra was sold by village producers to a network of middlemen, cooperatives, or ship owners, who on-sold to the CMB.

Table 5.21.1 Fiji and PNG cocoa prices, 1984–1986

Year	Fiji grower price (F$/tonne)	Fiji grower price (kina equivalent/ tonne)	PNG grower price (kina/ tonne)
1984	1,500	1,269	1,804
1985	1,500	1,364	1,890
1986	1,700	1,556	2,010

Source: ADB (2004:20).

The CMB performed satisfactorily for a number of years, but it became insolvent in the 1990s due to a combination of low international prices, deterioration in inter-island shipping, poor management and political interference. In response, the board was privatised in 2002 with the formation of the Kokonas Indastri Koporesen (KIK). The KIK was to form policy and regulate the industry, with marketing to be managed by the private sector. One result is that copra is no longer purchased in uneconomic, isolated places.

Growers continue to pay the same levies to the KIK as they did to the CMB, but the KIK does not provide the same services. The levies are high relative to the prices received, making the few services offered by the KIK very expensive. Copra does not have the same regulatory needs as more complex commodities such as coffee and cocoa and the involvement of a third party regulator adds unnecessary cost. There are no major research issues for copra that require the imposition of industry levies, and extension requirements are minimal.

Oil palm

Oil palm represents PNG's newest tree crop industry and is now the most important agricultural export earner (see Section 5.7). Oil palm is PNG's most successful agricultural industry in terms of efficiency and international competitiveness.

In contrast to coffee and cocoa, the oil palm industry is limited in both location and the number of processors and exporters. The industry is characterised by a few large companies that grow oil palm on nucleus plantations ('estates'). These companies operate central processing facilities and purchase fruit from hundreds of surrounding smallholders, whose leases require them to grow oil palm. All marketing functions are carried out by the estates. The nucleus plantation companies purchase smallholder fruit at prices determined by a formula (see footnote 5, Section 5.7). Although this structure brings considerable benefits to smallholder growers in the form of extension services and supply of inputs, they argue that the formula should be modified so that it is more favourable to them.

Sources

ADB (Asian Development Bank) (2004). Preparing the agriculture and rural development project, Papua New Guinea: Agricultural markets, marketing and rural enterprise development. ADB TA4055-PNG. Asian Development Bank.

CIC (Coffee Industry Corporation Ltd) (2003). Coffee Report No. 57. Coffee Industry Corporation Ltd, Goroka.

Coulter, H.K. (1992). *Commodity marketing institutions in the Pacific Islands: a case study of Papua New Guinea*. INA Discussion Paper No. 51. Institute of National Affairs and Pacific Islands Development Program, East-West Center, Port Moresby.

DAL (Department of Agriculture and Livestock) (2001). National Agriculture Development Strategy: Horizon 2002–2012. Department of Agriculture and Livestock, Konedobu.

International Bank for Reconstruction and Development (1965). *The Economic Development of the Territory of Papua and New Guinea*. The Johns Hopkins Press, Baltimore.

International Coffee Organization (2006). Historical data <http://www.ico.org/asp/select7.asp>. Accessed November 2007.

Kannapiran, C.A. and Fleming, E. (1999). *Competitiveness and comparative advantage of tree crop smallholdings in Papua New Guinea*. Working Paper Series in Agricultural and Resource Economics No. 99-10. Graduate School of Agricultural and Resource Economics, University of New England, Armidale, NSW.

McGregor, A. and Coulter, H. (1991). The institutional environment for private sector development in the agricultural sector. Pacific Islands Development Program, East-West Center, Honolulu, Hawaii.

McGregor, A., Lutulele, R. and Wapi, B. (2005). Papua New Guinea horticulture industry sector study. Secretariat of the Pacific Community, Suva, Fiji.

Wheeler, M. (2002). Papua New Guinea coffee: an integrated market development strategy. Coffee Industry Corporation Ltd, Goroka.

PART 6

Agricultural Development, Policies and Governance

Bryant Allen

6.1 Land tenure

Customary ownership

Around 97% of land in PNG is occupied and used under *customary tenure.* This section attempts to provide a basic understanding of customary land tenures. Custom is here defined as the long-established practices of the people of PNG. The principles of land tenure that arise from custom are not written down, but are maintained through time by human memory and everyday practice. Although custom is recognised legally in the laws and constitution of PNG, land used and occupied under customary tenure remains outside any formal state system of land administration. Until recently, this land could only be brought within a state system of land administration through a process called *alienation.* In this process, land was purchased by the state from its customary owners and leased to other users. When land was alienated, the customary owners legally lost all further interest in the land. This permanent loss of interest in land was always resisted by customary owners in PNG and today it is almost impossible to find landowning groups that will agree to land being alienated.

The first laws governing land registration were introduced to PNG in 1889. They provided for the registration of titles in 'alienated lands', or land that had been acquired by colonial administrations from customary owners. Until World War II, registered titles were only available to those who occupied alienated land.

After World War II, attempts were made to provide for registration of customary land. The Native Land Registration Ordinance was passed and a Native Land Commission was established to administer it. Around 1960 a Land Titles Commission was set up to hear claims and make decisions about ownership of customary land. In 1972 the Commission of Inquiry into Land Matters, with an all-PNG membership, was set up by the PNG Government to make recommendations on suitable land policies and laws for an independent PNG. The Commission's recommendations for customary land registration were never fully adopted but some legislation based on the inquiry was passed, including the Land Groups Incorporation Act and Land Disputes Settlement Act.

During the 1990s, under the influence of policies that argued that economic development depends on land registration, the World Bank attempted to persuade PNG to accept a Land Mobilisation Project that included the drafting of customary land registration laws. This project was to be trialled in East Sepik and East New Britain provinces. The laws were drafted but the trials were never carried out. In 2001, rumours began circulating that the World Bank was trying to take possession of customary land. This led to public protests and riots in which four people were shot by police.

In the absence of a suitable law for the registration of interests in customary land, the state has moved towards the option of temporary alienation. Under Section 11 of the Land Act 1962, the state can lease customary land from the recognised customary owners and then lease the land back to them as a legally incorporated body, or to another individual or body, designated by the owners. At the end of the lease, the land reverts to customary tenure unless a further lease is negotiated.

Landowning groups

The largest social groups, sometimes known as 'tribes' or 'language groups', occupy large areas that can be viewed as territorial domains. The rights to the territorial land may be held by the group as a whole, or by parts of the group, and are protected against infringements by agreements with neighbouring territorial groups, and if necessary by force. The size of territorial domains may vary from a few hectares to many square kilometres, and the groups that occupy them may number from a few hundred to thousands of people. The rights to occupy a territorial domain may be determined by kinship or by historical circumstance. Large landowning groups are often held together by the mutual advantages of shared labour and defence, as well as by constant references to kinship, tradition, history, culture and language.

Within a territory, smaller landowning groups that make up the territorial group are composed of people who say they are descended from a common ancestor. For this reason these groups are often known as 'descent' groups and in PNG they are often loosely called 'clans'. However, what determines clan membership can vary widely across PNG and the uncritical use of the term 'clan' can create considerable confusion and misunderstanding. Customary landowning groups are based on a dominant ethos of kinship in the broadest sense of that term, but legitimate membership of a landowning group is rarely determined only by biological descent.

The rights of landowning groups to the land they occupy are established through agreed-upon histories that describe how they came into being and how they came to occupy the land. In the case of disputes, it is often these histories that are disputed, rather than the boundaries of the land concerned. The particular historical events that endorse rights to occupy land are remembered by repeated storytelling. Particular individuals, usually older men, hold this knowledge and are relied upon to argue the historical precedents and principles, should disputes arise.

Landowning groups are commonly made up of a number of related extended families. In some places, the families may be the descendants of a number of ancestral brothers, the sons of the original founding ancestor. In other places the families may trace their descent through female ancestors. In yet other places, membership rules allow affiliation to be traced through descent from both male and female ancestors, such that the line of descent that an individual traces may jump back and forth between ancestral males and ancestral females. Intelligent, forceful men can argue their rights as landowners by either knowing these histories better than other men, or by being more skilled than others at manipulating the histories and presenting them more convincingly in public.

Importantly, some members of the landowning group will be individuals whose ancestors were incorporated into the group, by adoption for example, at some time in the past, perhaps hundreds of years ago. They will not be able trace a biological descent from the founding ancestor. Enough cases of non-descent incorporation into so-called 'descent' groups exist in PNG to argue that 'descent' group is an inappropriate term for a landholding group, even if the central ideology of most landowning groups is one of descent from a common ancestor. Adoption into a landholding group and change of group membership is a typical feature of PNG land tenure systems. The ability to transfer individuals serves to continually adjust the numbers of people to the amount of land available. It also ensures the survival of individuals from groups that have gone into decline because of the uncertainties of reproduction, mortality and warfare. However, adoption and change of membership of landowning groups is frequently poorly understood or not recognised at all by non-villagers who simplify rights to descent from a single male ancestor.

The means of determining rights to land within landowning groups vary from place to place and from group to group. Rights to land are determined by very local histories that are recorded only in the memories of the living. Within a single village, hundreds of individual pieces of land have particular histories. It is very difficult to predict the circumstances that will have led to any particular family owning any particular piece of land. Previous attempts to adjudicate land disputes in PNG have been forced to rely heavily on the merits of each particular case as argued by the disputants and local experts, and rarely on generalised principles of rightholding. An outsider wishing to determine the rights to a piece of land must understand the particular circumstances related to it. This can be a time consuming and tedious business, especially as some informants can be senile, confused and forgetful and some will present a case that favours their family's interests over those of other families. However, if membership of the landholding group is not properly investigated, great injustices can occur and major disputes can be generated.

All members of the landowning group are expected to meet certain social and economic obligations to other members. Through the exercise of these obligations and by daily social interaction, the rights to use land are maintained and strengthened. Group members who have been absent from the village for some time and who are presently not occupying or using land also often try to meet these obligations. They do so to ensure that, if need be, they or their children can return and exercise their rights to land in the future. Migrant parents are especially careful to show their children to their home-based siblings and cousins and to tell their children the locations and names of the pieces of land to which they have a legitimate right of use.

The land and its use

The landholdings of a particular landowning group may not be continuous, but may comprise a number of pieces of land scattered within a larger territory. The land claimed by families within the group's land may comprise a large number of very small areas of land. These individual pieces of land may be as small as 150 m^2. While land parcels may be used by an individual family, the rights to use them will probably not reside in one person, but in a number of close relatives, such as the brothers or cousins of the user. The rights held by different people in the same piece of land vary in nature and in degrees of importance. The use of the land for cultivation or another use may be decided from time to time by consultation within the family.

It is not necessary to be a customary rightholder to a piece of land to use it temporarily. A great deal of land is cultivated by a non-rightholder, with the permission of, or at the request of, the customary rightholder. When a person who does not have rights to a piece of land uses it without paying a rent, the use is known as *usufruct*. For many reasons, including shared labour arrangements or to ensure the maintenance of marriage, exchange and social relationships, the rightholders to a piece of land will invite non-rightholders to cultivate it. This is also how people who cannot establish rights to land within a group, such as in-marrying widows or recent migrants, can be given land on which to grow food and build houses during their lifetimes.

Usufructuary arrangements are usually limited to growing annual crops. People invited to use land to which they do not have rights are commonly prevented from planting trees, particularly export cash crop species like coffee or cocoa, but also food-producing trees such as breadfruit, *ton*, sago and coconut. Tree planting (and the burial of the dead) is usually interpreted as a public affirmation of right-holding. Even though much land is cultivated under usufructuary arrangements, it can be difficult for women and unmarried men to gain access to land on which to plant tree crops or to develop cash-earning business enterprises.

Change in customary tenure

Customary land tenure systems change over time. If, for example, population increase causes people to feel anxious about whether enough land will be available for their children or grandchildren, or if the land suddenly acquires a monetary value, as in the case of a logging or mining project, customary tenure systems may change quickly. Changes may include an insistence that group membership, and hence rights to use the group's land, is restricted to only male descendants of a putative ancestor;

the exclusion from the landowning group of the descendants of people who were adopted into the landowning group; or the ejection of people from land that they have used under usufruct arrangements for many years.

Some commentators on economic development in PNG assert that national economic development will be seriously constrained unless land is registered so that *individuals* can obtain legal title to land. These people argue that the economy will not grow until a market in land has been created. They contend that this will allow the most innovative and energetic individuals to accumulate land and so achieve economies of scale in farming, and enable landowners to raise capital by borrowing against their land as security. Whether or not these arguments have merit, the technical difficulties of registering individual titles in PNG should not be underestimated. The possibility that large numbers of people could lose their rights to land they are presently using should also be borne in mind. The present state land registration system struggles to administer the 3% of PNG's land that is alienated. A registry of customary land would have to be able to deal with thousands of changes of ownership annually, under circumstances where the customary owners will not be able to afford to travel to an administrative centre to register the changes. If land changes hands, but the changes of ownership are not registered, the disjuncture between the situation on the land and the records in the registry will soon grow to proportions that would render the system unworkable. The tasks of surveying the land, issuing titles and maintaining a land register, and of devising a new system of tenure that does not result in many people being disinherited, is presently beyond the competency of the PNG state.

Nevertheless, there is an urgent need, *in some places*, to survey the boundaries of land at landowner group level. There are circumstances in PNG where customary tenure systems are struggling to cope with rapid social or demographic changes occurring in the landholding group, or where urban development or other non-customary uses have spread onto customary land. There is a need to establish legally recognised surveys at a lower level of accuracy than is presently required by law. Legislation that defines membership of landowning groups and that allows land to be used by non-customary owners while the customary owners maintain an ongoing interest needs to be further developed. There is also an urgent need to teach more Papua New Guineans how to undertake the onerous and often tedious work required to properly establish the membership of customary landholding groups and the complex multiple rights that can be held by a number of individuals or groups in the same pieces of land.

Sources

Burton, J. (1991). Social mapping. In Larmour, P. (ed). *Customary Land Tenure: Registration and Decentralisation in Papua New Guinea.* IASER Monograph No. 29. National Research Institute, Boroko, Papua New Guinea. pp. 33–49.

Corin, B.E. (1981). Problems of ownership. In Walter, M.A.H.B. (ed). *What Do We Do About Plantations?* IASER Monograph No. 15. Institute of Applied Social and Economic Research, Boroko, Papua New Guinea. pp. 169–173.

Curtin, T., Holzknecht, H. and Larmour, P. (2003). *Land registration in Papua New Guinea: competing perspectives.* State, Society and Governance in Melanesia Discussion Paper No. 2003/1. Research School of Pacific and Asian Studies, The Australian National University, Canberra.

Fingleton, J.S. (1991). The East Sepik land legislation. In Larmour, P. (ed). *Customary Land Tenure: Registration and Decentralisation in Papua New Guinea.* IASER Monograph No. 29. National Research Institute, Boroko, Papua New Guinea. pp. 147–162.

Fingleton, J. (2004). Is Papua New Guinea viable *without* customary groups? *Pacific Economic Bulletin* 19(2):96–103.

Fingleton, J. (ed.) (2005). *Privatising Land in the Pacific: A Defence of Customary Tenures.* Discussion Paper Number 80. The Australia Institute, Canberra.

Gosarevski, S., Hughes, H. and Windybank, S. (2004). Is Papua New Guinea viable *with* cusomary land ownership? *Pacific Economic Bulletin* 19(3):133–136.

Holzknecht, H. (1999). Customary property rights and economic development. In van Meijl, T. and von Benda-Beckmann, F. (eds). *Property Rights and Economic Development: Land and Natural Resources in Southeast Asia and Oceania*. Kegan Paul International, London.

Hughes, H. (2004). *Can Papua New Guinea come back from the brink?* Issue Analysis No. 49. The Centre for Independent Studies, Sydney.

Jones, L.T. and McGavin, P.A. (2001). *Land Mobilisation in Papua New Guinea*. Asia Pacific Press, Canberra.

Lakau, A.A.L. (1997). Customary land tenure, customary landowners and the proposals for customary land reform in Papua New Guinea. *Anthropological Forum* 7(4):529–547.

Larmour, P. (ed.) (1991). *Customary Land Tenure: Registration and Decentralisation in Papua New Guinea*. IASER Monograph No. 29. National Research Institute, Boroko, Papua New Guinea.

Oliver, N.F. (2004). Comment on Helen Hughes, 'Can Papua New Guinea come back from the brink?' Unpublished paper. Cairns, Australia.

Simpson, S.R. (1971). Land problems in Papua-New Guinea. In Ward, M.W. (ed). *Land Tenure and Economic Development: Problems and Policies in Papua-New Guinea and Kenya*. New Guinea Research Bulletin No. 40. New Guinea Research Unit, The Australian National University, Port Moresby and Canberra. pp. 1–36.

Ward, A. (1983). The Commission of Inquiry into Land Matters 1973: choices, constraints and assumptions. *Melanesian Law Journal* 11:1–13.

6.2 Agriculture and gender

Gender refers to the distinction that all human societies make between 'male' and 'female' and how these distinctions are expressed in the everyday relationships between men and women. Melanesian societies express the relationships between men and women and the production, exchange and consumption of things in complex ways. These include the ownership and inheritance of land, the division of labour, the ability to exchange or distribute agricultural produce, and restrictions over whether, where and when, men and women can cultivate particular plants.

Debates over whether gender relations in Melanesia are exploitative of women are not discussed here. It is common for gender relations in PNG to be seen in terms of men dominating women. Print media, in particular, frequently report cases of women being disadvantaged in a range of everyday activities. Physical violence against women, including rape and other violent assaults, especially in urban areas, is reported daily. Some feminist scholars have argued that in pre-colonial societies, the view that women were disadvantaged is an oversimplification and a misinterpretation of how Papua New Guineans customarily perceived exchange, labour and the value of what was produced. They also argue that such a view fails to take into account complex domestic relationships of production at the household level. However, other scholars have argued that, in the societies they have studied, women were significantly disadvantaged.

Regardless of whether pre-colonial relations were exploitative or not, it is likely that contact with colonial institutions brought about changes in gender relations in PNG that worsened the position of women. Colonial officers were almost all male. Some older PNG women allege that members of the armed male police force regularly sexually assaulted village women. An indentured labour scheme was restricted to men and the introduction of cash cropping also favoured village men.

Gender and village production

Most agricultural output in PNG, other than that from plantations, is produced by a man *and* a woman, from land that has been accessed through complex family tenure arrangements (see Section 6.1) and with labour that is managed through generally harmonious domestic relationships. A highly successful man is usually married to a hard working and ambitious woman. Unmarried men or women are rarely outstanding agricultural producers or persons of importance in the village, unless they are aged. While men may denigrate or make fun of women's skills and knowledge in public, in private many express pride at their wife's critical contribution to the household economy.

The marriage that brings a productive family into being links it with the families of both the husband and the wife. The land that the family uses for agriculture is often obtained from the husband's family, his father, or his mother's brother, but it will not be unusual for some of the land used to belong to the wife's father or brother. The plants that they cultivate will have been brought into the marriage by both partners and they will continue to source seeds, cuttings and tubers from both their families. The families will exchange labour and food. The children from the marriage will, from time to time, reside for periods with their grandparents and uncles and aunts, as well as their parents.

In most parts of PNG, men and women work together in their gardens, sharing almost all of the labour and the contribution of knowledge and skills. Men, with help from other males, usually undertake the heavier work of clearing, fencing and preparing land for planting. Women remove slashed material, make heaps for burning, and do most of the planting, weeding and harvesting of food crops. In some places, women do not plant certain crops. For example, yams are usually planted and harvested by men, although some varieties are set aside for women to plant. Other crops, sweet potato for example, appear to have few restrictions on who can plant, weed or harvest them.

Examples of differences in the work done by men and women are illustrated in Tables 6.2.1, 6.2.2 and 6.2.3 with data from a village in Simbu Province, two villages in Southern Highlands Province, and a village in the Nomad area of Western Province. The marked differences between men's and women's labour contributions for certain agricultural activities is clear from these three studies.

In many PNG societies domesticated plants are believed to have spirits and must be spoken of and handled with respect. Furthermore, men perceive that women are closer to 'nature' than men are, because women menstruate monthly, in synchrony with the moon, and because they can bear children. Many rituals are performed to control natural forces, like the reproduction of plants and animals and to appease the attitudes of natural spirits towards humans. Likewise, many attributes of human reproduction (female genital secretions, menstrual blood, placenta) are thought to be detrimental to plants as well as to men, and there are many restrictions placed on the everyday behaviour of women – what they can plant and what they can do inside and outside the garden. For example, in many parts of PNG women never step over plants or food, they do not enter gardens if they are menstruating or have recently given birth and they usually do not urinate or defecate in or near gardens.

The contribution of women to agriculture in terms of physical labour expended is difficult to measure quantitatively. In low intensity systems, men and women spend about the same amount of time in gardens, with men doing the heavier work. However, where land use is more intensive, women contribute more labour. In more intensive systems, fields are cleared and fences rebuilt only once every five to ten years. Production becomes a continuous cycle of tilling, planting, weeding and harvesting, most of it done by women. In these systems, it is women who work continuously on the land, leaving men more time to spend negotiating exchanges, marriages and to become involved in cash cropping. Women also carry most of the food and firewood to the home and are responsible for raising and feeding pigs. These demands are additional to the work required to care for children and to prepare and cook food.

Gender and cash cropping

In most PNG villages, cash is earned by selling fresh food in local or urban markets (see Section 5.3) and by producing and selling export crops such as coffee, cocoa, copra and vanilla (see Sections 5.4, 5.5, 5.6 and 5.14). Women sellers significantly outnumber men in fresh food markets in PNG. However, men tend to be more involved in long-distance marketing, for example, the sale of highlands produce in Port Moresby, Lae or Madang food markets. A number of reasons have been suggested to explain the dominance of women in marketing fresh food: that women do not have the opportunity to earn cash from other activities, such as labouring or selling export crops; that women are not able to travel longer distances to urban and regional markets; that the role of women in carrying food to the village has been extended to carrying it to the local marketplace; that men see selling small amounts of food in markets as demeaning; and that men and women sell 'masculine' and 'feminine' crops, as determined by the local culture. All these possibilities remain speculative, or may apply in some places and times, but not others.

It is generally agreed that women earn less money selling food in markets (5–15% of the annual household income) than men do from selling export cash crops. However, the importance of fresh food marketing as a source of income is not in the absolute amounts earned but in the number of women participating in selling. The cash earned from fresh food sales accrues to adult women whose access to other income sources is limited and the contribution of women's income to the welfare of children is significant. It seems to be widely agreed that if women sell products that they have grown, the income earned is theirs to keep and spend as they see fit.

Women are also heavily involved in harvesting and processing export crops, such as coffee and cocoa, but are less involved in selling them. As a result, men receive the majority of the income from village cash cropping. There are, however, many ways in which money from the sale of cash crops is allocated within the household and this is not well documented. It is known that in some households, for example, members of the household, including older children, take it in turns to harvest, process and sell cash crops and they keep the income for themselves. Where unprocessed coffee cherry can be sold to an itinerant buyer on the roadside or unprocessed cocoa beans sold directly to fermentaries, women are likely to receive more of the earnings. This is usually only possible if husbands, fathers or brothers agree that women can have access to the trees, which are commonly viewed as belonging to men. In many places, the sons of a widow will give their mother access to some of the coffee or cocoa trees that they have inherited from their father, until her death. In many parts of PNG cash cropping was associated with cooperatives. Although most cooperatives are now defunct, it is of interest that the proportion of women shareholders was small.

Gender and settlement schemes

A number of formal settlement schemes exist in PNG, including the high-cost oil palm schemes (see Sections 5.7 and 6.7). The oil palm blocks have individual land titles but very few women possess a title to land in these schemes and few inherit the title to the land if their husband dies. This is despite women contributing as much as 40% of the labour on the blocks. Men also tend to control the income from palm fruit sales, which is paid directly into accounts in their name. The planners and implementers of the land settlement schemes were male administrators and they made no arrangements other than for the names of males to appear on the titles, or on the bank accounts associated with the palm fruit sales.

Women significantly outnumber men as sellers in fresh food markets in PNG. These images are from the Madang market in June 2006. Photos: Tracy Harwood.

A recent innovation (the Mama Lus Frut Scheme) that allows money for fruit sales to be paid directly into women's accounts has resulted in substantial increases in production and in the proportion of cash income received by women.

A similar pattern has been observed on low-cost settlement schemes, where the proportion of female titleholders is universally small. A number of schemes based on low-priced products such as copra or rubber have failed economically. It has been the settler women growing vegetables on the blocks and selling in local markets that has provided settler households with a cash income.

Under the Plantation Redistribution Scheme, land formally acquired (in the process of alienation – see Section 6.1) by foreign plantation owners was repurchased by the government and returned to customary owners. Evidence suggests that the outcomes of this scheme were less gender-biased than the settlement schemes. In redistributions on

Table 6.2.1 Division of labour by gender, Koge village, Simbu Province[a]

	Activity	Percentage contribution[b]	
		Men	Women
Activities carried out mainly by men	Fencing	98	2
	Courting	96	4
	Nut pandanus cultivation	91	9
	House construction	88	12
	Local Government Council	84	16
	Clearing	83	17
	Firewood (collecting and splitting)	81	19
	Coffee (maintaining and processing)	77	23
Activities carried out mainly by women	Sweet potato garden (weeding)	5	95
	Sweet potato garden (planting)	6	94
	Sweet potato garden (harvesting)	9	91
	Household (cooking, sweeping, childcare, etc.)	18	82
	Schoolwork	29	71
	Pig husbandry (feeding)	33	67
	Coffee (picking)	34	66
Activities carried out by both sexes	Ceremonies (various)	61	39
	Disputes	60	40
	Mourning	56	44
	Visiting and entertaining visitors	54	46
	Pig husbandry (moving, castrating, etc.)	54	46
	Vegetable garden (cultivation)	53	47
	Tools, equipment, clothing	50	50
	Marketing and purchasing	47	53
	Church	44	56
	Gambling	43	57
	Pig festivals (*bonagene*)	40	60

[a] Data were recorded over three 4-week periods (84 days) in 1972 and 1973.

[b] During this study, the labour contributions averaged 49.8 hours per week by men and 47.1 hours per week by women.

Source: Hide (1981:259).

the Gazelle Peninsula, East New Britain Province, 10–40% of block holders were women. However, this is an area where land is inherited through maternal relatives and granting women titles may have been one way families could gain access to multiple blocks.

Gender and plantations

Since before World War II, village men have travelled as indentured labourers to work on distant plantations, mainly from the north coast of mainland New Guinea to the islands of New Britain, New Ireland and Bougainville (see Section 1.4). It was rare for women to accompany the men and almost none were able to take an indenture. After the war, young men from these areas continued to travel informally to the Islands Region for plantation labour. The migration of men from villages caused changes in women's garden work in some places, with women taking up what had previously been men's work, but this was not a universal pattern.

Also after the war, the Highland Labour Scheme brought thousands of young highlands males to the Islands Region plantations on two-year contracts. Highlands women were excluded from the scheme. However, coffee plantations established in the highlands by foreign entrepreneurs began to employ women from nearby villages as casual labour, sometimes against the wishes of local men. It became common for women to be employed as seasonal casual labour, in picking and processing. Women also assisted employed men in picking, even though they themselves were not formally employed. One plantation manager is quoted as saying, 'I don't employ women, but they pick most of the coffee'. Although the evidence is not conclusive, it appears that women were paid lower rates than men for doing the same work.

Table 6.2.2 Division of labour by gender in two villages, Tari basin, Southern Highlands Province[a]

Activity	Time spent (hours per day)	
	Men	Women
Horticulture	1.66	3.95
Constructing ditches and pig sties	0.75	–
Pig rearing	0.32	0.32
Collecting wild plants, fishing and hunting	0.06	0.21
Total	2.79	4.48

[a] Data were recorded in Wenani village in 1993 and 1994 and in Heli village in 1994. A total of 77 adults in the two villages were studied over a 7-day period.

Source: Umezaki et al. (2002:135).

Table 6.2.3 Division of labour by gender for banana production, Gwaimasi village, Nomad area, Western Province[a]

Activity	Labour inputs (days per hectare)		
	Men	Women	Total
Clearing	10.8	12.7	23.5
Suckers[b]	25.5	15.6	41.1
Planting	5.3	13.0	18.3
Tree felling	35.0	0.0	35.0
Weeding	18.6	37.9	56.5
Total	95.2	79.2	174.4

[a] Data were recorded over a 15-month period in 1986–1987. Banana provides about half of the food energy in this community. Labour inputs were not recorded for harvesting bunches. Harvesting is shared by men and women, with women contributing more to day-to-day harvests, but men doing more work for large harvests.

[b] 'Suckers' refers to work collecting suckers from existing banana gardens and transporting them to new garden sites for planting.

Source: Dwyer and Minnegal (1993:11–13).

Summary

Village production depends on day-to-day domestic relationships between men and women. From daily work in producing food, women have an extensive knowledge of plants and the agricultural environment. They are highly skilled horticulturalists. Women contribute at least half of the labour to the production of food crops and cash crops from village gardens in PNG. Women therefore make a highly significant contribution, not only to household production, but also to the foreign earnings of PNG. Without their contribution, agricultural production would be substantially less than what it is. Women outnumber men as sellers in fresh food markets, but they receive less income from this activity than men receive from selling export cash crops. In more formal settings, such as plantations or settlement blocks, women are likely to be paid less or to be discriminated against in other ways.

Sources

Dwyer, P.D. and Minnegal, M. (1993). Banana production by Kubo people of the interior lowlands of Papua New Guinea. *Papua New Guinea Journal of Agriculture, Forestry and Fisheries* 36(1):1–21.

Hide, R.L. (1981). Aspects of pig production and use in colonial Sinasina, Papua New Guinea. PhD thesis. Columbia University, New York.

Hide, R.L. (1992). Unpublished papers on women in agriculture in PNG: Women and market trade; Women's participation in agriculture on smallholder land settlement schemes; Women's participation in estate agriculture; Other issues in women's participation in agriculture.

Jolly, M. (1994). *Women of the Place: Kastom, Colonialism, and Gender in Vanuatu.* Studies in Anthropology and History Volume 12. Harwood Academic Publishers, Chur, Switzerland.

Lewis, F. (2000). The "Lus frut mamas" scheme: developing economic opportunities for women in agriculture. Oil Palm Industry Corporation, Kimbe.

Modjeska, N. (1982). Production and inequality: perspectives from central New Guinea. In Strathern, A. (ed). *Inequality in New Guinea Highlands Societies.* Cambridge Papers in Social Anthropology No. 11. Cambridge University Press, Cambridge. pp. 50–108.

Queensland Department of Primary Industries, Papua New Guinea Department of Agriculture and Livestock and Papua New Guinea Coffee Industry Board (1987). *Papua New Guinea export tree crop study: volume 1 – largeholder coffee.* Australian Centre for International Agricultural Research, Port Moresby.

Sexton, L. (1986). *Mothers of Money, Daughters of Coffee: The Wok Meri Movement.* Studies in Cultural Anthropology No. 10. UMI Research Press, Ann Arbor, Michigan.

Sillitoe, P. (1985). Divide and no one rules: the implications of sexual divisions of labour in the Papua New Guinea highlands. *Man* (NS) 20:494–522.

Strathern, M. (1972). *Women in Between: Female Roles in a Male World: Mount Hagen, New Guinea.* Seminar Studies in Anthropology No. 2. Seminar Press, London.

Strathern, M. (1988). *The Gender of the Gift: Problems with Women and Problems with Society in Melanesia.* Studies in Melanesian Anthropology No. 6. University of California Press, Berkeley.

Umezaki, M., Yamauchi, T. and Ohtsuka, R. (2002). Time allocation to subsistence activities among the Huli in rural and urban Papua New Guinea. *Journal of Biosocial Science* 34(1):133–137.

Warry, W. (1986). Kafaina: female wealth and power in Chuave, Papua New Guinea. *Oceania* 57(1):4–21.

Yamauchi, T., Umezaki, M. and Ohtsuka, R. (2000). Energy expenditure, physical exertion and time allocation among Huli-speaking people in the Papua New Guinea highlands. *Annals of Human Biology* 27(6):571–585.

6.3 Policy making in the agricultural sector

Policies are political, management, financial and administrative mechanisms for achieving national or other goals. A policy is a plan to guide decision making and actions in order to achieve a stated goal. Policies must be distinguished clearly from laws, which are enforceable in the courts. Policies set out principles and guidelines, but they are not enforceable, except through acts of parliament that establish the principles of a policy as laws.

This section is concerned with agricultural policy in PNG.[1] Agriculture is a complex sector of the economy and involves many factors including the environment, land tenure, food security, export markets, domestic markets, prices of produce, costs of inputs, exchange rates, transport costs, subsidies, gender, education, research, information, labour and quarantine. This means firstly that many government policies impinge on agriculture and, secondly, that good policy making in agriculture is difficult. Thirdly, PNG is sometimes described as a 'weak state' because the government frequently does not have the capacity to enforce laws or to ensure that the many parts of a complex bureaucracy implement existing policies (see Section 6.4). Fourthly, because many present-day policies have evolved from the Australian colonial administration, they may reflect situations that have changed from the time they were developed. Lastly, agricultural policy in PNG is further complicated because it impacts directly on the welfare of the majority of the population. These are the people who live in rural areas and produce much of their own food and most of PNG's agricultural exports.

1 This section is based heavily on McKillop et al. (2007) and McKillop (1981).

Colonial policies and Independence

After World War II, Australian colonial agricultural policies had two main objectives: the first was to promote a commercial plantation sector and provide Australian settlers with the 'full benefits of modern agricultural scientific knowledge'; the second was to promote smallholder peasant proprietorship. This policy resulted in a Department of Agriculture, Stock and Fisheries (DASF)[2] that was a specialist, technically oriented organisation, modelled on Australian departments. DASF research work in tropical agriculture was internationally recognised and was focused on the export tree crops in which PNG had a comparative advantage: coffee, cocoa and coconut. The department was organised and structured by these commodities. In the late 1960s, oil palm was added to these crops.

2 This department has been known by different titles over the past 60 years. These are Department of Agriculture, Stock and Fisheries (1946–1975), Department of Primary Industry (1976–1985) and Department of Agriculture and Livestock (1986–present).

The colonial agricultural administration had a high level of autonomy to do what it thought best. Professional agriculturalists reported to generalist administrators in Canberra who knew little about the complexities of PNG agriculture, so there was little policy initiative from Australia. Headquarters officials in Port Moresby were isolated from the day-to-day administration of the districts and individual field officers had a high degree of latitude in formulating their own programs. Where they were highly skilled and motivated, this produced outstanding results and their skills and knowledge were passed on to their PNG understudies. Where they were not, the outcomes were sometimes poor.

In the transition to independence, policy making for PNG in Australia and in PNG was increasingly influenced by non-Australians with experience in the decolonisation of Africa and Asia, who criticised, for example, the strong policy emphasis on export cash cropping. They argued this would make an independent PNG less 'self-reliant', dependent on the vagaries of world commodity markets and at risk of not producing enough food. Within PNG, ministers of agriculture in the new PNG Government expressed views that heavy expenditure on research was irrelevant to the country's needs. Outsiders were also influential in persuading PNG's political leaders to adopt national ideas of equity and self reliance in the form of the Eight National Aims.

In the agricultural sector, the Eight Aims were expressed in a number of contradictory policy goals to achieve greater self-sufficiency in food and to replace the imports of rice and fresh food from Australia. In 1974 however, urban residents demonstrated in the streets against a rise in food prices and the threatened loss of imported rice. The policy response was to establish a Fresh Food Project with central marketing and storage facilities and the establishment of at least 50 large-scale farms to supply the fresh food needs of major urban centres. This briefly undermined the operations of private food-marketing organisations that were then emerging in the provinces and drove them out of business, before it also failed financially after incurring serious losses.

The basic policy issues of food production for urban markets remained unresolved. The goals remained confused between encouraging urban development through low food prices and promoting rural development by supporting higher prices for producers. The rhetoric of food self-sufficiency also brought political pressure to grow cereal crops, particularly rice, to replace imported rice from Australia. Despite many previous failures to produce rice economically in PNG, in the 1970s the cost-benefits and acceptability of cereals were investigated. Heavily subsidised machinery was provided for rice growing at Mekeo in Central Province, where rice growing remained unsuccessful (see Section 2.5).

On the positive side, new food crop cultivars that increased subsistence productivity were introduced. Rural households also improved food security by purchasing food with higher protein and energy content than local foods, using cash earned from export cash crop sales. Purchased food also evened out fluctuations in overall food supply caused by environmental variation.

Macro-economic policy

Macro-economic policies have had a major influence on the agricultural sector (see Section 4.1). From Independence to the late 1980s, the government followed a 'hard kina policy' where the value of the currency was held constant relative to the value of a number of international currencies.[3] Protective measures and subsidies were used to establish sugar and poultry industries during this period. Import bans or tariffs were partially imposed on many imported foods.

These policies inhibited investment and disadvantaged export cash cropping, but advantaged importers of food and other items. From 1989, PNG macro-economics was impacted by a series of severe shocks: natural disasters, external economic factors, internal conflict and poor fiscal management. The introduction of a complex and unmanageable provincial administrative structure; the 1989–1997 Bougainville civil war; the severe 1997–1998 drought; a World Bank/International Monetary

3 The international currencies were the Australian dollar, United States dollar, Japanese yen, German deutschmark, and British pound sterling.

Fund structural adjustment program; an inability to control expenditure; widespread corruption; and theft of public money brought the economy to the brink of failure. Foreign reserves were depleted and inflation doubled. Under World Bank pressure in 1994, the 'hard kina' policy was abandoned, the kina was floated and it fell in value from A$1.09 to A$0.42 by 2005 (Figure 4.1.1). The fall in the value of the currency advantaged rural cash crop exporters and domestic fresh food marketers, but disadvantaged urban dwellers who depended on wages, and rural people who could not access markets.

In 1990 the Medium Term Development Strategy (MTDS) was introduced. It sought to redefine the role and institutions of government to improve rural living standards. The key priorities were health, education, infrastructure and the private sector. Agriculture was placed in the private sector, which was to be the 'engine of growth' for productive enterprise. Government involvement in this sector was to be restricted to facilitating access to credit, research, extension, training services, and monitoring the commercial activities of government. However, agriculture has become a less important part of macro-economic policy with each succeeding MTDS plan, a result of its failure to become an 'engine of growth' (see Section 6.4).

Policy as public service politics

Since 2000, much agricultural policy making in PNG has reflected attempts by the Department of Agriculture and Livestock (DAL) to regain control of the agricultural sector, more than attempts to address agricultural production and marketing constraints. With the corporatisation of the export crop sectors, DAL's policy-making role became restricted to food security issues and so most attempts to gain influence were made here. Despite a 1989 White Paper on Agriculture that recognised that food supplies and the overall levels of nutrition were generally adequate, policy documents produced by DAL in 2000 claimed that PNG is a food-deficit country with poor food security. The *PNG National Food Security Policy 2000–2010* concluded that, 'the long-term sustainability of the national food security in PNG is precarious, based on the present trend of over-dependence on imported foods'.

An agricultural economist employed by DAL argued in a 1992 paper that rice production in PNG was uneconomic (see Section 2.5), and a 1993 PNG rice study showed that the cost of producing rice in PNG was more than twice the cost of importing rice. But because these conclusions were unpopular in the increasingly politicised culture of DAL, national officers were discouraged from undertaking rigorous assessments of policy issues, and policy making became increasingly unrealistic.

A submission by DAL in 1998 to the National Executive Council (PNG's cabinet) proposed the investment of K36 million over ten years for an intensive program to increase rice production. It put forward a 39-member DAL-led implementation team, new housing and vehicles for the team, training of an extension team, technicians and farmers, subsidised seed production and distribution, DAL-operated machinery pools, 1000 ha of irrigated land, and a further 4000 ha of mechanised rain-fed rice cultivation. The National Executive Council approved the establishment of a rice and grain authority and the allocation of K4 million to the rice program in 1999. The *PNG National Food Security Policy 2000–2010* proposed that PNG put some 100 000 ha of rain-fed lowland rice into production in order to produce some 150 000–170 000 tons of milled rice per year. This figure appears to be derived from the amount of rice PNG imported, but no analysis was provided as to how this target might be achieved.

Attempts to regain influence outside the food security area were also made. In August 2001, DAL launched a new agriculture development strategy called *Horizon 2002–2012*. It noted that most agricultural sector industries had declined or been stagnant since Independence and blamed, mostly correctly, inadequate infrastructure, inefficient government marketing services (copra, rubber), the failure of cooperative marketing (spices, honey) and ineffective research and extension services.

But the solutions contained in *Horizon 2002–2012* were often contradictory. The policy supported the principles of decentralisation and an efficient private sector as the key force in revitalising agriculture in the MTDS, but also proposed more centralised control of agriculture through legislation, increased government spending and a National Agriculture Development Plan. A cooperative model was proposed to attract private investment into projects on customary land, with the ownership of the projects to be restricted to the members of customary landowning groups, but another part of the plan recommended that the government 'secure land' for private investors. Marketing was to be privatised in the coconut industry, but DAL was to take over rubber marketing. An annual budget of K102–108 million was proposed. Not surprisingly, *Horizon 2002–2012* received a negative response from government and the private sector and it was seen as a blatant attempt by DAL to regain bureaucratic control over all aspects of agricultural policy and production in PNG, including the export tree crop sector.

The corporatisation policy has yielded mixed results. The PNG Oil Palm Research Association provides a cost-effective and technically well-managed research program for oil palm. The Oil Palm Industry Corporation has developed effective management and extension services (see Section 6.4). In the other tree crop industries, the results are less encouraging. Generous aid funding during the 1980s enabled the Coffee Industry Corporation (CIC) to rapidly expand its research facilities and extension services, with the latter employing some 300 field staff. When funding to these institutions was sharply reduced in 1997 following the end of donor support and general government budget constraints, a major cutback was required. Continual attempts by politicians to interfere in CIC affairs have also hindered the effectiveness of CIC operations.

Domestic food production has been successful, often despite agricultural sector policies. Domestically marketed food production appears to have kept pace with population growth and has accelerated rapidly since 1998 following the forced devaluation of the PNG currency and recovery from the 1997 drought and frosts, overcoming the constraints of poor governance and an unhelpful policy framework. Nevertheless, significant constraints continue to affect the supply of food to urban markets at competitive prices (see Section 5.3).

Non-government policies in agriculture

Community response to the breakdown of public sector service delivery has been a proliferation of non-government organisations (NGOs) and community-based organisations (CBOs). These groups are mainly voluntary organisations that seek to initiate local area rural development. Women's groups have been among the more successful and influential. Overall, however, NGOs and CBOs are weak in their analyses of genuine community need and in the identification, design, implementation and monitoring of projects that deliver real benefits to the community on a sustainable basis. In general, NGOs and CBOs are missing from all government-initiated agricultural policy.

Summary

The capacity of government to respond with appropriate policies to the challenges of the agricultural sector has been seriously eroded during the post-Independence period. The policies of 'corporatisation' and 'privatisation' have resulted in DAL losing responsibility for most projects, general extension, research and quarantine and for most research, development and extension functions for the coffee, copra, cocoa and oil palm industries. But rather than take up a policy-making role in support of the agricultural sector, DAL has wasted resources, and ultimately influence, by trying to regain its lost funcions.

As a result, DAL has produced policies that were based on misinformation and that recommend increased allocations of funds to DAL or to politically popular causes, like rice growing. Policy in the agricultural sector has thus lacked sound analysis of reality or of past outcomes and has been self-seeking and unrealistically optimistic. It is fortunate that an inability to implement these plans and resistance from other parts of the bureaucracy have prevented DAL from inflicting severe damage on the agricultural sector.

Sources

Bourke, R.M. (2001). An overview of food security in PNG. In Bourke, R.M., Allen, M.G. and Salisbury, J.G. (eds). *Food Security for Papua New Guinea. Proceedings of the Papua New Guinea Food and Nutrition 2000 Conference*. ACIAR Proceedings No. 99. Australian Centre for International Agricultural Research, Canberra. pp. 5–14.

Bourke, R.M. and Kesavan, V. (1982). Conference recommendations. In Bourke, R.M. and Kesavan, V. (eds). *Proceedings of the Second Papua New Guinea Food Crops Conference: Part One*. Department of Primary Industry, Port Moresby. pp. xi–xv.

DAL (Department of Agriculture and Livestock) (1989). White paper on agriculture: national policies for the development of the agricultural sub-sector. The Government of Papua New Guinea, Port Moresby.

DAL (Department of Agriculture and Livestock) (1996). Agriculture sector 5 years strategic development objectives, 1997–2001. Department of Agriculture and Livestock, Port Moresby.

DAL (Department of Agriculture and Livestock) (2000). Papua New Guinea National Food Security Policy 2000–2010. Department of Agriculture and Livestock, Port Moresby.

DAL (Department of Agriculture and Livestock) (2001). National Agriculture Development Strategy: Horizon 2002–2012. Department of Agriculture and Livestock, Port Moresby.

Gibson, J. (1992). *Rice self-sufficiency and the terms of trade (why rice is a good thing to import)*. Policy Working Paper No. 2. Programming, Planning and Budgeting Division, Department of Agriculture and Livestock, Port Moresby.

Gibson, J. (1993). *Trends in rice consumption*. Policy Working Paper No. 7. Programming, Planning and Budgeting Division, Department of Agriculture and Livestock, Port Moresby.

Kambori, V. (compiler) (2001). Conference summary and recommendations for policy and programs. In Bourke, R.M., Allen, M.G. and Salisbury, J.G. (eds). Food Security for Papua New Guinea. *Proceedings of the Papua New Guinea Food and Nutrition 2000 Conference*. ACIAR Proceedings No. 99. Australian Centre for International Agricultural Research, Canberra. pp. 873–882.

McKillop, R.F. (1981). Agricultural policy-making. In Ballard, J.A. (ed). *Policy-Making in a New State: Papua New Guinea 1972–77*. University of Queensland Press, St Lucia, Queensland. pp. 238–256.

McKillop, B., Bourke, R.M. and Kambori, V. (2007). Agriculture. In May, R.J. (ed). *Policy Making and Implementation: Studies from Papua New Guinea*. ANU E Press, Canberra. pp. 52–68.

Wilson, K. and Bourke, R.M. (1976). Recommendations. In Wilson, K. and Bourke, R.M. (eds). *1975 Papua New Guinea Food Crops Conference Proceedings*. Department of Primary Industry, Port Moresby. pp. 13–17.

6.4 Governance

'Governance' refers to the processes and systems by which organisations and societies govern themselves. Poor governance has been blamed for PNG's poor economic growth, deteriorating human development indicators, environmental degradation and inadequate law and order. Although these criticisms of PNG governance have been directed mainly at political governance, some of them also apply to the bodies that govern agriculture, including government departments and commodity boards.

In 2006 the World Bank developed the Worldwide Governance Indicators, which have been used to compare the standard of governance of all the countries in the world. The indicators are based on six measures:

- Voice and accountability – the extent to which a country's citizens are able to participate in selecting their government, as well as freedom of expression, freedom of association, and a free media.
- Political stability and absence of violence – the likelihood that the government will be destabilised or overthrown by unconstitutional or violent means, including domestic violence and terrorism.
- Government effectiveness – the quality of the public service and the degree of its independence from political pressures, the quality of policy formulation and implementation, and the credibility of the government's commitment to such policies.
- Regulatory quality – the ability of the government to formulate and implement sound policies and regulations that permit and promote private sector development.
- Rule of law – the extent to which agents have confidence in and abide by the rules of society, in particular the quality of contract enforcement, the police, and the courts, as well as the likelihood of crime and violence.
- Control of corruption – the extent to which public power is used for private gain, including petty and grand forms of corruption, as well as 'capture' of the state by elites and private interests.

Ranked on these indicators, relative to the rest of the world, PNG was in the 45th percentile[1] for voice and accountability, in the 24th percentile for political stability, and was at or below the 20th percentile for all the other indicators.

Many aspects of governance influence agriculture, forestry and fisheries production: the provision of health and education services; the law and justice sector; the building and maintenance of transport infrastructure (see Section 6.9); the management of the national economy (see Section 4.1) and local

[1] A percentile rank indicates the percentage of values that lie below any particular value. In this case, on the 45th percentile, 44% of all the countries in the world performed worse than PNG and 54% performed better. On the 20th percentile, 19% performed worse and 79% performed better. For details on how the index is constructed see World Bank (2006).

level government. This section, however, is primarily concerned with the constraints that poor governance places on agriculture. It focuses on the public bodies that provide services to agriculture, including regulation, quarantine, research and extension, and concludes that two critical issues affect the performance of these bodies. Firstly, an extremely complex and poorly understood law (known as the Organic Law) governs the relationships between national, provincial, district and local level governments. In these circumstances, even the most hardworking and honest administrator has difficulty promoting economic development. Secondly, the appointment of unqualified and incompetent political supporters, close family and friends to the managing boards of many of the bodies administering agriculture in PNG is widespread. This practice cripples these bodies and creates conditions where good policy development and effective administration is impossible.

Village agriculture

Most food in PNG is produced by customary landowning groups that occupy their land under constitutional rights. In reality, governments have no control over this land or the way in which food is produced from it (see Section 6.1). Food sold in local markets is also almost completely unregulated, except for the fairly arbitrary rules applied to marketplaces. Regulations are made about village agriculture (for example, the restraint of pigs) but they are rarely enforced. Villagers govern themselves with little external interference, yet most policies and plans designed to transform PNG agriculture do not take account of this reality.

Villagers also produce most of the export cash crops in PNG. Because these commodities are sent to international markets, they are subject to greater regulation. However, many of the regulations applying to the production of cash crops are not observed at the village level, and governing bodies have little capacity to enforce them. Once cash crops leave the village, their processing and marketing is controlled by laws, administered by commodity boards (see Section 5.21). The most important bodies are described on page 445 under National-level agencies.

Villagers are not well represented on organisations governing agriculture. In the colonial era, expatriate-dominated bodies, such as the Farmers & Settlers' Association in the highlands and the Planters' Association in the lowlands, had considerable influence on agricultural policy. Some members even travelled to Canberra a number of times to lobby the minister directly. The Planters' Association changed its name to the PNG Growers' Association in 1990 and became a lobby group for smallholders, as well as for the few remaining largeholders in the copra and cocoa industries (see page 447).

National, provincial and local government

PNG is a constitutional monarchy[2] and a parliamentary democracy. Four levels of government exist: national, provincial, district and local. The national parliament is a single legislature with members of parliament (MPs) elected for five-year terms. The prime minister is appointed and dismissed by the governor-general on the proposal of parliament. The National Executive Council, or PNG Cabinet, comprises ministers who are nominated by the prime minister. MPs are elected from 89 Open electorates and 20 Provincial electorates, sometimes called 'regional' electorates. Provincial electorates have the same boundaries as the provinces and the National Capital District. Provincial members often also serve as a provincial 'governor'. Each province has its own provincial assembly, comprising district MPs, managers and some local level government appointees. Eighty-five Open electorates have the same boundaries as the administrative districts (the other four Open electorates cover the cities of Port Moresby and Lae). MPs usually chair a number of district-level committees responsible for service delivery and planning. These committees include selected members of local level governments and administrators from within the district.

2 The Head of State is Queen Elizabeth II, represented in PNG by the Governor-General.

In order to win government for his[3] party, the prime minister must form a coalition from the fifteen or so unstable political parties usually represented in parliament.[4] This results in a great deal of 'pork barrel' politics,[5] including the granting of ministries to members of supportive parties and the sudden withdrawal of ministries to meet political contingencies. National ministries, as well as the seats on national commodity boards that national ministers can control, give access to monetary resources that can be diverted to local supporters or to personal use.[6] Similar behaviour occurs at the district, or Open electorate level, where MPs can use their District Support Grant funds (see Office of Rural Development, page 446) arbitrarily to gain and maintain the support of particular blocks of voters. This style of governance fails to address chronic problems that have been clearly identified by numerous investigations and also leads to the inefficient allocation of resources. Examples of these problems include the promotion of agricultural projects that will actually disadvantage village smallholders,[7] and appointments to agricultural institutions of people who are incompetent.

The relationship between the national and provincial governments is governed by the *Organic Law on Provincial Government and Local-Level Government 1995*, the outcome of a reform of the *Organic Law on Provincial Governments 1977*. The Organic Law specifies which functions are held by the different levels of government. The stated aim of the 1995 reforms was to improve the delivery of services to rural areas by decentralising them to district and local level governments. However, the most important outcome of the reforms appears to be a dislocation of administration and service delivery at all levels of government. A 2005 review of the decentralisation process found widespread confusion over administrative and financial accountability, responsibility for service delivery, the transfer and funding of functions, the separation of capital and recurrent expenditure, and taxation powers.

Section 108 of the Organic Law provides for the National Government to enact a 'National Planning Act', but this has not happened, leaving PNG without a coordinated national planning system. The Organic Law does provide for coordination of planning and budgeting between national and provincial governments through the Joint Provincial Planning and Budgeting Priorities Committee and between provincial and district governments through the Joint District Planning and Budgeting Priorities Committee. These committees are widely considered to have been a bottleneck in the funding process and to be controlled by MPs, who gained considerable

3 Only one of the 109 members of PNG's parliament is a woman, elected in 1997 and re-elected in 2002. From 1987 to 1997 there were no women members of parliament. Only three other women have been elected. No highlands woman has ever been elected to parliament. The central role of women in agriculture has largely been ignored by farmer bodies and the government institutions serving the agriculture sector. These institutions interact predominantly with men and, at the operational level, decisions about cash cropping and the sale of these products are dominated by men (see Section 6.2).

4 Party membership is unstable and political parties appear and disappear regularly between elections. Legislation was passed to prevent MPs from changing parties during the life of a parliament but it has been a difficult law to enforce. Parties and groups of MPs regularly change from government to opposition and back again.

5 'Pork barrel' politics describes government spending that is intended to benefit the constituents of a politician in return for their political support, either in the form of campaign contributions or votes.

6 One outcome is a struggle between ministers and senior public servants for control of the resources of government departments. The director of the PNG Growers' Association recently stated, 'The bad decisions and actions by some ministers are costing Papua New Guinea too much in terms of money, time and investments. Stakeholders of the cocoa, copra and other cash crop industries are not happy with the way ministers continue to violate established procedures to suit their own interests.' (Pacific Magazine Daily News, 14 January, 2005 <http://www.pacificmagazine.net/news/2005/01/14/png-punish-all-renegade-ministers-growers>.)

7 A recent example is a proposal to build a cocoa processing factory in Lae, well away from the main cocoa growing areas, supported by a 30% levy on smallholder cocoa growers. Despite strong lobbying against the proposal by growers' organisations, it was vigorously supported by some influential individuals and MPs.

powers under the 1995 reforms. The committees are preoccupied with spending the MPs' District Support Grants (DSGs) and not with coordinating activities between districts.[8] Many MPs only spend their DSG funds on new projects and not on district recurrent service delivery costs.

Other sources of funding leave most province and district health, education, agricultural extension and infrastructure maintenance programs chronically underfunded. Estimates of costs of service delivery and funding within provinces by the National Economic and Fiscal Commission in 2006 found that Sandaun Province receives only 20% of the funds required to deliver basic services; Manus 30%, Simbu, Central and East Sepik less than 40%; Milne Bay, Oro and Gulf less than 50%; Eastern Highlands 52%; Madang and Western Highlands less than 70%; East New Britain less than 80%; West New Britain less than 90%; and Morobe, Western, Enga, Southern Highlands, New Ireland and NCD more than 100%. The cost of services was not assessed for Bougainville Province.

The *Pacific 2020* review[9] of national agriculture development policies also found that major constraints affecting agricultural development in PNG have been the complex Organic Law; an ineffective extension system; the lack of an effective national agriculture development plan (see Section 6.3); the lack of transport infrastructure maintenance (see Section 6.9); ineffective communication and marketing facilities; an uncoordinated and inadequately staffed national agricultural research system; a lack of credit facilities; an acute lack of competency of staff in state institutions; and poor human resource training programs for the sector.

8 In some provinces (for example, East Sepik, Eastern Highlands and East New Britain), MPs have pooled their DSGs to repair roads or implement province-wide programs, but this does not appear to be a common practice.

9 *Pacific 2020* was an initiative by Australia (AusAID) to foster dialogue and debate in the Pacific, PNG and East Timor on how to accelerate economic growth. *Pacific 2020* was designed to provide practical policy guidance for these countries in nine critical growth areas: the productive sectors of agriculture, fisheries, forestry, mining and petroleum, and tourism, and the growth-enabling sectors of land, private investment, political governance, and employment and labour markets. *Pacific 2020* background papers can be accessed at <http://www.ausaid.gov.au/hottopics/pacific2020/papers.cfm>.

National-level agencies

Numerous national-level government agencies have roles in agricultural and rural industries. Key factors in determining their effectiveness is whether they have the resources and institutional linkages with other agencies and all levels of government to perform their specified functions and whether they have competent directors, managers and staff. This section examines the most important organisations related to agriculture.

Department of Agriculture and Livestock

The functions of the national Department of Agriculture and Livestock (DAL) include providing policy advice and sector coordination relating to agriculture and livestock (including advice on the application of agricultural legislation, administered by statutory bodies); promoting agricultural development; assisting provincial governments with the provision of extension; and preparing and implementing appropriate investment programs for major commodities and livestock. (See Section 6.3 for a brief history of the Department of Agriculture.)

In the 1970s DAL lost responsibility for extension services when they became a provincial function. Export tree crops research was transferred to specialised research institutions in the mid 1980s. During the 1990s, remaining research and quarantine functions held by DAL were moved into separate institutions, and commodity boards and corporations were given greater independence. DAL's role was narrowed to that described in the previous paragraph. However, the department struggled to adapt to its new role and wasted resources in trying to regain some of its lost functions. A 2004 review of DAL by the Asian Development Bank (ADB) found that DAL was without clear agricultural sector roles and did not have the capacity to plan or develop policy. The review argued that a National Agriculture Development Plan (NADP) 'owned' by all the stakeholders, including rural village people, was essential if agriculture was to bring about economic and social change in PNG. But the review stated that DAL was 'uncoordinated, was unsure of budget allocations and was poorly served by national and provincial financial information systems', and was unable to develop or implement such a plan. DAL produced an NADP (2002–2012) which was approved in principle by the National Executive Council.

Department of National Planning and Monitoring

The Department of National Planning and Monitoring (DNPM) is responsible for national strategic development policy, development planning and preparation of the development budget, aid coordination, and monitoring and evaluation. DNPM evolved from a pre-Independence National Planning Office (NPO). DNPM was disbanded in 1985 and its tasks split between the Department of the Prime Minister and the Department of Finance. The central planning and coordination functions and dominance of development activities previously carried out by the NPO disappeared during the next ten years. In 1995 these functions were included in a new DNPM, but the department has been reorganised four times in 10 years, which has created instability and lack of continuity. Part of the reorganisations involved the provincial and district coordination branch moving from DNPM to the Office of Rural Development, which caused the DNPM to lose touch with the provinces and resulted in the poor integration of agricultural policies into national development strategies. The role of DNPM also overlaps with a number of other departments, which causes confusion and inefficiency.

The Office of Rural Development

The Office of Rural Development (ORD) has the role of supporting provinces and districts in planning, implementing, monitoring and evaluating rural improvement programs. The ORD draws on funds from the Provincial Support Grant, the District Support Grant, the Social and Rural Development Program and the Targeted Community Development Program. ORD's role includes overseeing the allocation and spending of K1 million of provincial and district support grants per MP per year, K500 000 of which can be spent at the discretion of the MP.[10] Although the ORD is closer to the provinces and districts than the DNPM, it has been criticised for not collaborating sufficiently, or at all, with other government departments, the research organisations, or the universities, in the development of high quality district projects that can be supported by the MP's grants. The 2004 ADB review argued that 'projects submitted for funding … are not well designed, due to lack of skills in the provinces to support national members to formulate their projects'.

The funding of the districts (and provinces) is provided for under the Organic Law with decision making determined by the Joint District (and Provincial) Planning and Budgeting Priorities Committees, chaired by the MPs and comprising Local Level Government Area presidents and district managers (see page 444). Spending is widely considered to be unduly controlled by the MPs. In 2006 a new District Services Improvement Program (DSIP) was set up to complement the establishment of District Treasuries and to improve service delivery at the district level. The DSIP includes funds for 'agriculture'. However, confusion remains over which level of government is responsible for service provision. In addition, the amount of money provided does not cover the cost of even the most basic service provision (including agricultural extension) in most provinces.

National Agricultural Council

The National Agricultural Council (NAC) was established as a committee of national and provincial agriculture ministers, supported by an advisory council of national and provincial department and division heads. After the 1995 Organic Law reforms, its role changed to one of bringing together the chairs of the provincial agricultural committees (where they exist) and the provincial agricultural advisers and statutory body heads, with the national minister and the national secretary of DAL. The secretariat is provided by DAL. The mandate of the NAC is to

---

10 The District Support Grant is popularly known in PNG as 'the slush fund'.

review national research, training and skills building in agriculture and to report on this to the secretary of DAL. The full membership is large and unwieldy and lack of funds has meant the NAC rarely meets.

Agriculture Subcommittee of the Consultative Implementation and Monitoring Council

The Consultative Implementation and Monitoring Council (CIMC) was established by the National Executive Council and is chaired by the Minister for Planning and Implementation. The CIMC set up a number of sectoral committees, including the Agriculture Subcommittee, made up of representatives from the private sector commodity producers, civil society (including non-government organisations), DAL, DNPM and sectoral statutory bodies. The subcommittee operates independently of the government, under the direction of the CIMC. Its role is to ensure a dialogue between government, private enterprise and civil society. The subcommittee meets reasonably regularly. The ADB review suggested that, given the dormancy of the National Agricultural Council, the CIMC Agriculture Subcommittee should oversee the National Agriculture Development Plan and the agricultural sector program planning, budgeting and implementation, and monitoring and evaluation, for at present no body takes responsibility.

The Rural Industries Council and the PNG Growers' Association

The Rural Industries Council (RIC) mainly represents larger organisations and agricultural industry companies, although it is endeavouring to extend its representation more effectively, especially to growers. The chair and deputy chair come from the largest agricultural industry companies in PNG.[11] The RIC has 27 members, including a number of government departments and industry boards. It has an office in Port Moresby in association with the Institute of National Affairs.

A number of growers' associations exist in PNG, representing villagers (and in some cases largeholders) involved mainly in cash cropping. The Smallholder Coffee Growers' Association is the largest with around 14 000 members in 14 provinces. It is supported by the Coffee Industry Corporation (CIC) (see page 450) and has four members on CIC's board. The Palm Oil Producers' Association (POPA), on the other hand, is an association of large oil palm producers. Oil palm smallholders are represented by the Oil Palm Industry Corporation (OPIC) (see page 451). The PNG Growers' Association was formed out of the Planters' Association to bring together cocoa and copra growers. These organisations come together under the Rural Industries Council, which works to raise the profile of agriculture and give a stronger voice for policies to support the sector. Some branches of the growers' associations conduct field days and training activities.

National Agriculture Quarantine and Inspection Authority

The National Agriculture Quarantine and Inspection Authority (NAQIA) was created out of DAL in 1997. Its mandate is to protect the animals, plants and fish in PNG from exotic pests, diseases and weeds. It is also responsible for facilitating trade through export and import risk analysis and quality assurance systems. The head office is in Port Moresby and NAQIA operates in 15 seaports, 3 international airports and an international post office. It also maintains two animal and plant health laboratories to provide diagnostic and advisory services. Staff conduct meat inspections at five abattoirs.

The 2004 ADB report criticised NAQIA for charging high and unrealistic fees, overly restrictive rules imposed on the import of cultivars and seeds, and the slow issuing of import permits. However, the high risk that imported pests, diseases and exotic species could create severe economic damage to PNG's agricultural production means an effective and efficient quarantine service is critical to the future of agriculture in PNG.[12]

---

11 However, the present chair was formerly the Secretary of the Department of Primary Industry (now DAL) and was also previously the Secretary of the Prime Minister's Department.

12 The most recent economically damaging imported pest is a cocoa pod borer possibly contained in soil that was not removed from heavy logging equipment before it was brought into PNG from Malaysia in 2005.

Fresh Produce Development Agency

The Fresh Produce Development Agency (FPDA) was established in 1989, with assistance from New Zealand Aid, to develop a competitive and sustainable fruit and vegetable industry. FPDA compiles information on prices and quantities of fruit and vegetables in the main markets, provides extension and training on production, marketing and post-harvest handling, establishes contacts between sellers and buyers, supplies certified seed potato, assists in village commercial food processing initiatives, and assists women to engage in fresh food marketing. FPDA has had an important role in the re-establishment of the potato industry following an outbreak of blight.

The 2004 ADB report suggested that FPDA needs to make greater attempts to sell its services to the industry in order to become independent of government funding. The report also recommended that FPDA make greater efforts to collaborate with the National Agricultural Research Institute and to work more with small and medium producers.

Livestock Development Corporation

The Livestock Development Corporation (LDC) is a self-financing organisation that is responsible for the control of the slaughter and processing of livestock for retail sale, for the encouragement of the smallholder sector to increase the supply of poultry and breeding-age cattle, and the production of stockfeed. LDC has abattoirs in Central, Morobe and Eastern Highlands provinces. A fruit production project and a cashew nut nucleus estate project, both in Central Province, have been established using revenue from the abattoirs and from aid money.

However, the LDC board has recently been subject to numerous political appointments and its revenue depleted by excessive board and management spending. The Goroka piggery and the cashew nut projects are in financial difficulty and the abattoirs are said to be underfunded, poorly maintained and so far below public hygiene standards that they are in danger of closing. If these facilities close, all animals for slaughter will have to be transported to Ramu Agri-Industries' modern abattoir at Gusap in the Markham Valley. A chronic shortage of breeding-age poultry, day-old chicks and cattle for commercial smallholders exists in PNG, but LDC has seemingly been unable to respond to this need. The future of the LDC is in doubt.

Spice Industry Board

The Spice Industry Board (SIB) is responsible for regulating and collecting data on spice production and marketing, including vanilla, cardamom, pepper and turmeric. The board maintains a small office within DAL in Port Moresby. It has seven members, six of whom, including the chairman, are appointed by the Minister of DAL. SIB is poorly resourced and staff struggle to maintain basic information about production and trade. Licences to export spices are issued by the board on DAL's advice. Poor control over quality has recently damaged PNG's reputation as a vanilla producer (see Sections 5.14 and 5.21).

Rubber Board

A Rubber Board was established in the mid 1950s to regulate the export of rubber from PNG. The board is supposed to oversee inspections and hear appeals. It has five members, all appointed by the Minister of DAL. Considerable state investment has gone into rubber growing since the 1970s, but the schemes have been poorly managed by DAL and the Department of Lands.[13] The privately run Doa Plantations Ltd owned by Galley Reach Holdings and the Fly River Rubber Cooperative in Western Province produce most of PNG's rubber exports (see Section 5.11). An attempt to sell the government-owned Cape Rodney rubber factory in Central Province to private interests has been accompanied by alleged financial irregularity and a lack of transparency.

In late 2006, in an attempt to revitalise the rubber sector, an interim rubber board was formed to review the Rubber Act and to guide formation of a PNG Rubber Industry Corporation.

13 The Department of Lands was responsible for issuing land titles to the resettlement blocks. Many titles were never issued (see Section 6.7).

National Fisheries Authority

The National Fisheries Authority (NFA), established in 1998, is a non-commercial statutory authority owned by the government. The role of the NFA is to promote long-term sustainable development of PNG's marine resources. This includes ensuring that catch levels are such that maximum sustainable yield is achieved. Other roles include protection of entire marine ecosystems, preservation of biodiversity, minimising pollution and supporting village fishers. Major fisheries are managed under a national fisheries plan and local fisheries are also controlled nationally. The NFA trains staff in the National Fisheries College at Kavieng, New Ireland Province. A major review and restructure in 2000 and 2001 created an efficient and effective body, but recent political appointments to the board and management have been quickly followed by allegations of excessive licensing and other forms of malpractice.

PNG Forest Authority

The Papua New Guinea Forest Authority (PNGFA) was formed in 1993 as a statutory corporation to manage the national forest sector. Because forestry has been declared an area of national interest, control over forests has not been decentralised to the provinces under the 1995 Organic Law reforms. PNGFA comprises the National Forestry Board (NFB) and the National Forest Service (NFS). The NFB is made up of representatives of a number of government departments, provincial governments, women, forest owners, the forest industry and NGOs. The NFS works in all provinces.

PNG forestry policies and practices have been the focus of controversy, argument and allegations of corruption for some time (see Section 5.8). The most important issues are resource acquisition, allocation of licences to logging operators who do not comply with key requirements (like sustainable harvesting), poor monitoring of harvests, exports and enforcement of requirements, and the identification of landowners and lack of concern for their best interests. Aid donors, especially the World Bank, have attempted to place conditions on programs to force the PNG Government to comply with logging laws and control the damage being done to PNG's forest resources. Uncontrolled logging is leading to losses by forest-owning villagers and the national economy.

National agricultural plans

A failure to produce realistic national agricultural plans that deliver economic returns to PNG has contributed to the gradual marginalising of agriculture in national strategic plans (see Section 6.3). This has been accompanied by a reduction of public and aid donor resources to agriculture. This trend is reflected in the Medium Term Development Strategies (MTDSs) produced since 1990. The 1990–1994 MTDS emphasised the importance of agriculture as a driver of economic growth and rural development. It gave rise to a number of projects within DAL around the themes of 'strengthening extension, improved coordination between national and provincial government agriculture departments, encouraging private sector development, self sufficiency in food production, environmental protection, opportunities for women and youth and efficient use of scarce resources'. In the 1997–2002 plan, the emphasis had shifted to 'planning and fiscal discipline' and in the 2003–2007 plan to 'good governance; export-driven economic growth; and rural development, poverty reduction and empowerment through human resource development'. In the 2005–2010 plan agriculture was not an expenditure priority[14] and, in a speech on economic growth in May 2006, the Minister for Treasury mentioned agriculture only once, in the same sentence as tourism. The failure of agriculture to maintain prominence in plans for national economic development reflects a lack of capacity in DAL to devise sensible plans, a lack of ability by various institutions to implement plans (particularly after the confusion caused by the 1995 Organic Law reforms), the unrealistic and contradictory nature of many of the proposals put forward, and frustration in other parts of government and among aid donors at the seeming failure of the agriculture sector to deliver economic growth.

14 The 2005–2010 MTDS identifies priority areas for development. They are transport infrastructure, promotion of income-earning opportunities, education, health, and law and justice.

Agricultural research organisations and commodity boards

The International Food Policy Research Institute argues that every dollar invested in effective agricultural research results in a $6 increase in agricultural output and a $15 increase in economic growth. Effective agricultural research is vital to improve food security and cash income for rural Papua New Guineans. In the 1990s, the DAL research division was split into separate institutions and the commodity boards given greater powers (see page 445). Almost all agricultural research in PNG is now conducted by several statutory research organisations.[15] The most important are:

- National Agricultural Research Institute (NARI).
- Coffee Industry Corporation (CIC).
- Cocoa Coconut Institute of Papua New Guinea (CCI).
- Papua New Guinea Oil Palm Research Association (OPRA).

National Agricultural Research Institute

The National Agricultural Research Institute (NARI) was formed from the research division of DAL in 1997. It is a publicly funded, statutory research organisation that conducts applied and development-oriented research on food crops, emerging food crops, emerging cash crops, livestock, and resource management issues. The major targets are the smallholder, semi-subsistence, semi-commercial and commercial farmers. NARI manages six research stations: Keravat (East New Britain Province), Bubia and Labu (Morobe Province), Aiyura (Eastern Highlands Province), Laloki (Central Province) and Tambul (Western Highlands Province). Additionally, NARI owns the National Chemistry Laboratory and National Agricultural Insect Collection in Port Moresby, and has an office in Mount Hagen. NARI headquarters is at Bubia near Lae.

NARI has been well managed. It has had difficulty training and retaining high quality staff. Because NARI does not have access to funding from a levy on exports as do the export commodity-based research institutions, funding for capital equipment and maintenance are proportionately less than in other agencies. Aid projects, in particular the Australian Contribution to a National Agricultural Research System, have supported NARI for some years, but long-term sources need to be found to maintain an adequate income.[16]

Coffee Industry Corporation

Before 1991, coffee growing and exporting was governed by the Coffee Industry Board based at Goroka. Research on coffee was conducted by the Coffee Research Institute (set up in 1986) at Aiyura and extension to growers was the responsibility of the Coffee Development Agency. This last body was created after coffee rust appeared in PNG in 1986. In August 1991, the three organisations merged into the largely self-financing Coffee Industry Corporation Ltd (CIC).

The CIC has a broad range of powers, including buying and selling coffee, setting prices, registering and controlling exports, setting quality standards and controlling credit worthiness and capacity of market participants. CIC is unusual in that it is established under the Companies Act, but has been granted

---

15 Three private companies are also involved in agricultural research: Ramu Agri-Industries Ltd, New Britain Palm Oil Ltd and Trukai Industries Ltd. They conduct research on sugar cane, oil palm and rice production respectively. Other research is conducted at the University of Technology in Lae (Agriculture and Forestry departments) and University of Vudal in East New Britain Province. One non-government organisation, World Vision PNG, is involved in sweet potato research. The Republic of China (Taiwan) Technical Mission conducts research on introduced vegetables and rice. This group is based at Bubia near the NARI head office, but operates independently of PNG institutions and produces few technical reports. Some research on wheat and other crops was conducted by an agricultural mission from China near Kandep in Enga Province in the late 1990s.

16 In 1999 the Skate Government cut all funding to research organisations in PNG and suggested they should be abolished. NARI survived because of careful financial management and international assistance. Government funding has since been reinstated, but the danger remains that it can be arbitrarily cut again.

specific regulatory functions and powers by parliament. In practice, the CIC only applies its regulatory functions to setting guidelines, implementing firm quality control, and approving export contracts (and contract prices). The marketing of coffee is left in the hands of private companies licensed by the corporation. There is a risk that the board could become involved in the international marketing of coffee, and it has the power to do so, but the only occasion when the former Coffee Industry Board used this power (in the early 1980s), it failed badly. That experience provides a strong deterrent to using the powers again. CIC now has two divisions: the Research & Grower Services Division (made up of the Coffee Research Institute and Extension Services Division) and the Industry Operations Division. The CIC is well resourced (from an 8 toea/kg levy on green coffee beans).

Problems facing the CIC are how to:

- Improve the quality of village coffee and increase overall production.
- Ensure board members are competent, represent industry interests and have a good knowledge of the coffee industry.
- Handle increasing attempts at political interference in its powers and functions.

Cocoa and coconut institutions

The Cocoa Coconut Institute of Papua New Guinea (CCI) was formed in 2003 from the merger of the PNG Cocoa and Coconut Research Institute and PNG Cocoa and Coconut Extension Agency. CCI is owned jointly by two statutory bodies, the Cocoa Board of Papua New Guinea and Kokonas Indastri Koporesen (KIK), which fund the institute through levies on exports.

The Cocoa Board is responsible for the inspection of all export cocoa. It is funded by a K40/tonne levy on exported cocoa, some of which goes to supporting CCI. The board licenses around 5500 cocoa fermentaries and 14 cocoa exporters. Price competition at all stages of the marketing chain has kept marketing margins low, to the benefit of growers. However, the board has suffered from 'irregularities' in management in recent years.

The CCI is responsible for all cocoa and coconut research, development and extension in PNG. CCI has two active research stations, one at Tavilo in East New Britain Province and the other, the Stewart Research Station, in Madang Province. At a third research station, in Bougainville Province, operations are temporarily suspended. CCI owns 3234 ha of cocoa in eight plantations and two hybrid seed gardens. The plantations and seed gardens generate a significant proportion of CCI income.

The Kokonas Indastri Koporesen, based in Port Moresby, evolved out of the privatisation of the Copra Marketing Board trading functions in 2002. KIK's role is to contribute to policy and regulate the copra and coconut industry. Marketing is undertaken by the private sector, where an increasing proportion of exports is in the form of coconut oil, particularly from the Toboi mill in Rabaul. KIK provides some financial support for coconut research as well as funds for seed gardens. Growers feel strongly that KIK provides few benefits (see Section 5.21), while imposing high costs in the form of levies on producers.[17] The 2004 ADB report stated that it is 'difficult to justify the existence of KIK' and recommended that it be abolished.

KIK's former coconut mill in Madang, set up in the 1990s at exorbitant cost and managed by subsidiary company PNG Coconut Commodities (PNGCC), was sold to a New Zealand company exploring biofuels.

Oil palm organisations

The oil palm industry is governed by a number of organisations: the Oil Palm Industry Corporation (OPIC), Papua New Guinea Oil Palm Research Association (OPRA), the Oil Palm Growers' Association (OPGA), and the Papua New Guinea Palm Oil Producers' Association (POPA).

OPIC was established in 1992, as part of a reform of the oil palm industry in response to grower frustration over low prices, a then unsatisfactory pricing

17 Levies imposed by KIK were reduced in September 2006. The levy on copra was cut from K45 to K34 per tonne; that on copra oil from K60 to K44 per tonne; while the levy on coconut meal of K10 per tonne remained unchanged.

formula and declining government services. OPIC is funded by a levy on sales of fruit, matched by the oil palm companies. International aid funding has also provided significant financial support to the corporation. Funding is expected to continue under a proposed World Bank smallholder agriculture project. OPIC's main role is to provide extension services to smallholders in order to increase productivity, promote improved management, and enhance the wellbeing of producers. OPIC also liaises with government, the oil palm companies and other organisations involved in the industry. OPIC has five local planning committees, comprising representatives of smallholders, companies and the government, in five project areas.

OPRA is a non-profit research organisation with its headquarters at Dami Oil Palm Research Station in West New Britain Province, and another facility at Popondetta in Oro Province. OPRA was established by pooling research facilities of three companies[18] to bring together government, plantation companies and smallholders under a single research organisation. OPRA's main areas of research include agronomy, entomology, smallholder studies, and plant pathology. OPRA is funded by a levy on production (50 toea/tonne of fresh fruit bunch for smallholders and 80 toea/tonne of fresh fruit bunch for plantations), government funding and research grants. OPRA also provides technical support and training to smallholders, extension officers and plantation company officers. OPRA's research is highly regarded internationally.

POPA represents the joint interests of the milling companies. Each project area also has a growers' association which represents the interests of smallholders to the companies, OPIC, OPRA, and national and provincial governments. The chair of each oil palm growers' association sits on the board of OPIC. The extent of smallholder involvement in the associations varies between project areas and over time. At various times the associations have experienced problems with financial mismanagement resulting in members losing confidence in their organisations. For example, in 2000 the Hoskins growers' association suffered a significant loss of members after the association's funds were misappropriated. In Popondetta the association membership has been limited because the settlers believe that the organisation is dominated by local landowner interests. The distribution of financial benefits between the milling companies and the smallholders has been significantly adjusted in favour of smallholders in successive reviews in the 1990s and in 2000. However, smallholder advocates argue that the mills (or 'nucleus estates') have economic advantages over smallholders and the pricing formula fails to properly value customary land as well as heavily discounting smallholder labour.

Agricultural extension services

Before the establishment of provincial governments, agricultural extension services were conducted by DAL. In the late 1970s, responsibility moved to Divisions of Primary Industry within provincial departments. This arrangement ceased in 1995 when provincial governments were reformed. Agricultural extension staff now report to district managers at the district level. The outcome is that national staff are isolated from provincial extension services and province staff are isolated from district-level staff. Confused lines of authority, reporting and responsibility exist for extension staff. Consequently, most government agricultural extension services barely function and staff morale is severely degraded. This is exacerbated by confusion over financial responsibility and long delays in funds reaching districts. Regulations that specify responsibilities for funding and reporting on particular functions have never been implemented, and responsibilities remain confused and chaotic. Many district administrations are still ignorant of efforts to improve funding through the District Services Improvement Program (see page 446). Most government provincial radio stations, which once provided remote rural people with information, are no longer operating. Where radio stations are working, they are forced to charge fees for presenting programs because they receive inadequate funding.

---

18 Higaturu Oil Palms Ltd, New Britain Palm Oil Ltd and Hargy Oil Palms Ltd.

Some extension is conducted by some of the organisations described above, including Fresh Produce Development Agency, National Agricultural Research Institute, Coffee Industry Corporation, Cocoa Coconut Institute and Oil Palm Industry Corporation. The University of Vudal is involved in agricultural extension in the Islands Region. Some extension is also provided by organisations affiliated with churches.[19] Numerous locally based NGOs have some involvement in agricultural extension, but most are under-resourced and have limited technical capacity. The lack of a national coordinating NGO body means duplication and competition in relationships with government and funding agencies.

More extension effort is directed at rice production than for any other commodity. Staff of a number of NGOs, provincial Divisions of Primary Industry, Trukai Industries Ltd, and the Republic of China (Taiwan) Technical Mission are involved in promoting rice production. Despite these inputs and the handout of free equipment, seed and fertiliser, domestic rice production is negligible compared with that of the staple food crops (see Sections 2.2 and 2.5).

The ADB review in 2004 found that public sector extension services lacked direction, were very high cost and low benefit, and that the main research organisations did not have effective extension arms and failed to coordinate their efforts if they did extension at all. The review also found that linkages between research, new technology and extension needed to be greatly improved, communications in rural areas were extremely poor and needed urgent improvement, school curriculums did not emphasise agriculture, and the quality of most vocational training was poor.

Overall, agricultural extension and outreach is limited. With some exceptions, extension activity takes place near research stations, urban areas and main roads. Most rural villagers have little or no access to information about improved production and marketing of agricultural produce. A large unsatisfied demand for information exists on many aspects of food, cash crop and animal production and marketing, as well as for face-to-face contact with extension agents.

Summary

Governance in the agricultural sector of PNG occurs at national, provincial, district and local levels and also by organisations devoted to particular crops and commodities. Other sectors, for example health, education and transport, influence agricultural development. However, lack of coordination, widespread confusion and ignorance of laws and regulations that govern responsibilities for service delivery, funding and reporting cripple attempts to make agriculture a primary driver of national economic and local rural development. The chaos results partly from the provisions of the 1995 reforms to the Organic Law and partly from the instability of political parties. The 1995 reforms have introduced confusion about relationships and responsibilities between national, provincial, district and local level governments. Within political parties, positions on the boards of bodies that govern and promote agriculture are used as rewards to supporters, who may not be competent or who may not act in the best interests of rural smallholders. This situation occurs in other sectors that are critical to agriculture: infrastructure, transport, education and health, such that agricultural development is subject to multiple constraints.

The majority of rural dwellers produce their own food and house themselves, so their basic livelihoods are insulated from the chaos in the world beyond their villages. They also produce most of the

19 Lutheran Development Service is responsible for Yangpela Didiman, a self-help movement carrying out agricultural extension, using around 30 paid staff and up to 3700 volunteer 'farmer motivators', usually young men and women who have taken a 12-month training course. The Salvation Army provides services to remote villages in Eastern Highlands Province with support from AusAID's Incentive Fund, international NGOs, the Swiss Government and the United States Peace Corps. The Baptist Church has a small operation in Mount Hagen focused on promoting rabbits for meat in remote villages. CARITAS, the Catholic community development organisation, supports rural development initiatives in a number of provinces. The Anglican Church does similar work in Milne Bay and Oro provinces. The Seventh-Day Adventist Church has promoted vegetable marketing in Eastern Highlands and Bougainville provinces.

agricultural commodities exported from PNG, often despite significant disincentives from government policies and practices, and lack of services. On the basis of the past two decades, it seems unlikely agriculture will become the driver of economic growth that it could be, until at least some of these constraints of poor governance are removed.

Sources

ADB (Asian Development Bank) (2004). Preparing the agriculture and rural development project, Papua New Guinea: Agricultural markets, marketing and rural enterprise development. ADB TA4055-PNG. Asian Development Bank.

Anderson, T. (2006). Oil palm and small farmers in Papua New Guinea. Unpublished report. Australian Centre for Environmental Law and Community Rights, The University of Sydney, Sydney.

Department of National Planning and Monitoring (n.d.). Medium Term Development Strategy 2003–2007. Papua New Guinea Department of National Planning and Monitoring, Port Moresby.

Department of National Planning and Monitoring (2004). The Medium Term Development Strategy 2005–2010: our plan for economic and social advancement. Papua New Guinea Department of National Planning and Monitoring, Port Moresby.

Koczberski, G., Curry, G.N. and Gibson, K. (2001). *Improving productivity of the smallholder oil palm sector in Papua New Guinea: a socio-economic study of the Hoskins and Popondetta schemes*. Department of Human Geography, Research School of Pacific and Asian Studies, The Australian National University, Canberra.

Manning, M. (2005). Papua New Guinea thirty years on. *Pacific Economic Bulletin* 20(1):145–158.

McGregor, A. (2006). *Pacific 2020 background paper: agriculture*. Australian Agency for International Development, Canberra.

McKillop, B., Bourke, R.M. and Kambori, V. (2007). Agriculture. In May, R.J. (ed). *Policy Making and Implementation: Studies from Papua New Guinea*. ANU E Press, Canberra. pp. 52–68.

National Economic and Fiscal Commission (2006). Cost of services study: review of intergovernmental financing arrangements (RIGFA). National Economic and Fiscal Commission, Port Moresby.

National Planning Office (1996). Medium Term Development Strategy 1997–2002: a bridge into the 21st century. Papua New Guinea National Planning Office, Port Moresby.

Philemon, B. (2006). The PNG economy and constraints on the private sector: achieving sustainable economic growth in PNG. Speech to 22nd Australia – Papua New Guinea Business Forum, Cairns, Australia, 15 May 2006. Accessed October 2006 at <http://www.treasury.gov.pg/files/speech/Ministers_Speech_on_22nd_PNG&AUS_Business_Council.pdf>.

Pinstrup-Andersen, P., Lundberg, M. and Garrett, J.L. (1995). Foreign assistance to agriculture: a win-win proposition. Food policy report. The International Food Policy Research Institute, Washington D.C.

Saldanha, C. (2005). *Pacific 2020 background paper: political governance*. Australian Agency for International Development, Canberra.

Whimp, K. (2005). Comments on problems with the Organic Law on provincial governments and local-level governments. Unpublished paper.

World Bank (2006). *A decade of measuring the quality of governance. Governance matters 2006. Worldwide governance indicators*. The World Bank, Washington. <http://siteresources.worldbank.org/INTWBIGOVANTCOR/Resources/1740479-1150402582357/2661829-1158008871017/booklet_decade_of_measuring_governance.pdf>.

6.5 Agricultural surveys

Papua New Guinea's national strategic goals include economic growth and poverty alleviation in rural areas. In PNG, where over 80% of people live in rural areas, grow most of their own food and earn foreign earnings by growing cash crops, agricultural development will be a critical part of achieving these goals.

Agricultural development will depend on an effective, rational, planning process. Planning has at least two parts: the first is knowing, as realistically as possible, the present circumstances; the second part involves the development of a set of rational steps that will allow progress from the present circumstances to a desired position, at a given future time. Both parts of the planning process require up-to-date, accurate information about agriculture.

Knowledge about agriculture in PNG comes from three main sources:

- The national census, which every ten years collects information on the number of people in PNG as well as some other critical attributes of *all* of the population.
- Administrative or points-of-service information. This is information from the places where services are delivered. These provide information on the number of people serviced and their needs, and the types of facilities and the staff offering the services.
- Surveys that are conducted to collect information on particular aspects of agriculture.

This section is mainly concerned with surveys, and only briefly examines census and administrative information.

Surveys of smallholder agriculture

A relatively large number of surveys of various aspects of smallholder agriculture have been undertaken in PNG. Here they are reviewed briefly. The PNG Resource Information System (PNGRIS) and the Mapping Agricultural Systems of PNG Project (MASP) are described in greater detail in Section 1.15. Both are very important sources of national-level information for rural development. Most of the information in PNGRIS and MASP is still relevant for development, either because it does not become dated (for example, data on the physical environment) or because it was recorded not long ago (for example, the MASP data).

Survey of Indigenous Agriculture 1961–1962

The 1961–1962 Survey of Indigenous Agriculture was a national sample survey of 100 villages carried out by the Bureau of Statistics, the Department of Agriculture, Stock and Fisheries (DASF) and the Australian Bureau of Statistics. The sample frame was the annual village censuses carried out by Department of Native Affairs patrol officers, stratified by estimates of staple food, population size, population growth and agricultural environment.

The total sample included approximately 0.1% of the total population. Sample villages were visited three times at six-month intervals. An ancillary survey of coconuts was also undertaken.

Only basic summary statistics are available from this survey. Even these must be used with caution because of the large variation in the data. Nevertheless, this survey provided important baseline information. The original field datasheets from this survey appear to have been lost.

Department of Agriculture and Livestock surveys

Since the early 1960s, the Rural Statistics Section of the Department of Agriculture and Livestock (DAL) has produced statistics on the major cash crops grown in PNG. In the 1960s, these figures were based on regular annual counts that all agricultural field officers were supposed to carry out, and from regular sample surveys. In the 1970s the annual tree censuses were discontinued because it became obvious that it was impossible for all of the trees represented by the census totals to be counted properly every year in the time available.[1] As a replacement for the annual tree counts, from around 1975, the national department began arranging sample surveys in every province. The surveys involved counting all trees by age and condition. Planting density was estimated using a triangle and the total areas planted calculated. Food crops were listed but not counted. Observations of field enumeration suggest that trees were not actually counted, but growers were just asked to estimate how many trees they owned. This method of counting trees in PNG villages produces very unreliable data.

1 In the Maprik District in the 1970s, the figures in DASF files appeared to have been arrived at by multiplying the previous year's figures by 5%. However, some field surveys were conducted. In 1972 in one village where a DASF survey was observed, the Australian agricultural field officer asked all coffee growers to go to their gardens and pick one leaf off every tree. While the officer waited in the village, the villagers all went to the coffee garden nearest the village and picked handfuls of leaves from there. The officer then duly counted the leaves picked and entered the figures into the census forms.

Sample survey of smallholder coffee producers 1975

In 1973 a national sample survey of coffee smallholdings was carried out by the Rural Economics and Commodity Marketing Branch of DASF. A detailed report was produced following the 1973 survey. The survey was conducted because it was recognised that the annual censuses of tree crops conducted by DASF field officers (see above) were inaccurate. The sample frame was the list of villages censused every year by patrol officers, selected in proportion to the estimated tree numbers, population numbers and names on the 1972 electoral roll. In the selected villages five growers were selected at random from the electoral roll. The survey was carried out by six special teams, under close supervision.

The survey produced useful results, but was reported to have been unsuccessful in two provinces (Simbu and Western Highlands), where, compared to existing tree counts and production figures, a 'gross underestimate' of trees appeared to have occurred.

Economic survey of smallholder coffee producers 1976

In 1976 the Department of Primary Industry (DPI) contracted an ex-DPI officer (D. Anderson) to carry out another survey of smallholder coffee growers in the most important highlands coffee-producing provinces (Western Highlands, Simbu and Eastern Highlands). The survey was designed to complement the 1975 survey in the provinces where it had been unsuccessful. Thirteen villages were selected from each province on the basis of distance from a main centre, road access, the length of time coffee had been grown at the village and land availability. Agricultural students from the villages surveyed were used as enumerators. Trees were counted and selected growers interviewed about coffee stocks, costs of production and labour. Coffee-producing work was observed. The data from this survey are considered to be of high quality and a comprehensive report on methods and results was completed.

Coffee farming systems survey 1980

In 1980 a survey was conducted by the DPI Rural Statistics Section of coffee growing in Western Highlands and Simbu provinces. It was to be a 5-year follow-up to the 1975 and 1976 surveys. A sample was drawn from all villages censused, proportional to size. The enumerators were university students from the provinces being surveyed.

This survey was unsuccessful because the enumerators were not supervised in the field. The tree counts were significantly different from those of the 1975 and 1976 surveys. An assessment of what happened revealed that many enumerators did not visit the coffee gardens of the households they were allocated, but asked the growers to estimate tree numbers. They also had difficulty with the many qualitative judgments that the questionnaire required them to make.

Coffee farming systems research project

The coffee farming systems research project began in 1981. It used a number of innovative approaches to data collection at the village level. Two villages in the Asaro Valley of Eastern Highlands Province were studied intensively. Coffee trees were counted, air photographs were used to map land use, a soil survey was carried out and the coffee gardens examined by a coffee agronomist. Data on production, yields and income were collected by having literate youths from the villages monitor roadside sales and prices. The reporters also kept daily diaries of coffee producing and other village activities. The reporters were visited every two weeks, when they were paid, their data checked and new forms issued. During the first year of the project data quality was good, but during the second year, the project supervisor was posted elsewhere and data quality quickly became poor when the village reporters were not contacted for long periods of time.

Food crop market surveys

Numerous surveys of fresh food markets have been conducted in PNG over the past 55 years (see Section 5.3). The purpose of the survey and type of information collected varies, but the quantity and value of food sold was usually recorded. Information on surveys conducted between 1951 and 1982 is summarised in a review by Bourke (1986). His paper includes details of methods and the value and quantity of produce sold in Koki (Port Moresby); Aiyura, Goroka, Kainantu and Ukarumpa (Eastern Highlands); Kokopo, Rabaul and Vunapope (East New Britain); Kerema (Gulf); Lae (Morobe); Popondetta (Oro); Koge, Kavugara, Kimbe, Marakewa, Mosa and Talasea (West New Britain); and Mount Hagen (Western Highlands). In 1988 the Rural Statistics Section of DAL carried out surveys in food markets in Alotau, Port Moresby (Koki), Lae, Madang, Mount Hagen and Rabaul. The results were published between 1988 and 1992.

An important source of information on prices of fresh food over a long period is an ongoing survey of prices of 15 fresh foods conducted weekly in five urban markets (Port Moresby, Goroka, Lae, Madang and Rabaul). These surveys commenced in 1971 with recording done by local DPI staff. The data are used to generate the consumer price index (see Sections 4.2 and 4.3). The data are not published, but are available from the National Statistical Office (NSO) in Port Moresby. Staff of the Fresh Produce Development Agency (FPDA) record prices of about 50 fruit and vegetables in selected urban food markets and stores. Surveys are undertaken in Port Moresby, Lae, Madang, Goroka, Mount Hagen, Kokopo and Popondetta. Data are available from the FPDA in Goroka.

Rapid rural appraisal

A series of rapid rural appraisals (RRAs) were conducted in various locations between 1978 and the late 1990s. An RRA is a survey that is conducted by a number of experts who collaborate to study a well-defined problem from different viewpoints. The various disciplines they bring to the study may include, for example, land use mapping, human nutrition, tree crop agronomy and social systems. The term PRA (participatory rural appraisal) has tended to replace the term RRA in recent years, although the techniques usually differ little in practice.

In 1978 an RRA was conducted on the Nembi Plateau in Southern Highlands Province where high levels of child malnutrition had been identified. Other RRAs were subsequently done in the following areas: Amanab (Sandaun), Bougainville Province, Bubia (Morobe), Gazelle Peninsula (East New Britain), Gumine (Simbu), Jimi Valley (Western Highlands), Kanabea (Gulf), Okapa area (Eastern Highlands), Rabaraba (Milne Bay), upper Ramu Valley (Madang) and Wosera (East Sepik). A number of these RRAs were conducted in the early stages of the large rural development projects described in Section 6.8. Reports from most of these surveys were lodged in the DAL library in Port Moresby. The quality of the information is uneven, but the reports contain much useful information about certain locations and are a valuable resource.

Designing monitoring systems for smallholder agriculture in PNG 1988–1991

Between 1988 and 1991, the Department of Human Geography at The Australian National University (ANU) and the PNG Department of Agriculture and Livestock, funded by the Australian Centre for International Agricultural Research, produced 20 reports on methods of data collection on smallholder cocoa and coffee production in the highlands, the Gazelle Peninsula and East Sepik Province. The reports are available from the authors of this book at the ANU.

Other surveys

There have been many other surveys of smallholder agriculture over the past 60 years. The list below is a small sample to indicate the range of surveys conducted.

- In 1947, a survey of village agriculture in Manus Province (Conroy).
- In 1947, a survey of village agriculture at seven sites as part of the PNG nutrition survey expedition (Conroy and Bridgland) (see Section 6.6).
- In 1961, surveys of 'intensive agriculture' at Maprik (East Sepik Province), Enga Province and Simbu Province. These surveys produced more detailed reports than that from the 1962–1963 national agricultural census for these areas.
- In 1974, an economic survey of cocoa growing in PNG (Godyn).
- A survey of smallholder cocoa in Bougainville Province in 1976 (Coulter).
- In 1975, a survey of food production on Hoskins (West New Britain Province) oil palm blocks (and subsequent surveys) (Benjamin).
- A survey of yam growing in Central Province in 1986 (King).
- A survey of smallholder agriculture in Sandaun (West Sepik) Province in 1987.
- In 1991, a survey of farm management by coffee smallholders in Western Highlands and Southern Highlands provinces.
- In 1992, a survey of smallholder agriculture on the Gazelle Peninsula (Levett).
- Surveys on farming systems by DAL from 1993 (Woruba and Humphrey).

Censuses and administrative data

PNG National Census

The PNG National Census has been conducted approximately every ten years since 1971. Previous censuses in PNG have been conducted in 1966, 1971, 1980, 1990 and 2000 (see Section 1.1). The national census is a critical source of information for agricultural planning. In 2000, census units (CUs),[2] were geo-located and their locations recorded as latitudes and longitudes, which enables them to be mapped on digital maps by geographical information systems software (see Section 1.15). The ability to map CUs and display information about them on computerised maps is a very powerful analytical and planning tool.[3] The National Statistical Office sells census data and digital maps at the census unit and household level.

As well as collecting population counts, the census asks questions of all households included in the census (around one million households were counted in the 2000 census). The questions included:

2 Rural people in PNG live in a range of settlement types, which include scattered homesteads, hamlets and villages. Census units are often villages, or parts of villages, but they are sometimes just geographical points where, in colonial times, people were required to assemble to be censused. In the 2000 National Census, the average number of people censused at a census unit was 232.

3 There are a number of problems with the geo-located census units in the 2000 census. Most important is that around 40% of CUs in the 2000 census cannot be identified in the 1990 census. A similar problem exists between the 1980 and the 1990 censuses. Changes also occurred in higher-level units. Census Divisions were replaced with Local Level Government Areas, and district boundaries changed.

- A question on 'agricultural activity', which was defined as involvement in growing or raising any of the following either for cash or own use: cocoa, coffee, rubber, oil palm, coconut, betel nut, livestock (cattle, pigs, goats, sheep), poultry, food crops/vegetables/root crops, fish (including crabs and seashells), or any other major crop or livestock.
- A question on 'economic activity', which was defined as receiving money from or involvement in any of the following: selling food crops/cooked food at a market/roadside, selling fish (including crabs and seashells), selling meat at a market/roadside, selling manufactured items (such as carvings, bilums, mats), selling betel nut/mustard, running a Public Motor Vehicle, hiring boats, running a trade store, or any other major income-generating activity.

The national census, acknowledging problems with some of the data, remains a critical resource for agricultural and rural development planning. It provides standards for data collection, definitions, codes and identification of places (CUs). It also provides a national frame for sample surveys and the denominator in the calculation of rates that allows comparisons between different parts of the country.

Provincial Data System

The Provincial Data System (PDS) was set up in the 1970s by the then National Planning Office and Department of Decentralisation, to replace village registers and the Area Studies reports that had been written regularly by Australian patrol officers. The original PDS was compiled by hand; data were collected at the village level and entered into village registers. Summaries were passed up the administrative hierarchy and were combined into larger administrative units. There were problems with data aggregation resulting from misunderstandings of definitions and clerical errors during data entry. Disaggregated data was not available at levels above that at which it had been collected.

In 1978 the Bureau of Statistics attempted to computerise the PDS. By 1981 Rural Community Registers were published which showed all the CUs in a Census Division with their resident populations, absentee populations and services available to them. By 1983 Services Access Tables were prepared showing what services were accessible from CUs. A well-organised and presented PDS, using computer mapping methods, has the potential to be a critical planning tool.

Prior to the 1980 census, the PDS population registers were updated using the Department of District Administration patrol officers as enumerators. The patrol officers believed they should be conducting the national census, rather than the NSO census enumerators, and it is said they carried out a very good quality nationwide rural village census to show what they were capable of. The results were stored on a computer in the National Statistical Office. The present location of the data is not known.

Administrative data

The knowledge of where services are located, their capacity to provide a service and the numbers of people who access the service is fundamental to the effective administration of any bureaucracy. Information about health, education, transport and agricultural service provision should be available annually from the particular government departments providing the services. In PNG the production of information about service provision has become fragmented and dysfunctional in recent decades.[4] Confusion over responsibilities, jealousies and competition between national departments

4 For example, the national Department of Education does not have a list of all schools in PNG. Two lists of schools, one collected from the department and the other from the NSO in 2004, have only a 45% overlap. The national Department of Health has produced province and district health profiles, including digital maps of health service facilities, as part of its national health plan, but the digital data have not been made available to any departments outside health, including the departments responsible for national planning. The Department of Provincial and Local Government Affairs refused access to everyone outside the department to its district database. When food shortages occurred in 1997 and an attempt was made to use that database, it was found to have a whole province missing and to contain other errors that made it worthless. For many years the Department of Agriculture and Livestock refused to make the PNG Resource Information System (PNGRIS) available to other potential users (see Section 1.15).

and national and provincial administrations means that reliable information is not freely available for planners or analysts outside the department concerned.

Other relevant non-agricultural surveys

Rural household survey

In 1985 an attempt was made to carry out a rural household survey (RHS). The sample was drawn from the 1980 CUs. The RHS was to survey the populations of all households in a selected CU and to use this information to select households for a two-week long intensive survey of income and expenditure. The final survey had three forms: the *Household Form*, which collected information on dwellings, ownership of assets, and household composition; the *Census Unit Form*, which collected information on health and education facilities, drinking water supply, transport and communication facilities, commercial, agricultural and administrative activities, local assets, seasonal characteristics of the CU, land quality, land use, and trade store and commodity prices; and the *Household Census Form*, which collected demographic and occupational information of residents and visitors in each household in the CU. A household diary was used to record stocks at the beginning and end of the intensive survey period.

Two trial surveys were carried out in Milne Bay Province and at Cape Rodney in Central Province. The Milne Bay trial demonstrated that too much was being asked of the survey enumerators, especially the food garden surveys, which were difficult to carry out and were poorly done. The second trial, with a reduced number of questions, was said to be successful, but planning for the national census in 1990 overtook the implementation of a national RHS.

A successful urban household survey was completed between 1985 and 1987. Data from this survey have been used, for example, to examine the demand for fresh food that is produced in rural areas and sold in urban markets in PNG towns.

The World Bank funded a sample household income and expenditure survey in 1996 as part of a poverty assessment (the PNG Household Survey). The methods used in this survey are described in more detail in Section 6.10 on rural poverty, and in other publications.

Demographic and Health Survey

The Demographic and Health Survey (DHS) was carried out between 1993 and 1998. It contains information that is indirectly related to agriculture. The survey was a sample survey designed to provide reliable national-level information on fertility, infant mortality, child mortality, contraception and maternal and child health indicators. At the province level, the survey was designed to provide information on fertility, mortality, education, employment and housing.

The 1996 DHS was conducted in two phases. The first phase was a survey of 1250 households using the 1990 CUs as the sampling frame. The second phase was a survey of women aged 15–49 years in 250 households selected from the first phase sample. A report published in 1997 contains information on characteristics of the households, fertility, family planning, marriage, polygamy, infant mortality, child mortality, maternal and child health care, infant feed practices and HIV/AIDS knowledge and behaviour. A second DHS was planned for 2006.

Summary

Much information can be derived from previous rural surveys in PNG. Due to the uneven quality of data sets, some are of historical and baseline value only. Others have useful data that is still relevant for planning today.

The problems that occurred in many of the past surveys act as a reminder that conducting rural surveys in PNG is not easy. A great deal of thought must be put into the primary objectives of the survey and the questions to be used that will achieve these objectives. Common sense must be used when devising the questions: can they be understood in Tok Pisin or Motu; do people really have the information that is being sought; can the responses be related to previous surveys so that changes over time can be analysed? Questionnaires must be tested before a larger survey is done and, perhaps most importantly, surveys will almost certainly fail unless the enumerators are closely supervised in the field and the tasks they are assigned are within their capabilities. Time and funds must be allocated for entering, cleaning, compiling and analysing data. Assistance with statistical methods must be sought from well-qualified experts. If all these details are not given sufficient rigour, wrong conclusions can be drawn, or the original effort and money used to do the fieldwork wasted.

Sources

Allen, B.J. (1987). A review of smallholder data sources in PNG relevant to the export tree crops sector. In Barker, P., Bodman, P. and Remenyi, J. (eds). *Smallholders and Export Tree Crops in Papua New Guinea: Data Sources and Collection Methods*. IASER Special Publication No. 12. Institute of Applied Social and Economic Research in collaboration with the PNG Department of Primary Industry and the Australian Centre for International Agricultural Research, Boroko, Papua New Guinea. pp. 14–54.

Allen, B.J. (ed.) (1984). *Agricultural and nutritional studies on the Nembi Plateau, Southern Highlands*. Department of Geography Occasional Paper No. 4 (New series). University of Papua New Guinea and the Southern Highlands Rural Development Project, Mendi.

Anderson, D. (1977). *An economic survey of smallholder coffee producers – 1976*. Planning Economics Marketing Branch, Department of Primary Industry, Port Moresby.

Barker, P., Bodman, P. and Remenyi, J. (eds) (1987). *Smallholders and Export Tree Crops in Papua New Guinea: Data Sources and Collection Methods*. IASER Special Publication No. 12. Institute of Applied Social and Economic Research in collaboration with the PNG Department of Primary Industry and the Australian Centre for International Agricultural Research, Boroko, Papua New Guinea.

Benjamin, C. (1977). A survey of food gardens in the Hoskins Oil Palm Scheme. *Papua New Guinea Agricultural Journal* 28(2–4):57–71.

Bourke, R.M. (1986). Periodic markets in Papua New Guinea. *Pacific Viewpoint* 27(1):60–76.

Bureau of Statistics (1965). Report on intensive agricultural surveys in the Maprik sub-district 1961–4. Unpublished report. Territory of Papua and New Guinea, Port Moresby.

Carrad, B. (1981). The coffee farming systems research project. *Coffee Industry Board Research Newsletter* 1(1):7–10.

Collett, G. (1991). *Evaluation of methodologies to monitor smallholder coffee, cocoa and copra production*. Designing monitoring systems for smallholder agriculture in Papua New Guinea. Working Paper No. 19. Department of Human Geography, Research School of Pacific Studies, The Australian National University, Canberra.

Conroy, W.L. (1947). Report on agricultural and economic survey – Manus area. Department of Agriculture, Stock and Fisheries, Port Moresby.

Conroy, W.L. and Bridgland, L.A. (1950). Native agriculture in Papua–New Guinea. In Hipsley, E.H. and Clements, F.W. (eds). *Report of the New Guinea Nutrition Survey Expedition 1947*. A.H. Pettifer, Government Printer, Sydney. pp. 73–91.

Coulter, H. (1976). *Sample survey of smallholder cocoa, North Solomons*. Department of Primary Industry, Port Moresby.

DAL (Department of Agriculture and Livestock) (1992). Report on food crop market survey 1988: Rabaul market. Rural Statistics Section, Policy Programming and Budgeting Division, Department of Agriculture and Livestock, Port Moresby. (Published reports are also available for five other markets surveyed).

French, W. (ed.) (1984). *Rural Data Collection for Planning Purposes*. IASER Special Publication No. 8. Institute of Applied Social and Economic Research, Boroko, Papua New Guinea.

Gibson, J. (2000). The Papua New Guinea household survey. *The Australian Economic Review* 33(4):377–380.

Gibson, J. (2001). Food demand in the rural and urban sectors of PNG. In Bourke, R.M., Allen, M.G. and Salisbury, J.G. (eds). *Food Security for Papua New Guinea. Proceedings of the Papua New Guinea Food and Nutrition 2000 Conference*. ACIAR Proceedings No. 99. Australian Centre for International Agricultural Research, Canberra. pp. 45–53.

Gibson, J. (2001). The economic and nutritional importance of household food production in PNG. In Bourke, R.M., Allen, M.G. and Salisbury, J.G. (eds). *Food Security for Papua New Guinea. Proceedings of the Papua New Guinea Food and Nutrition 2000 Conference*. ACIAR Proceedings No. 99. Australian Centre for International Agricultural Research, Canberra. pp. 37–44.

Godyn, D.L. (1974). *An economic survey of cocoa in Papua New Guinea: Part I and II*. Department of Agriculture, Stock and Fisheries, Port Moresby.

Godyn, D.L. (1974). *An economic survey of cocoa in Papua New Guinea: Part III*. Department of Agriculture, Stock and Fisheries, Port Moresby.

Hayfield, S. (1982). The coffee farming systems research and the smallholders crop surveys. In Bourke, R.M. and Kesavan, V. (eds). *Proceedings of the Second Papua New Guinea Food Crops Conference: Part Two*. Department of Primary Industry, Port Moresby. pp. 289–293.

King, G.A. (1986). A survey of Central Province yam (*Dioscorea* spp.) gardens. Technical Report 86/7. Department of Primary Industry, Port Moresby.

Levett, M.P. (ed.) (1992). *A survey of the north-east lowlands of the Gazelle Peninsula of East New Britain Province with particular reference to the potential and acceptability of new agroforestry projects*. Research Bulletin No. 53. Department of Agriculture and Livestock, Port Moresby.

Ministry of Health (2000). Health Vision 2010: National Health Plan 2001–2010. Papua New Guinea Department of Health, Port Moresby.

Monitoring and Evaluation Unit (1987). Agricultural sample survey in West Sepik Province (1986). West Sepik Provincial Development Project, Vanimo.

Niangop, M., Palan, V., Wai, M. and Cox, E. (1977). A food garden and basic needs survey of the Gavien Rubber Resettlement Scheme. Bagi, East Sepik Province.

NSO (National Statistical Office of Papua New Guinea) (1997). Papua New Guinea Demographic and Health Survey 1996: National Report. National Statistical Office, Port Moresby.

Overfield, D. (1991). *The economics of smallholder coffee production: analysis, policy and future research directions*. Coffee Discussion Paper No. 6. Papua New Guinea Coffee Industry Board, Goroka.

Shadlow, J. (1984). Surveys in Papua New Guinea: a progress report and prospects for the future. In French, W. (ed). *Rural Data Collection for Planning Purposes*. IASER Special Publication No. 8. Institute of Applied Social and Economic Research, Boroko, Papua New Guinea. pp. 52–57.

Territory of Papua and New Guinea (1967). Report on intensive agricultural surveys in the Chimbu survey area, 1962–64. Port Moresby.

Territory of Papua and New Guinea (1967). Report on intensive agricultural surveys in the Wabag survey area, 1962–64. Port Moresby.

Walters, C.L. (1963). *Survey of Indigenous Agriculture and Ancillary Surveys, 1961–1962*. Territory of Papua and New Guinea Bureau of Statistics, Port Moresby.

Woruba, M. and Humphrey, W. (1993). *A survey report of cropping practices on atolls in Papua New Guinea*. Sustainable Agriculture Series Workshop Paper 1/93. Lowlands Agricultural Experiment Station, Department of Agriculture and Livestock, Keravat.

6.6 Nutrition surveys

The adequate nutrition of children in PNG is closely associated with the food production systems (or agriculture) where they live. Three surveys of the nutritional status of Papua New Guineans have been carried out: in 1947, in 1982–1983 and in 2005. The first and third surveys covered adults and children, while the second covered only children five years of age and under.

1947 New Guinea Nutrition Survey

The 1947 survey was carried out by staff from the Australian Department of Health and officers from the Department of Education and the Department of Agriculture, Stock and Fisheries of the provisional Papua and New Guinea administrations. The field party included a medical doctor, a parasitologist, a nutritionist, two biochemists, an agriculturalist, a sociologist, a dentist and a photographer. An intensive survey was carried out in five villages selected on the basis of their staple foods, ease of access and the absence of any major, sudden changes to the village economy. The villages surveyed were Busama (where the staple food was taro), on the coast 30 km south of Lae; Kaiapit (banana) in the Markham Valley 100 km west of Lae; Patep No. 2 (sweet potato and taro) in the Wampit Valley near the Lae-Wau road around 80 km from Lae; Kavitaria (yam) in the Trobriand Islands of Milne Bay; and Koravaki (sago) in the Purari River delta in Gulf Province.

The survey found that, in general, PNG village diets were adequate, but they were often badly balanced, with too many calories coming from carbohydrate and not enough from fat and protein. Protein levels were 'markedly inadequate'. Starchy vegetables (root crops and banana) comprised 86% by weight of diets, with meat and fish only 0.6%. Greens contributed 5.3%, fruit and vegetables 4.5%, and cereals and legumes 3.9%.[1] The intake of fat was notably low. Mineral and vitamin intakes were probably adequate. The lack of protein was especially concerning in children. However, few signs of clinical malnutrition were observed in adults, although children weighed less for their age than was satisfactory.

Recommendations included improved agriculture, although the survey found agricultural systems that produced food in such difficult environments were to be admired. Increased variety in food plants, including cereals and legumes (rice, millet, sorghum, peas, beans and peanuts), and increased use of animals for milk and meat, and fish, was recommended.

1982–1983 National Nutrition Survey

The National Nutrition Survey (NNS) of 1982–1983 measured the weight and length of a sample of approximately 30 000 children, selected from all

1 These figures add up to 100.3%. The error is in the source (Hipsley and Clements 1950:23).

districts and environments in PNG. It was conducted by provincial nutritionists in the Department of Health and the PNG Institute of Medical Research. Rather than comparing PNG children against international standards, the NNS compared them against mean PNG growth curves for three altitude classes (below 600 m, 600–1200 m and above 1200 m). The results are expressed in terms of the number of standard deviations that the weight and length of sample children, aged 18 months and 30 months, fell above or below the altitude mean (see Box 6.6).

The distribution by district of the estimated proportion (per cent) of children who were below the PNG standards for length-for-age (stunted) and weight-for-length (wasted) in 1982–1983 is shown in Figures 6.6.1 and 6.6.2, respectively. Highlands children were shorter and heavier than lowlands children. Children in the 600–1200 m altitude class were lighter than lowlands children and shorter than highlands children. These differences were found to be the outcome of differences in nutrient intake, birthweight and sickness, with all these factors being heavily influenced by environment.

A re-analysis in 1999 of NNS data from almost 16 000 children showed that variation in growth among children living in different environments was largely explained by differences in diet, although significant differences in relation to altitude, relief and rainfall patterns persisted. Other important determinants of child growth and nutrition were socioeconomic status, maternal education, marital status of the mother, and the father's occupation. Further analysis showed that the geographical variation in child growth (see Figures 6.6.1 and 6.6.2) was significantly associated with the environment, diet, socioeconomic conditions, agriculture and demography. The authors of this study (Mueller and Smith 1999) concluded that most of the factors which were found to be associated with child growth were related in one form or another to differences in local subsistence agriculture and that agriculture is probably the main determinant of child growth and nutrition patterns in PNG.

Since 1983 when the NNS was conducted, the district boundaries have been changed. The results presented in Table 6.6.1 are in the districts as they were at the 2000 National Census. They were recalculated from the original survey results by Dr Ivo Mueller, PNG Institute of Medical Research, Goroka. Because of this recalculation, the new means will not be as statistically reliable as the original means. Nevertheless, in the absence of any other reliable information, they are presented here as a reasonable approximation of the situation in 1982–1983, within the borders of the 2000 census districts.

Districts in which the greatest proportion of children are significantly below the standard weight-for-length and length-for-age (lightest and shortest) are listed in the top left corner of Table 6.6.1. Conversely, in the bottom right-hand corner are the districts where the proportion of children who are taller and heavier is greatest. In the other corners are districts where children are either tall and light, or heavy and short. In the lightest and shortest corner are the districts of Nuku (Sandaun Province), Ambunti-Dreikikir and Maprik (East Sepik Province), Esa'ala (Milne Bay Province), Middle Ramu and Sumkar (Madang Province) and North Fly (Western Province). Districts in which most children are among the heaviest and tallest in PNG include Central Bougainville (Bougainville Province), Gazelle (East New Britain Province), Goroka and Unggai-Bena (Eastern Highlands Province), Karimui-Nomane (Simbu Province),[2] North Wahgi (Western Highlands Province), Namatanai (New Ireland Province) and Talasea (West New Britain Province). These are all districts in which average cash income from cash cropping is high.

2005 National Micronutrient Survey

The National Micronutrient Survey was carried out between May and October 2005 by the PNG Department of Health, UNICEF PNG, and the University of Papua New Guinea to assess the

2 The inclusion of Karimui-Nomane district is a good illustration of how great care must be taken in the interpretation of information collected at the district level in PNG. Nomane children are highlands children who are in general heavier, shorter and more numerous than the children of Karimui, many of whom are seriously malnourished. The district has been created for administrative and political reasons and lumps together children of very different nutritional status.

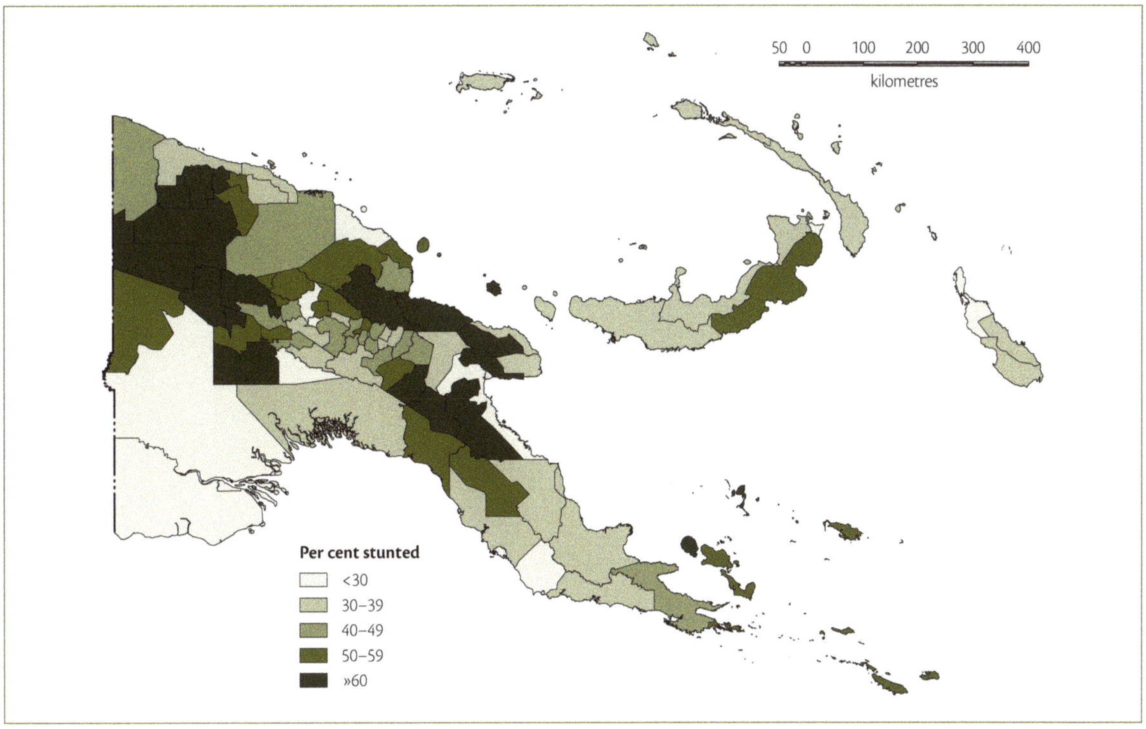

Figure 6.6.1 Proportion of children stunted, or below the PNG standard length-for-age. **Note:** The National Capital District was excluded from this analysis. Source: National Nutrition Survey 1982–1983 data recalculated by Dr Ivo Mueller into the redefined district boundaries of the 2000 National Census.

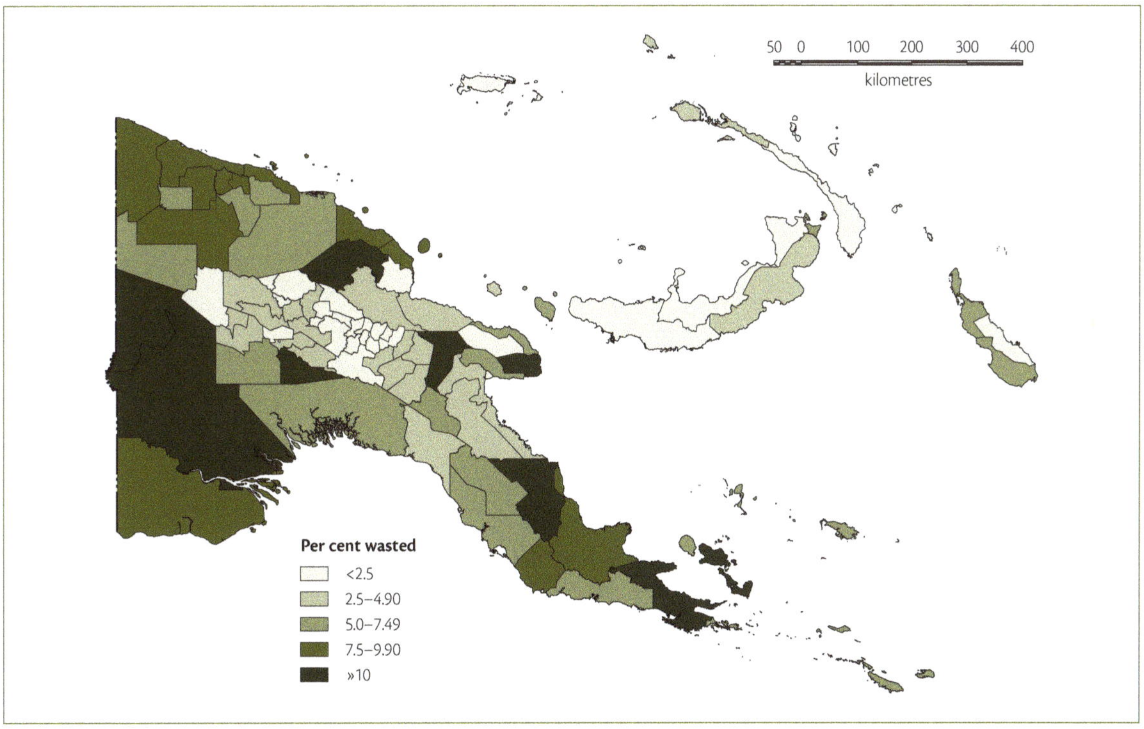

Figure 6.6.2 Proportion of children wasted, or below the PNG standard weight-for-length. **Note:** The National Capital District was excluded from this analysis. Source: National Nutrition Survey 1982–1983 data recalculated by Dr Ivo Mueller into the redefined district boundaries of the 2000 National Census.

Table 6.6.1 Classification of districts by age-independent length (stunted) and height-independent weight (wasted) scores

Per cent wasted[b]	Per cent stunted[a]				
	>= 60	**50 to 59**	**40 to 49**	**30 to 39**	**<30**
>= 10		Esa'ala Middle Ramu North Fly	Alotau	Finschhafen Markham Sohe	Middle Fly Kagua-Erave
7.5 to 9.9	Ambunti-Dreikikir Nuku	Maprik Sumkar	Vanimo-Green River	Aitape-Lumi Ijivitari Wewak	Bogia Rigo South Fly
5 to 7.49	Kiriwina-Goodenough Menyamya Nawae Nipa-Kutubu Telefomin	Goilala Samarai-Murua Wosera-Gaui	Angoram	Abau Kairuku-Hiri Kikori Rabaul South Bougainville Tewai-Siassi Yangoru-Saussia	Kokopo North Bougainville
2.5 to 4.9	Lagaip-Porgera Obura-Wonenara Rai Coast Tari Usino-Bundi	Kerema Komo-Margarima Okapa Pomio	Henganofi Imbonggu Kainantu Kandep Lufa Wapenamanda	Ialibu-Pangia Kandrian-Gloucester Kavieng Manus Tambul-Nebilyer	Huon Gulf Baiyer-Mul Hagen
<2.5	Koroba-Lake Kopiago Kabwum	Jimi Kompiam-Ambum Kundiawa Mendi Wabag	Angalimp-South Wahgi Chuave Daulo Gumine Kerowagi Madang Sina Sina–Yonggamugl	Central Bougainville Gazelle Goroka Karimui-Nomane North Wahgi Namatanai Talasea Unggai-Bena	

[a] Length (-for-age) score; expressed in terms of standard deviations from the national mean, independent of age.

[b] Weight (-for-length) score; expressed in terms of standard deviations from the national mean, independent of height.

Source: National Nutrition Survey 1982–1983 data recalculated by Dr Ivo Mueller into the redefined district boundaries of the 2000 National Census.

nutritional status of the sample population, to evaluate and better manage current health care practices, and to plan and implement new prevention programs and evaluate their impacts. The survey included preschool children aged six months to 59 months, non-pregnant women aged 15–49 years and men 18 years and older. The sample comprised 16 households each from 100 census units; all children in every selected household were surveyed and all non-pregnant women and all adult men from every second household. The census units were selected by PNGRIS environments (resource mapping units – see Section 1.15) and Region (see map on page xix). As well as anthropometry, the survey measured iodine intake, anaemia, and iron, vitamin A and protein deficiencies.

Preliminary results found that PNG children tend to be much shorter than the world average.[3] Just over one-third of children measured were below standard length-for-age (stunted) and 17% of these children were severely stunted. Children in Momase

3 In the draft report of the National Micronutrient Survey, the height-for-age z-score of all PNG children measured (n = 895) was –1.58z, compared to the international standard mean of zero.

Box 6.6 Measuring malnutrition in children

Malnutrition in children is measured indirectly, by measuring their physical growth. It is assumed that a child who is poorly nourished will grow more slowly than a child who is well nourished. Two measures are commonly used: *weight-for-age* and *length-for-age*. They can be combined into a third measure, *weight-for-length*.

The idea behind weight or length '-for-age' is that normally children grow taller and weigh more as they get older, and their weight and length is directly associated with their age. If large numbers of children whose ages are known accurately are weighed and measured, then age can be correlated with weight and length. Then the average weight and length for a given age can be calculated, as well as the variation (standard deviation – often expressed as a 'z-score') that occurs in weight and length among individual children at different ages. These means and standard deviations for weight and length can be graphed against age to produce *growth curves*. The growth curves can be used as standards against which individual children can be measured.

Individual children whose ages are known can be compared against a growth standard to see how far above or below the mean weight and length they are for their age. Low weight-for-age means a child is *wasted*, or is thin. This can be the result of short-term malnourishment, caused by a temporary problem with food supply or, for example, the death of the mother, or a recent sickness. Low length-for-age, or *stunting*, is usually a sign of chronic, or longer-term malnourishment. A number of different standard growth curves for children exist and care must be taken when comparing children from a particular population against an international standard. In PNG, where many highlands children are shorter than children elsewhere, *weight-for-length* is a better measure for comparing children across PNG.

In the 1982–1983 National Nutrition Survey, PNG children were compared against their own mean growth curves, calculated from the survey data. In this way, the slowest-growing children in PNG could be compared against PNG standards. Many children in PNG fall below international standards for weight and length for age, but they appear not to suffer greatly from this and grow into normal, healthy adults. However, at Tari in Southern Highlands Province, a study found that when a child became significantly wasted or stunted, the probability of them dying within the next 18 months was greatly increased (Heywood 1982).

Region and in rural areas were most likely to be stunted. Southern Region children were least likely to be stunted.

Just over 25% of children measured had low weight-for-age (wasted) and the overall mean z-score was −1.13z. Children in Momase Region and in rural areas were most likely to be underweight. Children in Highlands Region were least likely to be underweight.

Body mass index was calculated for 722 non-pregnant women and 787 men. More women and men were overweight than were underweight. Overweight men and women were most prevalent in Southern Region (including Port Moresby) and in urban areas. Men and women with higher education were more likely to be overweight.

Mean birthweight was 3.06 kg. The highest prevalence of low birthweights (<2.5 kg) occurred in Southern Region.

Overall, iodine intake was adequate but was lowest in Highlands Region; iodised salt use was also adequate, but was lowest in Southern Region. Almost 50% of all children measured were anaemic (low levels of haemoglobin in the blood), and 70% of children aged 6–11 months were anaemic. The proportion of anaemic children in Momase and Southern regions was particularly severe. The incidence of anaemia in non-pregnant women was also severe, with 40% of all women anaemic and 60% of women in Momase Region anaemic. Fewer men were anaemic (26%), with a greater proportion of rural men than urban men suffering from anaemia. Anaemia in PNG men is a moderate problem by World Health Organization standards, but it is more severe in Momase Region. Anaemia was found to be significantly associated with malaria, but not with hookworm infections.

Throughout PNG, vitamin A deficiency was found to be a moderate problem, but was severe in Momase Region and in urban areas. Vitamin A deficiency was significantly associated with malaria.

Sources

Department of Health of Papua New Guinea, UNICEF Papua New Guinea, University of Papua New Guinea and United States Centers for Disease Control and Prevention (2006). National Micronutrient Survey PNG 2005: draft summary report September 2006. Department of Health of Papua New Guinea.

Heywood, P. (1982). The functional significance of malnutrition – growth and prospective risk of death in the highlands of Papua New Guinea. *Journal of Food and Nutrition* 39(1):13–19.

Heywood, P.F. (1983). Growth and nutrition in Papua New Guinea. *Journal of Human Evolution* 12:133–143.

Heywood, P. (1987). Agricultural development, nutrition and health. *Papua New Guinea Medical Journal* 30:3–4.

Heywood, P., Singleton, N. and Ross, J. (1988). Nutritional status of young children – the 1982/83 National Nutrition Survey. *Papua New Guinea Medical Journal* 31(2):91–101.

Heywood, P.F. and Norgan, N.G. (1992). Human growth in Papua New Guinea. In Attenborough, R.D. and Alpers, M.P. (eds). *Human Biology in Papua New Guinea: The Small Cosmos*. Clarendon Press, Oxford. pp. 234–248.

Hipsley, E.H. and Clements, F.W. (eds) (1950). *Report of the New Guinea Nutrition Survey Expedition, 1947*. A.H. Pettifer, Government Printer, Sydney.

Mueller, I. (2001). The spatial pattern of child growth in PNG. In Bourke, R.M., Allen, M.G. and Salisbury, J.G. (eds). *Food Security for Papua New Guinea. Proceedings of the Papua New Guinea Food and Nutrition 2000 Conference*. ACIAR Proceedings No. 99. Australian Centre for International Agricultural Research, Canberra. pp. 414–431.

Mueller, I. and Smith, T.A. (1999). Patterns of child growth in Papua New Guinea and their relation to environmental, dietary and socioeconomic factors – further analyses of the 1982–1983 Papua New Guinea National Nutrition Survey. *Papua New Guinea Medical Journal* 42(3–4):94–113.

Mueller, I., Vounatsou, P., Allen, B.J. and Smith, T. (2001). Spatial patterns of child growth in Papua New Guinea and their relation to environment, diet, socio-economic status and subsistence activities. *Annals of Human Biology* 28(3):263–280.

6.7 Land settlement schemes[1]

From the 1950s to 1970s, land settlement schemes were a major part of the Australian administration's economic development strategy in PNG. The objectives of the schemes were to raise agricultural production and to demonstrate that individual titles to land were superior to customary tenure. Between 1952 and 1981, at least 130 land settlement schemes were started in every province in PNG except Western, Simbu and Madang provinces; 7203 blocks totalling 76 335 ha in area were made available over this period (Figure 6.7.1, Table A6.7.1).

The Commission of Inquiry into Land Matters in 1973 classified these schemes by the cost per settler. Using this method, three groups can be formed:

- Low cost schemes (less than K500 per family) – the government purchased land from landowners and gave it to land-short people to grow subsistence crops. The government had little or no further involvement.
- Medium cost schemes (K500 to K5000 per family) – the government planned the scheme layout, purchased and surveyed land, built roads, sometimes provided houses and water supply, and encouraged settlers to plant a single crop for sale, such as rubber.
- High cost schemes (K5000 or more per family) – highly planned and coordinated by the government, which purchased land, built roads, schools, health centres, port facilities and townships. Land titles contained covenants that forced people to grow a single crop for sale (oil palm).

Low cost schemes

Typical low cost schemes were the Gavien scheme north of Angoram and the Wosera (or Gawanga) resettlement scheme, both in East Sepik Province, or the Bakoiudu settlement in Central Province. The Wosera scheme involved the purchase of unoccupied land to the south-west of Wosera, owned by Gawanga people. The boundaries of the land were surveyed (but no internal subdivisions were made) and a road was built, by the villagers under government supervision, from Wosera to the land. The land was covered in tall secondary forest. After five years only 44 families had occupied the land, and these only on a part-time basis. After ten years, the road and bridges had been washed out.

The Gavien scheme, which started in 1967, was designed to provide agricultural land to Grass Country people who lived in swamps south of the Sepik River. It had a similar history to Wosera, with only eight families moving permanently to the 82 blocks provided by 1973. Then in 1977 it was upgraded as part of the East Sepik Rural Development Project (see Section 6.8) such that it became a medium cost scheme, with rubber as

1 Much of the information in this section is drawn from papers by David Hulme (1982, 1983).

the main cash crop. The Bakoiudu scheme was low cost to the government, but was sponsored by the Catholic Church, which subsidised rubber prices for many years to stop people leaving their blocks and returning to their scattered mountain villages.

Medium cost schemes

Like the low cost schemes, medium cost settlement schemes have performed disappointingly and have frequently required expensive redevelopment projects to rescue them from complete failure. The best-documented medium cost scheme is the Cape Rodney Resettlement Scheme south-east of Port Moresby. The scheme was begun in the 1950s as 14 plantations leased by Australians and 300 smallholder blocks. The plantations failed, leaving the smallholders to continue on their own. The scheme was not linked by road to the rest of the province and had no education or health services. What services existed were provided by missions. A 1971 report described Cape Rodney as a 'depressing shambles'. After nine years only 124 of the original 300 blocks had been occupied and only 300 ha of rubber had been planted of the 2129 ha of land available then. In the late 1970s the Magi Highway linked the scheme with Port Moresby. Throughout the 1980s and 1990s the scheme was the subject of a number of redevelopment attempts, but many of the settlers still have no formal title to their land and fresh food sales in Port Moresby remain the main source of their income. The rubber factory has been destroyed by fire twice (see Section 5.11).

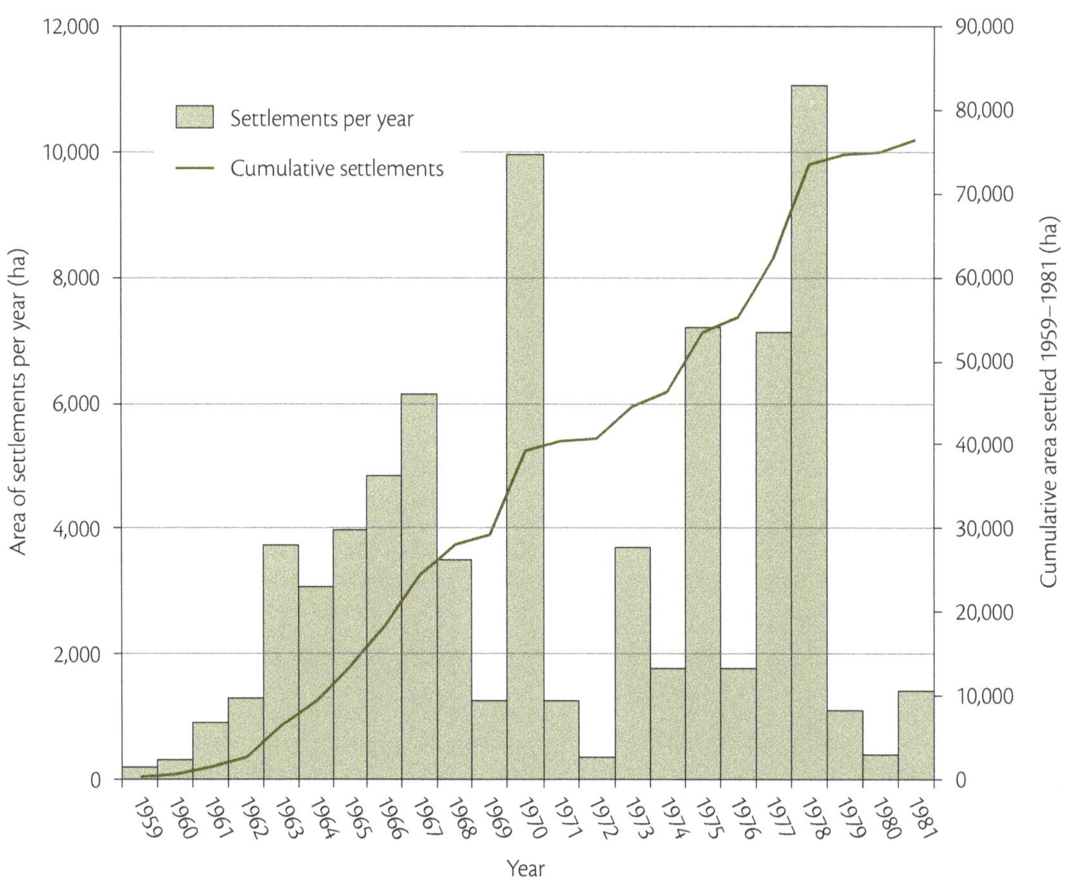

Figure 6.7.1 Area of land taken into land settlement schemes, 1959–1981. Source: Hulme (1982).

Elsewhere in PNG, medium cost settlement schemes have survived by abandoning original plans and growing fresh food for urban markets. This was the case at Situm and Gobari in Morobe Province. Settlers on some schemes in Western Highlands Province now grow vegetables instead of the tea that the scheme was designed for. In Oro Province a medium cost settlement scheme based on cocoa production that was abandoned because of pests and diseases was absorbed into high cost oil palm schemes in the 1980s. The medium cost schemes on the Gazelle Peninsula of New Britain, such as Vudal and Vunapaladin, are based on production of cocoa and fresh food and have been successful.

High cost schemes

All the high cost schemes in PNG are associated with oil palm (see Section 5.7) and take the form of a nucleus estate, surrounded by smallholders. They are established at Hoskins and Bialla in West New Britain Province, at Sangara in Oro Province, Hagita in Milne Bay Province and in New Ireland Province. At Hoskins, settlers came from East Sepik, Simbu and East New Britain provinces and initially local villagers were not involved. At Sangara, 50% of the settlers were from Oro Province, 18% from Morobe Province and 20% from Madang, Sandaun and East Sepik provinces. In Milne Bay and New Ireland, settlers came from within the provinces. The nucleus estates in Oro were established on abandoned plantations and smallholder resettlement blocks that previously grew cocoa, and on village land in New Ireland. The most recent land settlement schemes have been in the Bialla area of West New Britain Province, and these were established in the early to mid 1990s.

The oil palm schemes have been judged economically an outstanding success. The Hoskins scheme has recorded economic rates of return of 25%; settlers' incomes have been well above PNG means for smallholders; a high proportion of settler loans are repaid; and blocks are sold and are in high demand. The schemes generate significant export earnings and are responsible for a large local labour market.

Whether these high cost oil palm schemes have been a social success is more equivocal. Critics have pointed to inter-ethnic conflicts, reasonably severe in the beginnings of the Hoskins schemes; second and third generation settler underemployment and criminal activity; alcoholism; prostitution; and malnutrition among settler children. The schemes were also blamed for underdevelopment in the surrounding villages. Some of these criticisms have proven groundless, but a number remain serious problems. Vigorous attempts have been made at Hoskins to involve the local villagers in producing oil palm on village land, and the village oil palm production has expanded significantly in recent years. At Popondetta and Milne Bay the villagers have participated enthusiastically. David Hulme points out that the Hoskins schemes were groundbreaking in that they were the first attempt to create a mixed ethnic community on a large scale in PNG.

Other settlement schemes

Since the mid 1990s, few new land settlement schemes have been established.[2] However, the resettlement of refugees from natural and political disasters has resulted in the establishment of schemes, although these are not always formal settlements. In 1987, the East Awin Refugee Relocation Area was established east of Kiunga in Western Province for refugees from Papua Province of Indonesia. This settlement area was administered by the UN High Commissioner for Refugees and in 1992 had 3000 residents. Most of these settlers were not village farmers but were public servants, teachers, tradespeople and professionals and they required considerable support with food. The settlers were taught intensive farming methods and

2 Descriptions of agriculture on various schemes in the early to mid 1990s are given in Working Papers of the Mapping Agricultural Systems of PNG Project. For example, System 1412 in Allen et al. (2002) covers the Gavien scheme in East Sepik Province. See the Working Papers for Central Province (System 0308, Cape Rodney settlement); East New Britain Province (System 1808, various schemes on the Gazelle Peninsula); Gulf Province (System 0205, Murua settlement); and West New Britain Province (Systems 1916, 1917, 1918, various oil palm settlements).

they sold fresh food to Ok Tedi Mining Ltd and also sold paintings and carvings in Kiunga. In 2004, PNG issued birth certificates to over 4000 children born at East Awin and residency permits to all eligible adults, which enabled them to leave the settlement area if they wished.

After the eruption of volcanoes near Rabaul in 1994, people from villages near Rabaul were allocated land on former plantations in the Warangoi Valley and at Sigute between the Warangoi and Sigute rivers, East New Britain Province. Refugees from the 1998 tsunami in Sissano Lagoon area of Sandaun Province have been settled on land inland of the lagoon. The eruption of Manam Island, Madang Province, in 2005 and 2006 has resulted in the displacement of about 10 000 people from the island to former plantations in the Bogia area on the nearby mainland. Most Manam refugees had not been allocated land by late 2006, except for housing areas.

Summary

The low and medium cost settlement schemes in PNG are judged, in general, to have been only moderately successful or unsuccessful. On the other hand, the high cost schemes have been a marked success and have created considerable economic benefits. An important reason for this success has been the high price of palm oil, the commodity produced from the high cost schemes. In contrast, medium cost schemes based on rubber production have failed economically and socially, unless the settlers have had good access to urban markets and have been able to switch to fresh food production. The low and medium cost schemes were poorly planned and implemented and badly administered, an outcome of a critical shortage of well-qualified planners and experienced administrators. High cost schemes directly benefit a small number of people but may benefit large numbers of people indirectly by creating a strong agricultural export sector. Low and medium costs schemes, if they could be better planned and administered, would contribute to a broader improvement in rural living standards.

Sources

Allen, B.J., Hide, R.L., Bourke, R.M., Fritsch, D., Grau, R., Lowes, E., Nen, T., Nirsie, E., Risimeri, J. and Woruba, M. (2002). *East Sepik Province: Text Summaries, Maps, Code Lists and Village Identification*. Agricultural Systems of Papua New Guinea Working Paper No. 2. Revised edition. Land Management Group, Department of Human Geography, Research School of Pacific and Asian Studies, The Australian National University, Canberra.

Eng, J. (1980). A socio-economic survey of the Popondetta oil palm resettlement scheme blocks – Isevere and Egora, November–December 1979. Unpublished paper. Nutrition Section, Department of Health, Port Moresby.

French, W. and Nen, T. (1986). Benchline study of the Cape Rodney agriculture development project. Report to the Department of Finance and Planning. Institute of Applied Social and Economic Research, Port Moresby.

Gostin, O. (1986). *Cash Cropping, Catholicism and Change: Resettlement Among the Kuni of Papua*. Pacific Research Monograph No. 14. National Centre for Development Studies, The Australian National University, Canberra.

Hulme, D. (1982). Land settlement schemes in Papua New Guinea: an overview. *Land Reform, Land Settlement and Cooperatives* (1/2):21–42.

Hulme, D. (1983). Agricultural land settlement: the relationship between scheme costs and scheme performance. *Pacific Viewpoint* 24(1):69–79.

Hulme, D. (1983). An economic appraisal of the Hoskins Oil Palm Scheme. *Australian Geographer* 15:330–334.

Montgomery, D. (1979). A handbook of information for senior land settlement officers. Department of Lands, Surveys and Environment, Port Moresby.

Papua New Guinea Commission of Inquiry into Land Matters (1973). *Report of the Commission of Inquiry into Land Matters*. Government Printer, Port Moresby.

United Nations High Commissioner for Refugees (2004). *Global report: East Asia and the Pacific*. United Nations High Commissioner for Refugees, Geneva.

6.8 Rural development projects

Rural development projects (sometimes known as *integrated* rural development projects) arose from a belief that it was possible to bring about sweeping economic and social change in poor rural areas through the application of rational, scientific knowledge. In addition, hierarchical control over appraisal, planning, implementation and the use of funds was required. Rural development projects were an extension of a belief that the characteristics of mature capitalist economies (industrialisation, modern scientific agriculture, infrastructure and improved social services) would bring about political stability within developing countries.

The largest of the international development agencies, the World Bank, had been successful in rebuilding Europe after World War II, but when it applied the same practices to bring about change in poor rural countries it had little success. In 1973, the president of the bank, Robert McNamara, recognised the bank's failures. He announced a radical 'redirection' of effort, away from trickle-down theories to direct investment in poor rural areas in developing countries. Other developed country aid agencies around the world followed the World Bank's example and began to use rural development projects to attempt to bring about economic and social improvements in poor rural areas.

PNG has been the recipient of a number of integrated rural development projects. A full evaluation of the outcomes of these projects and why they succeeded or failed to achieve their goals in PNG has never been undertaken. Assessments that do exist were done either during the life of the project (a mid-term evaluation, for example) or immediately on completion. A brief review by Crittenden and Lea in 1989 of PNG projects found that it was not possible to make broad generalisations on the outcomes achieved. They also found that the argument that the projects had done very little to accelerate development in poor rural areas was an oversimplification. Nevertheless, while it acknowledged that in rural PNG 'everything is related to everything else', the 1989 review argued that, far from being integrated, project components were disparate, separated in space and benefited only particular groups of people. Projects were integrated only by being implemented in a single province under a single management organisation.

Rural development projects in PNG include the East Sepik Rural Development Project (1977–1984), Southern Highlands Rural Development Project (1978–1985), Enga Rural Development Project (1982–1987), West Sepik Provincial Development Project (1986–1991), South Simbu Rural Development Project (1986–1995), North Simbu Rural Development Project (1993–2002), Kandrian–Gloucester Integrated Development

Project (1993–1998) and Pomio–Bainings Rural Development Project (late 1980s to early 1990s). Only three of the largest projects are reviewed here.[1]

East Sepik Rural Development Project, 1977–1984

The East Sepik Rural Development Project was the first of the PNG integrated rural development projects. It was proposed as a means to achieve 'total development' through the delivery of a number of components that would address the problems faced by small farmers and the causes of rural poverty. It was also to deliver social services and meet the economic and infrastructure needs of the province. The components in the final plan were upgrades of existing activities, including the Gavien land settlement scheme (see Section 6.7), buffalo farming, inland fisheries, crop intensification, agricultural research and the construction of schools and an agricultural college. The total expenditure was US$14.87 million (around K97 million in 2005 values) provided in the form of a loan from the Asian Development Bank.

The Gavien settlement scheme was designed to benefit poor families who lived in swamps south of the Sepik River. The scheme required settlers to produce rubber, a cash crop with poor returns to labour in PNG, but it gave them access to urban markets for fresh food and to health and education services. Buffalo farming and inland fisheries failed. (A longer-term outcome was an Australian-funded aid project to shoot, from helicopters, buffalo that had escaped from the project, were infected with tuberculosis and were damaging village gardens.) The crop intensification program did not achieve many of its objectives, with the exception of the food and nutrition component. The agricultural college opened at Bainyik, but within ten years was closed due to lack of funds and today there is very little to be seen on the site.

The failures were blamed on poor planning; rushed implementation in order to qualify for the loan; a failure to relate the project to political, social and administrative realities in the province; overuse of foreign consultants (from a number of different companies, making control and coordination difficult); failure to train local staff; slowness in appointing project staff; severe administrative and procurement delays; carrying out research after rather than before the implementation of many agricultural subprojects; the use of completely new activities such as buffalo farming and fish ponds; frequent staff changes; poor monitoring; low economic returns to villagers of the proposed agricultural activities (for example, rubber and rice); and an extremely complex management structure.

Southern Highlands Rural Development Project, 1978–1985

The Southern Highlands Rural Development Project was started in response to a proposal to extend the Highlands Highway from Mendi to Tari in 1973. The new road gave 100 000 rural people access to domestic and international markets for the first time. The project was designed to help people take advantage of this, as well as to boost the internal rates of return on the road project and so justify it in economic terms. The main components of the project were to upgrade 138 km of road between Mendi and Tari; surface and upgrade 41 km of feeder roads to service proposed tea plantations; construct bridges and culverts on 900 km of feeder roads; establish 850 ha of tea, 1280 ha of coffee and 100 ha of cardamom, plus tea and coffee processing facilities; conduct food crop trials and extension work on food crops; and construct and operate secondary schools, a teachers' college, a nursing school and health subcentres. The food crops component was added

1 Also excluded are the more focused projects including Agriculture Quarantine and Inspection Project, Australian Contribution to a National Agricultural Research System, Bougainville Cocoa Rehabilitation Project, Cocoa Quality Project, Agriculture Support Services Project, East New Britain Smallholder Development Project, Smallholder Livestock Credit Project, Milne Bay Nucleus Estate and Smallholder Project, Cape Rodney Smallholder Development Project and the Popondetta Smallholder Oil Palm Development Project. Post-project reviews of these projects are difficult to locate.

to counter the perceived adverse effects of cash cropping on food crop production. The education and health facilities were planned to raise the level of services to an underdeveloped area. The project cost was around US$32 million (K180 million in 2005 values) and was funded by a loan from the World Bank.

A project completion report judged the construction of roads, schools and health centres as successful and they were seen to be having a positive effect. The food crop trials and extension component was judged as less than successful. The cash crop components were 'disastrous', with only 260 ha of tea planted, 67 ha of coffee (of which 10 ha survived) and 18 ha of cardamom. The large cash crop plantations were abandoned. The provincial government failed to take over the operation and management of the project components.

The main reasons for the failure of the non-infrastructure components of the project were said to be trying to start too many activities at once, such that project management, implementation capacity and the provincial government ability to absorb them was overwhelmed (provincial government was established in Southern Highlands Province after the project had begun). Further reasons were a failure to train provincial staff; inflexibility and a failure to respond to changing circumstances over time; a lack of commercial orientation in the cash crops components and 'incompetence' in managing this component; and a serious loss of knowledge caused by the failure of foreign staff to write reports on the outcome of their work before leaving the project.

Enga Yaaka Lasemana, 1982–1987

The Enga Rural Development Project grew out of the Enga Rural Development Study, which was instituted by the national government under the less developed areas sectoral program of the National Public Expenditure Plan. The Enga Provincial Government was responsible for the implementation of the project and the national government was only responsible for negotiating the loan and administering the funds. From over 200 possible projects arising from the Enga Rural Development Study, 40 were selected for the project. A final plan was produced in 1981 after appraisal by the World Bank. A loan was negotiated with the World Bank to provide funds from 1982.

The project included infrastructure (new office buildings, a remand centre, aid posts and health centres, road upgrading, hydro-electricity plants, sawmills, schools); a law and order component (Enga was the site of frequent inter-group fighting); community development; village food crop production and cash cropping systems (coffee, pyrethrum and cardamom); livestock management; vegetable marketing; industrial training; and improved province financial and management services.

Infrastructure and road upgrading was generally completed satisfactorily. However, the objectives of many of the other components either were not achieved, or failed to be sustained after the project was completed. An assessment of the project suggested that there were too many components; the project suffered from high staff turnover; excessive delays in recruiting staff; poor staff selection; poor training; lack of motivation; poor supervision; problems with securing land from customary owners; continued fighting and law and order problems that resulted in some infrastructure built by the project being destroyed; local political upheavals, sometimes associated with the fighting; personality clashes; and a loss of interest by the World Bank in the outcomes.

Summary

The Crittenden and Lea review of five PNG rural development projects recommended that future projects needed to be small and have a low budget; integrated into provincial planning structures; oriented towards training, institution-building and reform of local institutions; have local consent and commitment; have simple management and monitoring procedures; have low numbers of foreigners involved; and be properly evaluated. This review argued that the projects were not complete failures. The widely held perception that the projects failed was the result of comparing the outcomes with unrealistic objectives and with the manipulated cost-benefit analyses and inflated rhetoric that were used to justify their funding. More needs to be known about the long-term consequences of these projects, whether any parts of them were successful and the reasons for that success.

A workshop was held in Madang in 2004 on rural development 'projects' in PNG that had been successful. The findings of the workshop were published in the *Development Bulletin No. 67* in 2005. The successful projects were small, low budget and, above all, were focused on the development of rural communities and not on whole provinces. The participants concluded that successful rural development in PNG will be community-based and focused on the transformation of local polities and economies.

Sources

Carrad, B., Talyaga, K.K. and Lea, D.A.M. (1980). Developing the periphery: self-sufficient planning in the Enga Province, Papua New Guinea. *Proceedings of the Second International Symposium on Asian Studies, 1980*. Asian Research Service, Hong Kong. pp. 945–958.

Carrad, B., Lea, D.A.M. and Talyaga, K.K. (eds) (1982). *Enga: Foundations for Development. Volume 3 of Enga Yaaka Lasemana*. Department of Geography, University of New England, Armidale.

Crittenden, R., Eveno, J., Kimbu, R., Wane, T., Betitis, T. and Takahu, T. (1988). *Gavien and Kaupena: assessing the impacts of development projects in the East Sepik and Southern Highlands provinces*. Occasional Paper No. 8. Department of Geography, University of Papua New Guinea, Port Moresby.

Crittenden, R. and Lea, D.A.M. (1989). *Integrated Rural Development Programmes in Papua New Guinea: External Aid and Provincial Planning*. IASER Monograph No. 28. Institute of Applied Social and Economic Research and Department of Geography and Planning, University of New England, Boroko, Papua New Guinea.

French, W. (1984). IASER midterm evaluation of Enga Yaaka Lasemana: overview. Institute of Applied Social and Economic Research, Boroko, Papua New Guinea.

Hegarty, D. and Thomas, P. (eds) (2005). Effective development in Papua New Guinea: local initiatives and community innovation. *Development Bulletin No. 67*.

Hirschman, A.O. (1967). *Development Projects Observed*. The Brookings Institution, Washington DC.

Institute of Applied Social and Economic Research (1985). Southern Highlands Rural Development Project completion report. Institute of Applied Social and Economic Research, Boroko, Papua New Guinea.

World Bank (c. 1978). Annex 2. Papua New Guinea rural development project – Southern Highlands (Credit 841-PNG) project completion report. East Asia and Pacific Regional Office, World Bank, Washington.

6.9 Transport infrastructure

Infrastructure is usually understood to refer to roads, bridges, ports, airstrips, communication facilities, water supplies and dams, sewerage systems, and health and education facilities. Infrastructure is the physical structure that enables economic activities to take place, or the structures through which services are delivered. Infrastructure can be publicly or privately owned, but infrastructure in PNG remains almost all publicly owned, despite moves to privatise some port facilities and airports.

It is widely accepted that effective infrastructure is essential for economic growth and also makes a significant contribution to reducing poverty. Transport is critical for village access to markets, particularly to sell agricultural produce, and so for economic growth. Transport is also important for people to access health and education services. An inverse relationship exists between measures of poverty and access to markets, health and education services. Failure to maintain transport infrastructure, as has occurred in PNG in recent decades, constrains economic growth. This section describes the state of basic transport infrastructure in PNG, in particular roads, bridges, sea ports and airstrips.

Roads and bridges

The first major road in PNG was the Boluminski Highway, built on the east coast of New Ireland in 1910–1914. Prior to 1914, the German colonial administration built three roads on the Gazelle Peninsula and one each on Bougainville and New Hanover islands. During World War II there was a tremendous upsurge in road building activity, with both Allied and Japanese forces involved. This activity was continued after the war by the Australian administration. Many early roads were built with unpaid, local, hand labour and were little more than graded foot tracks. After World War II, roads for motorised vehicles were constructed all over the country, first by unpaid local labour and then with earthmoving equipment.

Many of the first roads in PNG were made through areas with the highest populations. This was partly because of the labour requirements for road construction and partly because colonial administrators sought to reach the greatest numbers of people for the lowest cost. Beginning in the 1950s, local road networks began to be linked together by trunk roads, constructed using machinery. These roads were built to enable the export of agricultural products, particularly from the highlands. The economic benefits of new roads were meant to cover the costs of construction and ongoing maintenance, and estimates of economic returns were supposed to be prepared before the roads were built. However, many roads were built without cost-benefit analyses, on a perception of reasonable returns and an understanding that regional development would not take place without a road.

The outcome of this period of road construction in PNG is that the majority of the population now lives close to a road (Table 6.9.1).[1] In 2000, 53% of the total population lived within 5 km of a national road and 70% lived within 15 km of a national road. A majority of people in PNG live within four hours walk of a national road. If district and rural roads are taken into consideration, a greater proportion of the population lives within 5 km of a road.

A number of provinces are particularly well served by roads (Figures 6.9.1, 6.9.2, Table A6.9.1). In Western Highlands and Simbu provinces, more than 70% of the populations live within 5 km of a national road; and in Southern Highlands, Central, Enga and East New Britain provinces more than 60% of the populations do so. Of the highlands provinces, only in Eastern Highlands do less than 50% of the population live within 5 km of a national road but, even here, only 30% are further than 15 km from a national road. In contrast, in Western and Milne Bay provinces, more than 70% of the populations live further than 15 km from a national road. More than 100 000 people live further than 15 km from a national road in each of Madang, Milne Bay, Eastern Highlands, Western and East Sepik provinces. In Morobe Province, more than 245 000 people (53% of the provincial population) live further than 15 km from a national road.

Table 6.9.1 Distance people live from a national road

Distance from a national road[a]	Population	%
Within 5 km	2,748,257	53
5 to 10 km	545,812	10
10 to 15 km	301,921	6
More than 15 km	1,594,796	31
Total population	5,190,786	100

[a] Distance is a straight-line distance and does not take into account the local terrain.

Sources: NSO (2002); PNG Road Asset Management System.

Residence near a road does not guarantee that the road will be trafficable all year round. Since about 1980, many roads and bridges have not been adequately maintained. As a result, many roads have deteriorated to the point where they are impassable when wet. When dry, they are so potholed and corrugated that damage is caused to vehicles travelling along them. Failure to maintain many bridges means that they have collapsed or must be crossed with care. Consequently, travel time and costs are greatly increased.

The PNG Road Asset Management System (RAMS) indicates that there were about 7300 km of national road in PNG in 2000 (Table 6.9.2).[2] As a proportion of total road length, 22% of all national roads in 2000 were in a 'poor' condition; 36% in a 'fair' condition; and 42% were in 'good' condition. In seven provinces (Simbu, Enga, Manus, Bougainville, East New Britain, Southern Highlands and Western Highlands), more than 75% of national roads were in 'poor' or 'fair' condition. In Eastern Highlands, Sandaun, Madang, Central and Gulf provinces, less than half of the national roads were in 'good' condition. Only in Oro and New Ireland provinces were more than 80% of national roads in 'good' condition (Figure 6.9.3, Table A6.9.2).

In 2000, two-thirds of national roads were gravel-surfaced and most of the remainder were sealed. Roughly equal proportions of both sealed and gravel-surfaced roads were in 'poor' or 'fair' condition (Table 6.9.2). Along the full length of roads, 'poor' sections of road are interspersed with 'fair' and

1 Roads in PNG are classed as national, district and rural. The national government is responsible for maintaining national roads. Provincial, district and local level governments maintain all other roads. The details of national roads have been entered into a management database (PNG Road Asset Management System – RAMS). This enables estimates to be made of the number of people living a given distance from a national road. This is not possible for other classes of roads.

2 The database on national roads contains many errors and omissions. There is little provincial-level or national-level data on length and condition of district and rural roads.

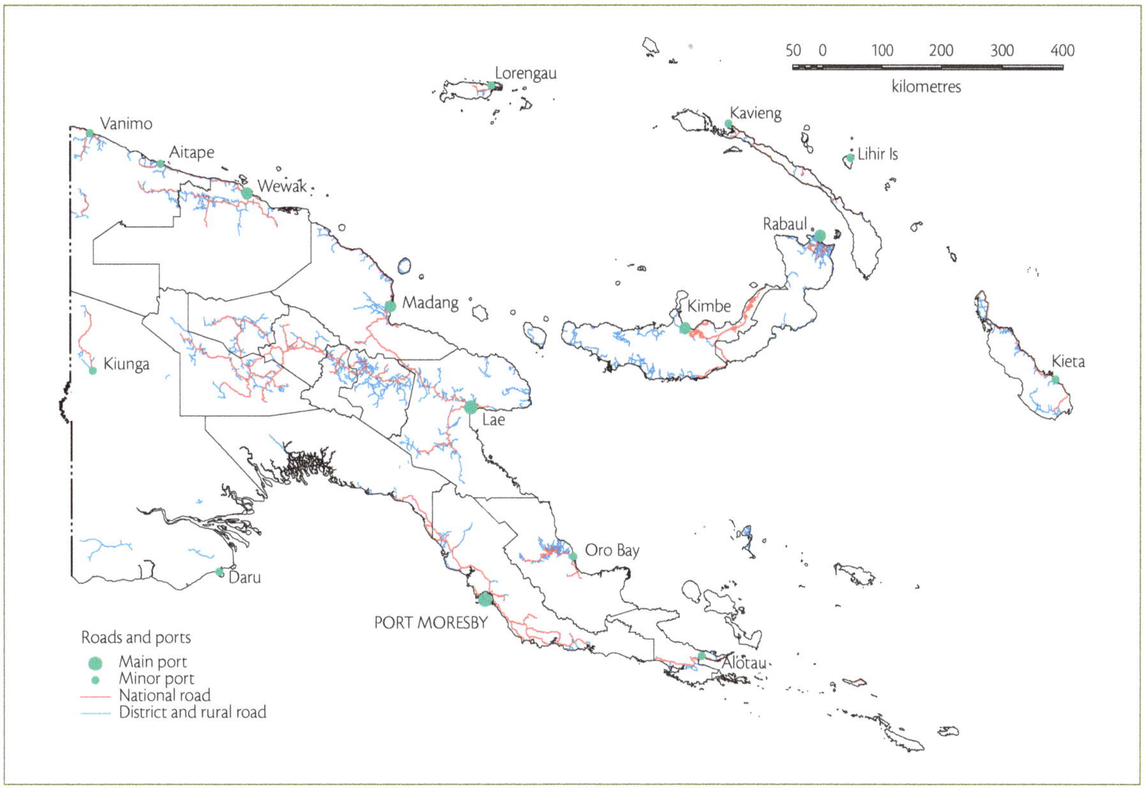

Figure 6.9.1 Roads and ports in PNG. Source: PNG Road Asset Management System.

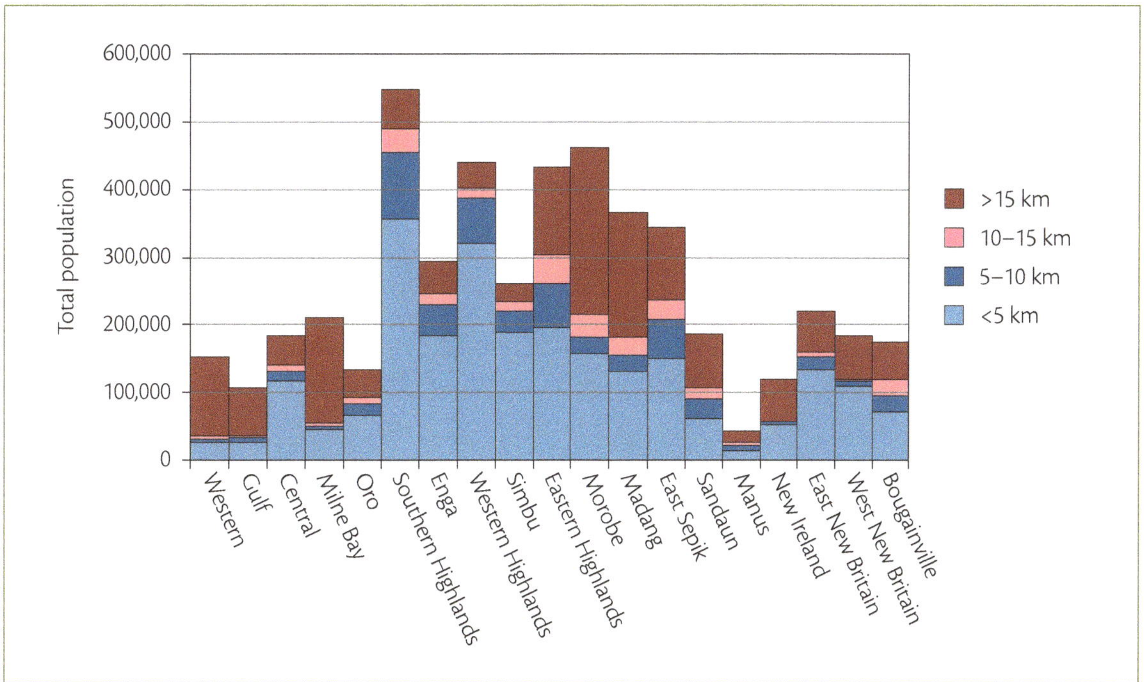

Figure 6.9.2 Total population (including urban and rural non-village populations) living within a given distance of a national road, by province. **Note:** Distance is a straight-line distance and does not take into account the local terrain. The urban populations of National Capital District and Lae City, which live within 5 km of a national road, have been excluded from this analysis (see Table A6.9.1). Sources: NSO (2002); PNG Road Asset Management System.

'good' sections. Thus a 'poor' section of road near an urban centre affects traffic originating along its total length, while impassable sections of road prevent the movement of traffic beyond that point. The condition of many district and rural roads is not known but, anecdotally, most are in very poor condition.

The Medium Term Development Strategy (see Sections 6.3 and 6.4) emphasises spending on transport infrastructure, road rehabilitation and road maintenance, but adequate spending on roads and bridges will require a radical departure from how the national budget has been allocated in the past. In 2002, the PNG Department of Transport estimated that it would require K120 million to bring the roads of PNG back to the condition they were in 1980, but only K10 million was allocated for road maintenance in that budget year. It was recently estimated that only around 40% of what is required to maintain

Table 6.9.2 Length of national roads by surface and road condition

Road surface	Road condition						Total road length	
	Poor	%	Fair	%	Good	%	(km)	(%)
Dirt	23	100.0	0	0.0	0	0.0	23	0.3
Gravel	1,346	27.3	1,542	31.3	2,044	41.4	4,932	67.3
Sealed	285	12.0	1,072	45.1	1,019	42.9	2,376	32.4
Total	1,654	22.5	2,614	35.7	3,063	41.8	7,331	100.0

Note: The distances in this table are slightly different to those in Table A6.9.2. These differences occur in the original RAMS document. These data refer only to national roads and do not include district or rural roads.

Sources: DWI (2001); PNG Road Asset Management System.

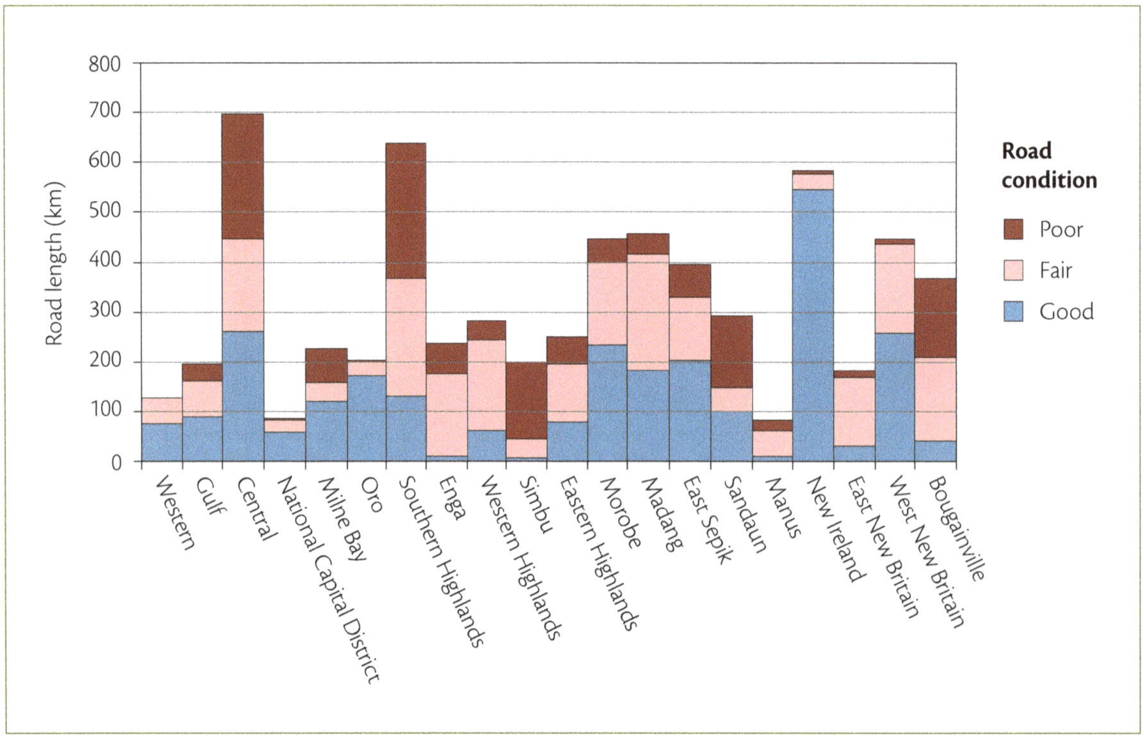

Figure 6.9.3 Length of national roads by road condition and province. Source: DWI (2001).

present main roads in a reasonable condition has been spent in an average year. The Highlands Highway is kept open by aid funds from AusAID, the Asian Development Bank, the World Bank, and mining companies.

The lack of maintenance of roads and bridges has a negative effect on marketing agricultural produce and on economic growth. It also makes it more difficult for rural people to access education and health services, as well as for wholesalers of food and other goods to transport these to where consumers can purchase them.

Sea ports

PNG sea ports provide a vital link between PNG and the world market, with more than 80% of exports shipped from ports. PNG has 22 declared international ports, but only 16 are operational. A number of publicly and privately operated minor wharves, jetties, and landings also exist around the country. In 2004, PNG ports handled slightly more than five million tons of general cargo and 195 000 twenty-foot equivalent units of containerised cargo.

The most important port in PNG is Lae, at the end of the Highlands Highway (Figure 6.9.1). Lae has no natural harbour; its port is constructed on land that is geologically unstable and the berths require continual dredging. In 2004, Lae port handled 51% of the general cargo and 56% of the containerised cargo that passed through PNG ports. About 50% of PNG exports and 90% of coffee exports are handled through Lae. The port also serves a major mine at Porgera in Enga Province, the gas and oil fields in Southern Highlands Province, and three gold mines in Morobe and Eastern Highlands provinces. Lae is also the main hub for coastal shipping to the smaller ports of Kimbe, Kavieng, Rabaul, Madang, Wewak and Kieta. Much of the sweet potato sold in Port Moresby's urban markets is grown in the highlands and shipped through Lae on coastal ships.

Port development and maintenance has not kept pace with the growth in cargo. Since 1995, container cargo at Lae has grown at 5% per year and general cargo at 2.5% per year. The current cargo volume and vessel types calling at Lae require at least four international berths with a total length of 800 m. Lae has only two berths suitable for international shipping with a total length of approximately 300 m, but the main wharf was being extended in late 2006. The storage and cargo marshalling areas are not sufficient to handle increasing cargo volumes and ship sizes. As a result, port congestion at Lae is frequent and imposes high costs on port users. Delays of 3–5 days have become common and cost international shipping companies about US$20 000 per day. Some companies are refusing to call at Lae and others are considering imposing congestion charges, which will translate into higher costs for importers and exporters (including village coffee growers). Port congestion also poses health and safety problems, and the issue of non-compliance with the International Shipping and Port Security Code. The wharf and shipboard facilities for handling fresh food are poor and much food is lost from damage inflicted during shipping from poor handling and unsuitable containers.

Port Moresby, despite being the capital city and having a magnificent harbour, is a less important sea port than Lae, particularly for agriculture. With no road connections to other regions in the country (Figure 6.9.1), exports from Port Moresby are mainly transhipped from coastal ships to international ships. Hence Port Moresby has a number of berths for coastal vessels but only one for overseas vessels. Copper concentrate from Ok Tedi mine is loaded on ships at Kiunga on the Fly River and transhipped to large international ore carriers in Port Moresby harbour. Crude oil from Kikori is loaded at the Kumul Platform in the Gulf of Papua. Some is shipped to the Napa Napa oil refinery in Port Moresby harbour where it is processed for domestic use and export.

Other main ports are:

- **Rabaul.** An important port for the export of cocoa and coconut products, with three international berths, one being a bulk coconut oil-loading berth. Rabaul is also an important coastal shipping focus. The 1994 volcanic eruptions did not damage the port facilities, although dredging is required.

- **Madang.** An excellent harbour, but with more difficult and less reliable road connection to the highlands and thus overshadowed as a port by Lae. However, if further port development at Lae proves technically too difficult, Madang could be developed to take over some of Lae's trade.
- **Wewak.** Serves East Sepik and Sandaun provinces via the Sepik Highway, but is restricted in the size of ships that can be handled by the depth of water at the single berth.
- **Kimbe.** On the north coast of West New Britain Province; is the main palm oil export point. In the last year, volcanoes to the east and west of Kimbe threatened to cause significant disruptions to this coastline, but a major eruption has not yet occurred.

Of the minor ports, Alotau, Kavieng and Oro Bay export palm oil, Lihir services the gold mine in New Ireland Province, and Kieta on Bougainville Island was the main port for the copper mine before the mine was closed by the civil war. Other minor ports include Aitape, Daru, Lorengau and Vanimo (Figure 6.9.1).

Maintenance on navigational aids has not been adequate and the situation has become critical. A joint venture project with the Australian Maritime Service began in 2004 to rehabilitate maritime navigation aids. The project will provide more lighthouses, day markers and buoys. The first phase began in Milne Bay and Port Moresby in 2004 and in the Islands Region in 2005.

People in the maritime provinces, such as Milne Bay, are especially dependent on marine transport. The infrequency, unreliability and slowness of domestic shipping in PNG is a major constraint to further agricultural and fisheries development as well as to delivery of services.

Airstrips

Aircraft have been particularly important in the development of PNG. Almost all inland centres began as small outposts established where it was possible to build a light aircraft landing strip. Light aircraft continue to be important where roads have not been constructed, mainly in mountainous country. In most cases, airstrips close once a road has been made to an area. About 490 airstrips existed in PNG in 2006; only 4% were sealed and about 80% were less than 900 m long (Table 6.9.3).

International airports offer opportunities for niche marketing of fruit and flowers to Australia and Asia.[3] International services operate from Port Moresby only, and all international air cargo has to be transhipped through Port Moresby, although international flights are possible through Alotau, Mount Hagen, Nadzab (Lae) and Rabaul. Other provincial centres with airstrips capable of handling Fokker F28 jets are Hoskins, Kavieng, Lorengau, Madang, Vanimo and Wewak. Twin engine aircraft, such as the Dash 8, fly into Daru, Goroka, Kerema, Kiunga, Kundiawa, Lihir Island, Mendi, Popondetta, Porgera, Tabubil, Tari and Wapenamanda. Daily flights occur from Port Moresby to some of these centres.

Of greater importance are the large numbers of light aircraft landing strips in remote and isolated areas, where they serve as the only means of access other than walking. Following the fall of the kina relative to the US dollar in 1997 (see Sections 4.1 and 4.2), the cost of fuel and air travel has increased sharply. This has resulted in the withdrawal or reduction

Table 6.9.3 Surface and length of airstrips

Surface	Length (m)	Number
Sealed	>2500	2
	1500–2500	13
	900–1500	5
Unsealed	1500–2500	13
	900–1500	57
	<900	402
Total airstrips		492

Source: Trevor Michie / <http://www.michie.net>.

[3] Air Niugini flies to Brisbane, Cairns, Honiara, Hong Kong, Manila, Nadi (Fiji), Singapore, Sydney and Tokyo. Another company, Airlines PNG, operates services from Port Moresby to Brisbane and Cairns.

of services by light aircraft operators, including missions, to a number of places where passenger traffic is light or where people cannot afford to pay the increased cost. Some coffee was transported by light aircraft to provincial centres, but the cost of air freight relative to the price of coffee is now generally too great for coffee to be moved by air economically.

A recent review of the PNG Air Transport Industry by PNG's Independent Consumer and Competition Commission concluded that PNG's airports are below minimum standards.[4] Lack of finance to maintain runways, terminal facilities and navigational aids; poor management through the Civil Aviation Authority (including an inefficient organisational structure and poor accounting practices); increasing fuel prices; decreasing numbers of customers; and law and order problems that contribute to increased operating costs for airlines are the main factors causing operating inefficiencies and posing safety concerns about the air industry.

Summary

PNG is relatively well provided with transport infrastructure. However, most roads, bridges, ports and airstrips were constructed prior to 1980 and in general have not been adequately maintained or developed to keep pace with increasing traffic and cargo. As a result, many national, district and rural roads are impassable when wet and in poor condition when dry. The most important sea port at Lae has frequent delays. The high cost of air travel and air cargo restricts the use of air transport for agricultural products, particularly perishable items such as vegetables and seafood. Sea and air navigational aids have also been allowed to deteriorate. Foreign aid is commonly the only source of funding keeping important roads, sea ports and airports in operation. The failure to maintain roads, bridges, sea ports and airstrips is a constraint to market agricultural produce and to economic growth.

4 The review was publicly released on 20 September 2006 and is available from the ICCC website at <http://www.iccc.gov.pg/home.htm>.

Sources

ADB (Asian Development Bank) (2006). Papua New Guinea: preparing the Lae Port Development Project–Tidal Basin Phase I. ADB Technical Assistance Report Project Number: 40037. Asian Development Bank, Manila.

Allen, B. and Lowe, M. (2005). Papua New Guinea roads priority study: review of national roads in Papua New Guinea. Australian Agency for International Development, Canberra.

Benediktsson, K. (2002). *Harvesting Development: The Construction of Fresh Food Markets in Papua New Guinea*. Nordic Institute of Asian Studies, Denmark.

Crotty, T.M. (1970). History of road development in the Territory of Papua and New Guinea. *Search* 1(5):239–243.

Doulman, D.J. (1977). Benefit cost analysis and income distribution in developing countries: a review of the Rigo Road proposal. Discussion Paper No. 28. Economics Department, University of Papua New Guinea, Port Moresby.

DWI (Government of Papua New Guinea Department of Works and Implementation) (2001). PNG national road network road statistics 2000. Government of Papua New Guinea Department of Works and Implementation, Port Moresby.

Finnish National Road Administration and Pacific Asia Consultants Pty Ltd (2000). RAMS: Road Asset Management System non technical users manual. ADB TA No. 3004-PNG. Government of PNG and Asian Development Bank, Port Moresby.

NSO (National Statistical Office of Papua New Guinea) (2002). Papua New Guinea 2000 Census: Final Figures. National Statistical Office of Papua New Guinea, Port Moresby.

Sinclair, J. (1978). *Wings of Gold: How the Aeroplane Developed New Guinea*. Pacific Publications, Sydney.

Ward, M.W. (1970). *The Rigo Road: a study of the economic effects of new road construction*. New Guinea Research Bulletin No. 33. New Guinea Research Unit, The Australian National University, Canberra and Port Moresby.

6.10 Rural poverty

People are said to be in poverty when they do not have the necessities of daily living. Poverty is often associated with need, hardship and lack of resources across a wide range of circumstances. Poverty may involve a lack of sufficient essential goods and services, or a lack of wealth (including capital, cash, material goods and natural resources). Poverty is often associated with social exclusion, dependency, and an inability to live what is understood as a 'normal' life: for example, being unable to raise a healthy family or to educate children.

Poverty may be *absolute* or *relative*. Absolute poverty is measured against a set of standards that are consistent over time and between countries; relative poverty is measured against other groups of people in a particular social context. Poverty is measured by indices of consumption or of income. The measure may be developed by identifying the cost of a minimum dietary level or 'basket of goods', or it may be based on a level of income set at a certain proportion of median household incomes.

It is sometimes said in PNG that because everyone has access to land through customary landowning groups (see Section 6.1), as long as people are prepared to work hard, there is no reason for anyone to be hungry or poor. Therefore, this argument goes, people who are in poverty have only themselves to blame. Leaving aside the likelihood that not everyone in PNG has access to customary land (for example unmarried youths, women, migrants), there is a great deal of evidence, much of it presented elsewhere in this book, that a greater proportion of poor people live in rural areas than in towns. They commonly live in remote locations that are either mountainous, with high rainfall, high levels of cloud cover and poor soils, or live on flood plains that are inundated regularly, or on small islands. These poorest people have limited access to basic services, including health and education services and, importantly, to markets where they can sell agricultural produce in exchange for cash (see Section 1.14). They also have poor access to information and knowledge about important matters like health care, nutrition and political developments. They cannot afford to buy foods to supplement diets that are low in protein and oils. They are often poorly represented politically and are 'invisible' from the main centres of the country.

Poverty in PNG

Since the 1960s, a number of studies have attempted to define 'less developed' areas using measures of food consumed, cash incomes earned, levels of health and education services accessed and numbers of public servants in a district. These studies did not use the term 'poverty', but the definitions of 'development' and 'disadvantage' used are similar to contemporary definitions of poverty. The areas these studies have revealed as being 'less developed' or 'disadvantaged' are summarised here as a set of maps (Figures 6.10.1 to 6.10.4). The type and quality of data varied between these studies, as did the definition of 'disadvantaged'.

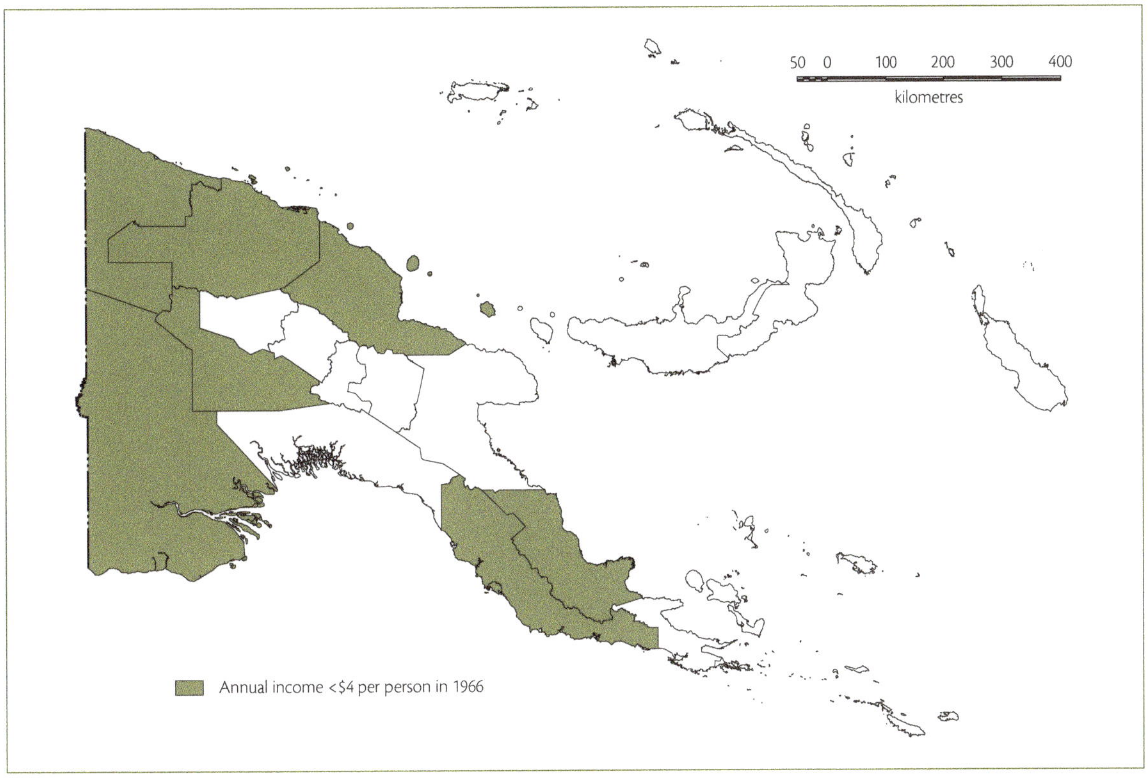

Figure 6.10.1 Provinces in which the average annual income from cash cropping was less than $4 per person in 1966. Source: Brookfield with Hart (1971:300)..

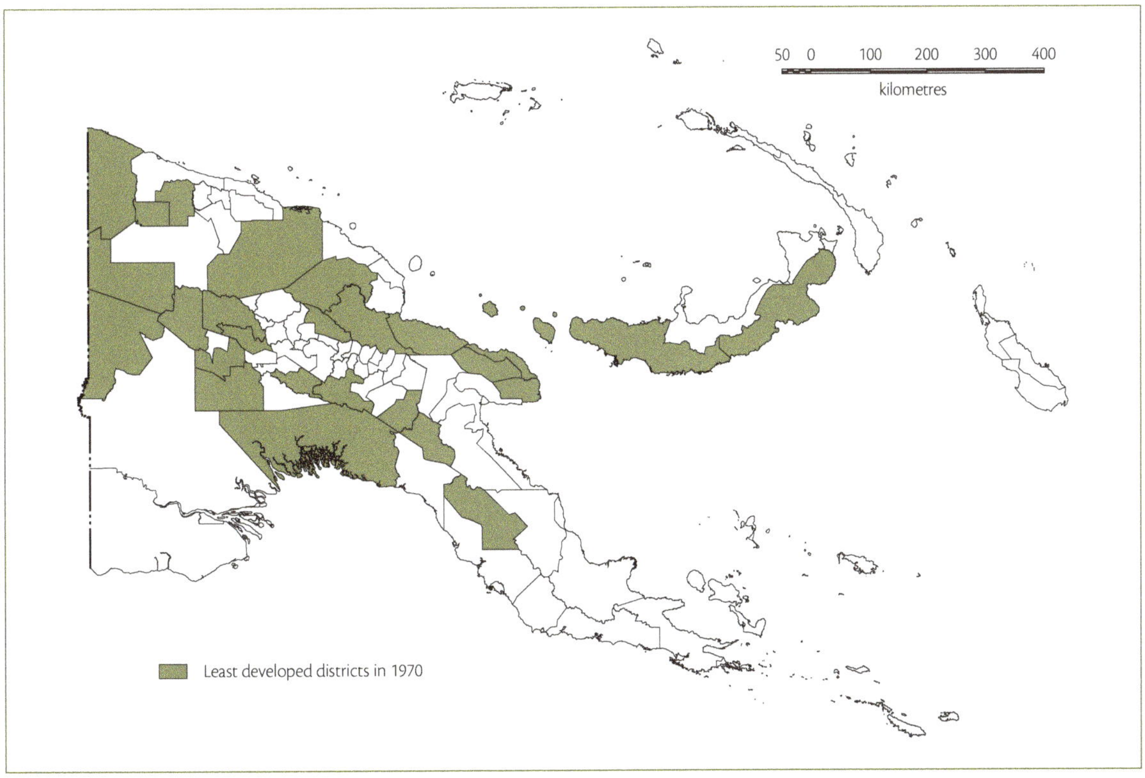

Figure 6.10.2 Least developed districts in 1970. Source: Wilson (1975, Groups 5 and 6).

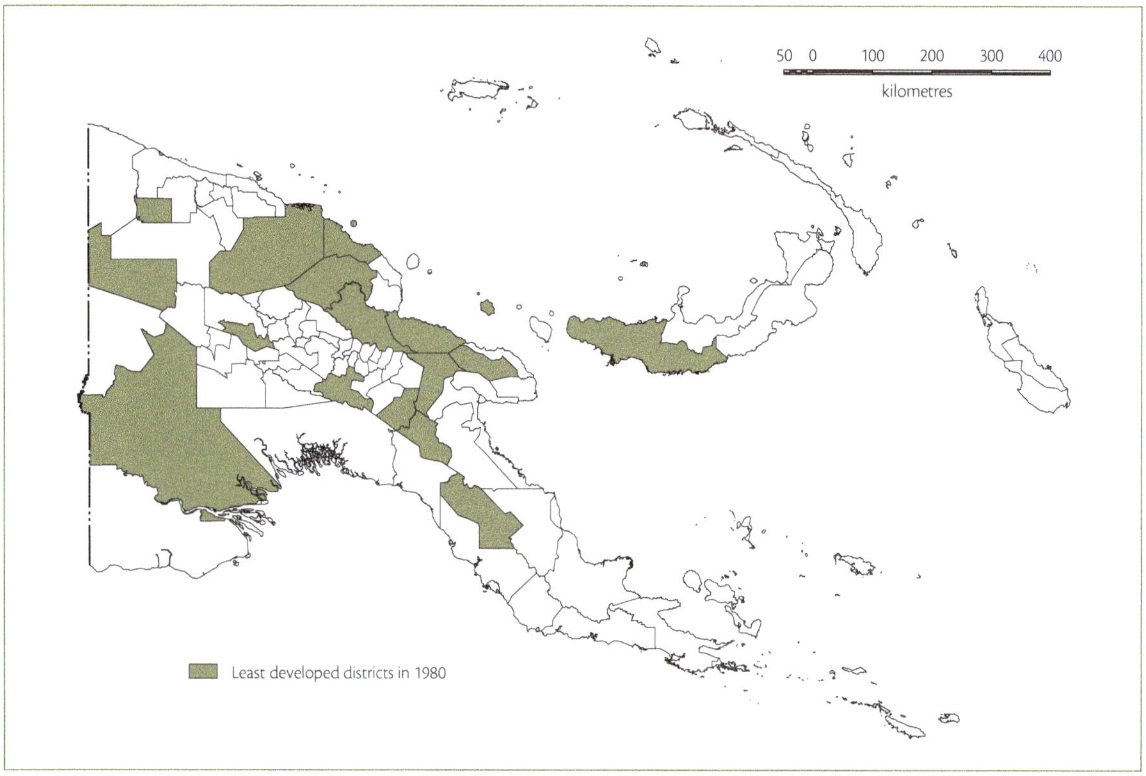

Figure 6.10.3 Least developed districts in 1980. Source: de Albuquerque and D'Sa (1986, Cluster 7).

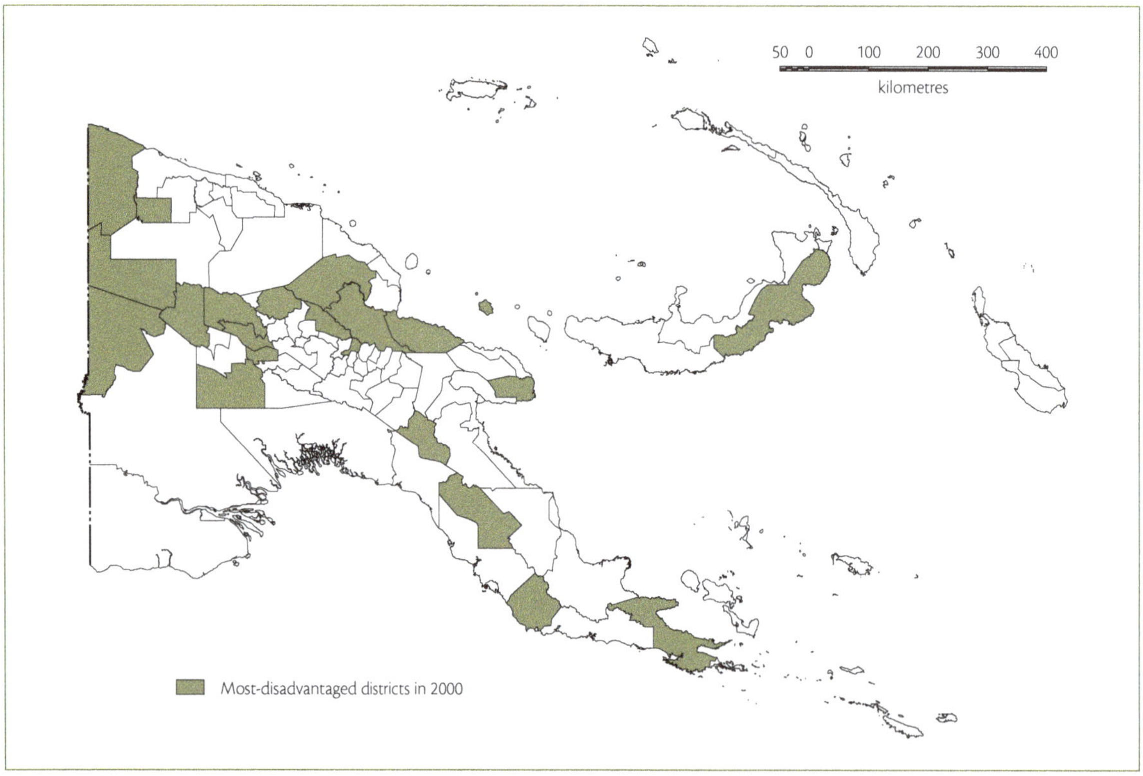

Figure 6.10.4 Most-disadvantaged districts in 2000. These worst 20 districts out of 85 are ranked by land potential, agricultural pressure, access to services, income from agriculture, and child malnutrition rate. Source: Hanson et al. (2001:310).

For example, in the *Papua New Guinea Rural Development Handbook* (2001:300), it was estimated that 18% of the rural population were 'strongly disadvantaged' or 'moderately disadvantaged'. This estimate of the proportion of people who are poor is not the same in all surveys. Nevertheless, the broad pattern of poverty in PNG was established by these four studies.

In 1996, the World Bank funded the PNG Household Survey, the first nationwide survey of consumption and living standards in PNG. The survey was based on a random sample of 1200 rural and urban households, selected by census unit from the 1990 census and stratified by environmental conditions (elevation and rainfall) and the level of agricultural development (from PNGRIS and MASP – see Section 1.15). The survey collected information about education, literacy, occupation, employment (but not income levels), dwelling characteristics, agricultural assets, and inputs. It also collected data on all food consumed and other frequent expenses. The expenditure estimates include the monetary value of subsistence production, gifts of food received, and food held in the house, as measures of consumption. The survey also collected information on infrequent expenditure and durable items owned. Poverty lines were established for the five regions of PNG – National Capital District, Southern, Highlands, Momase and Islands (see map on page xix). As a result, it was estimated that 42% of PNG's population lived in households in which the value of consumption per adult equivalent was below a poverty line. Of these households, 94% were in rural areas, making it clear that poverty in PNG is a rural problem. Statistical techniques were used to combine information from the sample household survey with the 2000 census and PNGRIS and MASP, to create disaggregated maps of predicted poverty in all districts in the country (with a known level of probable error) (Figure 6.10.5).

The location of poor places

The patterns of 'poverty' depicted in Figures 6.10.1 to 6.10.5 are broadly similar over the 40 years they cover. The poorest areas are located along the western end of PNG, along both sides

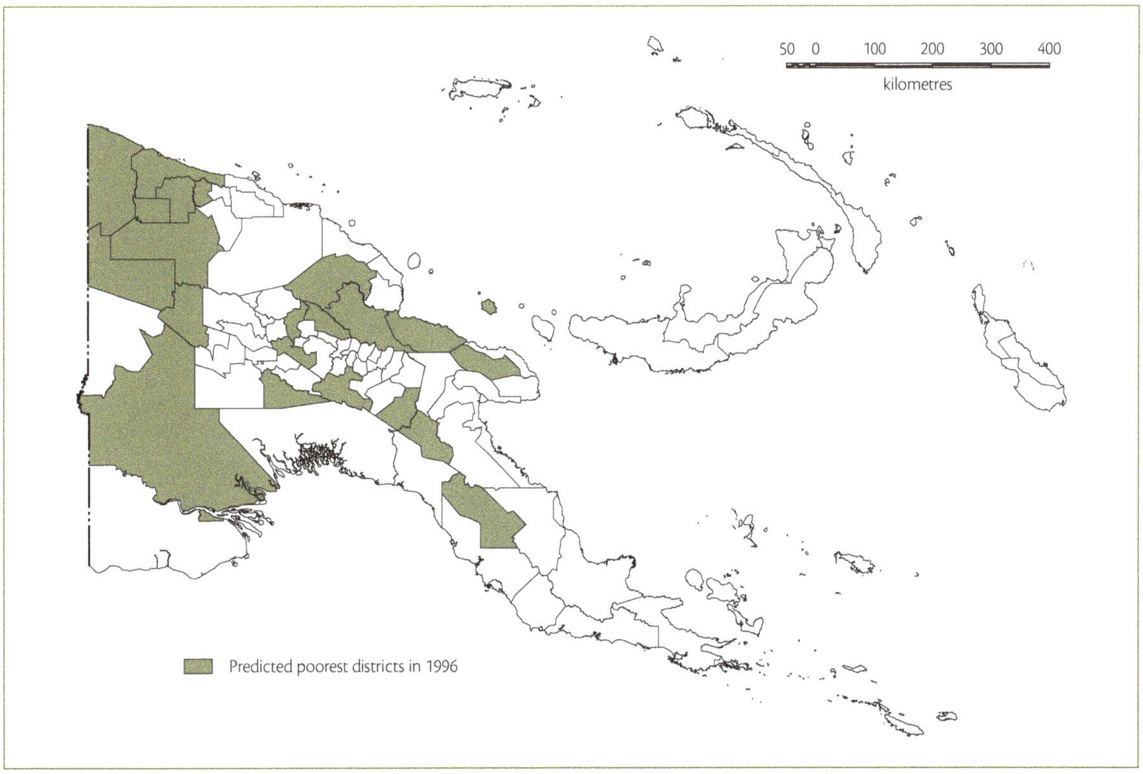

Figure 6.10.5 Predicted poorest districts in 1996 (% of district population; worst 19 districts out of 85).
Source: Gibson pers. comm. (2004).

of the highlands and down the length of the main mountain range, on the Finisterre and Sarawaget mountains and in inland New Britain. These areas are often located on mountains and are high and therefore relatively cold and wet, steep and subject to erosion, and are isolated from roads and urban centres. Some are also located in inland lowland areas, or on places that are flooded annually, or on small islands.

Places with these environmental conditions have probably always been 'poorer' than places with better environments. Evidence from nutrition surveys and other studies carried out in the 1950s and 1960s suggests that in many parts of PNG, *before* colonisation, subsistence food production and diets were such that people in many places were 'vulnerable' to higher rates of diseases and death. After PNG was colonised, these places did not attract investment in the form of infrastructure, plantations or village cash cropping. Where village cash cropping provided cash incomes, the 'vulnerable' pre-colonial diets were able to be supplemented with purchased, imported food that is high in protein, fats and oils. This has led to improved human nutrition, lower infant and child mortality, and higher population growth rates.

Colonisation resulted in the environmentally poorer areas of PNG being left behind, while the more favoured areas became 'developed'. Colonial administrators, planters and missionaries were drawn towards the higher-potential land, with its higher population densities and easier access, and it was here that 'development' occurred, in the form of infrastructure construction (towns, roads, wharves) and plantation agriculture. Villagers were able to take advantage of these developments and rapidly adopted cash cropping. These are now the districts with relatively high personal cash incomes from cash cropping.

For these reasons, the eradication of poverty will not be easy or simple. It will require separate policies for economic growth and for poverty eradication, which take into account the environmental realities of the poorer areas. A World Bank assessment of poverty in PNG published in 2004 found that poverty levels have probably worsened as a result of a severe contraction in the PNG economy. The assessment also found that education services have not improved, health services have deteriorated, and almost every socioeconomic indicator is significantly worse in rural areas. It urges the use of poverty mapping to allow the identification and specific targeting of the poorest places and the urgent restoration of education and health services in rural areas.

Sources

Allen, B.J., Bourke, R.M. and Gibson, J. (2005). Poor rural places in Papua New Guinea. *Asia Pacific Viewpoint* 46(2):201–217.

Brookfield, H.C. and Hart, D. (1971). *Melanesia: A Geographical Interpretation of an Island World.* Methuen & Co Ltd, London.

de Albuquerque, K. and D'Sa, E. (1986). *Spatial inequalities in Papua New Guinea: a district-level analysis.* IASER Discussion Paper No. 49. Institute of Applied Social and Economic Research, Boroko, Papua New Guinea.

Gibson, J. (2000). The Papua New Guinea household survey. *The Australian Economic Review* 33(4):377–380.

Gibson, J. and Rozelle, S. (2003). Poverty and access to roads in Papua New Guinea. *Economic Development and Cultural Change* 52(1):159–185.

Gibson, J., Datt, G., Allen, B., Hwang, V., Bourke, R.M. and Parajuli, D. (2005). Mapping poverty in rural Papua New Guinea. *Pacific Economic Bulletin* 20(1):27–43.

Hanson, L.W., Allen, B.J., Bourke, R.M. and McCarthy, T.J. (2001). *Papua New Guinea Rural Development Handbook.* Land Management Group, Department of Human Geography, Research School of Pacific and Asian Studies, The Australian National University, Canberra.

Heywood, P.F. and Hide, R.L. (1994). Nutritional effects of export crop production in Papua New Guinea: a review of the evidence. *Food and Nutrition Bulletin* 15(3):233–249.

Wilson, R.K. (1975). Socio-economic indicators applied to sub-districts of Papua New Guinea. *Yagl-Ambu* 2(1):71–87.

World Bank (2004). Papua New Guinea: poverty assessment. World Bank, Washington.

Appendix Tables

Appendix Table A1.2.1 Cultivated land area[a] by PNGRIS land use intensity class and province, 1975 (km^2)

Province	PNGRIS land use intensity class												Total area of cultivated land
	Extremely low and very low	%	Low	%	Moderate	%	High	%	Very high	%	Very high with tree crops	%	
Western	5,174	65.2	2,271	28.6	438	5.5	0	0.0	0	0.0	48	0.6	7,931
Gulf	2,919	76.8	546	14.4	321	8.4	0	0.0	0	0.0	15	0.4	3,801
Central	1,534	23.9	3,288	51.3	1,317	20.6	1	0.0	0	0.0	267	4.2	6,407
Milne Bay	1,711	30.1	3,408	59.9	405	7.1	92	1.6	0	0.0	75	1.3	5,691
Oro	2,219	52.1	1,044	24.5	742	17.4	149	3.5	0	0.0	104	2.4	4,258
Southern Highlands	983	13.8	2,419	34.0	2,002	28.1	1,576	22.2	132	1.9	0	0.0	7,112
Enga	469	12.5	318	8.5	1,419	37.9	1,257	33.5	286	7.6	0	0.0	3,749
Western Highlands	347	7.7	1,320	29.5	918	20.5	1,612	36.0	123	2.7	162	3.6	4,482
Simbu	286	11.4	503	20.0	479	19.0	175	7.0	1,072	42.6	0	0.0	2,515
Eastern Highlands	1,229	22.2	1,087	19.6	2,587	46.7	619	11.2	0	0.0	17	0.3	5,539
Morobe	5,938	48.5	4,757	38.8	1,236	10.1	132	1.1	0	0.0	182	1.5	12,245
Madang	9,870	61.5	4,762	29.7	1,007	6.3	148	0.9	0	0.0	259	1.6	16,046
East Sepik	6,122	68.1	1,697	18.9	667	7.4	370	4.1	98	1.1	38	0.4	8,992
Sandaun	5,181	63.1	2,330	28.4	596	7.3	83	1.0	0	0.0	21	0.3	8,211
Manus	1,544	88.4	136	7.8	27	1.5	0	0.0	0	0.0	40	2.3	1,747
New Ireland	3,100	68.4	992	21.9	25	0.6	0	0.0	0	0.0	414	9.1	4,531
East New Britain	1,625	43.1	1,093	29.0	321	8.5	0	0.0	0	0.0	733	19.4	3,772
West New Britain	4,747	83.6	522	9.2	45	0.8	23	0.4	0	0.0	339	6.0	5,676
Bougainville	2,625	50.9	1,622	31.5	739	14.3	0	0.0	0	0.0	167	3.2	5,153
Papua New Guinea	57,623	48.9	34,115	28.9	15,291	13.0	6,237	5.3	1,711	1.5	2,881	2.4	117,858

[a] Everywhere else in this book we use the term 'land used for agriculture' to refer to the CSIRO class 'cultivated land'.

Source: PNGRIS.

Appendix Table A1.2.2 Cultivated land area by MASP land use intensity class and province, 1995 (km²)

Province	Land use intensity class (R-value)										Total area of cultivated land
	Very low	%	Low	%	Medium	%	High	%	Very high	%	
Western	19,320	91.1	0	0.0	1,886	8.9	0	0.0	0	0.0	21,206
Gulf	9,070	99.2	70	0.8	0	0.0	0	0.0	0	0.0	9,140
Central	1,334	28.6	2,338	50.1	953	20.4	46	1.0	0	0.0	4,670
Milne Bay	3,438	77.2	995	22.3	19	0.4	0	0.0	0	0.0	4,452
Oro	1,848	60.4	1,213	39.6	0	0.0	0	0.0	0	0.0	3,061
Southern Highlands	1,577	32.0	1,035	21.0	1,755	35.7	264	5.4	290	5.9	4,921
Enga	1,085	43.8	307	12.4	154	6.2	0	0.0	931	37.6	2,477
Western Highlands	988	33.4	1,043	35.3	792	26.8	0	0.0	134	4.5	2,958
Simbu	640	36.7	690	39.6	412	23.7	0	0.0	0	0.0	1,742
Eastern Highlands	1,353	34.3	1,453	36.9	906	23.0	188	4.8	42	1.1	3,942
Morobe	3,302	41.7	3,115	39.4	1,496	18.9	0	0.0	2	0.0	7,915
Madang	8,583	78.1	2,264	20.6	137	1.2	0	0.0	0	0.0	10,984
East Sepik	11,608	70.2	4,887	29.6	34	0.2	0	0.0	0	0.0	16,529
Sandaun	10,195	100.0	0	0.0	0	0.0	0	0.0	0	0.0	10,195
Manus	1,128	95.2	54	4.6	3	0.2	0	0.0	0	0.0	1,185
New Ireland	2,022	71.4	798	28.2	13	0.5	0	0.0	0	0.0	2,833
East New Britain	1,632	71.1	239	10.4	0	0.0	424	18.5	0	0.0	2,295
West New Britain	3,463	87.9	417	10.6	60	1.5	0	0.0	0	0.0	3,940
Bougainville	585	17.1	2,814	82.4	4	0.1	0	0.0	11	0.3	3,414
Papua New Guinea	83,171	70.6	23,732	20.1	8,623	7.3	922	0.8	1,409	1.2	117,858

Source: MASP.

Appendix Table A1.2.3 Number of rural people living on cultivated land by MASP land use intensity class and province

Province	Land use intensity class (R-value)										Total rural population
	Very low	%	Low	%	Medium	%	High	%	Very high	%	
Western	95,067	88.2	0	0.0	12,770	11.8	0	0.0	0	0.0	107,837
Gulf	88,606	96.0	3,659	4.0	0	0.0	0	0.0	0	0.0	92,265
Central	27,625	17.6	83,860	53.4	45,573	29.0	0	0.0	0	0.0	157,058
Milne Bay	119,435	63.4	66,094	35.1	2,805	1.5	0	0.0	0	0.0	188,334
Oro	40,910	38.5	65,378	61.5	0	0.0	0	0.0	0	0.0	106,288
Southern Highlands	83,256	15.8	99,989	19.0	216,433	41.1	60,427	11.5	66,294	12.6	526,398
Enga	49,993	17.6	28,145	9.9	15,314	5.4	0	0.0	190,045	67.0	283,498
Western Highlands	47,700	12.9	121,744	32.8	178,943	48.2	0	0.0	22,627	6.1	371,014
Simbu	30,169	12.4	126,442	52.1	86,137	35.5	0	0.0	0	0.0	242,748
Eastern Highlands	81,044	20.6	148,142	37.7	109,194	27.8	39,117	9.9	15,922	4.0	393,418
Morobe	110,958	31.2	181,108	50.9	63,044	17.7	0	0.0	989	0.3	356,100
Madang	204,241	66.3	89,530	29.1	14,364	4.7	0	0.0	0	0.0	308,135
East Sepik	214,329	70.6	75,824	25.0	13,553	4.5	0	0.0	0	0.0	303,706
Sandaun	166,919	100.0	0	0.0	0	0.0	0	0.0	0	0.0	166,919
Manus	28,653	82.1	5,237	15.0	1,009	2.9	0	0.0	0	0.0	34,899
New Ireland	66,412	64.3	33,445	32.4	3,402	3.3	0	0.0	0	0.0	103,259
East New Britain	56,314	32.3	2,434	1.4	0	0.0	115,482	66.3	0	0.0	174,230
West New Britain	67,051	61.3	28,216	25.8	14,032	12.8	0	0.0	0	0.0	109,299
Bougainville	28,488	17.0	133,425	79.8	2,461	1.5	0	0.0	2,782	1.7	167,156
Papua New Guinea	1,607,170	38.3	1,292,672	30.8	779,034	18.6	215,025	5.1	298,660	7.1	4,192,561

Sources: NSO (2002); MASP.

Appendix Table A1.2.4 Cultivated land area by population density class and province, 1995 (km²)

Province	Population density class (persons/km²)										Total area of cultivated land
	Very low <10	%	Low 10–30	%	Medium 31–50	%	High 51–100	%	Very high >100	%	
Western	20,974	98.9	232	1.1	0	0.0	0	0.0	0	0.0	21,206
Gulf	8,866	97.0	235	2.6	36	0.4	0	0.0	3	0.0	9,140
Central	1,730	37.0	2,808	60.1	60	1.3	61	1.3	12	0.2	4,670
Milne Bay	1,299	29.2	2,640	59.3	161	3.6	349	7.8	4	0.1	4,452
Oro	1,034	33.8	2,027	66.2	0	0.0	0	0.0	0	0.0	3,061
Southern Highlands	592	12.0	2,262	46.0	1,617	32.9	450	9.1	0	0.0	4,921
Enga	116	4.7	960	38.8	876	35.4	308	12.4	217	8.7	2,477
Western Highlands	267	9.0	1,045	35.3	20	0.7	1,282	43.3	344	11.6	2,958
Simbu	91	5.2	474	27.2	35	2.0	569	32.6	573	32.9	1,742
Eastern Highlands	504	12.8	942	23.9	778	19.7	1,571	39.9	147	3.7	3,942
Morobe	1,504	19.0	5,276	66.7	828	10.5	305	3.8	2	0.0	7,915
Madang	7,181	65.4	3,277	29.8	238	2.2	287	2.6	2	0.0	10,984
East Sepik	14,140	85.5	1,075	6.5	693	4.2	565	3.4	56	0.3	16,529
Sandaun	7,641	74.9	2,411	23.6	79	0.8	64	0.6	0	0.0	10,195
Manus	675	57.0	466	39.3	24	2.1	11	0.9	9	0.7	1,185
New Ireland	1,052	37.1	1,736	61.3	0	0.0	34	1.2	11	0.4	2,833
East New Britain	711	31.0	915	39.9	130	5.7	494	21.5	45	2.0	2,295
West New Britain	2,504	63.5	1,172	29.8	22	0.6	192	4.9	49	1.3	3,940
Bougainville	575	16.8	2,799	82.0	0	0.0	37	1.1	4	0.1	3,414
Papua New Guinea	71,060	60.3	32,975	28.0	5,625	4.8	6,648	5.6	1,550	1.3	117,858

Note: The data in the MASP database was mainly recorded in 1990–1995. The population figures used to generate population densities are derived from the 2000 census.

Sources: NSO (2002); MASP.

Appendix Table A1.3.1 Total land area by population density class and province (km²)

Province	Population density class (persons/km²)												Total land area
	No population	%	Very low <10	%	Low 10–30	%	Medium 31–50	%	High 51–100	%	Very high >100	%	
Western	69,423	71.5	27,340	28.2	303	0.3	0	0.0	0	0.0	0	0.0	97,065
Gulf	21,773	64.3	11,712	34.6	310	0.9	47	0.1	0	0.0	4	0.0	33,847
Central	23,962	80.0	2,220	7.4	3,603	12.0	76	0.3	78	0.3	15	0.0	29,954
Milne Bay	8,174	57.9	1,616	11.4	3,696	26.2	200	1.4	434	3.1	5	0.0	14,125
Oro	18,603	82.6	1,319	5.9	2,588	11.5	0	0.0	0	0.0	0	0.0	22,510
Southern Highlands	19,327	75.2	764	3.0	2,941	11.4	2,086	8.1	580	2.3	0	0.0	25,698
Enga	8,617	72.8	151	1.3	1,249	10.6	1,139	9.6	401	3.4	282	2.4	11,839
Western Highlands	4,925	55.4	358	4.0	1,403	15.8	27	0.3	1,721	19.3	462	5.2	8,897
Simbu	3,710	61.6	121	2.0	629	10.4	46	0.8	755	12.5	761	12.6	6,022
Eastern Highlands	5,676	51.6	681	6.2	1,274	11.6	1,052	9.6	2,125	19.3	199	1.8	11,006
Morobe	23,160	69.1	1,777	5.3	7,247	21.6	979	2.9	360	1.1	3	0.0	33,525
Madang	13,800	48.0	7,764	27.0	6,599	23.0	257	0.9	310	1.1	2	0.0	28,732
East Sepik	21,252	48.6	15,317	35.0	5,728	13.1	751	1.7	612	1.4	61	0.1	43,720
Sandaun	22,495	62.5	10,129	28.1	3,196	8.9	105	0.3	85	0.2	0	0.0	36,010
Manus	402	19.2	967	46.1	666	31.8	35	1.7	15	0.7	13	0.6	2,098
New Ireland	7,253	75.4	1,501	15.6	703	7.3	13	0.1	115	1.2	30	0.3	9,615
East New Britain	12,168	80.5	712	4.7	1,558	10.3	130	0.9	495	3.3	45	0.3	15,109
West New Britain	12,622	60.8	3,020	14.6	4,982	24.0	0	0.0	97	0.5	31	0.1	20,753
Bougainville	4,703	50.4	779	8.3	3,792	40.6	0	0.0	50	0.5	6	0.1	9,329
Papua New Guinea	302,046	65.7	88,248	19.2	52,467	11.4	6,944	1.5	8,234	1.8	1,916	0.4	459,854

Note: The difference between the total land area and the area with no population is 157 808 km². This is greater than the 117 858 km² of 'land used for agriculture' (see Section 1.2) because areas in which sago is the staple crop are excluded from the 'land used for agriculture' as sago is not cultivated in gardens.

Sources: McAlpine and Quigley (c. 1995); NSO (2002); MASP.

Appendix Table A1.3.2 Rural population by population density class and province

Province	Population density class (persons/km^2)										Total rural population
	Very low <10	%	Low 10–30	%	Medium 31–50	%	High 51–100	%	Very high >100	%	
Western	94,951	88.1	0	0.0	12,886	11.9	0	0.0	0	0.0	107,837
Gulf	87,557	94.9	4,708	5.1	0	0.0	0	0.0	0	0.0	92,265
Central	27,677	17.6	83,841	53.4	45,540	29.0	0	0.0	0	0.0	157,058
Milne Bay	118,505	62.9	67,012	35.6	2,817	1.5	0	0.0	0	0.0	188,334
Oro	40,855	38.4	65,433	61.6	0	0.0	0	0.0	0	0.0	106,288
Southern Highlands	83,153	15.8	99,842	19.0	216,212	41.1	93,207	17.7	33,983	6.5	526,398
Enga	49,766	17.6	28,181	9.9	15,293	5.4	103,098	36.4	87,160	30.7	283,498
Western Highlands	47,660	12.8	121,782	32.8	178,900	48.2	0	0.0	22,672	6.1	371,014
Simbu	30,296	12.5	126,289	52.0	86,163	35.5	0	0.0	0	0.0	242,748
Eastern Highlands	81,324	20.7	147,984	37.6	108,862	27.7	39,255	10.0	15,993	4.1	393,418
Morobe	94,448	26.5	198,750	55.8	61,909	17.4	0	0.0	994	0.3	356,100
Madang	194,362	63.1	94,608	30.7	19,166	6.2	0	0.0	0	0.0	308,135
East Sepik	210,432	69.3	79,422	26.2	13,851	4.6	0	0.0	0	0.0	303,706
Sandaun	166,919	100.0	0	0.0	0	0.0	0	0.0	0	0.0	166,919
Manus	28,693	82.2	5,198	14.9	1,008	2.9	0	0.0	0	0.0	34,899
New Ireland	66,420	64.3	33,498	32.4	3,341	3.2	0	0.0	0	0.0	103,259
East New Britain	56,476	32.4	2,347	1.3	0	0.0	115,407	66.2	0	0.0	174,230
West New Britain	67,041	61.3	28,188	25.8	14,069	12.9	0	0.0	0	0.0	109,299
Bougainville	28,444	17.0	133,507	79.9	2,445	1.5	0	0.0	2,761	1.7	167,156
Papua New Guinea	1,606,774	38.3	1,336,765	31.9	766,724	18.3	333,793	8.0	148,504	3.5	4,192,561

Sources: NSO (2002); MASP.

Appendix Table A1.9.1 Distribution of USDA Soil Taxonomy soil orders by land area and province (km^2)

Province	Entisols	%	Histosols	%	Inceptisols and Andisols	%	Vertisols	%	Mollisols	%	Alfisols	%	Ultisols	%	Oxisols	%	Lakes	%	Total land area
Western	26,344	27.1	0	0.0	13,310	13.7	0	0.0	1,631	1.7	0	0.0	55,001	56.7	0	0.0	779	0.8	97,065
Gulf	12,244	36.2	616	1.8	14,234	42.1	0	0.0	5,534	16.4	135	0.4	1,084	3.2	0	0.0	0	0.0	33,847
Central	9,668	32.3	22	0.1	19,177	64.0	22	0.1	181	0.6	838	2.8	10	0.0	0	0.0	36	0.1	29,954
Milne Bay	2,170	15.4	33	0.2	9,503	67.3	0	0.0	1,692	12.0	630	4.5	26	0.2	71	0.5	0	0.0	14,125
Oro	6,104	27.1	1,213	5.4	13,673	60.7	0	0.0	1,346	6.0	135	0.6	39	0.2	0	0.0	0	0.0	22,510
Southern Highlands	5,265	20.5	297	1.2	14,021	54.6	0	0.0	6,057	23.6	0	0.0	0	0.0	0	0.0	58	0.2	25,698
Enga	2,578	21.8	1,705	14.4	6,683	56.4	0	0.0	873	7.4	0	0.0	0	0.0	0	0.0	0	0.0	11,839
Western Highlands	288	3.2	634	7.1	7,621	85.7	0	0.0	39	0.4	315	3.5	0	0.0	0	0.0	0	0.0	8,897
Simbu	50	0.8	376	6.2	5,052	83.9	0	0.0	544	9.0	0	0.0	0	0.0	0	0.0	0	0.0	6,022
Eastern Highlands	650	5.9	78	0.7	9,191	83.5	0	0.0	164	1.5	697	6.3	226	2.1	0	0.0	0	0.0	11,006
Morobe	11,586	34.6	0	0.0	18,276	54.5	0	0.0	3,316	9.9	333	1.0	0	0.0	0	0.0	14	0.0	33,525
Madang	7,659	26.7	0	0.0	16,290	56.7	80	0.3	2,366	8.2	1,932	6.7	334	1.2	0	0.0	71	0.2	28,732
East Sepik	19,072	43.6	2,891	6.6	13,122	30.0	153	0.3	703	1.6	3,991	9.1	3,584	8.2	0	0.0	204	0.5	43,720
Sandaun	9,724	27.0	14	0.0	21,604	60.0	0	0.0	2,116	5.9	343	1.0	2,209	6.1	0	0.0	0	0.0	36,010
Manus	311	14.8	0	0.0	1,589	75.7	0	0.0	174	8.3	0	0.0	24	1.1	0	0.0	0	0.0	2,098
New Ireland	513	5.3	0	0.0	4,464	46.4	0	0.0	1,931	20.1	2,707	28.2	0	0.0	0	0.0	0	0.0	9,615
East New Britain	1,236	8.2	0	0.0	8,191	54.2	0	0.0	5,295	35.0	387	2.6	0	0.0	0	0.0	0	0.0	15,109
West New Britain	3,367	16.2	0	0.0	17,035	82.1	0	0.0	230	1.1	64	0.3	0	0.0	0	0.0	57	0.3	20,753
Bougainville	1,728	18.5	696	7.5	5,869	62.9	0	0.0	173	1.9	863	9.3	0	0.0	0	0.0	0	0.0	9,329
Papua New Guinea	120,557	26.2	8,575	1.9	218,905	47.6	255	0.1	34,365	7.5	13,370	2.9	62,537	13.6	71	0.0	1,219	0.3	459,854

Source: PNGRIS.

Appendix Table A1.9.2 Distribution of USDA Soil Taxonomy soil orders by rural population and province

Province	Entisols	%	Histosols	%	Inceptisols and Andisols	%	Vertisols	%	Mollisols	%	Alfisols	%	Ultisols	%	Oxisols	%	Total rural population
Western	45,681	42.4	0	0.0	26,914	25.0	0	0.0	709	0.7	0	0.0	34,532	32.0	0	0.0	107,837
Gulf	48,128	52.2	2,360	2.6	29,768	32.3	0	0.0	11,647	12.6	226	0.2	136	0.1	0	0.0	92,265
Central	123,455	78.6	0	0.0	30,007	19.1	727	0.5	1,965	1.3	905	0.6	0	0.0	0	0.0	157,058
Milne Bay	36,397	19.3	484	0.3	96,352	51.2	0	0.0	44,667	23.7	9,488	5.0	0	0.0	947	0.5	188,334
Oro	41,947	39.5	1,526	1.4	56,996	53.6	0	0.0	5,773	5.4	46	0.0	0	0.0	0	0.0	106,288
Southern Highlands	91,234	17.3	20,750	3.9	386,061	73.3	0	0.0	28,354	5.4	0	0.0	0	0.0	0	0.0	526,398
Enga	27,874	9.8	9,393	3.3	237,827	83.9	0	0.0	8,404	3.0	0	0.0	0	0.0	0	0.0	283,498
Western Highlands	4,767	1.3	0	0.0	316,987	85.4	0	0.0	0	0.0	49,260	13.3	0	0.0	0	0.0	371,014
Simbu	0	0.0	387	0.2	236,718	97.5	0	0.0	5,643	2.3	0	0.0	0	0.0	0	0.0	242,748
Eastern Highlands	25,344	6.4	0	0.0	259,279	65.9	0	0.0	188	0.0	42,141	10.7	66,466	16.9	0	0.0	393,418
Morobe	163,405	45.9	0	0.0	128,601	36.1	0	0.0	50,348	14.1	13,745	3.9	0	0.0	0	0.0	356,100
Madang	92,304	30.0	0	0.0	168,824	54.8	3,167	1.0	32,704	10.6	10,920	3.5	215	0.1	0	0.0	308,135
East Sepik	73,181	24.1	23,668	7.8	35,418	11.7	0	0.0	49,539	16.3	99,675	32.8	22,226	7.3	0	0.0	303,706
Sandaun	27,426	16.4	0	0.0	112,726	67.5	0	0.0	18,614	11.2	1,361	0.8	6,793	4.1	0	0.0	166,919
Manus	5,471	15.7	0	0.0	13,767	39.4	0	0.0	15,661	44.9	0	0.0	0	0.0	0	0.0	34,899
New Ireland	8,503	8.2	0	0.0	15,040	14.6	0	0.0	6,035	5.8	73,681	71.4	0	0.0	0	0.0	103,259
East New Britain	23,021	13.2	0	0.0	114,319	65.6	0	0.0	20,550	11.8	16,340	9.4	0	0.0	0	0.0	174,230
West New Britain	8,962	8.2	0	0.0	95,301	87.2	0	0.0	1,432	1.3	3,604	3.3	0	0.0	0	0.0	109,299
Bougainville	6,981	4.2	1,724	1.0	113,108	67.7	0	0.0	16,148	9.7	29,196	17.5	0	0.0	0	0.0	167,156
Papua New Guinea	854,082	20.4	60,292	1.4	2,474,011	59.0	3,895	0.1	318,380	7.6	350,588	8.4	130,368	3.1	947	0.0	4,192,561

Sources: NSO (2002); PNGRIS.

Appendix Table A1.10.1 Total land area by landform and province (km^2)

Province	Mountains and hills	%	Volcanic landforms	%	Plains and plateaus	%	Floodplains	%	Raised coral reefs and littoral areas	%	Total land area
Western	8,637	8.9	3,676	3.8	58,520	60.3	23,547	24.3	2,685	2.8	97,065
Gulf	17,337	51.2	1,425	4.2	3,802	11.2	8,005	23.7	3,278	9.7	33,847
Central	23,336	77.9	635	2.1	236	0.8	4,503	15.0	1,244	4.2	29,954
Milne Bay	9,752	69.0	357	2.5	1,158	8.2	1,007	7.1	1,851	13.1	14,125
Oro	11,190	49.7	4,721	21.0	2,218	9.9	4,117	18.3	264	1.2	22,510
Southern Highlands	16,803	65.4	7,513	29.2	894	3.5	488	1.9	0	0.0	25,698
Enga	10,723	90.6	485	4.1	332	2.8	299	2.5	0	0.0	11,839
Western Highlands	6,075	68.3	1,832	20.6	625	7.0	365	4.1	0	0.0	8,897
Simbu	4,728	78.5	1,210	20.1	74	1.2	10	0.2	0	0.0	6,022
Eastern Highlands	9,950	90.4	405	3.7	512	4.7	139	1.3	0	0.0	11,006
Morobe	25,716	76.7	950	2.8	4,854	14.5	1,513	4.5	492	1.5	33,525
Madang	19,903	69.3	885	3.1	2,369	8.2	4,988	17.4	587	2.0	28,732
East Sepik	17,238	39.4	50	0.1	4,862	11.1	20,619	47.2	951	2.2	43,720
Sandaun	23,995	66.6	0	0.0	3,402	9.4	8,374	23.3	239	0.7	36,010
Manus	1,078	51.4	535	25.5	0	0.0	141	6.7	344	16.4	2,098
New Ireland	5,531	57.5	442	4.6	1,594	16.6	0	0.0	2,048	21.3	9,615
East New Britain	12,536	83.0	939	6.2	112	0.7	818	5.4	704	4.7	15,109
West New Britain	11,711	56.4	5,107	24.6	88	0.4	1,944	9.4	1,903	9.2	20,753
Bougainville	1,789	19.2	5,567	59.7	281	3.0	679	7.3	1,013	10.9	9,329
Papua New Guinea	238,028	51.8	36,734	8.0	85,933	18.7	81,556	17.7	17,603	3.8	459,854

Sources: McAlpine and Quigley (c. 1995); PNGRIS.

Appendix Table A1.10.2 Land area used for agriculture by landform and province (km^2)

Province	Mountains and hills	%	Volcanic landforms	%	Plains and plateaus	%	Floodplains	%	Raised coral reefs and littoral areas	%	Total land used for agriculture
Western	2,100	26.5	531	6.7	4,698	59.2	537	6.8	65	0.8	7,931
Gulf	2,852	75.0	37	1.0	168	4.4	588	15.5	156	4.1	3,801
Central	4,458	69.6	176	2.7	31	0.5	1,379	21.5	363	5.7	6,407
Milne Bay	3,378	59.4	254	4.5	545	9.6	683	12.0	831	14.6	5,691
Oro	1,015	23.8	1,966	46.2	553	13.0	677	15.9	47	1.1	4,258
Southern Highlands	4,433	62.3	2,293	32.2	276	3.9	110	1.5	0	0.0	7,112
Enga	3,356	89.5	84	2.2	233	6.2	76	2.0	0	0.0	3,749
Western Highlands	2,635	58.8	999	22.3	583	13.0	265	5.9	0	0.0	4,482
Simbu	2,152	85.6	298	11.8	55	2.2	10	0.4	0	0.0	2,515
Eastern Highlands	5,044	91.1	56	1.0	313	5.7	126	2.3	0	0.0	5,539
Morobe	9,338	76.3	317	2.6	1,968	16.1	431	3.5	191	1.6	12,245
Madang	12,261	76.4	379	2.4	1,467	9.1	1,412	8.8	527	3.3	16,046
East Sepik	5,857	65.1	47	0.5	1,011	11.2	1,939	21.6	138	1.5	8,992
Sandaun	6,268	76.3	0	0.0	673	8.2	1,133	13.8	137	1.7	8,211
Manus	1,002	57.4	511	29.3	0	0.0	40	2.3	194	11.1	1,747
New Ireland	1,993	44.0	397	8.8	627	13.8	0	0.0	1,514	33.4	4,531
East New Britain	2,188	58.0	739	19.6	22	0.6	379	10.0	444	11.8	3,772
West New Britain	2,499	44.0	1,337	23.6	31	0.5	409	7.2	1,400	24.7	5,676
Bougainville	1,033	20.0	3,121	60.6	145	2.8	83	1.6	771	15.0	5,153
Papua New Guinea	73,862	62.7	13,542	11.5	13,399	11.4	10,277	8.7	6,778	5.8	117,858

Sources: McAlpine and Quigley (c. 1995); PNGRIS.

Appendix Table A1.10.3 Total population (including urban and rural non-village populations) by landform and province

Province	Mountains and hills	%	Volcanic landforms	%	Plains and plateaus	%	Floodplains	%	Raised coral reefs and littoral areas	%	Total population
Western	29,801	19.4	6,928	4.5	52,505	34.2	49,909	32.6	14,161	9.2	153,304
Gulf	36,919	34.5	0	0.0	1,611	1.5	17,176	16.1	51,192	47.9	106,898
Central and NCD	330,789	75.5	7,763	1.8	514	0.1	49,339	11.3	49,736	11.4	438,141
Milne Bay	85,490	40.6	11,434	5.4	27,918	13.3	17,636	8.4	67,934	32.3	210,412
Oro	6,546	4.9	90,038	67.7	12,701	9.5	19,134	14.4	4,646	3.5	133,065
Southern Highlands	318,822	58.4	179,173	32.8	38,523	7.1	9,747	1.8	0	0.0	546,265
Enga	201,469	68.3	8,016	2.7	78,605	26.6	6,941	2.4	0	0.0	295,031
Western Highlands	140,946	32.0	157,054	35.7	91,126	20.7	50,899	11.6	0	0.0	440,025
Simbu	227,187	87.5	13,629	5.2	18,887	7.3	0	0.0	0	0.0	259,703
Eastern Highlands	319,956	73.9	2,410	0.6	84,490	19.5	26,116	6.0	0	0.0	432,972
Morobe	287,316	53.3	12,804	2.4	196,123	36.4	22,935	4.3	20,226	3.7	539,404
Madang	177,338	48.6	53,790	14.7	25,186	6.9	33,805	9.3	74,987	20.5	365,106
East Sepik	165,921	48.3	3,315	1.0	37,762	11.0	94,544	27.5	41,639	12.1	343,181
Sandaun	118,027	63.5	0	0.0	12,182	6.6	25,849	13.9	29,683	16.0	185,741
Manus	11,757	27.1	5,359	12.4	0	0.0	1,507	3.5	24,765	57.1	43,387
New Ireland	15,988	13.5	4,365	3.7	5,815	4.9	0	0.0	92,181	77.9	118,350
East New Britain	31,737	14.4	101,099	45.9	687	0.3	46,020	20.9	40,590	18.4	220,133
West New Britain	35,051	19.0	110,073	59.7	133	0.1	10,979	6.0	28,271	15.3	184,508
Bougainville	36,414	20.8	91,630	52.3	469	0.3	1,559	0.9	45,088	25.7	175,160
Papua New Guinea	2,577,475	49.7	858,880	16.5	685,237	13.2	484,095	9.3	585,099	11.3	5,190,786

Sources: McAlpine and Quigley (c. 1995); NSO (2002); PNGRIS.

Appendix Table A1.10.4 Total land area by altitude class and province (km^2)

Province	Altitude class (metres above sea level)												Total land area
	Lowlands 0–600	%	Intermediate 600–1200	%	Highlands 1200–1800	%	High altitude 1800–2400	%	Very high altitude 2400–2800	%	Uninhabited >2800	%	
Western	88,584	91.3	6,938	7.1	1,481	1.5	61	0.1	0	0.0	0	0.0	97,065
Gulf	30,890	91.3	2,419	7.1	517	1.5	21	0.1	0	0.0	0	0.0	33,847
Central	17,834	59.5	5,457	18.2	3,098	10.3	1,873	6.3	642	2.1	1,051	3.5	29,954
Milne Bay	11,348	80.3	1,729	12.2	706	5.0	307	2.2	13	0.1	22	0.2	14,125
Oro	13,023	57.9	5,124	22.8	2,054	9.1	1,418	6.3	20	0.1	872	3.9	22,510
Southern Highlands	3,660	14.2	7,854	30.6	7,821	30.4	3,840	14.9	1,095	4.3	1,427	5.6	25,698
Enga	283	2.4	1,142	9.6	2,242	18.9	2,111	17.8	1,968	16.6	4,092	34.6	11,839
Western Highlands	526	5.9	1,531	17.2	3,537	39.8	2,177	24.5	178	2.0	948	10.7	8,897
Simbu	131	2.2	2,160	35.9	1,903	31.6	1,255	20.8	146	2.4	427	7.1	6,022
Eastern Highlands	209	1.9	1,101	10.0	5,036	45.8	3,519	32.0	747	6.8	395	3.6	11,006
Morobe	10,133	30.2	8,952	26.7	7,298	21.8	5,141	15.3	872	2.6	1,130	3.4	33,525
Madang	19,945	69.4	4,719	16.4	1,920	6.7	1,429	5.0	163	0.6	555	1.9	28,732
East Sepik	38,987	89.2	3,054	7.0	974	2.2	526	1.2	29	0.1	149	0.3	43,720
Sandaun	24,921	69.2	5,330	14.8	3,317	9.2	1,650	4.6	403	1.1	389	1.1	36,010
Manus	2,098	100.0	0	0.0	0	0.0	0	0.0	0	0.0	0	0.0	2,098
New Ireland	18,213	87.8	2,347	11.3	164	0.8	29	0.1	0	0.0	0	0.0	20,753
East New Britain	8,870	58.7	5,760	38.1	478	3.2	1	0.0	0	0.0	0	0.0	15,109
West New Britain	6,808	70.8	2,320	24.1	486	5.1	0	0.0	0	0.0	0	0.0	9,615
Bougainville	7,381	79.1	1,565	16.8	383	4.1	0	0.0	0	0.0	0	0.0	9,329
Papua New Guinea	303,844	66.1	69,505	15.1	43,416	9.4	25,359	5.5	6,275	1.4	11,455	2.5	459,854

Sources: McAlpine and Quigley (c. 1995); PNGRIS.

Appendix Table A1.10.5 Land area used for agriculture by altitude class and province (km^2)

Province	Altitude class (metres above sea level)										Total land used for agriculture
	Lowlands 0–600	%	Intermediate 600–1200	%	Highlands 1200–1800	%	High altitude 1800–2400	%	Very high altitude 2400–2800	%	
Western	7,219	91.0	318	4.0	329	4.1	66	0.8	0	0.0	7,931
Gulf	2,547	67.0	760	20.0	478	12.6	16	0.4	0	0.0	3,801
Central	4,273	66.7	1,035	16.2	750	11.7	348	5.4	0	0.0	6,407
Milne Bay	5,208	91.5	359	6.3	103	1.8	20	0.4	0	0.0	5,691
Oro	3,533	83.0	710	16.7	14	0.3	1	0.0	0	0.0	4,258
Southern Highlands	219	3.1	1,299	18.3	3,878	54.5	1,582	22.2	133	1.9	7,112
Enga	26	0.7	465	12.4	1,085	28.9	985	26.3	1,188	31.7	3,749
Western Highlands	133	3.0	765	17.1	2,769	61.8	733	16.4	82	1.8	4,482
Simbu	24	1.0	484	19.2	1,045	41.5	875	34.8	88	3.5	2,515
Eastern Highlands	45	0.8	371	6.7	3,594	64.9	1,494	27.0	34	0.6	5,539
Morobe	3,941	32.2	4,502	36.8	3,134	25.6	636	5.2	32	0.3	12,245
Madang	12,662	78.9	2,450	15.3	752	4.7	162	1.0	19	0.1	16,046
East Sepik	8,417	93.6	468	5.2	99	1.1	8	0.1	0	0.0	8,992
Sandaun	6,214	75.7	1,021	12.4	760	9.3	199	2.4	16	0.2	8,211
Manus	1,747	100.0	0	0.0	0	0.0	0	0.0	0	0.0	1,747
New Ireland	4,496	99.2	35	0.8	0	0.0	0	0.0	0	0.0	4,531
East New Britain	3,306	87.7	464	12.3	2	0.1	0	0.0	0	0.0	3,772
West New Britain	4,902	86.4	743	13.1	31	0.5	0	0.0	0	0.0	5,676
Bougainville	4,616	89.6	515	10.0	22	0.4	0	0.0	0	0.0	5,153
Papua New Guinea	73,531	62.4	16,766	14.2	18,844	16.0	7,126	6.0	1,591	1.4	117,858

Sources: McAlpine and Quigley (c. 1995); PNGRIS.

Appendix Table A1.10.6 Total population (including urban and rural non-village populations) by altitude class and province

Province	Altitude class (metres above sea level)										Total population
	Lowlands 0–600	%	Intermediate 600–1200	%	Highlands 1200–1800	%	High altitude 1800–2400	%	Very high altitude 2400–2800	%	
Western	147,636	96.3	2,439	1.6	3,054	2.0	174	0.1	0	0.0	153,304
Gulf	93,699	87.7	7,347	6.9	4,986	4.7	866	0.8	0	0.0	106,898
Central and NCD	383,059	87.4	17,214	3.9	24,844	5.7	12,619	2.9	405	0.1	438,141
Milne Bay	192,880	91.7	12,756	6.1	4,776	2.3	0	0.0	0	0.0	210,412
Oro	115,574	86.9	16,875	12.7	615	0.5	0	0.0	0	0.0	133,065
Southern Highlands	5,223	1.0	33,099	6.1	330,120	60.4	166,548	30.5	11,275	2.1	546,265
Enga	566	0.2	11,214	3.8	93,944	31.8	100,249	34.0	89,058	30.2	295,031
Western Highlands	754	0.2	16,048	3.6	360,035	81.8	52,805	12.0	10,382	2.4	440,025
Simbu	117	0.0	11,654	4.5	144,554	55.7	101,633	39.1	1,744	0.7	259,703
Eastern Highlands	0	0.0	2,385	0.6	304,402	70.3	123,214	28.5	2,971	0.7	432,972
Morobe	233,425	43.3	131,050	24.3	148,832	27.6	22,989	4.3	3,108	0.6	539,404
Madang	282,971	77.5	39,069	10.7	33,149	9.1	8,828	2.4	1,089	0.3	365,106
East Sepik	342,924	99.9	171	0.0	87	0.0	0	0.0	0	0.0	343,181
Sandaun	148,802	80.1	18,531	10.0	17,405	9.4	1,003	0.5	0	0.0	185,741
Manus	43,387	100.0	0	0.0	0	0.0	0	0.0	0	0.0	43,387
New Ireland	118,241	99.9	0	0.0	109	0.1	0	0.0	0	0.0	118,350
East New Britain	210,043	95.4	10,090	4.6	0	0.0	0	0.0	0	0.0	220,133
West New Britain	172,660	93.6	11,848	6.4	0	0.0	0	0.0	0	0.0	184,508
Bougainville	162,558	92.8	12,472	7.1	130	0.1	0	0.0	0	0.0	175,160
Papua New Guinea	2,654,521	51.1	354,262	6.8	1,471,042	28.3	590,928	11.4	120,033	2.3	5,190,786

Sources: McAlpine and Quigley (c. 1995); NSO (2002); PNGRIS.

Appendix Table A1.10.7 Total population (including urban and rural non-village populations) by altitude range and province

Province	Altitude range (metres above sea level)															
	0–100	%	100–200	%	200–400	%	400–600	%	600–800	%	800–1000	%	1000–1200	%	1200–1400	%
Western	93,509	61.0	33,932	22.1	8,049	5.3	3,185	2.1	10,724	7.0	1,306	0.9	771	0.5	564	0.4
Gulf	75,392	70.5	1,245	1.2	1,025	1.0	215	0.2	2,383	2.2	3,390	3.2	4,554	4.3	6,834	6.4
Central	126,678	68.9	12,468	6.8	4,043	2.2	4,829	2.6	5,568	3.0	2,063	1.1	3,401	1.8	4,442	2.4
National Capital District	154,229	60.7	99,207	39.0	723	0.3	0	0.0	0	0.0	0	0.0	0	0.0	0	0.0
Milne Bay	180,143	85.6	10,124	4.8	5,483	2.6	3,409	1.6	3,151	1.5	2,191	1.0	3,280	1.6	2,079	1.0
Oro	52,589	39.5	31,713	23.8	13,305	10.0	16,733	12.6	6,690	5.0	9,676	7.3	1,702	1.3	226	0.2
Southern Highlands	0	0.0	0	0.0	1,038	0.2	1,742	0.3	4,787	0.9	8,028	1.5	5,128	0.9	7,723	1.4
Enga	0	0.0	0	0.0	0	0.0	110	0.0	316	0.1	490	0.2	2,077	0.7	4,942	1.7
Western Highlands	0	0.0	0	0.0	0	0.0	915	0.2	1,563	0.4	1,135	0.3	1,261	0.3	14,544	3.3
Simbu	0	0.0	0	0.0	0	0.0	194	0.1	113	0.0	1,275	0.5	5,271	2.0	6,793	2.6
Eastern Highlands	0	0.0	0	0.0	0	0.0	61	0.0	0	0.0	0	0.0	751	0.2	5,702	1.3
Morobe	218,312	40.5	39,790	7.4	9,340	1.7	25,150	4.7	24,593	4.6	28,210	5.2	25,596	4.7	45,769	8.5
Madang	164,809	45.1	62,986	17.3	36,447	10.0	27,798	7.6	10,218	2.8	4,574	1.3	8,327	2.3	10,171	2.8
East Sepik	179,688	52.4	56,285	16.4	75,906	22.1	26,640	7.8	3,417	1.0	439	0.1	56	0.0	170	0.0
Sandaun	61,610	33.2	23,637	12.7	20,431	11.0	36,063	19.4	14,017	7.5	7,792	4.2	778	0.4	860	0.5
Manus	37,543	86.5	1,069	2.5	4,425	10.2	349	0.8	0	0.0	0	0.0	0	0.0	0	0.0
New Ireland	111,556	94.3	3,777	3.2	1,605	1.4	414	0.4	0	0.0	211	0.2	787	0.7	0	0.0
East New Britain	97,886	44.5	51,317	23.3	36,677	16.7	22,521	10.2	4,846	2.2	3,999	1.8	2,577	1.2	218	0.1
West New Britain	139,801	75.8	33,001	17.9	6,762	3.7	4,286	2.3	657	0.4	0	0.0	0	0.0	0	0.0
Bougainville	100,333	57.3	26,974	15.4	14,092	8.0	18,161	10.4	8,333	4.8	3,115	1.8	2,419	1.4	1,465	0.8
Papua New Guinea	1,794,077	34.6	487,525	9.4	239,352	4.6	192,777	3.7	101,379	2.0	77,892	1.5	68,738	1.3	112,503	2.2

... continued

Appendix Table A1.10.7 (continued)

Province	Altitude range (metres above sea level)														Total population
	1400–1600	%	1600–1800	%	1800–2000	%	2000–2200	%	2200–2400	%	2400–2600	%	2600–2800	%	
Western	338	0.2	926	0.6	0	0.0	0	0.0	0	0.0	0	0.0	0	0.0	153,304
Gulf	8,960	8.4	1,646	1.5	1,254	1.2	0	0.0	0	0.0	0	0.0	0	0.0	106,898
Central	4,546	2.5	5,578	3.0	4,961	2.7	4,073	2.2	1,039	0.6	293	0.2	0	0.0	183,983
National Capital District	0	0.0	0	0.0	0	0.0	0	0.0	0	0.0	0	0.0	0	0.0	254,158
Milne Bay	553	0.3	0	0.0	0	0.0	0	0.0	0	0.0	0	0.0	0	0.0	210,412
Oro	345	0.3	53	0.0	32	0.0	0	0.0	0	0.0	0	0.0	0	0.0	133,065
Southern Highlands	26,603	4.9	92,047	16.9	207,142	37.9	114,019	20.9	56,124	10.3	14,777	2.7	7,106	1.3	546,265
Enga	9,744	3.3	13,758	4.7	42,597	14.4	55,660	18.9	68,133	23.1	51,023	17.3	46,181	15.7	295,031
Western Highlands	44,873	10.2	184,617	42.0	137,346	31.2	20,706	4.7	14,521	3.3	13,768	3.1	4,775	1.1	440,025
Simbu	10,404	4.0	72,375	27.9	79,046	30.4	51,780	19.9	21,276	8.2	8,298	3.2	2,878	1.1	259,703
Eastern Highlands	32,078	7.4	161,836	37.4	156,694	36.2	66,429	15.3	8,057	1.9	504	0.1	859	0.2	432,972
Morobe	46,190	8.6	40,235	7.5	19,310	3.6	9,328	1.7	4,786	0.9	2,289	0.4	506	0.1	539,404
Madang	13,327	3.7	10,819	3.0	11,566	3.2	2,989	0.8	1,077	0.3	0	0.0	0	0.0	365,106
East Sepik	579	0.2	0	0.0	0	0.0	0	0.0	0	0.0	0	0.0	0	0.0	343,181
Sandaun	4,731	2.5	4,478	2.4	2,441	1.3	7,290	3.9	1,006	0.5	607	0.3	0	0.0	185,741
Manus	0	0.0	0	0.0	0	0.0	0	0.0	0	0.0	0	0.0	0	0.0	43,387
New Ireland	0	0.0	0	0.0	0	0.0	0	0.0	0	0.0	0	0.0	0	0.0	118,350
East New Britain	91	0.0	0	0.0	0	0.0	0	0.0	0	0.0	0	0.0	0	0.0	220,133
West New Britain	0	0.0	0	0.0	0	0.0	0	0.0	0	0.0	0	0.0	0	0.0	184,508
Bougainville	267	0.2	0	0.0	0	0.0	0	0.0	0	0.0	0	0.0	0	0.0	175,160
Papua New Guinea	203,629	3.9	588,368	11.3	662,388	12.8	332,275	6.4	176,019	3.4	91,559	1.8	62,306	1.2	5,190,786

Sources: NSO (2002); digital 1:250 000 topographic map; PNGRIS.

Appendix Table A1.12.1 Total land area by land quality and province (km^2)

Province	Very low	%	Low	%	Moderate	%	High	%	Very high	%	Total land area
Western	31,400	32.3	60,682	62.5	4,984	5.1	0	0.0	0	0.0	97,065
Gulf	3,035	9.0	26,594	78.6	3,727	11.0	205	0.6	286	0.8	33,847
Central	10,469	34.9	12,076	40.3	5,525	18.4	1,365	4.6	519	1.7	29,954
Milne Bay	613	4.3	5,410	38.3	5,394	38.2	2,325	16.5	384	2.7	14,125
Oro	2,439	10.8	9,869	43.8	7,556	33.6	1,467	6.5	1,179	5.2	22,510
Southern Highlands	6,583	25.6	14,848	57.8	3,877	15.1	356	1.4	35	0.1	25,698
Enga	5,301	44.8	5,589	47.2	817	6.9	132	1.1	0	0.0	11,839
Western Highlands	980	11.0	4,695	52.8	2,106	23.7	427	4.8	689	7.7	8,897
Simbu	959	15.9	3,623	60.2	1,097	18.2	199	3.3	144	2.4	6,022
Eastern Highlands	2,952	26.8	5,071	46.1	2,261	20.5	414	3.8	308	2.8	11,006
Morobe	5,210	15.5	19,183	57.2	8,165	24.4	733	2.2	233	0.7	33,525
Madang	6,018	20.9	11,828	41.2	10,017	34.9	576	2.0	293	1.0	28,732
East Sepik	2,140	4.9	28,575	65.4	9,749	22.3	3,256	7.4	0	0.0	43,720
Sandaun	2,034	5.6	13,735	38.1	10,383	28.8	5,867	16.3	3,991	11.1	36,010
Manus	60	2.9	878	41.9	1,052	50.1	91	4.4	17	0.8	2,098
New Ireland	127	1.3	6,667	69.3	2,599	27.0	176	1.8	46	0.5	9,615
East New Britain	1,921	12.7	9,405	62.2	2,235	14.8	597	4.0	951	6.3	15,109
West New Britain	2,714	13.1	9,709	46.8	7,060	34.0	823	4.0	446	2.2	20,753
Bougainville	316	3.4	3,126	33.5	3,518	37.7	1,522	16.3	847	9.1	9,329
Papua New Guinea	85,270	18.5	251,563	54.7	92,121	20.0	20,532	4.5	10,368	2.3	459,854

Source: PNGRIS.

Appendix Table A1.12.2 Total population (including urban and rural non-village populations) by land quality and province

Province	Very low	%	Low	%	Moderate	%	High	%	Very high	%	Total population
Western	45,554	29.7	99,500	64.9	8,250	5.4	0	0.0	0	0.0	153,304
Gulf	15,543	14.5	66,856	62.5	15,591	14.6	278	0.3	8,629	8.1	106,898
Central and NCD	156,747	35.8	216,132	49.3	52,346	11.9	6,570	1.5	6,346	1.4	438,141
Milne Bay	17,794	8.5	68,615	32.6	88,995	42.3	29,528	14.0	5,480	2.6	210,412
Oro	925	0.7	35,781	26.9	46,027	34.6	12,884	9.7	37,447	28.1	133,065
Southern Highlands	71,327	13.1	271,214	49.6	139,518	25.5	52,152	9.5	12,054	2.2	546,265
Enga	54,209	18.4	196,820	66.7	44,002	14.9	0	0.0	0	0.0	295,031
Western Highlands	0	0.0	181,534	41.3	126,376	28.7	6,062	1.4	126,054	28.6	440,025
Simbu	32,637	12.6	123,779	47.7	38,785	14.9	30,116	11.6	34,386	13.2	259,703
Eastern Highlands	72,456	16.7	187,654	43.3	102,876	23.8	26,381	6.1	43,605	10.1	432,972
Morobe	52,351	9.7	294,824	54.7	168,108	31.2	19,684	3.6	4,438	0.8	539,404
Madang	32,266	8.8	139,411	38.2	139,856	38.3	10,875	3.0	42,697	11.7	365,106
East Sepik	4,329	1.3	98,457	28.7	124,942	36.4	115,453	33.6	0	0.0	343,181
Sandaun	1,438	0.8	43,507	23.4	44,991	24.2	54,646	29.4	41,159	22.2	185,741
Manus	6,417	14.8	11,820	27.2	20,373	47.0	4,014	9.3	764	1.8	43,387
New Ireland	5,831	4.9	53,099	44.9	40,377	34.1	16,117	13.6	2,926	2.5	118,350
East New Britain	12,960	5.9	38,300	17.4	26,601	12.1	41,651	18.9	100,621	45.7	220,133
West New Britain	3,887	2.1	91,484	49.6	74,196	40.2	10,190	5.5	4,751	2.6	184,508
Bougainville	1,907	1.1	20,388	11.6	77,623	44.3	51,582	29.4	23,660	13.5	175,160
Papua New Guinea	588,579	11.3	2,239,175	43.1	1,379,832	26.6	488,182	9.4	495,017	9.5	5,190,786

Sources: NSO (2002); PNGRIS.

Appendix Table A1.13.1 The usual and extreme altitudinal ranges of 22 crops in PNG

Common name	Scientific name	Economic product	Mean usual altitudinal range (m)	Extreme altitudinal range (m)	Number of observations/ standard deviation (m)	
					Usual minimum	Usual maximum
Arabica coffee[a]	*Coffea arabica*	export crop	700–2050	80–2380	15/150	30/90
Betel nut	*Areca catechu*	stimulant	0–1100	0–1390	–	17/140
Broad bean	*Vicia faba*	vegetable	2050–2650	1620–2760	4/90	4/80
Cassava	*Manihot esculenta*	staple food	0–1800	0–2210	–	32/120
Coastal pandanus	*Pandanus tectorius*	fruit	0–50	0–?		
Cocoa	*Theobroma cacao*	export crop	0–800	0–1390	–	7/140
Irish potato[b]	*Solanum tuberosum*	staple food	700–2750	0–2850	7/90	4/60
Kangkong	*Ipomoea aquatica*	vegetable	0–600	0–760	–	5/100
Kapok[c]	*Ceiba pentandra*	export crop	0–1250	0–1550	–	6/200
Karuka	*Pandanus julianettii*	edible nut	1800–2600	1450–2800	50/110	18/100
Pak choi	*Brassica rapa*	vegetable	0–2550	0–2800	–	7/160
Pao	*Barringtonia procera*	edible nut	0–500	0–620	–	4/90
Pawpaw	*Carica papaya*	fruit	0–1700	0–1950	–	30/100
Purple passionfruit	*Passiflora edulis f. edulis*	fruit	800–2300	700–2520	7/90	13/90
Pyrethrum	*Chrysanthemum cinerariaefolium*	export crop	2400–2800	1630–2850	8/20	5/50
Sis nut	*Pangium edule*	edible nut	0–1050	0–1380	–	11/120
Spring onion	*Allium cepa* var. *cepa*	vegetable	0–2700	0–2850	–	5/60
Swamp taro	*Cyrtosperma chamissonis*	staple food	0–50	0–?		
Sweet potato[d]	*Ipomoea batatas*	staple food	0–2700	0–2850	–	10/150
Tomato	*Lycopersicon esculentum*	vegetable	0–2250	0–2630	–	13/110
Watercress	*Nasturtium officinale*	vegetable	0–2900	0–3580	–	3/70
Winged bean[e]	*Psophocarpus tetragonolobus*	vegetable	0–1900	0–2070	–	28/90

[a] Plantation Arabica coffee occurs over a range of 930 m (Wau area) to 1820 m (Asaro Valley and Wahgi Valley).

[b] Irish potato grown by villagers for subsistence only is usually planted above 1900 m. Below 1900 m most crops are intended for both sale and subsistence. Smith (1977:189) records that he grew Irish potato on Mt Wilhelm at 3580 m.

[c] Kapok was introduced in early colonial times as a potential export crop, but the industry did not develop.

[d] The usual upper limit of sweet potato (2700 m) is derived from the 10 highest observations, not from all observations.

[e] The usual range from sea level to 1900 m is for all plantings of winged bean. This species is planted mainly for tuber production over the range 1200–1900 m.

Source: Bourke (1989).

Appendix Table A1.14.1 Rural population by accessibility class and province

Province	Accessibility class										Total rural population
	Very poor	%	Poor	%	Moderate	%	Good	%	Very good	%	
Western	7,488	6.9	16,265	15.1	78,282	72.6	5,802	5.4	0	0.0	107,837
Gulf	3,819	4.1	22,895	24.8	62,057	67.3	3,493	3.8	0	0.0	92,265
Central	7,005	4.5	10,375	6.6	47,107	30.0	85,963	54.7	6,609	4.2	157,058
Milne Bay	4,755	2.5	57,465	30.5	92,920	49.3	33,195	17.6	0	0.0	188,334
Oro	7,850	7.4	22,506	21.2	27,497	25.9	48,435	45.6	0	0.0	106,288
Southern Highlands	21,217	4.0	71,534	13.6	213,533	40.6	220,115	41.8	0	0.0	526,398
Enga	11,327	4.0	17,823	6.3	106,155	37.4	148,193	52.3	0	0.0	283,498
Western Highlands	916	0.2	48,617	13.1	21,362	5.8	211,662	57.0	88,457	23.8	371,014
Simbu	841	0.3	12,399	5.1	22,185	9.1	207,323	85.4	0	0.0	242,748
Eastern Highlands	6,247	1.6	17,403	4.4	57,991	14.7	267,456	68.0	44,321	11.3	393,418
Morobe	5,030	1.4	40,959	11.5	260,019	73.0	50,092	14.1	0	0.0	356,100
Madang	10,785	3.5	106,152	34.4	126,545	41.1	31,611	10.3	33,041	10.7	308,135
East Sepik	7,487	2.5	0	0.0	148,010	48.7	148,209	48.8	0	0.0	303,706
Sandaun	36,812	22.1	10,316	6.2	116,811	70.0	2,979	1.8	0	0.0	166,919
Manus	0	0.0	0	0.0	29,228	83.8	5,671	16.2	0	0.0	34,899
New Ireland	0	0.0	11,700	11.3	45,878	44.4	45,681	44.2	0	0.0	103,259
East New Britain	20,190	11.6	7,338	4.2	23,189	13.3	19,382	11.1	104,132	59.8	174,230
West New Britain	2,320	2.1	35,424	32.4	23,559	21.6	47,995	43.9	0	0.0	109,299
Bougainville	2,700	1.6	0	0.0	102,069	61.1	62,387	37.3	0	0.0	167,156
Papua New Guinea	156,791	3.7	509,171	12.1	1,604,395	38.3	1,645,644	39.3	276,560	6.6	4,192,561

Sources: NSO (2002); MASP.

Appendix Table A2.1.1 Estimated food consumption in PNG, 2006 and 1996

Food	Quantity in 2006[a]		Quantity in 1996 (kg/person/year)			Notes
	Tonnes/year	kg/person/year	PNG	Rural	Urban	
Staple foods						The 2006 source for staple foods is Table 2.2.1, extrapolated from 2000 to 2006 using a population growth rate of 2.76% per year.
Sweet potato	2,542,000	416	260	299	42	The 2006 figure for sweet potato consumption is 75% of the estimated production, as an estimated 25% of sweet potato in PNG is fed to pigs.
Banana	515,000	84	83	90	47	
Yam (all species)	322,000	53	28	31	9	
Cassava	321,000	52	25	27	9	
Taro	276,000	45	62	68	23	Data for taro and Chinese taro were combined in the 1996 Household Survey.
Chinese taro	267,000	44				
Sago	98,000	16	23	21	33	Sago has a higher food energy value than the other staple foods.
Irish potato	22,000	4	3	3	6	
Rice (local)	1,000	0.2				
Other garden foods						2006 estimates for other garden foods are crude and have a large error factor.
Sugar cane	372,000	61	35	40	10	
Coconut	120,000	20	44	42	51	The 2006 source is Table 2.2.1, extrapolated from 2000 to 2006 using a population growth rate of 2.76% per year.
Vegetables	500,000	81	72	78	34	Corn production, perhaps 50,000 tonnes/year, is included in the 2006 category.
Fruit	100,000	16	14	14	16	
Peanuts, other nuts	20,000	3	3	3	1	

... continued

Appendix Table A2.1.1 (continued)

Food	Quantity in 2006[a]		Quantity in 1996 (kg/person/year)			Notes
	Tonnes/year	kg/person/year	PNG	Rural	Urban	
Imported energy foods						2006 data for imported energy foods are from Table A2.7.1.
Rice	184,000	30	31	24	66	
Flour	107,000	18	7	5	14	Flour is milled in PNG from imported wheat.
Bread			3	1	13	
Biscuits			2	1	4	
Meat						2006 meat data are from Tables 2.6.3 and A2.9.1.
Pork	29,300	5	11	13	2	
Chicken	25,900	4	6	5	13	
Sheep meat	24,200	4	5	4	10	
Bush meat	21,000	3	5	5	3	
Beef, local and imported	7,300	1				The figure for tinned meat in the 1996 survey can be compared with the combined figures for beef and tinned meat and offal in the 2006 estimates.
Tinned meat and offal (imported)	3,600	0.6				
Tinned meat			2	1	7	
Fish and other seafood						2006 data for fish and other seafood are derived from Sections 2.10 and 5.9 and Table 2.1.2. Figures for fish and other seafood have a large error.
Fresh and smoked	50,000	8	10	8	21	
Local tuna	15,000	2				
Imported fish	10,800	2	3	2	7	

... continued

Appendix Table A2.1.1 (continued)

Food	Quantity in 2006[a]		Quantity in 1996 (kg/person/year)			Notes
	Tonnes/year	kg/person/year	PNG	Rural	Urban	
Other foods						The categories used for other foods differ somewhat between the two estimates.
Sugar	34,000	6	4	3	8	2006 data are from Section 5.10.
Imported fruit and vegetables	6,000	1				Data are from Table A2.8.1.
Milk and other dairy	5,400	1	0	0	1	2006 data are from Table 2.1.2.
Animal fat	6,200	1				Data are from Table 2.1.2.
Vegetable oils	5,500	1				Data are from Table 2.1.2.
Eggs	3,100	0.5				Data are from Table 2.6.3.
Butter, margarine, oil and dripping			1	1	3	
Minor dairy, cereal, eggs			1	0	3	
Miscellaneous						The foods classed under miscellaneous were recorded in the 1996 survey, but data are not available for 2006.
Snack foods			0	0	1	
Salt, spices, sauces			1	1	1	
Soft drink			4	3	12	
Beer			4	3	15	
Tea, coffee, milo			0	0	1	
Betel nut, lime, mustard			11	11	10	

[a] The quantities of foods consumed per person in 2006 are based on the estimated total population of 6 112 000 for that year.

Sources: 1996: Gibson (2001:41); 2006: various sections, this volume.

Appendix Table A2.2.1 Estimated production of banana, cassava, coconut, Irish potato, Queensland arrowroot, rice, sago and sweet potato in 2000, by province (tonnes/year)

Province	Banana	Cassava	Coconut	Irish potato	Queensland arrowroot	Rice	Sago	Sweet potato
Western	15,167	9,141	4,312	0	0	0	12,940	6,863
Gulf	10,537	3,892	2,857	171	54	0	10,369	20,308
Central	43,014	19,379	5,142	1,305	787	47	588	49,267
Milne Bay	23,591	51,485	9,795	444	590	4	1,676	43,831
Oro	9,866	10,403	1,549	0	0	48	1,624	53,309
Southern Highlands	15,121	12,626	449	4,920	0	0	2,405	619,561
Enga	6,052	4,215	0	6,485	0	0	104	340,745
Western Highlands	30,989	10,412	0	1,725	0	0	7	425,964
Simbu	10,334	2,482	0	593	0	0	166	294,708
Eastern Highlands	17,038	6,134	0	0	0	0	3	469,939
Morobe	63,311	12,111	5,704	2,455	0	1	572	194,695
Madang	41,729	16,612	14,091	661	0	0	5,288	77,746
East Sepik	37,100	5,086	17,806	0	0	0	23,484	26,175
Sandaun	18,322	883	8,133	0	0	0	16,711	25,036
Manus	1,389	5,715	1,956	0	0	0	4,575	4,477
New Ireland	8,288	15,652	5,612	0	0	0	1,797	38,891
East New Britain	55,587	33,900	8,847	0	0	0	0	42,642
West New Britain	11,885	39,855	5,101	0	0	0	222	45,103
Bougainville	17,176	11,911	9,575	0	0	307	431	92,591
Papua New Guinea	436,496	271,894	100,929	18,759	1,431	407	82,962	2,871,851

Source: Bourke and Vlassak (2004).

Appendix Table A2.2.2 Estimated production of various taro and yam species in 2000, by province (tonnes/year)

Province	*Colocasia* taro	Chinese taro	Taro (*Alocasia*)	Taro (*Amorphophallus*)	Swamp taro	Lesser yam (*D. esculenta*)	Greater yam (*D. alata*)	Yam (*D. nummularia*)	Aerial yam (*D. bulbifera*)	Yam (*D. pentaphylla*)
Western	5,066	2,165	0	108	0	7,794	1,910	0	0	0
Gulf	1,300	3,625	0	49	108	204	271	0	0	0
Central	5,796	6,715	310	504	0	10,462	7,109	0	197	0
Milne Bay	15,678	4,892	851	455	0	28,361	12,455	101	12	0
Oro	11,784	11,182	29	2	0	6,337	5,298	0	0	0
Southern Highlands	11,409	5,431	0	0	0	0	3,275	0	0	0
Enga	8,780	287	0	0	0	0	2,240	0	0	0
Western Highlands	9,212	1,586	0	0	0	567	3,143	0	0	0
Simbu	4,828	416	0	0	0	0	2,408	0	0	0
Eastern Highlands	11,160	15	0	0	0	6	5,792	0	0	0
Morobe	22,699	58,938	99	0	0	8,435	17,855	377	53	37
Madang	38,621	46,636	0	0	0	32,733	9,049	0	0	0
East Sepik	24,689	8,045	48	0	0	64,927	6,810	0	205	0
Sandaun	15,489	6,117	0	0	0	5,903	2,912	0	0	0
Manus	390	636	0	0	406	1,120	81	0	0	0
New Ireland	9,165	2,896	375	0	34	10,684	3,642	0	0	0
East New Britain	9,230	30,993	565	0	0	0	491	0	0	0
West New Britain	16,002	23,794	68	99	0	2,714	1,832	0	0	0
Bougainville	7,790	12,167	44	0	275	123	4,785	0	0	0
Papua New Guinea	229,088	226,536	2,389	1,217	823	180,370	91,358	478	467	37

Source: Bourke and Vlassak (2004).

Appendix Table A2.5.1 Estimated rice production, 1962–2000 (tonnes)

Year	Estimated production	Year	Estimated production	Year	Estimated production
1962	311	1975	1,268	1988	110
1963	–	1976	1,493	1989	340
1964	807	1977	–	1990	240
1965	–	1978	920	1991	940
1966	1,168	1979	760	1992	880
1967	–	1980	920	1993	340
1968	1,391	1981	610	1994	400
1969	1,779	1982	1,128	1995	300
1970	1,608	1983	380	1996	320
1971	2,206	1984	330	1997	60
1972	1,608	1985	1,143	1998	80
1973	1,008	1986	140	1999	180
1974	1,458	1987	80	2000	300

Note: Data from different sources do not always agree. For example, for the year 2000, Bourke estimated 410 tonnes (Table 2.2.1), whereas Blakeney and Clough (2001) estimated 300 tonnes.

Sources: 1962–1976: Hale (c. 1978); 1978–1990: DAL (1992:51); 1991–2000: Blakeney and Clough (2001).

Appendix Table A2.7.1 Volume of rice and wheat imports, 1961–2007 (tonnes)

Year	Rice	Wheat[a]	Year	Rice	Wheat[a]
1961	21,600	9,135	1985	127,600	69,000
1962	21,600	9,190	1986	169,982	80,000
1963	23,200	10,022	1987	132,058	77,000
1964	24,000	11,589	1988	117,300	69,500
1965	28,100	13,994	1989	130,155	85,000
1966	33,800	16,402	1990	120,810	93,500
1967	32,800	17,939	1991	122,970	86,500
1968	38,700	19,889	1992	131,180	103,000
1969	38,300	25,725	1993	146,960	116,000
1970	45,200	31,534	1994	156,070	120,500
1971	47,100	32,270	1995	156,100	110,000
1972	48,600	30,818	1996	162,670	107,000
1973	61,700	28,697	1997	207,690[b]	114,000
1974	41,700	30,804	1998	168,900	120,237
1975	55,500	32,503	1999	150,130	134,410
1976	55,100	33,345	2000	151,745	136,844
1977	72,449	38,067	2001	143,319	120,000
1978	77,642	42,844	2002	153,371	108,688
1979	81,277	46,769	2003	147,318	130,826
1980	88,924	48,129	2004	153,179	106,077
1981	88,099	44,675	2005	160,508	160,856
1982	93,524	44,283	2006	187,654	123,911
1983	92,633	50,262	2007	191,470	149,006
1984	102,774	56,164			

[a] Wheat import data have been converted from a June year to a calendar year.

[b] Rice imports were actually greater than this in 1997 because of imports from Asia for food relief following the 1997 drought and frosts (see Section 1.6). The figure given here is that imported by Trukai Industries Limited from Australia.

Sources: 1961–1999: Gibson (2001a: Appendix C); 2000–2007: Trukai Industries Limited, Port Moresby, and Australian Government Wheat Export Authority.

Appendix Table A2.8.1 Volume of fruit and vegetable imports from Australia and New Zealand, 1983–2003 (tonnes)

Year	Australia	New Zealand	Australia and New Zealand combined
1983	5,768	3,708	9,475
1984	4,763	3,541	8,304
1985	5,130	3,405	8,535
1986	5,967	3,265	9,232
1987	6,163	1,542	7,705
1988	6,844	1,202	8,046
1989	8,163	1,599	9,763
1990	7,177	2,023	9,200
1991	6,683	2,232	8,915
1992	6,654	1,504	8,158
1993	6,669	2,764	9,434
1994	7,029	2,019	9,048
1995	6,561	1,150	7,712
1996	7,547	1,428	8,975
1997	7,603	1,636	9,238
1998	6,660	1,749	8,409
1999	4,426	2,551	6,978
2000	3,448	3,125	6,573
2001	3,408	3,078	6,486
2002	3,483	2,957	6,440
2003	3,152	2,503	5,654

Sources: Australian Bureau of Statistics; Statistics New Zealand.

Appendix Table A2.9.1 Composition of meat imports from Australia and New Zealand, 1983–2003 (tonnes)

Year	Sheep meat	Beef	Offal	Tinned meat	Poultry	Pig meat	Other	Total
1983	7,780	5,187	1,270	9,371	539	326	342	24,815
1984	9,981	3,768	1,409	10,699	458	351	152	26,818
1985	11,481	5,038	1,564	11,448	577	267	320	30,695
1986	14,680	6,842	1,960	7,041	1,276	229	252	32,279
1987	19,405	10,326	3,182	2,176	1,034	638	149	36,910
1988	23,037	12,238	3,369	670	304	57	495	40,170
1989	23,235	11,930	2,522	712	386	209	476	39,470
1990	28,323	9,111	2,013	385	80	226	195	40,333
1991	31,228	9,635	2,782	361	83	209	310	44,609
1992	36,755	10,301	3,173	149	57	94	706	51,237
1993	41,963	11,894	3,150	97	37	60	331	57,532
1994	42,985	13,349	3,298	226	31	52	141	60,081
1995	38,618	11,017	2,750	159	25	37	122	52,728
1996	39,209	7,688	8,779	287	51	331	400	56,745
1997	39,774	4,677	9,532	373	17	35	397	54,805
1998	33,613	4,473	7,635	123	37	72	309	46,263
1999	31,447	3,201	7,727	189	3	46	441	43,055
2000	37,196	3,236	6,819	286	10	35	167	47,749
2001	27,311	4,512	3,405	264	26	41	34	35,593
2002	22,038	5,618	2,501	889	32	138	85	31,301
2003	23,158	4,457	1,985	937	57	243	32	30,868

Sources: Australian Bureau of Statistics; Statistics New Zealand.

Appendix Table A2.10.1 Value of fishery imports by country, 1981–2003 (US$)

Country	1981	1982	1983	1984	1985
Australia	559,870	589,085	794,113	1,062,726	586,245
Solomon Islands	–	–	–	–	–
Canada	33,608	27,351	39,035	31,810	18,283
Chile	–	–	–	247,303	–
China	178,344	249,783	476,609	399,242	205,121
Congo	–	–	–	–	–
Denmark	5,856	3,357	3,045	1,047	6,466
Fiji	–	–	–	–	–
France	2,989	2,216	–	1,080	–
(East) Germany	17,483	13,908	54,831	15,257	5,411
Hong Kong	66,338	99,162	27,783	69,654	17,371
Indonesia	–	2,588	–	933	1,386
Japan	31,621,443	24,264,220	24,162,512	22,718,711	23,267,482
North Korea	5,367	–	–	681	–
South Korea	23,242	28,525	134,315	282,381	60,583
Malaysia	38,986	40,063	18,771	4,779	2,578
Taiwan	29,496	22,118	2,311	927	2,652
Netherlands	1,083	734	1,456	–	1,046
New Zealand	1,070,872	1,428,273	1,528,425	1,722,479	1,647,772
Norway	22,827	19,489	20,014	6,504	1,231
Peru	1,255,223	552,666	17,660	175,296	–
Philippines	2,280	5,592	10,322	3,707	5,835
Portugal	3,075	3,664	4,622	2,167	3,123
Singapore	10,979	57,984	37,298	12,670	22,648
Slovenia	–	–	–	–	–
South Africa	714	–	–	–	–
Spain	1,856	2,556	1,922	–	–
Switzerland	–	–	–	–	–
Thailand	131,026	158,177	63,394	16,077	25,713
United Kingdom	37,008	15,135	16,052	13,618	40,470
United States	88,622	160,107	417,730	283,089	51,153
Other countries	17,183	2,652	11,560	2,310	1,159
Source country not specified	8,339	7,132	9,952	11,455	6,490
Total	**35,234,109**	**27,756,537**	**27,853,732**	**27,085,903**	**25,980,218**

... continued

Appendix Table A2.10.1 (continued)

Country	1986	1987	1988	1989	1990
Australia	514,914	1,248,560	1,169,452	1,082,954	916,231
Solomon Islands	–	–	–	–	–
Canada	14,156	15,585	8,763	17,203	7,501
Chile	3,175,651	3,115,182	4,042,505	3,023,073	3,818,579
China	1,161,099	2,072,036	2,748,277	1,853,119	1,748,482
Congo	–	–	–	–	–
Denmark	6,799	2,307	1,485	691	6,877
Fiji	–	–	–	–	65,117
France	538	1,031	–	28,377	1,299
(East) Germany	6,152	2,644	–	657	15,356
Hong Kong	96,933	238,732	300,264	203,420	126,135
Indonesia	–	–	835	705	6,265
Japan	29,626,667	28,378,301	33,738,736	31,460,807	27,342,298
North Korea	–	1,518	–	–	622
South Korea	28,120	54,917	381,005	58,723	66,729
Malaysia	4,178	9,398	1,423	705	–
Taiwan	–	1,307	67,861	199,206	–
Netherlands	1,695	–	–	–	–
New Zealand	1,428,819	1,341,123	798,758	746,418	791,003
Norway	–	–	–	1,674	4,563
Peru	–	–	–	–	–
Philippines	27,020	1,586	–	–	–
Portugal	2,419	1,030	774	–	–
Singapore	22,648	61,123	168,880	753,500	363,899
Slovenia	–	–	–	–	–
South Africa	–	2,830	–	–	–
Spain	14,653	–	32,664	6,066	–
Switzerland	–	–	–	1,989	–
Thailand	98,047	151,184	184,958	593,552	728,456
United Kingdom	6,482	2,429	12,161	20,565	35,370
United States	36,242	72,687	17,649	149,559	81,280
Other countries	631	54,654	1,616	136,681	66,004
Source country not specified	416,688	4,620	3,941	5,392	5,434
Total	**36,690,551**	**36,834,784**	**43,682,007**	**40,345,036**	**36,197,500**

... continued

Appendix Table A2.10.1 (continued)

Country	1998	2000	2001	2002	2003
Australia	489,166	289,111	326,819	490,056	135,960
Solomon Islands	703,488	181,735	–	–	883,757
Canada	–	–	805	772	–
Chile	802,654	29,844	47,628	90,395	69,381
China	25,169	7,606	–	11,320	21,784
Congo	–	–	94,308	15,075	99,212
Denmark	–	–	–	–	–
Fiji	151,182	18,050	11,218	–	44,723
France	–	–	–	–	–
(East) Germany	–	–	–	22,271	–
Hong Kong	36,244	32,604	–	–	–
Indonesia	35,130	41,001	10,030	4,539	–
Japan	6,761,395	15,338	9,961	–	35,533
North Korea	–	–	–	–	–
South Korea	317,052	–	765,597	–	555,759
Malaysia	19,526	14,336	7,835	8,722	390,592
Taiwan	49,535	–	127,268	265,818	258,274
Netherlands	–	–	–	–	–
New Zealand	733,381	379,450	418,187	482,184	766,120
Norway	–	–	–	–	–
Peru	–	20,361	58,649	–	–
Philippines	1,998,643	6,948	6,948	2,750	705,032
Portugal	733	–	–	–	–
Singapore	626,022	138,994	45,564	223,318	36,272
Slovenia	45,717	41,394	47,778	91,242	123,340
South Africa	–	134,305	314,929	331,185	48,769
Spain	–	–	–	–	–
Switzerland	–	107,996	22,589	2,801	95,960
Thailand	1,214,547	891,600	119,253	261,205	70,223
United Kingdom	4,193	1,148	–	–	–
United States	33,529	8,282,579	3,615,734	6,193,125	4,088,621
Other countries	412,615	34,183	3,688	81,296	15,412
Source country not specified	5,845	7,354	5,243	5,883	6,110
Total	**14,465,766**	**10,675,937**	**6,060,031**	**8,583,957**	**8,450,834**

Note: These data are based on datasets available for PNG fishery imports in the commodity trade statistics database compiled by the United Nations (UNComtrade). Data on PNG fishery imports are not contained in the UNComtrade figures for the years 1991–1997 and 1999.

Source: United Nations (2005).

Appendix Table A3.2.1 Rural population growing the 16 most important vegetables, by province

Province	Corn	%	Pumpkin tips	%	*Aibika*	%	Amaranthus	%	Highland *pitpit*	%	Lowland *pitpit*	%
Western	84,924	78.8	49,653	46.0	107,837	100.0	18,076	16.8	64,299	59.6	68,404	63.4
Gulf	68,019	73.7	74,389	80.6	82,425	89.3	50,108	54.3	39,117	42.4	92,265	100.0
Central	157,058	100.0	151,823	96.7	156,937	99.9	40,127	25.5	38,639	24.6	65,778	41.9
Milne Bay	172,675	91.7	172,328	91.5	188,334	100.0	2,877	1.5	12,371	6.6	151,708	80.6
Oro	106,288	100.0	106,288	100.0	106,288	100.0	105,116	98.9	18,000	16.9	106,288	100.0
Southern Highlands	514,363	97.7	388,568	73.8	59,851	11.4	336,164	63.9	526,398	100.0	59,851	11.4
Enga	257,212	90.7	119,444	42.1	149,433	52.7	55,219	19.5	272,450	96.1	40,359	14.2
Western Highlands	353,501	95.3	321,196	86.6	57,540	15.5	303,525	81.8	370,013	99.7	16,179	4.4
Simbu	242,748	100.0	151,200	62.3	19,157	7.9	200,304	82.5	242,748	100.0	18,460	7.6
Eastern Highlands	361,921	92.0	34,729	8.8	26,731	6.8	272,345	69.2	393,418	100.0	16,629	4.2
Morobe	315,855	88.7	228,403	64.1	320,149	89.9	278,165	78.1	147,655	41.5	320,194	89.9
Madang	308,135	100.0	305,972	99.3	277,341	90.0	220,087	71.4	51,601	16.7	276,505	89.7
East Sepik	301,345	99.2	201,346	66.3	301,876	99.4	293,081	96.5	531	0.2	269,731	88.8
Sandaun	166,919	100.0	136,074	81.5	166,919	100.0	153,802	92.1	49,817	29.8	141,827	85.0
Manus	25,430	72.9	34,899	100.0	34,899	100.0	27,456	78.7	–	–	22,450	64.3
New Ireland	87,466	84.7	93,301	90.4	103,259	100.0	77,368	74.9	–	–	86,594	83.9
East New Britain	166,643	95.6	159,400	91.5	174,230	100.0	21,295	12.2	16,969	9.7	174,230	100.0
West New Britain	100,704	92.1	105,623	96.6	109,299	100.0	65,274	59.7	3,757	3.4	108,569	99.3
Bougainville	166,472	99.6	167,156	100.0	167,156	100.0	–	–	–	–	161,441	96.6
Papua New Guinea	3,957,678	94.4	3,001,792	71.6	2,609,661	62.2	2,520,389	60.1	2,247,783	53.6	2,197,462	52.4

... continued

Appendix Table A3.2.1 (continued)

Province	Common bean	%	Cucumber	%	Rungia	%	Winged bean	%	*Tulip*	%	Snake bean	%
Western	8,622	8.0	66,570	61.7	6,575	6.1	10,287	9.5	100,183	92.9	18,214	16.9
Gulf	22,244	24.1	54,568	59.1	35,060	38.0	13,196	14.3	72,722	78.8	50,996	55.3
Central	28,527	18.2	129,943	82.7	1,338	0.9	4,247	2.7	56,913	36.2	128,237	81.6
Milne Bay	–	–	39,042	20.7	–	–	41,241	21.9	178,019	94.5	163,969	87.1
Oro	2,668	2.5	102,224	96.2	–	–	19,715	18.5	82,606	77.7	98,208	92.4
Southern Highlands	476,194	90.5	298,393	56.7	523,138	99.4	35,887	6.8	19,903	3.8	–	–
Enga	257,212	90.7	24,343	8.6	243,136	85.8	149,433	52.7	–	–	–	–
Western Highlands	354,868	95.6	31,609	8.5	370,013	99.7	306,895	82.7	58,574	15.8	–	–
Simbu	228,496	94.1	75,732	31.2	214,921	88.5	166,521	68.6	14,945	6.2	–	–
Eastern Highlands	349,316	88.8	190,783	48.5	311,427	79.2	354,002	90.0	153	0.0	153	0.0
Morobe	155,689	43.7	299,032	84.0	26,427	7.4	125,556	35.3	100,830	28.3	114,209	32.1
Madang	59,283	19.2	96,178	31.2	36,301	11.8	108,338	35.2	205,124	66.6	201,899	65.5
East Sepik	53,341	17.6	40,304	13.3	–	–	222,895	73.4	301,876	99.4	240,437	79.2
Sandaun	11,713	7.0	59,417	35.6	–	–	91,806	55.0	154,120	92.3	91,806	55.0
Manus	–	–	1,077	3.1	–	–	–	–	4,506	12.9	25,384	72.7
New Ireland	–	–	37,871	36.7	–	–	–	–	15,405	14.9	101,437	98.2
East New Britain	–	–	144,262	82.8	–	–	4,649	2.7	20,091	11.5	155,429	89.2
West New Britain	–	–	30,249	27.7	–	–	–	–	25,092	23.0	43,103	39.4
Bougainville	–	–	111,029	66.4	–	–	–	–	142,272	85.1	103,658	62.0
Papua New Guinea	2,008,173	47.9	1,832,626	43.7	1,768,336	42.2	1,654,668	39.5	1,553,334	37.0	1,537,139	36.7

... continued

Appendix Table A3.2.1 (continued)

Province	*Kumu musong* leaves	%	Peanut	%	Oenanthe	%	Cabbage	%	Total rural population
Western	7,510	7.0	12,558	11.6	–	–	–	–	107,837
Gulf	31,913	34.6	–	–	32,051	34.7	3,361	3.6	92,265
Central	40,691	25.9	79,854	50.8	22,823	14.5	22,934	14.6	157,058
Milne Bay	151,231	80.3	4,140	2.2	–	–	–	–	188,334
Oro	82,897	78.0	79,923	75.2	–	–	–	–	106,288
Southern Highlands	9,493	1.8	40,611	7.7	466,942	88.7	339,160	64.4	526,398
Enga	–	–	27,061	9.5	243,136	85.8	139,033	49.0	283,498
Western Highlands	39,994	10.8	256,999	69.3	250,984	67.6	112,970	30.4	371,014
Simbu	40,418	16.7	117,664	48.5	82,312	33.9	223,595	92.1	242,748
Eastern Highlands	–	–	267,633	68.0	201,191	51.1	54,885	14.0	393,418
Morobe	99,423	27.9	147,311	41.4	30,515	8.6	106,201	29.8	356,100
Madang	59,717	19.4	65,498	21.3	4,142	1.3	22,656	7.4	308,135
East Sepik	198,075	65.2	67,599	22.3	–	–	–	–	303,706
Sandaun	134,939	80.8	–	–	–	–	11,713	7.0	166,919
Manus	28,533	81.8	2,980	8.5	–	–	–	–	34,899
New Ireland	49,110	47.6	–	–	–	–	–	–	103,259
East New Britain	148,987	85.5	135,338	77.7	–	–	6,248	3.6	174,230
West New Britain	104,128	95.3	24,970	22.8	–	–	466	0.4	109,299
Bougainville	166,472	99.6	35,935	21.5	–	–	–	–	167,156
Papua New Guinea	1,393,531	33.2	1,366,074	32.6	1,334,096	31.8	1,043,222	24.9	4,192,561

Sources: NSO (2002), MASP.

Appendix Table A3.3.1 Rural population growing the 10 most important fruit, by province

Province	Pawpaw	%	*Marita* pandanus	%	Pineapple	%	Mango	%	Watermelon	%
Western	86,628	80.3	69,955	64.9	102,960	95.5	28,695	26.6	23,264	21.6
Gulf	74,754	81.0	41,264	44.7	84,589	91.7	51,692	56.0	51,924	56.3
Central	134,225	85.5	31,072	19.8	157,058	100.0	134,052	85.4	130,103	82.8
Milne Bay	186,246	98.9	53,863	28.6	150,155	79.7	186,246	98.9	180,419	95.8
Oro	103,613	97.5	85,744	80.7	106,288	100.0	103,613	97.5	103,627	97.5
Southern Highlands	87,000	16.5	269,253	51.2	97,626	18.5	–	–	11,005	2.1
Enga	40,359	14.2	40,359	14.2	27,061	9.5	–	–	–	–
Western Highlands	297,038	80.1	298,027	80.3	230,216	62.1	14,063	3.8	–	–
Simbu	131,650	54.2	242,748	100.0	14,945	6.2	–	–	–	–
Eastern Highlands	27,108	6.9	363,740	92.5	153	0.0	153	0.0	737	0.2
Morobe	216,274	60.7	335,979	94.3	235,369	66.1	226,540	63.6	83,826	23.5
Madang	282,599	91.7	206,728	67.1	219,550	71.3	275,281	89.3	111,695	36.2
East Sepik	268,105	88.3	264,556	87.1	272,839	89.8	272,839	89.8	32,346	10.7
Sandaun	141,827	85.0	153,941	92.2	92,124	55.2	91,806	55.0	–	–
Manus	25,384	72.7	–	–	19,847	56.9	30,436	87.2	5,537	15.9
New Ireland	103,259	100.0	–	–	90,219	87.4	97,314	94.2	44,527	43.1
East New Britain	160,754	92.3	–	–	135,338	77.7	174,230	100.0	117,313	67.3
West New Britain	108,825	99.6	8,247	7.5	65,840	60.2	109,299	100.0	32,246	29.5
Bougainville	167,156	100.0	–	–	95,744	57.3	166,472	99.6	161,441	96.6
Papua New Guinea	2,642,804	63.0	2,465,476	58.8	2,197,921	52.4	1,962,731	46.8	1,090,010	26.0

... continued

Appendix Table A3.3.1 (continued)

Province	*Ton*	%	Malay apple	%	Guava	%	Orange	%	Passionfuit	%	Total rural population
Western	–	–	5,431	5.0	14,278	13.2	19,327	17.9	–	–	107,837
Gulf	–	–	45,798	49.6	51,692	56.0	76,705	83.1	–	–	92,265
Central	–	–	84,175	53.6	11,435	7.3	66,740	42.5	51,361	32.7	157,058
Milne Bay	162	0.1	161,892	86.0	34,244	18.2	32,336	17.2	–	–	188,334
Oro	2,459	2.3	56,588	53.2	29,576	27.8	81,669	76.8	–	–	106,288
Southern Highlands	–	–	–	–	–	–	–	–	118,702	22.5	526,398
Enga	–	–	–	–	–	–	–	–	144,407	50.9	283,498
Western Highlands	–	–	–	–	–	–	17,546	4.7	21,154	5.7	371,014
Simbu	–	–	–	–	–	–	–	–	697	0.3	242,748
Eastern Highlands	–	–	–	–	–	–	–	–	32,296	8.2	393,418
Morobe	91,240	25.6	95,355	26.8	110,750	31.1	111,117	31.2	58,444	16.4	356,100
Madang	134,396	43.6	1,140	0.4	90,301	29.3	30,466	9.9	28,291	9.2	308,135
East Sepik	284,616	93.7	13,769	4.5	11,010	3.6	8,594	2.8	–	–	303,706
Sandaun	141,509	84.8	–	–	–	–	–	–	–	–	166,919
Manus	24,307	69.6	34,899	100.0	9,512	27.3	–	–	–	–	34,899
New Ireland	101,437	98.2	51,040	49.4	–	–	1,815	1.8	–	–	103,259
East New Britain	170,297	97.7	168,007	96.4	135,338	77.7	6,248	3.6	–	–	174,230
West New Britain	90,657	82.9	104,935	96.0	62,926	57.6	11,674	10.7	–	–	109,299
Bougainville	5,031	3.0	162,121	97.0	162,588	97.3	39,251	23.5	–	–	167,156
Papua New Guinea	1,046,111	25.0	985,150	23.5	723,650	17.3	503,488	12.0	455,352	10.9	4,192,561

Sources: NSO (2002), MASP.

Appendix Table A3.4.1 Rural population growing the 13 most important nuts, by province

Province	Breadfruit	%	*Karuka*, planted	%	*Karuka*, wild	%	*Galip* (*Canarium* spp.)	%	Polynesian chestnut (*aila*)	%	Sea almond (*talis*)	%	*Pao* (*Barringtonia procera*)	%
Western	107,280	99.5	762	0.7	762	0.7	5,752	5.3	–	–	6,320	5.9	–	–
Gulf	83,798	90.8	10,751	11.7	–	–	–	–	–	–	–	–	–	–
Central	138,535	88.2	22,410	14.3	28,758	18.3	–	–	19,210	12.2	6,340	4.0	–	–
Milne Bay	187,846	99.7	–	–	14,145	7.5	32,640	17.3	173,664	92.2	162,026	86.0	14,300	7.6
Oro	99,845	93.9	357	0.3	4,255	4.0	80,417	75.7	–	–	5,182	4.9	–	–
Southern Highlands	31,418	6.0	508,009	96.5	302,071	57.4	–	–	–	–	–	–	–	–
Enga	14,699	5.2	275,816	97.3	173,567	61.2	–	–	–	–	–	–	–	–
Western Highlands	57,617	15.5	364,729	98.3	351,336	94.7	–	–	–	–	–	–	–	–
Simbu	16,669	6.9	232,087	95.6	231,228	95.3	–	–	–	–	–	–	–	–
Eastern Highlands	8,789	2.2	373,213	94.9	106,010	26.9	–	–	–	–	–	–	–	–
Morobe	274,660	77.1	135,305	38.0	62,517	17.6	112,036	31.5	4,627	1.3	7,488	2.1	1,504	0.4
Madang	285,275	92.6	49,148	16.0	43,945	14.3	228,497	74.2	46,788	15.2	21,602	7.0	–	–
East Sepik	297,779	98.0	–	–	–	–	190,598	62.8	9,590	3.2	5,882	1.9	9,725	3.2
Sandaun	158,610	95.0	25,866	15.5	24,612	14.7	118,730	71.1	–	–	–	–	–	–
Manus	34,899	100.0	–	–	–	–	34,899	100.0	22,019	63.1	18,665	53.5	34,899	100.0
New Ireland	101,751	98.5	–	–	–	–	97,622	94.5	90,615	87.8	103,103	99.8	101,751	98.5
East New Britain	166,293	95.4	–	–	–	–	161,914	92.9	160,501	92.1	125,646	72.1	161,237	92.5
West New Britatin	106,659	97.6	–	–	–	–	87,951	80.5	102,241	93.5	73,376	67.1	45,101	41.3
Bougainville	163,119	97.6	–	–	–	–	163,119	97.6	7,494	4.5	10,361	6.2	163,119	97.6
Papua New Guinea	2,335,541	55.7	1,998,454	47.7	1,343,205	32.0	1,314,173	31.3	636,749	15.2	545,993	13.0	531,634	12.7

... continued

Appendix Table A3.4.1 (continued)

Province	*Okari* (*T. kaernbachii*)	%	*Okari* (*T. impediens*)	%	*Sis* (*solomon*)	%	Castanopsis	%	*Dausia*	%	*Tulip*	%	Total rural population
Western	43,919	40.7	–	–	–	–	762	0.7	–	–	–	–	107,837
Gulf	81,624	88.5	–	–	4,002	4.3	–	–	–	–	699	0.8	92,265
Central	118,215	75.3	–	–	1,319	0.8	510	0.3	6,340	4.0	10,990	7.0	157,058
Milne Bay	15,066	8.0	–	–	95,860	50.9	9,237	4.9	83,723	44.5	43,267	23.0	188,334
Oro	101,935	95.9	–	–	11,977	11.3	–	–	–	–	191	0.2	106,288
Southern Highlands	22,850	4.3	–	–	37,425	7.1	226,124	43.0	–	–	11,347	2.2	526,398
Enga	–	–	–	–	15,053	5.3	29,666	10.5	–	–	–	–	283,498
Western Highlands	1,577	0.4	–	–	2,797	0.8	1,523	0.4	–	–	–	–	371,014
Simbu	17,116	7.1	–	–	17,116	7.1	–	–	–	–	–	–	242,748
Eastern Highlands	–	–	–	–	3,769	1.0	–	–	–	–	–	–	393,418
Morobe	65,583	18.4	–	–	17,181	4.8	–	–	–	–	677	0.2	356,100
Madang	–	–	89,324	29.0	40,140	13.0	19,966	6.5	–	–	1,449	0.5	308,135
East Sepik	–	–	174,472	57.4	9,112	3.0	–	–	–	–	–	–	303,706
Sandaun	–	–	88,334	52.9	22,033	13.2	13,426	8.0	–	–	–	–	166,919
Manus	30,907	88.6	–	–	–	–	–	–	–	–	–	–	34,899
New Ireland	–	–	–	–	–	–	–	–	–	–	–	–	103,259
East New Britain	–	–	–	–	14,872	8.5	–	–	–	–	–	–	174,230
West New Britatin	2,817	2.6	–	–	31,019	28.4	–	–	–	–	–	–	109,299
Bougainville	–	–	–	–	–	–	–	–	–	–	–	–	167,156
Papua New Guinea	501,609	12.0	352,130	8.4	323,675	7.7	301,214	7.2	90,063	2.1	68,619	1.6	4,192,561

Sources: NSO (2002), MASP.

Appendix Table A3.5.1 Rural population growing the six most important stimulants, by province

Province	Tobacco	%	Betel nut	%	Lowland betel pepper	%	Highland betel pepper	%	Highland betel nut[a]	%	Kava	%	Total rural population
Western	102,400	95.0	60,971	56.5	60,971	56.5	–	–	–	–	31,529	29.2	107,837
Gulf	86,366	93.6	75,106	81.4	75,106	81.4	17,154	18.6	17,154	18.6	–	–	92,265
Central	152,277	97.0	105,687	67.3	105,687	67.3	343	0.2	24,504	15.6	–	–	157,058
Milne Bay	188,334	100.0	187,636	99.6	187,636	99.6	15,991	8.5	15,991	8.5	–	–	188,334
Oro	106,288	100.0	106,288	100.0	106,288	100.0	–	–	2,654	2.5	–	–	106,288
Southern Highlands	526,398	100.0	3,300	0.6	9,493	1.8	323,385	61.4	52,831	10.0	–	–	526,398
Enga	283,498	100.0	11,045	3.9	11,045	3.9	93,836	33.1	–	–	–	–	283,498
Western Highlands	371,014	100.0	41,028	11.1	14,156	3.8	60,115	16.2	17,546	4.7	–	–	371,014
Simbu	242,748	100.0	14,244	5.9	14,244	5.9	194,807	80.3	39,655	16.3	–	–	242,748
Eastern Highlands	393,418	100.0	153	0.0	153	0.0	388,653	98.8	293,730	74.7	–	–	393,418
Morobe	355,746	99.9	267,446	75.1	267,446	75.1	151,153	42.4	151,911	42.7	–	–	356,100
Madang	308,135	100.0	275,702	89.5	273,145	88.6	27,133	8.8	26,782	8.7	–	–	308,135
East Sepik	301,876	99.4	301,876	99.4	301,876	99.4	–	–	–	–	–	–	303,706
Sandaun	166,919	100.0	141,509	84.8	141,509	84.8	–	–	–	–	–	–	166,919
Manus	24,353	69.8	34,899	100.0	34,899	100.0	–	–	–	–	1,903	5.5	34,899
New Ireland	99,790	96.6	95,499	92.5	95,499	92.5	–	–	–	–	–	–	103,259
East New Britain	161,247	92.5	174,230	100.0	174,230	100.0	–	–	6,248	3.6	–	–	174,230
West New Britain	100,704	92.1	109,299	100.0	109,299	100.0	–	–	466	0.4	–	–	109,299
Bougainville	162,588	97.3	166,472	99.6	166,472	99.6	–	–	–	–	–	–	167,156
Papua New Guinea	4,134,099	98.6	2,172,390	51.8	2,149,154	51.3	1,272,570	30.4	649,472	15.5	33,432	0.8	4,192,561

[a] Data include highland betel nut (*Areca macrocalyx*) and other wild betel nut species.

Sources: NSO (2002), MASP.

Appendix Table A3.8.1 Rural population using fallows of given periods, by province

Province	1–4 years	%	5–15 years	%	More than 15 years	%	No long fallow	%	Total rural population
Western	5,592	5.2	25,884	24.0	76,360	70.8	0	0.0	107,837
Gulf	0	0.0	13,452	14.6	78,813	85.4	0	0.0	92,265
Central	16,650	10.6	85,674	54.5	54,734	34.8	0	0.0	157,058
Milne Bay	2,805	1.5	57,541	30.6	127,988	68.0	0	0.0	188,334
Oro	0	0.0	44,096	41.5	62,192	58.5	0	0.0	106,288
Southern Highlands	149,448	28.4	181,645	34.5	128,763	24.5	66,542	12.6	526,398
Enga	0	0.0	15,293	5.4	77,947	27.5	190,258	67.1	283,498
Western Highlands	0	0.0	345,480	93.1	2,862	0.8	22,672	6.1	371,014
Simbu	0	0.0	211,186	87.0	31,562	13.0	0	0.0	242,748
Eastern Highlands	148,117	37.6	48,453	12.3	180,855	46.0	15,993	4.1	393,418
Morobe	26,326	7.4	260,394	73.1	68,333	19.2	1,048	0.3	356,100
Madang	14,977	4.9	197,169	64.0	95,990	31.2	0	0.0	308,135
East Sepik	13,491	4.4	117,229	38.6	171,426	56.4	1,561	0.5	303,706
Sandaun	0	0.0	11,783	7.1	155,136	92.9	0	0.0	166,919
Manus	1,008	2.9	4,329	12.4	24,278	69.6	5,284	15.1	34,899
New Ireland	3,341	3.2	39,536	38.3	60,382	58.5	0	0.0	103,259
East New Britain	13,622	7.8	149,542	85.8	11,066	6.4	0	0.0	174,230
West New Britain	34,832	31.9	44,202	40.4	30,265	27.7	0	0.0	109,299
Bougainville	30,823	18.4	133,572	79.9	0	0.0	2,761	1.7	167,156
Papua New Guinea	461,031	11.0	1,986,460	47.4	1,438,952	34.3	306,118	7.3	4,192,561

Sources: NSO (2002); MASP.

Appendix Table A3.8.2 Rural population clearing different fallow vegetation, by province

Province	Short grass	%	Tall grass	%	Grass and woody regrowth	%	Low woody regrowth	%	Tall woody regrowth[a]	%	Savanna	%	Fallow not defined[b]	%	Total rural population
Western	12,349	11.5	3,210	3.0	6,780	6.3	6,876	6.4	66,335	61.5	12,287	11.4	0	0.0	107,837
Gulf	3,254	3.5	0	0.0	0	0.0	7,561	8.2	81,450	88.3	0	0.0	0	0.0	92,265
Central	46,764	29.8	8,230	5.2	7,193	4.6	37,494	23.9	53,983	34.4	3,394	2.2	0	0.0	157,058
Milne Bay	3,944	2.1	0	0.0	5,933	3.2	44,339	23.5	134,118	71.2	0	0.0	0	0.0	188,334
Oro	14,277	13.4	0	0.0	0	0.0	50,220	47.2	41,791	39.3	0	0.0	0	0.0	106,288
Southern Highlands	107,588	20.4	158,986	30.2	183,560	34.9	33,765	6.4	42,499	8.1	0	0.0	0	0.0	526,398
Enga	0	0.0	111,532	39.3	80,057	28.2	71,764	25.3	20,144	7.1	0	0.0	0	0.0	283,498
Western Highlands	1,474	0.4	262,116	70.6	48,697	13.1	41,613	11.2	17,114	4.6	0	0.0	0	0.0	371,014
Simbu	0	0.0	127,365	52.5	90,699	37.4	7,715	3.2	16,969	7.0	0	0.0	0	0.0	242,748
Eastern Highlands	162,485	41.3	27,044	6.9	117,861	30.0	58,368	14.8	27,660	7.0	0	0.0	0	0.0	393,418
Morobe	35,744	10.0	6,007	1.7	54,492	15.3	80,129	22.5	179,728	50.5	0	0.0	0	0.0	356,100
Madang	42,208	13.7	543	0.2	20,630	6.7	78,325	25.4	166,429	54.0	0	0.0	0	0.0	308,135
East Sepik	17,927	5.9	27,603	9.1	38,292	12.6	55,832	18.4	162,825	53.6	0	0.0	1,227	0.4	303,706
Sandaun	0	0.0	0	0.0	9,345	5.6	0	0.0	157,574	94.4	0	0.0	0	0.0	166,919
Manus	0	0.0	0	0.0	0	0.0	4,871	14.0	24,977	71.6	0	0.0	5,051	14.5	34,899
New Ireland	0	0.0	0	0.0	1,000	1.0	39,185	37.9	63,073	61.1	0	0.0	0	0.0	103,259
East New Britain	0	0.0	66,372	38.1	0	0.0	10,389	6.0	97,469	55.9	0	0.0	0	0.0	174,230
West New Britain	0	0.0	0	0.0	1,005	0.9	35,226	32.2	73,067	66.9	0	0.0	0	0.0	109,299
Bougainville	0	0.0	0	0.0	0	0.0	31,331	18.7	133,452	79.8	0	0.0	2,373	1.4	167,156
Papua New Guinea	448,015	10.7	799,006	19.1	665,545	15.9	695,003	16.6	1,560,659	37.2	15,681	0.4	8,651	0.2	4,192,561

[a] 'Tall woody regrowth' is the technical term for tall secondary forest.

[b] Agricultural systems with 'fallow not defined' occur in swamps at the mouth of the Sepik River and on some atolls.

Sources: NSO (2002), MASP.

Appendix Table A3.9.1 Rural population using soil retention barriers, by province

Province	Importance of soil retention barriers								Total rural population
	Minor	%	Important	%	Very important	%	Not used	%	
Western	4,325	4.0	0	0.0	0	0.0	103,512	96.0	107,837
Gulf	0	0.0	0	0.0	0	0.0	92,265	100.0	92,265
Central	31,025	19.8	0	0.0	869	0.6	125,164	79.7	157,058
Milne Bay	70,819	37.6	47,649	25.3	18,094	9.6	51,771	27.5	188,334
Oro	0	0.0	0	0.0	0	0.0	106,288	100.0	106,288
Southern Highlands	25,839	4.9	0	0.0	0	0.0	500,559	95.1	526,398
Enga	31,542	11.1	0	0.0	0	0.0	251,956	88.9	283,498
Western Highlands	89,057	24.0	16,374	4.4	0	0.0	265,583	71.6	371,014
Simbu	125,614	51.7	58,912	24.3	40,565	16.7	17,658	7.3	242,748
Eastern Highlands	98,246	25.0	10,931	2.8	5,376	1.4	278,865	70.9	393,418
Morobe	84,365	23.7	31,101	8.7	26,519	7.4	214,115	60.1	356,100
Madang	127,042	41.2	3,596	1.2	0	0.0	177,497	57.6	308,135
East Sepik	131,622	43.3	65,879	21.7	0	0.0	106,204	35.0	303,706
Sandaun	88,751	53.2	0	0.0	0	0.0	78,168	46.8	166,919
Manus	0	0.0	0	0.0	0	0.0	34,899	100.0	34,899
New Ireland	1,000	1.0	0	0.0	0	0.0	102,259	99.0	103,259
East New Britain	0	0.0	3,617	2.1	0	0.0	170,613	97.9	174,230
West New Britain	0	0.0	773	0.7	0	0.0	108,526	99.3	109,299
Bougainville	0	0.0	0	0.0	0	0.0	167,156	100.0	167,156
Papua New Guinea	909,247	21.7	238,832	5.7	91,422	2.2	2,953,059	70.4	4,192,561

Sources: NSO (2002); MASP.

Appendix Table A3.10.1 Rural population planting trees in fallows, by province

Province	Importance of planting trees in fallows								Total rural population
	Minor	%	Important	%	Very important	%	Not used	%	
Western	0	0.0	0	0.0	0	0.0	107,837	100.0	107,837
Gulf	0	0.0	0	0.0	0	0.0	92,265	100.0	92,265
Central	0	0.0	0	0.0	0	0.0	157,058	100.0	157,058
Milne Bay	0	0.0	764	0.4	0	0.0	187,570	99.6	188,334
Oro	0	0.0	0	0.0	0	0.0	106,288	100.0	106,288
Southern Highlands	316,948	60.2	0	0.0	0	0.0	209,450	39.8	526,398
Enga	203,280	71.7	0	0.0	0	0.0	80,218	28.3	283,498
Western Highlands	348,915	94.0	0	0.0	1,486	0.4	20,613	5.6	371,014
Simbu	127,658	52.6	85,036	35.0	9,394	3.9	20,660	8.5	242,748
Eastern Highlands	333,667	84.8	23,958	6.1	2,683	0.7	33,110	8.4	393,418
Morobe	4,023	1.1	7,355	2.1	0	0.0	344,722	96.8	356,100
Madang	12,593	4.1	0	0.0	16,555	5.4	278,987	90.5	308,135
East Sepik	0	0.0	0	0.0	0	0.0	303,706	100.0	303,706
Sandaun	6,341	3.8	4,514	2.7	0	0.0	156,064	93.5	166,919
Manus	0	0.0	0	0.0	0	0.0	34,899	100.0	34,899
New Ireland	0	0.0	0	0.0	0	0.0	103,259	100.0	103,259
East New Britain	0	0.0	0	0.0	0	0.0	174,230	100.0	174,230
West New Britain	0	0.0	0	0.0	0	0.0	109,299	100.0	109,299
Bougainville	0	0.0	0	0.0	0	0.0	167,156	100.0	167,156
Papua New Guinea	1,353,425	32.3	121,628	2.9	30,117	0.7	2,687,391	64.1	4,192,561

Sources: NSO (2002); MASP.

Appendix Table A3.10.2 Rural population planting legumes (peanut and winged bean) in rotations with root crops, by province

Province	Importance of planting legumes in rotations with root crops								Total rural population
	Minor	%	Important	%	Very important	%	Not used	%	
Western	3,149	2.9	0	0.0	0	0.0	104,688	97.1	107,837
Gulf	0	0.0	0	0.0	0	0.0	92,265	100.0	92,265
Central	5,860	3.7	0	0.0	1,589	1.0	149,609	95.3	157,058
Milne Bay	0	0.0	0	0.0	0	0.0	188,334	100.0	188,334
Oro	0	0.0	0	0.0	0	0.0	106,288	100.0	106,288
Southern Highlands	0	0.0	0	0.0	0	0.0	526,398	100.0	526,398
Enga	0	0.0	0	0.0	0	0.0	283,498	100.0	283,498
Western Highlands	48,203	13.0	245,138	66.1	0	0.0	77,674	20.9	371,014
Simbu	114,067	47.0	64,477	26.6	0	0.0	64,204	26.4	242,748
Eastern Highlands	108,580	27.6	47,545	12.1	166,830	42.4	70,463	17.9	393,418
Morobe	4,619	1.3	27,286	7.7	4,023	1.1	320,173	89.9	356,100
Madang	15,029	4.9	1,715	0.6	0	0.0	291,391	94.6	308,135
East Sepik	8,093	2.7	0	0.0	0	0.0	295,613	97.3	303,706
Sandaun	0	0.0	0	0.0	0	0.0	166,919	100.0	166,919
Manus	0	0.0	896	2.6	0	0.0	34,003	97.4	34,899
New Ireland	4,231	4.1	0	0.0	0	0.0	99,028	95.9	103,259
East New Britain	22,893	13.1	107,215	61.5	0	0.0	44,122	25.3	174,230
West New Britain	19,714	18.0	0	0.0	0	0.0	89,585	82.0	109,299
Bougainville	0	0.0	0	0.0	0	0.0	167,156	100.0	167,156
Papua New Guinea	354,438	8.5	494,271	11.8	172,442	4.1	3,171,410	75.6	4,192,561

Sources: NSO (2002); MASP.

Appendix Table A3.11.1 Rural population using small mounds in food production systems, by province

Province	Importance of small mounds in food production systems								Total rural population
	Minor	%	Important	%	Very important	%	Not used	%	
Western	19,508	18.1	8,761	8.1	10,333	9.6	69,235	64.2	107,837
Gulf	60,509	65.6	0	0.0	0	0.0	31,756	34.4	92,265
Central	8,718	5.6	105,102	66.9	40,516	25.8	2,721	1.7	157,058
Milne Bay	28,138	14.9	143,274	76.1	583	0.3	16,339	8.7	188,334
Oro	3,786	3.6	30,405	28.6	69,445	65.3	2,652	2.5	106,288
Southern Highlands	39,106	7.4	0	0.0	0	0.0	487,292	92.6	526,398
Enga	84,628	29.9	0	0.0	0	0.0	198,870	70.1	283,498
Western Highlands	16,505	4.4	1,535	0.4	40,793	11.0	312,181	84.1	371,014
Simbu	4,394	1.8	15,238	6.3	222,088	91.5	1,028	0.4	242,748
Eastern Highlands	10,208	2.6	44,051	11.2	332,901	84.6	6,258	1.6	393,418
Morobe	74,297	20.9	129,230	36.3	66,489	18.7	86,084	24.2	356,100
Madang	189,224	61.4	49,027	15.9	8,730	2.8	61,154	19.8	308,135
East Sepik	197,011	64.9	0	0.0	0	0.0	106,695	35.1	303,706
Sandaun	124,352	74.5	0	0.0	0	0.0	42,567	25.5	166,919
Manus	7,396	21.2	20,757	59.5	1,779	5.1	4,967	14.2	34,899
New Ireland	42,943	41.6	15,400	14.9	24,211	23.4	20,705	20.1	103,259
East New Britain	134,140	77.0	21,904	12.6	8,466	4.9	9,720	5.6	174,230
West New Britain	46,624	42.7	51,194	46.8	0	0.0	11,481	10.5	109,299
Bougainville	23,701	14.2	25,782	15.4	112,600	67.4	5,074	3.0	167,156
Papua New Guinea	1,115,188	26.6	661,661	15.8	938,933	22.4	1,476,779	35.2	4,192,561

Sources: NSO (2002); MASP.

Appendix Table A3.11.2 Rural population using medium-sized mounds in food production systems, by province

Province	Importance of medium-sized mounds in food production systems								Total rural population
	Minor	%	Important	%	Very important	%	Not used	%	
Western	12,755	11.8	0	0.0	0	0.0	95,082	88.2	107,837
Gulf	0	0.0	0	0.0	0	0.0	92,265	100.0	92,265
Central	0	0.0	0	0.0	399	0.3	156,659	99.7	157,058
Milne Bay	0	0.0	0	0.0	0	0.0	188,334	100.0	188,334
Oro	0	0.0	0	0.0	0	0.0	106,288	100.0	106,288
Southern Highlands	125,459	23.8	30,983	5.9	324,377	61.6	45,579	8.7	526,398
Enga	1,059	0.4	32,190	11.4	103,503	36.5	146,746	51.8	283,498
Western Highlands	19,078	5.1	46,668	12.6	2,419	0.7	302,849	81.6	371,014
Simbu	0	0.0	0	0.0	0	0.0	242,748	100.0	242,748
Eastern Highlands	0	0.0	0	0.0	0	0.0	393,418	100.0	393,418
Morobe	22,811	6.4	4,837	1.4	3,144	0.9	325,309	91.4	356,100
Madang	0	0.0	0	0.0	672	0.2	307,463	99.8	308,135
East Sepik	0	0.0	8,093	2.7	0	0.0	295,613	97.3	303,706
Sandaun	0	0.0	0	0.0	0	0.0	166,919	100.0	166,919
Manus	0	0.0	0	0.0	0	0.0	34,899	100.0	34,899
New Ireland	0	0.0	0	0.0	0	0.0	103,259	100.0	103,259
East New Britain	0	0.0	0	0.0	0	0.0	174,230	100.0	174,230
West New Britain	0	0.0	0	0.0	0	0.0	109,299	100.0	109,299
Bougainville	0	0.0	0	0.0	0	0.0	167,156	100.0	167,156
Papua New Guinea	181,162	4.3	122,770	2.9	434,514	10.4	3,454,115	82.4	4,192,561

Sources: NSO (2002); MASP.

Appendix Table A3.11.3 Rural population using large mounds in food production systems, by province

Province	Importance of large mounds in food production systems								Total rural population
	Minor	%	Important	%	Very important	%	Not used	%	
Western	0	0.0	0	0.0	0	0.0	107,837	100.0	107,837
Gulf	0	0.0	0	0.0	0	0.0	92,265	100.0	92,265
Central	0	0.0	0	0.0	0	0.0	157,058	100.0	157,058
Milne Bay	0	0.0	0	0.0	0	0.0	188,334	100.0	188,334
Oro	0	0.0	0	0.0	0	0.0	106,288	100.0	106,288
Southern Highlands	55,670	10.6	0	0.0	0	0.0	470,728	89.4	526,398
Enga	0	0.0	19,566	6.9	129,767	45.8	134,165	47.3	283,498
Western Highlands	0	0.0	47,316	12.8	19,429	5.2	304,269	82.0	371,014
Simbu	0	0.0	0	0.0	0	0.0	242,748	100.0	242,748
Eastern Highlands	0	0.0	0	0.0	0	0.0	393,418	100.0	393,418
Morobe	0	0.0	0	0.0	0	0.0	356,100	100.0	356,100
Madang	0	0.0	0	0.0	0	0.0	308,135	100.0	308,135
East Sepik	0	0.0	0	0.0	0	0.0	303,706	100.0	303,706
Sandaun	0	0.0	0	0.0	0	0.0	166,919	100.0	166,919
Manus	0	0.0	0	0.0	0	0.0	34,899	100.0	34,899
New Ireland	0	0.0	0	0.0	0	0.0	103,259	100.0	103,259
East New Britain	0	0.0	0	0.0	0	0.0	174,230	100.0	174,230
West New Britain	0	0.0	0	0.0	0	0.0	109,299	100.0	109,299
Bougainville	0	0.0	0	0.0	0	0.0	167,156	100.0	167,156
Papua New Guinea	55,670	1.3	66,882	1.6	149,196	3.6	3,920,813	93.5	4,192,561

Sources: NSO (2002); MASP.

Appendix Table A3.11.4 Rural population using square beds in food production systems, by province

Province	Importance of square beds in food production systems								Total rural population
	Minor	%	Important	%	Very important	%	Not used	%	
Western	5,613	5.2	0	0.0	0	0.0	102,224	94.8	107,837
Gulf	0	0.0	0	0.0	0	0.0	92,265	100.0	92,265
Central	0	0.0	0	0.0	0	0.0	157,058	100.0	157,058
Milne Bay	0	0.0	0	0.0	0	0.0	188,334	100.0	188,334
Oro	0	0.0	0	0.0	0	0.0	106,288	100.0	106,288
Southern Highlands	122,227	23.2	18,152	3.4	0	0.0	386,019	73.3	526,398
Enga	0	0.0	0	0.0	0	0.0	283,498	100.0	283,498
Western Highlands	46,668	12.6	0	0.0	245,138	66.1	79,209	21.3	371,014
Simbu	114,325	47.1	0	0.0	64,477	26.6	63,945	26.3	242,748
Eastern Highlands	0	0.0	0	0.0	0	0.0	393,418	100.0	393,418
Morobe	0	0.0	0	0.0	0	0.0	356,100	100.0	356,100
Madang	0	0.0	0	0.0	0	0.0	308,135	100.0	308,135
East Sepik	0	0.0	0	0.0	0	0.0	303,706	100.0	303,706
Sandaun	0	0.0	0	0.0	0	0.0	166,919	100.0	166,919
Manus	0	0.0	0	0.0	0	0.0	34,899	100.0	34,899
New Ireland	0	0.0	0	0.0	0	0.0	103,259	100.0	103,259
East New Britain	0	0.0	0	0.0	0	0.0	174,230	100.0	174,230
West New Britain	0	0.0	0	0.0	0	0.0	109,299	100.0	109,299
Bougainville	0	0.0	0	0.0	0	0.0	167,156	100.0	167,156
Papua New Guinea	288,832	6.9	18,152	0.4	309,615	7.4	3,575,962	85.3	4,192,561

Sources: NSO (2002); MASP.

Appendix Table A3.11.5 Rural population using long beds in food production systems, by province

Province	Importance of long beds in food production systems								Total rural population
	Minor	%	Important	%	Very important	%	Not used	%	
Western	8,825	8.2	0	0.0	16,025	14.9	82,987	77.0	107,837
Gulf	0	0.0	0	0.0	0	0.0	92,265	100.0	92,265
Central	0	0.0	0	0.0	1,589	1.0	155,469	99.0	157,058
Milne Bay	0	0.0	0	0.0	0	0.0	188,334	100.0	188,334
Oro	0	0.0	0	0.0	0	0.0	106,288	100.0	106,288
Southern Highlands	68,709	13.1	140,379	26.7	0	0.0	317,311	60.3	526,398
Enga	39,154	13.8	0	0.0	0	0.0	244,344	86.2	283,498
Western Highlands	264,215	71.2	0	0.0	0	0.0	106,799	28.8	371,014
Simbu	64,477	26.6	0	0.0	839	0.3	177,432	73.1	242,748
Eastern Highlands	93,694	23.8	21,855	5.6	237,465	60.4	40,403	10.3	393,418
Morobe	0	0.0	0	0.0	2,850	0.8	353,250	99.2	356,100
Madang	13,285	4.3	0	0.0	1,414	0.5	293,436	95.2	308,135
East Sepik	0	0.0	0	0.0	0	0.0	303,706	100.0	303,706
Sandaun	0	0.0	0	0.0	0	0.0	166,919	100.0	166,919
Manus	0	0.0	0	0.0	0	0.0	34,899	100.0	34,899
New Ireland	1,000	1.0	0	0.0	0	0.0	102,259	99.0	103,259
East New Britain	0	0.0	0	0.0	0	0.0	174,230	100.0	174,230
West New Britain	3,206	2.9	0	0.0	0	0.0	106,093	97.1	109,299
Bougainville	0	0.0	0	0.0	0	0.0	167,156	100.0	167,156
Papua New Guinea	556,566	13.3	162,234	3.9	260,182	6.2	3,213,579	76.6	4,192,561

Sources: NSO (2002); MASP.

Appendix Table A3.11.6 Rural population using green manure (compost) in food production systems, by province

Province	Importance of green manure (compost) in food production systems								Total rural population
	Minor	%	Important	%	Very important	%	Not used	%	
Western	0	0.0	0	0.0	0	0.0	107,837	100.0	107,837
Gulf	2,654	2.9	0	0.0	0	0.0	89,611	97.1	92,265
Central	0	0.0	0	0.0	0	0.0	157,058	100.0	157,058
Milne Bay	0	0.0	0	0.0	0	0.0	188,334	100.0	188,334
Oro	0	0.0	0	0.0	0	0.0	106,288	100.0	106,288
Southern Highlands	30,983	5.9	172,342	32.7	268,069	50.9	55,004	10.4	526,398
Enga	0	0.0	28,950	10.2	223,885	79.0	30,664	10.8	283,498
Western Highlands	0	0.0	0	0.0	68,165	18.4	302,849	81.6	371,014
Simbu	0	0.0	0	0.0	0	0.0	242,748	100.0	242,748
Eastern Highlands	0	0.0	0	0.0	0	0.0	393,418	100.0	393,418
Morobe	48,883	13.7	5,994	1.7	0	0.0	301,223	84.6	356,100
Madang	7,021	2.3	672	0.2	0	0.0	300,443	97.5	308,135
East Sepik	0	0.0	13,155	4.3	0	0.0	290,551	95.7	303,706
Sandaun	7,261	4.4	0	0.0	0	0.0	159,658	95.6	166,919
Manus	0	0.0	0	0.0	0	0.0	34,899	100.0	34,899
New Ireland	0	0.0	0	0.0	0	0.0	103,259	100.0	103,259
East New Britain	0	0.0	0	0.0	0	0.0	174,230	100.0	174,230
West New Britain	0	0.0	0	0.0	0	0.0	109,299	100.0	109,299
Bougainville	0	0.0	0	0.0	0	0.0	167,156	100.0	167,156
Papua New Guinea	96,802	2.3	221,112	5.3	560,119	13.4	3,314,528	79.1	4,192,561

Sources: NSO (2002); MASP.

Appendix Table A4.1.1 Exchange rate for one PNG kina against the Australian dollar and United States dollar, 1975–2007

Year	A$	US$
1975	1.0000	1.2571
1976	1.1335	1.2314
1977	1.1565	1.3200
1978	1.2629	1.4530
1979	1.3105	1.4488
1980	1.3154	1.5531
1981	1.3029	1.4695
1982	1.3636	1.3371
1983	1.2734	1.1422
1984	1.2785	1.0599
1985	1.4519	0.9877
1986	1.5669	1.0404
1987	1.5787	1.1384
1988	1.4181	1.2100
1989	1.4659	1.1633
1990	1.3616	1.0511
1991	1.3818	1.0498
1992	1.4708	1.0127
1993	1.5077	1.0190
1994	1.0927	0.8485
1995	1.0176	0.7545
1996	0.9653	0.7553
1997	0.9365	0.6971
1998	0.7708	0.4856
1999	0.6086	0.3922
2000	0.6256	0.3623
2001	0.5476	0.2976
2002	0.4740	0.2573
2003	0.4326	0.2816
2004	0.4222	0.3101
2005	0.4229	0.3224
2006	0.4342	0.3272
2007	0.4027	0.3375

Source: Bank of Papua New Guinea Quarterly Economic Bulletins.

Appendix Table A4.3.1 Average price per year of selected items in Port Moresby and Madang, 1971–2005[a]

Year	Sweet potato		Aibika		Rice		Kerosene	
	Port Moresby	Madang	Port Moresby	Madang	Port Moresby	Madang	Port Moresby	Madang
1971	13.0	6.6	18.8	9.5	22.3	22.1	8.8	7.7
1972	11.5	8.1	25.7	12.5	22.2	22.0	8.9	7.8
1973	14.8	7.2	27.7	10.1	26.1	25.3	8.6	7.9
1974	17.5	10.5	26.9	10.2	38.7	38.5	11.5	10.8
1975	22.9	11.3	26.3	12.2	40.2	41.3	13.3	13.2
1976	25.3	12.5	31.0	13.6	40.5	39.7	15.1	14.9
1977	22.0	13.2	33.5	13.2	38.2	38.5	16.9	16.1
1978	27.0	12.2	41.8	13.1	37.3	36.7	17.3	16.6
1979	30.3	13.9	46.6	17.4	39.0	37.6	17.5	17.5
1980	45.4	13.5	53.0	20.0	46.2	46.1	26.4	26.4
1981	50.4	15.9	45.6	18.1	49.1	48.3	29.5	29.5
1982	48.2	20.7	46.3	20.2	51.2	50.2	30.6	30.6
1983	62.1	18.6	47.3	22.6	46.2	45.7	32.2	32.2
1984	51.5	16.5	49.1	24.6	53.0	52.7	31.9	31.8
1985	47.6	20.4	60.9	19.6	57.9	57.9	34.6	34.8
1986	52.2	19.5	73.9	22.2	57.2	56.9	33.5	33.6
1987	57.8	24.9	63.8	30.3	52.8	53.3	30.7	32.9
1988	69.0	34.7	59.3	28.5	57.7	60.0	31.6	33.5
1989	57.3	35.0	67.5	23.0	61.5	64.1	32.4	32.4
1990	61.3	33.0	80.6	23.6	68.9	70.7	39.4	43.0
1991	58.0	32.3	78.0	23.3	71.8	73.0	44.4	44.9
1992	71.1	31.2	92.3	33.0	70.5	73.7	45.2	45.1
1993	71.9	35.0	88.1	35.4	69.8	72.2	45.4	45.2
1994	66.3	28.5	95.6	32.7	70.1	71.6	43.7	44.1
1995	73.2	32.2	80.7	28.5	84.4	87.7	48.2	47.8
1996	71.3	33.2	92.6	31.6	92.4	95.3	54.1	53.1
1997[b]	95.2	–	138.5	–	91.8	93.8	62.7	63.0
1998	90.3	39.6	95.3	36.7	137.6	134.3	60.5	59.5
1999	93.0	40.9	94.2	36.1	174.6	171.4	69.1	69.8
2000	107.2	36.0	140.6	46.9	180.8	188.6	98.3	98.0
2001	128.9	90.4	120.9	50.0	223.3	221.6	112.9	112.6
2002	129.7	112.3	147.3	105.1	275.8	276.3	112.1	112.5
2003	103.7	57.6	173.9	97.5	286.3	283.6	149.3	151.7
2004	125.6	45.1	184.0	91.8	279.5	269.8	165.8	168.7
2005	132.2	54.7	220.4	79.4	282.2	275.5	175.6	176.4

[a] Prices are toea per kilogram for sweet potato, aibika and rice, and toea per litre for kerosene.

[b] The gap in prices of fresh foods in 1997 occurred because there was very little or no fresh produce in many markets during the drought, hence it was not possible to record prices at that time.

Source: PNG National Statistical Office.

Appendix Table A5.1.1 Estimated annual cash income (kina) of the rural population from the six most important agricultural activities, by province, 1990–1995

Province	Arabica coffee	Fresh food	Cocoa	Betel nut	Copra	Oil palm	All agricultural activities[a]
Western	0	661,535	0	147,684	0	0	1,894,488
Gulf	161,224	462,605	5,550	1,387,117	0	0	2,472,386
Central	26,989	1,416,696	0	2,218,875	78,636	0	4,933,053
Milne Bay	72,804	1,231,478	24,754	1,260,083	387,539	106,444	4,506,198
Oro	69,599	1,375,756	285,706	422,873	0	3,335,491	6,130,723
Southern Highlands	2,373,297	2,363,636	0	0	0	0	6,755,959
Enga	3,136,665	2,949,277	0	0	0	0	7,638,569
Western Highlands	24,370,651	6,027,311	0	0	0	0	33,356,490
Simbu	9,808,452	2,421,425	0	0	0	0	12,897,846
Eastern Highlands	24,348,815	7,343,498	0	693,819	0	0	38,119,355
Morobe	1,790,880	5,664,761	232,776	5,636,345	207,066	0	14,423,894
Madang	778,406	2,525,167	2,166,737	2,845,524	3,796,524	0	13,101,744
East Sepik	0	1,522,015	2,788,349	587,425	12,561	0	8,138,594
Sandaun	0	745,184	417,974	257,470	0	0	2,253,345
Manus	0	619,750	90,359	519,716	632,225	0	2,306,037
New Ireland	0	1,336,543	1,067,613	795,538	3,659,143	104,942	7,605,285
East New Britain	0	3,139,511	9,793,408	2,680,497	3,091,375	0	19,999,025
West New Britain	0	765,931	504,600	388,887	1,346,606	2,600,276	6,032,606
Bougainville	0	1,564,656	4,859,606	374,176	3,248,555	0	10,742,165
Papua New Guinea	66,937,782	44,136,734	22,237,432	20,216,030	16,460,230	6,147,153	203,307,762

[a] See Table 5.1.1 for the list of agricultural activities.

Source: MASP.

Appendix Table A5.1.2 Estimated annual cash income of the rural population from agricultural activities, by income class and province, 1990–1995

Province	Income class (kina/person/year)												Total rural population[a]
	K0–20	%	K21–50	%	K51–100	%	K101–150	%	K151–200	%	K201–300	%	
Western	39,555	48.3	42,267	51.7	0	0.0	0	0.0	0	0.0	0	0.0	81,822
Gulf	17,637	25.2	36,778	52.6	0	0.0	15,526	22.2	0	0.0	0	0.0	69,940
Central	42,893	36.0	51,786	43.5	3,307	2.8	19,253	16.2	1,929	1.6	0	0.0	119,168
Milne Bay	51,444	36.0	70,208	49.1	21,245	14.9	0	0.0	0	0.0	0	0.0	142,896
Oro	14,024	17.4	29,407	36.5	181	0.2	37,035	45.9	0	0.0	0	0.0	80,646
Southern Highlands	342,582	85.8	56,880	14.2	0	0.0	0	0.0	0	0.0	0	0.0	399,461
Enga	72,044	33.5	71,869	33.4	71,203	33.1	0	0.0	0	0.0	0	0.0	215,116
Western Highlands	13,513	4.8	87,091	30.9	0	0.0	0	0.0	180,946	64.3	0	0.0	281,550
Simbu	48,703	26.4	80,380	43.6	529	0.3	0	0.0	54,579	29.6	0	0.0	184,191
Eastern Highlands	34,394	11.5	39,575	13.3	62,364	20.9	9,053	3.0	128,896	43.2	24,263	8.1	298,546
Morobe	143,345	53.1	30,470	11.3	42,336	15.7	2,607	1.0	50,576	18.7	858	0.3	270,192
Madang	93,143	39.8	50,583	21.6	37,411	16.0	15,574	6.7	37,089	15.9	0	0.0	233,801
East Sepik	69,020	30.0	115,026	49.9	42,475	18.4	3,927	1.7	0	0.0	0	0.0	230,448
Sandaun	116,876	92.3	9,778	7.7	0	0.0	0	0.0	0	0.0	0	0.0	126,654
Manus	0	0.0	3,799	14.3	5,680	21.5	17,000	64.2	0	0.0	0	0.0	26,479
New Ireland	13,140	16.8	24,598	31.4	3,190	4.1	18,952	24.2	18,534	23.6	0	0.0	78,414
East New Britain	26,003	19.7	3,528	2.7	0	0.0	16,681	12.6	0	0.0	86,014	65.1	132,225
West New Britain	36,982	44.6	0	0.0	1,799	2.2	44,149	53.2	0	0.0	0	0.0	82,930
Bougainville	516	0.4	61,009	48.1	8,872	7.0	56,438	44.5	0	0.0	0	0.0	126,835
Papua New Guinea	1,175,815	37.0	865,030	27.2	300,591	9.4	256,194	8.1	472,550	14.9	111,135	3.5	3,181,315

[a] Population figures are for estimated rural population in 1990 extrapolated from the 2000 census data, using the 1980 to 2000 growth rate.

Source: MASP.

Appendix Table A5.4.1 Coffee production by province, 1993–2006 (tonnes)

Province	1993	1994	1995	1996	1997	1998	1999	2000	2001	2002	2003	2004	2005	2006
Gulf	60	0	0	0	9	25	26	0	0	0	0	0	0	15
Central	60	0	0	0	0	0	105	22	6	13	16	0	1,079	29
Milne Bay	0	0	31	27	0	0	25	5	0	0	0	0	0	0
Oro	60	60	13	3	5	29	401	28	201	77	93	0	0	30
Southern Highlands	240	420	934	954	1,322	927	874	1,163	1,254	849	885	822	1,295	939
Enga	780	840	1,749	2,248	1,822	2,025	1,304	2,043	2,329	1,810	2,182	2,024	3,309	1,792
Western Highlands	30,600	29,640	27,393	29,972	28,501	34,141	35,136	29,902	27,792	30,039	27,843	25,683	32,375	21,217
Simbu	4,260	3,720	4,422	6,131	5,447	7,944	7,189	5,208	6,832	6,186	7,990	7,401	8,633	2,895
Eastern Highlands	22,320	24,480	19,237	23,229	18,814	23,764	24,286	19,664	20,712	18,783	26,099	24,102	20,144	20,603
Morobe	2,580	3,240	2,562	1,839	4,051	5,342	6,334	4,513	3,316	5,035	3,293	3,036	4,820	3,204
Madang	180	240	222	0	40	77	114	139	3	9	20	63	0	1
East Sepik	2,400	3,360	3,026	2,316	4,512	2,595	3,691	2,022	2,473	1,052	162	127	288	202
Papua New Guinea	63,540	66,000	59,588	66,719	64,524	76,869	79,484	64,709	64,918	63,853	68,583	63,259	71,944	50,927

Source: Coffee Industry Corporation Ltd.

Appendix Table A5.4.2 Volume and value of coffee exports, 1947–2006

Year	Volume (tonnes)	Value (K'000)	Year	Volume (tonnes)	Value (K'000)	Year	Volume (tonnes)	Value (K'000)
1947	48	8	1967	12,911	10,208	1987	64,840	134,643
1948	7	3	1968	18,267	14,320	1988	44,784	116,353
1949	17	7	1969	19,871	15,531	1989	84,307	140,347
1950	21	8	1970	25,376	20,182	1990	63,429	103,539
1951	33	17	1971	23,556	20,572	1991	47,206	79,418
1952	35	22	1972	27,748	20,457	1992	53,000	68,100
1953	47	62	1973	30,650	23,394	1993	62,800	100,500
1954	87	120	1974	32,196	28,847	1994	64,700	204,800
1955	107	150	1975	36,136	33,501	1995	55,100	214,500
1956	171	188	1976	37,091	42,225	1996	62,300	190,300
1957	274	367	1977	37,565	143,441	1997	59,200	325,900
1958	385	450	1978	45,801	107,225	1998	83,500	476,400
1959	1,088	912	1979	49,586	124,996	1999	79,200	417,100
1960	1,488	1,434	1980	51,006	118,643	2000	62,572	274,712
1961	2,294	2,212	1981	47,057	74,218	2001	65,743	247,791
1962	3,443	3,114	1982	41,014	77,780	2002	63,423	276,647
1963	4,845	4,028	1983	52,518	94,659	2003	68,830	298,256
1964	6,831	5,366	1984	50,771	113,317	2004	62,860	283,790
1965	8,687	7,295	1985	40,607	117,112	2005	71,979	471,009
1966	10,804	8,787	1986	53,069	203,624	2006	50,928	337,985

Sources: 1947–1976: Munnull and Densley (c. 1978); 1977–1991: DAL (1992); 1992–1999: Bank of PNG; 2000–2006: Coffee Industry Corporation Ltd.

Appendix Table A5.4.3 Coffee production by sector, 1985–2006 (tonnes)

Year	Smallholder	Plantation[a]	Total	Year	Smallholder	Plantation[a]	Total
1985	32,100	17,220	49,320	1996	49,500	15,600	65,100
1986	30,780	14,280	45,060	1997	52,260	12,240	64,500
1987	46,680	16,200	62,880	1998	64,560	16,140	80,700
1988	57,600	16,080	73,680	1999	67,800	11,940	79,740
1989	44,520	17,040	61,560	2000	53,040	11,640	64,680
1990	41,220	15,000	56,220	2001	55,860	9,060	64,920
1991	33,420	12,060	45,480	2002	54,300	9,540	63,840
1992	38,460	18,060	56,520	2003	56,430	16,818	73,248
1993	44,520	17,040	61,560	2004	53,505	7,995	61,500
1994	41,220	15,000	56,220	2005	63,750	11,250	75,000
1995	43,920	15,420	59,340	2006	42,787	8,126	50,913

[a] Plantation includes both plantations (30 ha or more under coffee) and coffee blocks (5–29 ha).

Source: Coffee Industry Corporation Ltd.

Appendix Table A5.4.4 Destination of coffee exports, 2000–2006

Destination	2000		2001		2002		2003	
	Volume (tonnes)	Value (K'000)	Volume (tonnes)	Value (K'000)	Volume (tonnes)	Value (K'000)	Volume (tonnes)	Value (K'000)
Australia	13,241.7	65,849.9	7,008.5	27,530.8	6,979.3	30,047.4	7,620.5	35,753.7
Belgium	277.2	649.1	433.1	1,358.9	617.0	3,518.4	503.5	2,733.7
Bulgaria	0.0	0.0	0.0	0.0	18.0	67.2	0.0	0.0
Canada	36.0	211.3	0.0	0.0	18.0	123.3	0.0	0.0
China	0.0	0.0	0.0	0.0	0.0	0.0	58.7	148.1
Czech Republic	0.0	0.0	0.0	0.0	0.0	0.0	21.6	79.1
Denmark	33.6	64.4	34.8	50.4	313.0	693.2	220.9	664.3
Fiji	20.2	163.9	30.7	172.9	38.7	288.4	39.2	303.8
Finland	0.0	0.0	294.0	1,194.7	378.0	1,723.3	108.0	487.6
France	140.4	687.6	258.1	1,171.5	154.8	1,042.1	187.2	1,380.3
Germany	32,808.6	138,076.2	32,741.9	120,547.2	35,382.1	145,615.1	36,113.9	138,240.4
Greece	0.0	0.0	0.0	0.0	0.0	0.0	0.0	0.0
Hong Kong	90.0	358.0	147.0	538.5	76.2	365.8	189.0	1,171.8
Hungary	0.0	0.0	72.0	226.1	0.0	0.0	0.0	0.0
Ireland	270.0	1,231.4	111.6	375.5	414.0	1,808.3	522.0	2,215.6
Israel	0.0	0.0	18.0	61.2	0.0	0.0	0.0	0.0
Italy	54.0	231.8	144.0	517.9	36.0	143.3	92.1	511.8
Japan	2,482.3	12,201.7	4,830.8	18,925.6	6,887.6	30,636.4	6,594.7	29,157.9
Malaysia	0.4	5.6	0.7	14.1	0.0	0.0	252.2	1,013.1
Netherlands	285.0	487.4	0.0	0.0	20.3	167.9	311.9	2,109.8
New Caledonia	274.2	1,120.0	310.1	1,322.7	351.6	1,688.3	233.4	1,213.4
New Zealand	659.4	3,646.6	803.5	4,108.8	698.0	3,967.6	522.5	3,707.2
North Korea	87.7	525.1	0.0	0.0	0.0	0.0	0.0	0.0
Norway	54.0	267.5	56.4	242.7	0.0	0.0	162.0	635.9
Oman	54.0	176.7	0.0	0.0	0.0	0.0	54.0	131.5
Philippines	0.0	0.0	0.0	0.0	0.7	13.7	0.0	0.0
Poland	1,213.8	4,886.2	979.5	3,526.4	615.0	2,365.3	574.2	2,072.9
Portugal	522.0	1,577.2	648.0	1,696.4	478.8	1,714.5	338.4	1,583.7
Russia	0.0	0.0	18.0	79.7	54.0	254.0	18.0	71.4
Singapore	124.9	469.7	81.2	398.7	62.1	347.7	92.1	491.5
Solomon Islands	0.5	4.2	0.1	0.5	0.0	0.5	0.4	5.1
South Africa	0.0	0.0	0.0	0.0	0.0	0.0	234.0	940.3
South Korea	0.0	0.0	465.8	1,916.3	61.3	270.8	0.0	0.0
Spain	162.0	232.2	451.2	783.1	36.0	99.5	18.0	51.3
Sweden	0.0	0.0	0.0	0.0	0.0	0.0	370.8	1,624.4
Switzerland	0.2	6.0	0.0	0.0	126.0	851.6	0.1	4.9
Syria	0.0	0.0	13.1	35.4	11.8	34.3	15.7	39.6
Tahiti	6.0	15.5	42.0	207.2	6.0	19.1	24.0	146.2
Taiwan	54.0	81.0	69.0	243.9	24.0	183.0	7.2	66.9
Tunisia	0.0	0.0	0.0	0.0	0.0	0.0	18.0	89.6
Ukraine	18.0	76.5	0.0	0.0	18.0	65.8	0.0	0.0
United Kingdom	3,693.0	16,355.4	2,256.0	8,506.6	503.2	2,561.8	225.0	1,177.6
United States	5,872.1	24,890.6	13,369.9	51,828.9	8,989.7	45,734.9	13,032.8	67,968.8
Vanuatu	0.4	9.8	0.0	0.0	0.0	0.0	0.0	0.0
Western Samoa	36.0	153.4	54.0	208.4	54.0	234.2	53.9	263.0
Total	62,571.6	274,712.0	65,743.0	247,790.9	63,423.3	276,646.7	68,830.1	298,256.4

... continued

Appendix Table A5.4.4 (continued)

Destination	2004		2005		2006	
	Volume (tonnes)	Value (K'000)	Volume (tonnes)	Value (K'000)	Volume (tonnes)	Value (K'000)
Australia	8,895.5	39,249.0	10,263.5	67,813.4	10,295.7	63,378.4
Belgium	553.2	3,380.7	2,520.6	17,105.5	162.0	1,344.0
Bulgaria	0.0	0.0	0.0	0.0	0.0	0.0
Canada	0.0	0.0	3.6	48.1	3.9	59.3
China	0.0	0.0	0.0	0.0	0.5	12.0
Czech Republic	0.0	0.0	0.0	0.0	0.0	0.0
Denmark	439.2	1,430.6	277.8	1,381.1	279.0	1,602.2
Fiji	39.6	338.9	21.5	206.8	20.5	187.4
Finland	345.6	1,567.9	0.0	0.0	0.0	0.0
France	342.0	1,929.3	435.6	2,654.2	90.0	789.4
Germany	31,549.4	129,376.9	34,631.9	214,363.2	20,262.7	126,515.8
Greece	0.0	0.0	0.0	0.0	40.8	271.5
Hong Kong	99.0	554.6	72.4	516.8	52.6	337.4
Hungary	0.0	0.0	0.0	0.0	0.0	0.0
Ireland	538.5	2,161.8	684.0	4,817.2	342.0	1,947.6
Israel	0.0	0.0	0.0	0.0	0.0	0.0
Italy	40.8	216.4	100.8	713.6	326.4	2,443.5
Japan	5,037.7	23,470.5	6,231.5	42,351.3	4,467.8	28,394.8
Malaysia	504.0	2,422.4	864.0	5,788.3	972.0	6,202.8
Netherlands	350.2	2,451.1	487.8	3,710.3	300.6	2,727.3
New Caledonia	234.2	1,298.1	195.4	1,515.9	193.2	1,626.7
New Zealand	666.9	3,856.7	769.4	5,773.8	698.3	5,419.3
North Korea	0.0	0.0	0.0	0.0	0.0	0.0
Norway	0.0	0.0	0.0	0.0	0.0	0.0
Oman	0.0	0.0	0.0	0.0	0.0	0.0
Philippines	0.0	0.0	0.0	0.0	0.0	0.0
Poland	0.0	0.0	36.0	207.6	0.0	0.0
Portugal	86.4	399.3	0.0	0.0	127.1	684.4
Russia	18.0	85.9	72.0	479.2	0.0	0.0
Singapore	140.2	720.7	100.1	486.7	59.6	319.1
Solomon Islands	1.1	13.1	0.4	5.8	1.2	23.3
South Africa	779.4	3,596.9	694.2	4,424.6	546.0	3,494.3
South Korea	454.9	1,904.8	1,921.8	11,497.3	651.7	4,175.0
Spain	108.0	452.8	262.8	1,746.6	252.0	1,522.2
Sweden	662.4	3,142.1	910.8	5,681.4	320.4	2,240.6
Switzerland	0.0	0.0	0.0	0.0	0.0	0.0
Syria	0.0	0.0	0.0	0.0	0.0	0.0
Tahiti	24.0	159.6	0.0	0.0	20.4	182.6
Taiwan	19.2	158.7	18.0	139.8	9.0	74.1
Tunisia	36.0	186.1	0.0	0.0	0.0	0.0
Ukraine	0.0	0.0	0.0	0.0	0.0	0.0
United Kingdom	231.0	1,312.2	216.0	1,653.9	163.2	1,407.9
United States	10,591.6	57,751.5	10,154.1	75,835.7	10,218.2	80,381.1
Vanuatu	0.0	0.0	0.0	0.0	0.0	0.0
Western Samoa	72.0	201.1	33.0	91.2	51.1	221.5
Total	62,860.0	283,789.8	71,978.8	471,009.1	50,928.0	337,985.5

Source: Coffee Industry Corporation Ltd.

Appendix Table A5.4.5 Smallholder coffee production and sales of food and alcohol in the Kainantu area, Eastern Highlands Province, 1982–1984

Month-Year	Coffee production (kg)	Food sales (kina)	Alcohol sales (kina)
Jan-82	49,858	106,212	27,100
Feb-82	48,370	105,920	25,700
Mar-82	106,291	126,825	32,042
Apr-82	81,079	147,651	35,250
May-82	173,445	151,206	55,282
Jun-82	265,978	159,907	–[a]
Jul-82	403,910	179,024	55,533
Aug-82	234,187	144,523	41,403
Sep-82	264,431	142,569	52,659
Oct-82	151,522	138,096	44,378
Nov-82	64,778	121,477	35,447
Dec-82	56,474	145,721	36,885
Jan-83	30,660	99,914	24,950
Feb-83	28,560	94,492	24,062
Mar-83	37,563	127,663	30,828
Apr-83	53,929	106,130	31,682
May-83	71,001	122,341	35,904
Jun-83	235,020	143,138	34,598
Jul-83	442,754	162,982	62,740
Aug-83	449,110	159,086	57,949
Sep-83	275,422	159,470	58,365
Oct-83	196,560	130,161	32,576
Nov-83	87,540	120,534	30,178
Dec-83	59,400	142,417	38,552
Jan-84	13,260	101,176	16,150
Feb-84	4,500	102,300	11,610
Mar-84	15,120	140,405	14,295
Apr-84	14,580	104,146	18,373
May-84	48,360	114,380	22,614
Jun-84	151,020	154,088	41,455
Jul-84	178,860	–	54,760
Aug-84	144,600	–	58,379
Sep-84	97,920	–	48,464
Oct-84	87,240	–	44,508
Nov-84	112,740	–	47,198
Dec-84	29,580	–	36,957

[a] Alcohol sales were banned in Eastern Highlands Province in June 1982 because of national elections.

Source: Bourke (1988:328–9).

Appendix Table A5.5.1 Volume and value of cocoa exports, 1948–2006

Year	Volume (tonnes)	Value (K'000)	Year	Volume (tonnes)	Value (K'000)
1948	114	34	1978	27,129	62,955
1949	200	73	1979	26,139	56,603
1950	207	62	1980	27,421	44,773
1951	341	179	1981	27,183	33,404
1952	477	296	1982	28,222	31,479
1953	639	350	1983	25,955	40,874
1954	727	499	1984	33,569	66,125
1955	1,106	1,051	1985	30,427	61,554
1956	1,290	722	1986	30,509	55,592
1957	2,125	924	1987	34,297	56,076
1958	2,589	1,657	1988	37,142	45,329
1959	4,253	2,984	1989	46,340	44,340
1960	5,870	3,343	1990	35,813	32,006
1961	7,291	3,333	1991	35,151	33,059
1962	10,014	3,965	1992	38,438	33,911
1963	14,071	5,911	1993	35,919	32,207
1964	15,640	6,843	1994	31,902	39,107
1965	20,146	7,049	1995	29,078	46,568
1966	16,781	4,435	1996	37,619	52,742
1967	21,592	9,545	1997	26,643	54,519
1968	24,385	11,794	1998	26,024	85,634
1969	27,738	16,060	1999	38,792	103,516
1970	23,725	15,549	2000	45,873	99,831
1971	27,839	13,643	2001	39,342	124,575
1972	29,262	11,021	2002	39,070	255,633
1973	22,048	11,175	2003	40,378	257,880
1974	28,742	23,338	2004	41,466	218,071
1975	35,498	40,067	2005	49,070	219,507
1976	30,431	28,645	2006	47,034	217,080
1977	29,435	86,371			

Sources: 1948–1978: Territory of Papua and New Guinea Bureau of Statistics; 1979–2006: Cocoa Board of PNG.

Appendix Table A5.5.2 Cocoa production by sector, 1970/71–2005/06[a] (tonnes)

Year	Smallholder	Plantation	Total
1970/71	8,182	17,895	26,077
1971/72	8,540	18,732	27,272
1972/73	6,817	15,541	22,358
1973/74	11,370	18,067	29,437
1974/75	12,646	22,396	35,042
1975/76	12,642	18,984	31,626
1976/77	12,954	13,662	26,616
1977/78	14,658	15,147	29,805
1978/79	16,843	13,283	30,126
1979/80	16,698	14,116	30,814
1980/81	16,082	12,047	28,129
1981/82	18,055	11,819	29,874
1982/83	20,076	9,323	29,399
1983/84	19,165	8,547	27,712
1984/85	21,170	9,617	30,787
1985/86	22,795	9,923	32,718
1986/87	21,427	10,431	31,858
1987/88	19,947	11,868	31,815
1988/89	29,482	18,349	47,831
1989/90	24,580	13,871	38,451
1990/91	18,633	13,331	31,964
1991/92	22,888	14,407	37,295
1992/93	22,628	13,105	35,733
1993/94	19,885	11,390	31,275
1994/95	20,796	9,730	30,526
1995/96	17,923	9,264	27,187
1996/97	16,997	7,584	24,581
1997/98	20,029	7,077	27,106
1998/99	26,357	6,542	32,899
1999/00	26,922	7,846	34,768
2000/01	20,861	4,999	25,860
2001/02	35,057	4,857	39,914
2002/03	36,126	5,145	41,271
2003/04	28,589	4,488	33,077
2004/05	40,155	4,626	44,781
2005/06	40,514	4,447	44,961

[a] Production is by cocoa year, October–September.

Sources: 1970/71–1975/76: Department of Primary Industry; 1976/77–2005/06: Cocoa Board of PNG.

Appendix Table A5.5.3 Cocoa production by province, 1978–2006[a] (tonnes)

Province	1978	1979	1980	1981	1982	1983	1984	1985	1986	1987	1988	1989	1990	1991	1992
Gulf	0	0	0	0	0	0	0	0	9	14	17	73	17	19	3
Central	44	45	55	78	85	90	58	45	71	124	70	137	31	35	20
Milne Bay	34	22	50	51	66	47	0	36	43	20	16	10	45	83	62
Oro	546	356	419	331	506	402	338	488	529	590	591	1,016	1,179	1,450	1,592
Morobe	176	130	134	122	156	128	143	195	206	226	350	559	398	459	440
Madang	1,829	1,466	2,145	1,843	1,624	1,188	1,628	1,637	1,908	1,479	2,114	2,176	2,496	2,669	3,339
East Sepik	661	502	906	1,319	1,510	1,518	2,219	2,158	2,039	1,524	1,615	1,914	2,320	2,363	2,661
Sandaun	39	29	29	26	28	16	416	41	130	111	271	264	302	360	344
Manus	24	36	19	26	47	22	48	58	87	83	85	105	70	38	48
New Ireland	2,201	1,929	1,397	1,565	960	755	880	920	934	897	850	1,663	1,513	1,614	2,506
East New Britain	10,416	8,548	8,197	8,553	7,875	7,712	10,428	6,299	10,835	11,019	11,568	21,863	18,064	18,913	19,970
West New Britain	1,035	980	651	841	610	712	849	567	803	738	631	1,662	946	816	941
Bougainville	14,250	13,514	13,547	17,180	13,160	15,654	15,958	16,430	14,544	14,133	14,766	18,615	7,530	3,146	5,369
Papua New Guinea	31,255	27,557	27,549	31,935	26,627	28,244	32,965	28,874	32,138	30,958	32,944	50,057	34,908	31,965	37,295

... continued

Appendix Table A5.5.3 (continued)

Province	1993	1994	1995	1996	1997	1998	1999	2000	2001	2002	2003	2004	2005	2006
Gulf	10	0	8	0	3	0	0	0	1	1	0	0	0	0
Central	22	13	8	2	5	1	0	0	0	0	0	0	0	0
Milne Bay	82	20	34	17	64	11	8	0	1	3	5	10	3	5
Oro	1,321	1,048	955	484	385	387	330	109	215	361	1,019	331	177	22
Morobe	421	432	412	80	831	441	509	203	539	840	1,157	831	1,141	779
Madang	2,892	2,703	2,380	2,249	2,006	2,309	2,532	1,731	2,801	2,045	4,443	3,826	3,877	3,181
East Sepik	2,590	2,793	2,289	1,776	1,863	1,427	1,709	1,176	1,468	4,125	3,291	1,426	3,676	3,438
Sandaun	400	468	380	160	303	205	410	312	203	153	599	545	756	892
Manus	59	25	33	12	21	44	7	13	13	21	10	7	3	1
New Ireland	2,816	2,543	5,880	2,391	2,251	1,580	2,263	1,295	1,105	1,512	1,191	1,380	984	1,185
East New Britain	18,739	15,790	14,509	16,237	11,966	15,912	20,527	20,522	15,003	23,882	16,920	18,241	20,626	19,027
West New Britain	939	661	987	1,160	775	836	739	722	435	769	783	763	698	803
Bougainville	5,442	4,779	5,109	2,619	4,108	3,699	3,481	4,073	5,447	9,995	11,525	6,881	11,559	13,071
Papua New Guinea	35,733	31,275	32,984	27,187	24,581	26,852	32,515	30,156	27,231	43,707	40,943	34,241	43,500	42,404

[a] Values for 1991–1999 are cocoa years (ie, 1991 = October 1990 – September 1991); all other values are for calendar years.

Source: Cocoa Board of PNG.

Appendix Table A5.5.4 Destination of cocoa exports, 1996–2006

Destination	1996		1997		1998		1999		2000		2001	
	Volume (tonnes)	Value (K'000)	Volume (tonnes)	Value (K'000)	Volume (tonnes)	Value (K'000)	Volume (tonnes)	Value (K'000)	Volume (tonnes)	Value (K'000)	Volume (tonnes)	Value (K'000)
Australia	4	6	11	22	10	18	8	22	15	34	18	47
Belgium	3,197	5,159	3,018	5,854	138	410	130	326	230	504	1,221	3,791
Germany	2,595	4,874	1,969	4,271	2,727	9,686	2,938	8,385	2,483	5,865	1,942	6,332
Indonesia	–	–	–	–	–	–	–	–	5,243	11,173	5,575	17,899
Malaysia	917	1,519	3,263	6,421	1,460	5,065	3,174	7,685	3,017	6,399	2,276	8,216
Netherlands	4,846	8,337	251	480	310	859	–	–	81	139	–	–
New Zealand	–	–	–	–	–	–	–	–	–	–	14	44
Philippines	1,300	2,288	1,875	3,897	2,512	8,288	–	–	2,505	5,473	800	2,780
Singapore	15,726	26,515	7,692	15,458	6,951	21,503	4,962	12,675	20,554	43,834	11,274	37,186
Spain	–	–	–	–	–	–	–	–	–	–	101	90
Thailand	–	–	–	–	–	–	–	–	2,730	6,094	423	1,476
United Kingdom	4,676	8,130	1,436	3,105	1,336	4,366	1,708	4,948	1,323	3,177	1,161	4,189
United States	4,258	6,983	6,594	14,183	9,040	30,192	19,553	52,862	7,691	17,139	14,536	42,524
Others	100	166	533	1,187	1,540	5,235	6,319	16,612	–	–	–	–
Total	37,619	63,977	26,642	54,878	26,024	85,622	38,792	103,515	45,872	99,831	39,341	124,574

... continued

Appendix Table A5.5.4 (continued)

Destination	2002		2003		2004		2005		2006	
	Volume (tonnes)	Value (K'000)	Volume (tonnes)	Value (K'000)	Volume (tonnes)	Value (K'000)	Volume (tonnes)	Value (K'000)	Volume (tonnes)	Value (K'000)
Australia	15	91	9	59	56	261	30	155	36	180
Belgium	2,793	18,205	260	1,412	179	919	1,231	5,826	8,017	37,620
China	–	–	–	–	–	–	–	–	210	844
France	–	–	–	–	25	207	–	–	38	318
Germany	1,854	11,014	1,315	9,622	1,533	9,155	1,389	7,240	2,137	11,627
Indonesia	6,575	42,394	4,246	29,116	7,991	40,936	6,056	27,229	5,015	24,057
Malaysia	–	–	1,799	9,872	4,697	23,617	12,654	56,097	7,342	34,473
Netherlands	–	–	–	–	–	–	–	–	130	725
New Zealand	6	51	10	61	10	50	15	73	28	143
Philippines	693	5,110	187	1,386	–	–	–	–	–	–
Singapore	8,410	53,252	12,942	74,245	6,900	31,898	10,129	42,486	7,305	31,536
Spain	43	59	–	–	–	–	–	–	–	–
Switzerland	–	–	–	–	–	–	–	–	108	656
Thailand	1,165	9,620	4,100	25,867	4,885	24,375	3,344	15,351	2,940	13,607
United Kingdom	3,523	26,765	1,329	9,396	1,704	9,495	1,605	8,288	618	3,704
United States	13,993	89,072	14,181	96,844	13,486	77,156	12,617	56,762	13,110	57,590
Total	39,070	255,633	40,378	257,880	41,466	218,069	49,070	219,507	47,034	217,080

Source: Cocoa Board of PNG.

Appendix Table A5.6.1 Volume and value of copra exports, 1922–2006

Year	Volume (tonnes)	Value (K'000)	Year	Volume (tonnes)	Value (K'000)	Year	Volume (tonnes)	Value (K'000)
1922	31,454	1,123	1951	76,694	7,690	1980	90,820	24,594
1923	39,136	1,464	1952	78,140	11,439	1981	99,390	19,315
1924	42,968	1,646	1953	77,619	12,490	1982	74,357	12,878
1925	47,669	1,978	1954	84,707	13,668	1983	78,711	23,981
1926	55,298	2,442	1955	80,860	13,051	1984	94,453	49,373
1927	58,072	2,073	1956	81,473	12,372	1985	104,650	33,922
1928	76,315	2,740	1957	80,287	11,297	1986	92,968	10,154
1929	74,085	2,296	1958	72,984	9,042	1987	84,138	15,113
1930	76,737	2,082	1959	63,037	11,508	1988	77,263	19,486
1931	72,890	1,621	1960	66,446	12,253	1989	72,647	16,110
1932	70,578	1,438	1961	77,753	10,227	1990	55,300	8,700
1933	69,830	1,267	1962	76,041	9,221	1991	44,000	5,200
1934	71,481	653	1963	73,663	9,436	1992	47,500	11,800
1935	65,861	838	1964	70,384	10,018	1993	59,000	14,200
1936	78,472	1,724	1965	74,745	12,409	1994	50,300	14,700
1937	91,453	2,846	1966	89,265	14,298	1995	64,200	27,400
1938	86,328	1,878	1967	75,313	9,994	1996	99,200	49,000
1939	84,029	1,572	1968	76,710	13,943	1997	90,300	47,200
1940	67,085	1,141	1969	95,251	14,804	1998	58,100	38,800
1941	59,781	652	1970	85,601	13,340	1999	63,500	66,500
1942	27,677	425	1971	92,002	14,207	2000	67,200	59,900
1943	–	–	1972	87,532	9,392	2001	46,400	15,500
1944	–	–	1973	79,810	8,629	2002	15,800	10,700
1945	12,816	766	1974	82,909	28,718	2003	8,400	6,500
1946	11,317	631	1975	95,455	28,841	2004	19,200	17,200
1947	9,368	582	1976	83,596	12,053	2005	22,300	17,300
1948	27,414	1,885	1977	88,932	23,219	2006	12,700	8,300
1949	47,127	3,929	1978	97,056	24,999			
1950	58,814	5,677	1979	90,880	38,162			

Sources: 1922–1951: Bureau of Agricultural Economics (1953); 1952–1970: Wheeler et al. (c. 1978); 1971–1978: Bank of PNG Quarterly Economic Bulletins; 1979–1989: DAL (1992); 1990–2006: Bank of PNG.

Appendix Table A5.6.2 Copra production by province, 1976–2005 (tonnes)

Province	1976	1977	1978	1979	1980	1981	1982	1983	1984	1985	1986	1987	1988	1989	1990
Western	30	18	8	22	28	15	51	28	0	0	0	1	5	1	0
Gulf	2,031	2,122	1,732	1,890	1,053	892	669	952	0	0	0	337	311	251	70
Central	5,074	4,260	4,661	5,086	2,868	4,486	2,873	3,094	4,204	5,224	4,000	2,113	1,809	1,219	1,339
Milne Bay	4,671	5,454	5,948	6,481	6,590	5,127	3,725	5,107	0	0	0	5,532	4,618	3,673	4,257
Oro	119	127	99	146	114	58	65	93	0	0	0	79	56	44	0
Morobe	1,508	1,505	1,676	2,366	2,121	2,043	1,845	1,908	3,470	4,450	3,587	2,454	2,119	1,950	1,806
Madang	18,919	19,333	19,715	21,438	19,494	19,542	26,808	19,813	22,525	24,568	25,618	24,369	18,713	20,295	25,997
East Sepik	1,519	1,941	1,251	2,553	2,317	2,203	1,817	2,289	3,717	3,179	3,059	2,272	1,700	1,468	2,669
Sandaun	229	267	47	471	793	846	983	899	0	0	0	541	464	734	0
Manus	1,337	1,030	1,454	1,566	1,370	1,250	1,135	1,342	1,888	2,184	757	682	703	588	380
New Ireland	39,421	16,720	25,767	24,223	23,923	24,137	21,089	17,870	16,589	21,700	17,765	17,574	15,984	12,438	11,426
East New Britain	31,845	35,621	40,855	42,712	40,717	41,055	41,206	41,321	58,597	58,923	55,538	46,924	44,762	46,254	47,908
West New Britain	7,186	8,687	11,102	12,639	12,373	13,221	15,267	14,658	13,421	16,400	14,615	16,198	12,637	14,150	15,349
Bougainville	22,049	22,532	26,114	23,773	26,584	24,991	20,877	20,859	26,658	30,782	25,835	24,698	27,128	26,756	1,801
Papua New Guinea	135,938	119,617	140,429	145,366	140,345	139,866	138,410	130,233	151,069	167,410	150,774	143,774	131,009	129,821	113,002

... continued

Appendix Table A5.6.2 (continued)

Province	1991	1992	1993	1994	1995	1996	1997	1998	1999	2000	2001	2002	2003	2004	2005
Western	0	0	0	0	0	0	0	0	0	0	0	0	0	0	0
Gulf	12	0	0	0	50	283	343	217	400	706	0	0	0	0	0
Central	613	832	698	470	1,114	1,189	1,389	1,404	1,025	1,274	63	12	0	0	0
Milne Bay	2,811	4,083	3,864	3,123	4,849	7,638	7,399	6,326	6,174	6,682	943	256	201	513	1,296
Oro	0	0	0	0	0	0	0	302	631	372	9	0	0	0	0
Morobe	773	1,413	1,557	1,182	1,133	2,437	2,121	2,007	3,191	3,162	985	620	0	1,024	752
Madang	20,811	23,925	25,826	16,624	24,515	33,480	28,748	27,014	28,654	29,029	20,366	16,088	9,391	24,958	20,138
East Sepik	1,892	2,403	3,227	3,055	3,538	3,987	3,611	4,082	4,844	4,017	1,264	293	0	1,624	664
Sandaun	0	0	0	0	0	0	0	0	0	0	0	0	0	0	0
Manus	147	378	602	683	918	1,063	878	929	1,606	1,724	115	68	0	277	170
New Ireland	8,874	14,316	15,739	13,876	16,309	18,632	17,479	12,547	17,491	19,250	6,423	13,483	14,003	10,740	12,490
East New Britain	38,796	44,253	45,110	35,634	37,736	46,843	43,651	42,378	46,462	41,357	21,601	30,229	41,563	43,966	46,276
West New Britain	10,839	14,881	15,775	14,772	17,537	19,975	18,997	15,862	20,898	19,560	11,231	7,263	9,260	8,340	7,908
Bougainville	1,984	10,212	10,900	11,410	14,342	13,453	15,231	12,558	19,931	15,458	949	2,888	4,236	14,231	10,161
Papua New Guinea	87,553	116,696	123,298	100,829	122,042	148,980	139,849	125,625	151,307	142,591	63,950	71,201	78,654	105,673	99,855

Sources: 1976–2000: Copra Marketing Board; 2001–2005: Kokonas Indastri Koporesen.

Appendix Table A5.6.3 Volume and value of copra oil exports, 1953–2006

Year	Volume (tonnes)	Value (K'000)	Year	Volume (tonnes)	Value (K'000)	Year	Volume (tonnes)	Value (K'000)
1953	3,626	946	1971	27,161	7,805	1989	34,697	15,179
1954	10,379	2,796	1972	26,519	5,880	1990	34,800	11,600
1955	8,799	2,431	1973	27,738	5,567	1991	33,200	12,800
1956	10,497	2,654	1974	26,594	13,762	1992	34,800	24,200
1957	11,521	2,632	1975	26,565	14,284	1993	45,500	19,600
1958	15,039	2,946	1976	27,550	7,853	1994	34,700	20,100
1959	21,887	5,509	1977	27,665	11,616	1995	33,100	29,700
1960	25,936	7,627	1978	27,608	11,706	1996	49,600	51,400
1961	20,757	4,722	1979	30,822	20,599	1997	48,600	51,100
1962	19,942	3,939	1980	34,142	16,610	1998	53,200	69,700
1963	24,021	4,668	1981	34,772	12,508	1999	50,300	95,800
1964	21,435	4,637	1982	37,593	12,110	2000	48,000	65,800
1965	25,945	6,781	1983	36,179	20,038	2001	27,100	27,300
1966	22,251	5,864	1984	40,867	40,421	2002	28,200	33,300
1967	23,554	5,181	1985	38,761	22,706	2003	47,700	67,400
1968	24,485	6,875	1986	41,109	10,887	2004	45,100	81,000
1969	20,894	5,772	1987	40,183	14,486	2005	54,400	93,700
1970	21,670	5,801	1988	36,246	17,456	2006	41,500	60,400

Sources: 1953–1970: Wheeler et al. (c. 1978); 1971–1978: Bank of PNG Quarterly Economic Bulletins; 1979–1989: DAL (1992); 1990–2006: Bank of PNG.

Appendix Table A5.6.4 Destination of copra exports, 1998–2005 (tonnes)

Destination	1998	1999	2000	2001[a]	2002	2003	2004	2005
Australia	0	3,716	3,442	0	0	0	0	1,799
Bangladesh	0	0	0	0	0	0	1,799	0
Europe[b]	23,043	22,804	39,298	12,346	9,202	4,916	10,721	19,057
India	0	0	0	0	0	0	0	105
Japan	24,370	29,329	27,165	8,982	3,332	383	0	0
Singapore	8,727	2,679	0	1,136	3,110	596	1,088	1,628
Solomon Islands	0	0	0	0	0	0	0	398
United Kingdom	0	0	0	0	0	0	7,542	0
Vanuatu	0	0	0	0	0	0	774	0
Total	56,140	58,528	69,905	22,464	15,644	5,895	21,925	22,988

[a] Less copra has been exported from 2001 onwards because more copra is being processed into copra oil at Madang and Toboi mills.

[b] 'Europe' is not defined in the data source for 1998–2003; in 2004–2005 exports were to Germany.

Source: Kokonas Indastri Koporesen.

Appendix Table A5.7.1 Volume and value of palm oil exports, 1972–2007

Year	Volume (tonnes)	Value (K'000)	Year	Volume (tonnes)	Value (K'000)
1972	3,409	515	1990	142,700	32,700
1973	7,939	1,148	1991	199,600	52,500
1974	8,734	2,685	1992	206,100	64,200
1975	18,438	6,786	1993	245,700	79,200
1976	27,262	6,776	1994	230,800	77,500
1977	24,532	8,582	1995	186,600	142,200
1978	28,413	10,483	1996	267,000	182,400
1979	34,527	14,441	1997	274,900	207,100
1980	33,347	11,956	1998	213,000	271,900
1981	44,031	14,223	1999	253,800	337,900
1982	76,715	21,655	2000	336,300	306,600
1983	77,940	23,740	2001	327,600	290,500
1984	129,900	75,700	2002	323,900	389,900
1985	123,800	61,600	2003	326,900	421,300
1986	129,000	28,300	2004	339,000	438,700
1987	97,300	23,900	2005	295,200	391,400
1988	102,600	32,900	2006	362,300	430,100
1989	131,700	38,300	2007[a]	368,300	672,200

[a] The 2007 figures are provisional.

Sources: 1972–1975: Territory of Papua and New Guinea Bureau of Statistics; 1976–1979: Bank of PNG Quarterly Economic Bulletins; 1980–1983: DAL (1992); 1984–2007: Bank of PNG.

Appendix Table A5.7.2 Oil palm production (fresh fruit bunch) by sector, 1985–2007 (tonnes)

Year	Smallholder	Plantation	Others	Total
1985	284,800	283,700		568,500
1986	286,200	280,700		566,900
1987	269,800	270,300		540,100
1988	246,200	254,300		500,500
1989	277,700	329,200		606,900
1990	248,100	350,100	60,200	658,400
1991	263,600	473,500	46,400	783,500
1992	261,800	612,200	18,100	892,100
1993	261,000	703,500	20,500	985,000
1994	263,100	723,800	1,400	988,300
1995	278,900	702,800	2,000	983,700
1996	309,511	865,923	21,338	1,196,772
1997	287,641	812,478		1,100,119
1998	395,800	705,500		1,101,300
1999	423,700	767,900	4,900	1,196,500
2000	531,263	1,086,821		1,618,084
2001	525,298	995,165		1,520,463
2002	554,608	1,042,890		1,597,498
2003	545,333	1,134,356		1,679,689
2004	586,910	1,183,850		1,770,760
2005	614,004	1,241,036		1,855,040
2006	656,119	1,355,719		2,011,838
2007	719,419	1,452,011		2,171,430

Sources: 1985–1995: Oil Palm Industry Corporation; 1996–1999: Palm Oil Producers Association (compiled by DAL); 2000–2007: Ian Orrell, Oil Palm Research Association.

Appendix Table A5.8.1 Log exports by province, 1991–2007 (m^3)

Province	1991	1992	1993	1994	1995	1996	1997	1998	1999
Western	247,314	273,510	315,808	250,458	245,136	234,960	449,136	250,311	342,962
Gulf	97,805	152,372	170,834	165,995	108,053	156,321	448,394	184,104	342,307
Central	126,285	107,756	125,629	77,458	22,558	73,537	46,046	27,763	36,688
Milne Bay	36,604	31,432	43,812	30,425	19,901	62,202	89,906	91,695	144,861
Oro	0	35,100	67,279	64,111	27,117	55,606	30,765	20,671	26,723
Morobe	103,768	111,961	111,705	62,220	49,521	61,326	88,054	55,039	50,438
Madang	0	0	23,220	43,455	35,023	77,661	90,114	12,404	5,491
East Sepik	2,141	0	4,952	16,178	12,487	6,568	14,771	8,315	7,307
Sandaun	68,694	76,432	179,320	249,472	283,360	329,112	352,815	145,901	197,215
Manus	22,002	34,143	113,931	83,019	18,241	39,953	58,287	56,389	77,664
New Ireland	75,746	155,658	153,925	61,026	109,946	58,196	61,964	23,545	42,997
East New Britain	84,203	165,039	296,642	270,367	286,404	417,649	196,655	243,960	303,088
West New Britain	434,346	679,123	1,119,968	1,359,803	973,315	1,077,628	1,079,281	496,469	406,112
Other	73,185	43,426	5,833	–	–	–	–	–	–
Papua New Guinea	1,372,093	1,865,952	2,732,858	2,733,987	2,191,062	2,650,719	3,006,188	1,616,566	1,983,853

... continued

Appendix Table A5.8.1 (continued)

Province	2000	2001	2002	2003	2004	2005	2006	2007
Western	423,302	347,594	458,984	563,020	335,904	312,028	339,802	236,360
Gulf	333,554	408,213	443,718	354,020	482,217	471,472	359,028	384,134
Central	59,412	62,805	104,013	75,741	52,002	7,745	45,774	100,544
Milne Bay	85,522	46,965	24,840	20,817	4,503	0	0	0
Oro	6,586	0	11,535	2,713	16,015	105,235	110,682	85,649
Morobe	44,574	41,409	30,557	19,418	24,078	36,333	97,778	98,852
Madang	0	0	2,479	18,443	0	34,914	37,218	18,071
East Sepik	5,561	6,759	0	0	0	57,664	127,621	120,082
Sandaun	221,904	134,158	144,638	195,140	176,454	148,884	232,076	413,629
Manus	60,522	42,391	43,016	31,557	46,060	38,728	41,834	35,989
New Ireland	31,667	14,780	34,043	30,221	33,746	61,213	102,393	117,083
East New Britain	332,145	191,314	198,715	282,342	255,479	259,072	454,412	304,356
West New Britain	387,778	259,832	357,012	421,776	585,677	749,128	694,606	920,361
Papua New Guinea	1,992,527	1,556,220	1,853,550	2,015,208	2,012,135	2,282,416	2,643,224	2,835,110

Source: SGS PNG Ltd.

Appendix Table A5.8.2 Log exports by province, 1991–2007 (kina)[a]

Province	1991	1992	1993	1994	1995	1996	1997	1998	1999
Western	17,940,150	22,544,666	51,980,730	48,736,251	43,850,484	42,063,138	75,848,288	34,300,414	67,698,673
Gulf	6,501,907	13,533,804	26,517,430	33,649,391	19,146,028	26,212,469	71,605,053	23,159,439	58,167,014
Central	8,694,037	8,320,747	20,484,842	13,819,465	3,555,824	12,656,718	6,848,997	3,079,778	5,714,468
Milne Bay	2,045,731	2,156,669	6,323,420	4,999,237	3,010,363	9,359,768	12,309,906	11,533,039	25,174,176
Oro	0	2,761,098	10,709,286	13,699,452	5,526,710	10,863,305	5,186,752	3,704,348	5,102,000
Morobe	6,951,661	9,345,370	22,052,482	12,995,979	9,541,088	10,553,126	14,561,935	7,720,929	9,874,794
Madang	0	0	3,938,198	7,423,606	5,175,506	12,940,859	15,121,242	1,161,599	610,898
East Sepik	114,204	0	601,109	3,088,812	1,960,603	992,865	1,874,356	580,090	1,431,883
Sandaun	8,046,210	8,346,507	33,849,893	49,563,696	61,902,253	70,067,773	68,253,321	28,573,599	43,961,006
Manus	1,951,503	3,668,009	27,293,847	19,876,388	3,821,478	8,291,695	12,427,903	8,956,097	20,223,559
New Ireland	5,573,538	12,878,377	24,576,045	11,345,847	20,300,119	11,433,155	11,875,012	2,586,289	7,421,474
East New Britain	6,224,159	13,594,344	48,809,278	49,611,098	49,886,390	73,029,468	29,122,888	29,759,535	57,892,046
West New Britain	29,855,507	53,683,269	171,591,970	241,957,748	163,707,601	179,023,143	171,115,651	62,968,499	82,285,067
Other	5,636,309	4,102,791	885,253	–	–	–	–	–	–
Papua New Guinea	99,534,914	154,935,650	449,613,782	510,766,970	391,384,447	467,487,482	496,151,304	218,083,655	385,557,058

... continued

Appendix Table A5.8.2 (continued)

Province	2000	2001	2002	2003	2004	2005	2006	2007
Western	82,741,270	66,659,999	98,019,320	109,872,779	67,946,108	63,453,933	67,320,647	47,362,562
Gulf	62,070,678	74,393,438	88,435,881	63,321,191	90,817,446	87,693,986	64,108,344	75,126,214
Central	8,466,926	10,274,548	18,943,421	12,742,536	9,976,127	4,545,791	10,424,840	19,213,881
Milne Bay	14,774,866	7,279,578	3,417,572	3,219,581	541,583	0	0	0
Oro	1,474,798	0	2,472,448	499,955	2,876,214	19,029,482	19,459,505	16,093,169
Morobe	9,373,135	6,217,634	4,844,623	2,788,873	4,214,373	6,032,329	15,583,192	15,853,376
Madang	0	0	497,127	2,794,726	0	5,395,239	6,559,117	2,870,827
East Sepik	793,648	822,981	0	0	0	13,475,236	26,358,083	27,922,727
Sandaun	48,184,626	24,809,499	31,740,253	36,294,489	34,499,023	31,408,030	45,858,376	101,249,271
Manus	15,540,520	9,676,609	11,979,774	8,080,612	11,463,205	9,841,647	9,841,689	8,438,072
New Ireland	4,309,600	1,793,289	5,672,012	4,416,886	5,748,520	8,862,658	16,854,785	20,482,423
East New Britain	60,917,316	34,450,091	38,822,065	53,026,940	55,358,636	47,306,694	84,263,776	59,035,736
West New Britain	75,096,816	46,712,232	62,088,224	72,498,211	100,221,422	132,835,112	123,677,411	176,940,466
Papua New Guinea	383,744,199	283,089,898	366,932,720	369,556,779	383,662,657	429,880,136	490,309,765	570,588,724

[a] 1991–1996 and 2004–2005 converted from US$ to kina using Bank of PNG exchange rates.

Source: SGS PNG Ltd.

Appendix Table A5.8.3 Volume and value of log exports, 1961–2007

Year	Volume (m^3)	Value (kina)	Year	Volume (m^3)	Value (kina)
1961	136,400		1985	1,052,600	53,578,921
1962	159,100		1986	1,279,600	67,746,059
1963	181,800		1987	1,312,600	93,167,164
1964	204,500		1988	1,174,700	77,260,826
1965	227,300		1989	1,259,200	84,354,251
1966	261,400		1990	1,002,000	68,539,340
1967	318,200		1991	1,374,883	99,751,254
1968	363,600		1992	1,869,210	155,182,520
1969	340,900		1993	2,732,860	449,613,783
1970	363,600		1994	2,761,036	513,476,634
1971	454,500		1995	2,191,064	391,384,449
1972	434,500		1996	2,650,717	467,487,482
1973	456,600		1997	3,006,188	496,151,305
1974	686,900		1998	1,616,568	218,083,655
1975	434,400		1999	1,983,853	385,572,125
1976	552,000	9,747,000	2000	1,992,528	383,746,252
1977	468,000	10,968,000	2001	1,556,219	283,089,897
1978	453,000	11,846,000	2002	1,853,550	367,043,758
1979	510,000	20,883,000	2003	2,015,208	369,556,781
1980	641,900	20,084,283	2004	2,012,136	355,660,116
1981	742,800	21,274,787	2005	2,282,414	405,370,831
1982	1,063,300	37,077,406	2006	2,638,297	489,463,821
1983	1,019,200	38,570,916	2007	2,835,110	570,588,726
1984	1,283,900	66,301,349			

Sources: 1961–1971: PNG Department of Forests; 1972–1979: Bank of PNG Quarterly Economic Bulletins; 1980–1990: PNG Forest Authority; 1991–2007: SGS PNG Ltd.

Appendix Table A5.8.4 Destination of log exports, 1991–2007 (m^3)

Destination	1991	1992	1993	1994	1995	1996	1997	1998	1999
China	47,252	36,179	69,054	55,045	46,183	57,228	104,418	157,862	328,890
Hong Kong	22,536	31,158	55,489	127,744	76,321	49,905	129,445	39,246	34,209
India	16,138	17,100	22,817	14,193	17,502	43,373	49,440	39,619	26,169
Indonesia	0	0	0	0	0	0	0	0	0
Japan	760,705	1,031,568	1,531,757	1,744,567	1,237,083	1,682,844	1,960,249	897,884	988,711
Malaysia	5,112	17,951	6,503	0	15,906	6,023	29,253	13,042	8,201
Philippines	9,600	41,436	74,002	45,075	117,731	268,760	196,016	147,008	136,229
Singapore	0	0	0	0	0	2,033	0	6,251	0
South Korea	492,654	654,700	748,771	715,856	607,946	486,518	421,219	281,053	416,680
Taiwan	11,707	22,641	56,287	47,944	58,610	33,034	88,373	24,435	41,049
Thailand	9,179	16,477	14,998	10,612	13,782	20,999	27,775	10,168	3,715
Vietnam	0	0	0	0	0	0	0	0	0
Other	–	–	153,182	–	–	–	–	–	–
Total	1,374,883	1,869,210	2,732,860	2,761,036	2,191,064	2,650,717	3,006,188	1,616,568	1,983,853

... continued

Appendix Table A5.8.4 (continued)

Destination	2000	2001	2002	2003	2004	2005	2006	2007
China	687,779	785,734	1,133,701	1,245,119	1,296,434	1,720,500	2,228,934	2,355,738
Hong Kong	53,183	46,831	12,945	54,185	7,918	0	0	0
India	6,481	0	0	0	19,673	27,286	45,646	95,405
Indonesia	0	8,345	37,855	0	0	0	12	5,050
Japan	745,612	418,235	418,486	406,806	339,969	215,875	193,145	152,820
Malaysia	13,274	4,192	0	0	0	0	0	7,238
Philippines	96,182	30,159	36,773	21,579	7,216	17,919	21,690	23,353
Singapore	0	0	0	0	2,730	0	0	0
South Korea	323,091	224,054	171,079	183,217	157,317	116,662	70,686	57,431
Taiwan	47,913	28,279	42,711	36,428	44,567	25,712	5,134	389
Thailand	8,558	0	0	0	0	12,105	8,959	6,553
Vietnam	10,455	10,390	0	67,874	136,312	146,355	64,091	131,133
Total	1,992,528	1,556,219	1,853,550	2,015,208	2,012,136	2,282,414	2,638,297	2,835,110

Source: SGS PNG Ltd.

Appendix Table A5.8.5 Destination of log exports, 1991–2006[a] (US$)

Destination	1991	1992	1993	1994	1995	1996
China	4,212,580	3,907,394	10,799,706	8,838,369	8,083,815	9,985,820
Hong Kong	3,172,531	5,160,710	10,208,909	25,459,001	17,457,936	11,262,097
India	2,674,476	2,399,253	4,697,083	3,119,903	3,798,896	9,673,679
Indonesia	0	0	0	0	0	0
Japan	60,043,718	91,591,007	279,517,862	288,975,785	173,568,346	230,146,889
Malaysia	358,758	1,159,345	764,957	0	1,798,256	632,498
Philippines	473,991	2,598,361	9,729,470	4,950,760	12,091,713	28,046,691
Singapore	0	0	0	0	0	447,174
South Korea	32,508,084	47,768,007	106,367,796	96,658,904	70,766,647	57,555,618
Taiwan	731,441	1,515,445	7,161,150	6,402,430	6,125,198	2,962,477
Thailand	543,287	1,053,817	1,671,854	1,279,774	1,608,760	2,380,351
Vietnam	0	0	0	0	0	0
Other	–	–	27,237,659	–	–	–
Total	104,718,866	157,153,339	458,156,446	435,684,926	295,299,567	353,093,294

... continued

Appendix Table A5.8.5 (continued)

Destination	2000	2001	2002	2003	2004	2005	2006
China	48,765,988	44,246,289	62,929,046	68,720,273	76,560,021	105,669,106	143,126,416
Hong Kong	4,442,856	2,791,332	791,442	2,957,013	417,316	0	0
India	777,668	0	0	0	2,543,413	2,637,206	5,364,123
Indonesia	0	444,831	2,002,607	0	0	0	718
Japan	56,972,759	26,030,620	22,209,475	21,359,766	19,446,139	13,265,985	11,621,455
Malaysia	834,889	239,830	0	0	0	0	0
Philippines	5,914,051	1,717,255	1,865,592	1,143,655	486,438	1,126,892	1,722,877
Singapore	0	0	0	0	158,314	0	0
South Korea	21,488,133	11,819,418	7,920,166	8,344,703	7,708,430	6,192,170	3,916,969
Taiwan	2,779,515	1,332,847	2,032,526	1,743,467	2,319,599	1,270,289	289,429
Thailand	722,062	0	0	0	0	675,421	524,516
Vietnam	1,007,174	778,718	0	4,765,637	9,334,121	7,756,289	3,717,073
Total	143,705,095	89,401,140	99,750,854	109,034,514	118,973,791	138,593,358	170,283,576

[a] Data were not available for 1997–1999.

Source: SGS PNG Ltd.

Appendix Table A5.9.1 Volume and value of marine product exports, 1977–2006

Year	Volume (tonnes)	Value (K'000)
1977	24,883	19,263
1978	46,831	24,851
1979	28,695	20,861
1980	34,729	31,739
1981	31,404	26,913
1982	4,139	8,288
1983	2,060	9,519
1984	3,133	8,173
1985	10,197	14,419
1986	2,030	11,303
1987	1,815	11,598
1988	1,536	8,945
1989	2,092	11,422
1990	2,164	11,322
1991	1,898	9,707
1992	3,580	9,300
1993	1,896	7,800
1994	1,443	10,300
1995	1,936	12,300
1996	4,508	25,880
1997	7,810	37,151
1998	38,147	135,636
1999	39,904	124,858
2000	50,048	158,212
2001	52,699	227,326
2002	57,610	307,863
2003	52,206	246,464
2004	42,899	253,330
2005	60,944	291,853
2006	43,708	261,022

Sources: 1977–1991: DAL (1992); 1992–2006: National Fisheries Authority.

Appendix Table A5.9.2 Volume of marine exports by product, 1992–2006 (tonnes)

Year	Tuna[a]					Shark[b]	Bêche-de-mer[c]	Shell[d]	Shrimp (frozen)	Fish[e]	Crab[f]	Lobster[g]	Total
		frozen	canned	fresh/chilled	dried/meal								
1992	1,514	–	–	–	–	296	656	286	757	32	0.4	38	3,579
1993	255	–	–	–	–	205	500	180	695	5	0.8	55	1,896
1994	177	–	–	–	–	98	209	229	601	25	7	97	1,443
1995	88	–	–	–	–	153	445	332	738	56	8	116	1,936
1996	2,287	–	–	–	–	162	596	225	725	418	5	90	4,508
1997	–	4,286	1,210	453	302	95	505	243	567	38	4	107	7,810
1998	–	28,322	5,184	590	1,011	867	679	333	946	97	3	115	38,147
1999	–	29,122	6,711	679	260	1,410	395	264	833	123	2	105	39,904
2000	–	32,977	10,299	1,197	1,690	1,804	608	246	934	191	4	98	50,048
2001	–	34,611	11,197	1,799	1,320	1,629	484	347	947	295	4	66	52,699
2002	–	33,908	13,886	3,229	3,120	1,451	389	420	787	312	–	108	57,610
2003	–	27,472	15,494	3,655	5,213	1,302	401	466	761	355	–	87	55,206
2004	–	17,504	16,377	2,310	3,521	1,464	488	399	620	121	4	91	42,900
2005	–	37,355	15,397	929	3,943	1,471	597	399	522	207	7	117	60,944
2006	–	16,627	16,079	1,565	6,142	1,684	666	453	286	139	9	55	43,708

[a] For the period 1992–1996, all categories of tuna are combined.

[b] Shark includes dried maw, fresh chilled, frozen and oil products and also dried and frozen sharkfin.

[c] See Kinch et al. (2007) for volume of bêche-de-mer exports for 1960–2006.

[d] The category 'shell' includes various shell products such as buttons and pearlshell.

[e] Fish includes canned, fresh chilled, frozen, live, maw, oil meat and unstated products.

[f] Crab includes cooked, frozen and live products. Crab is included with fish for 2002 and 2003.

[g] Lobster includes frozen and live products.

Source: National Fisheries Authority.

Appendix Table A5.9.3 Value of marine exports by product, 1996–2006 (K'000)

Year	Tuna					Shark[b]	Bêche-de-mer	Shell[c]	Shrimp (frozen)	Fish[d]	Crab[e]	Lobster[f]	Total
		frozen	canned	fresh/chilled	dried/meal								
1996	3,584[a]	–	–	–	–	431	7,872	3,300	7,612	491	47	2,542	25,880
1997	–	6,240	5,324	2,336	167	492	7,683	3,529	7,190	534	34	3,622	37,151
1998	–	55,234	23,946	3,480	507	4,371	16,893	5,312	19,150	741	30	5,972	135,636
1999	–	38,425	32,173	5,752	204	6,140	11,024	5,083	18,537	1,003	17	6,227	124,585
2000	–	36,942	48,728	14,120	1,075	6,641	16,321	4,979	22,100	993	58	6,255	158,212
2001	–	75,533	61,422	26,815	1,641	7,934	17,278	3,712	25,845	1,362	124	5,660	227,326
2002	–	117,102	94,576	29,854	1,748	6,633	17,344	4,218	24,100	1,644	–	10,644	307,863
2003	–	50,544	103,217	31,317	2,106	4,947	20,412	3,855	20,210	1,739	–	8,117	246,464
2004	–	37,488	122,402	33,275	4,520	6,205	23,049	3,439	13,417	1,388	92	8,054	253,330
2005	–	84,240	125,950	11,849	4,845	5,803	29,448	4,890	12,564	1,980	135	10,150	291,853
2006	–	40,576	126,224	21,464	9,473	6,352	36,715	7,308	6,980	1,735	169	4,026	261,022

[a] For 1996, all categories of tuna are combined.

[b] Shark includes dried maw, fresh chilled, frozen and oil products and also dried and frozen sharkfin.

[c] The category 'shell' includes various shell products such as buttons and pearlshell.

[d] Fish includes canned, fresh chilled, frozen, live, maw, oil meat and unstated products.

[e] Crab includes cooked, frozen and live products. Crab is included with fish for 2002 and 2003.

[f] Lobster includes frozen and live products.

Source: National Fisheries Authority.

Appendix Table A5.10.1 Commercial sugar harvest, production, domestic sales and exports, 1982–2007

Year	Area harvested (hectares)			Cane harvested (tonnes)			Sugar production (tonnes)	Domestic sales (tonnes)	Exports (tonnes)
	Estate	Outgrowers	Total	Estate	Outgrowers	Total			
1982	1,345	0	1,345	118,408	0	118,408	11,072	0	0
1983	4,551	87	4,638	356,113	6,877	362,990	33,320	26,249	0
1984	4,994	681	5,675	323,854	38,079	361,933	34,114	27,616	11,003
1985	4,855	818	5,673	245,051	49,746	294,797	30,050	26,625	0
1986	3,724	934	4,658	99,583	29,203	128,786	9,377	27,504	10,780
1987	3,800	794	4,594	243,070	43,327	286,397	22,107	28,493	6,519
1988	5,725	950	6,675	492,077	79,443	571,520	50,870	29,155	7,852
1989	5,958	919	6,877	316,589	47,634	364,223	30,410	29,384	14,844
1990	5,984	894	6,878	291,375	10,189	301,564	28,274	27,666	9,818
1991	6,371	705	7,076	303,729	32,699	336,428	29,804	25,989	7,765
1992	6,277	725	7,002	369,931	45,783	415,714	36,967	28,443	6,974
1993	6,129	712	6,841	307,730	32,331	340,061	32,203	30,020	0
1994	5,835	842	6,677	356,775	59,473	416,248	39,522	32,319	7,529
1995	6,032	797	6,830	371,930	55,411	427,341	35,162	30,803	7,435
1996	5,365	1,325	6,691	307,431	82,876	390,307	29,797	33,912	0
1997	5,815	1,731	7,546	308,538	88,598	397,136	39,279	36,982	7,012
1998	6,070	1,635	7,705	363,894	101,430	465,324	41,141	36,213	7,021
1999	5,954	2,193	8,147	380,113	139,471	519,584	44,521	35,543	6,960
2000	5,847	2,153	7,999	319,633	113,057	432,690	38,519	32,881	6,952
2001	5,333	2,498	7,831	311,421	143,155	454,576	42,438	32,590	8,241
2002	5,204	2,132	7,335	365,786	155,486	521,272	49,102	34,000	8,327
2003	5,066	1,941	7,007	349,092	147,499	496,591	46,490	32,000	10,350
2004	5,463	1,793	7,255	326,708	101,758	428,466	42,658	33,000	19,227
2005	6,258	1,710	7,968	368,256	101,113	469,369	44,074	37,000	7,823
2006	6,044	1,231	7,275	309,356	52,201	361,557	31,796	33,000	8,000
2007	6,642	1,219	7,862	321,991	49,802	371,793	33,516	n.a.	0

Source: Ramu Agri-Industries Ltd.

Appendix Table A5.11.1 Volume and value of rubber exports, 1936–2007[a]

Year	Volume (tonnes)	Value (kina)	Year	Volume (tonnes)	Value (kina)	Year	Volume (tonnes)	Value (kina)
1936	1,101		1960	4,470	3,000,394	1984	3,009	2,543,000
1937	1,183		1961	4,443	2,584,302	1985	5,019	3,668,000
1938	1,240		1962	4,755	2,421,444	1986	5,105	3,855,000
1939	1,315		1963	4,836	2,348,090	1987	4,252	3,397,000
1940	1,446		1964	5,088	2,452,062	1988	4,921	4,859,000
1941	1,481		1965	5,405	2,564,000	1989	4,905	3,359,000
1942	1,063		1966	5,474	2,576,000	1990	3,223	2,182,000
1943	403		1967	5,791	2,480,000	1991	2,772	1,835,000
1944	1,792		1968	5,690	1,956,000	1992	2,598	1,880,000
1945	1,416		1969	5,791	2,276,000	1993	3,438	2,494,000
1946	964		1970	6,350	2,797,000	1994	4,215	4,141,000
1947	958		1971	6,350	2,297,000	1995	5,396	8,946,000
1948	1,295	313,678	1972	5,944	1,995,000	1996	7,019	11,932,000
1949	1,217	291,946	1973	5,639	1,998,000	1997	4,586	6,709,000
1950	1,659	722,678	1974	6,147	3,563,000	1998	4,745	5,576,000
1951	1,983	1,604,354	1975	5,474	2,585,000	1999	4,425	5,651,000
1952	2,896	2,488,518	1976	4,956	2,653,000	2000	4,433	7,040,000
1953	2,958	1,472,146	1977	4,152	2,897,000	2001	4,133	7,465,732
1954	3,195	1,225,576	1978	4,135	2,630,000	2002	4,149	10,172,120
1955	3,650	1,907,246	1979	4,025	3,497,000	2003	4,298	12,520,100
1956	3,854	2,773,574	1980	4,027	3,751,000	2004	4,536	15,500,000
1957	4,030	2,297,084	1981	4,537	3,403,000	2005	5,231	17,985,000
1958	4,327	2,229,432	1982	2,337	1,406,000	2006	4,400	23,800,000
1959	4,450	2,303,742	1983	2,735	2,167,000	2007	4,100	22,900,000

[a] Figures for 1935–36 to 1946–47 are for imports of rubber into Australia from PNG. This was the main export destination for PNG rubber in this period. Value data were not available.

Sources: 1936–1947: Bureau of Agricultural Economics (1952:15); 1948–1976: Territory of Papua and New Guinea Bureau of Statistics; 1977–1984: DAL (1992); 1985–1999: DAL (2001); 2000–2005: DAL; 2006–2007: Bank of PNG.

Appendix Table A5.11.2 Rubber production by sector, 1985–2001 (tonnes)

Year	Smallholder	Plantation	Total
1985	1,296	3,723	5,019
1986	1,556	3,549	5,105
1987	1,401	2,851	4,252
1988	1,449	3,472	4,921
1989	1,610	3,442	5,052
1990	1,616	1,607	3,223
1991	1,659	1,113	2,772
1992	1,745	2,122	3,867
1993	2,615	2,233	4,848
1994	2,899	2,124	5,023
1995	4,368	2,142	6,510
1996	4,896	2,243	7,139
1997	2,676	1,910	4,586
1998	2,779	1,876	4,655
1999	2,322	2,103	4,425
2000	2,283	2,150	4,433
2001	2,033	2,100	4,133

Source: DAL.

Appendix Table A5.11.3 Destination of rubber exports, 2004–2005

Destination	2004		2005	
	Volume (tonnes)	Value (kina)	Volume (tonnes)	Value (kina)
Australia	1,332	5,393,793	416	1,529,847
Belgium	42	12,020	448	1,746,409
France	375	1,459,783	875	2,494,296
Germany	1,769	3,437,387	1,581	6,908,478
Italy	116	134,021	396	1,756,426
Netherlands	219	1,153,013	454	1,841,962
Spain	125	143,131	124	461,771
Total	3,977	11,733,148	4,294	16,739,189

Source: DAL.

Appendix Table A5.12.1 Volume and value of tea exports, 1963–2006

Year	Volume (tonnes)	Value (K'000)	Year	Volume (tonnes)	Value (K'000)
1963	19	8	1985	7,025	13,330
1964	29	25	1986	5,322	7,133
1965	18	17	1987	5,553	5,571
1966	11	11	1988	5,829	6,438
1967	4	3	1989	5,447	6,036
1968	38	42	1990	5,375	6,651
1969	298	297	1991	4,700	5,300
1970	689	645	1992	5,600	6,600
1971	1,169	1,094	1993	6,400	7,200
1972	1,803	1,500	1994	3,400	4,200
1973	2,791	2,048	1995	4,200	5,400
1974	3,965	2,601	1996	9,300	12,700
1975	4,489	3,866	1997	6,500	10,400
1976	4,871	3,978	1998	6,600	18,900
1977	6,192	9,068	1999	8,200	19,000
1978	6,979	7,833	2000	8,500	20,400
1979	6,978	7,982	2001	8,800	22,000
1980	7,915	8,507	2002	5,200	18,545
1981	6,959	7,131	2003	6,600	20,411
1982	6,475	6,682	2004	8,100	22,769
1983	7,234	10,391	2005	6,934	20,223
1984	7,530	19,058	2006	6,507	20,991

Sources: 1963–1976: Territory of Papua and New Guinea Bureau of Statistics; 1977–1990: DAL (1992); 1991–2004: Bank of PNG; 2005–2006: Carpenter Estates.

Appendix Table A5.12.2 Destination of tea exports, 1992–2006 (K'000)[a]

Destination	1992	1993	1994	1995	1996	1997	1998	1999	2000	2001	2002	2003	2004	2005	2006
Australia	3,840	3,792	2,377	2,992	7,036	5,824	10,546	9,880	10,404	6,789	5,448	1,013	776	860	839
Belgium	–	–	–	–	–	–	–	–	–	–	191	106	960	683	1,172
Canada	–	–	–	–	–	–	–	–	–	–	769	1,337	1,210	1,174	1,116
Chile	–	–	–	–	–	–	–	–	–	–	0	0	0	0	104
Fiji	–	–	–	–	–	–	–	–	–	–	0	100	65	63	33
Germany	–	–	–	–	–	–	–	–	–	–	1,385	1,597	2,669	1,861	3,089
India	–	–	–	–	–	–	–	–	–	–	1,326	2,207	2,440	2,087	1,939
Indonesia	–	–	–	–	–	–	–	–	–	–	1,020	1,625	2,032	1,818	980
Israel	–	–	–	–	–	–	–	–	–	–	0	34	32	0	0
Japan	–	–	–	–	–	–	–	–	–	–	200	140	0	37	0
Kenya	–	–	–	–	–	–	–	–	–	–	638	629	484	278	0
Malaysia	–	–	–	–	–	–	–	–	–	–	70	496	208	214	172
Netherlands	–	–	–	–	–	–	–	–	–	–	0	179	677	74	0
New Caledonia	–	–	–	–	–	–	–	–	–	–	6	6	6	0	0
New Zealand	395	268	269	270	635	186	813	756	612	–	55	92	133	520	589
Pakistan	221	157	168	194	508	279	662	950	1,020	–	0	527	159	258	347
Poland	–	–	–	–	–	–	–	–	–	–	619	1,058	1,529	0	0
Russia	–	–	–	–	–	–	–	–	–	–	1,413	3,499	4,165	5,925	6,507
Singapore	60	241	109	140	381	128	416	654	408	1,533	0	80	81	85	47
United Kingdom	420	1,014	202	356	889	837	1,493	1,140	1,836	6,570	2,555	3,597	3,625	3,389	2,043
United States	1,266	1,013	678	896	2,159	2,047	3,289	4,290	5,304	4,599	2,850	2,089	1,518	897	2,013
Others	310	697	397	551	1,092	1,099	1,682	1,330	816	2,409	–	–	–	–	–
Total	6,512	7,182	4,200	5,399	12,700	10,400	18,901	19,000	20,400	21,900	18,545	20,411	22,769	20,223	20,991

[a] 'Other' countries are not specified in the source data for 1992–2001.

Sources: 1992–2001: DAL (2001); 2002–2006: Carpenter Estates.

Appendix Table A5.13.1 Volume and value of balsa exports, 1996–2007

Year	Volume (m^3)	Value (US$)	Value (kina)
1996	2,706	1,186,788	1,571,280
1997	2,034	1,006,796	1,444,263
1998	3,570	1,648,549	3,394,870
1999	3,067	1,484,555	3,785,199
2000	2,668	1,268,310	3,500,718
2001	2,043	911,986	3,064,469
2002	2,700	926,260	3,599,922
2003	3,490	1,019,463	3,620,252
2004	3,755	1,168,339	3,767,620
2005	5,304	1,825,246	5,661,433
2006	8,197	2,852,079	8,716,623
2007	11,058	3,409,479	10,214,137

Sources: 1996–2002: ITTO ENB Balsa Industry Strengthening Project Phase 2; 2003–2007: PNG Forest Authority.

Appendix Table A5.13.2 Destination of balsa exports, 2000–2007

Destination	2000			2001			2002			2003		
	Volume (m^3)	Value (US$)	Value (kina)	Volume (m^3)	Value (US$)	Value (kina)	Volume (m^3)	Value (US$)	Value (kina)	Volume (m^3)	Value (US$)	Value (kina)
Australia	275	142,125	392,285	653	313,277	1,052,678	531	151,462	588,659	862	257,336	913,835
China	400	239,075	659,881	275	147,525	495,716	681	255,172	991,729	1,291	234,492	832,713
Germany	150	106,625	294,300	200	120,925	406,334	177	101,115	392,985	361	126,428	448,963
Hong Kong	0	0	0	54	15,460	51,949	54	17,032	66,195	64	23,038	81,811
Italy	0	0	0	577	205,703	691,206	552	193,023	750,187	30	138,903	493,263
Japan	0	0	0	3	435	1,462	26	8,970	34,862	50	17,250	61,257
Malaysia	0	0	0	1	1,136	3,817	9	5,338	20,746	4	1,156	4,105
New Zealand	0	0	0	0	0	0	51	10,804	41,990	27	6,148	21,832
Singapore	0	0	0	0	0	0	29	7,530	29,265	54	13,500	47,940
South Africa	0	0	0	0	0	0	27	14,850	57,715	18	5,760	20,455
Spain	0	0	0	25	8,750	29,402	107	43,283	168,220	243	87,885	312,092
Taiwan	50	20,375	56,238	25	8,375	28,142	0	90	350	58	21,261	75,501
Thailand	0	0	0	0	0	0	0	0	0	27	4,037	14,336
United Kingdom	1,793	760,110	2,098,013	230	90,400	303,763	404	107,521	417,882	401	82,269	292,148
Vietnam	0	0	0	0	0	0	52	10,070	39,137	0	0	0
Total	2,668	1,268,310	3,500,718	2,043	911,986	3,064,469	2,700	926,260	3,599,922	3,490	1,019,463	3,620,252

... continued

Appendix Table A5.13.2 (continued)

Destination	2004			2005			2006			2007		
	Volume (m^3)	Value (US$)	Value (kina)	Volume (m^3)	Value (US$)	Value (kina)	Volume (m^3)	Value (US$)	Value (kina)	Volume (m^3)	Value (US$)	Value (kina)
Australia	895	280,960	906,030	1,015	342,447	1,062,181	1,239	405,563	1,239,496	1,529	404,094	1,210,587
Bahrain	0	0	0	0	0	0	27	10,476	32,017	0	0	0
China	954	252,156	813,144	898	336,614	1,044,088	2,232	803,880	2,456,846	4,778	1,409,478	4,222,522
Germany	378	112,455	362,641	324	123,795	383,980	270	95,580	292,115	1,161	393,885	1,180,003
Hong Kong	0	0	0	83	30,747	95,369	54	22,140	67,665	215	85,070	254,853
India	0	0	0	617	234,266	726,632	2,890	1,024,072	3,129,804	2,163	706,848	2,117,579
Italy	30	10,000	32,248	0	0	0	0	0	0	0	0	0
Japan	81	27,496	88,668	162	60,957	189,073	108	38,610	118,001	135	52,110	156,111
Malaysia	3	806	2,599	4	1,325	4,110	4	1,396	4,267	4	1,188	3,559
New Zealand	29	8,497	27,401	52	13,255	41,114	53	18,787	57,417	78	28,350	84,931
South Africa	0	0	0	0	0	0	27	8,640	26,406	0	0	0
Spain	918	347,650	1,121,090	1,566	504,827	1,565,841	648	223,640	683,496	0	0	0
Taiwan	38	12,972	41,832	54	21,600	66,998	27	10,665	32,595	27	11,070	33,164
Thailand	27	10,935	35,263	54	21,870	67,835	27	11,070	33,833	0	0	0
United Kingdom	375	96,859	312,348	354	99,474	308,542	429	124,810	381,449	320	106,300	318,454
United States	0	0	0	0	0	0	0	0	0	432	141,966	425,303
Vietnam	27	7,553	24,357	121	34,069	105,673	162	52,750	161,216	216	69,120	207,070
Total	3,755	1,168,339	3,767,620	5,304	1,825,246	5,661,433	8,197	2,852,079	8,716,623	11,058	3,409,479	10,214,137

Sources: 2000–2002: ITTO ENB Balsa Industry Strengthening Project Phase 2; 2003–2007: PNG Forest Authority.

Appendix Table A5.15.1 Volume and value of chilli and cardamom exports, 1971–2000

Year	Chilli		Cardamom	
	Volume (kg)	Value (kina)	Volume (kg)	Value (kina)
1971	19,800	10,845	–	–
1972	36,800	27,762	–	–
1973	42,500	31,964	–	–
1974	145,600	106,192	660	2,686
1975	154,400	124,716	6,640	28,579
1976	161,598	118,696	11,764	49,453
1977	76,022	114,000	3,765	38,000
1978	190,243	407,000	9,950	94,000
1979	236,673	400,000	9,704	87,000
1980	244,491	222,000	6,358	48,000
1981	265,883	248,000	10,259	45,000
1982	38,216	63,000	49,491	227,000
1983	9,589	20,000	80,388	493,000
1984	32,785	85,000	186,635	1,946,000
1985	18,470	56,000	323,981	2,360,000
1986	24,292	163,000	387,268	1,794,000
1987	75,920	144,000	364,396	1,158,000
1988	43,685	94,000	185,846	506,000
1989	75,295	220,000	156,540	367,000
1990	57,141	159,000	98,104	364,000
1991	32,995	60,000	47,624	154,000
1992	22,250	–	73,240	–
1993	27,280	–	75,820	–
1994	–	–	–	–
1995	–	–	–	–
1996	17,752	50,459	32,003	175,800
1997	4,665	17,225	17,431	89,025
1998	4,051	27,331	7,755	39,548
1999	7,237	16,964	–	–
2000	9,970	58,328	–	–

Sources: 1971–1976: Wyatt (c. 1978); 1977–1991: DAL (1992); 1992–1993: Waisime (2000); 1996–2000: National Statistical Office of PNG (NSO data are increasingly unreliable from 1998 onwards).

Appendix Table A5.16.1 Purchases of dried pyrethrum flowers, 1963–2006

Year	Purchases (tonnes)	Year	Purchases (tonnes)	Year	Purchases (tonnes)
1963	1	1977	201	1991	290
1964	15	1978	150	1992	271
1965	190	1979	172	1993	273
1966	288	1980	205	1994[a]	?
1967	590	1981	252	1995–1999[a]	0
1968	502	1982	245	2000	24
1969	396	1983	279	2001	9
1970	461	1984	300	2002	25
1971	360	1985	352	2003	46
1972	311	1986	327	2004	40
1973	255	1987	256	2005	37
1974	316	1988	232	2006	26
1975	333	1989	265		
1976	253	1990	281		

[a] It is not known whether dried flowers were purchased in 1994, the year in which the extraction plant at Kagamuga closed. There were no purchases between 1995 and 1999.

Sources: 1963–1977: Anderson (c. 1978: Table 1) (data are for Australian financial years, that is, 1962–63 to 1976–77); 1978–1979: Carrad (1982:156); 1980–1993: DAL (1994); 2000–2006: Wakasa Mecksaene, Wabag (pers. comm.).

Appendix Table A5.16.2 Volume and value of pyrethrum extract exports, 1967–1991[a]

Year	Volume (kg)	Value (K'000)	Year	Volume (kg)	Value (K'000)
1967	25,800	390	1980	2,914	85
1968	27,100	417	1981	14,973	565
1969	20,300	313	1982	13,390	498
1970	21,600	332	1983	8,950	397
1971	18,600	286	1984	11,153	441
1972	14,200	227	1985	19,444	704
1973	11,800	192	1986	14,675	484
1974	14,400	217	1987	8,370	249
1975	14,800	190	1988	3,070	148
1976	12,800	194	1989[b]	–	–
1977	9,120	145	1990	6,760	489
1978	5,737	118	1991	9,200	682
1979	4,960	202			

[a] There have been some exports after 1991, according to National Statistical Office data. Small quantities of extract have been exported since 2003, including 4800 kg in 2006.

[b] Pyrethrum extract was not exported in 1989.

Sources: 1967–1976: Anderson (c. 1978: Table 2) (data are for Australian financial years, that is, 1966–67 to 1975–76); 1977–1991: DAL (1992).

Appendix Table A5.18.1 Value of crocodile skin exports, 1958–1968, 1977–1991 (kina)

Year	Value	Year	Value	Year	Value
1958	134,000	1977	814,000	1985	2,370,000
1959	89,000	1978	1,231,000	1986	2,430,000
1960	116,000	1979	1,610,000	1987	1,879,000
1961	241,000	1980	1,620,000	1988	876,000
1962	500,000	1981	1,320,000	1989	929,000
1963	661,000	1982	2,341,000	1990	902,000
1964	902,000	1983	936,000	1991	872,000
1965	857,000	1984	2,295,000		
1966	982,000				
1967	866,000				
1968	491,000				

Sources: 1958–1968: Downes (1971); 1977–1991: DAL (1992).

Appendix Table A5.18.2 Number of freshwater and saltwater crocodile skins exported, 1997–2005

Year	Freshwater crocodiles		Saltwater crocodiles		Total
	Wild animals	Farmed animals	Wild animals	Farmed animals	
1997	28,800	3,900	2,900	5,700	41,300
1998	10,700	5,800	2,600	7,800	26,900
1999	11,700	3,600	1,900	7,300	24,500
2000	13,900	2,300	1,600	6,800	24,600
2001	20,400	0	2,600	7,800	30,800
2002	18,500	0	2,300	7,100	27,900
2003	17,500	0	2,800	4,800	25,100
2004	22,700	0	2,600	7,300	32,600
2005	17,800	0	4,600	5,800	28,200

Source: Department of Environment and Conservation (2007).

Appendix Table A5.18.3 Estimated honey production in Eastern Highlands Province, 1975–2007 (tonnes)

Year	Production	Year	Production	Year	Production
1975	25	1986	120	1997	30
1976	30	1987	98	1998	25
1977	38	1988	50	1999	25
1978	41	1989	60	2000	30
1979	50	1990	79	2001	37
1980	60	1991	60	2002	40
1981	65	1992	56	2003	45
1982	75	1993	42	2004	40
1983	80	1994	42	2005	42
1984	85	1995	42	2006	44
1985	95	1996	40	2007	40

Note: Production data include an estimated 20 tonnes/year sold in the informal sector.

Sources: Hardie et al. (2005) and Tella Loie, Department of Primary Industry, Goroka.

Appendix Table A6.7.1 Land settlement schemes, 1959–1981

Province	Name of settlement area	No. of blocks	Approx. total area (ha)	Average block size (ha)	Recommended activity	Year of allocation
Central	Boregaina	22	185	8.4	rubber	1959
Oro	Girua	10	125	12.5	cocoa/oil palm	1960
East New Britain	Warangoi	33	208	6.3	cocoa	1960
Central	Block 7 Cape Rodney	23	208	9.0	coconut/rubber	1961
Central	Dom, Cape Rodney	30	336	11.2	coconut/rubber	1961
East New Britain	Ilugi	52	372	7.1	cocoa/coconut	1961
Morobe	Situm	22	210	9.5	coconut	1962
Morobe	Gobari	8	97	12.0	coconut	1962
East Sepik	Sowam 1	17	299	17.6	coconut	1962
West New Britain	Dagi	58	705	12.1	coconut	1962
Central	Bereina	30	367	12.2	subsistence	1963
Central	Mori 1 Cape Rodney	103	1111	10.8	rubber	1963
Oro	Sorovi Jumburu	55	658	12.0	cocoa/oil palm	1963
East New Britain	Keravat	36	343	9.5	cocoa/coconut	1963
East New Britain	Tavilo	36	318	8.8	cocoa	1963
West New Britain	Lavilelo	22	243	11	coconut	1963
West New Britain	Tabai Rikau	62	688	11.1	coconut	1963
Central	Mori 2 Cape Rodney	130	1575	12.1	rubber	1964
Oro	Wanigela	84	801	9.5	coconut	1964
Morobe	Bumbu	7	74	10.5	coconut/cattle	1964
Morobe	Umboi	18	142	7.9	coconut	1964
East Sepik	Karawura	3	53	17.6	coconut	1964
Manus	Silin	28	320	11.4	coconut	1964
Manus	Logenbrowa	20	121	6.0	coconut	1964
Morobe	Wau 3 & 4	63	441	7.0	market gardening	1965
Western Highlands	Nondugl	39	183	4.7	tea	1965
Western Highlands	Kondepina	57	245	4.3	tea	1965
Manus	Bokeles	5	82	16.4	coconut	1965
Manus	Putjoliu	1	5	5.0	coconut	1965
New Ireland	Balgai	48	68	1.4	subsistence	1965
East New Britain	Crewder Island	2	34	17.0	cocoa	1965
West New Britain	Silanga	224	1781	8	cocoa	1965
West New Britain	Uasilau	110	913	8.3	cocoa	1965
West New Britain	Pangalu	22	232	10.5	coconut	1965
Central	Kubuna 1	27	356	13.2	rubber	1966
Central	Bailebo	110	1113	10.1	rubber	1966

... continued

Appendix Table A6.7.1 (continued)

Province	Name of settlement area	No. of blocks	Approx. total area (ha)	Average block size (ha)	Recommended activity	Year of allocation
Central	Laloki 1	50	111	2.2	market gardening	1966
Oro	Saga-Amada	37	385	10.4	rubber/cocoa	1966
Oro	Savaia	23	255	11.0	rubber/cocoa	1966
Morobe	Wau 1	68	520	7.6	market gardening	1966
Morobe	Wau 2	18	814	45.0	market gardening	1966
Manus	Ndromalmal	6	82	13.6	coconut	1966
Manus	Haunel	26	227	8.7	coconut	1966
Manus	Salesia	20	198	9.9	coconut	1966
Manus	Malai Bay	17	209	12.2	coconut	1966
East New Britain	Sunam	17	120	7.0	cocoa	1966
West New Britain	Bulu	16	177	11	coconut	1966
Bougainville	Wianin-Teanin	16	155	9.7	cocoa/coconut	1966
Bougainville	Kotstawan	3	22	7.3	cocoa/coconut	1966
Central	Merani Estate	17	103	6.0	coconut	1960, 1966
Gulf	Kufala	8	66	8.2	rubber	1967
Morobe	Webiak	29	338	11.6	market gardening	1967
Manus	Beheni	10	87	8.7	coconut	1967
West New Britain	Kapore	130	911	7	oil palm	1967
West New Britain	Tamba	193	1158	6	oil palm	1967
West New Britain	Sale Malasi	241	2554	10.5	coconut	1967
Central	Anieurebau	7	93	13.3	coconut	1963 ,1967
Western Highlands	Kindeng West	214	862	4.0	tea	1966, 1967
Central	Baramata	5	60	12.0	coconut	
Oro	Girua Extension	23	214	9.3	cocoa/oil palm	1968
Morobe	Gobari Extension	12	135	11.2	coconut/cattle	1968
Morobe	Ali Walu	17	242	14.2	coconut/cattle	1968
Morobe	Wahu	15	191	12.7	coconut/cattle	1968
West New Britain	Sarakolok-Lakiemata	258	1801	7	oil palm	1968
Central	Kuriva	63	903	14.3	market gardening	1967, 1968
Southern Highlands	Pilibaru	5	68	13.6	mixed farming/ cattle	1969
West New Britain	Siki	27	310	11.5	cocoa/coconut	1969
Oro	Kokoda	26	266	10.2	rubber	1960, 1969
Gulf	Murua	53	606	11.4	rubber	1962, 1969
Central	Kubuna 2	78	1010	12.9	rubber	1970
Oro	Embi	14	1738	124.0	cattle	1970
Sandaun	Yalingi	55	660	12.0	coconut	1970

... continued

Appendix Table A6.7.1 (continued)

Province	Name of settlement area	No. of blocks	Approx. total area (ha)	Average block size (ha)	Recommended activity	Year of allocation
East New Britain	Vudal-Vunapaladin	175	2200	12.6	cocoa	1970
East New Britain	Mandress	96	1660	17.2	cocoa	1970
East New Britain	Japlik	44	710	16.0	cocoa	1970
West New Britain	Buvussi	321	1962	6.1	oil palm	1970
Western Highlands	Awalibu	3	178	59.0	mixed farming/ cattle	1971
Western Highlands [Enga]	Pawari	2	124	62.0	mixed farming/ cattle	1971
East New Britain	Nengmutka	40	600	15.0	cocoa	1971
West New Britain	Pota Galai	57	342	6	oil palm	1971
Oro	Butue	38	359	9.5	rubber/cocoa	1972
Oro	Kamondo-Pirive	27	258	9.5	rubber/cocoa	1973
East Sepik	Yauwosoru	19	107	5.6	coconut	1973
Sandaun	Sissano	7	62	8.8	coconut	1973
Western Highlands	Kagal 2	6	27	4.5	mixed farming/ cattle	1973
Western Highlands [Enga]	Paus	8	486	61.0	mixed farming/ cattle	1973
Manus	Saha	6	55	9.0	coconut	1973
Manus	Sopasopa	13	106	8.0	coconut	1973
Manus	Salmai	66	690	10.5	coconut	1973
West New Britain	Kavui	319	1905	6	oil palm	1973
Western Highlands	Koban	10	29	3.9	tea/vegetables	1974
Western Highlands	Rabe	10	36	3.6	tea/vegetables	1974
Western Highlands	Mt Ambra	5	478	95.6	mixed farming/ cattle	1974
Western Highlands	Pugmi Gaugwi	14	215	15.4	mixed farming/ cattle	1974
West New Britain	Galai	153	918	6	oil palm	1974
West New Britain	Lavilelo 2	12	108	9	cocoa/coconut	1974
East Sepik	Sowam 2	10	616	61.0	mixed farming/ cattle	1975
East Sepik	Chimbian-Timbunke	14	5194	371.0	cattle	1975
Sandaun	Marok	18	330	18.3	coconut	1975
West New Britain	Kavugara	157	1059	6.7	oil palm	1975
New Ireland	Danfu	22	206	9.4	cocoa	1976
West New Britain	Tiauru	225	1553	6.9	oil palm	1976
Oro	Isivini 2	144	1776	6.5	oil palm	1977
Western Highlands	Avi	119	555	4.7	tea	1977

... continued

Appendix Table A6.7.1 (continued)

Province	Name of settlement area	No. of blocks	Approx. total area (ha)	Average block size (ha)	Recommended activity	Year of allocation
Western Highlands	Madan	44	166	3.8	tea	1977
Western Highlands	Karpena	6	33	5.5	tea	1977
West New Britain	Wilelo 1 & 2	337	2202	6.5	oil palm	1977
West New Britain	Balimu	111	730	6.6	oil palm	1977
Oro	Ambogo	46	437	9.5	cocoa/oil palm	1960, 1977
Oro	Isivini 1	56	742	13.2	cocoa/oil palm	1961, 1977
Oro	Arehe	21	105	5.0	cocoa/oil palm	1962, 1977
Oro	North Sangara 1	15	164	10.9	cocoa/oil palm	1964, 1977
Oro	North Sangara 2	22	212	9.6	cocoa/oil palm	1964, 1977
Central	Laloki 2	3	75	25.0	mixed farming/ cattle	1978
Milne Bay	Wataluma	4	268	67.0	mixed farming/ cattle	1978
Oro	Igora	258	1707	6.6	oil palm	1978
Oro	Kepara-Pirive	51	482	9.5	rubber/cocoa	1978
East Sepik	Kanauki	17	4375	257.0	cattle	1978
New Ireland	Balgai 2	4	228	57.0	mixed farming/ cattle	1978
East New Britain	Kokopo	22	248	11.2	cocoa	1978
Oro	East Ambogo	36	397	11.0	cocoa/oil palm	1963, 1978
East Sepik	Gavien	234	3282	14.0	subsistence/ rubber	1967, 1975–78
Gulf	Beara	4	76	19.0	rubber	1979
Central	14 Mile	4	4	1.0	poultry/pigs	1979
Milne Bay	Giligili	7	68	9.7	mixed farming/ cattle	1979
Sandaun	Pes	84	958	11.4	subsistence	1973–79
West New Britain	Sale Sege	60	390	6.5	oil palm	1980
Central	Portion 393	70	455	6.5	rubber	1981
West New Britain	Salimalau	41	267	6.5	oil palm	1981
Oro	Afore	52	697	7.0	cardamom	1980, 1981

Note: This may not be a complete listing of all schemes. The spelling of some names varies from that used elsewhere, for example Japlik or Japalak in East New Britain Province.

The following oil palm schemes, some of which involved resettlement and some village blocks, were established between 1982 and 1995: at Tiauru, Lalopo, Silanga, Usailau, Salelubu/Malasi, Mamota, Kabaia and Soi in the Bialla and Kapiura areas of West New Britain Province; at Gurney, Hagita, Waigani, Naura and Sagarai in Milne Bay Province; and between Kapsu and Kabil, Lamasong and Lamerika, Ngavalus and Konos and Lavalai and Namasalong in New Ireland Province.

Source: Derived from Hulme (1982:38–42), who lists 130 land settlement schemes extracted from Montgomery (1979) and the Department of Lands, Survey and Environment File 145/55.

Appendix Table A6.9.1 Total population (including urban and rural non-village populations) living within a given distance of a national road, by province[a]

Province	<5km	%	5–10 km	%	10–15 km	%	>15 km	%	Total population
Western	26,425	17.2	5,649	3.7	4,540	3.0	116,690	76.1	153,304
Gulf	25,434	23.8	8,372	7.8	2,143	2.0	70,949	66.4	106,898
Central	115,973	63.0	15,483	8.4	8,442	4.6	44,085	24.0	183,983
National Capital District	254,158	100.0	0	0.0	0	0.0	0	0.0	254,158
Milne Bay	44,497	21.1	5,760	2.7	5,676	2.7	154,479	73.4	210,412
Oro	67,086	50.4	17,049	12.8	9,173	6.9	39,757	29.9	133,065
Southern Highlands	356,883	65.3	96,729	17.7	37,284	6.8	55,369	10.1	546,265
Enga	184,437	62.5	45,407	15.4	16,619	5.6	48,568	16.5	295,031
Western Highlands	321,354	73.0	65,769	14.9	15,551	3.5	37,351	8.5	440,025
Simbu	187,740	72.3	32,319	12.4	14,651	5.6	24,993	9.6	259,703
Eastern Highlands	196,591	45.4	62,922	14.5	42,984	9.9	130,475	30.1	432,972
Morobe	158,351	34.4	23,750	5.2	32,884	7.1	245,727	53.3	460,712
Lae City	78,692	100.0	0	0.0	0	0.0	0	0.0	78,692
Madang	132,502	36.3	22,539	6.2	26,460	7.2	183,605	50.3	365,106
East Sepik	150,180	43.8	56,977	16.6	29,207	8.5	106,817	31.1	343,181
Sandaun	62,248	33.5	29,010	15.6	16,875	9.1	77,608	41.8	185,741
Manus	15,413	35.5	5,244	12.1	4,607	10.6	18,123	41.8	43,387
New Ireland	53,351	45.1	3,250	2.7	1,725	1.5	60,024	50.7	118,350
East New Britain	134,870	61.3	18,253	8.3	7,885	3.6	59,125	26.9	220,133
West New Britain	110,035	59.6	7,130	3.9	1,694	0.9	65,649	35.6	184,508
Bougainville	72,037	41.1	24,200	13.8	23,521	13.4	55,402	31.6	175,160
Papua New Guinea	2,748,257	52.9	545,812	10.5	301,921	5.8	1,594,796	30.7	5,190,786

[a] Distance is a straight-line distance and does not take into account the local terrain.

Sources: NSO (2002); PNG Road Asset Management System.

Appendix Table A6.9.2 Length of national roads by road condition and province (km)

Province	Road condition						Total road length
	Poor	%	Fair	%	Good	%	
Western	0	0.0	50	39.4	77	60.6	127
Gulf	33	17.0	73	37.6	88	45.4	194
Central	249	35.8	187	26.9	260	37.4	696
National Capital District	6	6.9	22	25.3	59	67.8	87
Milne Bay	67	29.6	39	17.3	120	53.1	226
Oro	4	2.0	26	12.9	172	85.1	202
Southern Highlands	271	42.4	237	37.1	131	20.5	639
Enga	61	25.8	163	69.1	12	5.1	236
Western Highlands	35	12.5	182	65.0	63	22.5	280
Simbu	156	78.4	36	18.1	7	3.5	199
Eastern Highlands	56	22.4	115	46.0	79	31.6	250
Morobe	45	10.1	166	37.3	234	52.6	445
Madang	40	8.8	234	51.3	182	39.9	456
East Sepik	65	16.5	128	32.5	201	51.0	394
Sandaun	146	49.8	48	16.4	99	33.8	293
Manus	23	27.4	52	61.9	9	10.7	84
New Ireland	5	0.9	31	5.3	547	93.8	583
East New Britain	15	8.2	138	75.4	30	16.4	183
West New Britain	10	2.2	178	39.9	258	57.8	446
Bougainville	156	42.6	170	46.4	40	10.9	366
Papua New Guinea	1,443	22.6	2,275	35.6	2,668	41.8	6,386

Note: The distances in this table are slightly different to those in Table 6.9.2. These differences occur in the original RAMS document. These data refer only to national roads and do not include district or rural roads.

Source: DWI (2001).

Index

All localities mentioned in the text, no matter how fleetingly, are indexed. Nearly all localities appear on the map on pages xx–xxi. That map is not indexed here. A few minor localities have been excluded from the map due to space constraints. To help the reader find places on this map, the abbreviated province name follows the locality in the index.

The key to province abbreviations is:

WES	Western	ENG	Enga	SAN	Sandaun
GUL	Gulf	WHP	Western Highlands	MAN	Manus
CEN	Central	SIM	Simbu	ENB	East New Britain
NCD	National Capital District	EHP	Eastern Highlands	WNB	West New Britain
MBP	Milne Bay	MOR	Morobe	NIP	New Ireland
ORO	Oro	MAD	Madang	BOU	Bougainville
SHP	Southern Highlands	ESP	East Sepik		

Lower case letters following page numbers indicate information in places other than text, as follows:

b	box	fn	footnote	m	map
f	figure	t	table		

If a topic is mentioned in both the text and tables or figures on the same page, only the page is referenced, not the table or figure.

Italics are used for scientific names and for indigenous names for plants.

Authors of cited literature are not indexed.

www.ingramcontent.com/pod-product-compliance
Lightning Source LLC
LaVergne TN
LVHW072120210926
848482LV00028B/708